PREALGEBRA

SEVENTH EDITION

MARVIN L. BITTINGER

Indiana University Purdue University Indianapolis

DAVID J. ELLENBOGEN

Community College of Vermont

BARBARA L. JOHNSON

Indiana University Purdue University Indianapolis

PEARSON

Boston   Columbus   Hoboken   Indianapolis   New York   San Francisco
Amsterdam   Cape Town   Dubai   London   Madrid   Milan   Munich   Paris   Montréal   Toronto
Delhi   Mexico City   São Paulo   Sydney   Hong Kong   Seoul   Singapore   Taipei   Tokyo

Editorial Director	Christine Hoag
Executive Editor	Cathy Cantin
Editorial Assistant	Chase Hammond
Program Management Team Lead	Karen Wernholm
Program Manager	Patty Bergin
Project Management Team Lead	Christina Lepre
Project Manager	Ron Hampton
Composition	Lumina Datamatics
Production Services	Jane Hoover/Lifland et al., Bookmakers
Editorial Services	Martha K. Morong/Quadrata, Inc.
Art Editor and Photo Researcher	The Davis Group, Inc.
Associate Producer	Jonathan Wooding
Executive Content Manager	Rebecca Williams (MathXL)
Senior Content Developer	John Flanagan (TestGen)
Marketing Manager	Rachel Ross
Marketing Assistant	Kelly Cross
Procurement Specialist	Carol Melville
Text Designer	The Davis Group, Inc.
Associate Design Director	Andrea Nix
Program Design Lead	Barbara Atkinson
Cover Photograph	Getty

Photo and Text Credits
Credits appear on page vi.

Library of Congress Cataloging-in-Publication Data
Bittinger, Marvin L.
    Prealgebra / Marvin L. Bittinger, David J. Ellenbogen,
Barbara L. Johnson.—Seventh edition.
        pages cm
    Includes bibliographical references and index.
    ISBN: Student: 0-321-99715-8
    1. Arithmetic—Textbooks.
I. Johnson, Barbara L. (Barbara Loreen),   II. Title.
QA39.3.B58 2016
513′.14—dc23   2014035496

1 2 3 4 5 6 7 8 9 10—CRK—17 16 15 14

www.pearsonhighered.com

ISBN-13: 978-0-321-99715-9
ISBN-10: 0-321-99715-8

Contents

PHOTO CREDITS

TEXT CREDITS

Index of Applications

Preface

The Bittinger Program

Math hasn't changed, but students—and the way they learn it—have.

Prealgebra, 7th Edition, continues the Bittinger tradition of objective-based, guided learning, while integrating timely updates to the proven pedagogy. In this edition, there is a greater emphasis on guided learning and helping students get the most out of all of the course resources available with the Bittinger program, including new opportunities for mobile learning.

The program has expanded to include these comprehensive new teaching and learning resources: *MyMathGuide* **workbook**, **To-the-Point Objective Videos**, and enhanced, media-rich **MyMathLab** courses. Feedback from instructors and students motivated these and several other significant improvements: a new design to support guided learning, new figures and photos to help students visualize both concepts and applications, and many new and updated real-data applications to bring the math to life.

With so many resources available in so many formats, the trusted guidance of the Bittinger team on *what to do* and *when* will help today's math students stay on task. Students are encouraged to use *Your Guide to Success in Math*, a four-step learning path and checklist available on the handy reference card in the front of this text and in MyMathLab. The guide will help students identify the resources in the textbook, supplements, and MyMathLab that support *their* learning style, as they develop and retain the skills and conceptual understanding they need to succeed in this and future courses.

In this preface, a look at the key new *and* hallmark resources and features of the *Prealgebra* program—including the textbook/eText, video program, *MyMathGuide* workbook, and MyMathLab—is organized around *Your Guide to Success in Math*. This will help instructors direct students to the tools and resources that will help them most in a traditional lecture, hybrid, lab-based, or online environment.

NEW AND HALLMARK FEATURES IN RELATION TO Your Guide to Success in Math

STEP 1   Learn the Skills and Concepts

Students have several options for learning, reviewing, and practicing the math concepts and skills.

Textbook/eText

☐ **Skill to Review.** At the beginning of nearly every text section, *Skill to Review* offers a just-in-time review of a previously presented skill that relates to the new material in the section. Section and objective references are included for the student's convenience, and two practice exercises are provided for review and reinforcement.

☐ **Margin Exercises.** For each objective, problems labeled "Do Exercise . . ." give students frequent opportunities to solve exercises while they learn.

☐ *New!*   **Guided Solutions.** Nearly every section has *Guided Solution* margin exercises with fill-in blanks at key steps in the problem-solving process.

☐ *Enhanced!* **MyMathLab.** MyMathLab now includes *Active Learning Figures* for directed exploration of concepts; more problem types, including *Reading Checks* and *Guided Solutions*; and new, objective-based videos. (See pp. xvi–xix for a detailed description of the features of MyMathLab.)

☐ *New!* **Skills Checks.** In the Learning Path for Ready-to-Go MyMathLab, each chapter begins with a brief assessment of students' mastery of the prerequisite skills needed to learn the new material in the chapter. Based on the results of this pre-test, a personalized homework set is designed to help each student prepare for the chapter.

☐ *New!* **To-the-Point Objective Videos.** This is a comprehensive new program of objective-based, interactive videos that are incorporated into the Learning Path in MyMathLab and can be used hand-in-hand with the *MyMathGuide* workbook.

☐ *New!* **Interactive Your Turn Exercises.** For each objective in the videos, students solve exercises and receive instant feedback on their work.

☐ *New!* ***MyMathGuide: Notes, Practice, and Video Path.*** This is an objective-based workbook (available printed and in MyMathLab) for guided, hands-on learning. It offers vocabulary, skill, and concept review—along with problem-solving practice—with space to show work and write notes. Incorporated in the Learning Path in MyMathLab, it can be used together with the To-the-Point Objective Video program, instructor lectures, and the textbook.

STEP 2   Check Your Understanding

Throughout the program, students have frequent opportunities to check their work and confirm that they understand each skill and concept before moving on to the next topic.

☐ *New!* **Reading Checks.** At the beginning of each set of section exercises in the text, students demonstrate their grasp of the skills and concepts.

☐ **Translating for Success.** In the text and in MyMathLab, these activities offer students extra practice with the important first step of the process for solving applied problems.

STEP 3   Do Your Homework

Prealgebra, 7th Edition, has a wealth of proven and updated exercises. Pre-built assignments are available for instructors in MyMathLab, and they are pre-assigned and incorporated into the Learning Path in the Ready-to-Go course.

☐ **Skill Maintenance.** In each section, these exercises offer a thorough review of the math in the preceding text.

☐ **Synthesis Exercises.** To help build critical-thinking skills, these section exercises require students to use what they know and combine learning objectives from the current section with those from previous sections.

STEP 4   Review and Test Your Understanding

Students have a variety of resources to check their skills and understanding along the way and to help them prepare for tests.

☐ **Mid-Chapter Review.** Mid-way through each chapter, students work a set of exercises (*Concept Reinforcement, Guided Solutions, Mixed Review,* and *Understanding Through Discussion and Writing*) to confirm that they have grasped the skills and concepts covered in the first half before moving on to new material.

☐ **Summary and Review.** This resource provides an in-text opportunity for active learning and review for each chapter. *Vocabulary Reinforcement, Concept Reinforcement,* objective-based *Study Guide* (examples paired with similar exercises),

Review Exercises (including *Synthesis* problems), and *Understanding Through Discussion and Writing* are included in these comprehensive chapter reviews.

☐ **Chapter Test.** Chapter Tests offer students the opportunity for comprehensive review and reinforcement prior to taking their instructor's exam. **Chapter Test-Prep Videos** (in MyMathLab and on YouTube) show step-by-step solutions to the Chapter Tests.

☐ **Cumulative Review.** Following every chapter beginning with Chapter 2, a Cumulative Review revisits skills and concepts from all preceding chapters to help students retain previously learned material.

Study Skills

Developing solid time-management, note-taking, test-taking, and other study skills is key to student success in math courses (as well as professionally and personally). Instructors can direct students to related study skills resources as needed.

☐ *New!* **Student Study Reference.** This pull-out card at the front of the text is perforated, three-hole-punched, and binder-ready for convenient reference. It includes **Your Guide to Success in Math** course checklist, **Student Organizer**, and **At a Glance**, a list of key information and expressions for quick reference as students work exercises and review for tests.

☐ *New!* **Studying for Success.** Checklists of study skills—designed to ensure that students develop the skills they need to succeed in math, school, and life—are integrated throughout the text at the beginning of selected sections.

☐ *New!* **Study Skills Modules.** In MyMathLab, interactive modules address common areas of weakness, including time-management, test-taking, and note-taking skills. Additional modules support career-readiness.

Learning Math in Context

☐ *New!* **Applications.** Throughout the text in examples and exercises, real-data applications encourage students to see and interpret the mathematics that appears every day in the world around them. Applications that use real data are drawn from business and economics, life and physical sciences, medicine, technology, and areas of general interest such as sports and daily life. New applications include "Fastest-Growing Occupations" (p. 68), "Training Regimens" (p. 276), "Media Usage" (p. 457), and "*The Hobbit: An Unexpected Journey*" (p. 485). For a complete list of applications, please refer to the Index of Applications (p. vii).

BREAKTHROUGH
To improving results

MyMathLab
Ties the Complete Learning Program Together

MyMathLab® Online Course (access code required)

MyMathLab from Pearson is the world's leading online resource in mathematics, integrating interactive homework, assessment, and media in a flexible, easy-to-use format. MyMathLab delivers **proven results** in helping individual students succeed. It provides **engaging experiences** that personalize, stimulate, and measure learning for each student. And it comes from an **experienced partner** with educational expertise and an eye on the future.

MyMathLab for Developmental Mathematics

Prepared to go wherever you want to take your students.

Personalized Support for Students

Exercises: The homework and practice exercises in MyMathLab are correlated to the exercises in the textbook, and they regenerate algorithmically to give students unlimited opportunities for practice and mastery. The software offers immediate, helpful feedback when students enter incorrect answers.

Multimedia Learning Aids: Exercises include guided solutions, sample problems, animations, videos, and eText access for extra help at point of use.

Expert Tutoring: Although many students describe the whole of MyMathLab as "like having your own personal tutor," students using MyMathLab do have access to live tutoring from qualified math instructors.

To help students achieve mastery, MyMathLab can generate **personalized homework** based on individual performance on tests or quizzes. Personalized homework allows students to focus on topics they have not yet mastered.

The **Adaptive Study Plan** makes studying more efficient and effective for every student. Performance and activity are assessed continually in real time. The data and analytics are used to provide personalized content—reinforcing concepts that target each student's strengths and weaknesses.

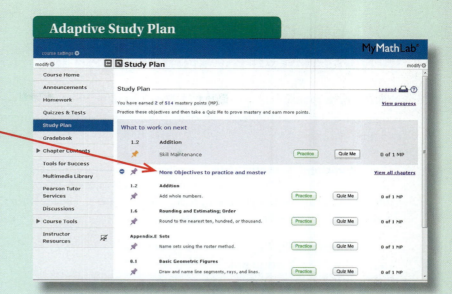

Flexible Design, Easy Start-Up, and Results for Instructors

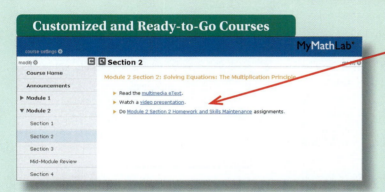

Instructors can modify the site navigation and insert their own directions on course-level landing pages; also, a **custom MyMathLab** course can be built that reorganizes and structures the course material by chapters, modules, units—whatever the need may be.

Ready-to-Go courses include preassigned homework, quizzes, and tests to make it even easier to get started. The Bittinger Ready-to-Go courses include new *Mid-Chapter Reviews* and *Reading Check Assignments*, plus a four-step Learning Path on each section-level landing page to help instructors direct students where to go and what resources to use.

The **comprehensive online gradebook** automatically tracks students' results on tests, quizzes, and homework and in the study plan. Instructors can use the gradebook to quickly intervene if students have trouble, or to provide positive feedback on a job well done. The data within MyMathLab is easily exported to a variety of spreadsheet programs, such as Microsoft Excel. Instructors can determine which points of data to export and then analyze the results to determine success.

New features, such as **Search/Email by Criteria**, make the gradebook a powerful tool for instructors. With this feature, instructors can easily communicate with both at-risk and successful students. They can search by score on specific assignments, by non-completion of assignments within a given time frame, by last login date, or by overall score.

Special Bittinger Resources
in MyMathLab for Students and Instructors

In addition to robust course delivery, MyMathLab offers the full Bittinger eText, additional Bittinger program features, and the entire set of instructor and student resources in one easy-to-access online location.

New! Four-Step Learning Path

Each of the section-level landing pages in the Ready-to-Go MyMathLab course includes a Learning Path that aligns with *Your Guide to Success in Math* to link students directly to the resources they should use when they need them. This also allows instructors to point students to the best resources to use at particular times.

New! Integrated Bittinger Video Program and *MyMathGuide* workbook

Bittinger Video Program*

The Video Program is available in MyMathLab and includes closed captioning and the following video types:

New! To-the-Point Objective Videos. These objective-based, interactive videos are incorporated into the Learning Path in MyMathLab and can be used along with the *MyMathGuide* workbook.

Chapter Test Prep Videos. The Chapter Test Prep Videos let students watch instructors work through step-by-step solutions to all the Chapter Test exercises from the textbook. Chapter Test Prep Videos are also available on YouTube™ (search using author name and book title).

New! *MyMathGuide: Notes, Practice, and Video Path* workbook*
(Printed Workbook ISBN: 978-0-321-92546-6)

This objective-based workbook for guided, hands-on learning offers vocabulary, skill, and concept review—along with problem-solving practice—with space to show work and write notes. Incorporated in the Learning Path in MyMathLab, *MyMathGuide* can be used together with the To-the-Point Objective Video program, instructor lectures, and the textbook. Instructors can assign To-the-Point Objective Videos in MyMathLab in conjunction with the *MyMathGuide* workbook.

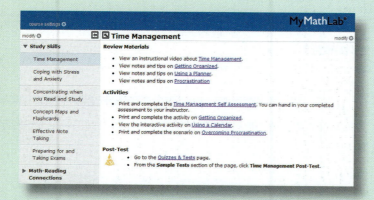

Study Skills Modules

In MyMathLab, interactive modules address common areas of weakness, including time-management, test-taking, and note-taking skills. Additional modules support career-readiness. Instructors can assign module material with a post-quiz.

Additional Resources in MyMathLab

For Students

Student's Solutions Manual*
(ISBN: 978-0-321-92007-9)
By Judy Penna

Contains completely worked-out annotated solutions for all the odd-numbered exercises in the text. Also includes fully worked-out annotated solutions for all the exercises (odd- and even-numbered) in the Mid-Chapter Reviews, the Summary and Reviews, the Chapter Tests, and the Cumulative Reviews.

For Instructors

Annotated Instructor's Edition**
(ISBN: 978-0-321-99715-9)

This version of the text includes answers to all exercises presented in the book, as well as helpful teaching tips.

Instructor's Resource Manual with Tests and Mini Lectures**
(download only)
By Laurie Hurley

This manual includes resources designed to help both new and experienced instructors with course preparation and classroom management. It contains support for media supplements, two multiple-choice tests per chapter, six free-response tests per chapter, and eight final exams.

Instructor's Solutions Manual**
(download only)
By Judy Penna

This manual contains detailed, worked-out solutions to all odd-numbered exercises and brief solutions to the even-numbered exercises in the exercise sets.

PowerPoint® Lecture Slides**
(download only)

Present key concepts and definitions from the text.

To learn more about how MyMathLab combines proven learning applications with powerful assessment, visit www.mymathlab.com or contact your Pearson representative.

---

*Printed supplements are also available for separate purchase through MyMathLab, MyPearsonStore.com, or other retail outlets. They can also be value-packed with a textbook or MyMathLab code.

**Available for download from the Instructor Resource Center (IRC) on www.pearsonhighered.com.

Acknowledgments

Our deepest appreciation to all of you who helped to shape this edition by reviewing and spending time with us on your campuses. In particular, we would like to thank the following reviewers:

Afsheen Akbar, *Bergen Community College*
Morgan Arnold, *Central Georgia Technical College*
Gurdial Arora, *Xavier University*
Donna Beatty, *Venture College*
Connie Buller, *Metropolitan Community College*
Erin Cooke, *Gwinnett Technical College*
Kay Davis, *Del Mar College*
Edward Dillon, *Century Community and Technical College*
Beverlee Drucker, *Northern Virginia Community College*
Sabine Eggleston, *Edison State College*
Dylan Faullin, *Dodge City Community College*
Anne Fischer, *Tulsa Community College Metro Campus*
Rebecca Gubitti, *Edison State College*
Shawna Haider, *Salt Lake Community College*
Exie Hall, *Del Mar College*
Nancy Hixson, *Columbia Southern University*
Stephanie Houdek, *St. Cloud Technical Institute*

Linda Kass, *Bergen Community College*
Chauncey Keaton, *Central Georgia Technical College*
Edith Lester, *Volunteer State Community College*
Dorothy Marshall, *Edison State College*
Kimberley McHale, *Heartland Community College*
Arda Melkonian, *Victor Valley College*
Christine Mirbaha, *Community College of Baltimore County–Dundalk*
Eli Nettles, *Nashville State Community College*
Louise Olshan, *County College of Morris*
Deborah Poetsch, *County College of Morris*
Nimisha Raval, *Central Georgia Technical College*
Chris Schultz, *Iowa State University*
Jane Serbousek, *Northern Virginia Community College*
Alexis Thurman, *County College of Morris*
Michelle Van Wagoner, *Nashville State Community College*
Melanie Walker, *Bergen Community College*

The endless hours of hard work by Martha Morong, Jane Hoover, and Geri Davis have led to products of which we are immensely proud. We also want to thank Judy Penna for accuracy checking pages, for writing the Student's and Instructor's Solutions Manuals, and for her strong leadership in the preparation of the printed supplements and video lectures. Other strong support has come from Laurie Hurley for the Instructor's Resource Manual and for accuracy checking pages. Michelle Lanosga assisted with applications research. We also wish to recognize Nelson Carter and Tom Atwater, who wrote video scripts.

In addition, a number of people at Pearson have contributed in special ways to the development and production of this textbook, including the Developmental Math team: Project Manager Ron Hampton, Senior Designer Barbara Atkinson, Editorial Assistant Chase Hammond, and Associate Media Producer Jonathan Wooding. Executive Editor Cathy Cantin and Marketing Manager Rachel Ross encouraged our vision and provided marketing insight.

CHAPTER 1

Whole Numbers

1.1 Standard Notation

OBJECTIVES

a Give the meaning of digits in standard notation.

b Convert from standard notation to expanded notation.

c Convert between standard notation and word names.

We study mathematics in order to be able to solve problems. In this section, we study how numbers are named. We begin with the concept of place value.

a PLACE VALUE

The numbers of jobs available in 2010 for several professions are shown in the following table.

PROFESSION	NUMBER OF JOBS, 2010
Registered nurses	2,737,400
Radiologic technologists	219,900
Radiation therapists	16,900

SOURCE: U.S. Department of Labor, Bureau of Labor Statistics

A **digit** is a number 0, 1, 2, 3, 4, 5, 6, 7, 8, or 9 that names a place-value location. For large numbers, digits are separated by commas into groups of three, called **periods**. Each period has a name: *ones, thousands, millions, billions, trillions,* and so on. To understand the number of jobs for registered nurses in the table above, we can use a **place-value chart**, as shown below.

PLACE-VALUE CHART																	
Periods →	Trillions			Billions			Millions			Thousands			Ones					
									2	7	3	7	4	0	0		
	Hundreds	Tens	Ones	Hundreds	Tens	Ones	Hundreds	Tens	Ones	Hundreds	Tens	Ones	Hundreds	Tens	Ones		

2 millions   737 thousands   400 ones

EXAMPLES In each of the following numbers, what does the digit 8 mean?

1. 27**8**,342   8 thousands
2. **8**72,342   8 hundred thousands
3. 2**8**,343,399,223   8 billions
4. 9**8**,413,099   8 millions
5. 632**8**   8 ones

Do Margin Exercises 1–6 (in the margin at right). ▶

What does the digit 2 mean in each number?

1. 526,555   2. 265,789

3. 42,789,654   4. 24,789,654

5. 8924   6. 5,643,201

EXAMPLE 6 *Charitable Organizations.* Since its founding in 1881 by Clara Barton, the American Red Cross has been the nation's best-known emergency response organization. As part of a worldwide organization, the American Red Cross also aids victims of devastating natural disasters. For the fiscal year ending June 2011, the total revenue of the American Red Cross was $3,452,960,387. What digit names the number of ten millions?
Source: charitynavigator.org

Ten millions

3,4**5**2,960,387

The digit 5 is in the ten millions place, so 5 names the number of ten millions.

Do Exercise 7. ▶

7. *Government Payroll.* In March 2011, the total payroll for all state employees in the United States was $19,971,861,990. What digit names the number of ten billions?
Source: *2011 Annual Survey of Public Employment and Payroll*

b CONVERTING FROM STANDARD NOTATION TO EXPANDED NOTATION

Heifer International is a charitable organization whose mission is to work with communities to end hunger and poverty and care for the earth by providing farm animals to impoverished families around the world. Consider the data in the following table.

GEOGRAPHICAL AREAS OF NEED	NUMBER OF FAMILIES ASSISTED DIRECTLY AND INDIRECTLY BY HEIFER INTERNATIONAL IN 2011
Africa	220,275
Americas	934,871
Asia, South Pacific	407,640
Central and Eastern Europe	344,945

SOURCE: *Heifer International 2011 Annual Report*

The number of families assisted in the Americas was 934,871. This number is expressed in **standard notation**. We write **expanded notation** for 934,871 as follows:

$$934,871 = 9 \text{ hundred thousands} + 3 \text{ ten thousands}$$
$$+ 4 \text{ thousands} + 8 \text{ hundreds}$$
$$+ 7 \text{ tens} + 1 \text{ one.}$$

EXAMPLE 7   Write expanded notation for 1815 ft, the height of the CN Tower in Toronto, Canada.

$$1815 = 1 \text{ thousand} + 8 \text{ hundreds} + 1 \text{ ten} + 5 \text{ ones}$$

EXAMPLE 8   Write expanded notation for 407,640, the number of families in Asia and the South Pacific assisted by Heifer International in 2011.

$$407,640 = 4 \text{ hundred thousands} + 0 \text{ ten thousands}$$
$$+ 7 \text{ thousands} + 6 \text{ hundreds} + 4 \text{ tens} + 0 \text{ ones}$$

or

$$4 \text{ hundred thousands} + 7 \text{ thousands} + 6 \text{ hundreds} + 4 \text{ tens}$$

◀ **Do Exercises 8–11.**

Write expanded notation.

8. 2718 mi, the length of the Congo River in Africa

$$2718 = 2 \boxed{} + 7 \boxed{}$$
$$+ \boxed{} \text{ ten} + \boxed{} \text{ ones}$$

9. 344,945, the number of families in Central and Eastern Europe assisted by Heifer International in 2011

10. 1670 ft, the height of the Taipei 101 Tower in Taiwan

11. 104,094 square miles, the area of Colorado

C   CONVERTING BETWEEN STANDARD NOTATION AND WORD NAMES

We often use **word names** for numbers. When we pronounce a number, we are speaking its word name. Russia won 82 medals in the 2012 Summer Olympics in London, Great Britain. A word name for 82 is "eighty-two." Word names for some two-digit numbers like 36, 51, and 72 use hyphens. Others like that for 17 use only one word, "seventeen."

2012 Summer Olympics Medal Count

COUNTRY	GOLD	SILVER	BRONZE	TOTAL
United States of America	46	29	29	104
People's Republic of China	38	27	23	88
Russia	24	26	32	82
Great Britain	29	17	19	65
Germany	11	19	14	44

SOURCE: espn.go.com

Answers

8. 2 thousands + 7 hundreds + 1 ten + 8 ones
9. 3 hundred thousands + 4 ten thousands + 4 thousands + 9 hundreds + 4 tens + 5 ones
10. 1 thousand + 6 hundreds + 7 tens + 0 ones, or 1 thousand + 6 hundreds + 7 tens
11. 1 hundred thousand + 0 ten thousands + 4 thousands + 0 hundreds + 9 tens + 4 ones, or 1 hundred thousand + 4 thousands + 9 tens + 4 ones

Guided Solution:
8. thousands, hundreds, 1, 8

EXAMPLES Write a word name.

9. 46, the number of gold medals won by the United States

Forty-six

10. 19, the number of silver medals won by Germany

Nineteen

11. 104, the total number of medals won by the United States

One hundred four

Do Exercises 12–14. ▶

For word names for larger numbers, we begin at the left with the largest period. The number named in the period is followed by the name of the period; then a comma is written and the next number and period are named. Note that the name of the ones period is not included in the word name for a whole number.

EXAMPLE 12 Write a word name for 46,605,314,732.

Forty-six billion,

six hundred five million,

three hundred fourteen thousand,

seven hundred thirty-two

The word "and" *should not* appear in word names for whole numbers. Although we commonly hear such expressions as "two hundred *and* one," the use of "and" is not, strictly speaking, correct in word names for whole numbers. For decimal notation, it is appropriate to use "and" for the decimal point. For example, 317.4 is read as "three hundred seventeen *and* four tenths."

Do Exercises 15–18. ▶

EXAMPLE 13 Write standard notation.

Five hundred six million,

three hundred forty-five thousand,

two hundred twelve

Standard notation is 506,345,212.

Do Exercise 19. ▶

Write a word name. (Refer to the chart on the previous page.)

12. 65, the total number of medals won by Great Britain

13. 14, the number of bronze medals won by Germany

14. 38, the number of gold medals won by the People's Republic of China

Write a word name.

15. 204

16. $44,640, the average annual wage for athletic trainers in the United States in 2012
Source: U.S. Bureau of Labor Statistics

GS **17.** 1,879,204

One _____ , eight hundred _____ thousand, two hundred _____

18. 7,052,428,785, the world population in 2012
Source: U.S. Census Bureau

19. Write standard notation.

Two hundred thirteen million, one hundred five thousand, three hundred twenty-nine

Answers

12. Sixty-five **13.** Fourteen
14. Thirty-eight **15.** Two hundred four
16. Forty-four thousand, six hundred forty
17. One million, eight hundred seventy-nine thousand, two hundred four **18.** Seven billion, fifty-two million, four hundred twenty-eight thousand, seven hundred eighty-five **19.** 213,105,329

Guided Solution:
17. Million, seventy-nine, four

✓ Reading Check

Complete each statement with the correct word from the following list.

digit   expanded   period   standard

RC1. In 983, the _____ 9 represents 9 hundreds.

RC2. In 615,702, the number 615 is in the thousands _____.

RC3. The phrase "3 hundreds + 2 tens + 9 ones" is _____ notation for 329.

RC4. The number 721 is written in _____ notation.

a   What does the digit 5 mean in each number?

1. 235,888

2. 253,777

3. 1,488,526

4. 500,736

Movie Receipts. The final movie of the Harry Potter series, *Harry Potter and the Deathly Hallows: Part II*, grossed $1,328,111,219 worldwide.

Source: Nash Information Services, LLC

What digit names the number of:

5. thousands?

6. millions?

7. ten millions?

8. hundred thousands?

b   Write expanded notation.

Stair-Climbing Races. The figure below shows the number of stairs in four buildings in which stair-climbing races are held. In Exercises 9–12, write expanded notation for the number of stairs in each race.

Stair-Climbing Races

2058	1776	1576	1081
International Towerthon, Kuala Lumpur, Malaysia	CN Tower Stair Climb, Toronto, Ontario, Canada	Empire State Building Run-Up, New York	Skytower Vertical Challenge, Auckland, New Zealand

SOURCE: towerrunning.com

9. 2058 steps in the International Towerthon, Kuala Lumpur, Malaysia

10. 1776 steps in the CN Tower Stair Climb, Toronto, Ontario, Canada

11. 1576 steps in the Empire State Building Run-Up, New York City, New York

12. 1081 steps in the Skytower Vertical Challenge, Auckland, New Zealand

13. 5702   **14.** 3097   **15.** 93,986   **16.** 38,453

Population. The table below shows the populations of four countries in 2012. In Exercises 17–20, write expanded notation for the population of the given country.

Four Most Populous Countries in the World

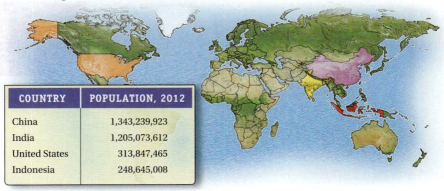

COUNTRY	POPULATION, 2012
China	1,343,239,923
India	1,205,073,612
United States	313,847,465
Indonesia	248,645,008

SOURCE: *CIA World Factbook*

17. 1,343,239,923 for China

18. 1,205,073,612 for India

19. 248,645,008 for Indonesia

20. 313,847,465 for the United States

C Write a word name.

21. 85   **22.** 48   **23.** 88,000   **24.** 45,987

25. 123,765   **26.** 111,013   **27.** 7,754,211,577   **28.** 43,550,651,808

29. *English Language Learners.* In the 2007–2008 academic year, there were 701,799 English language learners in Texas schools. Write a word name for 701,799.

Source: U.S. Department of Education

30. *College Football.* The 2012 Rose Bowl game was attended by 91,245 fans. Write a word name for 91,245.

Source: bizjournals.com

31. *Auto Racing.* Dario Franchitti, winner of the 2012 Indianapolis 500 auto race, won a prize of $2,474,280. Write a word name for 2,474,280.

Source: sbnation.com

32. *Busiest Airport.* In 2010, the world's busiest airport, Hartsfield-Jackson Atlanta International Airport, hosted 89,331,622 passengers. Write a word name for 89,331,622.

Source: Airports Council International

Write each number in standard notation.

33. Six hundred thirty-two thousand, eight hundred ninety-six

34. Three hundred fifty-four thousand, seven hundred two

35. Fifty thousand, three hundred twenty-four

36. Seventeen thousand, one hundred twelve

37. Two million, two hundred thirty-three thousand, eight hundred twelve

38. Nineteen million, six hundred ten thousand, four hundred thirty-nine

39. Eight billion

40. Seven hundred million

41. Forty million

42. Twenty-six billion

43. Thirty million, one hundred three

44. Two hundred thousand, seventeen

Write standard notation for the number in each sentence.

45. *Pacific Ocean.* The area of the Pacific Ocean is sixty-four million, one hundred eighty-six thousand square miles.

46. The average distance from the sun to Neptune is two billion, seven hundred ninety-three million miles.

Synthesis

To the student and the instructor: The Synthesis exercises found at the end of every exercise set challenge students to combine concepts or skills studied in the section or in preceding parts of the text. Exercises marked with a ▦ symbol are meant to be solved using a calculator.

47. How many whole numbers between 100 and 400 contain the digit 2 in their standard notation?

48. ▦ What is the largest number that you can name on your calculator? How many digits does that number have? How many periods?

Addition

a ADDITION OF WHOLE NUMBERS

To answer questions such as "How many?", "How much?", and "How tall?", we often use whole numbers. The set, or collection, of **whole numbers** is

> 0, 1, 2, 3, 4, 5, 6, 7, 8, 9, 10, 11, 12, ...

The set goes on indefinitely. There is no largest whole number, and the smallest whole number is 0. Each whole number can be named using various notations. The set 1, 2, 3, 4, 5, . . . , without 0, is called the set of **natural numbers**.

Addition of whole numbers corresponds to combining things.

We say that the **sum** of 3 and 2 is 5. The numbers added are called **addends**. The addition that corresponds to the figure above is

$$3 \quad + \quad 2 \quad = \quad 5.$$
Addend   Addend   Sum

To add whole numbers, we add the ones digits first, then the tens, then the hundreds, then the thousands, and so on.

EXAMPLE 1 Add: $878 + 995$.

Place values are lined up in columns.

$$\begin{array}{r} \overset{1}{8}\ 7\ 8 \\ +\ 9\ 9\ 5 \\ \hline 3 \end{array}$$

Add ones. We get 13 ones, or 1 ten + 3 ones. Write 3 in the ones column and 1 above the tens. This is called *carrying*, or *regrouping*.

$$\begin{array}{r} \overset{1}{8}\ \overset{1}{7}\ 8 \\ +\ 9\ 9\ 5 \\ \hline 7\ 3 \end{array}$$

Add tens. We get 17 tens, so we have 10 tens + 7 tens. This is also 1 hundred + 7 tens. Write 7 in the tens column and 1 above the hundreds.

$$\begin{array}{r} \overset{1}{8}\ \overset{1}{7}\ 8 \\ 9\ 9\ 5 \\ \hline 1\ 8\ 7\ 3 \end{array}$$

Add hundreds. We get 18 hundreds.

We show you these steps for explanation. You need write only this.

$$\begin{array}{r} \overset{1}{8}\ \overset{1}{7}\ 8 \\ +\ 9\ 9\ 5 \\ \hline 1\ 8\ 7\ 3 \end{array}$$

— Addends

← Sum

SKILL TO REVIEW

Objective 1.1a: Give the meaning of digits in standard notation.

In each of the following numbers, what does the digit 4 mean?

1. 8342
2. 14,976

Answers

Skill to Review:
1. 4 tens
2. 4 thousands

How do we perform an addition of three numbers, like $2 + 3 + 6$? We could do it by adding 3 and 6, and then 2. We can show this with parentheses:

$$2 + (3 + 6) = 2 + 9 = 11. \qquad \text{Parentheses tell what to do first.}$$

We could also add 2 and 3, and then 6:

$$(2 + 3) + 6 = 5 + 6 = 11.$$

Either way the result is 11. It does not matter how we group the numbers. This illustrates the **associative law of addition**, $a + (b + c) = (a + b) + c$. We can also add whole numbers in any order. That is, $2 + 3 = 3 + 2$. This illustrates the **commutative law of addition**, $a + b = b + a$. Together, the commutative and associative laws tell us that to add more than two numbers, we can use any order and grouping we wish. Adding 0 to a number does not change the number: $a + 0 = 0 + a = a$. That is, $6 + 0 = 0 + 6 = 6$, or $198 + 0 = 0 + 198 = 198$. We say that 0 is the **additive identity**.

EXAMPLE 2 Add: $391 + 1276 + 789 + 5498$.

Add ones. We get 24, so we have 2 tens + 4 ones. Write 4 in the ones column and 2 above the tens.

Add tens. We get 35 tens, so we have 30 tens + 5 tens. This is also 3 hundreds + 5 tens. Write 5 in the tens column and 3 above the hundreds.

Add hundreds. We get 19 hundreds, so we have 10 hundreds + 9 hundreds. This is also 1 thousand + 9 hundreds. Write 9 in the hundreds column and 1 above the thousands.

```
  1   3   2
      3   9   1
  1   2   7   6
      7   8   9
+ 5   4   9   8
  7   9   5   4
```

Add thousands. We get 7 thousands.

◀ **Do Exercises 1–4.**

Add.

1. $6203 + 3542$

2.

```
    7 9 6 8
  + 5 4 9 7
    _____
      7 9 6 8
  +   5 4 9 7
    _____
        , 4   5
```

3.
```
    9 8 0 4
  + 6 3 7 8
```

4.
```
    1 9 3 2
    6 7 2 3
    9 8 7 8
  + 8 9 4 1
```

Answers

1. 9745   **2.** 13,465   **3.** 16,182
4. 27,474

Guided Solution:
2.
```
    1 1 1
    7 9 6 8
  + 5 4 9 7
    _____
   13,4 6 5
```

b FINDING PERIMETER

Addition can be used when finding perimeter.

PERIMETER
The distance around an object is its **perimeter**.

EXAMPLE 3 Find the perimeter of the figure.

We add the lengths of the sides:

Perimeter = 29 in. + 19 in. + 37 in. + 44 in.

= 129 in.

The perimeter of the figure is 129 in. (inches).

Do Exercises 5 and 6. ▶

EXAMPLE 4 Lucas Oil Stadium in Indianapolis has a unique retractable roof. When the roof is opened (retracted) in good weather to create an open-air stadium, the opening approximates a rectangle 588 ft long and 300 ft wide. Find the perimeter of the opening.

Opposite sides of a rectangle have equal lengths, so this rectangle has two sides of length 588 ft and two sides of length 300 ft.

Perimeter = 588 ft + 300 ft + 588 ft + 300 ft

= 1776 ft

The perimeter of the opening is 1776 ft.

Do Exercise 7. ▶

Find the perimeter of each figure.

GS **5.**

Perimeter = 4 in. + 5 in. + ☐
+ 6 in. + 5 in.
= ☐ in.

6.

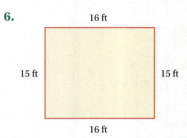

7. *Index Cards.* Two standard sizes for index cards are 3 in. by 5 in. and 5 in. by 8 in. Find the perimeter of each type of card.

For Extra Help

MyMathLab®

MathXL®   PRACTICE

 WATCH

READ

 REVIEW

✓ Reading Check

Complete each statement with the appropriate word or number from the following list. Not every choice will be used.

0	addends	law	product
1	factors	perimeter	sum

RC1. In the addition $5 + 2 = 7$, the numbers 5 and 2 are _____.

RC2. In the addition $5 + 2 = 7$, the number 7 is the _____.

RC3. The sum of _____ and any number a is a.

RC4. The distance around an object is its _____.

a Add.

1.
$$\begin{array}{r} 3\,6\,4 \\ +\quad 2\,3 \\ \hline \end{array}$$

2.
$$\begin{array}{r} 1\,5\,2\,1 \\ +\quad 3\,4\,8 \\ \hline \end{array}$$

3.
$$\begin{array}{r} 8\,6 \\ +\ 7\,8 \\ \hline \end{array}$$

4.
$$\begin{array}{r} 7\,3 \\ +\ 6\,9 \\ \hline \end{array}$$

5.
$$\begin{array}{r} 1\,7\,1\,6 \\ +\ 3\,4\,8\,2 \\ \hline \end{array}$$

6.
$$\begin{array}{r} 7\,5\,0\,3 \\ +\ 2\,6\,8\,3 \\ \hline \end{array}$$

7.
$$\begin{array}{r} 9\,9 \\ +\ \ 1 \\ \hline \end{array}$$

8.
$$\begin{array}{r} 9\,9\,9 \\ +\ \ 1\,1 \\ \hline \end{array}$$

9. $8113 + 390$

10. $271 + 3338$

11. $356 + 4910$

12. $280 + 34{,}902$

13. $3870 + 92 + 7 + 497$

14. $10{,}120 + 12{,}989 + 5738$

15.
$$\begin{array}{r} 4\,8\,2\,5 \\ +\ 1\,7\,8\,3 \\ \hline \end{array}$$

16.
$$\begin{array}{r} 3\,6\,5\,4 \\ +\ 2\,7\,0\,0 \\ \hline \end{array}$$

17.
$$\begin{array}{r} 2\,3{,}4\,4\,3 \\ +\ 1\,0{,}9\,8\,9 \\ \hline \end{array}$$

18.
$$\begin{array}{r} 4\,5{,}8\,7\,9 \\ +\ 2\,1{,}7\,8\,6 \\ \hline \end{array}$$

19.
$$\begin{array}{r} 7\,7{,}5\,4\,3 \\ +\ 2\,3{,}7\,6\,7 \\ \hline \end{array}$$

20.
$$\begin{array}{r} 9\,9{,}9\,9\,9 \\ +\quad\ \ 1\,1\,2 \\ \hline \end{array}$$

21.
$$\begin{array}{r} 4\,5 \\ 2\,5 \\ 3\,6 \\ 4\,4 \\ +\ 8\,0 \\ \hline \end{array}$$

22.
$$\begin{array}{r} 3\,8 \\ 2\,7 \\ 3\,2 \\ 1\,4 \\ +\ 7\,6 \\ \hline \end{array}$$

23.
$$\begin{array}{r} 1\,2{,}0\,7\,0 \\ 2{,}9\,5\,4 \\ +\ \ 3{,}4\,0\,0 \\ \hline \end{array}$$

24.
$$\begin{array}{r} 4\,2{,}4\,8\,7 \\ 8\,3{,}1\,4\,1 \\ +\ 3\,6{,}7\,1\,2 \\ \hline \end{array}$$

25.
$$\begin{array}{r} 4\,8\,3\,5 \\ 7\,2\,9 \\ 9\,2\,0\,4 \\ 8\,9\,8\,6 \\ +\ 7\,9\,3\,1 \\ \hline \end{array}$$

26.
$$\begin{array}{r} 9\,8\,9 \\ 5\,6\,6 \\ 8\,3\,4 \\ 9\,2\,0 \\ +\ 7\,0\,3 \\ \hline \end{array}$$

b Find the perimeter of each figure.

27.

50 yd
23 yd
19 yd
40 yd

28.

14 mi
13 mi
8 mi
22 mi
10 mi
47 mi

29.

402 ft
298 ft
196 ft
212 ft
100 ft
453 ft

30.

62 yd
39 yd
28 yd
46 yd
54 yd

31. Find the perimeter of a standard hockey rink.

85 ft
200 ft

32. In Major League Baseball, how far does a batter travel when circling the bases after hitting a home run?

90 ft
90 ft

Skill Maintenance

The exercises that follow begin an important feature called *Skill Maintenance exercises.* These exercises provide an ongoing review of topics previously covered in the book. You will see them in virtually every exercise set. It has been found that this kind of continuing review can significantly improve your performance on a final examination.

33. What does the digit 8 mean in 486,205? [1.1a]

34. The population of the world is projected to be 9,346,399,468 in 2050. Write a word name for 9,346,399,468. [1.1c]

Source: U.S. Census Bureau

Synthesis

35. A fast way to add all the numbers from 1 to 10 inclusive is to pair 1 with 9, 2 with 8, and so on. Use a similar approach to add all numbers from 1 to 100 inclusive.

1.3 Subtraction

OBJECTIVE

a Subtract whole numbers.

SKILL TO REVIEW

Objective 1.1a: Give the meaning of digits in standard notation.

Consider the number 328,974.

1. What digit names the number of hundreds?
2. What digit names the number of ones?

a SUBTRACTION OF WHOLE NUMBERS

Subtraction is finding the difference of two numbers. Suppose you purchase 6 tickets for a concert and give 2 to a friend.

6 tickets

Give 2 away   Keep 4

The subtraction that represents this situation is

$$6 \quad - \quad 2 \quad = \quad 4.$$

Minuend   Subtrahend   Difference

The **minuend** is the number from which another number is being subtracted. The **subtrahend** is the number being subtracted. The **difference** is the result of subtracting the subtrahend from the minuend.

In the subtraction above, note that the difference, 4, is the number we add to 2 to get 6. This illustrates the relationship between addition and subtraction and leads us to the following definition of subtraction.

> **SUBTRACTION**
>
> The difference $a - b$ is that unique whole number c for which $a = c + b$.

We see that $6 - 2 = 4$ because $4 + 2 = 6$.

To subtract whole numbers, we subtract the ones digits first, then the tens digits, then the hundreds, then the thousands, and so on.

EXAMPLE 1 Subtract: $9768 - 4320$.

$$
\begin{array}{r}
9\ 7\ 6\ \boxed{8} \\
-\ 4\ 3\ 2\ \boxed{0} \\
\hline
8
\end{array}
$$
Subtract ones.

$$
\begin{array}{r}
9\ 7\ \boxed{6}\ 8 \\
-\ 4\ 3\ \boxed{2}\ 0 \\
\hline
4\ 8
\end{array}
$$
Subtract tens.

$$
\begin{array}{r}
9\ \boxed{7}\ 6\ 8 \\
-\ 4\ \boxed{3}\ 2\ 0 \\
\hline
4\ 4\ 8
\end{array}
$$
Subtract hundreds.

We show these steps for explanation. You need write only this.

$$
\begin{array}{r}
\boxed{9}\ 7\ 6\ 8 \\
-\ \boxed{4}\ 3\ 2\ 0 \\
\hline
5\ 4\ 4\ 8
\end{array}
$$
Subtract thousands.

$$
\begin{array}{r}
9\ 7\ 6\ 8 \\
-\ 4\ 3\ 2\ 0 \\
\hline
5\ 4\ 4\ 8
\end{array}
$$

Answers

Skill to Review:
1. 9   2. 4

Because subtraction is defined in terms of addition, we can use addition to *check* subtraction.

Subtraction: *Check by Addition:*

$$\begin{array}{r} 9\ 7\ 6\ 8 \\ -\ 4\ 3\ 2\ 0 \\ \hline 5\ 4\ 4\ 8 \end{array}$$? $$\begin{array}{r} 5\ 4\ 4\ 8 \\ +\ 4\ 3\ 2\ 0 \\ \hline 9\ 7\ 6\ 8 \end{array}$$

Do Exercise 1. ▶

EXAMPLE 2 Subtract: $348 - 165$.
We have

$$\begin{array}{rcl} 3 \text{ hundreds} + 4 \text{ tens} + 8 \text{ ones} & = & 2 \text{ hundreds} + 14 \text{ tens} + 8 \text{ ones} \\ -\ 1 \text{ hundred}\ -\ 6 \text{ tens}\ -\ 5 \text{ ones} & = & -1 \text{ hundred}\ -\ 6 \text{ tens}\ -\ 5 \text{ ones} \\ & = & 1 \text{ hundred}\ +\ 8 \text{ tens}\ +\ 3 \text{ ones} \\ & = & 183. \end{array}$$

First, we subtract the ones.

$$\begin{array}{r} 3\ 4\ 8 \\ -\ 1\ 6\ 5 \\ \hline 3 \end{array}$$ Subtract ones.

We cannot subtract the tens because there is no whole number that when added to 6 gives 4. To complete the subtraction, we must *borrow* 1 hundred from 3 hundreds and regroup it with the 4 tens. Then we can do the subtraction 14 tens − 6 tens = 8 tens.

$$\begin{array}{r} \overset{2\ 14}{3\ 4}\ 8 \\ -\ 1\ 6\ 5 \\ \hline 3 \end{array}$$ Borrow one hundred. That is, 1 hundred = 10 tens, and 10 tens + 4 tens = 14 tens. Write 2 above the hundreds column and 14 above the tens.

$$\begin{array}{r} \overset{2\ 14}{3\ 4}\ 8 \\ -\ 1\ 6\ 5 \\ \hline 1\ 8\ 3 \end{array}$$ Subtract tens; subtract hundreds.

This is what you should write. → $$\begin{array}{r} \overset{2\ 14}{3\ 4}\ 8 \\ -\ 1\ 6\ 5 \\ \hline 1\ 8\ 3 \end{array}$$ *Check:* $$\begin{array}{r} \overset{1}{1}\ 8\ 3 \\ +\ 1\ 6\ 5 \\ \hline 3\ 4\ 8 \end{array}$$ ← The answer checks because this is the top number in the subtraction.

1. Subtract. Check by adding.

$$\begin{array}{r} 7\ 8\ 9\ 3 \\ -\ 4\ 0\ 9\ 2 \end{array}$$

$$\begin{array}{r} 7\ 8\ 9\ 3 \\ -\ 4\ 0\ 9\ 2 \\ \hline \boxed{}\ 8\ \boxed{}\ 1 \end{array}$$

Check:

$$\begin{array}{r} \boxed{} \\ +\ 4\ 0\ 9\ 2 \\ \hline \boxed{} \end{array}$$

CALCULATOR CORNER

Subtracting Whole Numbers
To subtract whole numbers on a calculator, we use the ☐− and ☐= keys.

EXERCISES Use a calculator to perform each subtraction. Check by adding.

1. $57 - 29$

2. $81 - 34$

3. $145 - 78$

4. $612 - 493$

5. $$\begin{array}{r} 4\ 9\ 7\ 6 \\ -\ 2\ 8\ 4\ 8 \end{array}$$

6. $$\begin{array}{r} 1\ 2,4\ 0\ 6 \\ -\ \ \ 9\ 8\ 1\ 3 \end{array}$$

Answer
1. 3801

Guided Solution:
1. 3, 0; 3801, 7893

EXAMPLE 3 Subtract: 6246 − 1879.

$$
\begin{array}{r}
\overset{\scriptstyle 3\ 16}{6\ 2\ \cancel{4}\ \cancel{6}} \\
-\ 1\ 8\ 7\ 9 \\
\hline
7
\end{array}
$$

We cannot subtract 9 ones from 6 ones, but we can
subtract 9 ones from 16 ones. We borrow 1 ten to get
16 ones.

$$
\begin{array}{r}
\overset{\scriptstyle 13}{} \\
\overset{\scriptstyle 1\ \cancel{3}\ 16}{6\ \cancel{2}\ \cancel{4}\ \cancel{6}} \\
-\ 1\ 8\ 7\ 9 \\
\hline
6\ 7
\end{array}
$$

We cannot subtract 7 tens from 3 tens, but we can
subtract 7 tens from 13 tens. We borrow 1 hundred
to get 13 tens.

$$
\begin{array}{r}
\overset{\scriptstyle 11\ 13}{} \\
\overset{\scriptstyle 5\ \cancel{1}\ \cancel{3}\ 16}{\cancel{6}\ \cancel{2}\ \cancel{4}\ \cancel{6}} \\
-\ 1\ 8\ 7\ 9 \\
\hline
4\ 3\ 6\ 7
\end{array}
$$

We cannot subtract 8 hundreds from 1 hundred, but we
can subtract 8 hundreds from 11 hundreds. We borrow
1 thousand to get 11 hundreds. Finally, we subtract the
thousands.

This is
what you
should
write.

The answer
checks because
this is the top
number in the
subtraction.

◀ **Do Exercises 2 and 3.**

EXAMPLE 4 Subtract: 902 − 477.

$$
\begin{array}{r}
\overset{\scriptstyle 8\ 9\ 12}{\cancel{9}\ \cancel{0}\ 2} \\
-\ 4\ 7\ 7 \\
\hline
4\ 2\ 5
\end{array}
$$

We cannot subtract 7 ones from 2 ones. We have
9 hundreds, or 90 tens. We borrow 1 ten to get 12 ones.
We then have 89 tens.

◀ **Do Exercises 4 and 5.**

EXAMPLE 5 Subtract: 8003 − 3667.

$$
\begin{array}{r}
\overset{\scriptstyle 7\ 9\ 9\ 13}{8\ \cancel{0}\ \cancel{0}\ \cancel{3}} \\
-\ 3\ 6\ 6\ 7 \\
\hline
4\ 3\ 3\ 6
\end{array}
$$

We have 8 thousands, or 800 tens.
We borrow 1 ten to get 13 ones. We then have 799 tens.

EXAMPLES

6. Subtract: 6000 − 3762.

$$
\begin{array}{r}
\overset{\scriptstyle 5\ 9\ 9\ 10}{\cancel{6}\ \cancel{0}\ \cancel{0}\ \cancel{0}} \\
-\ 3\ 7\ 6\ 2 \\
\hline
2\ 2\ 3\ 8
\end{array}
$$

7. Subtract: 6024 − 2968.

$$
\begin{array}{r}
\overset{\scriptstyle 11}{} \\
\overset{\scriptstyle 5\ 9\ \cancel{1}\ 14}{\cancel{6}\ \cancel{0}\ \cancel{2}\ \cancel{4}} \\
-\ 2\ 9\ 6\ 8 \\
\hline
3\ 0\ 5\ 6
\end{array}
$$

◀ **Do Exercises 6–9.**

Subtract. Check by adding.

2.
$$
\begin{array}{r}
8\ 6\ 8\ 6 \\
-\ 2\ 3\ 5\ 8 \\
\hline
\end{array}
$$

3.
$$
\begin{array}{r}
7\ 1\ 4\ 5 \\
-\ 2\ 3\ 9\ 8 \\
\hline
\end{array}
$$

Subtract.

4.
$$
\begin{array}{r}
7\ 0 \\
-\ 1\ 4 \\
\hline
\end{array}
$$

5.
$$
\begin{array}{r}
5\ 0\ 3 \\
-\ 2\ 9\ 8 \\
\hline
\end{array}
$$

$$
\begin{array}{r}
\overset{\scriptstyle \square\ \square\ 13}{5\ \cancel{0}\ \cancel{3}} \\
-\ 2\ 9\ 8 \\
\hline
\square\ 0\ \square
\end{array}
$$

Subtract.

6.
$$
\begin{array}{r}
7\ 0\ 0\ 7 \\
-\ 6\ 3\ 4\ 9 \\
\hline
\end{array}
$$

7.
$$
\begin{array}{r}
6\ 0\ 0\ 0 \\
-\ 3\ 1\ 4\ 9 \\
\hline
\end{array}
$$

8.
$$
\begin{array}{r}
9\ 0\ 3\ 5 \\
-\ 7\ 4\ 8\ 9 \\
\hline
\end{array}
$$

9.
$$
\begin{array}{r}
2\ 0\ 0\ 1 \\
-\ \ \ 1\ 2\ 4 \\
\hline
\end{array}
$$

Answers

2. 6328   **3.** 4747   **4.** 56   **5.** 205
6. 658   **7.** 2851   **8.** 1546   **9.** 1877

Guided Solution:

5.
$$
\begin{array}{r}
\overset{\scriptstyle 4\ 9\ 13}{5\ \cancel{0}\ \cancel{3}} \\
-\ 2\ 9\ 8 \\
\hline
2\ 0\ 5
\end{array}
$$

☑ Reading Check

Match each word or phrase from the following list with the indicated part of the subtraction sentence.

difference   minuend   subtraction symbol   subtrahend

RC1. A _____

RC2. B _____

RC3. C _____

RC4. D _____

Ⓐ   Ⓑ   Ⓒ   Ⓓ
↓   ↓   ↓   ↓
97 − 51 = 26

a   Subtract. Check by adding.

1. $\begin{array}{r} 6\ 5 \\ -\ 2\ 1 \\ \hline \end{array}$

2. $\begin{array}{r} 8\ 7 \\ -\ 3\ 4 \\ \hline \end{array}$

3. $\begin{array}{r} 8\ 6\ 6 \\ -\ 3\ 3\ 3 \\ \hline \end{array}$

4. $\begin{array}{r} 5\ 2\ 6 \\ -\ 3\ 2\ 3 \\ \hline \end{array}$

5. $86 - 47$

6. $73 - 28$

7. $51 - 37$

8. $64 - 19$

9. $\begin{array}{r} 5\ 6\ 3 \\ -\ 1\ 9\ 4 \\ \hline \end{array}$

10. $\begin{array}{r} 7\ 9\ 5 \\ -\ 3\ 9\ 8 \\ \hline \end{array}$

11. $\begin{array}{r} 3\ 9\ 1 \\ -\ 3\ 6\ 5 \\ \hline \end{array}$

12. $\begin{array}{r} 3\ 1\ 6 \\ -\ 2\ 4\ 7 \\ \hline \end{array}$

13. $981 - 747$

14. $887 - 698$

15. $683 - 266$

16. $342 - 217$

17. $\begin{array}{r} 7\ 7\ 6\ 9 \\ -\ 2\ 3\ 8\ 7 \\ \hline \end{array}$

18. $\begin{array}{r} 6\ 4\ 3\ 1 \\ -\ 2\ 8\ 9\ 6 \\ \hline \end{array}$

19. $\begin{array}{r} 4\ 5\ 1\ 2 \\ -\ 1\ 7\ 3\ 4 \\ \hline \end{array}$

20. $\begin{array}{r} 8\ 3\ 6\ 4 \\ -\ 5\ 3\ 7\ 5 \\ \hline \end{array}$

21. $5318 - 2249$

22. $9241 - 5643$

23. $3947 - 2858$

24. $7583 - 3641$

25.
$$\begin{array}{r} 12{,}647 \\ -4{,}899 \\ \hline \end{array}$$

26.
$$\begin{array}{r} 16{,}222 \\ -5{,}888 \\ \hline \end{array}$$

27.
$$\begin{array}{r} 51{,}342 \\ -47{,}198 \\ \hline \end{array}$$

28.
$$\begin{array}{r} 32{,}194 \\ -29{,}236 \\ \hline \end{array}$$

29.
$$\begin{array}{r} 80 \\ -24 \\ \hline \end{array}$$

30.
$$\begin{array}{r} 90 \\ -78 \\ \hline \end{array}$$

31.
$$\begin{array}{r} 690 \\ -236 \\ \hline \end{array}$$

32.
$$\begin{array}{r} 803 \\ -418 \\ \hline \end{array}$$

33.
$$\begin{array}{r} 6808 \\ -3059 \\ \hline \end{array}$$

34.
$$\begin{array}{r} 9405 \\ -258 \\ \hline \end{array}$$

35.
$$\begin{array}{r} 2300 \\ -109 \\ \hline \end{array}$$

36.
$$\begin{array}{r} 7500 \\ -3604 \\ \hline \end{array}$$

37. $90{,}237 - 47{,}209$

38. $84{,}703 - 298$

39. $101{,}734 - 5760$

40. $15{,}017 - 7809$

41.
$$\begin{array}{r} 6007 \\ -1589 \\ \hline \end{array}$$

42.
$$\begin{array}{r} 8003 \\ -599 \\ \hline \end{array}$$

43.
$$\begin{array}{r} 39{,}000 \\ -37{,}695 \\ \hline \end{array}$$

44.
$$\begin{array}{r} 17{,}000 \\ -11{,}598 \\ \hline \end{array}$$

45. $10{,}008 - 19$

46. $40{,}006 - 147$

47. $50{,}001 - 1984$

48. $30{,}004 - 6749$

Skill Maintenance

Add. [1.2a]

49. $567 + 778$

50. $901 + 23$

51. $12{,}885 + 9807$

52. $9909 + 1011$

53. Write a word name for 6,375,602. [1.1c]

54. Write expanded notation for 9103. [1.1b]

Synthesis

55. Fill in the missing digits to make the subtraction true:

$$9{,}\square 48{,}621 - 2{,}097{,}\square 81 = 7{,}251{,}140.$$

56. ▦ Subtract: $3{,}928{,}124 - 1{,}098{,}947.$

Multiplication

a MULTIPLICATION OF WHOLE NUMBERS

Repeated Addition

The multiplication 3 × 5 corresponds to this repeated addition.

We combine 3 sets of 5 $1 bills each.

The resulting set is a set of 15 $1 bills.

$$3 \times 5 = 5 + 5 + 5 = 15$$

3 addends; each is 5

The numbers that we multiply are called **factors**. The result of the multiplication is called a **product**.

$$3 \quad \times \quad 5 \quad = \quad 15$$

Factor   Factor   Product

Rectangular Arrays

Multiplications can also be thought of as rectangular arrays. Each of the following corresponds to the multiplication 3 × 5.

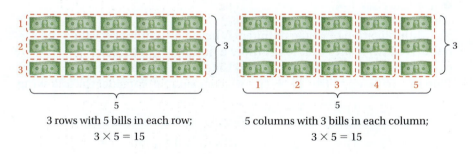

3 rows with 5 bills in each row;
3 × 5 = 15

5 columns with 3 bills in each column;
3 × 5 = 15

When you write a multiplication corresponding to a real-world situation, you should think of either a rectangular array or repeated addition. In some cases, it may help to think both ways.

We have used an "×" to denote multiplication. A dot " · " is also commonly used. (Use of the dot is attributed to the German mathematician Gottfried Wilhelm von Leibniz over three centuries ago.) Parentheses are also used to denote multiplication. For example,

$$3 \times 5 = 3 \cdot 5 = (3)(5) = 3(5) = 15.$$

Answers

Skill to Review:
1. 903   2. 6454

The product of 0 and any whole number is 0: $0 \cdot a = a \cdot 0 = 0$. For example, $0 \cdot 3 = 3 \cdot 0 = 0$. Multiplying a number by 1 does not change the number: $1 \cdot a = a \cdot 1 = a$. For example, $1 \cdot 3 = 3 \cdot 1 = 3$. We say that 1 is the **multiplicative identity**.

EXAMPLE 1 Multiply: 5×734.

We have

$$
\begin{array}{r}
7\ 3\ 4 \\
\times \qquad 5 \\
\hline
2\ 0 \\
1\ 5\ 0 \\
3\ 5\ 0\ 0 \\
\hline
3\ 6\ 7\ 0 \\
\end{array}
$$

← Multiply the 4 ones by 5: $5 \times 4 = 20$.
← Multiply the 3 tens by 5: $5 \times 30 = 150$.
← Multiply the 7 hundreds by 5: $5 \times 700 = 3500$.
← Add.

Instead of writing each product on a separate line, we can use a shorter form.

$$
\begin{array}{r}
\overset{2}{}\ \\
7\ \overset{}{3}\ 4 \\
\times \qquad 5 \\
\hline
0 \\
\end{array}
$$

Multiply the 4 ones by 5: $5 \cdot (4 \text{ ones}) = 20 \text{ ones}$ $= 2 \text{ tens} + 0 \text{ ones}$. Write 0 in the ones column and 2 above the tens.

$$
\begin{array}{r}
\overset{1}{7}\ \overset{2}{3}\ 4 \\
\times \qquad 5 \\
\hline
7\ 0 \\
\end{array}
$$

Multiply the 3 tens by 5 and add 2 tens: $5 \cdot (3 \text{ tens}) = 15 \text{ tens}$, $15 \text{ tens} + 2 \text{ tens} = 17 \text{ tens}$ $= 1 \text{ hundred} + 7 \text{ tens}$. Write 7 in the tens column and 1 above the hundreds.

$$
\begin{array}{r}
\overset{1}{7}\ \overset{2}{3}\ 4 \\
\times \qquad 5 \\
\hline
3\ 6\ 7\ 0 \\
\end{array}
$$

Multiply the 7 hundreds by 5 and add 1 hundred: $5 \cdot (7 \text{ hundreds}) = 35 \text{ hundreds}$, $35 \text{ hundreds} + 1 \text{ hundred} = 36 \text{ hundreds}$.

$$
\left.
\begin{array}{r}
\overset{1}{7}\ \overset{2}{3}\ 4 \\
\times \qquad 5 \\
\hline
3\ 6\ 7\ 0 \\
\end{array}
\right\}
$$
You should write only this.

Multiply.

1. $\begin{array}{r} 5\ 8 \\ \times\quad 2 \\ \hline \end{array}$

2. $\begin{array}{r} 3\ 7 \\ \times\quad 4 \\ \hline \end{array}$

3. $\begin{array}{r} 8\ 2\ 3 \\ \times\quad\ 6 \\ \hline \end{array}$

4. $\begin{array}{r} 1\ 3\ 4\ 8 \\ \times\qquad 5 \\ \hline \end{array}$ **GS**

$$
\begin{array}{r}
\square\ \overset{2}{}\ \square \\
1\ 3\ 4\ 8 \\
\times \qquad 5 \\
\hline
\square\quad 7\ \square\quad 0 \\
\end{array}
$$

◀ **Do Exercises 1–4.**

Multiplication of whole numbers is based on a property called the **distributive law**. It says that to multiply a number by a sum, $a \cdot (b + c)$, we can multiply each addend by a and then add like this: $(a \cdot b) + (a \cdot c)$. Thus, $a \cdot (b + c) = (a \cdot b) + (a \cdot c)$. For example, consider the following.

$$4 \cdot (2 + 3) = 4 \cdot 5 = 20$$
Adding first; then multiplying

$$4 \cdot (2 + 3) = (4 \cdot 2) + (4 \cdot 3) = 8 + 12 = 20$$
Multiplying first; then adding

The results are the same, so $4 \cdot (2 + 3) = (4 \cdot 2) + (4 \cdot 3)$.

Let's find the product 51×32. Since $32 = 2 + 30$, we can think of this product as

$$51 \times 32 = 51 \times (2 + 30) = (51 \times 2) + (51 \times 30).$$

That is, we multiply 51 by 2, then we multiply 51 by 30, and finally we add. We can write our work in columns.

Answers

1. 116 2. 148 3. 4938 4. 6740

Guided Solution:

4. $\begin{array}{r} \overset{1\ 2\ 4}{} \\ 1\ 3\ 4\ 8 \\ \times \qquad 5 \\ \hline 6\ 7\ 4\ 0 \end{array}$

```
    5 1
  × 3 2
  1 0 2      Multiplying by 2
  1 5 3 0    Multiplying by 30. (We write a 0 and then multiply 51 by 3.)
```

> You may have learned that such a 0 need not be written. You may omit it if you wish. If you do omit it, remember, when multiplying by tens, to start writing the answer in the tens place.

We add to obtain the product.

```
    5 1
  × 3 2
  1 0 2
  1 5 3 0
  1 6 3 2    Adding to obtain the product
```

EXAMPLE 2 Multiply: 457×683.

```
      5 2
    6 8 3
  × 4 5 7
  4 7 8 1        Multiplying 683 by 7
```

```
    4 1
    5 2
    6 8 3
  × 4 5 7
    4 7 8 1
  3 4 1 5 0      Multiplying 683 by 50
```

```
    3 1
    4 1
    5 2
      6 8 3
    × 4 5 7
      4 7 8 1
    3 4 1 5 0
  2 7 3 2 0 0    Multiplying 683 by 400. (We write 00 and then multiply 683 by 4.)
  3 1 2 ,1 3 1   Adding
```

Do Exercises 5–8.

EXAMPLE 3 Multiply: 306×274.

Note that $306 = 3$ hundreds $+ 6$ ones.

```
    2 7 4
  × 3 0 6
    1 6 4 4      Multiplying by 6
  8 2 2 0 0      Multiplying by 3 hundreds. (We write 00 and then multiply 274 by 3.)
  8 3 ,8 4 4     Adding
```

Do Exercises 9–12.

Multiply.

5.
```
  4 5
× 2 3
```
6. 48×63

7.
```
  7 4 6
×   6 2
```
8. 245×837

Multiply.

9.
```
  4 7 2
× 3 0 6
```
10. 408×704

11.
```
  2 3 4 4
× 6 0 0 5
```
12.
```
  1 0 0 6
×   7 0 3
```

Answers

5. 1035   6. 3024   7. 46,252   8. 205,065
9. 144,432   10. 287,232   11. 14,075,720
12. 707,218

Multiply.

13. 472
× 830

14. 2 3 4 4
× 7 4 0 0

15. 100×562

16. 1000×562

17. a) Find $23 \cdot 47$.

b) Find $47 \cdot 23$.

c) Compare your answers to parts (a) and (b).

Multiply.

18. $5 \cdot 2 \cdot 4$

19. $4 \cdot 2 \cdot 6$

EXAMPLE 4 Multiply: 360×274.

Note that $360 = 3$ hundreds $+ 6$ tens.

```
        2 7 4
     ×  3 6 0
      1 6 4 4 0
      8 2 2 0 0
      9 8,6 4 0
```

Multiplying by 6 tens. (We write 0 and then multiply 274 by 6.)

Multiplying by 3 hundreds. (We write 00 and then multiply 274 by 3.)

Adding

◀ **Do Exercises 13–16.**

When we multiply two numbers, we can change the order of the numbers without changing their product. For example, $3 \cdot 6 = 18$ and $6 \cdot 3 = 18$. This illustrates the **commutative law of multiplication:** $a \cdot b = b \cdot a$.

◀ **Do Exercise 17.**

To multiply three or more numbers, we group them so that we multiply two at a time. Consider $2 \cdot 3 \cdot 4$. We can group these numbers as $2 \cdot (3 \cdot 4)$ or as $(2 \cdot 3) \cdot 4$. The parentheses tell what to do first:

$2 \cdot (3 \cdot 4) = 2 \cdot (12) = 24.$ We multiply 3 and 4 and then that product by 2.

We can also multiply 2 and 3 and then that product by 4:

$(2 \cdot 3) \cdot 4 = (6) \cdot 4 = 24.$

Either way we get 24. It does not matter how we group the numbers. This illustrates the **associative law of multiplication:** $a \cdot (b \cdot c) = (a \cdot b) \cdot c$.

◀ **Do Exercises 18 and 19.**

b FINDING AREA

We can think of the area of a rectangular region as the number of square units needed to fill it. Here is a rectangle 4 cm (centimeters) long and 3 cm wide. It takes 12 square centimeters (sq cm) to fill it.

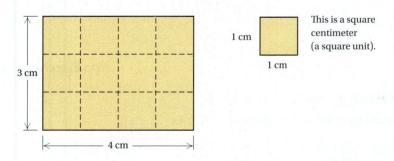

3 cm

4 cm

1 cm

1 cm

This is a square centimeter (a square unit).

In this case, we have a rectangular array of 3 rows, each of which contains 4 squares. The number of square units is given by $3 \cdot 4$, or 12. That is, $A = l \cdot w = 3\,\text{cm} \cdot 4\,\text{cm} = 12\,\text{sq cm}.$

Answers

13. 391,760   **14.** 17,345,600
15. 56,200   **16.** 562,000
17. (a) 1081; **(b)** 1081; **(c)** same
18. 40   **19.** 48

EXAMPLE 5 *Professional Pool Table.* The playing area of a pool table used in professional tournaments is 50 in. by 100 in. (There are 6-in. wide rails on the outside that are not included in the playing area.) Determine the playing area.

100 in.

50 in.

If we think of filling the rectangle with square inches, we have a rectangular array. The length $l = 100$ in. and the width $w = 50$ in. Thus the area A is given by the formula

$$A = l \cdot w = 100 \text{ in.} \cdot 50 \text{ in.} = 5000 \text{ sq in.}$$

Do Exercise 20. ▶

GS **20. Painting a Room.** Ben and Elizabeth plan to paint one wall of a bedroom in a dark accent color. The wall is a rectangle 12 ft long and 8 ft high. Determine the area of the wall.

$$A = l \cdot w$$
$$= 12 \text{ ft} \cdot \boxed{}$$
$$= \boxed{} \text{ sq ft}$$

Answer
20. 96 sq ft

Guided Solution:
20. 8 ft, 96

1.4 **Exercise Set**

☑ Reading Check

Complete each statement with the appropriate word or number from the following list. Not every choice will be used.

0	addends	product
1	factors	sum

RC1. In the multiplication $4 \times 3 = 12$, 4 and 3 are _____.

RC2. In the multiplication $4 \times 3 = 12$, 12 is the _____.

RC3. The product of _____ and any number a is 0.

RC4. The product of _____ and any number a is a.

1. $\begin{array}{r} 6\,5 \\ \times \quad 8 \\ \hline \end{array}$

2. $\begin{array}{r} 8\,7 \\ \times \quad 4 \\ \hline \end{array}$

3. $\begin{array}{r} 9\,4 \\ \times \quad 6 \\ \hline \end{array}$

4. $\begin{array}{r} 7\,6 \\ \times \quad 9 \\ \hline \end{array}$

5. $3 \cdot 509$

6. $7 \cdot 806$

7. $7(9229)$

8. $4(7867)$

9. $90(53)$

10. $60(78)$

11. $(47)(85)$

12. $(34)(87)$

13. $\begin{array}{r} 8\,7 \\ \times \;1\,0 \\ \hline \end{array}$

14. $\begin{array}{r} 2\,3\,4\,0 \\ \times \;1\,0\,0\,0 \\ \hline \end{array}$

15. $\begin{array}{r} 9\,6 \\ \times \;2\,0 \\ \hline \end{array}$

16. $\begin{array}{r} 8\,0\,0 \\ \times \;7\,0\,0 \\ \hline \end{array}$

17. $\begin{array}{r} 6\,4\,3 \\ \times \quad 7\,2 \\ \hline \end{array}$

18. $\begin{array}{r} 7\,7\,7 \\ \times \quad 7\,7 \\ \hline \end{array}$

19. $\begin{array}{r} 4\,4\,4 \\ \times \quad 3\,3 \\ \hline \end{array}$

20. $\begin{array}{r} 5\,4\,9 \\ \times \quad 8\,8 \\ \hline \end{array}$

21. $\begin{array}{r} 5\,6\,4 \\ \times \;4\,5\,8 \\ \hline \end{array}$

22. $\begin{array}{r} 4\,3\,2 \\ \times \;3\,7\,5 \\ \hline \end{array}$

23. $\begin{array}{r} 8\,5\,3 \\ \times \;9\,3\,6 \\ \hline \end{array}$

24. $\begin{array}{r} 3\,4\,6 \\ \times \;6\,5\,9 \\ \hline \end{array}$

25. $\begin{array}{r} 6\,4\,2\,8 \\ \times \;3\,2\,2\,4 \\ \hline \end{array}$

26. $\begin{array}{r} 8\,9\,2\,8 \\ \times \;3\,1\,7\,2 \\ \hline \end{array}$

27. $\begin{array}{r} 3\,4\,8\,2 \\ \times \quad 1\,0\,4 \\ \hline \end{array}$

28. $\begin{array}{r} 6\,4\,0\,8 \\ \times \;6\,0\,6\,4 \\ \hline \end{array}$

29. $\begin{array}{r} 8\,7\,6 \\ \times \;3\,4\,5 \\ \hline \end{array}$

30. $\begin{array}{r} 3\,5\,5 \\ \times \;2\,9\,9 \\ \hline \end{array}$

31. $\begin{array}{r} 7\,8\,8\,9 \\ \times \;6\,2\,2\,4 \\ \hline \end{array}$

32. $\begin{array}{r} 6\,5\,2\,1 \\ \times \;3\,4\,4\,9 \\ \hline \end{array}$

33.	5 6 0 8	34.	4 5 0 6	35.	5 0 0 6	36.	6 0 0 9
	× 4 5 0 0		× 7 8 0 0		× 4 0 0 8		× 2 0 0 3

b Find the area of each region.

37.

728 mi

728 mi

38.

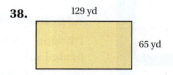

129 yd

65 yd

39. Find the area of the region formed by the base lines on a Major League Baseball diamond.

90 ft

90 ft

40. Find the area of a standard-sized hockey rink.

85 ft

200 ft

Skill Maintenance

Add.   [1.2a]

41.	4 9 0 8	**42.**	9 8 7 6
	5 6 6 7		8 7 6
	+ 2 1 1 0		7 6
			+ 6

Subtract.   [1.3a]

43.	9 8 7 6	**44.**	3 4 0 , 7 9 8
	− 9 8 7		− 8 6 , 6 7 9

45. What does the digit 4 mean in 9,482,157?   [1.1a]

46. What digit in 38,026 names the number of hundreds?   [1.1a]

47. Write expanded notation for 12,847.   [1.1b]

48. Write a word name for 7,432,000.   [1.1c]

Synthesis

49. ▦ An 18-story office building is box-shaped. Each floor measures 172 ft by 84 ft with a 20-ft by 35-ft rectangular area lost to an elevator and a stairwell. How much area is available as office space?

a DIVISION OF WHOLE NUMBERS

Repeated Subtraction

Division of whole numbers applies to two kinds of situations. The first is repeated subtraction. Suppose we have 20 doughnuts, and we want to find out how many sets of 5 there are. One way to do this is to repeatedly subtract sets of 5 as follows.

20 doughnuts

How many sets of 5 doughnuts each?

Since there are 4 sets of 5 doughnuts each, we have

$$20 \div 5 = 4.$$

Dividend   Divisor   Quotient

The division $20 \div 5$ is read "20 divided by 5." The **dividend** is 20, the **divisor** is 5, and the **quotient** is 4. We divide the *dividend* by the *divisor* to get the *quotient*. We can also express the division $20 \div 5 = 4$ as

$$\frac{20}{5} = 4 \quad \text{or} \quad 5\overline{)20}.$$

Rectangular Arrays

We can also think of division in terms of rectangular arrays. Consider again the 20 doughnuts and division by 5. We can arrange the doughnuts in a rectangular array with 5 rows and ask, "How many are in each row?"

We can also consider a rectangular array with 5 doughnuts in each column and ask, "How many columns are there?" The answer is still 4.

In each case, we are asking, "What do we multiply 5 by in order to get 20?"

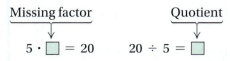

Missing factor   Quotient

$$5 \cdot \square = 20 \qquad 20 \div 5 = \square$$

This leads us to the following definition of division.

DIVISION

The quotient $a \div b$, where b is not 0, is that unique number c for which $a = b \cdot c$.

This definition shows the relation between division and multiplication. We see, for instance, that

$$20 \div 5 = 4 \text{ because } 20 = 5 \cdot 4.$$

This relation allows us to use multiplication to check division.

EXAMPLE 1   Divide. Check by multiplying.

a) $16 \div 8$ **b)** $\dfrac{36}{4}$ **c)** $7\overline{)56}$

We do so as follows.

a) $16 \div 8 = 2$ *Check*: $8 \cdot 2 = 16.$

b) $\dfrac{36}{4} = 9$ *Check*: $4 \cdot 9 = 36.$

c) $\overset{8}{7\overline{)56}}$ *Check*: $7 \cdot 8 = 56.$

Do Exercises 1–3. ▶

Divide. Check by multiplying.
1. $9\overline{)45}$

2. $27 \div 3$

3. $\dfrac{48}{6}$

Let's consider some basic properties of division.

DIVIDING BY 1

Any number divided by 1 is that same number: $a \div 1 = \dfrac{a}{1} = a.$

For example, $6 \div 1 = 6$ and $\dfrac{15}{1} = 15.$

DIVIDING A NUMBER BY ITSELF

Any nonzero number divided by itself is 1: $a \div a = \dfrac{a}{a} = 1, \quad a \neq 0.$

For example, $7 \div 7 = 1$ and $\dfrac{22}{22} = 1.$

Divide.
4. $\dfrac{9}{9}$ 5. $\dfrac{8}{1}$

6. $\dfrac{0}{12}$ 7. $\dfrac{28}{28}$

DIVIDENDS OF 0

Zero divided by any nonzero number is 0: $0 \div a = \dfrac{0}{a} = 0, \quad a \neq 0.$

For example, $0 \div 14 = 0$ and $\dfrac{0}{3} = 0.$

Do Exercises 4–7. ▶

Answers

1. 5   2. 9   3. 8   4. 1   5. 8   6. 0   7. 1

Division by 0 is not defined: $a \div 0$, or $\dfrac{a}{0}$, is **not defined**.

For example, $16 \div 0$, or $\dfrac{16}{0}$, is not defined.

---

8. Divide if possible: $0 \div 2$. If not possible, write "not defined."

$0 \div 2$ means ☐ divided by ☐.

Since zero divided by any nonzero number is 0, $0 \div 2 =$ ☐.

9. Divide if possible: $7 \div 0$. If not possible, write "not defined."

$7 \div 0$ means ☐ divided by ☐.

Since division by 0 is not defined, $7 \div 0$ is ☐.

Why can't we divide by 0? Suppose the number 4 could be divided by 0. Then if ☐ were the answer, we would have

$$4 \div 0 = \square,$$

and since 0 times any number is 0, we would have

$$4 = \square \cdot 0 = 0. \qquad \text{False!}$$

Thus, the only possible number that could be divided by 0 would be 0 itself. But such a division would give us any number we wish. For instance,

$$0 \div 0 = 8 \quad \text{because} \quad 0 = 8 \cdot 0;$$
$$0 \div 0 = 3 \quad \text{because} \quad 0 = 3 \cdot 0;$$
$$0 \div 0 = 7 \quad \text{because} \quad 0 = 7 \cdot 0.$$

All true!

We avoid the preceding difficulties by agreeing to exclude division by 0.

◀ **Do Exercises 8–9.**

Division with a Remainder

Suppose everyone in a group of 22 people wants to ride a roller coaster. If each car on the ride holds 6 people, the group will fill 3 cars and there will be 4 people left over.

We can think of this situation as the following division. The people left over are the **remainder**.

$$
\begin{array}{r}
3 \leftarrow \text{Quotient} \\
6\overline{)2\,2} \\
\underline{1\,8} \\
4 \leftarrow \text{Remainder}
\end{array}
$$

Answers
8. 0 **9.** Not defined
Guided Solutions:
8. 0, 2, 0
9. 7, 0, not defined

We express the result as

$$22 \div 6 = 3 \text{ R } 4.$$

Dividend　Divisor　Quotient　Remainder

Note that

Quotient $\cdot$ Divisor $+$ Remainder $=$ Dividend.

Thus we have

$3 \cdot 6 = 18$　　　Quotient $\cdot$ Divisor

and　$18 + 4 = 22.$　　Adding the remainder. The result is the dividend.

We now show a procedure for dividing whole numbers.

EXAMPLE 2　Divide and check: $4 \overline{)3457}$.

First, we try to divide the first digit of the dividend, 3, by the divisor, 4. Since $3 \div 4$ is not a whole number, we consider the first *two* digits of the dividend.

$$\begin{array}{r} 8 \\ 4\overline{)3\ 4\ 5\ 7} \\ 3\ 2 \\ \hline 2 \end{array}$$

Since $4 \cdot 8 = 32$ and 32 is smaller than 34, we write an 8 in the quotient above the 4. We also write 32 below 34 and subtract.

What if we had chosen a number other than 8 for the first digit of the quotient? Suppose we had used 7 instead of 8 and subtracted $4 \cdot 7$, or 28, from 34. The result would have been $34 - 28$, or 6. Because 6 is larger than the divisor, 4, we know that there is at least one more factor of 4 in 34, and thus 7 is too small. If we had used 9 instead of 8, then we would have tried to subtract $4 \cdot 9$, or 36, from 34. That difference is not a whole number, so we know 9 is too large. When we subtract, the difference must be smaller than the divisor.

Let's continue dividing.

$$\begin{array}{r} 8\ 6 \\ 4\overline{)3\ 4\ 5\ 7} \\ 3\ 2\ \downarrow \\ \hline 2\ 5 \\ 2\ 4 \\ \hline 1 \end{array}$$

Now we bring down the 5 in the dividend and consider $25 \div 4$. Since $4 \cdot 6 = 24$ and 24 is smaller than 25, we write 6 in the quotient above the 5. We also write 24 below 25 and subtract. The difference, 1, is smaller than the divisor, so we know that 6 is the correct choice.

$$\begin{array}{r} 8\ 6\ 4 \\ 4\overline{)3\ 4\ 5\ 7} \\ 3\ 2 \\ \hline 2\ 5 \\ 2\ 4 \\ \hline 1\ 7 \\ 1\ 6 \\ \hline 1 \end{array}$$ $\leftarrow$ The remainder is 1.

We bring down the 7 and consider $17 \div 4$. Since $4 \cdot 4 = 16$ and 16 is smaller than 17, we write 4 in the quotient above the 7. We also write 16 below 17 and subtract.

Check:　$864 \cdot 4 = 3456$ and $3456 + 1 = 3457.$

The answer is 864 R 1.

Do Exercises 10–12. ▶

Divide and check.

10. $3 \overline{)239}$

11. $5 \overline{)5864}$

12. $6 \overline{)3855}$

Answers

10. 79 R 2　　**11.** 1172 R 4

12. 642 R 3

EXAMPLE 3 Divide: 8904 ÷ 42.

Because 42 is close to 40, we think of the divisor as 40 when we make our choices of digits in the quotient.

```
        2
  4 2 ) 8 9 0 4    ← Think: 89 ÷ 40. We try 2. Multiply 42 · 2 and
        8 4 ↓         subtract. Then bring down the 0.
        ─────
          5 0
```

```
        2 1
  4 2 ) 8 9 0 4
        8 4
        ─────
          5 0        ← Think: 50 ÷ 40. We try 1. Multiply 42 · 1 and
          4 2 ↓         subtract. Then bring down the 4.
          ─────
            8 4
```

```
        2 1 2
  4 2 ) 8 9 0 4
        8 4
        ─────
          5 0
          4 2
          ─────
            8 4      ← Think: 84 ÷ 40. We try 2. Multiply 2 · 42 and subtract.
            8 4
            ───
              0
```

The remainder is 0, so the answer is 212.

◀ **Do Exercises 13 and 14.**

Divide.

13. $4\,5\,)\overline{6\,0\,3\,0}$

14. $5\,2\,)\overline{3\,2\,8\,8}$

.... **Caution!** ....

Be careful to keep the digits lined up correctly when you divide.

....

CALCULATOR CORNER

Dividing Whole Numbers   To divide whole numbers on a calculator, we use the ÷ and = keys.

When we enter 453 ÷ 15, the display reads [30.2]. Note that the result is not a whole number. This tells us that there is a remainder. The number 30.2 is expressed in decimal notation. The symbol "." is called a decimal point. Although it is possible to use the number to the right of the decimal point to find the remainder, we will not do so here.

EXERCISES   Use a calculator to perform each division.

1. $19)\overline{532}$

2. $7)\overline{861}$

3. 9367 ÷ 29

4. 12,276 ÷ 341

Answers

13. 134   **14.** 63 R 12

Zeros in Quotients

EXAMPLE 4 Divide: 6341 ÷ 7.

```
      9
   7 )6 3 4 1  ← Think: 63 ÷ 7 = 9. The first digit in the quotient is
     6 3 ↓           9. We do not write the 0 when we find 63 − 63. Bring
         4           down the 4.
```

```
      9 0
   7 )6 3 4 1
     6 3 ↓
         4 1  ← Think: 4 ÷ 7. If we subtract a group of 7's, such
                   as 7, 14, 21, etc., from 4, we do not get a whole
                   number, so the next digit in the quotient is 0.
                   Bring down the 1.
```

```
      9 0 5
   7 )6 3 4 1
     6 3
         4 1  ← Think: 41 ÷ 7. We try 5. Multiply 7 · 5
         3 5       and subtract.
           6  ← The remainder is 6.
```

The answer is 905 R 6.

Do Exercises 15 and 16. ▶

EXAMPLE 5 Divide: 8169 ÷ 34.

Because 34 is close to 30, we think of the divisor as 30 when we make our choices of digits in the quotient.

```
        2
  3 4 )8 1 6 9  ← Think: 81 ÷ 30. We try 2. Multiply 34 · 2 and
      6 8 ↓           subtract. Then bring down the 6.
      1 3 6
```

```
        2 4
  3 4 )8 1 6 9
      6 8
      1 3 6  ← Think: 136 ÷ 30. We try 4. Multiply 34 · 4 and
      1 3 6 ↓     subtract. The difference is 0, so we do not write it.
            9       Bring down the 9.
```

```
        2 4 0
  3 4 )8 1 6 9
      6 8
      1 3 6
      1 3 6 ↓     ┌ Think: 9 ÷ 34. If we subtract a group of 34's,
            9  ←──┤   such as 34 or 68, from 9, we do not get a whole
            0     └   number, so the last digit in the quotient is 0.
            9  ← The remainder is 9.
```

The answer is 240 R 9.

Do Exercises 17 and 18. ▶

TO THE INSTRUCTOR AND THE STUDENT

This section presents a review of division of whole numbers. Students who are successful should go on to Section 1.6. Those who have trouble should study developmental unit D near the back of this text and then repeat Section 1.5.

Divide.

15. 6)4 8 4 6

16. 7)7 6 1 6

Divide.

17. 2 7)9 7 2 4

18. 5 6)4 4, 8 4 7

Answers

15. 807 R 4   **16.** 1088

17. 360 R 4   **18.** 800 R 47

✓ Reading Check

Match each word from the following list with the indicated part of the division.

dividend   divisor   quotient   remainder

RC1. A _____

RC2. B _____

RC3. C _____

RC4. D _____

$$\begin{array}{r} 2\ 9 \leftarrow \text{\textcircled{A}} \\ \text{\textcircled{D}} \rightarrow 8\overline{)2\ 3\ 5} \leftarrow \text{\textcircled{B}} \\ \underline{1\ 6} \\ 7\ 5 \\ \underline{7\ 2} \\ 3 \leftarrow \text{\textcircled{C}} \end{array}$$

a   Divide, if possible. If not possible, write "not defined."

1. $72 \div 6$

2. $54 \div 9$

3. $\dfrac{23}{23}$

4. $\dfrac{37}{37}$

5. $22 \div 1$

6. $\dfrac{56}{1}$

7. $\dfrac{0}{7}$

8. $\dfrac{0}{32}$

9. $\dfrac{16}{0}$

10. $74 \div 0$

11. $\dfrac{48}{8}$

12. $\dfrac{20}{4}$

Divide.

13. $277 \div 5$

14. $699 \div 3$

15. $864 \div 8$

16. $869 \div 8$

17. $4\overline{)1\ 2\ 2\ 8}$

18. $3\overline{)2\ 1\ 2\ 4}$

19. $6\overline{)4\ 5\ 2\ 1}$

20. $9\overline{)9\ 1\ 1\ 0}$

21. $297 \div 4$

22. $389 \div 2$

23. $738 \div 8$

24. $881 \div 6$

25. $5 \overline{)8515}$ **26.** $3 \overline{)6027}$ **27.** $9 \overline{)8888}$ **28.** $8 \overline{)4139}$

29. $127{,}000 \div 10$ **30.** $127{,}000 \div 100$ **31.** $127{,}000 \div 1000$ **32.** $4260 \div 10$

33. $70 \overline{)3692}$ **34.** $20 \overline{)5798}$ **35.** $30 \overline{)875}$ **36.** $40 \overline{)987}$

37. $852 \div 21$ **38.** $942 \div 23$ **39.** $85 \overline{)7672}$ **40.** $54 \overline{)2729}$

41. $111 \overline{)3219}$ **42.** $102 \overline{)5612}$ **43.** $8 \overline{)843}$ **44.** $7 \overline{)749}$

45. $5 \overline{)8047}$ **46.** $9 \overline{)7273}$ **47.** $5 \overline{)5036}$ **48.** $7 \overline{)7074}$

49. $1058 \div 46$ **50.** $7242 \div 24$ **51.** $3425 \div 32$ **52.** $48 \overline{)4899}$

53. $24\overline{)8880}$ **54.** $36\overline{)7563}$ **55.** $28\overline{)17,067}$ **56.** $36\overline{)28,929}$

57. $80\overline{)24,320}$ **58.** $90\overline{)88,560}$ **59.** $285\overline{)999,999}$

60. $306\overline{)888,888}$ **61.** $456\overline{)3,679,920}$ **62.** $803\overline{)5,622,606}$

Skill Maintenance

Subtract. [1.3a]

63. $\begin{array}{r} 4\,9\,0\,8 \\ -\ 3\,6\,6\,7 \end{array}$ **64.** $\begin{array}{r} 8\,8,7\,7\,7 \\ -\ 2\,2,3\,3\,3 \end{array}$

Multiply. [1.4a]

65. $\begin{array}{r} 1\,9\,8 \\ \times\ 1\,0\,0 \end{array}$ **66.** $\begin{array}{r} 2\,6\,8 \\ \times\ 3\,5 \end{array}$

Use the following figure for Exercises 67 and 68.

11 ft

9 ft

67. Find the perimeter of the figure. [1.2b]

68. Find the area of the figure. [1.4b]

Synthesis

69. Complete the following table.

a	b	$a \cdot b$	$a + b$
	68	3672	
84			117
		32	12

70. Find a pair of factors whose product is 36 and:
   a) whose sum is 13.
   b) whose difference is 0.
   c) whose sum is 20.
   d) whose difference is 9.

71. A group of 1231 college students is going to use buses to take a field trip. Each bus can hold 42 students. How many buses are needed?

72. ▦ Fill in the missing digits to make the equation true:

$$34,584,132 \div 76\square = 4\square,386.$$

Mid-Chapter Review

Concept Reinforcement

Determine whether each statement is true or false.

_____ **1.** If $a - b = c$, then $b = a + c$. [1.3a]

_____ **2.** We can think of the multiplication 4×3 as a rectangular array containing 4 rows with 3 items in each row. [1.4a]

_____ **3.** We can think of the multiplication 4×3 as a rectangular array containing 3 columns with 4 items in each column. [1.4a]

_____ **4.** The product of two whole numbers is always greater than either of the factors. [1.4a]

_____ **5.** Zero divided by any nonzero number is 0. [1.5a]

_____ **6.** Any number divided by 1 is the number 1. [1.5a]

Guided Solutions

GS Fill in each blank with the number that creates a correct statement or solution.

7. Write a word name for 95,406,237. [1.1c]

8. Subtract: $604 - 497$. [1.3a]

Mixed Review

In each of the following numbers, what does the digit 6 mean? [1.1a]

9. 2698   **10.** 61,204   **11.** 146,237   **12.** 586

Consider the number 306,458,129. What digit names the number of: [1.1a]

13. tens?   **14.** millions?   **15.** ten thousands?   **16.** hundreds?

Write expanded notation. [1.1b]

17. 5602   **18.** 69,345

Write a word name. [1.1c]

19. 136   **20.** 64,325

Write standard notation. [1.1c]

21. Three hundred eight thousand, seven hundred sixteen

22. Four million, five hundred sixty-seven thousand, two hundred sixteen

Add. [1.2a]

23. $\begin{array}{r} 316 \\ + 482 \\ \hline \end{array}$

24. $\begin{array}{r} 593 \\ + 437 \\ \hline \end{array}$

25. $\begin{array}{r} 2638 \\ + 5284 \\ \hline \end{array}$

26. $\begin{array}{r} 4617 \\ 2436 \\ + 481 \\ \hline \end{array}$

Subtract. [1.3a]

27. $\begin{array}{r} 786 \\ - 321 \\ \hline \end{array}$

28. $\begin{array}{r} 624 \\ - 285 \\ \hline \end{array}$

29. $\begin{array}{r} 3602 \\ - 1748 \\ \hline \end{array}$

30. $\begin{array}{r} 5004 \\ - 676 \\ \hline \end{array}$

Multiply. [1.4a]

31. $\begin{array}{r} 36 \\ \times \ 6 \\ \hline \end{array}$

32. $\begin{array}{r} 567 \\ \times \ 28 \\ \hline \end{array}$

33. $\begin{array}{r} 407 \\ \times 325 \\ \hline \end{array}$

34. $\begin{array}{r} 9435 \\ \times \ 602 \\ \hline \end{array}$

Divide. [1.5a]

35. $4\overline{)1012}$

36. $38\overline{)4261}$

37. $60\overline{)1399}$

38. $56\overline{)8095}$

39. Find the perimeter of the figure (m stands for "meters"). [1.2b]

10 m
3 m
4 m
8 m

40. Find the area of the region. [1.4b]

2 in.
4 in.

Understanding Through Discussion and Writing

To the student and the instructor: The Discussion and Writing exercises are meant to be answered with one or more sentences. They can be discussed and answered collaboratively by the entire class or by small groups.

41. Explain in your own words what the associative law of addition means. [1.2a]

42. Is subtraction commutative? That is, is there a commutative law of subtraction? Why or why not? [1.3a]

43. Describe a situation that corresponds to each multiplication: 4 · $150; $4 · 150. [1.4a]

44. Suppose a student asserts that "0 ÷ 0 = 0 because nothing divided by nothing is nothing." Devise an explanation to persuade the student that the assertion is false. [1.5a]

Rounding and Estimating; Order

1.6

a   ROUNDING

We round numbers in various situations when we do not need an exact answer. For example, we might round to see if we are being charged the correct amount in a store. We might also round to check if an answer to a problem is reasonable or to check a calculation done by hand or on a calculator.

To understand how to round, we first look at some examples using the number line. The number line displays numbers at equally spaced intervals.

EXAMPLE 1   Round 47 to the nearest ten.

47 is between 40 and 50. Since 47 is closer to 50, we round up to 50.

EXAMPLE 2   Round 42 to the nearest ten.

42 is between 40 and 50. Since 42 is closer to 40, we round down to 40.

EXAMPLE 3   Round 45 to the nearest ten.

45 is halfway between 40 and 50. We could round 45 down to 40 or up to 50. We agree to round up to 50.

> When a number is halfway between rounding numbers, round up.

Do Margin Exercises 1–7. ▶

Based on these examples, we can state a rule for rounding whole numbers.

OBJECTIVES

a Round to the nearest ten, hundred, or thousand.

b Estimate sums, differences, products, and quotients by rounding.

c Use < or > for ☐ to write a true sentence in a situation like 6 ☐ 10.

SKILL TO REVIEW

Objective 1.1a: Give the meaning of digits in standard notation.

In the number 145,627, what digit names the number of:

1. Tens?

2. Thousands?

Round to the nearest ten.

1. 37

2. 52   **3.** 35

4. 73   **5.** 75

6. 88   **7.** 64

Answers

Skill to Review:
1. 2   **2.** 5

Margin Exercises:
1. 40   **2.** 50   **3.** 40   **4.** 70
5. 80   **6.** 90   **7.** 60

ROUNDING WHOLE NUMBERS

To round to a certain place:

a) Locate the digit in that place.

b) Consider the next digit to the right.

c) If the digit to the right is 5 or higher, round up. If the digit to the right is 4 or lower, round down.

d) Change all digits to the right of the rounding location to zeros.

EXAMPLE 4 Round 6485 to the nearest ten.

a) Locate the digit in the tens place, 8.

  6 4 **8** 5
      ↑

b) Consider the next digit to the right, 5.

  6 4 **8** **5**

c) Since that digit, 5, is 5 or higher, round 8 tens up to 9 tens.

d) Change all digits to the right of the tens digit to zeros.

  6 4 9 0 ← This is the answer.

◀ **Do Exercises 8–11.**

EXAMPLE 5 Round 6485 to the nearest hundred.

a) Locate the digit in the hundreds place, 4.

  6 **4** 8 5
    ↑

b) Consider the next digit to the right, 8.

  6 **4** **8** 5

c) Since that digit, 8, is 5 or higher, round 4 hundreds up to 5 hundreds.

d) Change all digits to the right of hundreds to zeros.

  6 5 0 0 ← This is the answer.

◀ **Do Exercises 12–15.**

EXAMPLE 6 Round 6485 to the nearest thousand.

a) Locate the digit in the thousands place, 6.

  6 4 8 5
  ↑

b) Consider the next digit to the right, 4.

  6 **4** 8 5

c) Since that digit, 4, is 4 or lower, round down, meaning that 6 thousands stays as 6 thousands.

d) Change all digits to the right of thousands to zeros.

  6 0 0 0 ← This is the answer.

◀ **Do Exercises 16–19.**

Round to the nearest ten.

8. 137   **9.** 473

10. 235   **11.** 285

Round to the nearest hundred.

12. 641   **13.** 759

14. 1871   **15.** 9325

·········· **Caution!** ··········

It is incorrect in Example 6 to round from the ones digit over, as follows:

  6485 → 6490 → 6500 → 7000.

Note that 6485 is closer to 6000 than it is to 7000.

Round to the nearest thousand.

16. 7896   **17.** 8459

18. 19,343   **19.** 68,500

Answers

8. 140   **9.** 470   **10.** 240   **11.** 290
12. 600   **13.** 800   **14.** 1900   **15.** 9300
16. 8000   **17.** 8000   **18.** 19,000   **19.** 69,000

Sometimes rounding involves changing more than one digit in a number.

EXAMPLE 7 Round 78,595 to the nearest ten.

a) Locate the digit in the tens place, 9.

7 8 , 5 9 5
      ↑

b) Consider the next digit to the right, 5.

7 8 , 5 9 5

c) Since that digit, 5, is 5 or higher, round 9 tens to 10 tens. We think of 10 tens as 1 hundred + 0 tens and increase the hundreds digit by 1, to get 6 hundreds + 0 tens. We then write 6 in the hundreds place and 0 in the tens place.

d) Change the digit to the right of the tens digit to zero.

7 8 , 6 0 0 ← This is the answer.

Note that if we round this number to the nearest hundred, we get the same answer.

Do Exercises 20 and 21. ▶

b ESTIMATING

Estimating can be done in many ways. In general, an estimate rounded to the nearest ten is more accurate than one rounded to the nearest hundred, and an estimate rounded to the nearest hundred is more accurate than one rounded to the nearest thousand, and so on.

EXAMPLE 8 Estimate this sum by first rounding to the nearest ten:

78 + 49 + 31 + 85.

We round each number to the nearest ten. Then we add.

```
   7 8       8 0
   4 9       5 0
   3 1       3 0
 + 8 5     + 9 0
           ───────
           2 5 0 ← Estimated answer
```

Do Exercises 22 and 23. ▶

EXAMPLE 9 Estimate the difference by first rounding to the nearest thousand: 9324 − 2849.

We have

```
   9 3 2 4       9 0 0 0
 − 2 8 4 9     − 3 0 0 0
               ─────────
               6 0 0 0 ← Estimated answer
```

Do Exercises 24 and 25. ▶

20. Round 48,968 to the nearest ten, hundred, and thousand.

21. Round 269,582 to the nearest ten, hundred, and thousand.

22. Estimate the sum by first rounding to the nearest ten. Show your work.

```
   7 4
   2 3
   3 5
 + 6 6
```

23. Estimate the sum by first rounding to the nearest hundred. Show your work.

```
   6 5 0
   6 8 5
   2 3 8
 + 1 6 8
```

24. Estimate the difference by first rounding to the nearest hundred. Show your work.

```
   9 2 8 5
 − 6 7 3 9
```

25. Estimate the difference by first rounding to the nearest thousand. Show your work.

```
   2 3,2 7 8
 − 1 1,6 9 8
```

Answers

20. 48,970; 49,000; 49,000
21. 269,580; 269,600; 270,000
22. 70 + 20 + 40 + 70 = 200
23. 700 + 700 + 200 + 200 = 1800
24. 9300 − 6700 = 2600
25. 23,000 − 12,000 = 11,000

In the sentence $7 - 5 = 2$, the equals sign indicates that $7 - 5$ is the *same* as 2. When we round to make an estimate, the outcome is rarely the same as the exact result. Thus we cannot use an equals sign when we round. Instead, we use the symbol $\approx$. This symbol means "**is approximately equal to**." In Example 9, for instance, we could have written

$$9324 - 2849 \approx 6000.$$

EXAMPLE 10 Estimate the following product by first rounding to the nearest ten and then to the nearest hundred: 683×457.

Nearest ten:

```
        6 8 0      683 ≈ 680
  ×     4 6 0      457 ≈ 460
    4 0 8 0 0
  2 7 2 0 0 0
  3 1 2,8 0 0
```

Nearest hundred:

```
        7 0 0      683 ≈ 700
  ×     5 0 0      457 ≈ 500
  3 5 0,0 0 0
```

Exact:

```
        6 8 3
  ×     4 5 7
      4 7 8 1
    3 4 1 5 0
  2 7 3 2 0 0
  3 1 2,1 3 1
```

We see that rounding to the nearest ten gives a better estimate than rounding to the nearest hundred.

◀ **Do Exercise 26.**

EXAMPLE 11 Estimate the following quotient by first rounding to the nearest ten and then to the nearest hundred: $12{,}238 \div 175$.

Nearest ten: *Nearest hundred:*

```
          6 8                            6 1
1 8 0)1 2,2 4 0        2 0 0)1 2,2 0 0
      1 0 8 0                      1 2 0 0
      1 4 4 0                          2 0 0
      1 4 4 0                          2 0 0
            0                                0
```

The exact answer is 69 R 163. Again we see that rounding to the nearest ten gives a better estimate than rounding to the nearest hundred.

◀ **Do Exercise 27.**

---

26. Estimate the product by first rounding to the nearest ten and then to the nearest hundred. Show your work. **GS**

```
    8 3 7
  × 2 4 5
```

Nearest ten:

```
          8 4 0
  ×
    4 2 0 0 0
          0 0 0
        ,0 0 0
```

Nearest hundred:

```
          8 0 0
  ×
      0,0 0 0
```

27. Estimate the quotient by first rounding to the nearest hundred. Show your work.

$$64{,}534 \div 349$$

Answers

26. $840 \times 250 = 210{,}000$;
  $800 \times 200 = 160{,}000$
27. $64{,}500 \div 300 = 215$

Guided Solution:
26. Nearest ten: 250, 1, 6, 8, 2, 1, 0;
  Nearest hundred: 200, 1, 6

The next two examples show how estimating can be used in making financial decisions.

EXAMPLE 12 *Tuition.* Ellen plans to take 12 credit hours of classes next semester. If she takes the courses on campus, the cost per credit hour is $248. Estimate, by rounding to the nearest ten, the total cost of tuition.

We have

$$
\begin{array}{r}
2\ 5\ 0 \\
\times\quad 1\ 0 \\
\hline
2\ 5\ 0\ 0.
\end{array}
$$

The tuition will cost about $2500.

Do Exercise 28. ▶

EXAMPLE 13 *Purchasing a New Car.* Jon and Joanna are shopping for a new car. They are considering buying a Ford Focus S sedan. The base price of the car is $16,200. A 6-speed automatic transmission package can be added to this, as well as several other options, as shown in the chart below. Jon and Joanna want to stay within a budget of $18,000.

Estimate, by rounding to the nearest hundred, the cost of the Focus with the automatic transmission package and all other options and determine whether this will fit within their budget.

FORD FOCUS S SEDAN	PRICE
Base price	$16,200
6-speed automatic transmission	$1,095
Remote start system (requires purchase of automatic transmission)	$445
SYNC basic (includes Bluetooth and USB input jacks)	$295
Cargo management	$115
Graphics package	$375
Exterior protection package	$245

SOURCE: motortrend.com

First, we list the base price of the car and then the cost of each of the options. We then round each number to the nearest hundred and add.

$$
\begin{array}{rr}
1\ 6,2\ 0\ 0 \qquad & 1\ 6,2\ 0\ 0 \\
1,0\ 9\ 5 \qquad & 1,1\ 0\ 0 \\
4\ 4\ 5 \qquad & 4\ 0\ 0 \\
2\ 9\ 5 \qquad & 3\ 0\ 0 \\
1\ 1\ 5 \qquad & 1\ 0\ 0 \\
3\ 7\ 5 \qquad & 4\ 0\ 0 \\
+\quad 2\ 4\ 5 \qquad & +\quad 2\ 0\ 0 \\
\hline
& 1\ 8,7\ 0\ 0 \leftarrow \text{Estimated cost}
\end{array}
$$

The estimated cost is $18,700. This exceeds Jon and Joanna's budget of $18,000, so they will have to forgo at least one option.

Do Exercises 29 and 30. ▶

28. *Tuition.* If Ellen takes courses online, the cost per credit hour is $198. Estimate, by rounding to the nearest ten, the total cost of 12 credit hours of classes.

Refer to the chart in Example 13 to do Margin Exercises 29 and 30.

29. By eliminating at least one option, determine how Jon and Joanna can buy a Focus S sedan and stay within their budget. Keep in mind that purchasing the remote-start system requires purchasing the automatic transmission.

30. Elizabeth and C.J. are also considering buying a Focus S sedan. Estimate, by rounding to the nearest hundred, the cost of this car with automatic transmission, SYNC basic, and exterior protection package.

Answers

28. $2000   **29.** Eliminate either the automatic transmission and the remote-start system or the remote-start system and the graphics package. There are other correct answers as well.   **30.** $17,800

C ORDER

A sentence like $8 + 5 = 13$ is called an **equation**. It is a *true* equation. The equation $4 + 8 = 11$ is a *false* equation.

A sentence like $7 < 11$ is called an **inequality**. The sentence $7 < 11$ is a *true* inequality. The sentence $23 > 69$ is a *false* inequality.

Some common **inequality symbols** follow.

INEQUALITY SYMBOLS
$<$ means   "is less than"
$>$ means   "is greater than"
$\neq$ means   "is not equal to"

We know that 2 is not the same as 5. We express this by the sentence $2 \neq 5$. We also know that 2 is less than 5. We symbolize this by the expression $2 < 5$. We can see this order on the number line: 2 is to the left of 5. The number 0 is the smallest whole number.

ORDER OF WHOLE NUMBERS

For any whole numbers a and b:

1. $a < b$ (read "a is less than b") is true when a is to the left of b on the number line.
2. $a > b$ (read "a is greater than b") is true when a is to the right of b on the number line.

Use $<$ or $>$ for ☐ to write a true sentence. Draw the number line if necessary.

31. 8 ☐ 12

> Since 8 is to the ▭ of 12 on the number line, 8 ▭ 12.

EXAMPLE 14 Use $<$ or $>$ for ☐ to write a true sentence: 7 ☐ 11.

Since 7 is to the left of 11 on the number line, $7 < 11$.

EXAMPLE 15 Use $<$ or $>$ for ☐ to write a true sentence: 92 ☐ 87.

Since 92 is to the right of 87 on the number line, $92 > 87$.

32. 12 ☐ 8

33. 76 ☐ 64

34. 64 ☐ 76

35. 217 ☐ 345

36. 345 ☐ 217

◀ **Do Exercises 31–36.**

Answers

31. $<$ **32.** $>$ **33.** $>$ **34.** $<$
35. $<$ **36.** $>$
Guided Solution:
31. left, $<$

✓ Reading Check

Determine whether each statement is true or false.

_____ **RC1.** When rounding to the nearest hundred, if the digit in the tens place is 5 or higher, we round up.

_____ **RC2.** When rounding 3500 to the nearest thousand, we should round down.

_____ **RC3.** An estimate made by rounding to the nearest thousand is more accurate than an estimate made by rounding to the nearest ten.

_____ **RC4.** Since 78 rounded to the nearest ten is 80, we can write 78 $\approx$ 80.

a Round to the nearest ten.

1. 48

2. 532

3. 463

4. 8945

5. 731

6. 54

7. 895

8. 798

Round to the nearest hundred.

9. 146

10. 874

11. 957

12. 650

13. 9079

14. 4645

15. 32,839

16. 198,402

Round to the nearest thousand.

17. 5876

18. 4500

19. 7500

20. 2001

21. 45,340

22. 735,562

23. 373,405

24. 6,713,255

b Estimate each sum or difference by first rounding to the nearest ten. Show your work.

25.
$$\begin{array}{r} 7\,8 \\ +\,9\,2 \\ \hline \end{array}$$

26.
$$\begin{array}{r} 6\,2 \\ 9\,7 \\ 4\,6 \\ +\,8\,1 \\ \hline \end{array}$$

27.
$$\begin{array}{r} 8\,0\,7\,4 \\ -\,2\,3\,4\,7 \\ \hline \end{array}$$

28.
$$\begin{array}{r} 6\,7\,3 \\ -\,2\,8 \\ \hline \end{array}$$

Estimate each sum by first rounding to the nearest ten. State if the given sum seems to be incorrect when compared to the estimate.

29.
$$\begin{array}{r} 4\,5 \\ 7\,7 \\ 2\,5 \\ +\,5\,6 \\ \hline 3\,4\,3 \end{array}$$

30.
$$\begin{array}{r} 4\,1 \\ 2\,1 \\ 5\,5 \\ +\,6\,0 \\ \hline 1\,7\,7 \end{array}$$

31.
$$\begin{array}{r} 6\,2\,2 \\ 7\,8 \\ 8\,1 \\ +\,1\,1\,1 \\ \hline 9\,3\,2 \end{array}$$

32.
$$\begin{array}{r} 8\,3\,6 \\ 3\,7\,4 \\ 7\,9\,4 \\ +\,9\,3\,8 \\ \hline 3\,9\,4\,7 \end{array}$$

Estimate each sum or difference by first rounding to the nearest hundred. Show your work.

33.   7 3 4 8
        + 9 2 4 7

34.   5 6 8
           4 7 2
           9 3 8
        + 4 0 2

35.   6 8 5 2
        − 1 7 4 8

36.   9 4 3 8
        − 2 7 8 7

Estimate each sum by first rounding to the nearest hundred. State if the given sum seems to be incorrect when compared to the estimate.

37.   2 1 6
             8 4
           7 4 5
        + 5 9 5
        ─────────
         1 6 4 0

38.   4 8 1
           7 0 2
           6 2 3
        + 1 0 4 3
        ─────────
         1 8 4 9

39.   7 5 0
           4 2 8
             6 3
        + 2 0 5
        ─────────
         1 4 4 6

40.   3 2 6
           2 7 5
           7 5 8
        + 9 4 3
        ─────────
         2 3 0 2

Estimate each sum or difference by first rounding to the nearest thousand. Show your work.

41.   9 6 4 3
           4 8 2 1
           8 9 4 3
        + 7 0 0 4

42.   7 6 4 8
           9 3 4 8
           7 8 4 2
        + 2 2 2 2

43.   9 2,1 4 9
        − 2 2,5 5 5

44.   8 4,8 9 0
        − 1 1,1 1 0

Estimate each product by first rounding to the nearest ten. Show your work.

45.   4 5
        × 6 7

46.   5 1
        × 7 8

47.   3 4
        × 2 9

48.   6 3
        × 5 4

Estimate each product by first rounding to the nearest hundred. Show your work.

49.   8 7 6
        × 3 4 5

50.   3 5 5
        × 2 9 9

51.   4 3 2
        × 1 9 9

52.   7 8 9
        × 4 3 4

Estimate each quotient by first rounding to the nearest ten. Show your work.

53. $347 \div 73$

54. $454 \div 87$

55. $8452 \div 46$

56. $1263 \div 29$

Estimate each quotient by first rounding to the nearest hundred. Show your work.

57. $1165 \div 236$ **58.** $3641 \div 571$ **59.** $8358 \div 295$ **60.** $32{,}854 \div 748$

Planning a Vacation. Most cruise ships offer a choice of rooms at varying prices, as well as additional packages and shore excursions at each port. The table below lists room prices for a seven-day Mediterranean cruise, as well as prices for several additional packages and excursions.

ROOMS	PRICE
Suite	$856
Balcony	$686
Ocean View	$586
Interior	$536
ADDITIONAL PACKAGES	**PRICE**
Spa	$115
Specialty Dining	$129
Beverage	$79
EXCURSIONS	**PRICE**
Athens, Greece; Private Tour	$289
Venice, Italy	$95
Istanbul, Turkey	$130
Pisa, Italy; Biking Tour	$199

61. Estimate the total price of a cruise with an ocean view room, a spa package, and an Istanbul excursion. Round each price to the nearest hundred dollars.

62. Estimate the total price of a cruise with a balcony room, no additional packages, a Venice excursion, and a biking tour in Pisa. Round each price to the nearest hundred dollars.

63. Antonio has a budget of $1000 for a Mediterranean cruise. He would like a balcony room, a specialty dining package, a private tour of Athens, and a tour of Venice. Estimate the total price of this cruise by rounding each price to the nearest hundred dollars. Can he afford his choices?

64. Alyssa has a budget of $1400 for a Mediterranean cruise. She is planning to book an interior room and would like to go on all the excursions listed. Estimate the total price of this cruise by rounding each price to the nearest hundred dollars. Does her budget cover her choices?

65. If you were going on a Mediterranean cruise and had a budget of $1500, what options would you choose? Decide on the options you would like and estimate the total price by rounding each price to the nearest hundred dollars. Could you afford all your chosen options?

66. If you were going on a Mediterranean cruise and had a budget of $1200, what options would you choose? Decide on the options you would like and estimate the total price by rounding each price to the nearest hundred dollars. Could you afford all your chosen options?

67. *Mortgage Payments.* To pay for their new home, Tim and Meribeth will make 360 payments of $751.55 each. In addition, they must add an escrow amount of $112.67 to each payment for insurance and taxes.

  a) Estimate the total amount they will pay by rounding the number of payments, the amount of each payment, and the escrow amount to the nearest ten.

  b) Estimate the total amount they will pay by rounding the number of payments, the amount of each payment, and the escrow amount to the nearest hundred.

68. *Conference Expenses.* The cost to attend a three-day teachers' conference is $245, and a hotel room costs $169 a night. One year, 489 teachers attended the conference, and 315 rooms were rented for two nights each.

  a) Estimate the total amount spent by the teachers by rounding the cost of attending the conference, the cost of a hotel room, the number of teachers, and the number of rooms to the nearest ten.

  b) Estimate the total amount spent by the teachers by rounding the cost of attending the conference, the cost of a hotel room, the number of teachers, and the number of rooms to the nearest hundred.

69. *Banquet Attendance.* Tickets to the annual awards banquet for the Riviera Swim Club cost $28 each. Ticket sales for the banquet totaled $2716. Estimate the number of people who attended the banquet by rounding the cost of a ticket to the nearest ten and the total sales to the nearest hundred.

70. *School Fundraiser.* For a school fundraiser, Charlotte sells trash bags at a price of $11 per roll. If her sales totaled $2211, estimate the number of rolls she sold by rounding the price per roll to the nearest ten and the total sales to the nearest hundred.

C Use < or > for ☐ to write a true sentence. Draw the number line if necessary.

71. 0 ☐ 17   **72.** 32 ☐ 0   **73.** 34 ☐ 12   **74.** 28 ☐ 18

75. 1000 ☐ 1001   **76.** 77 ☐ 117   **77.** 133 ☐ 132   **78.** 999 ☐ 997

79. 460 ☐ 17   **80.** 345 ☐ 456   **81.** 37 ☐ 11   **82.** 12 ☐ 32

New Book Titles. The number of new book titles published in the United States in each of three recent years is shown in the table below. Use this table to do Exercises 83 and 84.

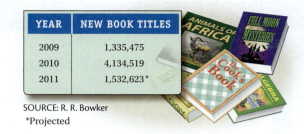

YEAR	NEW BOOK TITLES
2009	1,335,475
2010	4,134,519
2011	1,532,623*

SOURCE: R. R. Bowker
*Projected

83. Write an inequality to compare the numbers of new titles published in 2009 and in 2010.

84. Write an inequality to compare the numbers of new titles published in 2010 and in 2011.

85. *Public Schools.* The number of public schools in the United States increased from 97,382 in 2006 to 98,817 in 2010. Write an inequality to compare these numbers of schools.

Public Schools

SOURCE: National Center for Education Statistics

86. *Life Expectancy.* The life expectancy of a female in the United States in 2020 is predicted to be about 82 years and that of a male about 77 years. Write an inequality to compare these life expectancies.

Life Expectancy in the United States

SOURCE: U.S. Census Bureau

Skill Maintenance

Add. [1.2a]

87.
$$\begin{array}{r} 6\,7,7\,8\,9 \\ +\,1\,8,9\,6\,5 \\ \hline \end{array}$$

88.
$$\begin{array}{r} 9\,0\,0\,2 \\ +\,4\,5\,8\,7 \\ \hline \end{array}$$

Subtract. [1.3a]

89.
$$\begin{array}{r} 6\,7,7\,8\,9 \\ -\,1\,8,9\,6\,5 \\ \hline \end{array}$$

90.
$$\begin{array}{r} 9\,0\,0\,2 \\ -\,4\,5\,8\,7 \\ \hline \end{array}$$

Multiply. [1.4a]

91.
$$\begin{array}{r} 4\,6 \\ \times\,3\,7 \\ \hline \end{array}$$

92.
$$\begin{array}{r} 3\,0\,6 \\ \times\,\;5\,8 \\ \hline \end{array}$$

Divide. [1.5a]

93. $328 \div 6$

94. $4784 \div 23$

Synthesis

95.–98. Use a calculator to find the sums and the differences in each of Exercises 41–44. Then compare your answers with those found using estimation. Even when using a calculator it is possible to make an error if you press the wrong buttons, so it is a good idea to check by estimating.

Solving Equations

OBJECTIVES

a Solve simple equations by trial.

b Solve equations like $t + 28 = 54$, $28 \cdot x = 168$, and $98 \cdot 2 = y$.

SKILL TO REVIEW

Objective 1.5a: Divide whole numbers.

Divide.

1. $1008 \div 36$

2. $675 \div 15$

Find a number that makes each sentence true.

1. $8 = 1 + \square$

2. $\square + 2 = 7$

3. Determine whether 7 is a solution of $\square + 5 = 9$.

4. Determine whether 4 is a solution of $\square + 5 = 9$.

Solve by trial.

5. $n + 3 = 8$

6. $x - 2 = 8$

7. $45 \div 9 = y$

8. $10 + t = 32$

Answers

Skill to Review:
1. 28   **2.** 45

Margin Exercises:
1. 7   **2.** 5   **3.** No   **4.** Yes   **5.** 5
6. 10   **7.** 5   **8.** 22

a SOLUTIONS BY TRIAL

Let's find a number that we can put in the blank to make this sentence true:

$$9 = 3 + \square.$$

We are asking "9 is 3 plus what number?" The answer is 6.

$$9 = 3 + 6$$

◀ **Do Margin Exercises 1 and 2.**

A sentence with $=$ is called an **equation**. A **solution** of an equation is a number that makes the sentence true. Thus, 6 is a solution of

$$9 = 3 + \square \quad \text{because} \quad 9 = 3 + 6 \text{ is true.}$$

However, 7 is not a solution of

$$9 = 3 + \square \quad \text{because} \quad 9 = 3 + 7 \text{ is false.}$$

◀ **Do Exercises 3 and 4.**

We can use a letter in an equation instead of a blank:

$$9 = 3 + n.$$

We call n a **variable** because it can represent any number. If a replacement for a variable makes an equation true, the replacement is a solution of the equation.

SOLUTIONS OF AN EQUATION

A **solution of an equation** is a replacement for the variable that makes the equation true. When we find all the solutions, we say that we have **solved** the equation.

EXAMPLE 1 Solve $y + 12 = 27$ by trial.

We replace y with several numbers.

If we replace y with 13, we get a false equation: $13 + 12 = 27$.
If we replace y with 14, we get a false equation: $14 + 12 = 27$.
If we replace y with 15, we get a true equation: $15 + 12 = 27$.

No other replacement makes the equation true, so the solution is 15.

EXAMPLES Solve.

2. $7 + n = 22$
(7 plus what number is 22?)
The solution is 15.

3. $63 = 3 \cdot x$
(63 is 3 times what number?)
The solution is 21.

◀ **Do Exercises 5–8.**

b SOLVING EQUATIONS

We now begin to develop more efficient ways to solve certain equations. When an equation has a variable alone on one side and a calculation on the other side, we can find the solution by carrying out the calculation.

EXAMPLE 4 Solve: $x = 245 \times 34$.

To solve the equation, we carry out the calculation.

$$\begin{array}{r} 2\;4\;5 \\ \times\;\;3\;4 \\ \hline 9\;8\;0 \\ 7\;3\;5\;0 \\ \hline 8\;3\;3\;0 \end{array}$$

$$x = 245 \times 34$$
$$x = 8330$$

The solution is 8330.

Do Exercises 9–12. ▶

Look at the equation

$$x + 12 = 27.$$

We can get x alone by subtracting 12 *on both sides*. Thus,

$$x + 12 - 12 = 27 - 12 \qquad \text{Subtracting 12 on both sides}$$
$$x + 0 = 15 \qquad \text{Carrying out the subtraction}$$
$$x = 15.$$

SOLVING $x + a = b$

To solve $x + a = b$, subtract a on both sides.

If we can get an equation in a form with the variable alone on one side, we can "see" the solution.

EXAMPLE 5 Solve: $t + 28 = 54$.

We have

$$t + 28 = 54$$
$$t + 28 - 28 = 54 - 28 \qquad \text{Subtracting 28 on both sides}$$
$$t + 0 = 26$$
$$t = 26.$$

To check the answer, we substitute 26 for t in the original equation.

Check: $\begin{array}{r} t + 28 = 54 \\ \hline 26 + 28 \;?\; 54 \\ 54 \;\Big|\; \end{array}$ **TRUE**   Since $54 = 54$ is true, 26 checks.

The solution is 26.

Do Exercises 13 and 14. ▶

Solve.

9. $346 \times 65 = y$

10. $x = 2347 + 6675$

11. $4560 \div 8 = t$

12. $x = 6007 - 2346$

Solve. Be sure to check.

GS **13.** $x + 9 = 17$

$$x + 9 - \boxed{} = 17 - 9$$
$$x = \boxed{}$$

Check: $\begin{array}{r} x + 9 = 17 \\ \hline \boxed{} + 9 \;?\; 17 \\ \boxed{} \;\Big| \end{array}$

Since $17 = 17$ is $\boxed{}$, the answer checks. The solution is $\boxed{}$.

14. $77 = m + 32$

Answers

9. 22,490 **10.** 9022 **11.** 570 **12.** 3661
13. 8 **14.** 45

Guided Solution:
13. 9, 8, 8, 17; true; 8

EXAMPLE 6 Solve: $182 = 65 + n$.

We have

$$182 = 65 + n$$
$$182 - 65 = 65 + n - 65 \qquad \text{Subtracting 65 on both sides}$$
$$117 = 0 + n \qquad \text{65 plus } n \text{ minus 65 is } 0 + n.$$
$$117 = n.$$

Check:
$$\frac{182 = 65 + n}{182 \;?\; 65 + 117}$$
$$|\;182 \qquad \text{TRUE}$$

The solution is 117.

◀ **Do Exercise 15.**

15. Solve: $155 = t + 78$. Be sure to check.

EXAMPLE 7 Solve: $7381 + x = 8067$.

We have

$$7381 + x = 8067$$
$$7381 + x - 7381 = 8067 - 7381 \qquad \text{Subtracting 7381 on both sides}$$
$$x = 686.$$

Check:
$$\frac{7381 + x = 8067}{7381 + 686 \;?\; 8067}$$
$$8067 \;| \qquad \text{TRUE}$$

The solution is 686.

Solve. Be sure to check.

16. $4566 + x = 7877$

17. $8172 = h + 2058$

◀ **Do Exercises 16 and 17.**

We now learn to solve equations like $8 \cdot n = 96$. Look at

$$8 \cdot n = 96.$$

We can get n alone by dividing by 8 *on both sides*. Thus,

$$\frac{8 \cdot n}{8} = \frac{96}{8} \qquad \text{Dividing by 8 on both sides}$$
$$n = 12. \qquad \text{8 times } n \text{ divided by 8 is } n.$$

To check the answer, we substitute 12 for n in the original equation.

Check:
$$\frac{8 \cdot n = 96}{8 \cdot 12 \;?\; 96}$$
$$96 \;| \qquad \text{TRUE}$$

Since $96 = 96$ is a true equation, 12 is the solution of the equation.

SOLVING $a \cdot x = b$

To solve $a \cdot x = b$, divide by a on both sides.

EXAMPLE 8 Solve: $10 \cdot x = 240$.

We have

$$10 \cdot x = 240$$

$$\frac{10 \cdot x}{10} = \frac{240}{10} \qquad \text{Dividing by 10 on both sides}$$

$$x = 24.$$

Check: $\dfrac{10 \cdot x = 240}{10 \cdot 24 \,?\, 240}$

$\qquad\qquad 240 \mid \qquad$ TRUE

The solution is 24.

EXAMPLE 9 Solve: $5202 = 9 \cdot t$.

We have

$$5202 = 9 \cdot t$$

$$\frac{5202}{9} = \frac{9 \cdot t}{9} \qquad \text{Dividing by 9 on both sides}$$

$$578 = t.$$

Check: $\dfrac{5202 = 9 \cdot t}{5202 \,?\, 9 \cdot 578}$

$\qquad\qquad\qquad \mid 5202 \qquad$ TRUE

The solution is 578.

Do Exercises 18–20. ▶

EXAMPLE 10 Solve: $14 \cdot y = 1092$.

We have

$$14 \cdot y = 1092$$

$$\frac{14 \cdot y}{14} = \frac{1092}{14} \qquad \text{Dividing by 14 on both sides}$$

$$y = 78.$$

The check is left to the student. The solution is 78.

EXAMPLE 11 Solve: $n \cdot 56 = 4648$.

We have

$$n \cdot 56 = 4648$$

$$\frac{n \cdot 56}{56} = \frac{4648}{56} \qquad \text{Dividing by 56 on both sides}$$

$$n = 83.$$

The check is left to the student. The solution is 83.

Do Exercises 21 and 22. ▶

Solve. Be sure to check.

18. $8 \cdot x = 64$

GS **19.** $144 = 9 \cdot n$

$$\frac{144}{9} = \frac{9 \cdot n}{\boxed{}}$$

$$\boxed{} = n$$

Check: $\dfrac{144 = 9 \cdot n}{144 \,?\, 9 \cdot \boxed{}}$

$\qquad\qquad\quad \mid \boxed{}$

Since $144 = 144$ is $\boxed{}$,
the answer checks.
The solution is $\boxed{}$.

20. $5152 = 8 \cdot t$

Solve. Be sure to check.

21. $18 \cdot y = 1728$

22. $n \cdot 48 = 4512$

Answers

18. 8 **19.** 16 **20.** 644 **21.** 96 **22.** 94

Guided Solution:
19. 9, 16, 16, 144; true; 16

SECTION 1.7 Solving Equations **51**

1.7 Exercise Set

For Extra Help

MyMathLab®

 MathXL®
 PRACTICE WATCH
 READ
 REVIEW

✓ Reading Check

Match each word with its definition from the list on the right.

RC1. Equation _____

a) A replacement for the variable that makes an equation true

RC2. Solution _____

b) A letter that can represent any number

RC3. Solved _____

c) A sentence containing $=$

RC4. Variable _____

d) An equation for which we have found all solutions

a Solve by trial.

1. $x + 0 = 14$

2. $x - 7 = 18$

3. $y \cdot 17 = 0$

4. $56 \div m = 7$

b Solve. Be sure to check.

5. $x = 12{,}345 + 78{,}555$

6. $t = 5678 + 9034$

7. $908 - 458 = p$

8. $9007 - 5667 = m$

9. $16 \cdot 22 = y$

10. $34 \cdot 15 = z$

11. $t = 125 \div 5$

12. $w = 256 \div 16$

13. $13 + x = 42$

14. $15 + t = 22$

15. $12 = 12 + m$

16. $16 = t + 16$

17. $10 + x = 89$

18. $20 + x = 57$

19. $61 = 16 + y$

20. $53 = 17 + w$

21. $3 \cdot x = 24$

22. $6 \cdot x = 42$

23. $112 = n \cdot 8$

24. $162 = 9 \cdot m$

25. $3 \cdot m = 96$

26. $4 \cdot y = 96$

27. $715 = 5 \cdot z$

28. $741 = 3 \cdot t$

29. $8322 + 9281 = x$ **30.** $9281 - 8322 = y$ **31.** $47 + n = 84$ **32.** $56 + p = 92$

33. $45 \cdot 23 = x$ **34.** $23 \cdot 78 = y$ **35.** $x + 78 = 144$ **36.** $z + 67 = 133$

37. $6 \cdot p = 1944$ **38.** $4 \cdot w = 3404$ **39.** $567 + x = 902$ **40.** $438 + x = 807$

41. $234 \cdot 78 = y$ **42.** $10{,}534 \div 458 = q$ **43.** $18 \cdot x = 1872$ **44.** $19 \cdot x = 6080$

45. $40 \cdot x = 1800$ **46.** $20 \cdot x = 1500$ **47.** $2344 + y = 6400$ **48.** $9281 = 8322 + t$

49. $m = 7006 - 4159$ **50.** $n = 3004 - 1745$ **51.** $165 = 11 \cdot n$ **52.** $660 = 12 \cdot n$

53. $58 \cdot m = 11{,}890$ **54.** $233 \cdot x = 22{,}135$ **55.** $491 - 34 = y$ **56.** $512 - 63 = z$

Skill Maintenance

Divide. [1.5a]

57. $1283 \div 9$ **58.** $1278 \div 9$ **59.** $1\,7\,\overline{)\,5\,6\,7\,8}$ **60.** $1\,7\,\overline{)\,5\,6\,8\,9}$

Use $>$ or $<$ for ☐ to write a true sentence. [1.6c]

61. 123 ☐ 789 **62.** 342 ☐ 339 **63.** 688 ☐ 0 **64.** 0 ☐ 11

65. Round 6,375,602 to the nearest thousand. [1.6a]   **66.** Round 6,375,602 to the nearest ten. [1.6a]

Synthesis

Solve.

67. 🖩 $23{,}465 \cdot x = 8{,}142{,}355$ **68.** 🖩 $48{,}916 \cdot x = 14{,}332{,}388$

OBJECTIVE

a Solve applied problems involving addition, subtraction, multiplication, or division of whole numbers.

SKILL TO REVIEW

Objective 1.6b: Estimate sums, differences, products, and quotients by rounding.

Estimate each sum or difference by first rounding to the nearest thousand. Show your work.

1.
$$\begin{array}{r} 3\,6\,7{,}9\,8\,2 \\ +\quad 4\,3{,}4\,9\,5 \end{array}$$

2.
$$\begin{array}{r} 9\,2\,8\,7 \\ -\;3\,5\,0\,2 \end{array}$$

a A PROBLEM-SOLVING STRATEGY

One of the most important ways in which we use mathematics is as a tool in solving problems. To solve a problem, we use the following five-step strategy.

FIVE STEPS FOR PROBLEM SOLVING

1. **Familiarize** yourself with the problem situation.
2. **Translate** the problem to an equation using a variable.
3. **Solve** the equation.
4. **Check** to see whether your possible solution actually fits the problem situation and is thus really a solution of the problem.
5. **State** the answer clearly using a complete sentence and appropriate units.

The first of these five steps, becoming familiar with the problem, is probably the most important.

THE FAMILIARIZE STEP

- If the problem is presented in words, read and reread it carefully until you understand what you are being asked to find.
- Make a drawing, if it makes sense to do so.
- Write a list of the known facts and a list of what you wish to find out.
- Assign a letter, or *variable*, to the unknown.
- Organize the information in a chart or a table.
- Find further information. Look up a formula, consult a reference book or an expert in the field, or do research on the Internet.
- Guess or estimate the answer and check your guess or estimate.

Answers

Skill to Review:
1. $368{,}000 + 43{,}000 = 411{,}000$
2. $9000 - 4000 = 5000$

EXAMPLE 1 *Video Game Platforms.* The following table shows the total number of units of eight popular video game platforms sold world-wide, as of December 8, 2012. Three of these are made by PlayStation®. Find the total number of PlayStation units sold.

PLATFORM	GLOBAL SALES
Nintendo 3DS	682,396
Xbox 360	614,353
Nintendo Wii U	610,384
PlayStation® 3	576,565
Nintendo Wii	252,484
Nintendo DS	178,209
PlayStation® Vita	152,049
PlayStation® Portable	54,677

Source: www.vgchartz.com

1. **Familiarize.** First, we assign a letter, or variable, to the number we wish to find. We let $p =$ the total number of video game units sold by PlayStation. Since we are combining numbers, we will add.

2. **Translate.** We translate to an equation:

Number of PlayStation 3 units	plus	Number of PlayStation Vita units	plus	Number of PlayStation Portable units	is	Total number of PlayStation units
↓	↓	↓	↓	↓	↓	↓
576,565	+	152,049	+	54,677	=	p

3. **Solve.** We solve the equation by carrying out the addition.

$$576{,}565 + 152{,}049 + 54{,}677 = p$$
$$783{,}291 = p$$

$$\begin{array}{r} 576{,}565 \\ 152{,}049 \\ + \ \ 54{,}677 \\ \hline 783{,}291 \end{array}$$

4. **Check.** We check our result by rereading the original problem and seeing if 783,291 answers the question. Since we are looking for a total, we could repeat the addition calculation. We could also check whether the answer is reasonable. In this case, since the total is greater than any of the three separate sales numbers, the result seems reasonable. Another way to check is to estimate the expected result and compare the estimate with the calculated result. If we round each PlayStation sales number to the nearest ten thousand and add, we have

$$580{,}000 + 150{,}000 + 50{,}000 = 780{,}000.$$

Since $780{,}000 \approx 783{,}291$, our result again seems reasonable.

5. **State.** The total number of the three PlayStation units sold worldwide as of December 8, 2012, is 783,291.

Do Exercises 1–3. ▶

Refer to the table in Example 1 to do Margin Exercises 1–3.

1. Find the total number of Nintendo units sold.

2. Find the total number of units sold for the four most popular game platforms listed in the table.

3. Find the total number of units sold for all the game platforms listed in the table.

Answers

1. 1,723,473 units   **2.** 2,483,698 units
3. 3,121,117 units

EXAMPLE 2 *Travel Distance.* Abigail is driving from Indianapolis to Salt Lake City to attend a family reunion. The distance from Indianapolis to Salt Lake City is 1634 mi. In the first two days, she travels 1154 mi to Denver. How much farther must she travel?

1. **Familiarize.** We first make a drawing or at least visualize the situation. We let $d =$ the remaining distance to Salt Lake City.

2. **Translate.** We want to determine how many more miles Abigail must travel. We translate to an equation:

Distance already traveled	plus	Distance to go	is	Total distance of trip
↓	↓	↓	↓	↓
1154	+	d	=	1634.

3. **Solve.** To solve the equation, we subtract 1154 on both sides.

$$1154 + d = 1634$$
$$1154 + d - 1154 = 1634 - 1154$$
$$d = 480$$

$$\begin{array}{r} {\scriptstyle 5\ 13} \\ 1\ 6\ 3\ 4 \\ -\ 1\ 1\ 5\ 4 \\ \hline 4\ 8\ 0 \end{array}$$

4. **Check.** We check our answer of 480 mi in the original problem. This number should be less than the total distance, 1634 mi, and it is. We can add the distance traveled, 1154, and the distance left to go, 480: $1154 + 480 = 1634$. We can also estimate:

$$1634 - 1154 \approx 1600 - 1200$$
$$= 400 \approx 480.$$

The answer, 480 mi, checks.

5. **State.** Abigail must travel 480 mi farther to Salt Lake City.

◀ **Do Exercise 4.**

---

4. **Reading Assignment.** William **GS** has been assigned 234 pages of reading for his history class. He has read 86 pages. How many more pages does he have to read?

1. **Familiarize.** Let $p =$ the number of pages William still has to read.

2. **Translate.**

Pages already read	plus	Number of pages to read	is	Total number of pages
⏝	↓	⏝	↓	⏝
↓		↓		↓
86	+	☐	=	☐

3. **Solve.**

$$86 + p = 234$$
$$86 + p - \boxed{} = 234 - 86$$
$$p = \boxed{}$$

4. **Check.** If William reads 148 more pages, he will have read a total of $86 + 148$ pages, or ☐ pages.

5. **State.** William has ☐ more pages to read.

EXAMPLE 3 *Total Cost of Chairs.* What is the total cost of 6 Adirondack chairs if each one costs $169?

1. Familiarize. We make a drawing and let C = the cost of 6 chairs.

$169 $169 $169 $169 $169 $169

2. Translate. We translate to an equation:

Number of chairs	times	Cost of each chair	is	Total cost
6	×	169	=	C.

3. Solve. This sentence tells us what to do. We multiply.

$$6 \times 169 = C$$
$$1014 = C$$

$$\begin{array}{r} 1\,6\,9 \\ \times \quad 6 \\ \hline 1\,0\,1\,4 \end{array}$$

4. Check. We have an answer, 1014, that is greater than the cost of any one chair, which is reasonable. We can also check by estimating:

$$6 \times 169 \approx 6 \times 170 = 1020 \approx 1014.$$

The answer checks.

5. State. The total cost of 6 chairs is $1014.

Do Exercise 5. ▶

5. *Total Cost of Gas Grills.* What is the total cost of 14 gas grills, each with 520 sq in. of total cooking surface, if each one costs $398?

EXAMPLE 4 *Area of an Oriental Rug.* The dimensions of the oriental rug in the Fosters' front hallway are 42 in. by 66 in. What is the area of the rug?

1. Familiarize. We let A = the area of the rug and use the formula for the area of a rectangle, A = length · width = $l \cdot w$. Since we usually consider length to be larger than width, we will let l = 66 in. and w = 42 in.

42 in.

66 in.

2. Translate. We substitute in the formula:

$$A = l \cdot w = 66 \cdot 42.$$

3. Solve. We carry out the multiplication.

$$A = 66 \cdot 42$$
$$A = 2772$$

$$
\begin{array}{r}
6\ 6 \\
\times\ 4\ 2 \\
\hline
1\ 3\ 2 \\
2\ 6\ 4\ 0 \\
\hline
2\ 7\ 7\ 2
\end{array}
$$

4. Check. We can repeat the calculation. We can also round and estimate:

$$66 \times 42 \approx 70 \times 40 = 2800 \approx 2772.$$

The answer checks.

5. State. The area of the rug is 2772 sq in.

6. Bed Sheets. The dimensions of a flat sheet for a queen-size bed are 90 in. by 102 in. What is the area of the sheet?

◀ **Do Exercise 6.**

EXAMPLE 5 *Packages of Gum.* A candy company produces 3304 sticks of gum. How many 12-stick packages can be filled? How many sticks will be left over?

1. Familiarize. We make a drawing to visualize the situation and let $n =$ the number of 12-stick packages that can be filled. The problem can be considered as repeated subtraction, taking successive sets of 12 sticks and putting them into n packages.

12-stick packages

How many packages?   How many sticks are left over?

2. Translate. We translate to an equation:

Number of sticks	divided by	Number in each package	is	Number of packages
↓	↓	↓	↓	↓
3304	÷	12	=	n.

3. Solve. We solve the equation by carrying out the division.

$$3304 \div 12 = n$$
$$275 \text{ R } 4 = n$$

$$
\begin{array}{r}
2\ 7\ 5 \\
1\ 2\ {\overline{)\ 3\ 3\ 0\ 4}} \\
2\ 4 \\
\hline
9\ 0 \\
8\ 4 \\
\hline
6\ 4 \\
6\ 0 \\
\hline
4
\end{array}
$$

4. Check. We can check by multiplying the number of packages by 12 and adding the remainder, 4:

$$12 \cdot 275 = 3300,$$
$$3300 + 4 = 3304.$$

5. State. Thus, 275 twelve-stick packages of gum can be filled. There will be 4 sticks left over.

Do Exercise 7. ▶

EXAMPLE 6 *Automobile Mileage.* A 2013 Toyota Matrix gets 26 miles per gallon (mpg) in city driving. How many gallons will it use in 4758 mi of city driving?

Source: Toyota

1. Familiarize. We make a drawing and let g = the number of gallons of gasoline used in 4758 mi of city driving.

2. Translate. Repeated addition or multiplication applies here.

Number of miles per gallon	times	Number of gallons used	is	Number of miles driven
26	·	g	=	4758

3. Solve. To solve the equation, we divide by 26 on both sides.

$$26 \cdot g = 4758$$
$$\frac{26 \cdot g}{26} = \frac{4758}{26}$$
$$g = 183$$

```
        1 8 3
   2 6 ) 4 7 5 8
        2 6
        2 1 5
        2 0 8
            7 8
            7 8
             0
```

4. Check. To check, we multiply 183 by 26.

```
      1 8 3
    ×   2 6
    1 0 9 8
    3 6 6 0
    4 7 5 8
```

The answer checks.

5. State. The Toyota Matrix will use 183 gal of gasoline.

Do Exercise 8. ▶

7. *Packages of Gum.* The candy company in Example 5 also produces 6-stick packages. How many 6-stick packages can be filled with 2269 sticks of gum? How many sticks will be left over?

8. *Automobile Mileage.* A 2013 Toyota Matrix gets 32 miles per gallon (mpg) in highway driving. How many gallons will it use in 2528 mi of highway driving?

Source: Toyota

Answers

7. 378 packages with 1 stick left over
8. 79 gal

Multistep Problems

Sometimes we must use more than one operation to solve a problem, as in the following example.

EXAMPLE 7 *Weight Loss.* To lose one pound, you must burn about 3500 calories in excess of what you already burn doing your regular daily activities. The following chart shows how long a person must engage in several types of exercise in order to burn 100 calories. For how long would a person have to run at a brisk pace in order to lose one pound?

To burn 100 calories, you must:

- Run for 8 minutes at a brisk pace, or
- Swim for 2 minutes at a brisk pace, or
- Bicycle for 15 minutes at 9 mph, or
- Do aerobic exercises for 15 minutes, or
- Golf, walking, for 20 minutes, or
- Play tennis, singles, for 11 minutes

1. **Familiarize.** This is a multistep problem. We will first find how many hundreds are in 3500. This will tell us how many times a person must run for 8 min in order to lose one pound. Then we will find the total number of minutes required for the weight loss.

   We let $x =$ the number of hundreds in 3500 and $t =$ the time it takes to lose one pound.

2. **Translate.** We translate to two equations.

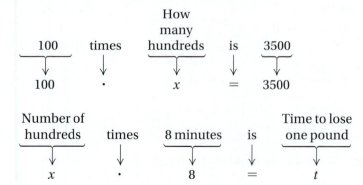

$$100 \cdot x = 3500$$

$$x \cdot 8 = t$$

3. **Solve.** We divide by 100 on both sides of the first equation to find x.

$$100 \cdot x = 3500$$

$$\frac{100 \cdot x}{100} = \frac{3500}{100}$$

$$x = 35$$

$$100 \overline{)3500}$$
$$\underline{3\ 0\ 0}$$
$$5\ 0\ 0$$
$$\underline{5\ 0\ 0}$$
$$0$$

Then we use the fact that $x = 35$ to find t.

$$x \cdot 8 = t$$

$$35 \cdot 8 = t$$

$$280 = t$$

$$\begin{array}{r} 3\ 5 \\ \times \quad 8 \\ \hline 2\ 8\ 0 \end{array}$$

4. **Check.** Suppose you run for 280 min. For every 8 min of running, you burn 100 calories. Since $280 \div 8 = 35$, there are 35 groups of 8 min in 280 min, so you will burn $35 \times 100 = 3500$ calories.

5. **State.** You must run for 280 min, or 4 hr 40 min, at a brisk pace in order to lose one pound.

Do Exercise 9. ▶

    The key words, phrases, and concepts in the following table are useful when translating the problems to equations.

Key Words, Phrases, and Concepts

ADDITION (+)	SUBTRACTION (−)	MULTIPLICATION (×)	DIVISION (÷)
add	subtract	multiply	divide
added to	subtracted from	multiplied by	divided by
sum	difference	product	quotient
total	minus	times	repeated subtraction
plus	less than	of	missing factor
more than	decreased by	repeated addition	finding equal quantities
increased by	take away	rectangular arrays	
	how much more		

The following tips are also helpful in problem solving.

PROBLEM-SOLVING TIPS
....

1. Look for patterns when solving problems.

2. When translating in mathematics, consider the dimensions of the variables and constants in the equation. The variables that represent length should all be in the same unit, those that represent money should all be in dollars or all in cents, and so on.

3. Make sure that units appear in the answer whenever appropriate and that you completely answer the original problem.

---

 9. **Weight Loss.** Use the information in Example 7 to determine how long an individual must swim at a brisk pace in order to lose one pound.

  1. **Familiarize.** Let $x =$ the number of hundreds in 3500. Let $t =$ the time it takes to lose one pound.

  2. **Translate.**

$$100 \cdot x = \boxed{}$$

$$x \cdot \boxed{} = t$$

  3. **Solve.** From Example 7, we know that $x = \boxed{}$.

$$x \cdot 2 = t$$

$$\boxed{} \cdot 2 = t$$

$$\boxed{} = t$$

  4. **Check.** Since $70 \div 2 = 35$, there are $\boxed{}$ groups of 2 min in 70 min. Thus, you will burn $35 \times 100 = \boxed{}$ calories.

  5. **State.** You must swim for $\boxed{}$ min, or 1 hr $\boxed{}$ min, in order to lose one pound.

Translating for Success

1. **Brick-Mason Expense.** A commercial contractor is building 30 two-unit condominiums in a retirement community. The brick-mason expense for each building is $10,860. What is the total cost of bricking the buildings?

2. **Heights.** Dean's sons are on the high school basketball team. Their heights are 73 in., 69 in., and 76 in. How much taller is the tallest son than the shortest son?

3. **Account Balance.** James has $423 in his checking account. Then he deposits $73 and uses his debit card for purchases of $76 and $69. How much is left in the account?

4. **Purchasing a Computer.** A computer is on sale for $423. Jenny has only $69. How much more does she need to buy the computer?

5. **Purchasing Coffee Makers.** Sara purchases 8 coffee makers for the newly remodeled bed-and-breakfast that she manages. If she pays $52 for each coffee maker, what is the total cost of her purchase?

The goal of these matching questions is to practice step (2), Translate, of the five-step problem-solving process. Translate each word problem to an equation and select a correct translation from equations A–O.

A. $8 \cdot 52 = n$

B. $69 \cdot n = 76$

C. $73 - 76 - 69 = n$

D. $423 + 73 - 76 - 69 = n$

E. $30 \cdot 10{,}860 = n$

F. $15 \cdot n = 195$

G. $69 + n = 423$

H. $n = 10{,}860 - 300$

I. $n = 423 \div 69$

J. $30 \cdot n = 10{,}860$

K. $15 \cdot 195 = n$

L. $n = 52 - 8$

M. $69 + n = 76$

N. $15 \div 195 = n$

O. $52 + n = 60$

Answers on page A-2

6. **Hourly Rate.** Miller Auto Repair charges $52 per hour for labor. Jackson Auto Care charges $60 per hour. How much more does Jackson charge than Miller?

7. **College Band.** A college band with 195 members marches in a 15-row formation in the homecoming halftime performance. How many members are in each row?

8. **Shoe Purchase.** A college football team purchases 15 pairs of shoes at $195 a pair. What is the total cost of this purchase?

9. **Loan Payment.** Kendra's uncle loans her $10,860, interest free, to buy a car. The loan is to be paid off in 30 payments. How much is each payment?

10. **College Enrollment.** At the beginning of the fall term, the total enrollment in Lakeview Community College was 10,860. By the end of the first two weeks, 300 students had withdrawn. How many students were then enrolled?

✓ Reading Check

List the steps of the problem-solving strategy in order, using the choices given below. The last step is already listed.

Check   Familiarize   Solve   Translate

RC1. 1. _____.

RC2. 2. _____.

RC3. 3. _____.

RC4. 4. _____.

5. State.

Towers Never Built. The buildings shown in the figure below were designed but never completed. Use the information to do Exercises 1–4.

Towers Never Built

SOURCE: http://en.wikipedia.org/wiki/Proposed_tall_buildings_and_structures

1. How much taller would the Aeropolis 2001 have been than the Nakheel Tower?

2. How much taller would the Grollo Tower have been than the Miglin-Beitler Skyneedle?

3. The Willis Tower (formerly the Sears Tower) is the tallest building in Chicago. If the Miglin-Beitler Skyneedle had been built, it would have been 551 ft higher than the Willis Tower. What is the height of the Willis Tower?

4. The Burj Khalifa is the tallest building in Dubai. If the Nakheel Tower had been built, it would have been 1883 ft higher than the Burj Khalifa. What is the height of the Burj Khalifa?

5. *Caffeine Content.* An 8-oz serving of Red Bull energy drink contains 76 milligrams of caffeine. An 8-oz serving of brewed coffee contains 19 more milligrams of caffeine than the energy drink. How many milligrams of caffeine does the 8-oz serving of coffee contain?

Source: The Mayo Clinic

6. *Caffeine Content.* Hershey's 6-oz milk chocolate almond bar contains 25 milligrams of caffeine. A 20-oz bottle of Coca-Cola has 32 more milligrams of caffeine than the Hershey bar. How many milligrams of caffeine does the 20-oz bottle of Coca-Cola have?

Source: *National Geographic,* "Caffeine," by T. R. Reid, January 2005

7. A carpenter drills 216 holes in a rectangular array to construct a pegboard. There are 12 holes in each row. How many rows are there?

8. Lou arranges 504 entries on a spreadsheet in a rectangular array that has 36 rows. How many entries are in each row?

9. *Olympics.* There were 302 events in the 2012 Summer Olympics in London, England. This was 259 more events than there were in the first modern Olympic games in Athens, Greece, in 1896. How many events were there in 1896?

Sources: *USA Today* research; infoplease.com

10. *Drilling Activity.* In 2011, there were 984 rotary rigs drilling for crude oil in the United States. This was 687 more rigs than were active in 2007. Find the number of active rotary oil rigs in 2007.

Source: Energy Information Administration

11. *Boundaries between Countries.* The boundary between mainland United States and Canada including the Great Lakes is 3987 mi long. The length of the boundary between the United States and Mexico is 1933 mi. How much longer is the Canadian border?

Source: U.S. Geological Survey

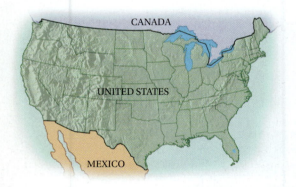

12. *Longest Rivers.* The longest river in the world is the Nile in Africa at about 4135 mi. The longest river in the United States is the Missouri–Mississippi at about 3860 mi. How much longer is the Nile?

13. *Pixels.* A high-definition television (HDTV) screen consists of small rectangular dots called *pixels*. How many pixels are there on a screen that has 1080 rows with 1920 pixels in each row?

Pixel

14. *Crossword Puzzle.* The *USA Today* crossword puzzle is a rectangle containing 15 rows with 15 squares in each row. How many squares does the puzzle have altogether?

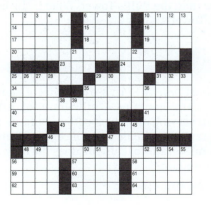

15. There are 24 hr in a day and 7 days in a week. How many hours are there in a week?

16. There are 60 min in an hour and 24 hr in a day. How many minutes are there in a day?

Housing Costs. The graph below shows the average monthly rent for a one-bedroom apartment in several cities in November 2012. Use this graph to do Exercises 17–22.

Average Rent for a One-Bedroom Apartment

SOURCE: rentjungle.com

17. How much higher is the average monthly rent in Seattle than in Dallas?

18. How much lower is the average monthly rent in Phoenix than in Atlanta?

19. Phil, Scott, and Julio plan to rent a one-bedroom apartment in Detroit immediately after graduation, sharing the rent equally. What average monthly rent can each of them expect to pay?

20. Maria and her sister Theresa plan to share a one-bedroom apartment in Dallas, dividing the monthly rent equally between them. About how much can each of them expect to pay?

21. On average, how much rent would a tenant pay for a one-bedroom apartment in Detroit during a 12-month period?

22. On average, how much rent would a tenant pay for a one-bedroom apartment in Seattle during a 6-month period?

23. *Colonial Population.* Before the establishment of the U.S. Census in 1790, it was estimated that the colonial population in 1780 was 2,780,400. This was an increase of 2,628,900 from the population in 1680. What was the colonial population in 1680?

Source: *Time Almanac*

24. *Interstate Speed Limits.* The speed limit for passenger cars on interstate highways in rural areas in Montana is 75 mph. This is 10 mph faster than the speed limit for trucks on the same roads. What is the speed limit for trucks?

25. *Yard-Sale Profit.* Ruth made $312 at her yard sale and divided the money equally among her four grandchildren. How much did each child receive?

26. *Paper Measures.* A quire of paper consists of 25 sheets, and a ream of paper consists of 500 sheets. How many quires are in a ream?

27. *Parking Rates.* The most expensive parking in the United States is found in midtown New York City, where the average rate is $585 per month. This is $545 per month more than in the city with the least expensive rate, Bakersfield, California. What is the average monthly parking rate in Bakersfield?

Source: Colliers International

28. *Trade Balance.* In 2011, international visitors spent $153,000,000,000 traveling in the United States, while Americans spent $110,200,000,000 traveling abroad. How much more was spent by visitors to the United States than by Americans traveling abroad?

Source: U.S. Office of Travel and Tourism Industries

29. *Refrigerator Purchase.* Gourmet Deli has a chain of 24 restaurants. It buys a commercial refrigerator for each store at a cost of $1019 each. Determine the total cost of the purchase.

30. *Microwave Purchase.* Each room in the new dorm at Bridgeway College has a small kitchen. To furnish the kitchens, the college buys 96 microwave ovens at $88 each. Determine the total cost of the purchase.

31. *"Seinfeld."* A local television station plans to air the 177 episodes of the long-running comedy series "Seinfeld." If the station airs 5 episodes per week, how many full weeks will pass before it must begin re-airing previously shown episodes? How many unaired episodes will be shown the following week before the previously aired episodes are rerun?

32. *"Everybody Loves Raymond."* The popular television comedy series "Everybody Loves Raymond" had 208 scripted episodes and 2 additional episodes consisting of clips from previous shows. A local television station plans to air the 208 scripted episodes, showing 5 episodes per week. How many full weeks will pass before it must begin re-airing episodes? How many unaired episodes will be shown the following week before the previously aired episodes are rerun?

33. *Crossword Puzzle.* The *Los Angeles Times* crossword puzzle is a rectangle containing 441 squares arranged in 21 rows. How many columns does the puzzle have?

34. *Mailing Labels.* A box of mailing labels contains 750 labels on 25 sheets. How many labels are on each sheet?

35. Automobile Mileage. The 2013 Hyundai Tucson GLS gets 30 miles per gallon (mpg) in highway driving. How many gallons will it use in 7080 mi of highway driving?

Source: Hyundai

36. Automobile Mileage. The 2013 Volkswagen Jetta (5 cylinder) gets 24 miles per gallon (mpg) in city driving. How many gallons will it use in 3960 mi of city driving?

Source: Volkswagen of America, Inc.

37. High School Court. The standard basketball court used by high school players has dimensions of 50 ft by 84 ft.
   a) What is its area?
   b) What is its perimeter?

84 ft

50 ft

38. College Court. The standard basketball court used by college players has dimensions of 50 ft by 94 ft.
   a) What is its area?
   b) What is its perimeter?
   c) How much greater is the area of a college court than a high school court? (See Exercise 37.)

94 ft

50 ft

39. Copies of this book are usually shipped from the warehouse in cartons containing 24 books each. How many cartons are needed to ship 1344 books?

40. The H. J. Heinz Company ships 16-oz bottles of ketchup in cartons containing 12 bottles each. How many cartons are needed to ship 528 bottles of ketchup?

41. Map Drawing. A map has a scale of 215 mi to the inch. How far apart *in reality* are two cities that are 3 in. apart on the map? How far apart *on the map* are two cities that, in reality, are 1075 mi apart?

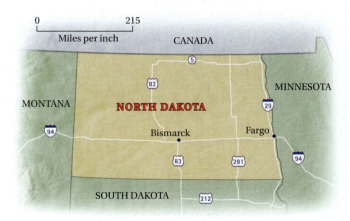

42. Map Drawing. A map has a scale of 288 mi to the inch. How far apart *on the map* are two cities that, in reality, are 2016 mi apart? How far apart *in reality* are two cities that are 8 in. apart on the map?

43. Loan Payments. Dana borrows $5928 for a used car. The loan is to be paid off in 24 equal monthly payments. How much is each payment (excluding interest)?

44. Home Improvement Loan. The Van Reken family borrows $7824 to build a detached garage next to their home. The loan is to be paid off in equal monthly payments of $163 (excluding interest). How many months will it take to pay off the loan?

Refer to the information in Example 7 to do Exercises 45 and 46.

45. For how long must you do aerobic exercises in order to lose one pound?

46. For how long must you bicycle at 9 mph in order to lose one pound?

New Jobs. Many of the fastest-growing occupations in the United States require education beyond a high school diploma. The following table lists some of these and gives the projected numbers of new jobs expected to be created between 2010 and 2020. Use the information in the table for Exercises 47 and 48.

New Jobs Created, 2010–2020

JOB	NUMBER
Registered nurse	711,900
Postsecondary teacher	305,700
Nursing aide	302,000
Elementary school teacher	248,800
Accountant	190,700
Licensed practical nurse	168,500

SOURCE: U.S. Bureau of Labor Statistics

47. The U.S. Bureau of Labor Statistics predicts that between 2010 and 2020, there will be 1,014,100 more new jobs created for registered nurses, nursing aides, and licensed practical nurses than there will be for physicians. How many new jobs will be created for physicians between 2010 and 2020?

48. The U.S. Bureau of Labor Statistics predicts that between 2010 and 2020, there will be 484,600 more new jobs created for postsecondary teachers and elementary teachers than there will be for secondary teachers. How many new jobs will be created for secondary teachers between 2010 and 2020?

49. Seating Configuration. The seats in the Boeing 737-500 airplanes in United Airlines' North American fleet are configured with 2 rows of 4 seats across in first class and 16 rows of 6 seats across in economy class. Determine the total seating capacity of one of these planes.

Source: United Airlines

Economy class: 16 rows of 6 seats First class: 2 rows of 4 seats

50. Seating Configuration. The seats in the Airbus 320 airplanes in United Airlines' North American fleet are configured with 3 rows of 4 seats across in first class and 21 rows of 6 seats across in economy class. Determine the total seating capacity of one of these planes.

Source: United Airlines

Economy class: 21 rows of 6 seats First class: 3 rows of 4 seats

51. Elena buys 5 video games at $64 each and pays for them with $10 bills. How many $10 bills does it take?

52. Pedro buys 5 video games at $64 each and pays for them with $20 bills. How many $20 bills does it take?

53. The balance in Meg's bank account is $568. She uses her debit card for purchases of $46, $87, and $129. Then she deposits $94 in the account after returning a textbook. How much is left in her account?

54. The balance in Dylan's bank account is $749. He uses his debit card for purchases of $34 and $65. Then he makes a deposit of $123 from his paycheck. What is the new balance?

55. *Bones in the Hands and Feet.* There are 27 bones in each human hand and 26 bones in each human foot. How many bones are there in all in the hands and feet?

56. An office for adjunct instructors at a community college has 6 bookshelves, each of which is 3 ft wide. The office is moved to a new location that has dimensions of 16 ft by 21 ft. Is it possible for the bookshelves to be put side by side on the 16-ft wall?

Skill Maintenance

57. Add: [1.2a]

```
  6 2 5 4
  1 5 3 7
+   4 8 2
```

58. Subtract: [1.3a]

```
  9 6 0 2
- 1 8 4 3
```

59. Multiply: [1.4a]

```
  3 4 0 5
× 2 3 7
```

60. Divide: [1.5a]

$$32\overline{)4708}$$

61. Find the perimeter of the figure. [1.2b]

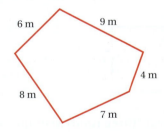

62. Find the area of the region. [1.4b]

211 ft

46 ft

63. Estimate 238×596 by rounding to the nearest hundred. [1.6b]

64. Solve: $x + 15 = 81$. [1.7b]

Synthesis

65. 🖩 *Speed of Light.* Light travels about 186,000 miles per second (mi/sec) in a vacuum such as in outer space. In ice it travels about 142,000 mi/sec, and in glass it travels about 109,000 mi/sec. In 18 sec, how many more miles will light travel in a vacuum than in ice? than in glass?

66. Carney Community College has 1200 students. Each instructor teaches 4 classes, and each student takes 5 classes. There are 30 students and 1 instructor in each classroom. How many instructors are there at Carney Community College?

OBJECTIVES

a Write exponential notation for products such as 4 · 4 · 4.

b Evaluate exponential notation.

c Simplify expressions using the rules for order of operations.

d Remove parentheses within parentheses.

SKILL TO REVIEW

Objective 1.4a: Multiply whole numbers.

Multiply.

1. 5 × 5 × 5

2. 2 × 2 × 2 × 2 × 2

a WRITING EXPONENTIAL NOTATION

Consider the product 3 · 3 · 3 · 3. Such products occur often enough that mathematicians have found it convenient to create a shorter notation, called **exponential notation**, for them. For example,

We read exponential notation as follows.

NOTATION	WORD DESCRIPTION
3^4	"three to the fourth power," or "the fourth power of three"
5^3	"five cubed," or "the cube of five," or "five to the third power," or "the third power of five"
7^2	"seven squared," or "the square of seven," or "seven to the second power," or "the second power of seven"

The wording "seven squared" for 7^2 is derived from the fact that a square with side s has area A given by $A = s^2$.

An expression like $3 \cdot 5^2$ is read "three times five squared," or "three times the square of five."

EXAMPLE 1 Write exponential notation for 10 · 10 · 10 · 10 · 10.

Exponential notation is 10^5.   5 is the *exponent.*
10 is the *base.*

EXAMPLE 2 Write exponential notation for 2 · 2 · 2.

Exponential notation is 2^3.

◀ Do Margin Exercises 1–4.

Write exponential notation.

1. 5 · 5 · 5 · 5

2. 5 · 5 · 5 · 5 · 5 · 5

3. 10 · 10

4. 10 · 10 · 10 · 10

Answers

Skill to Review:
1. 125 **2.** 32

Margin Exercises:
1. 5^4 **2.** 5^6 **3.** 10^2 **4.** 10^4

b EVALUATING EXPONENTIAL NOTATION

We evaluate exponential notation by rewriting it as a product and then computing the product.

EXAMPLE 3 Evaluate: 10^3.
$$10^3 = 10 \cdot 10 \cdot 10 = 1000$$

············· **Caution!** ···········

10^3 does not mean $10 \cdot 3$.

GS **5.** Evaluate: 10^4.

$10^4 = \boxed{} \cdot \boxed{} \cdot \boxed{} \cdot \boxed{} = \boxed{}$

EXAMPLE 4 Evaluate: 5^4.
$$5^4 = 5 \cdot 5 \cdot 5 \cdot 5 = 625$$

Evaluate.

6. 10^2 **7.** 8^3

Do Exercises 5–8. ▶

8. 2^5

c SIMPLIFYING EXPRESSIONS

Suppose we have a calculation like the following:

$$3 + 4 \cdot 8.$$

How do we find the answer? Do we add 3 to 4 and then multiply by 8, or do we multiply 4 by 8 and then add 3? In the first case, the answer is 56. In the second, the answer is 35. We agree to compute as in the second case:

$$3 + 4 \cdot 8 = 3 + 32 = 35.$$

The following rules are an agreement regarding the order in which we perform operations. These are the rules that computers and most scientific calculators use to do computations.

CALCULATOR CORNER

Exponential Notation

Many calculators have a $\boxed{y^x}$ or $\boxed{\wedge}$ key for raising a base to a power. To find 16^3, for example, we press $\boxed{1}$ $\boxed{6}$ $\boxed{y^x}$ $\boxed{3}$ $\boxed{=}$ or $\boxed{1}$ $\boxed{6}$ $\boxed{\wedge}$ $\boxed{3}$ $\boxed{=}$. The result is 4096.

EXERCISES Use a calculator to find each of the following.

1. 3^5
2. 5^6
3. 12^4
4. 2^{11}

RULES FOR ORDER OF OPERATIONS

1. Do all calculations within parentheses ( ), brackets [ ], or braces { } before operations outside.
2. Evaluate all exponential expressions.
3. Do all multiplications and divisions in order from left to right.
4. Do all additions and subtractions in order from left to right.

EXAMPLE 5 Simplify: $16 \div 8 \cdot 2$.

There are no parentheses or exponents, so we begin with the third step.

$\left. \begin{array}{l} 16 \div 8 \cdot 2 = 2 \cdot 2 \\ \qquad\qquad = 4 \end{array} \right\}$ Doing all multiplications and divisions in order from left to right

EXAMPLE 6 Simplify: $7 \cdot 14 - (12 + 18)$.

$$\begin{aligned} 7 \cdot 14 - (12 + 18) &= 7 \cdot 14 - 30 \\ &= 98 - 30 \\ &= 68 \end{aligned}$$

Carrying out operations inside parentheses

Doing all multiplications and divisions

Doing all additions and subtractions

Do Exercises 9–12. ▶

Simplify.

9. $93 - 14 \cdot 3$

10. $104 \div 4 + 4$

11. $25 \cdot 26 - (56 + 10)$

12. $75 \div 5 + (83 - 14)$

Answers

5. 10,000   6. 100   7. 512   8. 32   9. 51
10. 30   11. 584   12. 84

Guided Solution:
5. 10, 10, 10, 10, 10,000

Simplify and compare.

13. $64 \div (32 \div 2)$ and
$(64 \div 32) \div 2$

14. $(28 + 13) + 11$ and
$28 + (13 + 11)$

15. Simplify: **GS**

$9 \times 4 - (20 + 4) \div 8 - (6 - 2)$.

$9 \times 4 - (20 + 4) \div 8 - (6 - 2)$

$\quad = 9 \times 4 - \boxed{} \div 8 - \boxed{}$

$\quad = \boxed{} - 24 \div 8 - 4$

$\quad = 36 - \boxed{} - 4$

$\quad = \boxed{} - 4$

$\quad = \boxed{}$

Simplify.

16. $5 \cdot 5 \cdot 5 + 26 \cdot 71$
$- (16 + 25 \cdot 3)$

17. $30 \div 5 \cdot 2 + 10 \cdot 20 + 8 \cdot 8$
$- 23$

18. $95 - 2 \cdot 2 \cdot 2 \cdot 5 \div (24 - 4)$

Simplify.

19. $5^3 + 26 \cdot 71 - (16 + 25 \cdot 3)$

20. $(1 + 3)^3 + 10 \cdot 20 + 8^2 - 23$

21. $81 - 3^2 \cdot 2 \div (12 - 9)$

EXAMPLE 7   Simplify and compare: $23 - (10 - 9)$ and $(23 - 10) - 9$.
We have

$$23 - (10 - 9) = 23 - 1 = 22;$$
$$(23 - 10) - 9 = 13 - 9 = 4.$$

We can see that $23 - (10 - 9)$ and $(23 - 10) - 9$ represent different numbers. Thus subtraction is not associative.

◀ **Do Exercises 13 and 14.**

EXAMPLE 8   Simplify: $7 \cdot 2 - (12 + 0) \div 3 - (5 - 2)$.

$7 \cdot 2 - (12 + 0) \div 3 - (5 - 2)$

$\quad = 7 \cdot 2 - 12 \div 3 - 3$ Carrying out operations inside parentheses

$\quad = 14 - 4 - 3$ Doing all multiplications and divisions in order from left to right

$\left. \begin{array}{l} = 10 - 3 \\ = 7 \end{array} \right\}$ Doing all additions and subtractions in order from left to right

◀ **Do Exercise 15.**

EXAMPLE 9   Simplify: $15 \div 3 \cdot 2 \div (10 - 8)$.

$15 \div 3 \cdot 2 \div (10 - 8)$

$\quad = 15 \div 3 \cdot 2 \div 2$ Carrying out operations inside parentheses

$\left. \begin{array}{l} = 5 \cdot 2 \div 2 \\ = 10 \div 2 \\ = 5 \end{array} \right\}$ Doing all multiplications and divisions in order from left to right

◀ **Do Exercises 16–18.**

EXAMPLE 10   Simplify: $4^2 \div (10 - 9 + 1)^3 \cdot 3 - 5$.

$4^2 \div (10 - 9 + 1)^3 \cdot 3 - 5$

$\quad = 4^2 \div (1 + 1)^3 \cdot 3 - 5$ Subtracting inside parentheses

$\quad = 4^2 \div 2^3 \cdot 3 - 5$ Adding inside parentheses

$\quad = 16 \div 8 \cdot 3 - 5$ Evaluating exponential expressions

$\left. \begin{array}{l} = 2 \cdot 3 - 5 \\ = 6 - 5 \end{array} \right\}$ Doing all multiplications and divisions in order from left to right

$\quad = 1$ Subtracting

◀ **Do Exercises 19–21.**

Answers

13. 4; 1   **14.** 52; 52   **15.** 29   **16.** 1880
17. 253   **18.** 93   **19.** 1880   **20.** 305
21. 75

Guided Solution:
15. 24, 4, 36, 3, 33, 29

EXAMPLE 11   Simplify: $2^9 \div 2^6 \cdot 2^3$.

$2^9 \div 2^6 \cdot 2^3 = 512 \div 64 \cdot 8$ Since there are no parentheses, we evaluate the exponential expressions.

$\left. \begin{array}{l} = 8 \cdot 8 \\ = 64 \end{array} \right\}$ Doing all multiplications and divisions in order from left to right

Do Exercise 22. ▶

22. Simplify: $2^3 \cdot 2^8 \div 2^9$.

Averages

In order to find the average of a set of numbers, we use addition and then division. For example, the average of 2, 3, 6, and 9 is found as follows.

The number of addends is 4.

$$\text{Average} = \frac{2 + 3 + 6 + 9}{4} = \frac{20}{4} = 5$$

Divide by 4.

The fraction bar acts as a grouping symbol, so

$$\frac{2 + 3 + 6 + 9}{4} \text{ is equivalent to } (2 + 3 + 6 + 9) \div 4.$$

Thus we are using order of operations when we compute an average.

AVERAGE

The **average** of a set of numbers is the sum of the numbers divided by the number of addends.

EXAMPLE 12   *National Parks.*   Since 1995, four national parks have been established in the United States. The sizes of these parks are shown in the figure below. Determine the average size of the four parks.

U.S. National Parks

Size (in acres)

- Black Canyon of the Gunnison National Park, Colorado, 1999 — 32,950
- Cuyahoga Valley National Park, Ohio, 2000 — 32,860
- Congaree National Park, South Carolina, 2003 — 26,550
- Great Sand Dunes National Park and Preserve, Colorado, 2004 — 42,980

SOURCE: us-national-parks.findthedata.org

Answer

22. 4

23. *Average Number of Career Hits.* The numbers of career hits of five Hall of Fame baseball players are given in the graph below. Find the average number of career hits of all five.

Career Hits

SOURCES: Associated Press; Major League Baseball

Simplify.

24. $9 \times 5 + \{6 \div [14 - (5 + 3)]\}$

25. $[18 - (2 + 7) \div 3]$
$\quad - (31 - 10 \times 2)$ **GS**

$= [18 - \boxed{} \div 3] - (31 - 10 \times 2)$
$= [18 - \boxed{}] - (31 - \boxed{})$
$= \boxed{} - \boxed{}$
$= \boxed{}$

The average is given by

$$\frac{32{,}950 + 32{,}860 + 26{,}550 + 42{,}980}{4} = \frac{135{,}340}{4} = 33{,}835.$$

The average size of the four national parks is 33,835 acres.

◀ **Do Exercise 23.**

d REMOVING PARENTHESES WITHIN PARENTHESES

When parentheses occur within parentheses, we can make them different shapes, such as [ ] (called "brackets") and { } (called "braces"). All of these have the same meaning. When parentheses occur within parentheses, computations in the innermost ones are to be done first.

EXAMPLE 13 Simplify: $[25 - (4 + 3) \cdot 3] \div (11 - 7)$.

$[25 - (4 + 3) \cdot 3] \div (11 - 7)$

$= [25 - 7 \cdot 3] \div (11 - 7)$ Doing the calculations in the innermost parentheses first

$= [25 - 21] \div (11 - 7)$ Doing the multiplication in the brackets

$= 4 \div 4$ Subtracting

$= 1$ Dividing

EXAMPLE 14 Simplify: $16 \div 2 + \{40 - [13 - (4 + 2)]\}$.

$16 \div 2 + \{40 - [13 - (4 + 2)]\}$

$= 16 \div 2 + \{40 - [13 - 6]\}$ Doing the calculations in the innermost parentheses first

$= 16 \div 2 + \{40 - 7\}$ Again, doing the calculations in the innermost brackets

$= 16 \div 2 + 33$ Subtracting inside the braces

$= 8 + 33$ Dividing

$= 41$ Adding

◀ **Do Exercises 24 and 25.**

Answers

23. 3118 hits   **24.** 46   **25.** 4

Guided Solution:
25. 9, 3, 20, 15, 11, 4

✓ Reading Check

Complete each statement by choosing the correct word or number from below the blank.

RC1. In the expression 5^3, the number 3 is the _____.
base/exponent

RC2. The expression 9^2 can be read "nine _____."
cubed/squared

RC3. To calculate $10 - 4 \cdot 2$, we perform the _____ first.
multiplication/subtraction

RC4. To find the average of 7, 8, and 9, we add the numbers and divide the sum by _____.
2/3

a Write exponential notation.

1. $3 \cdot 3 \cdot 3 \cdot 3$

2. $2 \cdot 2 \cdot 2 \cdot 2 \cdot 2$

3. $5 \cdot 5$

4. $13 \cdot 13 \cdot 13$

5. $7 \cdot 7 \cdot 7 \cdot 7 \cdot 7$

6. $9 \cdot 9$

7. $10 \cdot 10 \cdot 10$

8. $1 \cdot 1 \cdot 1 \cdot 1$

b Evaluate.

9. 7^2

10. 5^3

11. 9^3

12. 8^2

13. 12^4

14. 10^5

15. 3^5

16. 2^6

c Simplify.

17. $12 + (6 + 4)$

18. $(12 + 6) + 18$

19. $52 - (40 - 8)$

20. $(52 - 40) - 8$

21. $1000 \div (100 \div 10)$

22. $(1000 \div 100) \div 10$

23. $(256 \div 64) \div 4$

24. $256 \div (64 \div 4)$

25. $(2 + 5)^2$

26. $2^2 + 5^2$

27. $(11 - 8)^2 - (18 - 16)^2$

28. $(32 - 27)^3 + (19 + 1)^3$

29. $16 \cdot 24 + 50$

30. $23 + 18 \cdot 20$

31. $83 - 7 \cdot 6$

32. $10 \cdot 7 - 4$

33. $10 \cdot 10 - 3 \cdot 4$

34. $90 - 5 \cdot 5 \cdot 2$

35. $4^3 \div 8 - 4$

36. $8^2 - 8 \cdot 2$

37. $17 \cdot 20 - (17 + 20)$

38. $1000 \div 25 - (15 + 5)$

39. $6 \cdot 10 - 4 \cdot 10$

40. $3 \cdot 8 + 5 \cdot 8$

41. $300 \div 5 + 10$

42. $144 \div 4 - 2$

43. $3 \cdot (2 + 8)^2 - 5 \cdot (4 - 3)^2$

44. $7 \cdot (10 - 3)^2 - 2 \cdot (3 + 1)^2$

45. $4^2 + 8^2 \div 2^2$

46. $6^2 - 3^4 \div 3^3$

47. $10^3 - 10 \cdot 6 - (4 + 5 \cdot 6)$

48. $7^2 + 20 \cdot 4 - (28 + 9 \cdot 2)$

49. $6 \cdot 11 - (7 + 3) \div 5 - (6 - 4)$

50. $8 \times 9 - (12 - 8) \div 4 - (10 - 7)$

51. $120 - 3^3 \cdot 4 \div (5 \cdot 6 - 6 \cdot 4)$

52. $80 - 2^4 \cdot 15 \div (7 \cdot 5 - 45 \div 3)$

53. $2^3 \cdot 2^8 \div 2^6$

54. $2^7 \div 2^5 \cdot 2^4 \div 2^2$

55. Find the average of $64, $97, and $121.

56. Find the average of four test grades of 86, 92, 80, and 78.

57. Find the average of 320, 128, 276, and 880.

58. Find the average of $1025, $775, $2062, $942, and $3721.

d Simplify.

59. $8 \times 13 + \{42 \div [18 - (6 + 5)]\}$

60. $72 \div 6 - \{2 \times [9 - (4 \times 2)]\}$

61. $[14 - (3 + 5) \div 2] - [18 \div (8 - 2)]$

62. $[92 \times (6 - 4) \div 8] + [7 \times (8 - 3)]$

63. $(82 - 14) \times [(10 + 45 \div 5) - (6 \cdot 6 - 5 \cdot 5)]$

64. $(18 \div 2) \cdot \{[(9 \cdot 9 - 1) \div 2] - [5 \cdot 20 - (7 \cdot 9 - 2)]\}$

65. $4 \times \{(200 - 50 \div 5) - [(35 \div 7) \cdot (35 \div 7) - 4 \times 3]\}$

66. $15(23 - 4 \cdot 2)^3 \div (3 \cdot 25)$

67. $\{[18 - 2 \cdot 6] - [40 \div (17 - 9)]\} + \{48 - 13 \times 3 + [(50 - 7 \cdot 5) + 2]\}$

68. $(19 - 2^4)^5 - (141 \div 47)^2$

Skill Maintenance

Solve. [1.7b]

69. $x + 341 = 793$

70. $4197 + x = 5032$

71. $7 \cdot x = 91$

72. $1554 = 42 \cdot y$

73. $6000 = 1102 + t$

74. $10,000 = 100 \cdot t$

Solve. [1.8a]

75. *Colorado.* The state of Colorado is roughly the shape of a rectangle that is 273 mi by 382 mi. What is its area?

76. On a long four-day trip, a family bought the following amounts of gasoline for their motor home: 23 gal, 24 gal, 26 gal, and 25 gal. How much gasoline did they buy in all?

Synthesis

Each of the answers in Exercises 77–79 is incorrect. First find the correct answer. Then place as many parentheses as needed in the expression in order to make the incorrect answer correct.

77. $1 + 5 \cdot 4 + 3 = 36$

78. $12 \div 4 + 2 \cdot 3 - 2 = 2$

79. $12 \div 4 + 2 \cdot 3 - 2 = 4$

80. Use one occurrence each of 1, 2, 3, 4, 5, 6, 7, 8, and 9, in order, and any of the symbols $+, -, \cdot, \div,$ and $(\ )$ to represent 100.

Vocabulary Reinforcement

In each of Exercises 1–8, fill in the blank with the correct term from the given list. Some of the choices may not be used and some may be used more than once.

1. The distance around an object is its _____. [1.2b]

2. The _____ is the number from which another number is being subtracted. [1.3a]

3. For large numbers, _____ are separated by commas into groups of three, called _____. [1.1a]

4. In the sentence $28 \div 7 = 4$, the _____ is 28. [1.5a]

5. In the sentence $10 \times 1000 = 10{,}000$, 10 and 1000 are called _____ and 10,000 is called the _____. [1.4a]

6. The number 0 is called the _____ identity. [1.2a]

7. The sentence $3 \times (6 \times 2) = (3 \times 6) \times 2$ illustrates the _____ law of multiplication. [1.4a]

8. We can use the following statement to check division: quotient · _____ + _____ = _____. [1.5a]

associative
commutative
addends
factors
area
perimeter
minuend
subtrahend
product
digits
periods
additive
multiplicative
dividend
quotient
remainder
divisor

Concept Reinforcement

Determine whether each statement is true or false.

_____ 1. $a > b$ is true when a is to the right of b on the number line. [1.6c]

_____ 2. Any nonzero number divided by itself is 1. [1.5a]

_____ 3. For any whole number a, $a \div 0 = 0$. [1.5a]

_____ 4. Every equation is true. [1.7a]

_____ 5. The rules for order of operations tell us to multiply and divide before adding and subtracting. [1.9c]

_____ 6. The average of three numbers is the middle number. [1.9c]

Study Guide

Objective 1.1a Give the meaning of digits in standard notation.

Example What does the digit 7 mean in 2,379,465?

$$2,3\ 7\ 9,4\ 6\ 5$$

7 means 7 ten thousands.

Practice Exercise

1. What does the digit 2 mean in 432,079?

Objective 1.2a Add whole numbers.

Example Add: 7368 + 3547.

```
    1 1
  7 3 6 8
+ 3 5 4 7
1 0,9 1 5
```

Practice Exercise

2. Add: 36,047 + 29,255.

Objective 1.3a Subtract whole numbers.

Example Subtract: 8045 − 2897.

```
        13
    7 9 3 15
  8 0 4 5
− 2 8 9 7
  5 1 4 8
```

Practice Exercise

3. Subtract: 4805 − 1568.

Objective 1.4a Multiply whole numbers.

Example Multiply: 57 × 315.

```
        2
    1   3
    3 1 5
×     5 7
  2 2 0 5   ← 315 × 7
1 5 7 5 0   ← 315 × 50
1 7,9 5 5
```

Practice Exercise

4. Multiply: 329 × 684.

Objective 1.5a Divide whole numbers.

Example Divide: 6463 ÷ 26.

```
        2 4 8
2 6 ) 6 4 6 3
      5 2
      1 2 6
      1 0 4
        2 2 3
        2 0 8
          1 5
```

The answer is 248 R 15.

Practice Exercise

5. Divide: 8519 ÷ 27.

Objective 1.6a Round to the nearest ten, hundred, or thousand.

Example Round to the nearest thousand.

6 4 7 1
↑

The digit 6 is in the thousands place. We consider the next digit to the right. Since the digit, 4, is 4 or lower, we round down, meaning that 6 thousands stays as 6 thousands. Change all digits to the right of the thousands digit to zeros. The answer is 6000.

Practice Exercise

6. Round 36,468 to the nearest thousand.

Objective 1.6c Use < or > for ☐ to write a true sentence in a situation like 6 ☐ 10.

Example Use < or > for ☐ to write a true sentence:

34 ☐ 29.

Since 34 is to the right of 29 on the number line,

34 > 29.

Practice Exercise

7. Use < or > for ☐ to write a true sentence:
   78 ☐ 81.

Objective 1.7b Solve equations like $t + 28 = 54$, $28 \cdot x = 168$, and $98 \cdot 2 = y$.

Example Solve: $y + 12 = 27$.

$$y + 12 = 27$$
$$y + 12 - 12 = 27 - 12$$
$$y + 0 = 15$$
$$y = 15$$

The solution is 15.

Practice Exercise

8. Solve: $24 \cdot x = 864$.

Objective 1.9b Evaluate exponential notation.

Example Evaluate: 5^4.

$$5^4 = 5 \cdot 5 \cdot 5 \cdot 5 = 625$$

Practice Exercise

9. Evaluate: 6^3.

Review Exercises

The review exercises that follow are for practice. Answers are given at the back of the book. If you miss an exercise, restudy the objective indicated in red next to the exercise or on the direction line that precedes it.

1. What does the digit 8 mean in 4,678,952? [1.1a]

2. In 13,768,940, what digit tells the number of millions? [1.1a]

Write expanded notation. [1.1b]

3. 2793

4. 56,078

5. 4,007,101

Write a word name. [1.1c]

6. 67,819

7. 2,781,427

Write standard notation. [1.1c]

8. Four hundred seventy-six thousand, five hundred eighty-eight

9. *Subway Ridership.* Ridership on the New York City Subway system totaled one billion, six hundred forty million in 2011.

Source: Metropolitan Transit Authority

Add. [1.2a]

10. $7304 + 6968$

11. $27{,}609 + 38{,}415$

12. $2703 + 4125 + 6004 + 8956$

13. $\begin{array}{r} 9\,1{,}4\,2\,6 \\ +\quad 7{,}4\,9\,5 \\ \hline \end{array}$

Subtract. [1.3a]

14. $8045 - 2897$

15. $9001 - 7312$

16. $6003 - 3729$

17. $\begin{array}{r} 3\,7{,}4\,0\,5 \\ -\,1\,9{,}6\,4\,8 \\ \hline \end{array}$

Multiply. [1.4a]

18. $17{,}000 \cdot 300$

19. $7846 \cdot 800$

20. $726 \cdot 698$

21. $587 \cdot 47$

22. $\begin{array}{r} 8\,3\,0\,5 \\ \times\quad 6\,4\,2 \\ \hline \end{array}$

Divide. [1.5a]

23. $63 \div 5$

24. $80 \div 16$

25. $7\,\overline{)\,6\,3\,9\,4}$

26. $3073 \div 8$

27. $6\,0\,\overline{)\,2\,8\,6}$

28. $4266 \div 79$

29. $3\,8\,\overline{)\,1\,7{,}1\,7\,6}$

30. $1\,4\,\overline{)\,7\,0{,}1\,1\,2}$

31. $52{,}668 \div 12$

Round 345,759 to the nearest: [1.6a]

32. Hundred.

33. Ten.

34. Thousand.

35. Hundred thousand.

Use $<$ or $>$ for ☐ to write a true sentence. [1.6c]

36. 67 ☐ 56

37. 1 ☐ 23

Estimate each sum, difference, or product by first rounding to the nearest hundred. Show your work. [1.6b]

38. $41{,}348 + 19{,}749$

39. $38{,}652 - 24{,}549$

40. $396 \cdot 748$

Solve. [1.7b]

41. $46 \cdot n = 368$

42. $47 + x = 92$

43. $1 \cdot y = 58$

44. $24 = x + 24$

45. Write exponential notation: $4 \cdot 4 \cdot 4$. [1.9a]

Evaluate. [1.9b]

46. 10^4

47. 6^2

Simplify. [1.9c, d]

48. $8 \cdot 6 + 17$

49. $10 \cdot 24 - (18 + 2) \div 4 - (9 - 7)$

50. $(80 \div 16) \times [(20 - 56 \div 8) + (8 \cdot 8 - 5 \cdot 5)]$

51. Find the average of 157, 170, and 168. [1.9c]

Solve. [1.8a]

52. *Computer Purchase.* Natasha has $196 and wants to buy a computer for $698. How much more does she need?

53. Toni has $406 in her checking account. She is paid $78 for a part-time job and deposits that in her checking account. How much is then in her account?

54. *Lincoln-Head Pennies.* In 1909, the first Lincoln-head pennies were minted. Seventy-three years later, these pennies were first minted with a decreased copper content. In what year was the copper content reduced?

55. A beverage company packed 228 cans of soda into 12-can cartons. How many cartons were filled?

56. An apartment builder bought 13 gas stoves at $425 each and 13 refrigerators at $620 each. What was the total cost?

57. An apple farmer keeps bees in her orchard to help pollinate the apple blossoms. The bees from an average beehive can pollinate 30 surrounding trees during one growing season. A farmer has 420 trees. How many beehives does she need to pollinate all of them?

Source: Jordan Orchards, Westminster, PA

58. *Olympic Trampoline.* Shown below is an Olympic trampoline. Determine the area and the perimeter of the trampoline. [1.2b], [1.4b]

Source: International Trampoline Industry Association, Inc.

59. A chemist has 2753 mL of alcohol. How many 20-mL beakers can be filled? How much will be left over?

60. A family budgeted $7825 a year for food and clothing and $2860 for entertainment. The yearly income of the family was $38,283. How much of this income remained after these two allotments?

61. Simplify: $7 + (4 + 3)^2$. [1.9c]
A. 32   B. 56
C. 151   D. 196

62. Simplify: $7 + 4^2 + 3^2$. [1.9c]
A. 32   B. 56
C. 130   D. 196

63. $[46 - (4 - 2) \cdot 5] \div 2 + 4$ [1.9d]
A. 6   B. 20
C. 114   D. 22

Synthesis

64. ▦ Determine the missing digit d. [1.4a]

$$\begin{array}{r} 9\,d \\ \times\ \ d\,2 \\ \hline 8\,0\,3\,6 \end{array}$$

65. ▦ Determine the missing digits a and b. [1.5a]

$$\begin{array}{r} 9\,a\,1 \\ 2\,b\,1\,)\overline{\,2\,3\,6,4\,2\,1} \end{array}$$

66. A mining company estimates that a crew must tunnel 2000 ft into a mountain to reach a deposit of copper ore. Each day, the crew tunnels about 500 ft. Each night, about 200 ft of loose rocks roll back into the tunnel. How many days will it take the mining company to reach the copper deposit? [1.8a]

Understanding Through Discussion and Writing

1. Is subtraction associative? Why or why not? [1.3a]

2. Explain how estimating and rounding can be useful when shopping for groceries. [1.6b]

3. Write a problem for a classmate to solve. Design the problem so that the solution is "The driver still has 329 mi to travel." [1.8a]

4. Consider the expressions $9 - (4 \cdot 2)$ and $(3 \cdot 4)^2$. Are the parentheses necessary in each case? Explain. [1.9c]

CHAPTER

1   Test

For Extra Help

For step-by-step test solutions, access the Chapter Test Prep Videos in MyMathLab® or on YouTube (search "BittingerPrealgebra" and click on Channels).

1. In the number 546,789, which digit tells the number of hundred thousands?

2. Write expanded notation: 8843.

3. Write a word name: 38,403,277.

Add.

4.
$$\begin{array}{r} 6811 \\ +\ 3178 \end{array}$$

5.
$$\begin{array}{r} 45{,}889 \\ +\ 17{,}902 \end{array}$$

6.
$$\begin{array}{r} 1239 \\ 843 \\ 301 \\ +\ \ 782 \end{array}$$

7.
$$\begin{array}{r} 6203 \\ +\ 4312 \end{array}$$

Subtract.

8.
$$\begin{array}{r} 7983 \\ -\ 4353 \end{array}$$

9.
$$\begin{array}{r} 2974 \\ -\ 1935 \end{array}$$

10.
$$\begin{array}{r} 8907 \\ -\ 2059 \end{array}$$

11.
$$\begin{array}{r} 23{,}067 \\ -\ 17{,}892 \end{array}$$

Multiply.

12.
$$\begin{array}{r} 4568 \\ \times\ \ \ \ \ \ 9 \end{array}$$

13.
$$\begin{array}{r} 8876 \\ \times\ \ \ 600 \end{array}$$

14.
$$\begin{array}{r} 65 \\ \times\ 37 \end{array}$$

15.
$$\begin{array}{r} 678 \\ \times\ 788 \end{array}$$

Divide.

16. $15 \div 4$

17. $420 \div 6$

18. $89 \overline{)8633}$

19. $44 \overline{)35{,}428}$

Round 34,528 to the nearest:

20. Thousand.

21. Ten.

22. Hundred.

Estimate each sum, difference, or product by first rounding to the nearest hundred. Show your work.

23.
$$\begin{array}{r} 23{,}649 \\ +\ 54{,}746 \end{array}$$

24.
$$\begin{array}{r} 54{,}751 \\ -\ 23{,}649 \end{array}$$

25.
$$\begin{array}{r} 824 \\ \times\ 489 \end{array}$$

Use < or > for ☐ to write a true sentence.

26. 34 ☐ 17

27. 117 ☐ 157

Solve.

28. $28 + x = 74$

29. $169 \div 13 = n$

30. $38 \cdot y = 532$

31. $381 = 0 + a$

Solve.

32. *Calorie Content.* An 8-oz serving of whole milk contains 146 calories. This is 63 calories more than the number of calories in an 8-oz serving of skim milk. How many calories are in an 8-oz serving of skim milk?

Source: *American Journal of Clinical Nutrition*

34. *Largest States.* The following table lists the five largest states in terms of their land area. Find the total land area of these states.

STATE	AREA (in square miles)
Alaska	571,951
Texas	261,797
California	155,959
Montana	145,552
New Mexico	121,356

Sources: U.S. Department of Commerce; U.S. Census Bureau

36. *Hostess Ding Dongs®.* Hostess packages its Ding Dong snack cakes in 12-packs. How many 12-packs can it fill with 22,231 cakes? How many will be left over?

38. Write exponential notation: $12 \cdot 12 \cdot 12 \cdot 12$.

Evaluate.

39. 7^3

Simplify.

41. $35 - 1 \cdot 28 \div 4 + 3$

42. $10^2 - 2^2 \div 2$

43. $(25 - 15) \div 5$

44. $2^4 + 24 \div 12$

45. $8 \times \{(20 - 11) \cdot [(12 + 48) \div 6 - (9 - 2)]\}$

46. Find the average of 97, 99, 87, and 89.

A. 93   **B.** 124   **C.** 186   **D.** 372

Synthesis

47. An open cardboard container is 8 in. wide, 12 in. long, and 6 in. high. How many square inches of cardboard are used?

49. Cara spends $229 a month to repay her student loan. If she has already paid $9160 on the 10-year loan, how many payments remain?

33. A box contains 5000 staples. How many staplers can be filled from the box if each stapler holds 250 staples?

35. *Pool Tables.* The Bradford™ pool table made by Brunswick Billiards comes in three sizes of playing area, 50 in. by 100 in., 44 in. by 88 in., and 38 in. by 76 in.

Source: Brunswick Billiards

a) Determine the perimeter and the playing area of each table.
b) By how much does the area of the largest table exceed the area of the smallest table?

37. *Office Supplies.* Morgan manages the office of a small graphics firm. He buys 3 black inkjet cartridges at $15 each and 2 photo inkjet cartridges at $25 each. How much does the purchase cost?

40. 10^5

48. Use trials to find the single-digit number a for which

$$359 - 46 + a \div 3 \times 25 - 7^2 = 339.$$

Introduction to Integers and Algebraic Expressions

2.1   Integers and the Number Line

OBJECTIVES

a State the integer that corresponds to a real-world situation.

b Form a true sentence using $<$ or $>$.

c Find the absolute value of any integer.

d Find the opposite of any integer.

SKILL TO REVIEW

Objective 1.6c: Use $<$ or $>$ for ☐ to write a true sentence in a situation like 6 ☐ 10.

Use $<$ or $>$ for ☐ to write a true sentence.

1. 0 ☐ 10   **2.** 51 ☐ 15

In this section, we extend the set of whole numbers to form the set of *integers*. You have probably already used negative numbers. For example, the outside temperature could drop to *negative five* degrees, or a credit card statement could indicate activity of *negative forty-eight* dollars.

To describe integers, we start with the whole numbers, 0, 1, 2, 3, and so on. For each number 1, 2, 3, and so on, we obtain a new number that is the same number of units to the left of zero on the number line.

For the number 1, there is the *opposite* number -1 (negative 1).

For the number 2, there is the *opposite* number -2 (negative 2).

For the number 3, there is the *opposite* number -3 (negative 3), and so on.

The **integers** consist of the whole numbers and these new numbers. We picture them on the number line as follows.

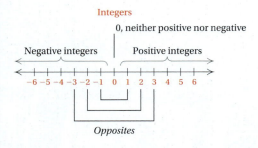

We call the integers to the left of zero **negative integers**. Those to the right of zero are called **positive integers**. Zero is neither positive nor negative. We call -1 and 1 **opposites** of each other. Similarly, -2 and 2 are opposites, -3 and 3 are opposites, -100 and 100 are opposites, and 0 is its own opposite.

INTEGERS

The **integers**: . . . , $-5, -4, -3, -2, -1, 0, 1, 2, 3, 4, 5,$ . . .

---

a INTEGERS AND THE REAL WORLD

Integers correspond to many real-world problems and situations. The following examples will help you get ready to translate problem situations to mathematical language.

Answers

Skill to Review:

1. $<$ 2. $>$

EXAMPLE 1 Tell which integer corresponds to this situation: The temperature is 4 degrees below zero.

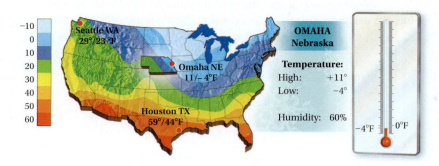

The integer −4 corresponds to the situation. The temperature is −4°. ■

EXAMPLE 2 *Stock Price Change.* Tell which integers correspond to this situation: Hal owns a stock whose price decreased from $27 per share to $11 per share over a recent time period. He owns another stock whose price increased from $20 per share to $22 per share over the same time period.

The integer −16 corresponds to the decrease in the value of the first stock. The integer 2 represents the increase in the value of the second stock.

Do Exercises 1–5. ▶

b ORDER ON THE NUMBER LINE

Numbers are written in order on the number line, increasing as we move from left to right. For any two numbers on the line, the one to the left is *less than* the one to the right.

Since the symbol < means "is less than," the sentence −5 < 9 is read "−5 is less than 9." The symbol > means "is greater than," so the sentence −4 > −8 is read "−4 is greater than −8."

EXAMPLES Use either < or > for ☐ to form a true sentence.

3. −9 ☐ 2   Since −9 is to the left of 2, we have −9 < 2.
4. 4 ☐ −6   Since 4 is to the right of −6, we have 4 > −6.
5. −8 ☐ −1   Since −8 is to the left of −1, we have −8 < −1.

Do Exercises 6–9. ▶

c ABSOLUTE VALUE

From the number line, we see that some integers, like 4 and −4, are the same distance from zero. We call the distance of a number from zero the **absolute value** of the number. Since distance is always a nonnegative number, absolute value is always nonnegative.

Tell which integers correspond to each situation.

1. *High and Low Temperatures.* As of 2010, the highest recorded temperature in Illinois was 117°F on July 14, 1954, in East St. Louis. The lowest recorded temperature in Illinois was 36°F below zero on January 5, 1999, in Congerville.
   Source: National Climate Data Center, NESDIS, NOAA, U.S. Dept. of Commerce

2. *Stock Decrease.* The price of a stock decreased from $41 per share to $38 per share over a recent period.

3. At 10 sec before liftoff, ignition occurs. At 148 sec after liftoff, the first stage is detached from the rocket.

4. The halfback gained 8 yd on first down. The quarterback was sacked for a 5-yd loss on second down.

5. A submarine dived 120 ft, rose 50 ft, and then dived 80 ft.

Use either < or > for ☐ to write a true sentence.

6. −3 ☐ 7

7. −8 ☐ −5

8. 7 ☐ −10

9. −4 ☐ −20

Answers

1. 117; −36   **2.** The integer −3 corresponds to the decrease in the value of the stock.   **3.** −10; 148   **4.** 8; −5   **5.** −120; 50; −80   **6.** <   **7.** <   **8.** >   **9.** >

ABSOLUTE VALUE

The **absolute value** of a number is its distance from zero on the number line. We use the symbol $|x|$ to represent the absolute value of a number x.

The distance of -4 from 0 is 4.   The distance of 4 from 0 is 4.
The absolute value of -4 is 4.   The absolute value of 4 is 4.

$|-4|=4$ $|4|=4$

-4 0   4

4 units   4 units

To find the absolute value of a number:

a) If a number is negative, its absolute value is its opposite.

b) If a number is positive or zero, its absolute value is the same as the number.

Find the absolute value.

10. $|18|$ **GS**

The distance of 18 from 0 is ___ , so $|18|$ = ___ .

11. $|-9|$ **12.** $|-29|$

13. $|52|$

EXAMPLES Find the absolute value of each number.

6. $|-3|$ The distance of -3 from 0 is 3, so $|-3| = 3$.

7. $|25|$ The distance of 25 from 0 is 25, so $|25| = 25$.

8. $|0|$ The distance of 0 from 0 is 0, so $|0| = 0$.

◀ **Do Exercises 10–13.**

---

d OPPOSITES

Given a number on one side of 0 on the number line, we can get a number on the other side by *reflecting* the number across zero. For example, the *reflection* of 2 is -2. We can read -2 as "negative 2" or "the opposite of 2."

NOTATION FOR OPPOSITES

The **opposite** of a number x is written $-x$ (read "the opposite of x").

The opposite of a number is also called its *additive inverse*.

EXAMPLE 9 If x is -3, find $-x$.

To find the opposite of x when x is -3, we reflect -3 to the other side of 0.

When $x = -3$, $-x = -(-3)$. We have $-(-3) = 3$. The opposite of -3 is 3.

Answers

10. 18 **11.** 9 **12.** 29 **13.** 52

Guided Solution:
10. 18, 18

When we replace a variable with a number and find the value of an expression we say that we are **evaluating** the expression.

EXAMPLE 10   Evaluate $-x$ when x is 4.

To find the opposite of x when x is 4, we reflect 4 to the other side of 0.

We have $-(4) = -4$. The opposite of 4 is -4.

EXAMPLE 11   Evaluate $-x$ when x is 0.

$-x = 0$ when x is 0.   The opposite of 0 is 0; $-0 = 0$.

Do Exercises 14–16. ▶

A negative number is sometimes said to have a negative *sign*. A positive number is said to have a positive sign, even though it is rarely written. Thus, for example, -7 has a negative sign and 23 has a positive sign. Replacing a number with its opposite, or additive inverse, is sometimes called *changing the sign*.

EXAMPLES   Change the sign (find the opposite, or additive inverse) of each number.

12. -6 $-(-6) = 6$ **13.** -10 $-(-10) = 10$

14. 0 $-(0) = 0$ **15.** 14 $-(14) = -14$

Do Exercises 17–20. ▶

EXAMPLE 16   If x is 2, find $-(-x)$.

If $x = 2$, then $-(-x) = -(-2) = 2$.   The opposite of the opposite of 2 is 2.

EXAMPLE 17   Evaluate $-(-x)$ for $x = -4$.

If $x = -4$, then $-(-x) = -(-(-4)) = -(4) = -4$.   The opposite of the opposite of -4 is -4.

When we change a number's sign twice, we return to the original number.

Do Exercises 21 and 22. ▶

It is important not to confuse parentheses with absolute-value symbols.

EXAMPLE 18   Evaluate $-|-x|$ for $x = 2$.

If $x = 2$, then $-|-x| = -|-2| = -2$.   The absolute value of -2 is 2.

Note that $-(-2) = 2$, whereas $-|-2| = -2$.

Do Exercises 23 and 24. ▶

In each case, draw a number line, if necessary.

14. Find $-x$ when x is 1.

15. Find $-x$ when x is 0.

16. Evaluate $-x$ when x is -2.

Change the sign. (Find the opposite, or additive inverse.)

17. -4 **18.** -13

19. 39   **20.** 0

21. If x is 7, find $-(-x)$.

GS **22.** Evaluate $-(-x)$ for $x = -2$.

$$-(-x) = -(-())$$
$$= -() = $$

23. Find $-|-7|$.

24. Find $-|-39|$.

Answers

14. -1 **15.** 0   **16.** 2   **17.** 4   **18.** 13
19. -39 **20.** 0   **21.** 7   **22.** -2 **23.** -7
24. -39

Guided Solution:
22. $-2, 2, -2$

☑ Reading Check

Use the number line below, on which the letters name numbers, for Exercises RC1–RC8.

Determine whether each statement is true or false.

_____ **RC1.** $K < B$ _____ **RC2.** $H < B$ _____ **RC3.** $E < C$ _____ **RC4.** $J > D$

_____ **RC5.** $|K| = 4$ _____ **RC6.** $|H| = |B|$ _____ **RC7.** $A = -F$ _____ **RC8.** $-G = H$

a Tell which integers correspond to each situation.

1. At tax time, Janine received an $820 refund while David owed $541.

2. A student deposited her tax refund of $750 in a savings account. Two weeks later, she withdrew $125 to pay technology fees.

3. *Temperature Extremes.* The highest temperature ever created on Earth was 950,000,000°F. The lowest temperature ever created was approximately 460°F below zero.

Source: *The Guinness Book of Records*

4. *Shipwreck.* There are numerous shipwrecks to explore near Bermuda. One of the most frequently visited wrecks is *L'Herminie*, a French warship that sank in 1837. This wreck is 35 ft below the surface.

Source: www./10best.com/interests/adventure/scuba-diving-in-pirate-territory/

5. *Empire State Building.* The Empire State Building has a total height above ground level, including the lightning rod at the top, of 1454 ft. The foundation depth is 55 ft below ground level.

Source: www.empirestatebuildingfacts.com

6. *Extreme Climate.* Verkhoyansk, a river port in northeast Siberia, has the most extreme climate on the planet. Its average monthly winter temperature is 59°F below zero, and its average monthly summer temperature is 57°F.

Source: *The Guinness Book of Records*

b Use either $<$ or $>$ for ☐ to form a true sentence.

7. -8 ☐ 0 **8.** 7 ☐ 0 **9.** 9 ☐ 0 **10.** -7 ☐ 0 **11.** 8 ☐ -8

12. 6 ☐ -6 **13.** -6 ☐ -4 **14.** -1 ☐ -7 **15.** -8 ☐ -5 **16.** -5 ☐ -3

17. -13 ☐ -9 **18.** -5 ☐ -11 **19.** -3 ☐ -4 **20.** -6 ☐ -5

c Find the absolute value.

21. $|57|$ **22.** $|11|$ **23.** $|0|$ **24.** $|-4|$ **25.** $|-24|$

26. $|-36|$ **27.** $|53|$ **28.** $|54|$ **29.** $|-8|$ **30.** $|-79|$

d Find $-x$ when x is each of the following.

31. -7 **32.** -6 **33.** 7 **34.** 6 **35.** 0 **36.** -1

Change the sign. (Find the opposite, or additive inverse.)

37. -21 **38.** -67 **39.** 53 **40.** 0 **41.** -1 **42.** 16

Evaluate $-(-x)$ when x is each of the following.

43. 7 **44.** -8 **45.** -9 **46.** 3 **47.** -17 **48.** -19

49. 23 **50.** 0 **51.** -1 **52.** 73 **53.** 85 **54.** -37

Evaluate $-|-x|$ when x is each of the following.

55. 345 **56.** 729 **57.** 0 **58.** 1 **59.** -8 **60.** -3

Skill Maintenance

61. Add: $327 + 498$.   [1.2a]

62. Evaluate: 5^3.   [1.9b]

63. Multiply: $209 \cdot 34$.   [1.4a]

64. Solve: $300 \cdot x = 1200$.   [1.7b]

65. Simplify: $7(9 - 3)$.   [1.9c]

66. Simplify: $9^2 - 3[2 + (10 - 8)]$.   [1.9d]

Synthesis

Use either $<$, $>$, or $=$ for ☐ to write a true sentence.

67. $|-5|$ ☐ $|-2|$

68. $|4|$ ☐ $|-7|$

69. $|-8|$ ☐ $|8|$

Solve. Consider only integer replacements.

70. $|x| = 7$

71. $|x| < 2$

72. Simplify $-(-x)$, $-(-(-x))$, and $-(-(-(-x)))$.

73. List these integers in order from least to greatest.
2^{10}, -5, $|-6|$, 4, $|3|$, -100, 0, 2^7, 7^2, 10^2

OBJECTIVE

 a Add integers without using the number line.

Add using the number line.

1. $1 + (-4)$

2. $-3 + (-2)$

3. $-3 + 7$

4. $-5 + 5$

For each illustration, write a corresponding addition sentence.

5.

6.

7.

a ADDITION

To explain addition of integers, we can use the number line. Once our understanding is developed, we will streamline our approach.

Addition on the Number Line

To find $a + b$, we start at a and then move according to b.

a) If b is positive, we move from a to the right.

b) If b is negative, we move from a to the left.

c) If b is 0, we stay at a.

EXAMPLE 1 Add: $2 + (-5)$.

$2 + (-5) = -3$

EXAMPLE 2 Add: $-1 + (-3)$.

$-1 + (-3) = -4$

EXAMPLE 3 Add: $-4 + 9$.

$-4 + 9 = 5$

◄ **Do Exercises 1–7.**

You may have noticed a pattern in Example 2 and Margin Exercises 2 and 6. When two negative integers are added, the result is negative.

ADDING NEGATIVE INTEGERS

To add two negative integers, add their absolute values and change the sign (making the answer negative).

Answers

1. -3 **2.** -5 **3.** 4 **4.** 0
5. $4 + (-5) = -1$ **6.** $-2 + (-4) = -6$
7. $-3 + 8 = 5$

EXAMPLES Add.

4. $-5 + (-7) = -12$ *Think*: Add the absolute values: $5 + 7 = 12$.
Make the answer negative, -12.

5. $-8 + (-2) = -10$

Do Exercises 8–11. ▶

When the number 0 is added to any number, that number remains unchanged. For this reason, the number 0 is referred to as the **additive identity**.

EXAMPLES Add.

6. $-4 + 0 = -4$ **7.** $0 + (-9) = -9$ **8.** $17 + 0 = 17$

Do Exercises 12–14. ▶

When we add a positive integer and a negative integer, as in Examples 1 and 3, the sign of the number with the greater absolute value is the sign of the answer.

ADDING POSITIVE AND NEGATIVE INTEGERS

To add a positive integer and a negative integer, find the difference of their absolute values.

a) If the negative integer has the greater absolute value, the answer is negative.

b) If the positive integer has the greater absolute value, the answer is positive.

c) If the integers have the same absolute value, the answer is 0.

EXAMPLES Add.

9. $3 + (-5) = -2$ *Think*: The absolute values are 3 and 5. The difference is 2. Since the negative number has the larger absolute value, the answer is *negative*, -2.

10. $11 + (-8) = 3$ *Think*: The absolute values are 11 and 8. The difference is 3. The positive number has the larger absolute value, so the answer is *positive*, 3.

11. $-7 + 4 = -3$
12. $-6 + 10 = 4$
13. $9 + (-9) = 0$

Do Exercises 15–19. ▶

Sometimes $-a$ is referred to as the **additive inverse** of a because adding any number to its additive inverse always results in the additive identity, 0.

ADDING OPPOSITES

For any integer a,
$$a + (-a) = -a + a = 0.$$

(The sum of any number and its additive inverse, or opposite, is 0.)

Add. Do not use the number line except as a check.

8. $-5 + (-6)$

9. $-9 + (-3)$

10. $-20 + (-14)$

11. $-11 + (-11)$

Add.

12. $0 + (-17)$

13. $49 + 0$

14. $-56 + 0$

Add, using the number line only as a check.

15. $-4 + 6$

16. $-7 + 3$

17. $5 + (-7)$

18. $10 + (-7)$

GS **19.** $-12 + 12$

     -12 and 12 have the same value. The answer is ▭.

Answers

8. -11 **9.** -12 **10.** -34 **11.** -22
12. -17 **13.** 49 **14.** -56 **15.** 2
16. -4 **17.** -2 **18.** 3 **19.** 0

Guided Solution:
19. absolute, 0

Add, using the number line only as a check.

20. $5 + (-5)$

21. $-6 + 6$

22. $89 + (-89)$

Add.

23. $-5 + (-10)$

24. $18 + (-11)$

25. $-13 + 13$

26. $-20 + 7$

Add. **GS**

27. $-15 + (-5) + 25 + (-9) + 10 + (-14)$

Add the positive numbers:

$25 + \boxed{} = \boxed{}$.

Add the negative numbers:

$-15 + (-5) + (-9) + (\boxed{}) = \boxed{}$.

Finally, add the results:

$35 + (\boxed{}) = \boxed{}$.

Answers

20. 0   **21.** 0   **22.** 0   **23.** -15 **24.** 7
25. 0   **26.** -13 **27.** -8

Guided Solution:
27. $10, 35, -14, -43, -43, -8$

EXAMPLES Add.

14. $-8 + 8 = 0$

15. $14 + (-14) = 0$

◄ **Do Exercises 20–22.**

In summary, to add integers, look first at the signs of the numbers you are adding. This tells you whether you should add or subtract to find the sum.

RULES FOR ADDITION OF INTEGERS

1. *Positive numbers*: Add the same way as you add arithmetic numbers. The answer is positive.
2. *Negative numbers*: Add absolute values. The answer is negative.
3. *A positive and a negative number*: Subtract absolute values.
   a) If the positive number has the greater absolute value, the answer is positive.
   b) If the negative number has the greater absolute value, the answer is negative.
   c) If the numbers have the same absolute value, they are additive inverses and the answer is 0.
4. *One number is zero*: The sum is the other number.

EXAMPLES Add.

16. $-12 + (-7) = -19$ Two negative numbers; add absolute values. The answer is negative.

17. $-20 + 36 = 16$ A positive and a negative number; subtract absolute values. The answer is positive.

◄ **Do Exercises 23–26.**

Suppose we wish to add several numbers, some positive and some negative, as in $15 + (-2) + 7 + 14 + (-5) + (-12)$. Because of the commutative and associative laws for addition, we can group the positive numbers together and the negative numbers together and add them separately. Then we add the two results.

EXAMPLE 18 Add: $15 + (-2) + 7 + 14 + (-5) + (-12)$.

First add the positive numbers: $15 + 7 + 14 = 36$.
Then add the negative numbers: $-2 + (-5) + (-12) = -19$.
Finally, add the results: $36 + (-19) = 17$.

We can also add in any other order we wish, say, from left to right:

$$
\begin{aligned}
15 + (-2) + 7 + 14 + (-5) + (-12) &= 13 + 7 + 14 + (-5) + (-12) \\
&= 20 + 14 + (-5) + (-12) \\
&= 34 + (-5) + (-12) \\
&= 29 + (-12) \\
&= 17.
\end{aligned}
$$

◄ **Do Exercise 27.**

☑ Reading Check

Fill in each blank with either "left" or "right" so that the statement describes the steps for adding the given numbers on the number line.

RC1. To add $7 + 2$, start at 7 and then move 2 units _____. The sum is 9.

RC2. To add $-3 + (-5)$, start at -3 and then move 5 units _____. The sum is -8.

RC3. To add $4 + (-6)$, start at 4 and then move 6 units _____. The sum is -2.

RC4. To add $-8 + 3$, start at -8 and then move 3 units _____. The sum is -5.

a   Add, using the number line.

1. $-7 + 2$ **2.** $1 + (-5)$ **3.** $-9 + 5$ **4.** $8 + (-3)$ **5.** $-3 + 9$

6. $9 + (-9)$ **7.** $-7 + 7$ **8.** $-8 + (-5)$ **9.** $-3 + (-1)$ **10.** $-2 + (-9)$

Add. Use the number line only as a check.

11. $-3 + (-9)$ **12.** $-3 + (-7)$ **13.** $-6 + (-5)$ **14.** $-10 + (-14)$

15. $-15 + 0$ **16.** $0 + (-11)$ **17.** $0 + 42$ **18.** $27 + 0$

19. $9 + (-4)$ **20.** $-7 + 8$ **21.** $-10 + 6$ **22.** $6 + (-13)$

23. $5 + (-5)$ **24.** $10 + (-10)$ **25.** $-2 + 2$ **26.** $-3 + 3$

27. $-4 + (-5)$ **28.** $10 + (-12)$ **29.** $13 + (-6)$ **30.** $-3 + 14$

31. $-25 + 25$ **32.** $40 + (-40)$ **33.** $63 + (-18)$ **34.** $85 + (-65)$

35. $-11 + 8$ **36.** $0 + (-34)$ **37.** $-19 + 19$ **38.** $-10 + 3$

39. $-16 + 6$ **40.** $-15 + 5$ **41.** $-17 + (-7)$ **42.** $-15 + (-5)$

43. $11 + (-16)$ **44.** $-8 + 14$ **45.** $-15 + (-6)$ **46.** $-8 + 8$

47. $-15 + (-15)$ **48.** $-25 + (-25)$ **49.** $-11 + 17$ **50.** $19 + (-19)$

51. $-15 + (-7) + 1$ **52.** $23 + (-5) + 4$ **53.** $30 + (-10) + 5$ **54.** $40 + (-8) + 5$

55. $-23 + (-9) + 15$ **56.** $-25 + 25 + (-9)$

57. $40 + (-40) + 6$ **58.** $63 + (-18) + 12$

59. $12 + (-65) + (-12)$ **60.** $-35 + (-63) + 35$

61. $-24 + (-37) + (-19) + (-45) + (-35)$ **62.** $75 + (-14) + (-17) + (-5)$

63. $28 + (-44) + 17 + 31 + (-94)$ **64.** $27 + (-54) + (-32) + 65 + 46$

65. $-19 + 73 + (-23) + 19 + (-73)$ **66.** $35 + (-51) + 29 + 51 + (-35)$

Skill Maintenance

67. Add: $587 + 6094$.   [1.2a]

68. Subtract: $3046 - 2973$.   [1.3a]

69. Write in expanded notation: $39{,}417$.   [1.1b]

70. Multiply: $42 \cdot 56$.   [1.4a]

71. Divide: $288 \div 9$.   [1.5a]

72. Round to the nearest ten: 3496.   [1.6a]

Synthesis

Add.

73. $-|27| + (-|-13|)$

74. $|-32| + (-|15|)$

75. ▦ $-3496 + (-2987)$

76. ▦ $497 + (-3028)$

77. For what numbers x is $-x$ positive?

78. For what numbers x is $-x$ negative?

Tell whether each sum is positive, negative, or zero.

79. If n is positive and m is negative, then $-n + m$ is
_____.

80. If $n = m$ and n is negative, then $-n + (-m)$ is
_____.

81. If n is negative and m is less than n, then $n + m$ is
_____.

82. If n is positive and m is greater than n, then $n + m$ is
_____.

Subtraction of Integers

a SUBTRACTION

OBJECTIVES

a Subtract integers, and simplify combinations of additions and subtractions.

b Solve applied problems involving addition and subtraction of integers.

We now consider subtraction of integers. To find the difference $a - b$, we look for a number to add to b that gives us a.

THE DIFFERENCE

The difference $a - b$ is the number that when added to b gives a.

For example, $45 - 17 = 28$ because $28 + 17 = 45$. Let's consider an example in which the answer is a negative number.

EXAMPLE 1 Subtract: $5 - 8$.

Think: $5 - 8$ is the number that when added to 8 gives 5. What number can we add to 8 to get 5? The number must be negative. The number is -3:

$$5 - 8 = -3.$$

That is, $5 - 8 = -3$ because $8 + (-3) = 5$.

Do Exercises 1–3. ▶

The definition of $a - b$ above does not always provide the most efficient way to subtract. To understand a faster way to subtract, consider finding $5 - 8$ using the number line. We start at 5. Then we move 8 units to the *left* to do the subtracting. Note that this is the same as adding the opposite of 8, or -8, to 5.

Move 8 units to the left. Start at 5.

$5 - 8 = -3$

Look for a pattern in the following table.

SUBTRACTIONS	ADDING AN OPPOSITE
$5 - 8 = -3$	$5 + (-8) = -3$
$-6 - 4 = -10$	$-6 + (-4) = -10$
$-7 - (-10) = 3$	$-7 + 10 = 3$
$-7 - (-2) = -5$	$-7 + 2 = -5$

Do Exercises 4–7. ▶

Perhaps you have noticed that we can subtract by adding the opposite of the number being subtracted. This can always be done.

Subtract.

1. $-6 - 4$
   Think: What number can be added to 4 to get -6:
   $\square + 4 = -6$?

2. $-7 - (-10)$
   Think: What number can be added to -10 to get -7:
   $\square + (-10) = -7$?

3. $-7 - (-2)$
   Think: What number can be added to -2 to get -7:
   $\square + (-2) = -7$?

Complete the addition and compare with the subtraction.

4. $4 - 6 = -2$;
   $4 + (-6) = $ _____

5. $-3 - 8 = -11$;
   $-3 + (-8) = $ _____

6. $-5 - (-9) = 4$;
   $-5 + 9 = $ _____

7. $-5 - (-3) = -2$;
   $-5 + 3 = $ _____

Answers

1. -10 2. 3 3. -5 4. -2 5. -11
6. 4 7. -2

SUBTRACTING BY ADDING THE OPPOSITE

To subtract, add the opposite, or additive inverse, of the number being subtracted:

$$a - b = a + (-b).$$

This is the method generally used for quick subtraction of integers.

EXAMPLES Write each subtraction as a corresponding addition. Then write the equation in words.

2. $-12 - 30$

$-12 - 30 = -12 + (-30)$ Adding the opposite of 30

Negative twelve minus thirty is negative twelve plus negative thirty.

3. $-20 - (-17)$

$-20 - (-17) = -20 + 17$ Adding the opposite of -17

Negative twenty minus negative seventeen is negative twenty plus seventeen.

◀ **Do Exercises 8–10.**

EXAMPLES Subtract.

4. $2 - 6 = 2 + (-6)$ The opposite of 6 is -6. We change the subtraction to addition and add the opposite. Instead of subtracting 6, we add -6.

$= -4$

5. $4 - (-9) = 4 + 9$ The opposite of -9 is 9. We change the subtraction to addition and add the opposite. Instead of subtracting -9, we add 9.

$= 13$

6. $-4 - 8 = -4 + (-8)$ We change the subtraction to addition and add the opposite. Instead of subtracting 8, we add -8.

$= -12$

7. $10 - 7 = 10 + (-7)$ We change the subtraction to addition and add the opposite. Instead of subtracting 7, we add -7.

$= 3$

8. $-4 - (-9) = -4 + 9$ Instead of subtracting -9, we add 9.

$= 5$ To check, note that $5 + (-9) = -4$.

9. $-7 - (-3) = -7 + 3$ Instead of subtracting -3, we add 3.

$= -4$ *Check*: $-4 + (-3) = -7.$

◀ **Do Exercises 11–16.**

Write each subtraction as a corresponding addition. Then write the equation in words.

8. $3 - 10$

9. $-12 - (-9)$

10. $-12 - 10$

Subtract.

11. $2 - 8$

$= 2 + \boxed{} = \boxed{}$

12. $-6 - 10$

13. $13 - 8$

14. $-7 - (-9)$

15. $-8 - (-2)$

16. $5 - (-8)$

Answers

8. $3 - 10 = 3 + (-10)$; three minus ten is three plus negative ten.   **9.** $-12 - (-9) = -12 + 9$; negative twelve minus negative nine is negative twelve plus nine.   **10.** $-12 - 10 = -12 + (-10)$; negative twelve minus ten is negative twelve plus negative ten.   **11.** -6 **12.** -16 **13.** 5   **14.** 2   **15.** -6 **16.** 13

Guided Solution:
11. (-8), -6

When several additions and subtractions occur together, we can make them all additions. The commutative law for addition can then be used.

EXAMPLE 10   Simplify: $-3 - (-5) - 9 + 4 - (-6)$.

$$-3 - (-5) - 9 + 4 - (-6) = -3 + 5 + (-9) + 4 + 6 \quad \text{Adding opposites}$$

$$= -3 + (-9) + 5 + 4 + 6 \quad \text{Using a commutative law}$$

$$= -12 + 15$$

$$= 3$$

Do Exercises 17 and 18. ▶

Simplify.

GS **17.** $-6 - (-2) - (-4) - 12 + 3$

$\quad = -6 + 2 + \boxed{} + (-12) + 3$

$\quad = -6 + (-12) + 2 + \boxed{} + 3$

$\quad = -18 + \boxed{}$

$\quad = \boxed{}$

18. $9 - (-6) + 7 - 9 - 8 - (-20)$

b APPLICATIONS AND PROBLEM SOLVING

We use addition and subtraction of real numbers to solve a variety of applied problems.

EXAMPLE 11   *Surface Temperatures on Mars.*   Surface temperatures on Mars vary from $-128°C$ during the polar night to $27°C$ at the equator at midday when Mars is at the point in its orbit closest to the sun. Find the difference between the highest value and the lowest value in this temperature range.

Source: Mars Institute

We let $D =$ the difference in the temperatures. Then the problem translates to the following subtraction:

Difference in temperature is Highest temperature minus Lowest temperature

$$D \quad = \quad 27 \quad - \quad (-128).$$

We then solve the equation: $D = 27 - (-128) = 27 + 128 = 155$.

The difference in the temperatures is $155°C$.

Do Exercise 19. ▶

19. *Temperature Extremes.*   The highest temperature ever recorded in the United States was $134°F$ in Greenland Ranch, California, on July 10, 1913. The lowest temperature ever recorded was $-80°F$ in Prospect Creek, Alaska, on January 23, 1971. How much higher was the temperature in Greenland Ranch than the temperature in Prospect Creek?

Source: National Oceanographic and Atmospheric Administration

Answers

17. -9 **18.** 25 **19.** $214°F$

Guided Solution:
17. $4, 4, 9, -9$

☑ **Reading Check**

Match each expression with the expression from the list at the right that names the same number.

RC1. $18 - 6$ _____

RC2. $-18 - (-6)$ _____

RC3. $-18 - 6$ _____

RC4. $18 - (-6)$ _____

a) $18 + 6$

b) $-18 + 6$

c) $18 + (-6)$

d) $-18 + (-6)$

a Subtract.

1. $3 - 7$

2. $5 - 10$

3. $0 - 7$

4. $0 - 8$

5. $-8 - (-2)$

6. $-6 - (-8)$

7. $-10 - (-10)$

8. $-8 - (-8)$

9. $12 - 16$

10. $14 - 19$

11. $20 - 27$

12. $26 - 7$

13. $-9 - (-3)$

14. $-6 - (-9)$

15. $-11 - (-11)$

16. $-14 - (-14)$

17. $7 - 7$

18. $9 - 9$

19. $7 - (-7)$

20. $4 - (-4)$

21. $8 - (-3)$

22. $-7 - 4$

23. $-6 - 8$

24. $6 - (-10)$

25. $-4 - (-9)$

26. $-14 - 2$

27. $2 - 9$

28. $2 - 8$

29. $-6 - (-5)$

30. $-4 - (-3)$

31. $8 - (-10)$

32. $5 - (-6)$

33. $0 - 10$

34. $0 - 23$

35. $-5 - (-2)$

36. $-3 - (-1)$

37. $-7 - 14$

38. $-9 - 16$

39. $0 - (-5)$

40. $0 - (-1)$

41. $-8 - 0$

42. $-9 - 0$

43. $7 - (-5)$

44. $7 - (-4)$

45. $6 - 25$

46. $18 - 63$

47. $-42 - 26$

48. $-18 - 63$

49. $-72 - 9$ **50.** $-49 - 3$ **51.** $24 - (-92)$ **52.** $48 - (-73)$

53. $-50 - (-50)$ **54.** $-70 - (-70)$ **55.** $-30 - (-85)$ **56.** $-25 - (-15)$

Simplify.

57. $7 - (-5) + 4 - (-3)$ **58.** $-5 - (-8) + 3 - (-7)$ **59.** $-31 + (-28) - (-14) - 17$

60. $-43 - (-19) - (-21) + 25$ **61.** $-34 - 28 + (-33) - 44$ **62.** $39 + (-88) - 29 - (-83)$

63. $-93 - (-84) - 41 - (-56)$ **64.** $84 + (-99) + 44 - (-18) - 43$ **65.** $-5 - (-30) + 30 + 40 - (-12)$

66. $14 - (-50) + 20 - (-32)$ **67.** $132 - (-21) + 45 - (-21)$ **68.** $81 - (-20) - 14 - (-50) + 53$

b   Solve.

69. *"Flipping" Houses.* Buying run-down houses, fixing them up, and reselling them is referred to as "flipping" houses. Charlie and Sophia flipped four houses in a recent year. The profits and losses are shown in the following bar graph. Find the sum of the profits and losses.

Flipping Houses

70. *Elevations in Asia.* The elevation of the highest point in Asia, Mt. Everest, on the border between Nepal and Tibet, is 29,035 ft. The lowest elevation, at the Dead Sea, between Israel and Jordan, is -1348 ft. What is the difference in the elevations of the two locations?

Dead Sea -1348 ft   Mt. Everest 29,035 ft

71. *Difference in Elevation.* The highest elevation in Japan is 3776 m above sea level at Fujiyama. The lowest elevation in Japan is 4 m below sea level at Hachirogata. Find the difference in the elevations.

Source: *The CIA World Factbook 2012*

72. *Copy-Center Account.* Rachel's copy-center bill for July was $327. She made a payment of $200 and then made $48 worth of copies in August. How much did she then owe on her account?

73. Through exercise, Rod went from 8 lb above his "ideal" body weight to 9 lb below it. How many pounds did Rod lose?

74. Laura has a charge of $477 on her credit card, but she then returns a sweater that cost $129. How much does she now owe on her credit card?

75. *Temperature Changes.* One day the temperature in Lawrence, Kansas, is 32° at 6:00 A.M. It rises 15° by noon, but falls 50° by midnight after a cold front moves in. What is the final temperature?

76. *Stock Price Changes.* On a recent day, the price of a stock opened at $61. It rose $5, dropped $7, and rose $4. Find the price of the stock at the end of the day.

77. *Points per Game Differential.* A basketball team's point differential is the difference between that team's score in a game and its opponent's score. The points per game differential is the sum of a team's point differentials divided by the number of games it has played. In January 2014, the Indiana Pacers had a points per game differential of +9, and the Milwaukee Bucks had a points per game differential of −9. How much higher was the Pacers' points per game differential?

78. *Golf.* As a result of coaching, Cedric's average golf score improved from 3 over par to 2 under. By how many strokes did his score change?

79. *Offshore Oil.* In 1998, the elevation of the world's deepwater drilling record was −7718 ft. In 2013, the deepwater drilling record was 2693 ft deeper. What was the elevation of the deepwater drilling record in 2013?

Source: www.deepwater.com

80. *Oceanography.* The deepest point in the Pacific Ocean is the Marianas Trench, with a depth of 11,033 m. The deepest point in the Atlantic Ocean is the Puerto Rico Trench, with a depth of 8648 m. What is the difference in the elevation of the two trenches?

81. *Toll Roads.* The E-Z Pass program allows drivers in the Northeast to travel certain toll roads without having to stop to pay. Instead, a transponder attached to the vehicle is scanned as the vehicle rolls through a toll booth. Recently the Ramones began a trip to New York City with a balance of $12 in their E-Z Pass account. They accumulated $15 in tolls on their trip, and because they overspent their balance, the Ramones had to pay $80 in administrative fees. By how much were the Ramones in debt as a result of their travel on toll roads?

Source: State of New Jersey

82. *Toll Roads.* The Murrays began a trip with $13 in their E-Z Pass account (see Exercise 81). They accumulated $20 in tolls and had to pay $80 in administrative fees. By how much were the Murrays in debt as a result of their travel on toll roads?

Skill Maintenance

Evaluate.

83. 4^3 [1.9b]

84. $68 \cdot 72$ [1.4a]

85. 1^7 [1.9b]

86. $143 \cdot 29$ [1.4a]

87. How many 12-oz cans of soda can be filled with 96 oz of soda?   [1.8a]

88. A case of soda contains 24 bottles. If each bottle contains 12 oz, how many ounces of soda are in the case?   [1.8a]

Simplify.   [1.9c]

89. $5 + 4^2 + 2 \cdot 7$

90. $45 \div (2^2 + 11)$

91. $(9 + 7)(9 - 7)$

92. $(13 - 2)(13 + 2)$

Synthesis

Subtract.

93. ▦ $123,907 - 433,789$

94. ▦ $23,011 - (-60,432)$

For Exercises 95–100, tell whether each statement is true or false for all integers a and b. If false, give an example to show why.

95. $a - 0 = 0 - a$

96. $0 - a = a$

97. If $a \neq b$, then $a - b \neq 0$.

98. If $a = -b$, then $a + b = 0$.

99. If $a + b = 0$, then a and b are opposites.

100. If $a - b = 0$, then $a = -b$.

101. If $a - 54$ is -37, find the value of a.

102. If $x - 48$ is -15, find the value of x.

103. Maureen kept track of the weekly changes in the stock market over a period of 5 weeks. By how many points (pts) had the market risen or fallen over this time?

WEEK 1	WEEK 2	WEEK 3	WEEK 4	WEEK 5
Down 13 pts	Down 16 pts	Up 36 pts	Down 11 pts	Up 19 pts

104. *Blackjack Counting System.*   Players of the casino game of blackjack make use of many card-counting systems in which a negative count means that a player has an advantage. One such system, called *High–Low*, was developed by Harvey Dubner in 1963. Each card already showing counts as $-1, 0,$ or 1 as follows:

2, 3, 4, 5, 6   count as $+1$;

7, 8, 9   count as 0;

10, J, Q, K, A   count as -1.

Source: Patterson, Jerry L. *Casino Gambling.* New York: Perigee, 1982

a) Find the total count on this sequence of cards:

K, A, 2, 4, 5, 10, J, 8, Q, K, 5.

b) Does the player have an advantage?

a MULTIPLICATION

Multiplication of integers is like multiplication of whole numbers. The difference is that we must determine whether the answer is positive or negative.

Multiplication of a Positive Integer and a Negative Integer

To see how to multiply a positive integer and a negative integer, consider the following pattern.

This number decreases by 1 each time.

$$4 \cdot 5 = 20$$
$$3 \cdot 5 = 15$$
$$2 \cdot 5 = 10$$
$$1 \cdot 5 = 5$$
$$0 \cdot 5 = 0$$
$$-1 \cdot 5 = -5$$
$$-2 \cdot 5 = -10$$
$$-3 \cdot 5 = -15$$

This number decreases by 5 each time.

◀ **Do Exercise 1.**

According to this pattern, it looks as though the product of a negative integer and a positive integer is negative. To confirm this, use repeated addition:

$$-1 \cdot 5 = 5 \cdot (-1) = -1 + (-1) + (-1) + (-1) + (-1) = -5$$
$$-2 \cdot 5 = 5 \cdot (-2) = -2 + (-2) + (-2) + (-2) + (-2) = -10$$
$$-3 \cdot 5 = 5 \cdot (-3) = -3 + (-3) + (-3) + (-3) + (-3) = -15$$

MULTIPLYING A POSITIVE INTEGER AND A NEGATIVE INTEGER

....

To multiply a positive integer and a negative integer, multiply their absolute values. The answer is negative.

EXAMPLES Multiply.

1. $8(-5) = -40$ **2.** $50(-1) = -50$ **3.** $-7 \cdot 6 = -42$

◀ **Do Exercises 2–4.**

Multiplication of Two Negative Integers

How do we multiply two negative integers? Again we look for a pattern.

This number decreases by 1 each time.

$$4 \cdot (-5) = -20$$
$$3 \cdot (-5) = -15$$
$$2 \cdot (-5) = -10$$
$$1 \cdot (-5) = -5$$
$$0 \cdot (-5) = 0$$
$$-1 \cdot (-5) = 5$$
$$-2 \cdot (-5) = 10$$
$$-3 \cdot (-5) = 15$$

This number increases by 5 each time.

◀ **Do Exercise 5.**

---

1. Complete, as in the example in the text.

$$4 \cdot 10 = 40$$
$$3 \cdot 10 = 30$$
$$2 \cdot 10 =$$
$$1 \cdot 10 =$$
$$0 \cdot 10 =$$
$$-1 \cdot 10 =$$
$$-2 \cdot 10 =$$
$$-3 \cdot 10 =$$

Multiply.

2. $-3 \cdot 6$ **3.** $20 \cdot (-5)$

4. $12(-1)$

5. Complete, as in the example in the text.

$$3 \cdot (-10) = -30$$
$$2 \cdot (-10) = -20$$
$$1 \cdot (-10) =$$
$$0 \cdot (-10) =$$
$$-1 \cdot (-10) =$$
$$-2 \cdot (-10) =$$
$$-3 \cdot (-10) =$$

Answers

1. 20; 10; 0; −10; −20; −30   **2.** −18   **3.** −100
4. −12   **5.** −10; 0; 10; 20; 30

According to the pattern, the product of two negative integers is positive. This leads to the second part of the rule for multiplying integers.

> ### MULTIPLYING TWO NEGATIVE INTEGERS
>
> To multiply two negative integers, multiply their absolute values. The answer is positive.

EXAMPLES Multiply.

4. $(-2)(-4) = 8$ **5.** $(-10)(-7) = 70$

Do Exercises 6–8. ▶

The following is another way to state the rules for multiplication.

> To multiply two nonzero integers:
>
> **a)** Multiply the absolute values.
> **b)** If the signs are the same, the answer is positive.
> **c)** If the signs are different, the answer is negative.

EXAMPLES Multiply.

6. $(-3)(-5) = 15$ **7.** $8(-10) = -80$

Do Exercises 9–12. ▶

Multiplication by 0

No matter how many times 0 is added to itself, the answer is 0.

> The product of 0 and any integer is 0:
> $$a \cdot 0 = 0.$$

EXAMPLES Multiply.

8. $-19 \cdot 0 = 0$ **9.** $0(-7) = 0$

Do Exercises 13 and 14. ▶

b MULTIPLICATION OF MORE THAN TWO INTEGERS

When multiplying more than two integers, we can choose order and grouping as we please, using the commutative and associative laws.

EXAMPLES Multiply.

10. a) $-8 \cdot 2(-3) = -16(-3)$ Multiplying the first two numbers
$\qquad\qquad\qquad = 48$ Multiplying the results

   b) $-8 \cdot 2(-3) = 24 \cdot 2$ Multiplying the negative numbers
$\qquad\qquad\qquad = 48$ The result is the same as above.

Multiply.

6. $(-3)(-4)$

7. $-9(-5)$

8. $(-1)(-6)$

Multiply.

 9. $3(-6)$

Multiply absolute values:
$$3 \cdot 6 = \boxed{}.$$
The signs are different, so the answer is $\boxed{}$.
$$3(-6) = \boxed{}$$

10. $(-6)(-5)$

11. $(-1)(-50)$

12. $(-8) \cdot 11$

Multiply.

13. $0(-5)$

14. $-23 \cdot 0$

Answers

6. 12   7. 45   8. 6   9. −18   10. 30
11. 50   12. −88   13. 0   14. 0

Guided Solution:
9. 18, negative, −18

Multiply.

15. $5 \cdot (-3) \cdot 2$

16. $-2 \cdot (-5) \cdot (-4) \cdot (-3)$

17. $(-4)(-5)(-2)(-3)(-1)$ **GS**

$$= \boxed{} \cdot 6 \cdot (-1)$$

$$= \boxed{} \cdot (-1)$$

$$= \boxed{}$$

18. $(-1)(-1)(-2)(-3)(-1)(-1)$

11. $7(-1)(-4)(-2) = (-7)8$ Multiplying the first two numbers and the last two numbers

$$= -56$$

12. $-5 \cdot (-2) \cdot (-3) \cdot (-6) = 10 \cdot 18$ Each pair of negative numbers gives a positive product.

$$= 180$$

13. $(-3)(-5)(-2)(-3)(-1) = 15 \cdot 6 \cdot (-1)$

$$= 90(-1) = -90$$

We can see the following pattern in the results of Examples 11–13.

> The product of an even number of negative integers is positive.
> The product of an odd number of negative integers is negative.

◀ **Do Exercises 15–18.**

Powers of Integers

A positive number raised to any power is positive. When a negative number is raised to a power, the sign of the result depends upon whether the exponent is even or odd.

EXAMPLES   Simplify.

14. $(-7)^2 = (-7)(-7) = 49$ The result is positive.

15. $(-4)^3 = (-4)(-4)(-4)$

$$= 16(-4)$$

$$= -64$$ The result is negative.

16. $(-3)^4 = (-3)(-3)(-3)(-3)$

$$= 9 \cdot 9$$

$$= 81$$ The result is positive.

17. $(-2)^5 = (-2)(-2)(-2)(-2)(-2)$

$$= 4 \cdot 4 \cdot (-2)$$

$$= 16(-2)$$

$$= -32$$ The result is negative.

Perhaps you noted the following.

> When a negative number is raised to an even exponent, the result is positive.
> When a negative number is raised to an odd exponent, the result is negative.

Simplify.

19. $(-2)^3$ **20.** $(-9)^2$

21. $(-1)^9$ **22.** 2^5

◀ **Do Exercises 19–22.**

When an integer is multiplied by -1, the result is the opposite of that integer.

> For any integer a,
> $$-1 \cdot a = -a.$$

EXAMPLE 18 Simplify: -7^2.

In the expression -7^2, the base is 7, not -7. We can regard -7^2 as $-1 \cdot 7^2$:

$$-7^2 = -1 \cdot 7^2 \qquad \text{\textcolor{red}{The rules for order of operations tell us to square first: } } 7^2 = 7 \cdot 7.$$

$$= -1 \cdot 49$$

$$= -49.$$

Compare Examples 14 and 18 and note that $(-7)^2 \neq -7^2$.

Do Exercises 23 and 24. ▶

23. Simplify: -5^2.

24. Simplify: $(-5)^2$.

Answers

23. -25 **24.** 25

CALCULATOR CORNER

Exponential Notation   When using a calculator to calculate expressions like $(-39)^4$ or -39^4, it is important to use the correct sequence of keystrokes.

Calculators with $\boxed{+/-}$ **key:** To calculate $(-39)^4$, we press $\boxed{3}\,\boxed{9}\,\boxed{+/-}\,\boxed{x^y}\,\boxed{4}\,\boxed{=}$.

To calculate -39^4, we must first raise 39 to the power 4. Then the sign of the result must be changed. This can be done with the keystrokes $\boxed{3}\,\boxed{9}\,\boxed{x^y}\,\boxed{4}\,\boxed{=}\,\boxed{+/-}$, or by multiplying 39^4 by -1 using the keystrokes $\boxed{3}\,\boxed{9}\,\boxed{x^y}\,\boxed{4}\,\boxed{=}\,\boxed{\times}\,\boxed{1}\,\boxed{+/-}\,\boxed{=}$.

Calculators with $\boxed{(-)}$ **key:** On some calculators, the $\boxed{(-)}$ key is pressed before a number to indicate that the number is negative. This is similar to the way the expression is written on paper. With these calculators, $(-39)^4$ is found by pressing $\boxed{(}\,\boxed{(-)}\,\boxed{3}\,\boxed{9}\,\boxed{)}\,\boxed{\wedge}\,\boxed{4}\,\boxed{\text{ENTER}}$, and -39^4 is found by pressing $\boxed{(-)}\,\boxed{3}\,\boxed{9}\,\boxed{\wedge}\,\boxed{4}\,\boxed{\text{ENTER}}$.

You can either experiment or consult a user's manual if you are unsure of the proper keystrokes for your calculator.

EXERCISES   Use a calculator to determine each of the following.

1. $(-23)^6$

2. $(-17)^5$

3. $(-104)^3$

4. $(-4)^{10}$

5. -9^6

6. -7^6

7. -6^5

8. -3^9

2.4 Exercise Set

✓ Reading Check

Fill in each blank with either "positive" or "negative."

RC1. To multiply a positive integer and a negative integer, we multiply their absolute values. The answer is _____.

RC2. To multiply two negative integers, we multiply their absolute values. The answer is _____.

RC3. The product of an even number of negative integers is _____.

RC4. The product of an odd number of negative integers is _____.

a Multiply.

1. $-2 \cdot 8$

2. $-7 \cdot 3$

3. $10 \cdot (-6)$

4. $12 \cdot (-2)$

5. $8 \cdot (-6)$

6. $8 \cdot (-3)$

7. $-10 \cdot 3$

8. $-9 \cdot 8$

9. $-3(-5)$

10. $-8 \cdot (-2)$

11. $-9 \cdot (-2)$

12. $(-8)(-9)$

13. $(-6)\,(-7)$

14. $-8 \cdot (-3)$

15. $-10(-2)$

16. $-9(-8)$

17. $12(-10)$

18. $15(-8)$

19. $-23 \cdot 0$

20. $-38 \cdot 0$

21. $(-72)(-1)$

22. $41(-3)$

23. $(-20)17$

24. $(-1)(-43)$

25. $-8(-50)$

26. $(-25) \cdot 8$

27. $0(-14)$

28. $0(-38)$

b Multiply.

29. $3 \cdot (-8) \cdot (-1)$

30. $(-7) \cdot (-4) \cdot (-1)$

31. $7(-4)(-3)5$

32. $9(-2)(-6)7$

33. $-2(-5)(-7)$

34. $(-2)(-5)(-3)(-5)$

35. $(-5)(-2)(-3)(-1)$

36. $-6(-5)(-9)$

37. $(-15)(-29) \cdot 0 \cdot 8$

38. $19(-7)(-8) \cdot 0 \cdot 6$

39. $(-7)(-1)(7)(-6)$

40. $(-5)6(-4)5$

Simplify.

41. $(-6)^2$

42. $(-8)^2$

43. $(-5)^3$

44. $(-2)^4$

45. $(-10)^4$

46. $(-1)^5$

47. -2^4

48. $(-2)^6$

49. $(-3)^5$

50. -10^4

51. $(-1)^{12}$

52. $(-1)^{13}$

53. -11^2

54. -2^6

55. -4^3

56. -2^5

Skill Maintenance

57. Round 532,451 to the nearest hundred.   [1.6a]

58. Write standard notation for sixty million.   [1.1c]

59. Divide: $2880 \div 36$.   [1.5a]

60. Multiply: 75×34.   [1.4a]

61. Simplify: $10 - 2^3 + 6 \div 2$.   [1.9c]

62. Simplify: $2 \cdot 5^2 - 3 \cdot 2^3 \div (3 + 3^2)$.   [1.9c]

63. A rectangular rug measures 5 ft by 8 ft. What is the area of the rug?   [1.4b], [1.8a]

64. How many 12-egg cartons can be filled with 2880 eggs?   [1.8a]

65. A ferry can accommodate 12 cars and 53 cars are waiting. How many trips will be required to ferry them all?   [1.8a]

66. An elevator can hold 16 people and 50 people are waiting to go up. How many trips will be required to transport all of them?   [1.8a]

Synthesis

Simplify.

67. $(-3)^5(-1)^{379}$

68 $(-2)^3 \cdot [(-1)^{29}]^{46}$

69. $-9^4 + (-9)^4$

70. $-5^2(-1)^{29}$

71. $|(-2)^5 + 3^2| - (3 - 7)^2$

72. $|-12(-3)^2 - 5^3 - 6^2 - (-5)^2|$

73. ▦ -47^2

74. ▦ -53^2

75. ▦ $(-19)^4$

76. ▦ $(-23)^4$

77. ▦ $(73 - 86)^3$

78. ▦ $(-49 + 34)^3$

79. ▦ $-935(238 - 243)^3$

80. ▦ $(-17)^4(129 - 133)^5$

81. Jo had a balance of $68 in her account and wrote seven checks for $13 each. What was her balance after writing the checks?

82. After diving 95 m below the surface, a diver rises at a rate of 7 meters per minute for 9 min. What is the diver's new elevation?

83. What must be true of m and n if $[(-5)^m]^n$ is to be **(a)** negative? **(b)** positive?

84. What must be true of m and n if $-mn$ is to be **(a)** positive? **(b)** zero? **(c)** negative?

SKILL TO REVIEW

Objective 1.9c: Simplify expressions using the rules for order of operations.

Simplify.

1. $5^2 - 2(10 - 3)$

2. $2[21 - (11 - 3)]$

We now consider division of integers. Because of the way in which division is defined, its rules are similar to those for multiplication.

a   DIVISION OF INTEGERS

> **DIVISION**
>
> The quotient $\dfrac{a}{b}$ (or $a \div b$, or a/b) is the number, if there is one, that when multiplied by b gives a.

Let's use the definition to divide integers.

EXAMPLES   Divide, if possible. Check each answer.

1. $14 \div (-7) = -2$ *Think:* What number multiplied by -7 gives 14? The number is -2. *Check:* $(-2)(-7) = 14$.

2. $\dfrac{-32}{-4} = 8$ *Think:* What number multiplied by -4 gives -32? The number is 8. *Check:* $8(-4) = -32$.

3. $-21 \div 7 = -3$ *Think:* What number multiplied by 7 gives -21? The number is -3. *Check:* $(-3) \cdot 7 = -21$.

4. $\dfrac{0}{-5} = 0$ *Think:* What number multiplied by -5 gives 0? The number is 0. *Check:* $0(-5) = 0$.

5. $\dfrac{-5}{0}$ is **not defined**.   *Think:* What number multiplied by 0 gives -5? There is no such number because the product of 0 and *any* number is 0.

The rules for determining the sign of a quotient are the same as those for determining the sign of a product. We state them together.

Divide.

1. $6 \div (-2)$
Think: What number multiplied by -2 gives 6?

2. $\dfrac{-15}{-3}$
Think: What number multiplied by -3 gives -15?

3. $-24 \div 8$
Think: What number multiplied by 8 gives -24?

4. $\dfrac{0}{-4}$ **5.** $\dfrac{30}{-5}$

6. $\dfrac{-45}{9}$

> To multiply or divide two integers:
>
> **a)** Multiply or divide the absolute values.
>
> **b)** If the signs are the same, the answer is positive.
>
> **c)** If the signs are different, the answer is negative.

◀ **Do Margin Exercises 1–6.**

Dividing by 0

Recall that, in general, $a \div b$ and $b \div a$ are different numbers. In Example 4, we divided *into* 0. In Example 5, we attempted to divide *by* 0. Since any number times 0 gives 0, not -5, we say that $-5 \div 0$ is **not defined** or is **undefined**. Also, since *any* number times 0 gives 0, $0 \div 0$ is also not defined.

Answers

Skill to Review:
1. 11   **2.** 26

Margin Exercises:
1. -3 **2.** 5   **3.** -3 **4.** 0
5. -6 **6.** -5

EXCLUDING DIVISION BY 0

Division by 0 is not defined:

$a \div 0$, or $\dfrac{a}{0}$, is undefined for all real numbers a.

Dividing 0 by Other Numbers

Note that $0 \div 8 = 0$ because $0 = 0 \cdot 8$.

DIVIDENDS OF 0

Zero divided by any nonzero real number is 0:

$\dfrac{0}{a} = 0, \quad a \neq 0.$

EXAMPLES   Divide.

6. $0 \div (-6) = 0$

7. $\dfrac{0}{12} = 0$

8. $\dfrac{-3}{0}$ is undefined.

Do Exercises 7–9. ▶

Divide, if possible.

7. $34 \div 0$

8. $0 \div (-4)$

9. $-52 \div 0$

b RULES FOR ORDER OF OPERATIONS

When more than one operation appears in a calculation or problem, we apply the rules for order of operations.

> **RULES FOR ORDER OF OPERATIONS**
>
> 1. Do all calculations within grouping symbols, including parentheses, brackets, braces, and absolute-value symbols, and within numerators or denominators.
> 2. Evaluate all exponential expressions.
> 3. Do all multiplications and divisions in order from left to right.
> 4. Do all additions and subtractions in order from left to right.

EXAMPLE 9   Simplify: $17 - 10 \div 2 \cdot 4$.

There are no grouping symbols or exponents, so we begin with the third rule.

$$17 - 10 \div 2 \cdot 4 = 17 - 5 \cdot 4$$
$$= 17 - 20$$
$$= -3$$

Carrying out all multiplications and divisions in order from left to right

Answers

7. Undefined   **8.** 0   **9.** Undefined

EXAMPLES Simplify.

10. $|(-2)^3 \div 4| - 5(-2)$

We first simplify within the absolute-value symbols.

$$
\begin{aligned}
|(-2)^3 \div 4| - 5(-2) &= |-8 \div 4| - 5(-2) && (-2)(-2)(-2) = -8 \\
&= |-2| - 5(-2) && \text{Dividing} \\
&= 2 - 5(-2) && |-2| = 2 \\
&= 2 - (-10) && \text{Multiplying} \\
&= 12 && \text{Subtracting;} \\
& && 2 - (-10) = 2 + 10
\end{aligned}
$$

11. $2^4 + 51 \cdot 4 - (37 + 23 \cdot 2)$

$$
\begin{aligned}
2^4 &+ 51 \cdot 4 - (37 + 23 \cdot 2) \\
&= 2^4 + 51 \cdot 4 - (37 + 46) && \text{Carrying out all operations inside parentheses first, following the rules for order of operations within the parentheses} \\
&= 2^4 + 51 \cdot 4 - 83 && \text{Adding inside parentheses} \\
&= 16 + 51 \cdot 4 - 83 && \text{Evaluating exponential expressions} \\
&= 16 + 204 - 83 && \text{Doing all multiplications} \\
&= 220 - 83 && \text{Doing all additions and subtractions in order from left to right} \\
&= 137
\end{aligned}
$$

Always regard a fraction bar as a grouping symbol. It separates any calculations in the numerator from those in the denominator.

EXAMPLE 12 Simplify: $\dfrac{5 - (-3)^2}{8 - 10}$.

$$
\begin{aligned}
\frac{5 - (-3)^2}{8 - 10} &= \frac{5 - 9}{-2} && \text{Calculating within the numerator and within the denominator: } (-3)^2 = (-3)(-3) = 9, \\
&= \frac{-4}{-2} && 8 - 10 = -2, \text{ and } 5 - 9 = -4 \\
&= 2 && \text{Dividing}
\end{aligned}
$$

◄ **Do Exercises 10–13.**

Simplify.

10. $5 - (-7)(-3)^2$

11. $(-2) \cdot |3 - 2^2| + 5$ **GS**

$\quad = (-2) \cdot |3 - \boxed{}| + 5$

$\quad = (-2) \cdot |\boxed{}| + 5$

$\quad = (-2) \cdot \boxed{} + 5$

$\quad = \boxed{} + 5$

$\quad = \boxed{}$

12. $52 \cdot 5 + 5^3 - (4^2 - 48 \div 4)$

13. $\dfrac{(-5)(-9)}{1 - 2 \cdot 2}$

Answers

10. 68 **11.** 3 **12.** 381 **13.** −15

Guided Solution:

11. 4, −1, 1, −2, 3

---

37. 4 · (6 + 8)(8 + 4)

✓ Reading Check

Tell whether each division results in 0 or is undefined.

RC1. $0 \div 17$ **RC2.** $(-3) \div 0$ **RC3.** $\dfrac{132}{0}$ **RC4.**

a Divide, if possible, and check each answer by multiplying. If an answer is undefined, state so.

1. $36 \div (-6)$

2. $\dfrac{42}{-7}$

3. $\dfrac{26}{-2}$

4. $24 \div (-12)$

5. $\dfrac{-16}{8}$

6. $-22 \div (-2)$

7. $\dfrac{-48}{-12}$

8. $-72 \div (-9)$

9. $\dfrac{-72}{8}$

10. $\dfrac{-50}{25}$

11. $-100 \div (-50)$

12. $\dfrac{-400}{8}$

13. $-108 \div 9$

14. $\dfrac{-128}{8}$

15. $\dfrac{200}{-25}$

16. $-651 \div (-31)$

17. $\dfrac{-56}{0}$

18. $\dfrac{0}{-5}$

19. $\dfrac{88}{-11}$

20. $\dfrac{-145}{-5}$

21. $-\dfrac{276}{12}$

22. $-\dfrac{217}{7}$

23. $\dfrac{0}{-2}$

24. $\dfrac{-13}{0}$

25. $\dfrac{19}{-1}$

26. $\dfrac{-17}{1}$

27. $-41 \div 1$

28. $23 \div (-1)$

b Simplify, if possible. If an answer is undefined, state so.

29. $8 - 2 \cdot 3 - 9$

30. $8 - (2 \cdot 3 - 9)$

31. $(8 - 2 \cdot 3) - 9$

32. $(8 - 2)(3 - 9)$

33. $16 \cdot (-24) + 50$

34. $10 \cdot 20 - 15 \cdot 24$

35. $40 - 3^2 - 2^3$

36. $2^4 + 2^2 - 10$

38. $4^3 + 10 \cdot 20 + 8^2 - 23$ **39.** $4 \cdot 5 - 2 \cdot 6 + 4$ **40.** $5^3 + 4 \cdot 9 - (8 + 9 \cdot 3)$

41. $1 - (-2)^2 \cdot 3 \div 6$ **42.** $-6 + (-3)^2 + 6 \div (-2)$ **43.** $18 - (-3)^3 - 3^2 \cdot 5$ **44.** $9 - (-2)^3 - 50 \div 2$

45. $\dfrac{9^2 - 1}{1 - 3^2}$ **46.** $\dfrac{100 - 6^2}{(-5)^2 - 3^2}$ **47.** $8(-7) + 6(-5)$ **48.** $10(-5) \div 1(-1)$

49. $20 \div 5(-3) + 3$ **50.** $14 \div 2(-6) + 7$ **51.** $18 - 0(3^2 - 5^2 \cdot 7 - 4)$ **52.** $9 \cdot 0 \div 5 \cdot 4$

53. $-4^2 + 6$ **54.** $-5^2 + 7$ **55.** $-8^2 - 3$ **56.** $-9^2 - 11$

57. $4 \cdot 5^2 \div 10$ **58.** $(2 - 5)^2 \div (-9)$ **59.** $(3 - 8)^2 \div (-1)$ **60.** $3 - 3^2$

61. $12 - 20^3$ **62.** $20 + 4^3 \div (-8)$ **63.** $2 \times 10^3 - 5000$ **64.** $-7(3^4) + 18$

65. $6[9 - (3 - 4)]$ **66.** $8[(6 - 13) - 11]$ **67.** $-1000 \div (-100) \div 10$ **68.** $256 + (-32) \div (-4)$

69. $-7 - 3[-80 \div (2 - 10)]$ **70.** $-1 - 5[3 - (7 - 4^2)]$ **71.** $-2[3 - (7 - 9)^3]$ **72.** $-10[(2 - 8)^2 - 6]$

73. $8 - |7 - 9| \cdot 3$ **74.** $|8 - 7 - 9| \cdot 2 + 1$ **75.** $9 - |7 - 3^2|$ **76.** $9 - |5 - 7|^3$

77. $\dfrac{6^3 - 7 \cdot 3^4 - 2^5 \cdot 9}{(1 - 2^3)^3 + 7^3}$ **78.** $\dfrac{6 \div 2 \cdot 4^2 - 7^2 + 1}{(7 - 4)^3 - 2 \cdot 5 - 4}$ **79.** $\dfrac{2 \cdot 3^2 \div (3^2 - (2 + 1))}{5^2 - 6^2 - 2^2(-3)}$ **80.** $\dfrac{5 \cdot 6^2 \div (2^2 \cdot 5) - 7^2}{3^2 - 4^2 - (-2)^3 - 2}$

81. $\dfrac{(-5)^3 + 17}{10(2 - 6) - 2(5 + 2)}$ **82.** $\dfrac{(3 - 5)^2 - (7 - 13)}{(2 - 5)3 + 2 \cdot 4}$ **83.** $\dfrac{2 \cdot 4^3 - 4 \cdot 32}{19^3 - 17^4}$ **84.** $\dfrac{-16 \cdot 28 \div 2^2}{5 \cdot 25 - 5^3}$

Skill Maintenance

85. Fabrikant Fine Diamonds ran a 4-in. by 7-in. advertisement in the *New York Times*. Find the area of the ad. [1.4b], [1.8a]

86. A hotel has 4 floors with 62 rooms on each floor. How many rooms are there in the hotel? [1.8a]

87. Cindi's Ford Focus gets 32 mpg (miles per gallon). How many gallons will it take to travel 384 mi? [1.8a]

88. Craig's Chevy Blazer gets 14 mpg. How many gallons will it take to travel 378 mi? [1.8a]

89. A 7-oz bag of tortilla chips contains 1050 calories. How many calories are in a 1-oz serving? [1.8a]

90. A 7-oz bag of tortilla chips contains 8 g (grams) of fat per ounce. How many grams of fat are in a carton containing 12 bags of chips? [1.8a]

91. There are 18 sticks in a large pack of Trident gum. If 4 people share the pack equally, how many whole pieces will each person receive? How many extra pieces will remain? [1.8a]

92. A bag of Ricola throat lozenges contains 24 cough drops. If 5 people share the bag equally, how many lozenges will each person receive? How many extra lozenges will remain? [1.8a]

Synthesis

Simplify, if possible.

93. $\dfrac{9 - 3^2}{2 \cdot 4^2 - 5^2 \cdot 9 + 8^2 \cdot 7}$

94. $\dfrac{7^3 \cdot 9 - 6^2 \cdot 8 + 4^3 \cdot 6}{5^2 - 25}$

95. $\dfrac{(25 - 4^2)^3}{17^2 - 16^2} \cdot ((-6)^2 - 6^2)$

96. $\dfrac{(7 - 8)^{37}}{7^2 - 8^2} \cdot (98 - 7^2 \cdot 2)$

97. ▦ $\dfrac{19 - 17^2}{13^2 - 34}$

98. ▦ $\dfrac{195 + (-15)^3}{195 - 7 \cdot 5^2}$

99. ▦ $28^2 - 36^2/4^2 + 17^2$

100. ▦ $9^3 - 36^3/12^2 + 9^2$

101. ▦ Write down the keystrokes needed to calculate $\dfrac{15^2 - 5^3}{3^2 + 4^2}$.

102. ▦ Write down the keystrokes needed to calculate $\dfrac{16^2 - 24 \cdot 23}{3 \cdot 4 + 5^2}$.

103. Evaluate the expression for which the keystrokes are as follows: [4] [−] [1] [0] [÷] [2] [+] [6].

104. Evaluate the expression for which the keystrokes are [4] [−] [1] [6] [÷] [(] [2] [+] [6] [)].

Determine the sign of each expression if *m* is negative and *n* is positive.

105. $\dfrac{-n}{m}$

106. $\dfrac{-n}{-m}$

107. $-\left(\dfrac{-n}{m}\right)$

108. $-\left(\dfrac{n}{-m}\right)$

109. $-\left(\dfrac{-n}{-m}\right)$

Mid-Chapter Review

Concept Reinforcement

Determine whether each statement is true or false.

_____ **1.** Every integer is either positive or negative. [2.1a]

_____ **2.** If $a > b$, then a lies to the left of b on the number line. [2.1b]

_____ **3.** The absolute value of a number is always nonnegative. [2.1c]

Guided Solutions

 Fill in each blank with the number that creates a correct statement or solution.

4. Evaluate $-x$ and $-(-x)$ when $x = -4$. [2.1d]

$-x = -(\boxed{}) = \boxed{}$;

$-(-x) = -(-(\boxed{})) = -(\boxed{}) = \boxed{}$

Subtract. [2.3a]

5. $5 - 13 = 5 + (\boxed{}) = \boxed{}$

6. $-6 - (-7) = -6 + \boxed{} = \boxed{}$

Mixed Review

7. State the integers that correspond to this situation.

Jerilyn deposited $450 in her checking account. Later that week she used her debit card for a $79 purchase. [2.1a]

8. Change the sign of 9. [2.1d]

Use either $<$ or $>$ for $\square$ to write a true sentence. [2.1b]

9. $-6 \ \square\ 6$

10. $-5 \ \square\ -3$

11. $-10 \ \square\ 0$

12. $-20 \ \square\ -30$

Find the absolute value. [2.1c]

13. $|38|$

14. $|-18|$

15. $|0|$

16. $|-12|$

Find the opposite, or additive inverse, of the number. [2.1d]

17. -56

18. 3

19. 0

20. -49

21. Find $-x$ when x is -19. [2.1d]

22. Evaluate $-(-x)$ when x is 23. [2.1d]

Compute and simplify. [2.2a], [2.3a], [2.4a], [2.5a], [2.5b]

23. $7 + (-9)$

24. $-6 + (-10)$

25. $36 + (-36)$

26. $-8 + (-9)$

27. $-9 + 10$

28. $19 + (-17)$

29. $2 - 28$

30. $-8 - (-4)$

31. $-3 - 10$

32. $5 - (-11)$

33. $0 - (-6)$

34. $12 - 24$

35. $-12 \cdot 3$

36. $6(-9)$

37. $(-13)(-2)$

38. $(-2)(-41)$

39. $(-9)^2$

40. -9^2

41. $-75 \div (-3)$

42. $-20 \div 4$

43. $17 - (-25) + 15 - (-18)$

44. $-9 + (-3) + 16 - (-10)$

45. $(-7)(-2)(-1)(-3)$

46. $3 - 6 \cdot 5 - 11$

47. $-5^2 + 6[1 - (3 - 4)]$

48. $\dfrac{6^2 - 3(5 - 9)}{7^2 - (-5)^2}$

Solve. [2.3b]

49. *Temperature Change.* In a chemistry lab, Ben works with a substance whose initial temperature is 25°C. During an experiment, the temperature falls to −8°C. Find the difference between the two temperatures.

50. *Stock Price Change.* The price of a stock opened at $56. During the day, it dropped $6, then rose $2, and dropped $8. Find the value of the stock at the end of the day.

Understanding Through Discussion and Writing

51. A student states "−45 is bigger than −21." What mistake do you think the student is making? [2.1b]

52. Is subtraction of integers associative? Why or why not? [2.3a]

53. Explain in your own words why the sum of two negative numbers is always negative. [2.2a]

54. If a negative number is subtracted from a positive number, will the result always be positive? Why or why not? [2.3a]

2.6   Introduction to Algebra and Expressions

OBJECTIVES

a Evaluate an algebraic expression by substitution.

b Use the distributive law to find equivalent expressions.

SKILL TO REVIEW

Objective 1.9b: Evaluate exponential notation.

Evaluate.

1. 8^2 **2.** 5^4

In this section, we will write *equivalent expressions* by making use of the *distributive law*.

a   EVALUATING ALGEBRAIC EXPRESSIONS

In arithmetic, we work with expressions such as

$$37 + 86, \quad 7 \cdot 8, \quad 19 - 7, \quad \text{and} \quad \frac{3}{8}.$$

In algebra, we use both numbers and letters and work with *algebraic expressions* such as

$$x + 86, \quad 7 \cdot t, \quad 19 - y, \quad \text{and} \quad \frac{a}{b}.$$

When a letter can stand for various numbers, we call the letter a **variable**. A number or a letter that stands for just one number is called a **constant**. Let $b =$ your birth year. Then b is a constant. Let $a =$ your age. Then a is a variable since the value of a changes from year to year.

An **algebraic expression** consists of variables, constants, numerals, and operation signs. When we replace a variable with a number, we say that we are **substituting** for the variable. Carrying out the **operations** of addition, subtraction, and so on, is called **evaluating the expression**.

EXAMPLE 1   Evaluate $x + y$ for $x = 37$ and $y = 29$.

We substitute 37 for x and 29 for y and carry out the addition:

$$x + y = 37 + 29 = 66.$$

The number 66 is called the **value** of the expression.

Algebraic expressions involving multiplication can be written in several ways. For example, "8 times a" can be written as $8 \times a$, $8 \cdot a$, $8(a)$, or simply $8a$. Two letters written together without an operation symbol, such as ab, also indicate multiplication.

1. Evaluate $a + b$ for $a = 38$ and $b = 26$.

2. Evaluate $x - y$ for $x = 57$ and $y = 29$.

3. Evaluate $5t$ for $t = -14$.

EXAMPLE 2   Evaluate $3y$ for $y = -14$.

$$3y = 3(-14) = -42 \qquad \text{Parentheses are required here.}$$

◄ **Do Margin Exercises 1–3.**

Algebraic expressions involving division can also be written several ways. For example, "8 divided by t" can be written as $8 \div t$, $8/t$, or $\frac{8}{t}$.

Answers

Skill to Review:
1. 64   **2.** 625

Margin Exercises:
1. 64   **2.** 28   **3.** −70

EXAMPLE 3 Evaluate $\dfrac{a}{b}$ and $\dfrac{-a}{-b}$ for $a = 35$ and $b = 7$.

We substitute 35 for a and 7 for b:

$$\frac{a}{b} = \frac{35}{7} = 5; \qquad \frac{-a}{-b} = \frac{-35}{-7} = 5.$$

EXAMPLE 4 Evaluate $-\dfrac{a}{b}$, $\dfrac{-a}{b}$, and $\dfrac{a}{-b}$ for $a = 15$ and $b = 3$.

We substitute 15 for a and 3 for b:

$$-\frac{a}{b} = -\frac{15}{3} = -5; \qquad \frac{-a}{b} = \frac{-15}{3} = -5; \qquad \frac{a}{-b} = \frac{15}{-3} = -5.$$

Examples 3 and 4 illustrate the following.

$\dfrac{-a}{-b}$ and $\dfrac{a}{b}$ represent the same number.

$-\dfrac{a}{b}$, $\dfrac{-a}{b}$, and $\dfrac{a}{-b}$ all represent the same number.

Do Exercises 4–7. ▶

EXAMPLE 5 Evaluate $\dfrac{9C}{5} + 32$ for $C = 20$.

This expression can be used to find the Fahrenheit temperature that corresponds to 20 degrees Celsius.

$$\frac{9C}{5} + 32 = \frac{9 \cdot 20}{5} + 32 = \frac{180}{5} + 32 = 36 + 32 = 68.$$

Do Exercise 8. ▶

EXAMPLE 6 Evaluate $5x^2$ for $x = 3$ and $x = -3$.

The rules for order of operations specify that the replacement for x be squared. That result is then multiplied by 5.

$$\left.\begin{array}{l} 5x^2 = 5(3)^2 = 5(9) = 45; \\ 5x^2 = 5(-3)^2 = 5(9) = 45 \end{array}\right\}$$

You can always use parentheses when substituting. They are usually necessary when substituting a negative number.

Do Exercises 9 and 10. ▶

EXAMPLE 7 Evaluate $(-x)^2$ and $-x^2$ for $x = 7$.

$$(-x)^2 = (-7)^2 = (-7)(-7) = 49 \qquad \text{Substitute 7 for } x. \text{ Then evaluate the power.}$$

To evaluate $-x^2$, we again substitute 7 for x. We must recall that taking the opposite of a number is the same as multiplying that number by -1.

$$\begin{aligned} -x^2 &= -1 \cdot x^2 & -a = -1 \cdot a \\ -7^2 &= -1 \cdot 7^2 & \text{Substituting 7 for } x \\ &= -1 \cdot 49 = -49 & \text{Using the rules for order of operations} \end{aligned}$$

Do Exercises 11–13. ▶

For each expression, find two equivalent expressions with the negative sign in different places.

GS 4. $\dfrac{-6}{x}$

$$\frac{-6}{x} = \frac{\boxed{}}{x} = \frac{6}{\boxed{}}$$

5. $-\dfrac{m}{n}$ 6. $\dfrac{r}{-4}$

7. Evaluate $\dfrac{a}{-b}$, $\dfrac{-a}{b}$, and $-\dfrac{a}{b}$ for $a = 28$ and $b = 4$.

8. Find the Fahrenheit temperature that corresponds to 10 degrees Celsius. (See Example 5.)

9. Evaluate $3x^2$ for $x = 4$ and $x = -4$.

10. Evaluate a^4 for $a = 3$ and $a = -3$.

11. Evaluate $(-x)^2$ and $-x^2$ for $x = 3$.

12. Evaluate $(-x)^2$ and $-x^2$ for $x = 2$.

13. Evaluate x^5 for $x = 2$ and $x = -2$.

Answers

4. $-\dfrac{6}{x}; \dfrac{6}{-x}$ 5. $\dfrac{-m}{n}; \dfrac{m}{-n}$ 6. $\dfrac{-r}{4}; -\dfrac{r}{4}$
7. $-7; -7; -7$ 8. 50   9. 48; 48   10. 81; 81
11. 9; -9 12. 4; -4 13. 32; -32

Guided Solution:
4. $-, -x$

Examples 6 and 7 illustrate the following.

$$(-x)^2 = (x)^2$$
$$(-x)^2 \neq -x^2$$

b EQUIVALENT EXPRESSIONS AND THE DISTRIBUTIVE LAW

Some pairs of algebraic expressions are *equivalent*.

EXAMPLE 8 Evaluate $x + x$ and $2x$ for $x = 3$ and $x = -5$.

We substitute 3 for x in $x + x$ and again in $2x$:

$$x + x = 3 + 3 = 6; \qquad 2x = 2 \cdot 3 = 6.$$

Next we repeat the procedure, substituting -5 for x:

$$x + x = -5 + (-5) = -10; \qquad 2x = 2(-5) = -10.$$

The results can be shown in a table. It appears that $x + x$ and $2x$ represent the same number.

	$x + x$	$2x$
$x = 3$	6	6
$x = -5$	-10	-10

◀ Do Exercises 14 and 15.

Example 8 suggests that $x + x$ and $2x$ represent the same number for any replacement of x. This is in fact true, so we can say that $x + x$ and $2x$ are **equivalent expressions**.

EQUIVALENT EXPRESSIONS

Two expressions that have the same value for all allowable replacements are called **equivalent**.

In Examples 3 and 7 we saw that the expressions $\dfrac{-a}{-b}$ and $\dfrac{a}{b}$ are equivalent but that the expressions $(-x)^2$ and $-x^2$ are *not* equivalent.

An important concept, known as the **distributive law**, is useful for finding equivalent algebraic expressions. The distributive law involves two operations: multiplication and either addition or subtraction.

To understand how the distributive law works, consider the following:

$$
\begin{array}{r}
4\,5 \\
\times\ 7 \\
\hline
3\,5 \quad \leftarrow \text{This is } 7 \cdot 5. \\
2\,8\,0 \quad \leftarrow \text{This is } 7 \cdot 40. \\
\hline
3\,1\,5 \quad \leftarrow \text{This is the sum } 7 \cdot 40 + 7 \cdot 5.
\end{array}
$$

To carry out the multiplication, we actually added two products. That is,

$$7 \cdot 45 = 7(40 + 5) = 7 \cdot 40 + 7 \cdot 5.$$

The distributive law says that if we want to multiply a sum of several numbers by a number, we can either add within the grouping symbols and then multiply, or multiply each of the terms separately and then add.

Complete each table by evaluating each expression for the given values.

14.

	$3x + 2x$	$5x$
$x = 4$		
$x = -2$		
$x = 0$		

15.

	$4x - x$	$3x$
$x = 2$		
$x = -2$		
$x = 0$		

Answers

14. 20, 20; $-10, -10$; 0, 0
15. 6, 6; $-6, -6$; 0, 0

THE DISTRIBUTIVE LAW
....

For any numbers a, b, and c,

$$a(b + c) = ab + ac.$$

EXAMPLE 9 Evaluate $a(b + c)$ and $ab + ac$ for $a = 3$, $b = 4$, and $c = 2$.

We have

$$a(b + c) = 3(4 + 2) = 3 \cdot 6 = 18$$

and $\quad ab + ac = 3 \cdot 4 + 3 \cdot 2 = 12 + 6 = 18.$ ∎

The parentheses in the statement of the distributive law tell us to multiply both b and c by a. Without the parentheses, we would have $ab + c$.

EXAMPLE 10 Use the distributive law to write an expression equivalent to $2(l + w)$.

$$2(l + w) = 2 \cdot l + 2 \cdot w \qquad \text{Note that the } + \text{ sign between } l \text{ and } w$$
$$\text{now appears between } 2 \cdot l \text{ and } 2 \cdot w.$$
$$= 2l + 2w. \qquad \text{Try to go directly to this step.}$$

Do Exercises 16 and 17. ▶

Since subtraction can be regarded as addition of the opposite, it follows that the distributive law is true for subtraction as well as addition.

EXAMPLE 11 Use the distributive law to write an expression equivalent to each of the following:

a) $9(x - 5)$; **b)** $(a - 7)b$; **c)** $-4(x - 2y + 3z)$; **d)** $-(2x - 3y)$.

a) $9(x - 5) = 9x - 9(5)$
$$\qquad = 9x - 45 \qquad \text{Try to go directly to this step.}$$

b) $(a - 7)b = b(a - 7) \qquad \text{Using a commutative law}$
$$\qquad = b \cdot a - b \cdot 7 \qquad \text{Using the distributive law}$$
$$\qquad = ab - 7b \qquad \text{Using a commutative law to write } ba \text{ alpha-}$$
$$\text{betically and } b \cdot 7 \text{ with the constant first}$$

c) $-4(x - 2y + 3z) = -4 \cdot x - (-4)(2y) + (-4)(3z) \qquad \text{Using the}$
$$\text{distributive law}$$

$$\qquad\qquad = -4x - (-4 \cdot 2)y + (-4 \cdot 3)z \qquad \text{Using an}$$
$$\text{associative law}$$
$$\text{(twice)}$$

$$\qquad\qquad = -4x - (-8y) + (-12z)$$
$$\qquad\qquad = -4x + 8y - 12z$$

d) $-(2x - 3y) = -1(2x - 3y) \qquad \text{The opposite of a number is the}$
$$\text{same as multiplying by } -1.$$

$$\qquad = -1 \cdot 2x - (-1)(3y) \qquad \text{Using the distributive law}$$
$$\qquad = -2x - (-3y) \qquad \text{Using an associative law (twice)}$$
$$\qquad = -2x + 3y$$

Do Exercises 18–22. ▶

Use the distributive law to write an equivalent expression.

16. $5(a + b)$

(GS) 17. $6(x + y + z)$
$$= 6 \cdot \boxed{} + 6 \cdot \boxed{} + 6 \cdot \boxed{}$$
$$= 6x + \boxed{} + 6z$$

Use the distributive law to write an equivalent expression.

18. $4(x - y)$

19. $3(a - b + c)$

20. $(m - 4)6$

21. $-8(2a - b + 3c)$

22. $-(c - 8d)$

Answers

16. $5a + 5b$ **17.** $6x + 6y + 6z$ **18.** $4x - 4y$
19. $3a - 3b + 3c$ **20.** $6m - 24$
21. $-16a + 8b - 24c$ **22.** $-c + 8d$

Guided Solution:
17. $x, y, z, 6y$

☑ **Reading Check**

Classify each algebraic expression as involving either multiplication or division.

RC1. $3/q$ _____

RC2. $3q$ _____

RC3. $3 \cdot q$ _____

RC4. $\dfrac{3}{q}$ _____

a Evaluate.

1. $10n$, for $n = 2$
(The cost, in cents, of sending 2 text messages)

2. $99n$, for $n = 2$
(The cost, in cents, of downloading 2 songs)

3. $\dfrac{x}{y}$, for $x = 6$ and $y = -3$

4. $\dfrac{m}{n}$, for $m = 18$ and $n = 2$

5. $\dfrac{2d}{c}$, for $c = 6$ and $d = 3$

6. $\dfrac{5y}{z}$, for $y = 15$ and $z = -25$

7. $\dfrac{72}{r}$, for $r = 4$
(The approximate doubling time, in years, for an investment earning 4% interest per year)

8. $\dfrac{72}{i}$, for $i = 2$
(The approximate doubling time, in years, for an investment earning 2% interest per year)

9. $3 - 5 \cdot x$, for $x = 2$

10. $9 - 2 \cdot x$, for $x = 5$

11. $2l + 2w$, for $l = 3$ and $w = 4$
(The perimeter, in feet, of a 3-ft by 4-ft rectangle)

12. $3(a + b)$, for $a = 2$ and $b = 4$

13. $2(l + w)$, for $l = 3$ and $w = 4$
(The perimeter, in feet, of a 3-ft by 4-ft rectangle)

14. $3a + 3b$, for $a = 2$ and $b = 4$

15. $7a - 7b$, for $a = -1$ and $b = 2$

16. $4x - 4y$, for $x = -5$ and $y = 1$

17. $7(a - b)$, for $a = -1$ and $b = 2$

18. $4(x - y)$, for $x = -5$ and $y = 1$

19. $16t^2$, for $t = 5$
(The distance, in feet, that an object falls in 5 sec)

20. $\dfrac{49t^2}{10}$, for $t = 10$
(The distance, in meters, that an object falls in 10 sec)

21. $a + (b - a)^2$, for $a = 6$ and $b = 10$

22. $(x + y)^2$, for $x = 2$ and $y = 10$

23. $9a + 9b$, for $a = 13$ and $b = -13$

24. $8x + 8y$, for $x = 17$ and $y = -17$

25. $\dfrac{n^2 - n}{2}$, for $n = 9$
(For determining the number of handshakes possible among 9 people)

26. $\dfrac{5(F - 32)}{9}$, for $F = 50$
(For converting 50 degrees Fahrenheit to degrees Celsius)

27. $1 - x^2$, for $x = -2$

28. $4 - y^2$, for $y = -1$

29. $m^2 - n^2$, for $m = 6$ and $n = 5$

30. $a^2 + b^2$, for $a = 3$ and $b = 4$

31. $a^3 - a^2$, for $a = -10$

32. $x^2 - x^3$, for $x = -10$

For each expression, write two equivalent expressions with the negative sign in different places.

33. $-\dfrac{5}{t}$

34. $\dfrac{7}{-x}$

35. $\dfrac{-n}{b}$

36. $-\dfrac{3}{r}$

37. $\dfrac{9}{-p}$

38. $\dfrac{-u}{5}$

39. $\dfrac{-14}{w}$

40. $\dfrac{-23}{m}$

Evaluate $\dfrac{-a}{b}$, $\dfrac{a}{-b}$, and $-\dfrac{a}{b}$ for the given values.

41. $a = 45, b = 9$

42. $a = 40, b = 2$

43. $a = 81, b = 3$

44. $a = 56, b = 7$

Evaluate.

45. $(-3x)^2$ and $-3x^2$, for $x = 2$

46. $(-2x)^2$ and $-2x^2$, for $x = 3$

47. $5x^2$, for $x = 3$ and $x = -3$

48. $2x^2$, for $x = 5$ and $x = -5$

49. x^3, for $x = 6$ and $x = -6$

50. x^6, for $x = 2$ and $x = -2$

51. x^8, for $x = 1$ and $x = -1$

52. x^5, for $x = 3$ and $x = -3$

53. a^5, for $a = 2$ and $a = -2$

54. a^7, for $a = 1$ and $a = -1$

b   Use the distributive law to write an equivalent expression.

55. $5(a + b)$

56. $7(x + y)$

57. $4(x + 1)$

58. $6(a + 1)$

59. $2(b + 5)$

60. $3(x - 6)$

61. $7(1 - t)$

62. $4(1 - y)$

63. $6(5x - 2)$

64. $9(6m - 7)$

65. $8(x + 7 + 6y)$

66. $4(5x + 8 + 3p)$

67. $-7(y - 2)$

68. $-9(y - 7)$

69. $(x + 2)3$

70. $(x + 4)2$

71. $-4(x - 3y - 2z)$

72. $8(2x - 5y - 8z)$

73. $8(a - 3b + c)$

74. $-6(a + 2b - c)$

75. $4(x - 3y - 7z)$

76. $5(9x - y + 8z)$

77. $(4a - 5b + c - 2d)5$

78. $(9a - 4b + 3c - d)7$

79. $-1(3m + 2n)$

80. $-1(6a + 7b)$

81. $-1(2a - 3b + 4)$

82. $-1(7x - 8y - 9)$

83. $-(x - y - z)$

84. $-(a - b - c)$

124   CHAPTER 2   Introduction to Integers and Algebraic Expressions

Skill Maintenance

85. Write a word name for 23,043,921.   [1.1c]

86. Multiply: $17 \cdot 53$.   [1.4a]

87. Estimate by rounding to the nearest ten. Show your work.   [1.6b]

$$\begin{array}{r} 5\ 2\ 8\ 3 \\ -\ 2\ 4\ 7\ 5 \\ \hline \end{array}$$

88. Divide: $2982 \div 3$.   [1.5a]

89. On March 9, it snowed 12 in., but on March 10, the sun melted 7 in. How much snow remained?   [1.8a]

90. For Tania's graduation party, her husband ordered three buckets of chicken wings at $12 apiece and 3 trays of nachos at $9 a tray. How much did he pay for the wings and nachos?   [1.8a]

Synthesis

91. A car's catalytic converter works most efficiently after it is heated to about 370°C. To what Fahrenheit temperature does this correspond? (*Hint*: See Example 5.)

92. Evaluate $\dfrac{9C}{5} + 32$ for $C = 10$ and for $C = 20$. (See Example 5.) When the Celsius temperature is doubled, is the corresponding Fahrenheit temperature also doubled?

Evaluate.

93. ▦ $a - b^3 + 17a$, for $a = 19$ and $b = -16$

94. ▦ $x^2 - 23y + y^3$, for $x = 18$ and $y = -21$

95. ▦ $r^3 + r^2t - rt^2$, for $r = -9$ and $t = 7$

96. ▦ $a^3b - a^2b^2 + ab^3$, for $a = -8$ and $b = -6$

97. $a^{1996} - a^{1997}$, for $a = -1$

98. $x^{1493} - x^{1492}$, for $x = -1$

99. $(m^3 - mn)^m$, for $m = 4$ and $n = 6$

100. $5a^{3a-4}$, for $a = 2$

Replace each blank with $\boxed{+}$, $\boxed{-}$, $\boxed{\times}$, or $\boxed{\div}$ to make each statement true.

101. ▦ $-32 \,\square\, (88 \,\square\, 29) = -1888$

102. ▦ $59 \,\square\, 17 \,\square\, 59 \,\square\, 8 = 1475$

Classify each statement as true or false. If false, write an example showing why.

103. For any choice of x, $x^2 = (-x)^2$.

104. For any choice of x, $x^3 = -x^3$.

105. For any choice of x, $x^6 + x^4 = (-x)^6 + (-x)^4$.

106. For any choice of x, $(-3x)^2 = 9x^2$.

Like Terms and Perimeter

OBJECTIVES

a Combine like terms.

b Determine the perimeter of a polygon.

One common way in which equivalent expressions are formed is by *combining like terms*.

a COMBINING LIKE TERMS

A **term** is a number, a variable, a product of numbers and/or variables, or a quotient of numbers and/or variables. Terms are separated by addition signs. If there are subtraction signs, we can find an equivalent expression that uses addition signs.

EXAMPLE 1 What are the terms of $3xy - 4y + \dfrac{2}{z}$?

$$3xy - 4y + \frac{2}{z} = 3xy + (-4y) + \frac{2}{z} \qquad \text{Separating parts with } + \text{ signs}$$

The terms are $3xy$, $-4y$, and $\dfrac{2}{z}$.

What are the terms of each expression?

1. $5x - 4y + 3$

2. $-4y - 2x + \dfrac{x}{y}$

◀ **Do Exercises 1 and 2.**

Terms in which the variable factors are exactly the same, such as $9x$ and $-4x$, are called **like**, or **similar**, **terms**. For example, $3y^2$ and $7y^2$ are like terms, but $5x$ and $6x^2$ are not. Constants, like 7 and 3, are also like terms.

EXAMPLES Identify the like terms.

2. $7x + 5x^2 + 2x + 8 + 5x^3 + 1$

$7x$ and $2x$ are like terms;   8 and 1 are like terms.

3. $5ab + a^3 - a^2b - 2ab + 7a^3$

$5ab$ and $-2ab$ are like terms; a^3 and $7a^3$ are like terms.

Identify the like terms.

3. $9a^3 + 4ab + a^3 + 3ab + 7$

4. $3xy - 5x^2 + y^2 - 4xy + y$

◀ **Do Exercises 3 and 4.**

When an algebraic expression contains like terms, an equivalent expression can be formed by **combining**, or **collecting**, **like terms**. To combine like terms, we use the distributive law.

EXAMPLE 4 Combine like terms to form an equivalent expression.

a) $4x + 3x$ **b)** $6mn - 7mn$

c) $7y - 2 - 6y + 5$ **d)** $2a^5 + 9ab + 3 + a^5 - 7 - 4ab$

a) $4x + 3x = (4 + 3)x$ Using the distributive law (in "reverse")

           $= 7x$ We usually go directly to this step.

b) $6mn - 7mn = (6 - 7)mn$ Try to do this mentally.

                $= -1mn$, or simply $-mn$

c) $7y - 2 - 6y + 5 = 7y + (-2) + (-6y) + 5$ Rewriting as addition

                    $= 7y + (-6y) + (-2) + 5$ Using a commutative law

                  $= 1y + 3$, or simply $y + 3$

Answers

1. $5x$; $-4y$; 3   **2.** $-4y$; $-2x$; $\dfrac{x}{y}$

3. $9a^3$ and a^3; $4ab$ and $3ab$ **4.** $3xy$ and $-4xy$

d) $2a^5 + 9ab + 3 + a^5 - 7 - 4ab$

$\quad = 2a^5 + 9ab + 3 + a^5 + (-7) + (-4ab)$

$\quad = 2a^5 + a^5 + 9ab + (-4ab) + 3 + (-7)$ Rearranging terms

$\quad = 3a^5 + 5ab + (-4)$ Think of a^5 as $1a^5$; $2a^5 + a^5 = 3a^5$

$\quad = 3a^5 + 5ab - 4$

Do Exercises 5–7. ▶

b   PERIMETER

PERIMETER OF A POLYGON

A **polygon** is a closed geometric figure with three or more sides. The **perimeter** of a polygon is the distance around it, or the sum of the lengths of its sides.

EXAMPLE 5   Find the perimeter of this polygon.

> A polygon with five sides is called a *pentagon*.

We add the lengths of all sides. Since all the units are the same, we are effectively combining like terms.

$\quad$ Perimeter $= 6\,\text{m} + 5\,\text{m} + 4\,\text{m} + 5\,\text{m} + 9\,\text{m}$

$\qquad\qquad\quad = (6 + 5 + 4 + 5 + 9)\,\text{m}$ Using the distributive law

$\qquad\qquad\quad = 29\,\text{m}$ Try to go directly to this step.

·············· **Caution!** ··············

When units of measurement are given in the statement of a problem, as in Example 5, the solution should also contain units of measurement.

Do Exercises 8 and 9. ▶

A **rectangle** is a polygon with four sides and four 90° angles. Opposite sides of a rectangle have the same measure. The symbol ⌐ or ⌐ indicates a 90° angle. A 90° angle is often referred to as a **right angle**.

EXAMPLE 6   Find the perimeter of a rectangle that is 3 cm by 4 cm.

$\quad$ Perimeter $= 3\,\text{cm} + 3\,\text{cm} + 4\,\text{cm} + 4\,\text{cm}$

$\qquad\qquad\quad = (3 + 3 + 4 + 4)\,\text{cm}$

$\qquad\qquad\quad = 14\,\text{cm}$

Do Exercise 10. ▶

Combine like terms to form an equivalent expression.

5. $2a + 7a$

6. $5x^2 + 9 - 4x^2 + 3$

GS **7.** $4m - 2n^2 + 5 + n^2 + m - 9$

The like terms are

$\qquad 4m$ and $\boxed{}$,

$\qquad -2n^2$ and $\boxed{}$,

$\qquad 5$ and $\boxed{}$.

$4m + (-2n^2) + 5 + n^2 + m + (-9)$

$\quad = 4m + m + (-2n^2) + \boxed{}$
$\qquad\qquad\qquad + 5 + (-9)$

$\quad = 5m + (\boxed{}) + (-4)$

$\quad = 5m - \boxed{} - 4$

Find the perimeter of each polygon.

8.

9.

10. Find the perimeter of a rectangle that is 4 cm by 2 cm.

Answers

5. $9a$ **6.** $x^2 + 12$ **7.** $5m - n^2 - 4$

8. 26 cm   **9.** 35 mm   **10.** 12 cm

Guided Solution:
7. $m, n^2, -9, n^2, -n^2, n^2$

The perimeter of the rectangle in Example 6 is $2 \cdot 3\,\text{cm} + 2 \cdot 4\,\text{cm}$, or equivalently $2(3\,\text{cm} + 4\,\text{cm})$. This can be generalized, as follows.

PERIMETER OF A RECTANGLE

The **perimeter P of a rectangle** of length l and width w is given by

$$P = 2l + 2w, \quad \text{or} \quad P = 2 \cdot (l + w).$$

EXAMPLE 7 A common door size is 7 ft by 3 ft. Find the perimeter of such a door.

$$
\begin{aligned}
P &= 2l + 2w && \text{We could also use } P = 2(l + w). \\
&= 2 \cdot 7\,\text{ft} + 2 \cdot 3\,\text{ft} \\
&= (2 \cdot 7)\,\text{ft} + (2 \cdot 3)\,\text{ft} && \text{Try to do this mentally.} \\
&= 14\,\text{ft} + 6\,\text{ft} \\
&= 20\,\text{ft} && \text{Combining like terms}
\end{aligned}
$$

The perimeter of the door is 20 ft.

11. Find the perimeter of a 4-ft by 8-ft sheet of plywood.

◀ **Do Exercise 11.**

A **square** is a rectangle in which all sides have the same length.

12. Find the perimeter of a square with sides of length 10 km.

EXAMPLE 8 Find the perimeter of a square with sides of length 9 mm.

$$
\begin{aligned}
P &= 9\,\text{mm} + 9\,\text{mm} + 9\,\text{mm} + 9\,\text{mm} \\
&= (9 + 9 + 9 + 9)\,\text{mm} && \text{Note that} \\
& && 9 + 9 + 9 + 9 = 4 \cdot 9. \\
&= 36\,\text{mm}
\end{aligned}
$$

◀ **Do Exercise 12.**

PERIMETER OF A SQUARE

The **perimeter P of a square** is four times s, the length of a side:

$$
\begin{aligned}
P &= s + s + s + s \\
&= 4s.
\end{aligned}
$$

---

13. Find the perimeter of a square book with sides of length 9 in.

$$
\begin{aligned}
P &= 4s \\
&= 4 \cdot \boxed{}\ \text{in.} \\
&= \boxed{}\ \text{in.}
\end{aligned}
$$

EXAMPLE 9 Find the perimeter of a square room with sides of length 12 ft.

$$
\begin{aligned}
P &= 4s \\
&= 4 \cdot 12\,\text{ft} \\
&= 48\,\text{ft}
\end{aligned}
$$

The perimeter of the room is 48 ft.

◀ **Do Exercise 13.**

Answers

11. 24 ft **12.** 40 km **13.** 36 in.

Guided Solution:
13. 9, 36

Translating for Success

1. **Wood Costs.** It costs $8 for the wood for each birdhouse that Annette builds. If she used $120 worth of wood, how many birdhouses did she build?

2. **Elevation.** Genine started hiking a 10-mi trail at an elevation that was 150 ft below sea level. At the end of the trail, she was 75 ft above sea level. How many feet higher was she at the end of the trail than at the beginning?

3. **Community Service.** In order to fulfill the requirements for a sociology class, Glen must log 120 hr of community service. So far, he has spent 75 hr volunteering at a youth center. How many more hours must he serve?

4. **Disaster Relief.** Each package that is prepared for a disaster relief effort contains 15 meal bars. How many packages can be filled from a donation of 750 bars?

5. **Perimeter.** A rectangular building lot is 75 ft wide and 150 ft long. What is the perimeter of the lot?

The goal of these matching questions is to practice step (2), Translate, of the five-step problem-solving process. Translate each word problem to an equation and select a correct translation from equations A–O.

A. $75 + x = 120$

B. $15 \div 750 = x$

C. $-10 - 15 = x$

D. $8 \cdot 120 = x$

E. $150 - 75 = x$

F. $15 \cdot 750 = x$

G. $75 - (-150) = x$

H. $2 \cdot 150 + 2 \cdot 75 = x$

I. $-10 - (-15) = x$

J. $8 \cdot x = 120$

K. $750 \div 15 = x$

L. $15 - (-10) = x$

M. $75 + 120 = x$

N. $75 - 150 = x$

O. $75 = 120 + x$

Answers on page A-4

6. **Account Balance.** Lorenzo had $75 in his checking account. He then wrote a check for $150. What was the balance in his account?

7. **Laptop Computers.** Great Graphics purchased a laptop computer for each of its 15 employees. If each laptop cost $750, how much did the computers cost?

8. **Basketball.** A basketball team scored 75 points in one game. In the next game, the team scored a record 120 points. How many points did the team score in the two games?

9. **Pizza Sales.** A youth club sold 120 pizzas for a fundraiser. If each pizza sold for $8, how much money was taken in?

10. **Temperature.** The temperature in Fairbanks was $-10°$ at 6:00 P.M. and fell another $15°$ by midnight. What was the temperature at midnight?

✓ Reading Check

Complete each statement with the correct word from the following list. A word may be used more than once or not at all.

closed   perimeter   rectangle
open   polygon   square

RC1. A polygon is a(n) _____ figure with three or more sides.

RC2. The distance around a polygon is its _____.

RC3. The formula $P = 2l + 2w$ gives the _____ of a rectangle.

RC4. The perimeter of a(n) _____ is given by the formula $P = 4s$.

a List the terms of each expression.

1. $2a + 5b - 7c$

2. $4x - 6y + 7z$

3. $mn - 6n + 8$

4. $7rs - s - 5$

5. $3x^2y - 4y^2 - 2z^3$

6. $4a^3b + ab^2 - 9b^3$

Combine like terms to form an equivalent expression.

7. $5x + 9x$

8. $9a + 7a$

9. $10a - 15a$

10. $-17x + x$

11. $2x + 6y + x$

12. $3t - y + 7t$

13. $27a + 70 - 40a - 8$

14. $42x - 6 - x + 2$

15. $9 + 5t + 7y - t - y - 13$

16. $8 - 4a + 9b + 5a - 3b - 15$

17. $a + 3b + 5a - 2 + b$

18. $x + 7y + 5 - 2y + 3x$

19. $-8 + 11a - 5b - 10a - 7b + 7$

20. $8x - 5x + 6 + 3y - y - 4$

21. $8x^2 + 3y - x^2$

22. $8y^3 - 3z + 4y^3$

23. $11x^4 + 2y^3 - 4x^4 - y^3$

24. $13a^5 + 9b^4 - 2a^5 - 4b^4$

25. $9a^2 - 4a + a - 3a^2$

26. $3a^2 + 7a^3 - a^2 + 5 + a^3$

27. $x^3 - 5x^2 + 2x^3 - 3x^2 + 4$

28. $9xy + 4y^2 - 2xy + 2y^2 - 1$

29. $9x^3y + 4xy^3 - 5xy^3 + 3xy$

30. $8a^2b - 3ab^2 - 7a^2b + 2ab$

31. $3a^6 - b^4 + 2a^6b^4 - 7a^6 - 2b^4$

32. $3x^4 - 2y^4 + 8x^4y^4 - 7x^4 + y^4$

b Find the perimeter of each polygon.

33.
3 ft, 2 ft

34.
5 in., 5 in.

35.
Each side 7 km

36.
4 mm, 6 mm, 7 mm

37.
3 m, 1 m, 1 m, 3 m

38.
4 m, 4 m, 5 m, 4 m, 1 m

Tennis Court. A tennis court contains many rectangles. Use the diagram of a regulation tennis court to calculate the perimeters in Exercises 39–42.

39. The perimeter of a singles court

40. The perimeter of a doubles court

41. The perimeter of the rectangle formed by the service lines and the singles sidelines

42. The perimeter of the rectangle formed by a service line, a baseline, and the singles sidelines

43. Find the perimeter of a rectangular 8-ft by 10-ft bedroom.

44. Find the perimeter of a rectangular 3-ft by 4-ft doghouse.

45. Find the perimeter of a checkerboard that is 14 in. on each side.

46. Find the perimeter of a square skylight that is 2 m on each side.

47. Find the perimeter of a square frame that is 65 cm on each side.

48. Find the perimeter of a square garden that is 12 yd on each side.

49. Find the perimeter of a 12-ft by 20-ft rectangular deck.

50. Find the perimeter of a 40-ft by 35-ft rectangular backyard.

Skill Maintenance

51. A box of Shaw's Corn Flakes contains 510 grams (g) of corn flakes. A serving of corn flakes weighs 30 g. How many servings are in one box? [1.8a]

52. Estimate the difference by rounding to the nearest ten. [1.6b]

$$\begin{array}{r} 7\ 0\ 4 \\ -\ 4\ 8\ 6 \\ \hline \end{array}$$

Solve. [1.7b]

53. $25 = t + 9$

54. $19 = x + 6$

55. $45 = 3x$

56. $2t = 50$

57. $25 + n = 400$

58. $25 \cdot n = 400$

Synthesis

Simplify. (Multiply and then combine like terms.)

59. $5(x + 3) + 2(x - 7)$

60. $3(a - 7) + 7(a + 4)$

61. $2(3 - 4a) + 5(a - 7)$

62. $7(2 - 5x) + 3(x - 8)$

63. $-5(2 + 3x + 4y) + 7(2x - y)$

64. $3(4 - 2x) + 5(9x - 3y + 1)$

65. In order to save energy, Andrea plans to run a bead of caulk sealant around 3 exterior doors and 13 windows. Each window measures 3 ft by 4 ft, each door measures 3 ft by 7 ft, and there is no need to caulk the bottom of each door. If each cartridge of caulk seals 56 ft and costs $9, how much will it cost Andrea to seal the windows and doors?

66. Jacie is adding a border to small rectangular bulletin boards that are 5 ft by 7 ft and to larger bulletin boards that are 7 ft by 9 ft. If the border material costs $5 per yard, how much will Jacie spend to do 4 small bulletin boards and 3 large bulletin boards?

67. 🖩 A square wooden rack is used to store the 15 numbered pool balls as well as the cue ball. If a pool ball has a diameter of 57 mm, find the inside perimeter of the storage rack.

68. A rectangular box is used to store six Christmas ornaments. Find the perimeter of such a box if each ornament has a diameter of 72 mm.

Solving Equations

A *solution* of an equation is a replacement for the variable that makes the equation true. To *solve* an equation means to find all of its solutions.

a THE ADDITION PRINCIPLE

The solution of the equation $x = 10$ is 10. The solution of the equation $x + 5 = 15$ is also 10, since $10 + 5 = 15$ is true. Because their solutions are identical, $x = 10$ and $x + 5 = 15$ are said to be **equivalent equations**.

EQUIVALENT EQUATIONS

Equations with the same solutions are called **equivalent equations**.

It is important to be able to distinguish between equivalent *expressions* and equivalent *equations*.

- $6a$ and $4a + 2a$ are equivalent *expressions* because, for any replacement of a, both expressions represent the same number.
- $3x = 15$ and $4x = 20$ are equivalent *equations* because any solution of one equation is also a solution of the other equation.

EXAMPLE 1 Classify each pair as either equivalent equations or equivalent expressions:

a) $5x + 1$; $2x - 4 + 3x + 5$ **b)** $x = -7$; $x + 2 = -5$.

a) First note that these are expressions, not equations. To see if they are equivalent, we combine like terms in the second expression:

$$2x - 4 + 3x + 5 = (2 + 3)x + (-4 + 5) \qquad \text{Regrouping and using the distributive law}$$

$$= 5x + 1.$$

We see that $2x - 4 + 3x + 5$ and $5x + 1$ are *equivalent expressions*.

b) Both $x = -7$ and $x + 2 = -5$ are equations. The solution of $x = -7$ is -7. We substitute to see if -7 is also the solution of $x + 2 = -5$:

$$x + 2 = -5$$
$$-7 + 2 = -5 \quad \text{TRUE}$$

Since $x = -7$ and $x + 2 = -5$ have the same solution, they are *equivalent equations*.

Do Margin Exercises 1 and 2. ▶

We now begin to consider principles that enable us to start with one equation and create an equivalent equation similar to $x = 15$, for which the solution is obvious. One such principle, the *addition principle*, tells us that we can add the same number to both sides of an equation without changing the solutions of the equation.

OBJECTIVES

a Use the addition principle to solve equations.

b Use the division principle to solve equations.

c Decide which principle should be used to solve an equation.

d Solve equations that require use of both the addition principle and the division principle.

SKILL TO REVIEW

Objective 1.7b: Solve equations like $t + 28 = 54$, $28 \cdot x = 168$, and $98 \cdot 2 = y$.

Solve.
  1. $x + 13 = 87$
  2. $17 \cdot y = 357$

Classify each pair as equivalent expressions or equivalent equations.

  1. $a - 5 = -3$; $a = 2$

  2. $a - 9 + 6a$; $7a - 9$

Answers

THE ADDITION PRINCIPLE

For any numbers a, b, and c,

$$a = b \quad \text{is equivalent to} \quad a + c = b + c.$$

EXAMPLE 2 Solve: $x - 7 = -2$.

We have

$$x - 7 = -2$$
$$x - 7 + 7 = -2 + 7 \qquad \text{Using the addition principle:}$$
$$\text{adding 7 to both sides}$$
$$x + 0 = 5 \qquad \text{Adding 7 "undoes" the subtraction of 7.}$$
$$x = 5. \qquad \text{This equation has the same solution as}$$
$$x - 7 = -2.$$

The solution appears to be 5. To check, we use the original equation.

Check:
$$\frac{x - 7 = -2}{5 - 7 \; ? \; -2}$$
$$-2 \; | \qquad \text{TRUE}$$

The solution is 5.

Solve.

3. $x - 5 = 19$

4. $x - 9 = -12$

◀ **Do Exercises 3 and 4.**

We can subtract by adding the opposite of the number being subtracted. Because of this, the addition principle allows us to subtract the same number from both sides of an equation.

EXAMPLE 3 Solve: $23 = t + 7$.

We have

$$23 = t + 7$$
$$23 - 7 = t + 7 - 7 \qquad \text{Using the addition principle to add } -7$$
$$\text{or to subtract 7 on both sides}$$
$$16 = t + 0 \qquad \text{Subtracting 7 "undoes" the addition of 7.}$$
$$16 = t. \qquad \text{The solution of } 23 = t + 7 \text{ is also 16.}$$

The solution is 16. The check is left to the student.

To visualize the addition principle, think of a jeweler's balance. When both sides of the balance hold equal amounts of weight, the balance is level. If weight is added or removed, equally, on both sides, the balance remains level.

Solve.

5. $42 = x + 17$

6. $a + 8 = -6$

◀ **Do Exercises 5 and 6.**

b THE DIVISION PRINCIPLE

We can solve $8n = 96$ by dividing both sides by 8:

$$8 \cdot n = 96$$

$$\frac{8 \cdot n}{8} = \frac{96}{8} \qquad \text{Dividing both sides by 8}$$

$$n = 12. \qquad \text{8 times } n \text{, divided by 8, is } n.\ 96 \div 8 \text{ is 12.}$$

Both $8n = 96$ and $n = 12$ have the solution 12. We can divide both sides of an equation by any nonzero number in order to find an equivalent equation.

THE DIVISION PRINCIPLE

For any numbers a, b, and c ($c \neq 0$),

$$a = b \quad \text{is equivalent to} \quad \frac{a}{c} = \frac{b}{c}.$$

After we have discussed multiplication of fractions, we can use an equivalent form of this principle: the multiplication principle.

EXAMPLE 4   Solve: $9x = 63$.

We have

$$9x = 63$$

$$\frac{9x}{9} = \frac{63}{9} \qquad \text{Using the division principle to divide both sides by 9}$$

$$x = 7.$$

Check:
$$\begin{array}{c|c} 9x = 63 & \text{Checking in the original equation} \\ \hline 9 \cdot 7 \;?\; 63 & \\ 63 & \text{TRUE} \end{array}$$

The solution is 7.

Do Exercises 7 and 8. ▶

EXAMPLE 5   Solve: $48 = -8n$.

To undo multiplication by -8, we use the division principle:

$$48 = -8n$$

$$\frac{48}{-8} = \frac{-8n}{-8} \qquad \text{Dividing both sides by } -8$$

$$-6 = n.$$

Check:
$$\begin{array}{c|c} 48 = -8n & \\ \hline 48 \;?\; -8(-6) & \\ 48 & \text{TRUE} \end{array}$$

The solution is -6.

Do Exercises 9 and 10. ▶

Solve.

7. $7x = 42$

8. $-24 = 3t$

Solve.

9. $63 = -7n$

10. $-6x = 72$

Answers

7. 6   8. −8   9. −9   10. −12

Our goal in equation solving is to have the variable by itself on one side of the equation. In an equation like $-x = 7$, the variable is not by itself—there is a negative sign in front of x. We can solve an equation like $-x = 7$ by dividing or by multiplying both sides of the equation by -1.

EXAMPLE 6 Solve: $-x = 7$.

One way to solve this equation is to note that $-x = -1 \cdot x$. Then we can divide both sides by -1:

$$-x = 7$$
$$-1 \cdot x = 7$$
$$\frac{-1 \cdot x}{-1} = \frac{7}{-1} \qquad \text{Using the division principle}$$
$$x = -7.$$

Check: $\dfrac{-x = 7}{-(-7) \; ? \; 7}$ Be sure to check in the original equation.
$ 7 \;|$ **TRUE**

The solution is -7.

Another way to solve the equation in Example 6 is to remember that when an expression is multiplied or divided by -1, its sign is changed. Here we multiply on both sides by -1 to change the sign of $-x$:

$$-x = 7$$
$$(-1)(-x) = (-1) \cdot 7 \qquad \text{Multiplying both sides by } -1$$
$$x = -7. \qquad \text{Note that } (-1)(-x) \text{ is the same as } (-1)(-1)x.$$

◀ **Do Exercises 11 and 12.**

Solve.

11. $-x = 23$

12. $-t = -3$

$\boxed{} \cdot t = -3$
$\dfrac{-1 \cdot t}{\boxed{}} = \dfrac{-3}{\boxed{}}$
$t = \boxed{}$

GS

C SELECTING THE CORRECT APPROACH

You can determine which principle should be used to solve a particular equation by thinking of undoing an operation.

EXAMPLE 7 Solve: $39 = -3 + t$.

Note that -3 is added to t. To undo addition of -3, we subtract -3 or simply add 3 on both sides:

$$3 + 39 = 3 + (-3) + t \qquad \text{Using the addition principle}$$
$$42 = 0 + t$$
$$42 = t.$$

Check: $\dfrac{39 = -3 + t}{39 \; ? \; -3 + 42}$
$ 39 \qquad$ **TRUE**

The solution is 42.

It is important to distinguish between addition of a negative number, as in $39 = -3 + t$, and multiplication by a negative number, as in $39 = -3t$. We undo addition by subtracting on both sides, and we undo multiplication by dividing on both sides.

Answers

11. -23 **12.** 3

Guided Solution:
12. $-1, -1, -1, 3$

EXAMPLE 8 Solve: $39 = -3t$.

Here t is multiplied by -3. To undo multiplication by -3, we divide by -3 on both sides:

$$39 = -3t$$

$$\frac{39}{-3} = \frac{-3t}{-3} \qquad \text{Using the division principle}$$

$$-13 = t.$$

Check:
$$\begin{array}{c} 39 = -3t \\ \hline 39 \ ? \ -3(-13) \\ \quad | \quad 39 \qquad \text{TRUE} \end{array}$$

The solution is -13.

Do Exercises 13–15. ▶

Solve.

13. $-2x = -52$

14. $-2 + x = -52$

15. $x \cdot 7 = -28$

d USING THE PRINCIPLES TOGETHER

Suppose we want to determine whether 7 is the solution of $5x - 8 = 27$. To check, we replace x with 7 and simplify.

Check:
$$\begin{array}{c} 5x - 8 = 27 \\ \hline 5 \cdot 7 - 8 \ ? \ 27 \\ 35 - 8 \quad | \\ 27 \quad | \qquad \text{TRUE} \end{array}$$

This shows that 7 is the solution.

Do Exercises 16 and 17. ▶

16. Determine whether -9 is the solution of $7x + 8 = -55$.

17. Determine whether -6 is the solution of $4x + 3 = -25$.

In the check above, note that the rules for order of operations require that we multiply before we subtract (or add). Thus, to evaluate $5x - 8$,

we *select* a value: $\quad x$

then *multiply* by 5: $\quad 5x$

and then *subtract* 8: $\quad 5x - 8.$

In Example 9, which follows, these steps are reversed to solve for x:

we *add* 8: $\quad 5x - 8 + 8$

then *divide* by 5: $\quad \dfrac{5x}{5}$

and then *isolate* x: $\quad x.$

In general, the *last* step performed when calculating is the *first* step to be reversed when finding a solution.

When x is by itself on one side of an equation, we say that it is *isolated*. Our goal in the remaining examples in this section will be to isolate the term containing the variable (using the addition principle) and then to isolate the variable itself (using the division principle).

EXAMPLE 9 Solve: $5x - 8 = 27$.

We first note that the term containing x is $5x$. To isolate $5x$, we add 8 on both sides:

$$5x - 8 = 27$$
$$5x - 8 + 8 = 27 + 8 \qquad \text{Using the addition principle}$$
$$5x + 0 = 35 \qquad \text{Try to do this step mentally.}$$
$$5x = 35. \qquad \text{We have isolated } 5x.$$

Next, we isolate x by dividing by 5 on both sides:

$$5x = 35$$
$$\frac{5x}{5} = \frac{35}{5} \qquad \text{Using the division principle}$$
$$1x = 7 \qquad \text{Try to do this step mentally.}$$
$$x = 7. \qquad \text{We have isolated } x.$$

The check was performed on the previous page. The solution is 7.

◀ **Do Exercise 18.**

EXAMPLE 10 Solve: $38 = -9t + 2$.

We first isolate $-9t$ by subtracting 2 on both sides:

$$38 = -9t + 2$$
$$38 - 2 = -9t + 2 - 2 \qquad \text{Subtracting 2 (or adding } -2) \text{ on both sides}$$
$$36 = -9t + 0 \qquad \text{Try to do this step mentally.}$$
$$36 = -9t.$$

Now that we have isolated $-9t$ on one side of the equation, we can divide by -9 to isolate t:

$$36 = -9t$$
$$\frac{36}{-9} = \frac{-9t}{-9} \qquad \text{Dividing both sides by } -9$$
$$-4 = t. \qquad \text{Simplifying}$$

Check:
$$\begin{array}{c} 38 = -9t + 2 \\ \hline 38 \ ? \ -9 \cdot (-4) + 2 \\ 36 + 2 \\ 38 \qquad \text{TRUE} \end{array}$$

The solution is -4.

◀ **Do Exercise 19.**

---

18. Solve: $2x - 9 = 43$.

$$2x - 9 = 43$$
$$2x - 9 + \boxed{} = 43 + \boxed{}$$
$$2x = \boxed{}$$
$$\frac{2x}{\boxed{}} = \frac{52}{\boxed{}}$$
$$x = \boxed{}$$

19. Solve: $-3n + 2 = 47$.

✓ Reading Check

Match each equation with the correct first step for solving it from the list on the right.

RC1. $7x = 28$ _____

RC2. $x - 7 = 13$ _____

RC3. $-7x = -14$ _____

RC4. $x + 7 = 1$ _____

a) Add 7 to both sides.

b) Add -7 to both sides.

c) Divide both sides by 7.

d) Divide both sides by -7.

a Classify each pair as either equivalent expressions or equivalent equations.

1. $x = -1;\ x + 5 = 4$

2. $4x + 1;\ 6 + 4x - 5$

3. $7a - 3;\ 13 + 7a - 16$

4. $t = -2;\ 5t = -10$

5. $4r + 3;\ 9 - r + 5r - 6$

6. $2r - 7;\ r - 10 + r + 3$

7. $x - 9 = 8;\ x = 20 - 3$

8. $t + 4 = 19;\ t = 9 + 6$

9. $3(t + 2);\ 5 + 3t + 1$

10. $2x = -14;\ x - 2 = -9$

11. $x + 4 = -8;\ 2x = -24$

12. $4(x - 7);\ 3x - 28 + x$

Solve.

13. $x - 6 = -9$

14. $x - 5 = -7$

15. $x - 4 = -12$

16. $x - 7 = 5$

17. $a + 7 = 25$

18. $x + 9 = -3$

19. $-8 = n + 7$

20. $38 = a + 12$

21. $24 = t - 8$

22. $-9 = x + 3$

23. $-12 = x + 5$

24. $17 = n - 6$

25. $-5 + a = 12$

26. $3 = 17 + x$

27. $-8 = -8 + t$

28. $-7 + t = -7$

b Solve.

29. $6x = 60$

30. $-8t = 40$

31. $-3t = 42$

32. $3x = 24$

33. $-7n = -35$

34. $64 = -2t$

35. $0 = 6x$

36. $-5n = -65$

37. $55 = -5t$

38. $-x = 83$

39. $-x = 56$

40. $-2x = 0$

41. $n(-4) = -48$

42. $-x = -475$

43. $-x = -390$

44. $n(-7) = 42$

c Solve.

45. $t - 6 = -2$

46. $3t = -45$

47. $6x = -54$

48. $x + 9 = -15$

49. $15 = -x$

50. $-13 = x - 4$

51. $-21 = x + 5$

52. $-42 = -x$

53. $35 = -7t$

54. $7 + t = -18$

55. $-17x = 68$

56. $-34 = x - 10$

57. $12 + t = -160$

58. $-48 = t(-12)$

59. $-27 = x - 23$

60. $-135 = -9t$

d Solve.

61. $5x - 1 = 34$

62. $7x - 3 = 25$

63. $4t + 2 = 14$

64. $3t + 5 = 26$

65. $6a + 1 = -17$

66. $8a + 3 = -37$

67. $2x - 9 = -23$

68. $3x - 5 = -35$

69. $-2x + 1 = 17$

70. $-4t + 3 = -17$

71. $-8t - 3 = -67$

72. $-7x - 4 = -46$

73. $-x + 9 = -15$

74. $-x - 6 = 8$

75. $7 = 2x - 5$

76. $9 = 4x - 7$

77. $13 = 3 + 2x$

78. $33 = 5 - 4x$

79. $13 = 5 - x$

80. $12 = 7 - x$

Skill Maintenance

Simplify.   [1.9c]

81. $5 + 3 \cdot 2^3$

82. $(9 - 7)^4 - 3^2$

83. $12 \div 3 \cdot 2$

84. $27 \div 3(2 + 1)$

85. $15 - 3 \cdot 2 + 7$

86. $30 - 4^2 \div 8 \cdot 2$

87. $3(8 - 6) - (9 - 4)$

88. $2(10 - 6)^2 - (120 \div 6 - 3 \cdot 4)$

Synthesis

Solve.

89. $2x - 7x = -40$

90. $9 + x - 5 = 23$

91. $2x - 7 + x = 5 - 12$

92. $3 - 6x + 5 = 2(4)$

93. $n + n = -2 - 3 \cdot 6 \div 2 + 1$

94. $10 \div 5 \cdot 2 + 3 = 5t - 4t$

95. $17 - 3^2 = 4 + t - 5^2$

96. $(-9)^2 = 2^3t + (3 \cdot 6 + 1)t$

97. $(-7)^2 - 5 = t + 4^3$

98. ▦ $(-42)^3 = 14^2t$

99. ▦ $x - (19)^3 = -18^3$

100. ▦ $23^2 = x + 22^2$

101. ▦ $35^3 = -125t$

102. ▦ $248 = 24 - 32x$

103. ▦ $529 - 143x = -1902$

Vocabulary Reinforcement

Complete each statement with the appropriate word or phrase from the column on the right. Some of the choices may not be used.

1. The _____ are . . . , $-3, -2, -1, 0, 1, 2, 3, ...$ [2.1a]

2. The _____ value of a number is its distance from zero on the number line. [2.1c]

3. The number 3 is the _____ of -3. [2.1d]

4. Division by 0 is _____. [2.5a]

5. When we replace a variable with a number, we say that we are _____ for the variable. [2.6a]

6. A letter that stands for just one number is called a _____. [2.6a]

7. The _____ states that for any numbers a, b, and c, $a(b + c) = ab + ac$. [2.6b]

8. The _____ for solving equations states that for any real numbers a, b, and c, $a = b$ is equivalent to $a + c = b + c$. [2.8a]

9. Equations with the same solutions are called _____ equations. [2.8a]

constant

variable

opposite

equivalent

distributive law

integers

substituting

absolute

not defined

zero

addition principle

division principle

distributive law

Concept Reinforcement

Determine whether each statement is true or false.

_____ **1.** The opposite of the opposite of a number is the original number. [2.1d]

_____ **2.** The product of an even number of negative numbers is positive. [2.4b]

_____ **3.** The expression $2(x + 3)$ is equivalent to the expression $2 \cdot x + 3$. [2.6b]

_____ **4.** $3y$ and $3y^2$ are like terms. [2.7a]

Study Guide

Objective 2.1c   Find the absolute value of any integer.

Example   Find the absolute value.

a) $|-97|$ **b)** $|35|$

a) The number is negative, so we write the opposite.

$$|-97| = 97$$

b) The number is positive, so the absolute value is the same as the number.

$$|35| = 35$$

Practice Exercise

1. Find the absolute value.

a) $|-17|$ **b)** $|300|$

Objective 2.2a Add integers without using the number line.

Example Add without using the number line: $-15 + 9$.

To add a negative number and a positive number, subtract the absolute values: $15 - 9 = 6$. The negative number has the larger absolute value, so the answer is negative.

$$-15 + 9 = -6$$

Example Add without using the number line: $-8 + (-9)$.

To add two negative numbers, add the absolute values: $8 + 9 = 17$. The answer is negative.

$$-8 + (-9) = -17$$

Practice Exercise

2. Add without using the number line: $37 + (-16)$.

Objective 2.3a Subtract integers, and simplify combinations of additions and subtractions.

Example Subtract: $8 - 12$.

We add the opposite of the number being subtracted.
$$8 - 12 = 8 + (-12)$$
$$= -4$$

Practice Exercise

3. Subtract: $6 - (-8)$.

Objective 2.4a Multiply integers.

Example Multiply: $-6(-4)$.

The signs are the same, so the answer is positive.
$$-6(-4) = 24$$

Example Multiply: $-4(3)$.

The signs are different, so the answer is negative.
$$-4(3) = -12$$

Practice Exercise

4. Multiply: $6(-15)$.

Objective 2.5a Divide integers.

Example Divide: $-36 \div (-4)$.

The signs are the same, so the answer is positive.
$$-36 \div (-4) = 9$$

Example Divide: $-30 \div 10$.

The signs are different, so the answer is negative.
$$-30 \div 10 = -3$$

Practice Exercise

5. Divide: $99 \div (-9)$.

Objective 2.5b Use the rules for order of operations with integers.

Example Simplify: $3^2 - 24 \div 2 - (4 + 2 \cdot 8)$.

$$3^2 - 24 \div 2 - (4 + 2 \cdot 8)$$
$$= 3^2 - 24 \div 2 - (4 + 16)$$
$$= 3^2 - 24 \div 2 - 20$$
$$= 9 - 24 \div 2 - 20$$
$$= 9 - 12 - 20$$
$$= -3 - 20$$
$$= -23$$

Practice Exercise

6. Simplify: $4 - 8^2 \div (10 - 6)$.

Objective 2.6b Use the distributive law to find equivalent expressions.

Example Use the distributive law to write an expression equivalent to $-2(3x - 4)$.

$$-2(3x - 4) = -2(3x) - (-2)(4)$$
$$= -6x - (-8)$$
$$= -6x + 8$$

Practice Exercise

7. Use the distributive law to write an expression equivalent to $5(6x - 8y - z)$.

Objective 2.7a Combine like terms.

Example Combine like terms to form an expression equivalent to $6 - 30x + 12x - 7$.

$$6 - 30x + 12x - 7 = 6 + (-30x) + 12x + (-7)$$
$$= -30x + 12x + 6 + (-7)$$
$$= -18x - 1$$

Practice Exercise

8. Combine like terms to form an expression equivalent to $8a - b + 9a - 6b$.

Objective 2.8d Solve equations that require use of both the addition principle and the division principle.

Example Solve: $-3x + 1 = 16$.

$$-3x + 1 = 16$$
$$-3x + 1 - 1 = 16 - 1$$
$$-3x = 15$$
$$\frac{-3x}{-3} = \frac{15}{-3}$$
$$x = -5$$

The solution is -5.

Practice Exercise

9. Solve: $-19 = 5x + 11$.

Review Exercises

1. Tell which integers correspond to this situation:
   David has a debt of $45 and Joe has $72 in his savings account.   [2.1a]

Use either $<$ or $>$ for ☐ to form a true statement.   [2.1b]
2. $0 \; \square \; -5$ 3. $-7 \; \square \; 6$ 4. $-4 \; \square \; -19$

Find the absolute value.   [2.1c]
5. $|-39|$ 6. $|23|$ 7. $|0|$

8. Find $-x$ when $x = -72$.   [2.1d]

9. Find $-(-x)$ when $x = 59$.   [2.1d]

Compute and simplify.
10. $-14 + 5$ [2.2a]

11. $-5 + (-6)$ [2.2a]

12. $14 + (-8)$ [2.2a]

13. $0 + (-24)$ [2.2a]

14. $17 - 29$ [2.3a]

15. $9 - (-14)$ [2.3a]

16. $-8 - (-7)$ [2.3a]

17. $-3 - (-3)$ [2.3a]

18. $-3 + 7 + (-8)$ [2.3a]

19. $8 - (-9) - 7 + 2$ [2.3a]

20. $-23 \cdot (-4)$ [2.4a]

21. $7(-12)$ [2.4a]

22. $2(-4)(-5)(-1)$ [2.4b]

23. $15 \div (-5)$ [2.5a]

24. $\dfrac{-55}{11}$ [2.5a]

25. $\dfrac{0}{7}$ [2.5a]

Simplify.   [2.5b]
26. $625 \div (-25) \div 5$

27. $-16 \div 4 - 30 \div (-5)$

28. $9[(7 - 14) - 13]$

29. $(-3)|4 - 3^2| - 5$

30. $[-12(-3) - 2^3] - (-9)(-10)$

31. Evaluate $3a + b$ for $a = 4$ and $b = -5$.   [2.6a]

32. Evaluate $\dfrac{-x}{y}, \dfrac{x}{-y}$, and $-\dfrac{x}{y}$ for $x = 30$ and $y = 5$.   [2.6a]

Use the distributive law to write an equivalent expression.   [2.6b]

33. $4(5x + 9)$ **34.** $3(2a - 4b + 5)$

35. $-10(2x + y)$

Combine like terms.   [2.7a]

36. $5a + 12a$ **37.** $-7x + 13x$

38. $9m + 14 - 12m - 8$

39. Find the perimeter of a rectangular frame that is 8 in. by 10 in.   [2.7b]

40. Find the perimeter of a square pane of glass that is 25 cm on each side.   [2.7b]

Solve.   [2.8a, b, c, d]

41. $x - 9 = -17$ **42.** $-4t = 36$

43. $13 = -x$ **44.** $56 = 6x - 10$

45. $-x + 3 = -12$ **46.** $18 = 4 - 2x$

Solve.

47. On the first, second, and third downs, a football team had these gains and losses: 5-yd gain, 12-yd loss, and 15-yd gain, respectively. Find the total gain (or loss). [2.3b]

48. Kaleb's total assets are $170. He borrows $300. What are his total assets now? [2.3b]

49. Evaluate $-|-x|$ when $x = -10$. [2.1c], [2.1d]
   A. -10 **B.** -20
   C. 100 **D.** 10

50. Simplify: $-3 \cdot 4 - 12 \div 4$. [2.5b]
   A. -16 **B.** -15
   C. 0 **D.** 6

Synthesis

51. The sum of two numbers is 800. The difference is 6. Find the numbers. [2.3b]

52. The following are examples of consecutive integers: $4, 5, 6, 7, 8; -13, -12, -11, -10$. [2.2a], [2.4b]
   a) Express the number 8 as the sum of 16 consecutive integers.
   b) Find the product of the 16 consecutive integers in part (a).

Simplify. [2.5b]

53. 🔲 $87 \div 3 \cdot 29^3 - (-6)^6 + 1957$

54. 🔲 $1969 + (-8)^5 - 17 \cdot 15^3$

55. 🔲 $\dfrac{113 - 17^3}{15 + 8^3 - 507}$

56. For what values of x will $8 + x^3$ be negative? [2.6a]

57. For what values of x is $|x| > x$? [2.1b, c]

Understanding Through Discussion and Writing

1. What rule have we developed that would tell you the sign of $(-7)^8$ and $(-7)^{11}$ without doing the computations? Explain. [2.4b]

2. Does $-x$ always represent a negative number? Why or why not? [2.1d]

3. Jake enters $18/2 \cdot 3$ on his calculator and expects the result to be 3. What mistake is he making? [2.5b]

4. Does $-x^2$ always represent a negative number? Why or why not? [2.6a]

For Extra Help For step-by-step test solutions, access the Chapter Test Prep Videos in MyMathLab® or on YouTube (search "BittingerPrealgebra" and click on Channels).

1. Tell which integers correspond to this situation: The Tee Shop sold 542 fewer muscle shirts than expected in January and 307 more than expected in February.

2. Use either $<$ or $>$ for $\square$ to form a true statement.
$$-14 \;\square\; -21$$

3. Find the absolute value: $|-739|$.

4. Find $-(-x)$ when $x = -19$.

Compute and simplify.

5. $6 + (-17)$

6. $-9 + (-12)$

7. $-8 + 17$

8. $0 - 12$

9. $7 - 22$

10. $-5 - 19$

11. $-8 - (-27)$

12. $31 - (-3) - 5 + 9$

13. $(-4)^3$

14. $27(-10)$

15. $-9 \cdot 0$

16. $-72 \div (-9)$

17. $\dfrac{-56}{7}$

18. $8 \div 2 \cdot 2 - 3^2$

19. $29 - (3 - 5)^2$

20. *Antarctica Highs and Lows.* The continent of Antarctica, which lies in the southern hemisphere, experiences winter in July. The average high temperature is $-67°F$ and the average low temperature is $-81°F$. How much higher is the average high than the average low?

Source: National Climatic Data Center

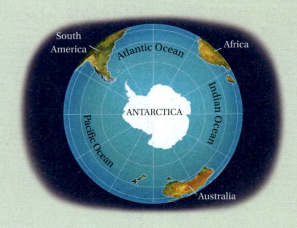

21. Evaluate $\dfrac{a - b}{6}$ for $a = -8$ and $b = 10$.

22. Use the distributive law to write an expression equivalent to $7(2x + 3y - 1)$.

23. Combine like terms.
$9x - 14 - 5x - 3$

24. Find the perimeter of a square garden that is 5 ft on each side.

Solve.

25. $-7x = -35$

26. $a + 9 = -3$

27. $95 = -x$

28. $3t - 7 = 5$

29. Use the distributive law to write an expression equivalent to $-2(n - 6m)$.
A. $-2n - 6m$ **B.** $-2n - 12m$
C. $-2n + 12m$ **D.** $2n + 12m$

Synthesis

30. Monty plans to attach trim around the doorway and along the base of the walls in a 12-ft by 14-ft room. If the doorway is 3 ft by 7 ft, how many feet of trim are needed? (Only three sides of a doorway get trim.)

Simplify.

31. $9 - 5[x + 2(3 - 4x)] + 14$

32. $15x + 3(2x - 7) - 9(4 + 5x)$

33. ⊞ $49 \cdot 14^3 \div 7^4 + 1926^2 \div 6^2$

34. ⊞ $3487 - 16 \div 4 \cdot 4 \div 2^8 \cdot 14^4$

1. Write standard notation for the number written in words in the following sentence: In 2012, Washington State University Master Gardeners taught two hundred twenty thousand, eight hundred sixty-three first-time gardeners.

   Source: *WSU Extension Master Gardener Report, 2012*

2. Write a word name for 5,380,001,437.

Add.

3.
$$\begin{array}{r} 1\,5{,}8\,9\,2 \\ +\ 2{,}9\,3\,5 \\ \hline \end{array}$$

4.
$$\begin{array}{r} 7\,9\,8\,9 \\ 7\,8\,9 \\ +\ \ \ \ 7\,9 \\ \hline \end{array}$$

5. $-16 + 72$

6. $-30 + (-72)$

Subtract.

7.
$$\begin{array}{r} 8\,2\,7\,6 \\ -\ \ \ 4\,3\,0 \\ \hline \end{array}$$

8.
$$\begin{array}{r} 3\,0\,0\,6 \\ -\ \ \ 5\,7\,8 \\ \hline \end{array}$$

9. $7 - (-7)$

10. $-18 - 25$

Multiply.

11.
$$\begin{array}{r} 6\,2\,1 \\ \times\ \ \ 2\,7 \\ \hline \end{array}$$

12.
$$\begin{array}{r} 2\,5\,0\,5 \\ \times\ 3\,3\,0\,0 \\ \hline \end{array}$$

13. $43 \cdot (-8)$

14. $-12(-6)$

Divide.

15. $6\,3 \overline{)\,6\,5\,5\,2}$

16. $6\,2 \overline{)\,3\,8\,4\,4}$

17. $0 \div (-67)$

18. $60 \div (-12)$

19. Round 427,931 to the nearest thousand.

20. Round 5309 to the nearest hundred.

Estimate each sum or product by rounding to the nearest hundred. Show your work.

21.
$$\begin{array}{r} 7\,4\,9{,}5\,5\,9 \\ +\ 3\,0\,1{,}3\,6\,2 \\ \hline \end{array}$$

22.
$$\begin{array}{r} 7\,4\,9 \\ \times\ 5\,3\,1 \\ \hline \end{array}$$

23. Use $<$ or $>$ for ☐ to form a true sentence:
    -26 ☐ 19.

24. Find the absolute value: $|-279|$.

Simplify.

25. $35 - 25 \div 5 + 2 \times 3$

26. $\{17 - [8 - (5 - 2 \times 2)]\} \div (3 + 12 \div 6)$

27. $10 \div 1(-5) - 6^2$

28. 5^3

29. Evaluate $\dfrac{x - y}{5}$ for $x = 11$ and $y = -4$.

30. Evaluate $7x^2$ for $x = -2$.

Use the distributive law to write an equivalent expression.

31. $-2(x + 5)$

32. $6(3x - 2y + 4)$

Simplify.

33. $12 + (-14)$ **34.** $(-2)^3$

35. $23 - 38$ **36.** $-64 \div (-2)$

37. $-12 - (-25)$ **38.** $(-2)(-3)(-5)$

39. $3 - (-8) + 2 - (-3)$ **40.** $16 \div 2(-8) + 7$

Solve.

41. $x + 8 = 35$ **42.** $-12t = 36$

43. $6 - x = -9$ **44.** $-39 = 4x - 7$

Solve.

45. In the movie *Little Big Man,* Dustin Hoffman plays a character who ages from 17 to 121. This represents the greatest age range depicted by one actor in one film. How many years did Hoffman's character age?
Source: Guinness Book of World Records

46. Seven of the ten largest hotels in the world are in Las Vegas. Of these, the four largest are the MGM Grand, the Luxor, the Venetian, and the Excalibur. These have 5690 rooms, 4408 rooms, 4027 rooms, and 4008 rooms, respectively. What is the total number of rooms in these four hotels?
Source: insidervlv.com/hotelslargestworld.html

47. Surface temperatures on the Moon vary from $-184°$C to $101°$C. Find the difference between the highest value and the lowest value in this temperature range.
Source: northern-stars.com/solar_system_info.htm

48. Eastside Appliance sells a refrigerator for $900 and $30 tax with no delivery charge. Westside Appliance sells the same model for $860 and $28 tax plus a $25 delivery charge. Which is the better buy?

49. Write an equivalent expression by combining like terms: $7x - 9 + 3x - 5$.

Synthesis

50. A soft-drink distributor has 166 loose cans of cola. The distributor wishes to form as many 24-can cases as possible and then, with any remaining cans, as many six-packs as possible. How many cases will be filled? How many six-packs? How many loose cans will remain?

51. Simplify: $a - \{3a - [4a - (2a - 4a)]\}$.

52. 🖩 Simplify: $37 \cdot 64 \div 4^2 \cdot 2 - (7^3 - (-4)^5)$.

53. Find two solutions of $5|x| - 2 = 13$.

Fraction Notation: Multiplication and Division

3.1   Multiples and Divisibility

In this section, we discuss *multiples* and *divisibility* in order to be able to simplify fractions like $\frac{117}{225}$.

OBJECTIVES

a Find some multiples of a number, and determine whether a number is divisible by another number.

b Test to see if a number is divisible by 2, 3, 5, 6, 9, or 10.

SKILL TO REVIEW

Objective 1.5a: Divide whole numbers.

Divide.
1. $329 \div 8$
2. $23 \overline{)1081}$

1. Show that each of the numbers 5, 45, and 100 is a multiple of 5.

2. Show that each of the numbers 10, 60, and 110 is a multiple of 10.

3. Multiply by 1, 2, 3, and so on, to find ten multiples of 5.

a MULTIPLES

A **multiple** of a number is a product of that number and an integer. For example, some multiples of 2 are

$2 \quad$ (because $2 = 1 \cdot 2$);
$4 \quad$ (because $4 = 2 \cdot 2$);
$6 \quad$ (because $6 = 3 \cdot 2$);
$8 \quad$ (because $8 = 4 \cdot 2$).

We can also find multiples of 2 by counting by twos: 2, 4, 6, 8, and so on.

EXAMPLE 1   Show that each of the numbers 3, 6, 9, and 15 is a multiple of 3.

$$3 = 1 \cdot 3; \qquad 6 = 2 \cdot 3; \qquad 9 = 3 \cdot 3; \qquad 15 = 5 \cdot 3.$$

◀ **Do Margin Exercises 1 and 2.**

EXAMPLE 2   Multiply by 1, 2, 3, and so on, to find eight multiples of six.

$1 \cdot 6 = 6$	$5 \cdot 6 = 30$
$2 \cdot 6 = 12$	$6 \cdot 6 = 36$
$3 \cdot 6 = 18$	$7 \cdot 6 = 42$
$4 \cdot 6 = 24$	$8 \cdot 6 = 48$

◀ **Do Exercise 3.**

DIVISIBILITY

A number b is said to be **divisible** by another number a if b is a multiple of a.

Thus,

6 is divisible by 2 because 6 is a multiple of 2 ($6 = 3 \cdot 2$), and
100 is divisible by 25 because 100 is a multiple of 25 ($100 = 4 \cdot 25$).

Answers

Skill to Review:
1. 41 R 1   2. 47

Margin Exercises:
1. $5 = 1 \cdot 5; 45 = 9 \cdot 5; 100 = 20 \cdot 5$
2. $10 = 1 \cdot 10; 60 = 6 \cdot 10; 110 = 11 \cdot 10$
3. 5, 10, 15, 20, 25, 30, 35, 40, 45, 50

Saying that *b* is divisible by *a* means that if we divide *b* by *a*, the remainder is 0. When this happens, we sometimes say that *a* divides *b* "evenly."

EXAMPLE 3 Determine **(a)** whether 45 is divisible by 9 and **(b)** whether 45 is divisible by 4.

a) We divide 45 by 9.

```
      5
   9)45
     45
  ─────
      0 ← Remainder is 0.
```

Because the remainder is 0, 45 is divisible by 9.

b) We divide 45 by 4.

```
     11
   4)45
     4
  ────
      5
      4
  ────
      1 ← Not 0
```

Since the remainder is not 0, 45 is not divisible by 4.

Do Exercises 4–6. ▶

b TESTS FOR DIVISIBILITY

We now look at quick ways of checking for divisibility by 2, 3, 5, 6, 9, and 10 without actually performing long division. Tests do exist for divisibility by 4, 7, and 8, but they can be as difficult to perform as the actual long division.

To test for divisibility by 2, 5, and 10, we examine the ones digit.

Divisibility by 2

All even numbers are divisible by 2.

> **BY 2**
>
> A number is **divisible by 2** (is *even*) if it has a ones digit of 0, 2, 4, 6, or 8 (that is, it has an even ones digit).

To see why this test works, start counting by twos: 2, 4, 6, 8, 10, 12, 14, 16, 18, 20, 22,.... Note that the ones digit will always be 0, 2, 4, 6, or 8, no matter how high we count.

EXAMPLES Determine whether each number is divisible by 2.

4. 35**5** *is not* divisible by 2 because **5** is not even.
5. 478**6** *is* divisible by 2 because **6** is even.
6. 899**0** *is* divisible by 2 because **0** is even.
7. 426**1** *is not* divisible by 2 because **1** is not even.

Do Exercises 7–10. ▶

GS 4. Determine whether 16 is divisible by 2.

```
      8
   ☐)16
     16
    ──
     ☐
```

Since the remainder is ☐, 16 ___ divisible by 2.
                      is/is not

5. Determine whether 125 is divisible by 5.

6. Determine whether 125 is divisible by 6.

Determine whether each number is divisible by 2.

7. 84   **8.** 59

9. 998   **10.** 2225

Answers

4. Yes   **5.** Yes   **6.** No   **7.** Yes
8. No   **9.** Yes   **10.** No

Guided Solution:
4. 2, 0; 0, is

Divisibility by 5

To determine the test for divisibility by 5, we start counting by fives: 5, 10, 15, 20, 25, 30, 35, ... Note that the ones digit will always be 5 or 0, no matter how high we count.

> **BY 5**
>
> A number is **divisible by 5** if its ones digit is 0 or 5.

EXAMPLES Determine whether each number is divisible by 5.

8. 220 *is* divisible by 5 because the ones digit is 0.

9. 475 *is* divisible by 5 because the ones digit is 5.

10. 6514 *is not* divisible by 5 because the ones digit is neither 0 nor 5.

◄ **Do Exercises 11–14.**

Determine whether each number is divisible by 5.

11. 5780   **12.** 3427

13. 34,678   **14.** 7775

Divisibility by 10

> **BY 10**
>
> A number is **divisible by 10** if its ones digit is 0.

We know that this test works because the product of 10 and *any* number has a ones digit of 0.

EXAMPLES Determine whether each number is divisible by 10.

11. 3440 *is* divisible by 10 because the ones digit is 0.

12. 3447 *is not* divisible by 10 because the ones digit is not 0.

◄ **Do Exercises 15–18.**

Determine whether each number is divisible by 10.

15. 305   **16.** 847

17. 300   **18.** 8760

Divisibility by 3

To test for divisibility by 3, we examine the sum of a number's digits.

Determine whether each number is divisible by 3.

19. 111   **20.** 1111

21. 309

> **BY 3**
>
> A number is **divisible by 3** if the sum of its digits is divisible by 3.

An explanation of why this test works is outlined in Exercise 69 at the end of this section.

22. 17,216 **(GS)**

Add the digits:

$1 + 7 + \boxed{} + 1 + 6 = \boxed{}$.

Since 17 $\underset{\text{is/is not}}{\boxed{}}$ divisible by 3,

the number 17,216 $\underset{\text{is/is not}}{\boxed{}}$ divisible by 3.

EXAMPLES Determine whether each number is divisible by 3.

13. 18 $1 + 8 = 9$

14. 93 $9 + 3 = 12$ Each *is* divisible by 3 because the sum of its digits *is* divisible by 3.

15. 201 $2 + 0 + 1 = 3$

16. 256 $2 + 5 + 6 = 13$ The sum of the digits, 13, *is not* divisible by 3, so 256 *is not* divisible by 3.

◄ **Do Exercises 19–22.**

Divisibility by 9

The test for divisibility by 9 is similar to the test for divisibility by 3.

> ### BY 9
>
> A number is **divisible by 9** if the sum of its digits is divisible by 9.

EXAMPLES Determine whether each number is divisible by 9.

17. 6984

Because $6 + 9 + 8 + 4 = 27$ and 27 is divisible by 9, 6984 *is* divisible by 9.

18. 322

Because $3 + 2 + 2 = 7$ and 7 is not divisible by 9, 322 *is not* divisible by 9.

Do Exercises 23–26. ▶

Determine whether each number is divisible by 9.

23. 16   **24.** 117

25. 309   **26.** 29,223

Divisibility by 6

A number divisible by 6 is a multiple of 6. But $6 = 2 \cdot 3$, so the number is also a multiple of 2 and 3. Since 2 and 3 have no factors in common, a number is divisible by 6 if it is divisible by 2 *and* by 3.

> ### BY 6
>
> A number is **divisible by 6** if its ones digit is 0, 2, 4, 6, or 8 (is even) and the sum of its digits is divisible by 3.

EXAMPLES Determine whether each number is divisible by 6.

19. 720

Because 720 is even, it is divisible by 2. Also, $7 + 2 + 0 = 9$ and 9 is divisible by 3, so 720 is divisible by 3. Thus, 720 *is* divisible by 6.

$$720 \quad 7 + 2 + 0 = 9$$

Even   Divisible by 3

20. 73

73 *is not* divisible by 6 because it is not even.

21. 256

Although 256 is even, it *is not* divisible by 6 because the sum of its digits, $2 + 5 + 6$, or 13, is not divisible by 3.

Do Exercises 27–30. ▶

Determine whether each number is divisible by 6.

27. 420   **28.** 106

29. 321   **30.** 444

Answers

23. No **24.** Yes **25.** No **26.** Yes
27. Yes **28.** No **29.** No **30.** Yes

✓ Reading Check

Match the beginning of each divisibility test with the appropriate ending from the list at the right.

RC1. A number is divisible by 2 if _____.

RC2. A number is divisible by 3 if _____.

RC3. A number is divisible by 5 if _____.

RC4. A number is divisible by 6 if _____.

RC5. A number is divisible by 9 if _____.

RC6. A number is divisible by 10 if _____.

a) the sum of its digits is divisible by 3

b) the sum of its digits is divisible by 9

c) it has an even ones digit

d) its ones digit is 0 or 5

e) its ones digit is 0

f) it has an even ones digit and the sum of its digits is divisible by 3

---

a Multiply by 1, 2, 3, and so on, to find ten multiples of each number.

1. 7

2. 4

3. 20

4. 50

5. 3

6. 8

7. 12

8. 15

9. 10

10. 11

11. 25

12. 100

13. Determine whether 83 is divisible by 3.

14. Determine whether 29 is divisible by 2.

15. Determine whether 525 is divisible by 7.

16. Determine whether 346 is divisible by 8.

17. Determine whether 8127 is divisible by 9.

18. Determine whether 4144 is divisible by 4.

---

b For Exercises 19–30, answer "Yes" or "No" and give a reason based on the tests for divisibility.

19. Determine whether 84 is divisible by 3.

20. Determine whether 467 is divisible by 9.

21. Determine whether 5553 is divisible by 5.

22. Determine whether 2004 is divisible by 6.

23. Determine whether 671,500 is divisible by 10.

24. Determine whether 6120 is divisible by 5.

25. Determine whether 1773 is divisible by 9.

26. Determine whether 3286 is divisible by 3.

27. Determine whether 21,687 is divisible by 2.

28. Determine whether 64,091 is divisible by 10.

29. Determine whether 32,109 is divisible by 6.

30. Determine whether 9840 is divisible by 2.

For Exercises 31–38, test each number for divisibility by 2, 3, 5, 6, 9, and 10.

31. 6825

32. 12,600

33. 119,117

34. 2916

35. 127,575

36. 25,088

37. 9360

38. 143,507

To answer Exercises 39–44, consider the following numbers. Use the tests for divisibility.

46	300	85	256
224	36	711	8064
19	45,270	13,251	1867
555	4444	254,765	21,568

39. Which of the above are divisible by 3?

40. Which of the above are divisible by 2?

41. Which of the above are divisible by 10?

42. Which of the above are divisible by 5?

43. Which of the above are divisible by 6?

44. Which of the above are divisible by 9?

To answer Exercises 45–50, consider the following numbers.

56	200	75	35
324	42	812	402
784	501	2345	111,111
55,555	3009	2001	1005

45. Which of the above are divisible by 2?

46. Which of the above are divisible by 3?

47. Which of the above are divisible by 5?

48. Which of the above are divisible by 10?

49. Which of the above are divisible by 9?

50. Which of the above are divisible by 6?

Skill Maintenance

Solve.

51. $16 \cdot t = 848$ [1.7b], [2.8b]

52. $m + 9 = 14$ [1.7b], [2.8a]

53. $23 + x = 15$ [1.7b], [2.8a]

54. $24 \cdot m = -576$ [1.7b], [2.8b]

Solve.   [1.8a]

55. Marty's automobile has a 5-speed transmission and gets 33 mpg in city driving. How many gallons of gas will it use in 1485 mi of city driving?

56. There are 60 min in 1 hr. How many minutes are there in 72 hr?

Evaluate.   [1.9b], [2.4b]

57. 5^3 **58.** $(-2)^4$

Write in exponential notation.   [1.9a]

59. $9 \cdot 9 \cdot 9$ **60.** $3 \cdot 3 \cdot 3 \cdot 3 \cdot 3 \cdot 3$

Synthesis

61. ▦ Find the largest five-digit number that is divisible by 47.

62. ▦ Find the largest six-digit number that is divisible by 53.

Find the smallest number that is simultaneously a multiple of the given numbers.

63. 2, 3, and 5

64. 3, 5, and 7

65. 6, 10, and 14

66. ▦ 17, 43, and 85

67. 30, 70, and 120

68. 25, 100, and 175

69. To help see why the tests for division by 3 and 9 work, note that any four-digit number *abcd* can be rewritten as $1000 \cdot a + 100 \cdot b + 10 \cdot c + d$, or $999a + 99b + 9c + a + b + c + d$.

   a) Explain why $999a + 99b + 9c$ is divisible by both 9 and 3 for all choices of *a, b, c,* and *d.*

   b) Explain why the four-digit number *abcd* is divisible by 9 if $a + b + c + d$ is divisible by 9 and is divisible by 3 if $a + b + c + d$ is divisible by 3.

70. A passenger in a taxicab asks for the cab number. The driver says abruptly, "Sure—it's the smallest multiple of 11 that, when divided by 2, 3, 4, 5, or 6, has a remainder of 1." What is the number?

Following are the tests for divisibility by 4, 8, 7, and 11. Use these for Exercises 71 and 72.

NUMBER	TEST	TEST 23,904,328 FOR DIVISIBILITY
4	A number is divisible by 4 if the number named by its last two digits is divisible by 4.	The number named by the last two digits is 28. Since 28 is divisible by 4, 23,904,328 is divisible by 4.
8	A number is divisible by 8 if the number named by its last three digits is divisible by 8.	The number named by the last three digits is 328. Since 328 is divisible by 8, 23,904,328 is divisible by 8.
7	Divide the number into groups of three digits, starting at the right. Start with the group at the right, subtract the next group to the left, add the next group to the left, and so on. If the resulting number is divisible by 7, the original number is divisible by 7.	The number is already divided into groups of three by commas. Calculate $328 - 904 + 23 = -553$. Since -553 is divisible by 7, 23,904,328 is divisible by 7.
11	Find the sum of the odd-numbered digits (the 1st digit plus the 3rd plus the 5th and so on). Find the sum of the even-numbered digits. Subtract these sums. If the difference is divisible by 11, then the original number is divisible by 11.	Add the odd-numbered digits: $2 + 9 + 4 + 2 = 17$. Add the even-numbered digits: $3 + 0 + 3 + 8 = 14$. Subtract: $17 - 14 = 3$. Since 3 is *not* divisible by 11, 23,904,328 is not divisible by 11.

71. Test 332,986,412 for divisibility by 4, 8, 7, and 11.

72. Test 6,637,105,860 for divisibility by 4, 8, 7, and 11.

Factorizations

When we express a number as a product, we say that we have *factored* the original number. The numbers in the product are called *factors*. Thus "factor" can be used as either a noun or a verb. The ability to factor is an important skill needed for a solid understanding of fractions.

OBJECTIVES

a Find the factors of a number.

b Given a number from 1 to 100, tell whether it is prime, composite, or neither.

c Find the prime factorization of a composite number.

a FACTORS AND FACTORIZATIONS

From the equation $3 \cdot 4 = 12$, we can say that 3 and 4 are **factors** of 12. Since $12 = 12 \cdot 1$, we know that 12 and 1 are also factors of 12.

SKILL TO REVIEW

Objective 1.9a: Write exponential notation for products such as $4 \cdot 4 \cdot 4$.

Write in exponential notation.

1. $2 \cdot 2 \cdot 2$ **2.** $5 \cdot 5 \cdot 5 \cdot 5$

FACTORS AND FACTORIZATIONS

- In the product $a \cdot b$, a and b are called **factors**.
- A number c is a **factor** of a if a is divisible by c.
- A **factorization** of a expresses a as a product of two or more numbers.

Note that saying that 3 and 4 are factors of 12 is the same as saying that 12 is a multiple of 3 and that 12 is a multiple of 4. Each of the following gives a factorization of 12.

$12 = 4 \cdot 3$ ← This factorization shows that 4 and 3 are factors of 12.

$12 = 12 \cdot 1$ ← This factorization shows that 12 and 1 are factors of 12.

$12 = 6 \cdot 2$ ← This factorization shows that 6 and 2 are factors of 12.

$12 = 2 \cdot 3 \cdot 2$ ← This factorization shows that 2 and 3 are factors of 12.

Thus, 1, 2, 3, 4, 6, and 12 are all factors of 12. Note that since $n = n \cdot 1$, every number has a factorization, and every number has itself and 1 as factors.

Find all the factors of each number.

1. 10 **2.** 62 **3.** 24

EXAMPLE 1 Find all the factors of 18.

Beginning at 1, we check all positive integers to see if they are factors of 18. If they are, we write the factorization. We stop when we have already included the next integer in a factorization.

1 is a factor of every number.	$1 \cdot 18$
2 is a factor of 18.	$2 \cdot 9$
3 is a factor of 18.	$3 \cdot 6$
4 is *not* a factor of 18.	
5 is *not* a factor of 18.	

 4. 45

1 is a factor of 45. $1 \cdot 45$
2 ▢ a factor of 45.
3 ▢ a factor of 45. $3 \cdot$ ▢
4 ▢ a factor of 45.
5 ▢ a factor of 45. $5 \cdot$ ▢
6 is not a factor of 45.
7 is not a factor of 45.
8 is not a factor of 45.

Factors of 45: 1, 3, 5, ▢ , ▢ , 45.

The next integer is 6, but we have already listed 6 as a factor in the product $3 \cdot 6$. We need check no additional numbers, because any integer greater than 6 must be paired with a factor less than 6.

We now write the factors of 18 beginning with 1, going down the list of factorizations writing each first factor, then up the list of factorizations writing each second factor:

1, 2, 3, 6, 9, 18.

Do Margin Exercises 1–4. ▶

Answers

Skill to Review:
1. 2^3 **2.** 5^4

Margin Exercises:
1. 1, 2, 5, 10 **2.** 1, 2, 31, 62
3. 1, 2, 3, 4, 6, 8, 12, 24 **4.** 1, 3, 5, 9, 15, 45

Guided Solution:
4. is not; is, 15; is not; is, 9; 9, 15

b PRIME AND COMPOSITE NUMBERS

> ### PRIME AND COMPOSITE NUMBERS
>
> A natural number that has exactly two different factors, only itself and 1, is called a **prime number**.
>
> - The number 1 is *not* prime.
> - A natural number, other than 1, that is not prime is **composite**.

EXAMPLE 2 Determine which of the numbers 1, 2, 7, 8, 9, 11, 18, 27, 39, 43, 56, 59, and 77 are prime, which are composite, and which are neither.

The number 1 is not prime. It does not have *two* different factors.

The number 2 is prime. It has only the factors 2 and 1.

The numbers 7, 11, 43, and 59 are prime. Each has only two factors, itself and 1.

The number 8 is not prime. It has the factors 1, 2, 4, and 8 and is composite.

The numbers 9, 18, 27, 39, 56, and 77 are composite. Each has more than two factors.

Thus we have

Prime: 2, 7, 11, 43, 59;

Composite: 8, 9, 18, 27, 39, 56, 77;

Neither: 1.

We can make several observations about prime numbers.

- Because 0 is not a natural number, it is neither prime nor composite.
- The number 1 is not prime because it does not have two different factors.
- The number 2 is the smallest prime and the only even prime, since 2 is a factor of all even numbers.
- To determine whether an odd number is prime, check divisibility by prime numbers beginning with 3 and 5. If you reach a point where the quotient is less than the divisor and none of the primes up to that point are factors, the number you are checking is prime.

5. Classify each number as prime, composite, or neither.
1, 2, 6, 12, 13, 23, 41, 65, 73, 99

◀ **Do Exercise 5.**

The following is a table of the prime numbers from 2 to 157. Although you need not memorize the entire list, remembering at least the first nine or ten is important.

> ### A TABLE OF PRIMES FROM 2 TO 157
>
> 2, 3, 5, 7, 11, 13, 17, 19, 23, 29, 31, 37, 41, 43, 47, 53, 59, 61, 67, 71, 73, 79, 83, 89, 97, 101, 103, 107, 109, 113, 127, 131, 137, 139, 149, 151, 157

There are infinitely many prime numbers. It takes extensive computer operations to determine whether very large numbers are prime. In 2014, the largest known prime was $2^{57885161} - 1$. This number has over 17 million digits!

(**Source:** primes.utm.edu/largest.html#largest.)

Answer

5. 2, 13, 23, 41, 73 are prime; 6, 12, 65, 99 are composite; 1 is neither

c PRIME FACTORIZATIONS

When we factor a composite number into a product of primes, we find the **prime factorization** of the number. We can do this by making a series of successive divisions or by using a *factor tree*.

To use division, we consider the prime numbers 2, 3, 5, 7, 11, 13, and so on, and determine whether a given number is divisible by the primes.

EXAMPLE 3 Find the prime factorization of 39.

We check for divisibility by the first prime, 2. Since 39 is not even, 2 is not a factor of 39. Since the sum of the digits in 39 is 12 and 12 is divisible by 3, we know that 39 is divisible by 3. We then perform the division.

$$\begin{array}{r} 13 \\ 3{\overline{)39}} \end{array} \quad R = 0 \qquad \text{A remainder of 0 confirms that 3 is a factor of 39.}$$

Because 13 is prime, we can now write the prime factorization:

$$39 = 3 \cdot 13.$$

EXAMPLE 4 Find the prime factorization of 220.

We consider the first prime, 2. Since 220 is even, it must have 2 as a factor.

$$\begin{array}{r} 110 \\ 2{\overline{)220}} \end{array} \quad R = 0 \qquad 220 = 2 \cdot 110$$

Because 110 is also even, we divide again by 2.

$$\begin{array}{r} 55 \\ 2{\overline{)110}} \end{array} \quad R = 0 \qquad 220 = 2 \cdot 2 \cdot 55$$

Since 55 is odd, it is not divisible by 2. We move to the next prime, 3. The sum of the digits is 10 and 10 is not divisible by 3, so we move to the next prime, 5. Since the ones digit of 55 is 5, we divide by 5.

$$\begin{array}{r} 11 \\ 5{\overline{)55}} \end{array} \quad R = 0 \qquad 220 = 2 \cdot 2 \cdot 5 \cdot 11$$

Because 11 is prime, we are finished. The prime factorization is

$$220 = 2 \cdot 2 \cdot 5 \cdot 11.$$

We abbreviate our procedure as follows.

$$\begin{array}{r} 11 \\ 5{\overline{)55}} \\ 2{\overline{)110}} \\ 2{\overline{)220}} \end{array}$$

$$220 = 2 \cdot 2 \cdot 5 \cdot 11$$

A factorization like $2 \cdot 2 \cdot 5 \cdot 11$ could also be expressed as $5 \cdot 2 \cdot 2 \cdot 11$ or $2 \cdot 5 \cdot 11 \cdot 2$ or $2^2 \cdot 5 \cdot 11$ or $11 \cdot 2^2 \cdot 5$. The prime factors are the same in each case. For this reason, we agree that any of these may be considered "the" prime factorization of 220.

> Every composite number has just one (unique) prime factorization.

This result is sometimes called the Fundamental Theorem of Arithmetic.

EXAMPLE 5 Find the prime factorization of 187.

We check for divisibility by 2, 3, and 5, and find that 187 is not divisible by any of these numbers. The next prime number, 7, does not divide 187 evenly. However, when we divide by 11, the remainder is 0, so 11 is a factor of 187.

$$11\overline{)187}^{17} \qquad \text{We can write } 187 = 11 \cdot 17.$$

Because 17 is prime, we can factor no further. The complete factorization is

$$187 = 11 \cdot 17. \qquad \text{All factors are prime.}$$

EXAMPLE 6 Find the prime factorization of 72.

$$3\overline{)9}^{3} \qquad \leftarrow \text{3 is prime, so we stop dividing.}$$
$$2\overline{)18}$$
$$2\overline{)36}$$
$$2\overline{)72} \qquad \leftarrow \text{Begin here and work upward.}$$
$$72 = 2 \cdot 2 \cdot 2 \cdot 3 \cdot 3$$

Another way to find the prime factorization of 72 is to use a **factor tree** as follows. Begin by determining any factorization you can, and then continue factoring until all of the factors are prime numbers. Each of the following trees gives the same prime factorization.

EXAMPLE 7 Find the prime factorization of 130.

We can use a string of divisions or a factor tree.

$$5\overline{)65}^{13}$$
$$2\overline{)130}$$

$$130 = 2 \cdot 5 \cdot 13$$

◀ **Do Exercises 6–13.**

Finding a number's prime factorization can be quite challenging, especially when the prime factors themselves are large. This difficulty is used worldwide as a way of securing transactions over the Internet.

Find the prime factorization of each number.

6. 6   **7.** 12

8. 45

9. 98

$$\square\overline{)49}$$
$$\square\overline{)98}$$
$$98 = \square \cdot \square \cdot \square$$

10. 126   **11.** 144

12. 91   **13.** 1925

Answers

6. 2 · 3   **7.** 2 · 2 · 3   **8.** 3 · 3 · 5
9. 2 · 7 · 7   **10.** 2 · 3 · 3 · 7
11. 2 · 2 · 2 · 2 · 3 · 3   **12.** 7 · 13
13. 5 · 5 · 7 · 11

Guided Solution:

9. $7\overline{)49}^{7}$; 2, 7, 7
  $2\overline{)98}$

✓ Reading Check

Determine whether each statement is true or false.

_____ **RC1.** One factorization of 20 is 4 · 5.

_____ **RC2.** One factor of 15 is 30.

_____ **RC3.** A prime number has exactly two different factors.

_____ **RC4.** The smallest prime number is 1.

_____ **RC5.** The prime factorization of 30 is 3 · 10.

a Determine whether the second number is a factor of the first.

1. 52; 14   **2.** 52; 13   **3.** 625; 25   **4.** 680; 16

List all the factors of each number.

5. 18   **6.** 16   **7.** 54   **8.** 48

9. 9   **10.** 4   **11.** 13   **12.** 11

13. 98   **14.** 100   **15.** 255   **16.** 120

b Classify each number as prime, composite, or neither.

17. 19   **18.** 24   **19.** 22   **20.** 31

21. 48   **22.** 43   **23.** 53   **24.** 54

25. 1   **26.** 2   **27.** 81   **28.** 37

29. 47   **30.** 51   **31.** 29   **32.** 49

c Find the prime factorization of each number.

33. 27   **34.** 16   **35.** 14   **36.** 15   **37.** 80

38. 32   **39.** 25   **40.** 75   **41.** 62   **42.** 169

43. 100   **44.** 110   **45.** 143   **46.** 40   **47.** 121

48. 170   **49.** 273   **50.** 675   **51.** 175   **52.** 196

53. 209   **54.** 217   **55.** 1200   **56.** 1800   **57.** 693

58. 675   **59.** 2884   **60.** 484   **61.** 1122   **62.** 6435

Skill Maintenance

Multiply.

63. $-2 \cdot 13$ [2.4a]   **64.** $(-8)(-32)$ [2.4a]

Add.

65. $-17 + 25$ [2.2a]   **66.** $-9 + (-14)$ [2.2a]

Divide.

67. $53 \div 53$ [1.5a]   **68.** $-98 \div 1$ [2.5a]   **69.** $0 \div 22$ [1.5a]   **70.** $0 \div (-42)$ [2.5a]

Synthesis

Find the prime factorization of each number.

71. ▦ 136,097   **72.** ▦ 102,971   **73.** ▦ 473,073,361

74. Describe an arrangement of 54 objects that corresponds to the factorization $54 = 6 \times 9$.

75. Describe an arrangement of 24 objects that corresponds to the factorization $24 = 2 \cdot 3 \cdot 4$.

76. Two numbers are **relatively prime** if there is no prime number that is a factor of both numbers. For example, 10 and 21 are relatively prime but 15 and 18 are not. List five pairs of composite numbers that are relatively prime.

77. *Factors and Sums.* In the table below, the top number in each column can be factored in such a way that the sum of the factors is the bottom number in the column. For example, in the first column, 56 has been factored as $7 \cdot 8$, and $7 + 8 = 15$, the bottom number. (Such thinking will be important in understanding the meaning of a factor and in algebra.) Fill in the missing numbers in the table.

PRODUCT	56	63	36	72	140	96	48	168	110	90	432	63
FACTOR	7											
FACTOR	8											
SUM	15	16	20	38	24	20	14	29	21	19	42	24

The study of arithmetic begins with the set of whole numbers. We also need to be able to use fractional parts of numbers such as halves and thirds.

Households in auto-dependent locations spend about $\frac{1}{4}$ of their income on transportation costs, while location-efficient households (those with easy access to public transportation) can hold transportation costs to $\frac{1}{10}$ of their income.

Auto-Dependent Households   **Location-Efficient Households**

$\frac{1}{4}$ Transportation $\frac{3}{4}$ Remaining $\frac{1}{10}$ Transportation $\frac{9}{10}$ Remaining

SOURCE: Based on data from U.S. Department of Transportation, Federal Highway Administration

a IDENTIFYING NUMERATORS AND DENOMINATORS

Expressions like those below are written in **fraction notation**. The top number is called the **numerator** and the bottom number is called the **denominator**.

$$\frac{1}{2}, \quad \frac{7}{8}, \quad \frac{-8}{5}, \quad \frac{x}{y}, \quad -\frac{4}{25}, \quad \frac{2a}{7b}$$

EXAMPLE 1 Identify the numerator and the denominator.

$$\frac{7}{8} \begin{array}{l} \leftarrow \text{Numerator} \\ \leftarrow \text{Denominator} \end{array}$$

Do Margin Exercises 1–4. ▶

Fractions as a Partition of an Object Divided into Equal Parts

Consider a candy bar divided into 5 equal sections. If you eat 2 sections, you have eaten $\frac{2}{5}$ of the candy bar. The denominator 5 tells us the unit, $\frac{1}{5}$. The numerator 2 tells us the number of equal parts we are considering, 2.

For each fraction, identify the numerator and the denominator.

1. About $\frac{1}{5}$ of people age 5 and older in the United States speak a language other than English at home.

   Source: 2010 American Community Survey

2. About $\frac{4}{5}$ of the parts of a Toyota Camry were produced in the United States.

   Source: "Made in America: Which Car Creates the Most Jobs?" by David Muir and Sharyn Alfonsi, on abcnews.go.com

3. $\dfrac{5a}{7m}$ 4. $\dfrac{-22}{3}$

Answers

Skill to Review:
1. 1   2. 50

Margin Exercises:
1. Numerator: 1; denominator: 5
2. Numerator: 4; denominator: 5
3. Numerator: $5a$; denominator: $7m$
4. Numerator: -22; denominator: 3

EXAMPLE 2 What part is shaded?

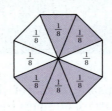

There are 8 equal parts. This tells us the unit, $\frac{1}{8}$. The *denominator* is 8. We have 5 of the units shaded. This tells us the *numerator,* 5. Thus,

$5 \leftarrow$ 5 units are shaded.
$\overline{8} \leftarrow$ The unit is $\frac{1}{8}$.

is shaded.

EXAMPLE 3 What part of the measuring cup is filled?

The measuring cup is divided into 3 parts of the same size, and 2 of them are filled. This is $2 \cdot \frac{1}{3}$, or $\frac{2}{3}$. Thus, $\frac{2}{3}$ (read *two-thirds*) of the cup is filled.

◀ **Do Exercises 5–8.**

The markings on a ruler use fractions.

EXAMPLE 4 What part of an inch is indicated?

16 equal spaces

Each inch on the ruler shown above is divided into 16 equal parts. The marked section extends to the 11th mark. Thus, $\frac{11}{16}$ (read *eleven-sixteenths*) of an inch is indicated.

◀ **Do Exercise 9.**

What part is shaded?

5.

6. 1 mile

7.

1 gallon

8.

9. What part of an inch is indicated?

Answers

5. $\frac{5}{6}$ **6.** $\frac{1}{3}$ **7.** $\frac{3}{4}$ **8.** $\frac{8}{15}$ **9.** $\frac{15}{16}$

Fractions greater than or equal to 1, such as $\frac{24}{24}$, $\frac{10}{3}$, and $\frac{5}{4}$, correspond to situations like the following.

EXAMPLE 5   What part is shaded?

a)

The rectangle is divided into 24 equal parts. Thus the unit is $\frac{1}{24}$. The denominator is 24. All 24 equal parts are shaded. This tells us that the numerator is 24. Thus, $\frac{24}{24}$ is shaded.

b)
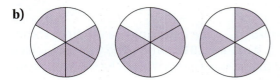

Each circle is divided into 6 parts. Thus the unit is $\frac{1}{6}$. The denominator is 6. We see that 11 of the equal units are shaded. This tells us that the numerator is 11. Thus, $\frac{11}{6}$ is shaded.

EXAMPLE 6   *Ice-Cream Roll-up Cake.*   What part of an ice-cream roll-up cake is shaded?

3 ice cream roll-up cakes

Each cake is divided into 6 equal slices. The unit is $\frac{1}{6}$. The *denominator* is 6. We see that 13 of the slices are shaded. This tells us that the *numerator* is 13. Thus, $\frac{13}{6}$ is shaded.

Do Exercises 10–12. ▶

Fractions larger than or equal to 1, such as $\frac{13}{6}$ or $\frac{9}{9}$, are sometimes referred to as "improper" fractions. We will not use this terminology because notation such as $\frac{27}{8}$, $\frac{11}{3}$, or $\frac{4}{4}$ is quite "proper" and very common in algebra.

Fractions as Ratios

A **ratio** is a quotient of two quantities. We can express a ratio with fraction notation. (We will consider ratios in more detail in Chapter 7.)

What part is shaded?

10.

11.

GS **12.**

Each gallon is divided into ☐ equal parts.

The unit is $\dfrac{1}{\square}$.

There are ☐ equal units shaded.

The part that is shaded is $\dfrac{\square}{\square}$.

Answers

10. $\dfrac{15}{15}$ 11. $\dfrac{8}{5}$ 12. $\dfrac{7}{4}$

Guided Solution:

12. $4, 4, 7, \dfrac{7}{4}$

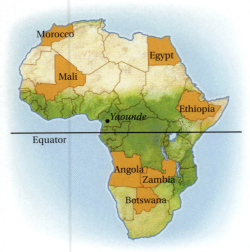

Morocco
Egypt
Mali
Ethiopia
Yaounde
Equator
Angola
Zambia
Botswana

13. What part of the set of countries in Example 7 is west of Yaounde?

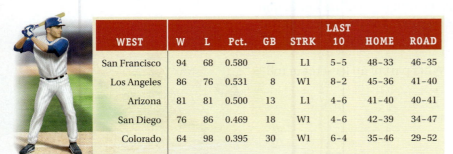

WEST	W	L	Pct.	GB	STRK	LAST 10	HOME	ROAD
San Francisco	94	68	0.580	—	L1	5–5	48–33	46–35
Los Angeles	86	76	0.531	8	W1	8–2	45–36	41–40
Arizona	81	81	0.500	13	L1	4–6	41–40	40–41
San Diego	76	86	0.469	18	W1	4–6	42–39	34–47
Colorado	64	98	0.395	30	W1	6–4	35–46	29–52

SOURCE: espn.go.com

14. *Baseball Standings.* Refer to the table in Example 8. The Arizona Diamondbacks finished third in the National League West in 2012. Find the ratio of Diamondback wins to losses, wins to total games, and losses to total games.

EXAMPLE 7 *Countries of Africa.* What part of this group of countries is north of the equator? south of the equator?

Angola   Mali
Botswana   Morocco
Egypt   Zambia
Ethiopia

There are 7 countries in the set, and 4 of them, Egypt, Ethiopia, Mali, and Morocco, are north of the equator. Thus, 4 of 7, or $\frac{4}{7}$, are north of the equator. The 3 remaining countries are south of the equator. Thus, $\frac{3}{7}$ are south of the equator.

◀ **Do Exercise 13.**

EXAMPLE 8 *Baseball Standings.* The following are the final standings in the National League West for 2012, when the San Francisco Giants won the division. Find the ratio of Giants wins to losses, wins to total games, and losses to total games.

The Giants won 94 games and lost 68 games. They played a total of 94 + 68, or 162 games. Thus we have the following.

The ratio of wins to losses is $\frac{94}{68}$.

The ratio of wins to total games is $\frac{94}{162}$.

The ratio of losses to total games is $\frac{68}{162}$.

◀ **Do Exercise 14.**

b SOME FRACTION NOTATION FOR WHOLE NUMBERS

Fraction Notation for 1

The number 1 corresponds to situations like those shown here. If we divide an object into n parts and take n of them, we get all of the object (1 whole object).

Since a negative number divided by a negative number is a positive number, we state the following for all nonzero integers.

1
$\frac{2}{2}$
$\frac{4}{4}$

THE NUMBER 1 IN FRACTION NOTATION

$\frac{n}{n} = 1$, for any integer n that is not 0.

Answers

13. $\frac{2}{7}$ 14. $\frac{81}{81}, \frac{81}{162}, \frac{81}{162}$

EXAMPLES Simplify. Assume any variables are nonzero.

9. $\dfrac{5}{5} = 1$ **10.** $\dfrac{-9}{-9} = 1$ **11.** $\dfrac{23x}{23x} = 1$

Do Exercises 15–20. ▶

Fraction Notation for 0

Consider the fraction $\frac{0}{4}$. This corresponds to dividing an object into 4 parts and taking none of them. We get 0.

THE NUMBER 0 IN FRACTION NOTATION

$\dfrac{0}{n} = 0$,   for any integer n that is not 0.

EXAMPLES Simplify.

12. $\dfrac{0}{1} = 0$ **13.** $\dfrac{0}{9} = 0$

14. $\dfrac{0}{-23} = 0$ **15.** $\dfrac{0}{5a} = 0$, assuming that $a \neq 0$

Fraction notation with a denominator of 0, such as $n/0$, does not represent a number because we cannot speak of an object being divided into *zero* parts. (If it is not divided at all, then we say it is undivided and remains in one part.)

A DENOMINATOR OF 0

$\dfrac{n}{0}$ is not defined. We say that $\dfrac{n}{0}$ is *undefined*.

Do Exercises 21–26. ▶

Other Integers

Consider the fraction $\frac{4}{1}$. This corresponds to taking 4 objects and dividing each into 1 part. (In other words, we do not divide them.) We have 4 objects.

ANY INTEGER IN FRACTION NOTATION

Any integer divided by 1 is the original integer. That is,

$\dfrac{n}{1} = n$,   for any integer n.

EXAMPLES Simplify.

16. $\dfrac{2}{1} = 2$ **17.** $\dfrac{-9}{1} = -9$ **18.** $\dfrac{3x}{1} = 3x$

Do Exercises 27–30. ▶

Simplify.

15. $\dfrac{1}{1}$ **16.** $\dfrac{4}{4}$

17. $\dfrac{a}{a}$ **18.** $\dfrac{-100}{-100}$

19. $\dfrac{2347}{2347}$ **20.** $\dfrac{54n}{54n}$

Simplify, if possible. Assume that $x \neq 0$.

21. $\dfrac{0}{1}$ **22.** $\dfrac{0}{-6}$

23. $\dfrac{0}{4x}$

 GS 24. $\dfrac{4-4}{567} = \dfrac{\square}{567} = \square$

25. $\dfrac{15}{0}$ **26.** $\dfrac{-4}{3-3}$

Simplify.

27. $\dfrac{8}{1}$ **28.** $\dfrac{-10}{1}$

29. $\dfrac{-346}{1}$ **30.** $\dfrac{24-1}{23-22}$

Answers

15. 1   16. 1   17. 1   18. 1   19. 1
20. 1   21. 0   22. 0   23. 0   24. 0
25. Undefined   26. Undefined   27. 8
28. −10   29. −346   30. 23

Guided Solution:
24. 0, 0

✓ Reading Check

Match each expression with the appropriate description or value from the list at the right.

RC1. The 3 in $\frac{3}{4}$ _____   **a)** a denominator

RC2. The 4 in $\frac{3}{4}$ _____   **b)** a numerator

RC3. The fraction $\frac{3}{4}$ _____   **c)** a ratio

RC4. $\frac{0}{1}$ _____   **d)** n

RC5. $\frac{n}{0}$ _____   **e)** 0

RC6. $\frac{n}{1}$ _____   **f)** undefined

a   Identify the numerator and the denominator.

1. $\frac{3}{4}$ **2.** $\frac{-9}{10}$ **3.** $\frac{7}{-9}$ **4.** $\frac{15}{8}$ **5.** $\frac{2x}{3y}$ **6.** $\frac{9a}{2b}$

What part of each object or set of objects is shaded?

7.

1 acre

8.

1 square inch

9.

1 yard

10. 1 mile

11.

12.

1 year

13.

1 pie

14.

15.

16.

17. 1 gold bar

2 gold bars

18.
1 spool

19.
1 quart

2 quarts

20. 1 foot

3 feet

21.
1 window

22.
1 quart

23.

24.

25.

26.

What part of an inch is indicated?

27.

28.

29.

30.

For each of Exercises 31–34, give fraction notation for the amount of gas **(a)** in the tank and **(b)** used from a full tank.

31.

32.

33.

34.

35. For the following set of animals, what is the ratio of:
   a) puppies to the total number of animals?
   b) puppies to kittens?
   c) kittens to the total number of animals?
   d) kittens to puppies?

36. For the following set of sports equipment, what is the ratio of:
   a) basketballs to footballs?
   b) footballs to basketballs?
   c) basketballs to the total number of balls?
   d) total number of balls to basketballs?

37. Bryce delivers car parts to auto service centers. On Thursday he had 15 deliveries scheduled. By noon he had delivered only 4 orders. What is the ratio of:
   a) orders delivered to total number of orders?
   b) orders delivered to orders not delivered?
   c) orders not delivered to total number of orders?

38. *Gas Mileage.* A Volkswagen Passat TDI® SE will travel 473 mi on 11 gal of gasoline in highway driving. What is the ratio of:
   a) miles driven to gasoline used?
   b) gasoline used to miles driven?

   Source: vw.com

For Exercises 39 and 40, use the following bar graph, which shows the number of registered nurses per 100,000 residents in each of twelve states or districts.

Registered Nurses per 100,000 Residents

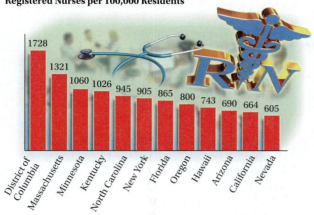

SOURCES: www.statehealthfacts.org; Kaiser Family Foundation, 2011

39. What is the ratio of registered nurses to 100,000 residents in the given state or district?

a) Minnesota   b) Hawaii
c) Florida   d) Kentucky
e) New York   f) District of Columbia

40. What is the ratio of registered nurses to 100,000 residents in the given state?

a) North Carolina   b) California
c) Massachusetts   d) Nevada
e) Arizona   f) Oregon

For Exercises 41 and 42, use the following set of states, as illustrated on the map.

Alabama	Nebraska	West Virginia
Arkansas	South Dakota	Wisconsin
Illinois		

41. What part of this group of states is east of the Mississippi River?

42. What part of this group of states is north of Nashville, Tennessee?

b   Simplify. Assume that all variables are nonzero.

43. $\dfrac{0}{8}$

44. $\dfrac{8}{8}$

45. $\dfrac{8-1}{9-8}$

46. $\dfrac{16}{1}$

47. $\dfrac{20}{20}$

48. $\dfrac{-20}{1}$

49. $\dfrac{-45}{-45}$

50. $\dfrac{11-1}{10-9}$

51. $\dfrac{0}{-238}$

52. $\dfrac{19x}{19x}$

53. $\dfrac{19x}{1}$

54. $\dfrac{0}{-16}$

55. $\dfrac{13t}{13t}$

56. $\dfrac{-27m}{1}$

57. $\dfrac{-87}{1}$

58. $\dfrac{-98}{-98}$

59. $\dfrac{0}{2a}$

60. $\dfrac{0}{18}$

61. $\dfrac{52}{0}$

62. $\dfrac{8-8}{1247}$

63. $\dfrac{7n}{1}$

64. $\dfrac{1317}{0}$

65. $\dfrac{5}{6-6}$

66. $\dfrac{13}{10-10}$

Skill Maintenance

Multiply.

67. $-7(30)$ [2.4a]

68. $23 \cdot (-14)$ [2.4a]

69. $(-71)(-12)0$ [2.4b]

70. $32(-29)0$ [2.4b]

Solve.   [1.8a]

71. *Museums.*   At the end of 2011, there were 3415 museums in China. By the end of 2012, the number of museums had increased to 3866. How many museums opened in China in 2012?
Source: China Museums Association

72. *Gas Mileage.*   The 2014 Subaru XV Cross gets 33 mpg in highway driving. How many gallons will it use in 2343 mi of highway driving?
Source: www.kbb.com

Synthesis

What part of each object is shaded?

73.

74.

75.

76.

Shade or mark each figure to show $\frac{3}{5}$.

77.

78.

79.

80.

81. The year 2013 began on a Tuesday. What fractional part of the days in 2013 were Wednesdays?

82. The year 2014 began on a Wednesday. What fractional part of the days in 2014 were Wednesdays?

83. The surface of Earth is 3 parts water and 1 part land. What fractional part of Earth is water? land?

84. A couple had 3 sons, each of whom had 3 daughters. If each daughter gave birth to 3 sons, what fractional part of the couple's descendants is female?

Multiplication and Applications

a MULTIPLICATION BY AN INTEGER

We can find $3 \cdot \frac{1}{4}$ by thinking of repeated addition. We add three $\frac{1}{4}$'s.

$\frac{1}{4}$ acre $\frac{1}{4}$ acre $\frac{1}{4}$ acre $\frac{3}{4}$ acre

We see that $3 \cdot \frac{1}{4} = \frac{1}{4} + \frac{1}{4} + \frac{1}{4} = \frac{3}{4}$.

Do Margin Exercises 1 and 2. ▶

To multiply a fraction by an integer,

a) multiply the top number (the numerator) by the integer and

$$6 \cdot \frac{4}{5} = \frac{6 \cdot 4}{5} = \frac{24}{5}$$

b) keep the same denominator.

EXAMPLES Multiply.

1. $5 \times \frac{3}{8} = \underbrace{\frac{5 \times 3}{8}} = \frac{15}{8}$ We generally replace the $\times$ symbol with $\cdot$.

Skip this step when you feel comfortable doing so.

2. $\frac{2}{5} \cdot 13 = \frac{2 \cdot 13}{5} = \frac{26}{5}$

3. $-10 \cdot \frac{1}{3} = \frac{-10}{3}$, or $-\frac{10}{3}$ Recall that $\frac{-a}{b} = -\frac{a}{b}$.

4. $a \cdot \frac{4}{7} = \frac{4a}{7}$ Recall that $a \cdot 4 = 4 \cdot a$.

Do Exercises 3–6. ▶

OBJECTIVES

a Multiply an integer and a fraction.

b Multiply using fraction notation.

c Solve problems involving multiplication of fractions.

SKILL TO REVIEW

Objective 1.4a: Multiply whole numbers.

Multiply.

1. $24 \cdot 17$ **2.** $5(13)$

---

1. Find $2 \cdot \frac{1}{3}$.

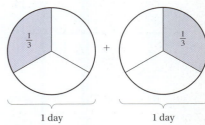

1 day   1 day

2. Find $5 \cdot \frac{1}{8}$.

Multiply.

3. $7 \times \frac{2}{3}$ **4.** $(-11) \cdot \frac{3}{10}$

5. $34 \cdot \frac{2}{5}$ **6.** $x \cdot \frac{4}{9}$

Answers

Skill to Review:
1. 408   **2.** 65

Margin Exercises:
1. $\frac{2}{3}$ **2.** $\frac{5}{8}$ **3.** $\frac{14}{3}$ **4.** $-\frac{33}{10}$, or $\frac{-33}{10}$
5. $\frac{68}{5}$ **6.** $\frac{4x}{9}$

---

7. Draw diagrams like the ones at right to illustrate $\frac{1}{4}$ and $\frac{1}{2} \cdot \frac{1}{4}$.

$\frac{1}{4}$ $\frac{1}{2} \cdot \frac{1}{4} = \frac{1}{8}$

To illustrate the meaning of an expression like $\frac{1}{2} \cdot \frac{1}{3}$, we first represent $\frac{1}{3}$ and then shade half of that region. Note that $\frac{1}{2} \cdot \frac{1}{3}$ is the same as $\frac{1}{2}$ of $\frac{1}{3}$.

$\frac{1}{3}$ $\frac{1}{2} \cdot \frac{1}{3} = \frac{1}{6}$

◀ **Do Exercise 7.**

To visualize $\frac{2}{5} \cdot \frac{3}{4}$, we first represent $\frac{3}{4}$. This is shown by the shading on the left below. Next, we divide the shaded area into 5 equal parts (using horizontal lines) and take 2 of them. That is shown by the darker shading on the right below. Note that $\frac{2}{5} \cdot \frac{3}{4}$ is the same as $\frac{2}{5}$ of $\frac{3}{4}$.

   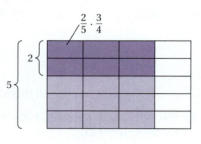

The entire object has now been divided into 20 parts, and we have shaded 6 of them twice. Thus,

$$\frac{2}{5} \cdot \frac{3}{4} = \frac{6}{20}. \quad \begin{array}{l} \leftarrow \text{This is the product of the numerators.} \\ \leftarrow \text{This is the product of the denominators.} \end{array}$$

◀ **Do Exercise 8.**

Notice that the product of two fractions is the product of the numerators over the product of the denominators.

8. Draw diagrams like the ones at right to illustrate $\frac{1}{3}$ and $\frac{4}{5} \cdot \frac{1}{3}$.

$\frac{1}{3}$ $\frac{4}{5} \cdot \frac{1}{3} = \frac{4}{15}$

To multiply a fraction by a fraction,

a) multiply the numerators and

$$\frac{9}{7} \cdot \frac{3}{4} = \frac{9 \cdot 3}{7 \cdot 4} = \frac{27}{28}$$

b) multiply the denominators.

EXAMPLES Multiply.

5. $\dfrac{5}{6} \cdot \dfrac{7}{4} = \dfrac{5 \cdot 7}{6 \cdot 4} = \dfrac{35}{24}$

> Skip this step when you feel comfortable doing so.

6. $\dfrac{3}{5} \cdot \dfrac{7}{8} = \dfrac{3 \cdot 7}{5 \cdot 8} = \dfrac{21}{40}$

Answers

7.

8.

7. $\left(-\dfrac{4}{x}\right)\left(-\dfrac{y}{7}\right) = \dfrac{4y}{7x}$ A negative times a negative is positive.

8. $(-6)\left(-\dfrac{4}{5}\right) = \dfrac{-6}{1} \cdot \dfrac{-4}{5} = \dfrac{24}{5}$ We can always write n as $\dfrac{n}{1}$.

Do Exercises 9–12. ▶

C APPLICATIONS AND PROBLEM SOLVING

Many problems that can be solved by multiplying fractions can be thought of in terms of rectangular arrays.

EXAMPLE 9 A real estate developer owns a plot of land and plans to use $\frac{4}{5}$ of the plot for a small strip mall and parking lot. Of this, $\frac{2}{3}$ will be needed for the parking lot. What part of the plot will be used for parking?

1. **Familiarize.** We first make a drawing to help familiarize ourselves with the problem. The land may not be rectangular, but we can think of it as a rectangle. The strip mall, including the parking lot, uses $\frac{4}{5}$ of the plot. We shade $\frac{4}{5}$ as shown on the left below. The parking lot alone uses $\frac{2}{3}$ of the part we just shaded. We shade that as shown on the right below.

2. **Translate.** We let $n =$ the part of the plot that is used for parking. We are taking "two-thirds of four-fifths." The word "of" corresponds to multiplication. Thus the following multiplication sentence corresponds to the situation:

$$\frac{2}{3} \cdot \frac{4}{5} = n.$$

3. **Solve.** The number sentence tells us what to do. We multiply:

$$\frac{2}{3} \cdot \frac{4}{5} = \frac{2 \cdot 4}{3 \cdot 5} = \frac{8}{15}.$$

Thus, $\dfrac{8}{15} = n$.

4. **Check.** We can do a partial check by noting that the answer is a fraction less than 1, which we expect since the developer is using only part of the original plot of land. Thus, $\frac{8}{15}$ is a reasonable answer. We can also check this in the figure above, where we see that 8 of 15 parts represent the parking lot.

5. **State.** The parking lot takes up $\frac{8}{15}$ of the plot of land.

Do Exercise 13. ▶

 Multiply.

9. $\dfrac{3}{8} \cdot \dfrac{5}{7} = \dfrac{3 \cdot 5}{8 \cdot \boxed{}}$

$= \dfrac{\boxed{}}{\boxed{}}$

10. $\dfrac{4}{3} \cdot \dfrac{8}{5}$

11. $\left(-\dfrac{3}{10}\right)\left(-\dfrac{1}{10}\right)$

12. $(-7)\dfrac{a}{b}$

13. A developer plans to set aside $\frac{3}{4}$ of the land in a housing development as open (undeveloped) space. Of this, $\frac{1}{2}$ will be green (natural) space. What part of the land will be green space?

Answers

9. $\dfrac{15}{56}$ **10.** $\dfrac{32}{15}$ **11.** $\dfrac{3}{100}$

12. $-\dfrac{7a}{b}$, or $\dfrac{-7a}{b}$ **13.** $\dfrac{3}{8}$

Guided Solution:

9. $7, \dfrac{15}{56}$

14. *Area of a Ceramic Tile.* The length of a rectangular ceramic tile inlaid on a countertop is $\frac{4}{9}$ ft. The width is $\frac{2}{9}$ ft. What is the area of the tile?

15. Of the students at Overton Junior College, $\frac{1}{8}$ participate in sports and $\frac{3}{5}$ of these play football. What fractional part of the students play football?

Answers

14. $\frac{8}{81}$ ft²   **15.** $\frac{3}{40}$

The area of a rectangular region is found by multiplying length by width. That is true whether length and width are whole numbers or not. Remember, the area of a rectangular region is given by the formula

$$A = l \cdot w \quad (Area = length \cdot width).$$

EXAMPLE 10 *Area of a Cranberry Bog.* The length of a rectangular cranberry bog is $\frac{9}{16}$ mi. The width is $\frac{3}{8}$ mi. What is the area of the bog?

1. **Familiarize.** Recall that area is length times width. We let $A =$ the area of the cranberry bog.

2. **Translate.** Next, we translate:

Area	is	Length	times	Width
A	$=$	$\frac{9}{16}$	$\times$	$\frac{3}{8}$.

3. **Solve.** The sentence tells us what to do. We multiply:

$$A = \frac{9}{16} \cdot \frac{3}{8} = \frac{9 \cdot 3}{16 \cdot 8} = \frac{27}{128}.$$

4. **Check.** We check by repeating the calculation. This is left to the student.

5. **State.** The area is $\frac{27}{128}$ square mile (mi²).

◀ **Do Exercise 14.**

EXAMPLE 11 A recipe for oatmeal chocolate chip cookies calls for $\frac{3}{4}$ cup of rolled oats. Monica is making $\frac{1}{2}$ of the recipe. How much oats should she use?

1. **Familiarize.** We first make a drawing or at least visualize the situation. We let $n =$ the amount of oats that Monica should use.

$\frac{3}{4}$ cup in the recipe $\frac{1}{2} \cdot \frac{3}{4}$ cup in $\frac{1}{2}$ the recipe

2. **Translate.** We are finding $\frac{1}{2}$ of $\frac{3}{4}$, so the multiplication sentence $\frac{1}{2} \cdot \frac{3}{4} = n$ corresponds to the situation.

3. **Solve.** We carry out the multiplication:

$$\frac{1}{2} \cdot \frac{3}{4} = \frac{1 \cdot 3}{2 \cdot 4} = \frac{3}{8}.$$

Thus, $\frac{3}{8} = n$.

4. **Check.** We check by repeating the calculation. This is left to the student.

5. **State.** Monica should use $\frac{3}{8}$ cup of oats.

◀ **Do Exercise 15.**

✓ Reading Check

Determine whether each statement is true or false.

_____ **RC1.** Multiplying $\frac{1}{2} \cdot \frac{3}{5}$ is the same as finding $\frac{1}{2}$ of $\frac{3}{5}$.

_____ **RC2.** When we multiply two fractions, the new numerator is the product of the numerators in the two fractions.

_____ **RC3.** The whole number 6 can be written $\frac{6}{1}$.

_____ **RC4.** The product of two fractions can be smaller than either of the two fractions.

a   Multiply.

1. $3 \cdot \frac{1}{8}$

2. $2 \cdot \frac{1}{5}$

3. $(-5) \times \frac{1}{6}$

4. $(-4) \times \frac{1}{7}$

5. $\frac{2}{3} \cdot 7$

6. $\frac{2}{5} \cdot 7$

7. $(-1)\frac{7}{9}$

8. $(-1)\frac{4}{11}$

9. $\frac{5}{6} \cdot x$

10. $\left(-\frac{7}{8}\right)y$

11. $\frac{2}{5}(-3)$

12. $\frac{3}{5}(-4)$

13. $a \cdot \frac{2}{7}$

14. $b \cdot \frac{3}{8}$

15. $-17 \times \frac{m}{6}$

16. $\frac{n}{7} \cdot 30$

17. $-3 \cdot \frac{-2}{5}$

18. $-4 \cdot \frac{-5}{7}$

19. $-\frac{2}{7}(-x)$

20. $-\frac{3}{4}(-a)$

b   Multiply.

21. $\frac{2}{5} \cdot \frac{2}{3}$

22. $\frac{3}{4} \cdot \frac{3}{5}$

23. $\left(-\frac{1}{4}\right) \times \frac{1}{10}$

24. $\left(-\frac{1}{3}\right) \times \frac{1}{10}$

25. $\frac{2}{3} \times \frac{1}{5}$

26. $\frac{3}{5} \times \frac{1}{5}$

27. $\frac{2}{y} \cdot \frac{x}{9}$

28. $\left(-\frac{3}{4}\right)\left(-\frac{3}{5}\right)$

29. $\left(-\frac{3}{4}\right)\left(-\frac{3}{4}\right)$

30. $\frac{3}{b} \cdot \frac{a}{7}$

31. $\frac{2}{3} \cdot \frac{7}{13}$

32. $\frac{3}{11} \cdot \frac{4}{5}$

33. $\frac{1}{10}\left(\frac{-3}{5}\right)$

34. $\frac{3}{10}\left(\frac{-7}{5}\right)$

35. $\frac{7}{8} \cdot \frac{a}{8}$

36. $\frac{4}{5} \cdot \frac{7}{x}$

37. $\frac{1}{y} \cdot 100$

38. $\frac{b}{10} \cdot 13$

39. $\frac{-21}{4} \cdot \frac{7}{5}$

40. $\frac{-8}{3} \cdot \frac{20}{9}$

c   Solve.

41. *Hair Bows.* It takes $\frac{5}{3}$ yd of ribbon to make a hair bow. How much ribbon is needed to make 8 bows?

42. *Gasoline Can Capacity.* A gasoline can holds $\frac{5}{2}$ gal. How much will the can hold when it is $\frac{1}{2}$ full?

43. *Basketball: High School to Pro.* One of 35 high school basketball players plays college basketball. One of 75 college players plays professional basketball. What fractional part of high school basketball players play professional basketball?

Source: National Basketball Association

44. *Football: High School to Pro.* One of 42 high school football players plays college football. One of 85 college players plays professional football. What fractional part of high school football players play professional football?

Source: National Football League

45. *Serving of Cheesecake.* At the Cheesecake Factory, a piece of cheesecake is $\frac{1}{12}$ of a cheesecake. How much of the cheesecake is $\frac{1}{2}$ piece?

Source: The Cheesecake Factory

46. *Tossed Salad.* The recipe for a tossed salad calls for $\frac{3}{4}$ cup of sliced almonds. How much is needed to make $\frac{1}{2}$ of the recipe?

47. *Floor Tiling.* The floor of a room is being covered with tile. An area $\frac{3}{5}$ of the length and $\frac{3}{4}$ of the width is covered. What fraction of the floor has been tiled?

48. *Table Top Size.* A rectangular table top measures $\frac{4}{5}$ m long by $\frac{3}{5}$ m wide. What is its area?

Skill Maintenance

Divide. [1.5a], [2.5a]

49. $7140 \div 35$

50. $32{,}200 \div 46$

51. $-65 \div (-5)$

52. $540 \div (-6)$

Simplify. [1.9c]

53. $8 \cdot 12 - (63 \div 9 + 13 \cdot 3)$

54. $(10 - 3)^4 + 10^3 \cdot 4 - 10 \div 5$

Synthesis

Multiply. Write each answer using fraction notation.

55. ▦ $\dfrac{341}{517} \cdot \dfrac{209}{349}$

56. ▦ $\left(\dfrac{57}{61}\right)^3$

57. $\left(\dfrac{2}{5}\right)^3\left(-\dfrac{7}{9}\right)$

58. $\left(-\dfrac{1}{2}\right)^5\left(\dfrac{3}{5}\right)$

59. A chain saw holds $\frac{1}{5}$ gal of fuel. Chain saw fuel is $\frac{1}{16}$ two-cycle oil and $\frac{15}{16}$ unleaded gasoline. How much two-cycle oil is in a freshly filled chain saw?

60. Evaluate $-\frac{2}{3}xy$ for $x = \frac{2}{5}$ and $y = -\frac{1}{7}$.

Simplifying

a MULTIPLYING BY 1

Recall the following:

$$1 = \frac{1}{1} = \frac{2}{2} = \frac{3}{3} = \frac{4}{4} = \frac{-13}{-13} = \frac{45}{45} = \frac{100}{100} = \frac{n}{n}.$$

$$1 = \qquad = \qquad = \qquad$$
$$1 \qquad \frac{2}{2} \qquad \frac{3}{3} \qquad \frac{4}{4}$$

Recall that for any whole number a, we have $1 \cdot a = a \cdot 1 = a$. Since any nonzero number divided by itself is 1, we can state the multiplicative identity using fraction notation.

MULTIPLICATIVE IDENTITY FOR FRACTIONS

When we multiply a number by 1, we get the same number:

$$a = a \cdot 1 = a \cdot \frac{n}{n}.$$

For example, $\dfrac{3}{5} = \dfrac{3}{5} \cdot 1 = \dfrac{3}{5} \cdot \dfrac{4}{4} = \dfrac{12}{20}.$

Since $\frac{3}{5} = \frac{12}{20}$, we know that $\frac{3}{5}$ and $\frac{12}{20}$ are two names for the same number. We also say that $\frac{3}{5}$ and $\frac{12}{20}$ are **equivalent fractions**.

$$\frac{3}{5} = \frac{12}{20}$$

Do Margin Exercises 1–4. ▶

Suppose we want to rename $\frac{2}{3}$, using a denominator of 15. We can multiply by 1 to find a number equivalent to $\frac{2}{3}$. We choose $\frac{5}{5}$ for 1 because $15 \div 3 = 5$.

$$\frac{2}{3} = \frac{2}{3} \cdot \frac{5}{5} = \frac{2 \cdot 5}{3 \cdot 5} = \frac{10}{15}.$$

EXAMPLE 1 Find a number equivalent to $\frac{1}{4}$ with a denominator of 24.

Since $24 \div 4 = 6$, we multiply by 1, using $\frac{6}{6}$:

$$\frac{1}{4} = \frac{1}{4} \cdot \frac{6}{6} = \frac{1 \cdot 6}{4 \cdot 6} = \frac{6}{24}.$$

OBJECTIVES

a Multiply by 1 to find an equivalent expression using a different denominator.

b Simplify fraction notation.

c Test to determine whether two fractions are equivalent.

SKILL TO REVIEW

Objective 3.2c: Find the prime factorization of a composite number.

Find the prime factorization.
1. 84
2. 2250

Multiply.

1. $\dfrac{1}{2} \cdot \dfrac{8}{8}$ 2. $\dfrac{3}{7} \cdot \dfrac{a}{a}$

3. $-\dfrac{8}{25} \cdot \dfrac{4}{4}$ 4. $\dfrac{8}{3}\left(\dfrac{-2}{-2}\right)$

Answers

Skill to Review:
1. $2 \cdot 2 \cdot 3 \cdot 7$ 2. $2 \cdot 3 \cdot 3 \cdot 5 \cdot 5 \cdot 5$

Margin Exercises:
1. $\dfrac{8}{16}$ 2. $\dfrac{3a}{7a}$ 3. $-\dfrac{32}{100}$ 4. $\dfrac{-16}{-6}$

Find an equivalent expression for each number, using the denominator indicated. Use multiplication by 1.

5. $\dfrac{4}{3} = \dfrac{?}{15}$

$$\dfrac{4}{3} = \dfrac{4}{3} \cdot \dfrac{\square}{\square}$$

$$= \dfrac{4 \cdot 5}{3 \cdot 5}$$

$$= \dfrac{\square}{15}$$

6. $\dfrac{3}{4} = \dfrac{?}{-24}$ **7.** $\dfrac{9}{10} = \dfrac{?}{10x}$

8. $\dfrac{15}{1} = \dfrac{?}{2}$ **9.** $\dfrac{-8}{7} = \dfrac{?}{49}$

10. $\dfrac{1}{-2} = \dfrac{?}{-6}$

EXAMPLE 2 Find a number equivalent to $\frac{2}{5}$ with a denominator of -35.

Since $-35 \div 5 = -7$, we multiply by 1, using $\frac{-7}{-7}$:

$$\dfrac{2}{5} = \dfrac{2}{5}\left(\dfrac{-7}{-7}\right) = \dfrac{2(-7)}{5(-7)} = \dfrac{-14}{-35}.$$

EXAMPLE 3 Find an expression equivalent to $\frac{9}{8}$ with a denominator of $8a$.

Since $8a \div 8 = a$, we multiply by 1, using $\dfrac{a}{a}$:

$$\dfrac{9}{8} \cdot \dfrac{a}{a} = \dfrac{9a}{8a}.$$

◀ **Do Exercises 5–10.**

b SIMPLIFYING FRACTION NOTATION

All of the following are names for three-fourths:

$$\dfrac{3}{4}, \quad \dfrac{-3}{-4}, \quad \dfrac{6}{8}, \quad \dfrac{30}{40}, \quad \dfrac{-15}{-20}.$$

We say that $\frac{3}{4}$ is **simplest** because it has the smallest positive denominator. Note that 3 and 4 have no factor in common other than 1.

To simplify fraction notation, we reverse the process of multiplying by 1:

$$\dfrac{12}{18} = \dfrac{2 \cdot 6}{3 \cdot 6} \begin{matrix}\leftarrow \text{Factoring the numerator} \\ \leftarrow \text{Factoring the denominator}\end{matrix}$$

$$= \dfrac{2}{3} \cdot \dfrac{6}{6} \quad \text{Factoring the fraction}$$

$$= \dfrac{2}{3} \cdot 1 \quad \dfrac{6}{6} = 1$$

$$= \dfrac{2}{3} \quad \text{Removing the factor 1: } \dfrac{2}{3} \cdot 1 = \dfrac{2}{3}$$

> ### SIMPLIFYING FRACTION NOTATION
>
> 1. Factor the numerator and factor the denominator.
> 2. Identify any common factors in the numerator and denominator.
> 3. Use the common factors to remove a factor equal to 1.

EXAMPLES Simplify.

4. $\dfrac{-8}{20} = \dfrac{-2 \cdot 4}{5 \cdot 4} = \dfrac{-2}{5} \cdot \dfrac{4}{4} = \dfrac{-2}{5}$ Removing a factor equal to 1: $\dfrac{4}{4} = 1$

5. $\dfrac{2}{6} = \dfrac{1 \cdot 2}{3 \cdot 2} = \dfrac{1}{3} \cdot \dfrac{2}{2} = \dfrac{1}{3}$

Writing 1 allows for pairing of factors in the numerator and the denominator.

6. $\dfrac{30}{6} = \dfrac{5 \cdot 6}{1 \cdot 6} = \dfrac{5}{1} \cdot \dfrac{6}{6} = \dfrac{5}{1} = 5$ ◀

We could also simplify $\frac{30}{6}$ by doing the division $30 \div 6$. That is, $\frac{30}{6} = 30 \div 6 = 5$.

7. $-\dfrac{15}{10} = -\dfrac{3 \cdot 5}{2 \cdot 5} = -\dfrac{3}{2} \cdot \dfrac{5}{5} = -\dfrac{3}{2}$ Removing a factor equal to 1: $\dfrac{5}{5} = 1$

Answers

5. $\dfrac{20}{15}$ **6.** $\dfrac{-18}{-24}$ **7.** $\dfrac{9x}{10x}$

8. $\dfrac{30}{2}$ **9.** $\dfrac{-56}{49}$ **10.** $\dfrac{3}{-6}$

Guided Solution:

5. $\dfrac{5}{5}$, 20

8. $\dfrac{4x}{15x} = \dfrac{4 \cdot x}{15 \cdot x} = \dfrac{4}{15} \cdot \dfrac{x}{x} = \dfrac{4}{15}$ Removing a factor equal to 1: $\dfrac{x}{x} = 1$

Do Exercises 11–16. ▶

The use of prime factorizations can be helpful for simplifying.

EXAMPLE 9 Simplify: $\dfrac{90}{84}$.

$$\dfrac{90}{84} = \dfrac{2 \cdot 3 \cdot 3 \cdot 5}{2 \cdot 2 \cdot 3 \cdot 7} \quad \text{Factoring the numerator and the denominator into primes}$$

$$= \dfrac{2 \cdot 3 \cdot 3 \cdot 5}{2 \cdot 3 \cdot 2 \cdot 7} \quad \text{Changing the order so that like primes are above and below each other}$$

$$= \dfrac{2}{2} \cdot \dfrac{3}{3} \cdot \dfrac{3 \cdot 5}{2 \cdot 7} \quad \text{Factoring the fraction}$$

$$= \dfrac{3 \cdot 5}{2 \cdot 7} \quad \text{Removing factors of 1}$$

$$= \dfrac{15}{14}$$

EXAMPLE 10 Simplify: $\dfrac{105}{135}$.

Since both 105 and 135 end in 5, we know that 5 is a factor of both the numerator and the denominator:

$$\dfrac{105}{135} = \dfrac{21 \cdot 5}{27 \cdot 5} = \dfrac{21}{27} \cdot \dfrac{5}{5} = \dfrac{21}{27}.$$ To find the 21, we divided 105 by 5.
To find the 27, we divided 135 by 5.

A fraction is not "simplified" if common factors of the numerator and the denominator remain. Because 21 and 27 are both divisible by 3, we must simplify further:

$$\dfrac{105}{135} = \dfrac{21}{27} = \dfrac{7 \cdot 3}{9 \cdot 3} = \dfrac{7}{9} \cdot \dfrac{3}{3} = \dfrac{7}{9}.$$ To find the 7, we divided 21 by 3.
To find the 9, we divided 27 by 3.

EXAMPLE 11 Simplify: $\dfrac{322}{434}$.

Since 322 and 434 are both even, we know that 2 is a common factor:

$$\dfrac{322}{434} = \dfrac{2 \cdot 161}{2 \cdot 217} = \dfrac{2}{2} \cdot \dfrac{161}{217} = \dfrac{161}{217} \quad \text{Removing a factor equal to 1: } \dfrac{2}{2} = 1$$

$$= \dfrac{7 \cdot 23}{7 \cdot 31} = \dfrac{7}{7} \cdot \dfrac{23}{31} = \dfrac{23}{31}. \quad \text{7 is also a common factor; removing a factor equal to 1}$$

We found the common factor 7 by focusing first on 161. After determining that 7 is a factor of 161, we checked to see if 7 is also a factor of 217.

Do Exercises 17–23. ▶

Simplify.

11. $\dfrac{8}{14}$ **12.** $\dfrac{-10}{12}$

13. $\dfrac{40}{8}$ **14.** $\dfrac{4a}{3a}$

15. $-\dfrac{50}{30}$ **16.** $\dfrac{x}{3x}$

Simplify.

17. $\dfrac{-35}{40}$ **18.** $\dfrac{801}{702}$

19. $\dfrac{24}{21}$ **20.** $\dfrac{429}{561}$

21. $\dfrac{280}{960}$ **22.** $\dfrac{1332}{2880}$

23. Simplify each fraction in this circle graph.

What College Graduates Will Do If They Are Not Offered a Full-Time Job in Their First-Choice Field

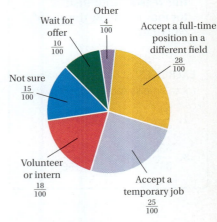

SOURCE: Yahoo Hotjobs survey

Answers

11. $\dfrac{4}{7}$ **12.** $\dfrac{-5}{6}$ **13.** 5 **14.** $\dfrac{4}{3}$ **15.** $-\dfrac{5}{3}$

16. $\dfrac{1}{3}$ **17.** $\dfrac{-7}{8}$ **18.** $\dfrac{89}{78}$ **19.** $\dfrac{8}{7}$ **20.** $\dfrac{13}{17}$

21. $\dfrac{7}{24}$ **22.** $\dfrac{37}{80}$ **23.** $\dfrac{28}{100} = \dfrac{7}{25}$; $\dfrac{25}{100} = \dfrac{1}{4}$;

$\dfrac{18}{100} = \dfrac{9}{50}$; $\dfrac{15}{100} = \dfrac{3}{20}$; $\dfrac{10}{100} = \dfrac{1}{10}$; $\dfrac{4}{100} = \dfrac{1}{25}$

Canceling

Canceling is a shortcut that you may have used for removing a factor that equals 1 when working with fraction notation. With concern, we mention it as a possibility for speeding up your work. Canceling may be done only when removing common factors in numerators and denominators. Each common factor allows us to remove a factor equal to 1 in a product.

Generally, slashes are used to indicate factors equal to 1 that have been removed. For instance, Example 11 might have been done faster as follows:

$$\frac{322}{434} = \frac{2 \cdot 161}{2 \cdot 217} \qquad \text{Factoring the numerator and the denominator}$$

$$= \frac{\cancel{2} \cdot 161}{\cancel{2} \cdot 217} \qquad \begin{array}{l}\text{When a factor equal to 1 is noted,}\\ \text{it is "canceled" as shown: } \frac{2}{2} = 1.\end{array}$$

$$= \frac{161}{217} = \frac{7 \cdot 23}{7 \cdot 31} = \frac{23}{31}.$$

Caution!

The problem with canceling is that it is often applied incorrectly in situations like the following:

$$\frac{\cancel{2} + 3}{\cancel{2}} = 3; \qquad \frac{\cancel{4} + 1}{\cancel{4} + 2} = \frac{1}{2}; \qquad \frac{15}{54} = \frac{1}{4}.$$

$$\quad\text{Wrong!}\qquad\qquad\text{Wrong!}\qquad\qquad\text{Wrong!}$$

The correct answers are

$$\frac{2 + 3}{2} = \frac{5}{2}; \qquad \frac{4 + 1}{4 + 2} = \frac{5}{6}; \qquad \frac{15}{54} = \frac{\cancel{3} \cdot 5}{\cancel{3} \cdot 18} = \frac{5}{18}.$$

In each incorrect case, the numbers canceled did not form a factor equal to 1. Factors are parts of products. For example, in $2 \cdot 3$, the numbers 2 and 3 are factors, but in $2 + 3$, 2 and 3 are terms, not factors.

- **If you cannot factor, do not cancel! If in doubt, do not cancel!**
- **Only factors can be canceled, and factors are never separated by + or − signs.**

C | A TEST FOR EQUALITY

When denominators are the same, we say that fractions have a **common denominator**. One way to compare fractions like $\frac{2}{4}$ and $\frac{3}{6}$ is to find a common denominator and compare numerators. We can multiply each fraction by 1, using the other denominator to write 1.

The denominator is 6.

$$\frac{3}{6} = \frac{3}{6} \cdot \frac{4}{4} = \frac{3 \cdot 4}{6 \cdot 4} = \frac{12}{24}$$

$$\frac{2}{4} = \frac{2}{4} \cdot \frac{6}{6} = \frac{2 \cdot 6}{4 \cdot 6} = \frac{12}{24} \quad \text{Both denominators are 24.}$$

The denominator is 4.

Because $\dfrac{12}{24} = \dfrac{12}{24}$ is true, it follows that $\dfrac{3}{6} = \dfrac{2}{4}$ is also true.

The "key" to the above work is that $3 \cdot 4$ and $2 \cdot 6$ are equal. Had these products differed, we would have shown that $\frac{3}{6}$ and $\frac{2}{4}$ were *not* equal.

A TEST FOR EQUALITY

Two fractions are equal if their cross products are equal.

We multiply these two numbers: $3 \cdot 4 = 12$.

We multiply these two numbers: $6 \cdot 2 = 12$.

$$\frac{3}{6} \; \square \; \frac{2}{4}$$

We call $3 \cdot 4$ and $6 \cdot 2$ **cross products**. Since the cross products are the same—that is, $3 \cdot 4 = 6 \cdot 2$—we know that $\frac{3}{6} = \frac{2}{4}$.

In the sentence $a \neq b$, the symbol $\neq$ means "is not equal to."

EXAMPLE 12 Use $=$ or $\neq$ for $\square$ to write a true sentence:

$$\frac{6}{7} \; \square \; \frac{7}{8}.$$

We multiply these two numbers: $6 \cdot 8 = 48$.

We multiply these two numbers: $7 \cdot 7 = 49$.

$$\frac{6}{7} \; \square \; \frac{7}{8}$$

Because $48 \neq 49$, $\frac{6}{7}$ and $\frac{7}{8}$ do not name the same number. Thus, $\frac{6}{7} \neq \frac{7}{8}$. ■

EXAMPLE 13 Use $=$ or $\neq$ for $\square$ to write a true sentence:

$$\frac{6}{10} \; \square \; \frac{3}{5}.$$

We multiply these two numbers: $6 \cdot 5 = 30$.

We multiply these two numbers: $10 \cdot 3 = 30$.

$$\frac{6}{10} \; \square \; \frac{3}{5}$$

Because the cross products are the same, we have $\frac{6}{10} = \frac{3}{5}$. ■

Remembering that $\frac{-a}{b}$, $\frac{a}{-b}$, and $-\frac{a}{b}$ all represent the same number can be helpful when checking for equality.

EXAMPLE 14 Use $=$ or $\neq$ for $\square$ to write a true sentence:

$$\frac{-6}{8} \; \square \; -\frac{9}{12}.$$

We rewrite $-\frac{9}{12}$ as $\frac{-9}{12}$ and then check cross products:

$$-6 \cdot 12 = -72 \qquad\qquad 8(-9) = -72$$

$$\frac{-6}{8} \; \square \; \frac{-9}{12}$$

Because the cross products are the same, we have $\frac{-6}{8} = -\frac{9}{12}$.

Do Exercises 24–26. ▶

Use $=$ or $\neq$ for $\square$ to write a true sentence.

24. $\dfrac{2}{6} \; \square \; \dfrac{3}{9}$

GS **25.** $\dfrac{2}{3} \; \square \; \dfrac{14}{20}$

$$2 \cdot \boxed{} = 40 \qquad 3 \cdot \boxed{} = 42$$

$$\frac{2}{3} \; \square \; \frac{14}{20}$$

Since $40 \neq 42$, $\dfrac{2}{3} \; \square \; \dfrac{14}{20}$.

26. $-\dfrac{10}{15} \; \square \; \dfrac{8}{-12}$

Answers

24. $=$ **25.** $\neq$ **26.** $=$

Guided Solution:
25. 20, 14; $\neq$

✓ Reading Check

Complete each statement with the appropriate the word from the following list.

common   cross   equivalent   simplify

RC1. _____ fractions name the same number.

RC2. To _____ a fraction, we find a fraction that names the same number and that has a numerator and a denominator with no common factor.

RC3. The fractions $\frac{2}{7}$ and $\frac{4}{7}$ have a _____ denominator.

RC4. Two fractions are equal if their _____ products are equal.

a   Find an equivalent expression for each number, using the denominator indicated. Use multiplication by 1.

1. $\frac{1}{2} = \frac{?}{10}$

2. $\frac{1}{6} = \frac{?}{12}$

3. $\frac{3}{4} = \frac{?}{-48}$

4. $\frac{2}{9} = \frac{?}{-18}$

5. $\frac{7}{10} = \frac{?}{50}$

6. $\frac{3}{8} = \frac{?}{48}$

7. $\frac{11}{5} = \frac{?}{5t}$

8. $\frac{5}{3} = \frac{?}{3a}$

9. $\frac{5}{1} = \frac{?}{4}$

10. $\frac{7}{1} = \frac{?}{5}$

11. $-\frac{17}{18} = -\frac{?}{54}$

12. $-\frac{11}{16} = -\frac{?}{256}$

13. $\frac{3}{-8} = \frac{?}{-40}$

14. $\frac{7}{-8} = \frac{?}{-32}$

15. $\frac{-7}{22} = \frac{?}{132}$

16. $\frac{-10}{21} = \frac{?}{126}$

17. $\frac{1}{8} = \frac{?}{8x}$

18. $\frac{1}{3} = \frac{?}{3a}$

19. $\frac{-10}{7} = \frac{?}{7a}$

20. $\frac{-4}{3} = \frac{?}{3n}$

21. $\frac{4}{9} = \frac{?}{9ab}$

22. $\frac{8}{11} = \frac{?}{11xy}$

23. $\frac{4}{9} = \frac{?}{27b}$

24. $\frac{8}{11} = \frac{?}{55y}$

b   Simplify.

25. $\frac{2}{4}$

26. $\frac{3}{6}$

27. $-\frac{6}{9}$

28. $\frac{-9}{12}$

29. $\frac{10}{25}$

30. $\frac{8}{10}$

31. $\frac{24}{8}$

32. $\frac{36}{9}$

33. $\frac{27}{36}$

34. $\frac{30}{40}$

35. $-\frac{24}{14}$

36. $-\frac{16}{10}$

37. $\frac{3n}{4n}$

38. $\frac{7x}{8x}$

39. $\frac{-17}{51}$

40. $-\frac{13}{26}$

41. $\frac{-100}{20}$

42. $\frac{-150}{25}$

43. $\dfrac{420}{480}$ **44.** $\dfrac{180}{240}$ **45.** $\dfrac{-540}{810}$ **46.** $\dfrac{-1000}{1080}$ **47.** $\dfrac{12x}{30x}$ **48.** $\dfrac{54n}{90n}$

49. $\dfrac{153}{136}$ **50.** $\dfrac{117}{91}$ **51.** $\dfrac{132}{143}$ **52.** $\dfrac{91}{259}$ **53.** $\dfrac{221}{247}$ **54.** $\dfrac{299}{403}$

55. $\dfrac{3ab}{8ab}$ **56.** $\dfrac{6xy}{7xy}$ **57.** $\dfrac{9xy}{6x}$ **58.** $\dfrac{10ab}{15a}$ **59.** $\dfrac{-18a}{20ab}$ **60.** $\dfrac{-19x}{38xy}$

c   Use = or ≠ for ☐ to write a true sentence.

61. $\dfrac{3}{4}$ ☐ $\dfrac{9}{12}$ **62.** $\dfrac{4}{8}$ ☐ $\dfrac{3}{6}$ **63.** $\dfrac{1}{5}$ ☐ $\dfrac{2}{9}$ **64.** $\dfrac{1}{4}$ ☐ $\dfrac{2}{9}$

65. $\dfrac{3}{8}$ ☐ $\dfrac{6}{16}$ **66.** $\dfrac{2}{6}$ ☐ $\dfrac{6}{18}$ **67.** $\dfrac{2}{5}$ ☐ $\dfrac{2}{7}$ **68.** $\dfrac{3}{10}$ ☐ $\dfrac{3}{11}$

69. $\dfrac{-3}{10}$ ☐ $\dfrac{-4}{12}$ **70.** $\dfrac{-2}{9}$ ☐ $\dfrac{-8}{36}$ **71.** $-\dfrac{12}{9}$ ☐ $\dfrac{-8}{6}$ **72.** $\dfrac{-8}{7}$ ☐ $-\dfrac{16}{14}$

73. $\dfrac{5}{-2}$ ☐ $-\dfrac{17}{7}$ **74.** $-\dfrac{10}{3}$ ☐ $\dfrac{24}{-7}$ **75.** $\dfrac{305}{145}$ ☐ $\dfrac{122}{58}$ **76.** $\dfrac{425}{165}$ ☐ $\dfrac{130}{66}$

Skill Maintenance

Use < or > for ☐ to write a true sentence.   [1.6c]

77. 0 ☐ 23 **78.** 34 ☐ 43 **79.** 124 ☐ 98 **80.** 999 ☐ 1001

Solve.   [1.7b]

81. $5280 = 1760 + t$ **82.** $10{,}947 = 123 \cdot y$ **83.** $8797 = y + 2299$ **84.** $x \cdot 74 = 6290$

Synthesis

Simplify. Use a list of prime numbers (see p. 160).

85. $\dfrac{391}{667}$ **86.** $\dfrac{209ab}{247ac}$ **87.** $-\dfrac{1073x}{555y}$ **88.** ▦ $\dfrac{3473}{3197}$

89. Sociologists have found that 4 of 10 people are shy. Write fraction notation for **(a)** the part of the population that is shy and **(b)** the part that is not shy. Simplify.

90. Sociologists estimate that 3 of 20 people are left-handed. In a crowd of 460 people, how many would you expect to be left-handed?

91. ▦ *Batting Averages.* For the 2013 season, Michael Cuddyer of the Colorado Rockies won the National League batting title with 162 hits in 489 times at bat. Miguel Cabrera of the Detroit Tigers won the American League title with 193 hits in 555 times at bat. Did they have the same fraction of hits in times at bat (batting average)? Why or why not?

Source: Major League Baseball

92. ▦ On a test with 82 questions, Taylor got 63 correct. On another test with 100 questions, she got 77 correct. Did she get the same portion of each test correct? Why or why not?

Concept Reinforcement

Determine whether each statement is true or false.

_____ **1.** A number a is divisible by another number b if b is factor of a. [3.2a]

_____ **2.** If a number is not divisible by 6, then it is not divisible by 3. [3.1b]

_____ **3.** The fraction $\dfrac{9}{4}$ is equal to the fraction $\dfrac{13}{6}$. [3.5c]

_____ **4.** The number 1 is not prime. [3.2b]

Guided Solutions

 Fill in each blank with the number that creates a correct statement or solution.

5. $\dfrac{25}{\boxed{}} = 1$ [3.3b]

6. $\dfrac{\boxed{}}{9} = 0$ [3.3b]

7. $\dfrac{8}{\boxed{}} = 8$ [3.3b]

8. $\dfrac{6}{13} = \dfrac{\boxed{}}{39}$ [3.5a]

9. Simplify: $\dfrac{70}{225}$. [3.5b]

$$\dfrac{70}{225} = \dfrac{2 \cdot \boxed{} \cdot 7}{\boxed{} \cdot 3 \cdot 5 \cdot \boxed{}}$$
Factoring the numerator
Factoring the denominator

$$= \dfrac{5}{5} \cdot \dfrac{\boxed{} \cdot 7}{3 \cdot \boxed{} \cdot 5}$$
Factoring the fraction

$$= \boxed{} \cdot \dfrac{\boxed{}}{45}$$
$\dfrac{5}{5} = 1$

$$= \dfrac{\boxed{}}{\boxed{}}$$
Removing the factor 1

Mixed Review

To answer Exercises 10–14, consider the following numbers. [3.1b]

84	132	594	350
300	500	120	14,850
17,576	180	1125	504
224	351	495	1632

10. Which of the above are divisible by 2 but not by 10?

11. Which of the above are divisible by 4 but not by 8?

12. Which of the above are divisible by 4 but not by 6?

13. Which of the above are divisible by 3 but not by 9?

14. Which of the above are divisible by 4, 5, and 6?

Determine whether each number is prime, composite, or neither. [3.2b]

15. 61

16. 2

17. 91

18. 1

Find all the factors of each composite number. Then find the prime factorization of the number. [3.2a], [3.2c]

19. 160

20. 222

21. 98

22. 315

What part of each object or set of objects is shaded? [3.3a]

23.

24.

Multiply. [3.4a], [3.4b]

25. $7 \cdot \dfrac{1}{9}$

26. $\dfrac{4}{15} \cdot \dfrac{2}{3}$

27. $\dfrac{5}{11}(-8)$

Simplify. [3.3b], [3.5b]

28. $\dfrac{24}{60}$

29. $\dfrac{220n}{60n}$

30. $\dfrac{17x}{17x}$

31. $\dfrac{0}{-23}$

32. $\dfrac{54}{186}$

33. $\dfrac{-36}{20}$

34. $\dfrac{75}{630}$

35. $\dfrac{315}{435}$

36. $\dfrac{14}{0}$

Use $=$ or $\neq$ for ☐ to write a true sentence. [3.5c]

37. $\dfrac{3}{-7}$ ☐ $\dfrac{-39}{91}$

38. $\dfrac{19}{3}$ ☐ $\dfrac{95}{18}$

39. *Job Applications.* Of every 200 online job applications started, only 25 are reviewed by a hiring manager. What is the ratio of applications reviewed to applications started? [3.3a], [3.5b]

Source: Talent Function Group LLC, in "Your Résumé vs. Oblivion," wsj.com, 1/24/12

40. *Area of an Ice-Skating Rink.* The length of a rectangular ice-skating rink in the atrium of a shopping mall is $\frac{7}{100}$ mi. The width is $\frac{3}{100}$ mi. What is the area of the rink? [3.4c]

Understanding Through Discussion and Writing

41. Explain a method for finding a composite number that contains exactly two factors other than itself and 1. [3.2b]

42. Which of the years from 2000 to 2020, if any, also happen to be prime numbers? Explain at least two ways in which you might go about solving this problem. [3.2b]

43. Explain in your own words when it *is* possible to "cancel" and when it *is not* possible to "cancel." [3.5b]

44. Can fraction notation be simplified if the numerator and the denominator are two different prime numbers? Why or why not? [3.5b]

3.6   Multiplying, Simplifying, and More with Area

OBJECTIVES

a Multiply and simplify using fraction notation.

b Solve applied problems involving multiplication of fractions.

SKILL TO REVIEW

Objective 3.1b: Test to see if a number is divisible by 2, 3, 5, 6, 9, or 10.

Determine whether each number is divisible by 9.

1. 486   **2.** 129

a   SIMPLIFYING WHEN MULTIPLYING

It is often possible to simplify after we multiply. To make such simplifying easier, it is usually best not to carry out the products in the numerator and the denominator immediately, but to factor and simplify first. Consider the product

$$\frac{5}{6} \cdot \frac{14}{15}.$$

We proceed as follows:

$$\frac{5}{6} \cdot \frac{14}{15} = \frac{5 \cdot 14}{6 \cdot 15}$$ We write the products in the numerator and the denominator, but we do not yet carry out the multiplication.

$$= \frac{5 \cdot 2 \cdot 7}{2 \cdot 3 \cdot 5 \cdot 3}$$ Factoring and identifying common factors

$$= \frac{5 \cdot 2}{5 \cdot 2} \cdot \frac{7}{3 \cdot 3}$$ Factoring the fraction

$$= 1 \cdot \frac{7}{3 \cdot 3}$$

$$= \frac{7}{3 \cdot 3}$$ Removing a factor equal to 1: $\frac{5 \cdot 2}{5 \cdot 2} = 1$

$$= \frac{7}{9}.$$

> To multiply and simplify:
>
> **a)** Write the products in the numerator and the denominator, but do not carry out the multiplication.
> **b)** Factor the numerator and the denominator.
> **c)** Factor the fraction to remove a factor equal to 1, if possible.
> **d)** Carry out the remaining products.

EXAMPLES   Multiply and simplify.

1. $\dfrac{2}{3} \cdot \dfrac{9}{4} = \dfrac{2 \cdot 9}{3 \cdot 4} = \dfrac{2 \cdot 3 \cdot 3}{3 \cdot 2 \cdot 2} = \dfrac{2 \cdot 3}{2 \cdot 3} \cdot \dfrac{3}{2} = 1 \cdot \dfrac{3}{2} = \dfrac{3}{2}$

Answers

Skill to Review:
1. Yes **2.** No

2. $\dfrac{6}{7} \cdot \dfrac{-5}{3} = \dfrac{3 \cdot 2 \cdot (-5)}{7 \cdot 3}$ Note that 3 is a common factor of 6 and 3.

$\qquad = \dfrac{3}{3} \cdot \dfrac{2(-5)}{7} = \dfrac{-10}{7}$, or $-\dfrac{10}{7}$ Removing a factor equal to 1: $\dfrac{3}{3} = 1$

3. $\dfrac{10}{21} \cdot \dfrac{14a}{15} = \dfrac{5 \cdot 2 \cdot 7 \cdot 2 \cdot a}{7 \cdot 3 \cdot 5 \cdot 3}$ Note that 5 is a common factor of 10 and 15.
Note that 7 is a common factor of 21 and 14a.

$\qquad = \dfrac{5 \cdot 7}{5 \cdot 7} \cdot \dfrac{2 \cdot 2 \cdot a}{3 \cdot 3}$

$\qquad = \dfrac{4a}{9}$ }   Removing a factor equal to 1: $\dfrac{5 \cdot 7}{5 \cdot 7} = 1$

4. $32 \cdot \dfrac{7}{8} = \dfrac{32}{1} \cdot \dfrac{7}{8} = \dfrac{8 \cdot 4 \cdot 7}{8 \cdot 1}$ Note that 8 is a common factor of 32 and 8.

$\qquad = \dfrac{8}{8} \cdot \dfrac{4 \cdot 7}{1} = 28$ Removing a factor equal to 1: $\dfrac{8}{8} = 1$

·· **Caution!** ··

Canceling can be used as follows for these examples.

1. $\dfrac{2}{3} \cdot \dfrac{9}{4} = \dfrac{2 \cdot \cancel{3} \cdot 3}{\cancel{3} \cdot 2 \cdot 2} = \dfrac{3}{2}$ Removing a factor equal to 1: $\dfrac{2 \cdot 3}{2 \cdot 3} = 1$

2. $\dfrac{6}{7} \cdot \dfrac{-5}{3} = \dfrac{\cancel{3} \cdot 2(-5)}{7 \cdot \cancel{3}} = \dfrac{-10}{7}$ Removing a factor equal to 1: $\dfrac{3}{3} = 1$

3. $\dfrac{10}{21} \cdot \dfrac{14a}{15} = \dfrac{\cancel{5} \cdot 2 \cdot \cancel{7} \cdot 2 \cdot a}{\cancel{7} \cdot 3 \cdot \cancel{5} \cdot 3} = \dfrac{4a}{9}$ Removing a factor equal to 1: $\dfrac{5 \cdot 7}{5 \cdot 7} = 1$

4. $32 \cdot \dfrac{7}{8} = \dfrac{\cancel{8} \cdot 4 \cdot 7}{\cancel{8} \cdot 1} = 28$ Removing a factor equal to 1: $\dfrac{8}{8} = 1$

Remember, only factors can be canceled!

··

Do Exercises 1–4.

Multiply and simplify.

GS 1. $\dfrac{2}{3} \cdot \dfrac{7}{8} = \dfrac{2 \cdot 7}{3 \cdot \square}$

$\qquad = \dfrac{2 \cdot 7}{3 \cdot 2 \cdot 2 \cdot \square}$

$\qquad = \dfrac{2}{\square} \cdot \dfrac{7}{3 \cdot 2 \cdot 2}$

$\qquad = \dfrac{\square}{} \cdot \dfrac{7}{3 \cdot 2 \cdot 2}$

$\qquad = \dfrac{7}{\square}$

2. $\dfrac{4}{5} \cdot \dfrac{-5}{12}$ **3.** $16 \cdot \dfrac{3}{8}$

4. $\dfrac{5}{2x} \cdot 4$

b   SOLVING PROBLEMS

EXAMPLE 5 *Landscaping.* Celina's Landscaping uses $\frac{2}{3}$ lb of peat moss when planting a rosebush. How much will be needed to plant 21 rosebushes?

1. Familiarize. We let $n =$ the number of pounds of peat moss needed. Each rosebush requires $\frac{2}{3}$ lb of peat moss, so repeated addition, or multiplication, applies.

2. Translate. The problem translates to the following equation:

$$n = 21 \cdot \dfrac{2}{3}.$$

3. Solve. To solve the equation, we carry out the multiplication:

$$n = 21 \cdot \dfrac{2}{3} = \dfrac{21}{1} \cdot \dfrac{2}{3} = \dfrac{21 \cdot 2}{1 \cdot 3} \qquad \text{Multiplying}$$

$$= \dfrac{3 \cdot 7 \cdot 2}{1 \cdot 3} = \dfrac{3}{3} \cdot \dfrac{7 \cdot 2}{1} = 14. \qquad \text{Removing the factor } \tfrac{3}{3} \text{ and simplifying}$$

Answers

1. $\dfrac{7}{12}$ **2.** $-\dfrac{1}{3}$ **3.** 6   **4.** $\dfrac{10}{x}$

Guided Solution:
1. 8, 2, 2, 1, 12

5. *Candy.* Chocolate Delight sells $\frac{4}{5}$-lb boxes of truffles. How many pounds of truffles will be needed to fill 85 boxes?

85 boxes

$\frac{4}{5}$ pound of truffles in each box

4. Check. We check by repeating the calculation. (This is left to the student.) We can also ask if the answer seems reasonable. We are putting less than a pound of peat moss on each bush, so the answer should be less than 21. Since 14 is less than 21, we have a partial check.

A second partial check can be performed using the units:

$$21 \text{ bushes} \cdot \frac{2}{3} \text{ pounds per bush}$$

$$= 21 \cdot \frac{2}{3} \cdot \text{bushes} \cdot \frac{\text{pounds}}{\text{bush}}$$

$$= 14 \text{ pounds}.$$

Since the resulting unit is pounds, we have another partial check.

5. State. Celina's Landscaping will need 14 lb of peat moss to plant 21 rosebushes.

Do Exercise 5.

Area

We multiply to find the area of a triangle. Consider a triangle with a base of length b and a height of h. A rectangle can be formed by splitting and inverting a copy of this triangle.

The rectangle's area, $b \cdot h$, is exactly twice the area of the triangle. We have the following result.

AREA OF A TRIANGLE

The **area *A* of a triangle** is half the length of the base b times the height h:

$$A = \frac{1}{2} \cdot b \cdot h.$$

EXAMPLE 6 Find the area of this triangle.

$$A = \frac{1}{2} \cdot b \cdot h$$

$$= \frac{1}{2} \cdot 9 \text{ yd} \cdot 6 \text{ yd}$$

$$= \frac{9 \cdot 6}{2} \text{yd}^2$$

$$= 27 \text{ yd}^2, \text{ or } 27 \text{ square yards}$$

.... **Caution!** ....

Use square units for units of area.

....

Answer

5. 68 lb

EXAMPLE 7   Find the area of this triangle.

$$A = \frac{1}{2} \cdot b \cdot h$$

$$= \frac{1}{2} \cdot \frac{10}{3} \text{cm} \cdot 4 \text{ cm}$$

$$= \frac{1 \cdot 10 \cdot 4}{2 \cdot 3} \text{cm}^2$$

$$= \frac{1 \cdot 2 \cdot 5 \cdot 4}{2 \cdot 3} \text{cm}^2 \qquad \text{Removing a factor equal to 1: } \frac{2}{2} = 1$$

$$= \frac{20}{3} \text{cm}^2$$

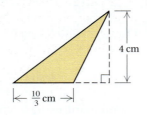

4 cm

$\frac{10}{3}$ cm

Do Exercises 6 and 7.

EXAMPLE 8   Find the area of this kite.

8 in.   8 in.   8 in.   8 in.

27 in.   27 in.

1. **Familiarize.**   We look for figures with areas we can calculate using area formulas that we already know. We let $K =$ the kite's area.

2. **Translate.**   The kite consists of two triangles, each with a base of 27 in. and a height of 8 in. We can apply the formula $A = \frac{1}{2} \cdot b \cdot h$ for the area of a triangle and then multiply by 2.

Rephrase:	Kite's area	is	twice	Area of long triangle
Translate:	K	$=$	2	$\cdot \quad \frac{1}{2}(27 \text{ in.}) \cdot (8 \text{ in.})$

3. **Solve.**   We have

$$K = 2 \cdot \frac{1}{2} \cdot (27 \text{ in.}) \cdot (8 \text{ in.})$$

$$= 1 \cdot 27 \text{ in.} \cdot 8 \text{ in.} = 216 \text{ in}^2.$$

4. **Check.**   We can check by repeating the calculations. The unit, in², is appropriate for area.

5. **State.**   The area of the kite is 216 in².

Do Exercise 8.

Find the area.

6.

12 m

16 m

GS 7.

$\frac{12}{5}$ cm

11 cm

$$A = \frac{1}{2} \cdot b \cdot h$$

$$= \frac{1}{2} \cdot 11 \text{ cm} \cdot \square \text{ cm}$$

$$= \frac{1 \cdot 11 \cdot \square}{2 \cdot 5} \text{cm}^2$$

$$= \frac{1 \cdot 11 \cdot 2 \cdot 2 \cdot \square}{2 \cdot 5} \text{cm}^2$$

$$= \frac{\square}{5} \text{cm}^2$$

8. Find the area. (*Hint*: The figure is made up of a rectangle and a triangle.)

8 in.

10 in.   15 in.

Answers

6. 96 m²   **7.** $\frac{66}{5}$ cm²

8. Rectangle: $(10 \text{ in.}) \cdot (8 \text{ in.}) = 80 \text{ in}^2$;

triangle: $\frac{1}{2}(8 \text{ in.}) \cdot (5 \text{ in.}) = 20 \text{ in}^2$;

$80 \text{ in}^2 + 20 \text{ in}^2 = 100 \text{ in}^2$

Guided Solution:

7. $\frac{12}{5}$, 12, 3, 66

✓ Reading Check

Complete each step in the process for multiplying and simplifying using fraction notation.

RC1. **a)** Write the _____ in the numerator and the denominator, but do not carry out the multiplication.

RC2. **b)** _____ the numerator and the denominator.

RC3. **c)** Factor the fraction to remove a factor equal to _____, if possible.

RC4. **d)** _____ the remaining products.

a Multiply. Don't forget to simplify, if possible.

1. $\dfrac{2}{3} \cdot \dfrac{1}{2}$

2. $\dfrac{4}{5} \cdot \dfrac{1}{4}$

3. $\dfrac{7}{8} \cdot \dfrac{-1}{7}$

4. $\dfrac{5}{6} \cdot \dfrac{-1}{5}$

5. $\dfrac{2}{3} \cdot \dfrac{6}{7}$

6. $\dfrac{2}{5} \cdot \dfrac{3}{10}$

7. $\dfrac{2}{9} \cdot \dfrac{3}{10}$

8. $\dfrac{3}{5} \cdot \dfrac{10}{9}$

9. $\dfrac{9}{-5} \cdot \dfrac{12}{8}$

10. $\dfrac{16}{-15} \cdot \dfrac{5}{4}$

11. $\dfrac{5x}{9} \cdot \dfrac{4}{5}$

12. $\dfrac{25}{4a} \cdot \dfrac{4}{3}$

13. $9 \cdot \dfrac{1}{9}$

14. $4 \cdot \dfrac{1}{4}$

15. $\dfrac{7}{10} \cdot \dfrac{10}{7}$

16. $\dfrac{8}{9} \cdot \dfrac{9}{8}$

17. $\dfrac{1}{4} \cdot 12$

18. $\dfrac{1}{6} \cdot 12$

19. $21 \cdot \dfrac{1}{3}$

20. $18 \cdot \dfrac{1}{2}$

21. $-16\left(-\dfrac{3}{4}\right)$

22. $-24\left(-\dfrac{5}{6}\right)$

23. $\dfrac{3}{8} \cdot 8a$

24. $\dfrac{2}{9} \cdot 9x$

25. $\left(-\dfrac{3}{8}\right)\left(-\dfrac{8}{3}\right)$

26. $\left(-\dfrac{7}{9}\right)\left(-\dfrac{9}{7}\right)$

27. $\dfrac{a}{b} \cdot \dfrac{b}{a}$

28. $\dfrac{n}{m} \cdot \dfrac{m}{n}$

29. $\dfrac{4}{10} \cdot \dfrac{5}{10}$

30. $\dfrac{11}{24} \cdot \dfrac{3}{5}$

31. $\dfrac{8}{10} \cdot \dfrac{45}{100}$

32. $\dfrac{3}{10} \cdot \dfrac{8}{10}$

33. $\frac{1}{6} \cdot 360n$

34. $\frac{1}{3} \cdot 12y$

35. $20\left(\dfrac{1}{-6}\right)$

36. $35\left(\dfrac{1}{-10}\right)$

37. $-8x \cdot \dfrac{1}{-8x}$

38. $-5a \cdot \dfrac{1}{-5a}$

39. $\dfrac{2x}{9} \cdot \dfrac{27}{2x}$

40. $\dfrac{10a}{3} \cdot \dfrac{3}{5a}$

41. $\dfrac{7}{10} \cdot \dfrac{34}{150}$

42. $\dfrac{15}{22} \cdot \dfrac{4}{7}$

43. $\dfrac{36}{85} \cdot \dfrac{25}{-99}$

44. $\dfrac{-70}{45} \cdot \dfrac{50}{49}$

45. $\dfrac{-98}{99} \cdot \dfrac{27a}{175a}$

46. $\dfrac{70}{-49} \cdot \dfrac{63}{300x}$

47. $\dfrac{110}{33} \cdot \dfrac{-24}{25x}$

48. $\dfrac{-19}{130} \cdot \dfrac{65}{38x}$

49. $\left(-\dfrac{11}{24}\right)\dfrac{3}{5}$

50. $\left(-\dfrac{15}{22}\right)\dfrac{4}{7}$

51. $\dfrac{10a}{21} \cdot \dfrac{3}{8b}$

52. $\dfrac{17}{21y} \cdot \dfrac{3x}{5}$

 Solve.

The *pitch* of a screw is the distance between its threads. With each complete rotation, the screw goes in or out a distance equal to its pitch. Use this information to answer Exercises 53 and 54.

⟩Pitch = p in.

Each rotation moves the screw in or out p in.

53. The pitch of a screw is $\frac{1}{16}$ in. How far will it go into a piece of oak when it is turned 10 complete rotations clockwise?

54. The pitch of a screw is $\frac{3}{32}$ in. How far will it come out of a piece of plywood when it is turned 10 complete rotations counterclockwise?

55. World Silver Supply. The total world supply of silver for new uses in 2011 was about 1040 million ounces. Of this, approximately $\frac{1}{4}$ was scrap silver being reused or repurposed. How many ounces of silver for new uses was supplied as scrap silver?

Source: Based on information from the *World Silver Survey 2012*

56. Substitute Teaching. After Jack completes 60 hr of teacher training in college, he can earn $75 for working a full day as a substitute teacher. How much will he receive for working $\frac{3}{5}$ of a day?

57. Mailing-List Changes. The United States Postal Service estimates that $\frac{4}{25}$ of the addresses on a mailing list will change in one year. A business has a mailing list of 3000 people. After one year, how many addresses on that list will be incorrect?

Source: Based on information from usps.com

58. Shy People. Sociologists have determined that $\frac{2}{5}$ of the people in the world are shy. A sales manager has 650 applicants for a sales position that requires an outgoing personality. How many of these people might be shy?

59. A recipe for piecrust calls for $\frac{2}{3}$ cup of flour. A chef is making $\frac{1}{2}$ of the recipe. How much flour should the chef use?

60. Of the students in the freshman class, $\frac{4}{5}$ have digital cameras; $\frac{1}{4}$ of these students also join the college photography club. What fraction of the students in the freshman class join the photography club?

61. Assessed Value. A house worth $154,000 is assessed for $\frac{3}{4}$ of its value. What is the assessed value of the house?

62. Roxanne's tuition was $4600. A loan was obtained for $\frac{3}{4}$ of the tuition. How much was the loan?

63. Map Scaling. On a map, 1 in. represents 240 mi. How much does $\frac{2}{3}$ in. represent?

64. Map Scaling. On a map, 1 in. represents 120 mi. How much does $\frac{3}{4}$ in. represent?

65. *Household Budgets.* A family has an annual income of $42,000. Of this, $\frac{1}{5}$ is spent for food, $\frac{1}{4}$ for housing, $\frac{1}{10}$ for clothing, $\frac{1}{14}$ for savings, $\frac{1}{5}$ for taxes, and the rest for other expenses. How much is spent for each?

66. *Household Budgets.* A family has an annual income of $28,140. Of this, $\frac{1}{5}$ is spent for food, $\frac{1}{4}$ for housing, $\frac{1}{10}$ for clothing, $\frac{1}{14}$ for savings, $\frac{1}{5}$ for taxes, and the rest for other expenses. How much is spent for each?

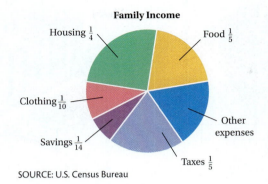

Family Income

Housing $\frac{1}{4}$ Food $\frac{1}{5}$

Clothing $\frac{1}{10}$

Savings $\frac{1}{14}$

Taxes $\frac{1}{5}$

Other expenses

SOURCE: U.S. Census Bureau

Find the area.

67.

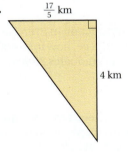

8 in.

15 in.

68.

18 yd

18 yd

69.

$\frac{7}{2}$ mm

5 mm

70.

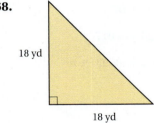

$\frac{17}{5}$ km

4 km

71.

$\frac{7}{2}$ m

$\frac{9}{2}$ m

72.

$\frac{8}{3}$ yd

$\frac{7}{4}$ yd

73.

10 mi

8 mi

13 mi

74.

15 cm

30 cm

30 cm

75. *Quilting.* Michelle is designing a quilt using hexagons. Each hexagon is made using six kite-shaped pieces of cloth, as shown below. Find the area of one of the kite-shaped pieces.

2 in.

$\frac{7}{8}$ in.

76. *Construction.* Find the total area of the sides and ends of the town office building shown. Do not subtract for any windows, doors, or steps.

25 ft

11 ft

75 ft

50 ft

Dumont Town Offices

Skill Maintenance

Solve. [1.7b]

77. $48 \cdot t = 1680$

78. $74 \cdot x = 6290$

79. $3125 = 25 \cdot t$

80. $2880 = 24 \cdot y$

81. $t + 28 = 5017$

82. $456 + x = 9002$

83. $8797 = y + 98$

84. $10{,}000 = 3593 + m$

Synthesis

Simplify. Use the list of prime numbers on p. 160.

85. ▦ $\dfrac{201}{535} \cdot \dfrac{4601}{6499}$

86. ▦ $\dfrac{667}{899} \cdot \dfrac{558}{621}$

87. *College Profile.* Of students entering a college, $\frac{7}{8}$ have completed high school and $\frac{2}{3}$ are older than 20. If $\frac{1}{7}$ of all students are left-handed, what fraction of students entering the college are left-handed high school graduates over the age of 20?

88. *College Profile.* Refer to the information in Exercise 87. If 480 students are entering the college, how many of them are left-handed high school graduates 20 years old or younger?

89. ▦ *Manufacturing.* A candy box is triangular at each end, as shown below. Find the surface area of the box.

30 mm

PRISM CANDY
FRUIT FLAVORS

30 mm

26 mm

140 mm

30 mm

90. ▦ *Painting.* Shoreline Painting needs to determine the surface area of an octagonal steeple. Find the total area, if the dimensions are as shown below.

15 ft

6 ft

4 ft

Reciprocals and Division

a RECIPROCALS

OBJECTIVES

a Find the reciprocal of a number.

b Divide and simplify using fraction notation.

Each of the following products is 1:

$$8 \cdot \frac{1}{8} = \frac{8}{8} = 1; \qquad \frac{-2}{3} \cdot \frac{3}{-2} = \frac{-6}{-6} = 1.$$

RECIPROCALS

If the product of two numbers is 1, we say that they are **reciprocals** of each other.* To find the reciprocal of a fraction, interchange the numerator and the denominator.

$$\text{Number: } \frac{3}{4} \longrightarrow \text{Reciprocal: } \frac{4}{3}$$

SKILL TO REVIEW

Objective 2.5a: Divide integers.

Divide.

1. $\dfrac{26}{-2}$ 2. $\dfrac{-100}{-25}$

EXAMPLES Find the reciprocal.

1. The reciprocal of $\dfrac{4}{5}$ is $\dfrac{5}{4}$. $\qquad \dfrac{4}{5} \cdot \dfrac{5}{4} = \dfrac{20}{20} = 1$

2. The reciprocal of $\dfrac{a}{b}$ is $\dfrac{b}{a}$. $\qquad \dfrac{a}{b} \cdot \dfrac{b}{a} = \dfrac{ab}{ba} = 1$

3. The reciprocal of 8 is $\dfrac{1}{8}$. $\qquad$ Think of 8 as $\dfrac{8}{1}$: $\dfrac{8}{1} \cdot \dfrac{1}{8} = \dfrac{8}{8} = 1.$

4. The reciprocal of $\dfrac{1}{3}$ is 3. $\qquad \dfrac{1}{3} \cdot 3 = \dfrac{3}{3} = 1$

5. The reciprocal of $-\dfrac{5}{9}$ is $-\dfrac{9}{5}$. $\qquad$ Negative numbers have negative reciprocals: $\left(-\dfrac{5}{9}\right)\left(-\dfrac{9}{5}\right) = \dfrac{45}{45} = 1.$

Do Margin Exercises 1–5. ▶

Find the reciprocal.

1. $\dfrac{2}{5}$ 2. $\dfrac{-6}{x}$

3. 9   4. $\dfrac{1}{5}$

5. $-\dfrac{3}{10}$

Does 0 have a reciprocal? If it did, it would have to be a number x such that $0 \cdot x = 1$. But 0 times any number is 0. Thus we have the following.

0 HAS NO RECIPROCAL

The number 0, or $\dfrac{0}{n}$, has no reciprocal. $\left(\text{Recall that } \dfrac{n}{0} \text{ is not defined.}\right)$

b DIVISION

Consider the division $\frac{3}{4} \div \frac{1}{8}$. We are asking how many $\frac{1}{8}$'s are in $\frac{3}{4}$. From the figure at the right, we see that there are six $\frac{1}{8}$'s in $\frac{3}{4}$. Thus,

$$\frac{3}{4} \div \frac{1}{8} = 6.$$

*A reciprocal is also called a *multiplicative inverse*.

Answers

Skill to Review:
1. -13 2. 4

Margin Exercises:
1. $\dfrac{5}{2}$ 2. $\dfrac{x}{-6}$ 3. $\dfrac{1}{9}$ 4. 5   5. $-\dfrac{10}{3}$

We can check this by multiplying:

$$6 \cdot \frac{1}{8} = \frac{6}{1} \cdot \frac{1}{8} = \frac{6}{8} = \frac{2 \cdot 3}{2 \cdot 4} = \frac{2}{2} \cdot \frac{3}{4} = \frac{3}{4}.$$

Here is a faster way to do this division:

$$\frac{3}{4} \div \frac{1}{8} = \frac{3}{4} \cdot \frac{8}{1} = \frac{3 \cdot 8}{4 \cdot 1} = \frac{24}{4} = 6. \qquad \text{Multiplying by the reciprocal of the divisor}$$

DIVISION OF FRACTIONS

To divide by a fraction, multiply by its reciprocal:

$$\frac{a}{b} \div \frac{c}{d} = \frac{a}{b} \cdot \frac{d}{c}. \qquad \text{Multiply by the reciprocal of the divisor.}$$

EXAMPLES Divide and simplify.

6. $\dfrac{5}{6} \div \dfrac{2}{3} = \dfrac{5}{6} \cdot \dfrac{3}{2}$ Multiplying by the reciprocal of the divisor

$$= \frac{5 \cdot 3}{3 \cdot 2 \cdot 2} \qquad \text{Factoring and identifying a common factor}$$

$$= \frac{3}{3} \cdot \frac{5}{2 \cdot 2} \qquad \text{Removing a factor equal to 1: } \frac{3}{3} = 1$$

$$= \frac{5}{4}$$

7. $\dfrac{-3}{5} \div \dfrac{1}{2} = \dfrac{-3}{5} \cdot 2$ The reciprocal of $\dfrac{1}{2}$ is 2.

$$= \frac{-3 \cdot 2}{5} = \frac{-6}{5}, \text{ or } -\frac{6}{5}$$

8. $\dfrac{2a}{5} \div 7 = \dfrac{2a}{5} \cdot \dfrac{1}{7}$ The reciprocal of 7 is $\dfrac{1}{7}$.

$$= \frac{2a \cdot 1}{5 \cdot 7} = \frac{2a}{35}$$

9. $\dfrac{\frac{7}{10}}{-\frac{14}{15}} = \dfrac{7}{10} \div \left(-\dfrac{14}{15} \right)$ The fraction bar indicates division.

$$= \frac{7}{10} \cdot \left(-\frac{15}{14} \right) \qquad \text{Multiplying by the reciprocal of the divisor}$$

$$= \frac{7 \cdot 5(-3)}{2 \cdot 5 \cdot 7 \cdot 2} \qquad \text{Factoring and identifying common factors}$$

$$= \frac{7 \cdot 5}{7 \cdot 5} \cdot \frac{-3}{4} \qquad \text{Removing a factor equal to 1: } \frac{7 \cdot 5}{7 \cdot 5} = 1$$

$$= \frac{-3}{4}, \text{ or } -\frac{3}{4}$$

Canceling can be used as follows for Examples 6 and 9.

6. $\dfrac{5}{6} \div \dfrac{2}{3} = \dfrac{5}{6} \cdot \dfrac{3}{2} = \dfrac{5 \cdot 3}{6 \cdot 2} = \dfrac{5 \cdot \cancel{3}}{\cancel{3} \cdot 2 \cdot 2} = \dfrac{5}{2 \cdot 2} = \dfrac{5}{4}$

Removing a factor equal to 1: $\dfrac{3}{3} = 1$

9. $\dfrac{7}{10} \div \left(-\dfrac{14}{15}\right) = \dfrac{7}{10} \cdot \left(-\dfrac{15}{14}\right) = \dfrac{\cancel{7} \cdot \cancel{5}(-3)}{2 \cdot \cancel{5} \cdot \cancel{7} \cdot 2} = \dfrac{-3}{4}$, or $-\dfrac{3}{4}$

Removing a factor equal to 1: $\dfrac{7 \cdot 5}{7 \cdot 5} = 1$

Remember, if you can't factor, you can't cancel!

....

Do Exercises 6–10. ▶

Why do we multiply by a reciprocal when dividing? To see why, consider $\frac{2}{3} \div \frac{7}{5}$. We will multiply by 1 to find an equivalent expression. To write 1, we use $(5/7)/(5/7)$, since $\frac{5}{7}$ is the reciprocal of $\frac{7}{5}$.

$\dfrac{2}{3} \div \dfrac{7}{5} = \dfrac{\frac{2}{3}}{\frac{7}{5}}$ Writing fraction notation for the division

$= \dfrac{\frac{2}{3}}{\frac{7}{5}} \cdot 1$ Multiplying by 1

$= \dfrac{\frac{2}{3}}{\frac{7}{5}} \cdot \dfrac{\frac{5}{7}}{\frac{5}{7}}$ Multiplying by 1; $\frac{5}{7}$ is the reciprocal of $\frac{7}{5}$ and $\frac{\frac{5}{7}}{\frac{5}{7}} = 1$.

$= \dfrac{\frac{2}{3} \cdot \frac{5}{7}}{\frac{7}{5} \cdot \frac{5}{7}}$ Multiplying the numerators and the denominators

$= \dfrac{\frac{2}{3} \cdot \frac{5}{7}}{1} = \dfrac{2}{3} \cdot \dfrac{5}{7} = \dfrac{10}{21}$

After we multiplied, we got 1 for the denominator. The numerator (in color) shows the multiplication by the reciprocal of $\frac{7}{5}$.

Thus,

$\dfrac{2}{3} \div \dfrac{7}{5} = \dfrac{2}{3} \cdot \dfrac{5}{7} = \dfrac{10}{21}.$

Expressions of the form $\dfrac{\frac{a}{b}}{\frac{c}{d}}$ are examples of *complex fractions*:

$\dfrac{\frac{a}{b}}{\frac{c}{d}} = \dfrac{a}{b} \div \dfrac{c}{d}.$

Do Exercise 11. ▶

Divide and simplify.

6. $\dfrac{6}{7} \div \dfrac{3}{4} = \dfrac{6}{7} \cdot \dfrac{\boxed{}}{\boxed{}}$

$= \dfrac{6 \cdot 4}{7 \cdot 3}$

$= \dfrac{2 \cdot 3 \cdot 2 \cdot \boxed{}}{7 \cdot 3}$

$= \dfrac{3}{\boxed{}} \cdot \dfrac{2 \cdot 2 \cdot 2}{7}$

$= \dfrac{2 \cdot 2 \cdot 2}{7}$

$= \dfrac{\boxed{}}{7}$

7. $\left(-\dfrac{2}{3}\right) \div \dfrac{1}{4}$ **8.** $\dfrac{4}{5} \div 8$

9. $60 \div \dfrac{3a}{5}$ **10.** $\dfrac{\frac{-6}{7}}{\frac{3}{5}}$

11. To remember *why* fractions are divided as they are, multiply by 1 to perform the following division, using the reciprocal of $\frac{4}{5}$ to write 1.

$\dfrac{\frac{6}{7}}{\frac{4}{5}}$

Answers

6. $\dfrac{8}{7}$ **7.** $-\dfrac{8}{3}$ **8.** $\dfrac{1}{10}$ **9.** $\dfrac{100}{a}$

10. $\dfrac{-10}{7}$, or $-\dfrac{10}{7}$ **11.** $\dfrac{15}{14}$

Guided Solution:

6. $\dfrac{4}{3}$, 2, 3, 8

✓ Reading Check

Determine whether each statement is true or false.

_____ **RC1.** The numbers $\frac{1}{7}$ and 7 are reciprocals.

_____ **RC2.** The number 1 has no reciprocal.

_____ **RC3.** To divide by a fraction, we multiply by its reciprocal.

_____ **RC4.** We can divide by $\frac{1}{2}$ by multiplying by 2.

a Find the reciprocal.

1. $\frac{7}{3}$

2. $\frac{6}{5}$

3. 9

4. 3

5. $\frac{1}{7}$

6. $\frac{1}{4}$

7. $-\frac{8}{9}$

8. $-\frac{12}{5}$

9. $\frac{a}{c}$

10. $\frac{x}{y}$

11. $\frac{-3n}{m}$

12. $\frac{8t}{-7r}$

13. $\frac{8}{-15}$

14. $\frac{-6}{25}$

15. $7m$

16. $5n$

17. $\frac{1}{4a}$

18. $\frac{1}{9t}$

19. $-\frac{1}{3z}$

20. $-\frac{1}{2x}$

b Divide. Don't forget to simplify when possible. Assume that all variables are nonzero.

21. $\frac{3}{7} \div \frac{3}{4}$

22. $\frac{2}{3} \div \frac{3}{4}$

23. $\frac{3}{5} \div \frac{9}{4}$

24. $\frac{6}{7} \div \frac{3}{5}$

25. $\frac{4}{3} \div \frac{1}{3}$

26. $\frac{10}{9} \div \frac{1}{2}$

27. $\left(-\frac{1}{3}\right) \div \frac{1}{6}$

28. $\left(-\frac{1}{4}\right) \div \frac{1}{5}$

29. $\left(-\frac{10}{21}\right) \div \left(-\frac{2}{15}\right)$

30. $-\frac{15}{28} \div \left(-\frac{9}{20}\right)$

31. $\frac{3}{8} \div 3$

32. $\frac{5}{6} \div 5$

33. $\frac{12}{7} \div 16$

34. $\frac{18}{5} \div 27$

35. $(-12) \div \frac{3}{2}$

36. $(-24) \div \frac{3}{8}$

37. $\dfrac{x}{8} \div \dfrac{1}{4}$ **38.** $\dfrac{3}{4} \div \dfrac{2}{y}$ **39.** $\dfrac{2}{3} \div (6x)$ **40.** $\dfrac{12}{5} \div (4x)$

41. $28 \div \dfrac{4}{5a}$ **42.** $40 \div \dfrac{2}{3m}$ **43.** $\left(-\dfrac{5}{8}\right) \div \left(-\dfrac{5}{8}\right)$ **44.** $\left(-\dfrac{2}{5}\right) \div \left(-\dfrac{2}{5}\right)$

45. $\dfrac{-8}{15} \div \dfrac{4}{5}$ **46.** $\dfrac{6}{-13} \div \dfrac{3}{26}$ **47.** $\dfrac{77}{64} \div \dfrac{49}{18}$ **48.** $\dfrac{81}{42} \div \dfrac{33}{56}$

49. $120a \div \dfrac{45}{14}$ **50.** $360n \div \dfrac{27n}{8}$ **51.** $\dfrac{\frac{2}{5}}{\frac{3}{7}}$ **52.** $\dfrac{\frac{5}{6}}{\frac{2}{7}}$

53. $\dfrac{-\frac{7}{20}}{-\frac{8}{5}}$ **54.** $\dfrac{-\frac{8}{21}}{-\frac{6}{5}}$ **55.** $\dfrac{-\frac{15}{8}}{\frac{9}{10}}$ **56.** $\dfrac{-\frac{27}{10}}{\frac{21}{20}}$

Skill Maintenance

57. Evaluate $-(-x)$ when $x = -9$.   [2.1d]

58. Evaluate $-x^2$ when $x = -9$.   [2.6a]

59. Evaluate $3x^2$ when $x = 5$.   [2.6a]

60. Evaluate $(-x)^2$ when $x = -9$.   [2.6a]

61. Evaluate $\dfrac{-a}{b}$ for $a = 20$ and $b = -10$.   [2.6a]

62. Evaluate $-\dfrac{a}{b}$ for $a = 20$ and $b = -10$.   [2.6a]

Use the distributive law to write an equivalent expression.   [2.6b]

63. $3(2x + 3y - 1)$

64. $-4(a - 3b + 2c)$

Synthesis

Simplify.

65. $\left(\dfrac{4}{15} \div \dfrac{2}{25}\right)^2$ **66.** $\left(\dfrac{9}{10} \div \dfrac{12}{25}\right)^2$ **67.** $\left(\dfrac{9}{10} \div \dfrac{2}{5} \div \dfrac{3}{8}\right)^2$

68. $\dfrac{\left(-\frac{3}{7}\right)^2 \div \frac{12}{5}}{\left(\frac{-2}{9}\right)\left(\frac{9}{2}\right)}$ **69.** $\left(\dfrac{14}{15} \div \dfrac{49}{65} \cdot \dfrac{77}{260}\right)^2$ **70.** $\left(\dfrac{10}{9}\right)^2 \div \dfrac{35}{27} \cdot \dfrac{49}{44}$

Simplify. Use the list of prime numbers on p. 160.

71. $\dfrac{711}{1957} \div \dfrac{10,033}{13,081}$ **72.** $\dfrac{8633}{7387} \div \dfrac{485}{581}$ **73.** $\dfrac{451}{289} \div \dfrac{123}{340}$ **74.** $\dfrac{530}{490} \div \dfrac{1060}{980}$

OBJECTIVES

a Use the multiplication principle to solve equations.

b Solve problems by using the multiplication principle.

SKILL TO REVIEW

Objective 2.8b: Use the division principle to solve equations.

Solve.

1. $-8t = 32$ **2.** $-x = -9$

Solve.

1. $\dfrac{2}{3}x = 8$ **GS**

$$\dfrac{\square}{\square} \cdot \dfrac{2}{3}x = \dfrac{\square}{\square} \cdot 8$$

$$x = \dfrac{3 \cdot 8}{2}$$

$$x = \dfrac{3 \cdot 2 \cdot \square}{2}$$

$$x = \square$$

2. $\dfrac{2}{7}a = -6$

Answers

Skill to Review:
1. -4 **2.** 9

Margin Exercises:
1. 12 **2.** -21

Guided Solution:
1. $\dfrac{3}{2}, \dfrac{3}{2}, 1, 4, 12$

With fraction notation, we can solve equations like $a \cdot x = b$ by using multiplication.

a THE MULTIPLICATION PRINCIPLE

To divide by a fraction, we multiply by the reciprocal of that fraction. This suggests that we restate the division principle in its more common form—the multiplication principle.

> **THE MULTIPLICATION PRINCIPLE**
>
> For any numbers a, b, and c, with $c \neq 0$,
>
> $$a = b \quad \text{is equivalent to} \quad a \cdot c = b \cdot c.$$

EXAMPLE 1 Solve: $\dfrac{3}{4}x = 15$.

We multiply both sides of the equation by the reciprocal of $\dfrac{3}{4}$.

$$\dfrac{3}{4}x = 15$$

$$\dfrac{4}{3} \cdot \dfrac{3}{4}x = \dfrac{4}{3} \cdot 15 \qquad \text{Using the multiplication principle; note that } \dfrac{4}{3} \text{ is the reciprocal of } \dfrac{3}{4}.$$

$$\left(\dfrac{4}{3} \cdot \dfrac{3}{4}\right)x = \dfrac{4 \cdot 15}{3} \qquad \text{Using an associative law; try to do this mentally.}$$

$$1x = 20 \qquad \text{Multiplying; note that } \dfrac{4 \cdot 15}{3} = \dfrac{4 \cdot 3 \cdot 5}{3}.$$

$$x = 20 \qquad \text{Remember that } 1x \text{ is } x.$$

To confirm that 20 is the solution, we perform a check.

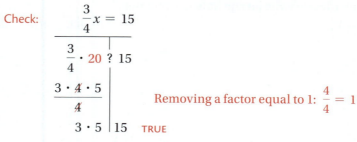

Check:
$$\dfrac{3}{4}x = 15$$
$$\dfrac{3}{4} \cdot 20 \; ? \; 15$$
$$\dfrac{3 \cdot 4 \cdot 5}{4}$$
$$3 \cdot 5 \;\big|\; 15 \quad \text{TRUE}$$

Removing a factor equal to 1: $\dfrac{4}{4} = 1$

The solution is 20.

◀ **Do Margin Exercises 1 and 2.**

In an expression like $\dfrac{3}{4}x$, the constant factor—in this case, $\dfrac{3}{4}$—is called the **coefficient**. In Example 1, we multiplied on both sides by $\dfrac{4}{3}$, the reciprocal of the coefficient of x. After we carried out the multiplication, the coefficient of x was 1. Note that using the multiplication principle to multiply by $\dfrac{4}{3}$ on both sides is the same as using the division principle to divide by $\dfrac{3}{4}$ on both sides.

EXAMPLE 2 Solve: $5a = -\dfrac{7}{3}$.

We have

$$5a = -\frac{7}{3}$$

$$\frac{1}{5} \cdot 5a = \frac{1}{5} \cdot \left(-\frac{7}{3}\right) \qquad \text{Multiplying both sides by } \tfrac{1}{5}, \text{ the reciprocal of 5}$$

$$a = -\frac{1 \cdot 7}{5 \cdot 3} = -\frac{7}{15}.$$

We leave the check to the student. The solution is $-\frac{7}{15}$.

EXAMPLE 3 Solve: $\dfrac{10}{3} = -\dfrac{4}{9}x$.

We have

$$\frac{10}{3} = -\frac{4}{9}x$$

$$-\frac{9}{4} \cdot \frac{10}{3} = -\frac{9}{4} \cdot \left(-\frac{4}{9}\right)x \qquad \text{Multiplying both sides by } -\tfrac{9}{4}, \text{ the reciprocal of } -\tfrac{4}{9}$$

$$-\frac{3 \cdot 3 \cdot 2 \cdot 5}{2 \cdot 2 \cdot 3} = x$$

$$-\frac{15}{2} = x. \qquad \text{Removing a factor equal to 1: } \frac{3 \cdot 2}{2 \cdot 3} = 1$$

We leave the check to the student. The solution is $-\frac{15}{2}$.

Do Exercises 3 and 4. ▶

Solve.

3. $-\dfrac{9}{8} = 4x$

GS **4.** $-\dfrac{6}{7}a = \dfrac{9}{14}$

$$-\frac{\square}{\square} \cdot \left(-\frac{6}{7}a\right) = -\frac{\square}{\square} \cdot \frac{9}{14}$$

$$\square = -\frac{7 \cdot 3 \cdot 3}{2 \cdot 3 \cdot 2 \cdot 7}$$

$$a = -\frac{3}{\square}$$

b APPLICATIONS AND PROBLEM SOLVING

EXAMPLE 4 *Doses of an Antibiotic.* How many doses, each containing $\frac{15}{4}$ milliliters (mL), can be obtained from a bottle of a children's antibiotic that contains 60 mL?

1. **Familiarize.** We make a drawing, as shown to the right, and let $n =$ the number of doses in the bottle.

2. **Translate.** The problem can be translated to the equation

$$\frac{15}{4} \cdot n = 60.$$

3. **Solve.** To solve the equation, we use the multiplication principle.

$$\frac{4}{15} \cdot \frac{15}{4} \cdot n = \frac{4}{15} \cdot 60 \qquad \text{Multiplying both sides by } \tfrac{4}{15}$$

$$1n = \frac{4 \cdot 60}{15}$$

$$n = \frac{2 \cdot 2 \cdot 2 \cdot 2 \cdot 3 \cdot 5}{1 \cdot 3 \cdot 5} = \frac{3 \cdot 5}{3 \cdot 5} \cdot \frac{2 \cdot 2 \cdot 2 \cdot 2}{1} = 16$$

$\frac{15}{4}$ milliliter in each dose

n doses in the bottle

Answers

3. $-\dfrac{9}{32}$ 4. $-\dfrac{3}{4}$

Guided Solution:

4. $\dfrac{7}{6}, \dfrac{7}{6}, a, 4$

5. Each loop in a spring uses $\frac{21}{8}$ in. of wire. How many loops can be made from 210 in. of wire?

6. *Servings of Cereal.* A box contains 12 cups of cereal. The nutrition label on the box states that the serving size is $\frac{3}{4}$ cup. How many servings of cereal are in the box?

4. Check. We check by multiplying the number of doses by the size of the dose: $16 \cdot \dfrac{15}{4} = 60$. Note too that

$$\text{doses} \cdot \frac{\text{mL}}{\text{dose}} = \text{mL},$$

so the units also check.

5. State. There are 16 doses of the antibiotic in the 60-mL bottle.

◀ **Do Exercises 5 and 6.**

EXAMPLE 5 *Bicycle Paths.* The city of Indianapolis has adopted the *Indianapolis Bicycle Master Plan* as a strategy for creating an environment where bicycling is a safe, practical, and enjoyable transportation choice. After the city finished constructing 60 mi of bike paths and on-road bike lanes, the master plan was $\frac{3}{10}$ complete. What is the total number of miles of bicycling surface that the city of Indianapolis plans to construct?

Source: *Indianapolis Bicycle Master Plan,* June 2012

1. Familiarize. We ask: "60 mi is $\frac{3}{10}$ of what length?" We make a drawing or at least visualize the problem. We let $b =$ the total number of miles of bicycling surface in the master plan.

2. Translate. We translate to an equation:

Fraction completed	of	Total miles planned	is	Amount completed
↓	↓	↓	↓	↓
$\dfrac{3}{10}$	$\cdot$	b	$=$	$60.$

7. *Sales Trip.* John Penna sells soybean seeds to seed companies. After he had driven 210 mi, $\frac{5}{6}$ of his sales trip was completed. How long was the total trip?

3. Solve. We divide by $\frac{3}{10}$ on both sides and carry out the division:

$$b = 60 \div \frac{3}{10} = \frac{60}{1} \cdot \frac{10}{3} = \frac{60 \cdot 10}{1 \cdot 3} = \frac{3 \cdot 20 \cdot 10}{1 \cdot 3} = \frac{3}{3} \cdot \frac{20 \cdot 10}{1} = 200.$$

4. Check. We determine whether $\frac{3}{10}$ of 200 is 60: $\frac{3}{10} \cdot 200 = 60$. The answer, 200, checks.

5. State. The *Indianapolis Bicycle Master Plan* calls for 200 mi of bicycling surface.

◀ **Do Exercise 7.**

Answers

5. 80 loops   **6.** 16 servings   **7.** 252 mi

Translating for Success

1. **Valentine Boxes.** Jane's Fudge Shop is preparing Valentine boxes. How many pounds of fudge will be needed to fill 80 boxes if each box contains $\frac{5}{16}$ lb?

2. **Gallons of Gasoline.** On the third day of a business trip, a sales representative used $\frac{4}{5}$ of a tank of gasoline. If the tank is a 20-gal tank, how many gallons were used on the third day?

3. **Purchasing a Shirt.** Tom received $36 for his birthday. If he spends $\frac{3}{4}$ of the gift on a new shirt, what is the cost of the shirt?

4. **Checkbook Balance.** The balance in Sam's checking account is $1456. He writes a check for $28 and makes a deposit of $52. What is the new balance?

5. **Valentine Boxes.** Jane's Fudge Shop prepared 80 lb of fudge for Valentine boxes. If each box contains $\frac{5}{16}$ lb, how many boxes can be filled?

The goal of these matching questions is to practice step (2), Translate, of the five-step problem-solving process. Translate each problem to an equation and select a correct translation from equations A–O.

A. $x = \frac{3}{4} \cdot 36$

B. $28 \cdot x = 52$

C. $x = 80 \cdot \frac{5}{16}$

D. $x = 1456 \div 28$

E. $\frac{5}{4} \cdot x = 20$

F. $20 = \frac{4}{5} \cdot x$

G. $x = 12 \cdot 28$

H. $x = \frac{4}{5} \cdot 20$

I. $\frac{3}{4} \cdot x = 36$

J. $x = 1456 - 52 - 28$

K. $x \div 28 = 1456$

L. $x = 52 - 28$

M. $x = 52 \cdot 28$

N. $x = 1456 - 28 + 52$

O. $\frac{5}{16} \cdot x = 80$

Answers on page A-6

---

6. **Gasoline Tank.** A gasoline tank contains 20 gal when it is $\frac{4}{5}$ full. How many gallons can it hold when full?

7. **Knitting a Scarf.** It takes Rachel 36 hr to knit a scarf. She can knit only $\frac{3}{4}$ hr per day because she is taking 16 hr of college classes. How many days will it take her to knit the scarf?

8. **Bicycle Trip.** On a recent 52-mi bicycle trip, David stopped to make a cell-phone call after completing 28 mi. How many more miles did he bicycle after the call?

9. **Crème de Menthe Thins.** Andes Candies L.P. makes Crème de Menthe Thins. How many 28-piece packages can be filled with 1456 pieces?

10. **Cereal Donations.** The Williams family donates 28 boxes of cereal weekly to the local Family in Crisis Center. How many boxes does this family donate in one year?

✓ **Reading Check**

From the list on the right, choose the fraction to use in solving each equation by multiplying on both sides.

RC1. $\dfrac{3}{4}x = 10$ _____

a) $\dfrac{3}{4}$

RC2. $-\dfrac{4}{9}x = \dfrac{3}{4}$ _____

b) $\dfrac{4}{3}$

RC3. $-7 = \dfrac{9}{4}x$ _____

c) $-\dfrac{9}{4}$

RC4. $\dfrac{4}{9} = \dfrac{4}{3}x$ _____

d) $\dfrac{4}{9}$

a Use the multiplication principle to solve each equation. Don't forget to check!

1. $\dfrac{4}{5}x = 12$

2. $\dfrac{4}{3}x = 20$

3. $\dfrac{7}{3}a = 21$

4. $\dfrac{4}{5}a = 24$

5. $\dfrac{2}{9}y = -10$

6. $\dfrac{3}{8}y = -21$

7. $6t = \dfrac{12}{17}$

8. $3t = \dfrac{15}{14}$

9. $\dfrac{1}{4}x = \dfrac{3}{5}$

10. $\dfrac{1}{6}x = \dfrac{2}{7}$

11. $\dfrac{3}{2}t = -\dfrac{8}{7}$

12. $\dfrac{4}{3}t = -\dfrac{5}{2}$

13. $\dfrac{4}{5} = -10a$

14. $\dfrac{6}{5} = -12a$

15. $x \cdot \dfrac{9}{5} = \dfrac{3}{10}$

16. $x \cdot \dfrac{10}{3} = \dfrac{8}{15}$

17. $-\dfrac{1}{10}x = 8$

18. $-\dfrac{1}{11}x = -5$

19. $a \cdot \dfrac{9}{7} = -\dfrac{3}{14}$

20. $a\left(-\dfrac{9}{4}\right) = -\dfrac{3}{10}$

21. $-x = \dfrac{7}{13}$

22. $-x = \dfrac{7}{11}$

23. $-x = -\dfrac{27}{31}$

24. $-x = -\dfrac{35}{39}$

25. $7t = 6$

26. $-6t = 1$

27. $-24 = -10a$

28. $-18 = -20a$

29. $-\dfrac{14}{9} = \dfrac{10}{3}t$

30. $-\dfrac{15}{7} = \dfrac{3}{2}t$

31. $n \cdot \dfrac{4}{15} = \dfrac{12}{25}$

32. $n \cdot \dfrac{5}{16} = \dfrac{15}{14}$

33. $-\dfrac{7}{20}x = -\dfrac{21}{10}$

34. $-\dfrac{7}{15}x = -\dfrac{21}{10}$

35. $-\dfrac{25}{17} = -\dfrac{35}{34}a$

36. $-\dfrac{49}{45} = -\dfrac{28}{27}a$

  Solve.

37. *Extension Cords.* An electrical supplier sells rolls of SJO 14-3 cable to a company that makes extension cords. It takes $\frac{7}{3}$ ft of cable to make each cord. How many extension cords can be made with a roll of cable containing 2240 ft of cable?

38. Benny uses $\frac{2}{5}$ gram (g) of toothpaste each time he brushes his teeth. If Benny buys a 30-g tube, how many times will he be able to brush his teeth?

39. *Sewing.* A pair of basketball shorts requires $\frac{3}{4}$ yd of nylon. How many pairs of shorts can be made from 24 yd of nylon?

40. *Sewing.* A child's baseball shirt requires $\frac{5}{6}$ yd of fabric. How many shirts can be made from 25 yd of fabric?

41. How many $\frac{2}{3}$-cup sugar bowls can be filled from 16 cups of sugar?

42. For a party, Kyrsten makes an 8-ft submarine sandwich. If one serving is $\frac{2}{3}$ ft, how many servings does Kyrsten's sub contain?

43. A bucket had 12 L of water in it when it was $\frac{3}{4}$ full. How much could it hold when full?

44. A tank had 20 L of gasoline in it when it was $\frac{4}{5}$ full. How much could it hold when full?

45. Packaging. The South Shore Co-op prepackages cheddar cheese in $\frac{3}{4}$-lb packages. How many packages can be made from a 15-lb wheel of cheese?

46. Meal Planning. Ian purchased 6 lb of cold cuts for a luncheon. If Ian is to allow $\frac{3}{8}$ lb per person, how many people can attend the luncheon?

47. Art Supplies. The Ferristown School District purchased $\frac{3}{4}$ T (ton) of clay. The clay is to be shared equally among the district's 6 art departments. How much will each art department receive?

48. Gardening. The Bingham community garden is to be split into 16 equally sized plots. If the garden occupies $\frac{3}{4}$ acre of land, how large will each plot be?

49. Honey. A worker bee will produce $\frac{1}{12}$ tsp (teaspoon) of honey in her lifetime. How many worker bees does it take to produce $\frac{3}{4}$ tsp of honey?

Source: www.pbs.org/wgbh/nova/bees/buzz.html

50. Gardening. Large quantities of soil, gravel, or mulch are normally sold by the *yard* (yd). Although technically the unit for volume is *cubic yard* (yd³), in this context only the word *yard* is used. Green Season Gardening uses about $\frac{2}{3}$ yd of bark mulch per customer every spring. How many customers can they accommodate with one 30-yd load of bark mulch?

51. Writing. On Monday, Ayesha wrote 450 words of an essay that was due on Friday. She then calculated that she had written $\frac{3}{16}$ of the minimum number of words required. What was the minimum number of words required for the essay?

52. Landscaping. As part of a semester project for an urban planning course, Fridrik planted 240 perennials in the median strip of an avenue. If Fridrik planted $\frac{5}{32}$ of the total number of perennials planted, how many perennials were planted in all?

53. Running. Chad runs 12 mi three days a week. In bad weather, he runs on an indoor track that is $\frac{3}{8}$ mi long. How many laps must he complete in order to run 12 mi?

54. Herbal Tea. At Perfect Tea, Donna fills each bag of Sereni-Tea with $\frac{3}{5}$ g (gram) of chamomile. If she begins with 51 g of chamomile, how many tea bags can she fill?

55. Yoshi Teramoto sells tools to hardware stores. After driving 180 kilometers (km), he has completed $\frac{5}{8}$ of a sales trip. How long is the total trip? How many kilometers are left to drive?

56. A piece of coaxial cable $\frac{4}{5}$ meter (m) long is to be cut into 8 pieces of the same length. What is the length of each piece?

Pitch of a Screw. The pitch of a screw is the distance between its threads. With each complete rotation, the screw goes in or out a distance equal to its pitch. Use this information to do Exercises 57 and 58.

57. After a screw has been turned 8 complete rotations, it is extended $\frac{1}{2}$ in. into a piece of wallboard. What is the pitch of the screw?

58. The pitch of a screw is $\frac{3}{32}$ in. How many complete rotations are necessary to drive the screw $\frac{3}{4}$ in. into a piece of pine wood?

Skill Maintenance

Simplify.

59. $-23 + 49$ [2.2a]

60. $-69 + 27$ [2.2a]

61. $-38 - 29$ [2.3a]

62. $-47 - 18$ [2.3a]

63. $36 \div (-3)^2 \times (7 - 2)$ [2.5b]

64. $(-37 - 12 + 1) \div (-2)^3$ [2.5b]

Form an equivalent expression by combining like terms.   [2.7a]

65. $13x + 4x$

66. $9a - 5a$

67. $2a + 3 + 5a$

68. $3x - 7 + x$

Synthesis

Solve.

69. $2x - 7x = -\dfrac{10}{9}$

70. $\left(-\dfrac{4}{7}\right)^2 = \left(\dfrac{2^3 - 9}{3}\right)^3 x$

Solve using the five-step problem-solving approach.

71. A package of coffee beans weighed $\frac{21}{32}$ lb when it was $\frac{3}{4}$ full. How much could the package hold when completely filled?

72. After swimming $\frac{2}{7}$ mi, Katie had swum $\frac{3}{4}$ of the race. How long a race was Katie competing in?

73. A block of Swiss cheese is 12 in. long. How many slices will it yield if half of the brick is cut by a slicer set for $\frac{3}{32}$-in. slices and half is cut by a slicer set for $\frac{5}{32}$-in. slices?

74. If $\frac{1}{3}$ of a number is $\frac{1}{4}$, what is $\frac{1}{2}$ of the number?

75. See Exercise 49. There are 3 teaspoons in a tablespoon and 4 tablespoons in $\frac{1}{4}$ cup. How many worker bees does it take to produce $\frac{1}{2}$ cup of honey?

76. ⊞ See Exercise 50. If each customer of Green Season Gardening uses $\frac{3}{4}$ yd of bark mulch and Green Season charges each customer $45 for the mulch, how much will Green Season receive from a 25-yd load of mulch?

Vocabulary Reinforcement

Fill in each blank with the correct term from the list at the right. Some of the choices may not be used.

1. For any number a, $a \cdot 1 = a$. The number 1 is the _____ identity.   [3.5a]

2. In the product $10 \cdot \frac{3}{4}$, 10 and $\frac{3}{4}$ are called _____.   [3.2a]

3. A natural number that has exactly two different factors, only itself and 1, is called a(n) _____ number.   [3.2b]

4. In the fraction $\frac{4}{17}$, we call 17 the _____.   [3.3a]

5. Since $\frac{2}{5}$ and $\frac{6}{15}$ are two names for the same number, we say that $\frac{2}{5}$ and $\frac{6}{15}$ are _____ fractions.   [3.5a]

6. The product of 6 and $\frac{1}{6}$ is 1. We say that 6 and $\frac{1}{6}$ are _____.   [3.7a]

7. Since $20 = 4 \cdot 5$, we say that $4 \cdot 5$ is a _____ of 20.   [3.2a]

8. Since $20 = 4 \cdot 5$, we say that 20 is a _____ of 5.   [3.1a]

equivalent
additive
multiplicative
reciprocals
factors
prime
composite
numerator
denominator
factorization
variables
multiple

Concept Reinforcement

Determine whether each statement is true or false.

_____ 1. For any natural number n, $\frac{n}{n} > \frac{0}{n}$.   [3.3b]

_____ 2. A number is divisible by 10 if its ones digit is 0 or 5.   [3.1b]

_____ 3. If a number is divisible by 9, then it is also divisible by 3.   [3.1b]

_____ 4. The fraction $\dfrac{-6}{8}$ is equivalent to the fraction $\dfrac{15}{-20}$.   [3.5c]

Study Guide

Objective 3.2a   Find the factors of a number.

Example   Find the factors of 84.

We find as many "two-factor" factorizations as we can.

$1 \cdot 84$ $4 \cdot 21$

$2 \cdot 42$ $6 \cdot 14$

$3 \cdot 28$ $7 \cdot 12$ ⟵ Since 8, 9, 10, and 11 are not factors, we are finished.

The factors are 1, 2, 3, 4, 6, 7, 12, 14, 21, 28, 42, and 84.

Practice Exercise

1. Find the factors of 104.

Objective 3.2c Find the prime factorization of a composite number.

Example Find the prime factorization of 84.

$$
\begin{array}{r}
7 \\
3\overline{)21} \\
2\overline{)42} \\
2\overline{)84}
\end{array}
$$

84
/ \
2   42
    / \
2   2   21
        / \
2   2   3   7

Thus, $84 = 2 \cdot 2 \cdot 3 \cdot 7$.

Practice Exercise

2. Find the prime factorization of 104.

Objective 3.3b Simplify fraction notation like n/n to 1, $0/n$ to 0, and $n/1$ to n.

Example Simplify $\dfrac{6}{6}, \dfrac{0}{6}$, and $\dfrac{6}{1}$.

$$\frac{6}{6} = 1, \qquad \frac{0}{6} = 0, \qquad \frac{6}{1} = 6$$

Practice Exercise

3. Simplify $\dfrac{0}{18}, \dfrac{18}{18}$, and $\dfrac{18}{1}$.

Objective 3.5b Simplify fraction notation.

Example Simplify: $\dfrac{315}{1650}$.

Using the test for divisibility by 5, we see that both the numerator and the denominator are divisible by 5:

$$\frac{315}{1650} = \frac{5 \cdot 63}{5 \cdot 330} = \frac{5}{5} \cdot \frac{63}{330} = 1 \cdot \frac{63}{330}$$

$$= \frac{63}{330} = \frac{3 \cdot 21}{3 \cdot 110} = \frac{3}{3} \cdot \frac{21}{110} = 1 \cdot \frac{21}{110} = \frac{21}{110}.$$

Practice Exercise

4. Simplify: $\dfrac{100}{280}$.

Objective 3.5c Test to determine whether two fractions are equivalent.

Example Use $=$ or $\neq$ for $\square$ to write a true sentence:

$$\frac{10}{54} \;\square\; \frac{15}{81}.$$

We find the cross products. We multiply 10 and 81: $10 \cdot 81 = 810$. Then we multiply 54 and 15: $54 \cdot 15 = 810$. Because the cross products are the same, we have

$$\frac{10}{54} = \frac{15}{81}.$$

If the cross products had been different, the fractions would not be equal.

Practice Exercise

5. Use $=$ or $\neq$ for $\square$ to write a true sentence:

$$\frac{8}{48} \;\square\; \frac{6}{44}.$$

Objective 3.6a Multiply and simplify using fraction notation.

Example Multiply and simplify: $\dfrac{7}{16} \cdot \dfrac{40}{49}$.

$$\frac{7}{16} \cdot \frac{40}{49} = \frac{7 \cdot 40}{16 \cdot 49} = \frac{7 \cdot 2 \cdot 2 \cdot 2 \cdot 5}{2 \cdot 2 \cdot 2 \cdot 2 \cdot 7 \cdot 7}$$

$$= \frac{2 \cdot 2 \cdot 2 \cdot 7}{2 \cdot 2 \cdot 2 \cdot 7} \cdot \frac{5}{2 \cdot 7} = 1 \cdot \frac{5}{14} = \frac{5}{14}$$

Practice Exercise

6. Multiply and simplify: $\dfrac{80}{3} \cdot \dfrac{21}{72}$.

Objective 3.7b Divide and simplify using fraction notation.

Example Divide and simplify: $\dfrac{9}{20} \div \dfrac{18}{25}$.

$$\dfrac{9}{20} \div \dfrac{18}{25} = \dfrac{9}{20} \cdot \dfrac{25}{18} = \dfrac{9 \cdot 25}{20 \cdot 18} = \dfrac{3 \cdot 3 \cdot 5 \cdot 5}{2 \cdot 2 \cdot 5 \cdot 2 \cdot 3 \cdot 3}$$

$$= \dfrac{3 \cdot 3 \cdot 5}{3 \cdot 3 \cdot 5} \cdot \dfrac{5}{2 \cdot 2 \cdot 2} = 1 \cdot \dfrac{5}{8} = \dfrac{5}{8}$$

Practice Exercise

7. Divide and simplify: $\dfrac{9}{4} \div \dfrac{45}{14}$.

Objective 3.8b Solve problems by using the multiplication principle.

Example A rental car had 18 gal of gasoline when its gas tank was $\frac{6}{7}$ full. How much could the tank hold when full?

The equation that corresponds to the situation is

$$\dfrac{6}{7} \cdot g = 18.$$

We multiply both sides by $\frac{7}{6}$:

$$\dfrac{7}{6} \cdot \dfrac{6}{7} \cdot g = \dfrac{7}{6} \cdot 18$$

$$g = \dfrac{7}{6} \cdot \dfrac{18}{1} = \dfrac{7 \cdot 3 \cdot \cancel{6}}{\cancel{6} \cdot 1} = 21.$$

The rental car can hold 21 gal of gasoline.

Practice Exercise

8. A flower vase has $\frac{7}{4}$ cups of water in it when it is $\frac{3}{4}$ full. How much can it hold when full?

Review Exercises

1. Multiply by 1, 2, 3, and so on, to find ten multiples of 8. [3.1a]

Use the tests for divisibility to answer Exercises 2–6. [3.1b]

2. Determine whether 3920 is divisible by 6.

3. Determine whether 68,537 is divisible by 3.

4. Determine whether 673 is divisible by 5.

5. Determine whether 4936 is divisible by 2.

6. Determine whether 5238 is divisible by 9.

Find all the factors of each number. [3.2a]

7. 60   8. 176

Classify each number as prime, composite, or neither. [3.2b]

9. 37   10. 1   11. 91

Find the prime factorization of each number. [3.2c]

12. 70   13. 72

14. 45   15. 150

16. 648   17. 1200

18. Identify the numerator and the denominator of $\dfrac{9}{7}$. [3.3a]

What part is shaded? [3.3a]

19.   20.

21. For a committee in the United States Senate that consists of 3 Democrats and 5 Republicans, what is the ratio of: [3.3a]

  a) Democrats to Republicans?
  b) Republicans to Democrats?
  c) Democrats to the total number of members of the committee?

Simplify, if possible. Assume that all variables are nonzero.

22. $\dfrac{0}{6}$ [3.3b]   **23.** $\dfrac{74}{74}$ [3.3b]   **24.** $\dfrac{48}{1}$ [3.3b]

25. $\dfrac{7x}{7x}$ [3.3b]   **26.** $-\dfrac{10}{15}$ [3.5b]   **27.** $\dfrac{7}{28}$ [3.5b]

28. $\dfrac{-42}{42}$ [3.5b]   **29.** $\dfrac{9m}{12m}$ [3.5b]   **30.** $\dfrac{-12}{-30}$ [3.5b]

31. $\dfrac{-27}{0}$ [3.3b]   **32.** $\dfrac{140}{490}$ [3.5b]   **33.** $\dfrac{288}{2025}$ [3.5b]

Find an equivalent expression for each number, using the denominator indicated. Use multiplication by 1. [3.5a]

34. $\dfrac{5}{7} = \dfrac{?}{21}$ **35.** $\dfrac{-6}{11} = \dfrac{?}{55}$

36. Simplify, if possible, the fractions on this circle graph. [3.5b]

Museums in the United States

Science $\dfrac{15}{100}$

History $\dfrac{38}{100}$

Art $\dfrac{24}{100}$

Other $\dfrac{23}{100}$

Use = or ≠ for ☐ to write a true sentence. [3.5c]

37. $\dfrac{3}{5}$ ☐ $\dfrac{4}{6}$ **38.** $\dfrac{4}{-7}$ ☐ $\dfrac{-8}{14}$

39. $\dfrac{4}{5}$ ☐ $\dfrac{5}{6}$ **40.** $\dfrac{4}{3}$ ☐ $\dfrac{28}{21}$

Find the reciprocal of each number. [3.7a]

41. $\dfrac{2}{13}$ **42.** -7

43. $\dfrac{1}{8}$ **44.** $\dfrac{3x}{5y}$

Perform the indicated operation and, if possible, simplify.

45. $\dfrac{2}{9} \cdot \dfrac{7}{5}$ [3.4b]   **46.** $\dfrac{3}{x} \cdot \dfrac{y}{7}$ [3.4b]

47. $\dfrac{3}{4} \cdot \dfrac{8}{9}$ [3.6a]   **48.** $-10 \cdot \dfrac{7}{5}$ [3.6a]

49. $\dfrac{11}{3} \cdot \dfrac{30}{77}$ [3.6a]   **50.** $\dfrac{4a}{7} \cdot \dfrac{7}{4a}$ [3.6a]

51. $\dfrac{6}{5} \cdot 20x$ [3.6a]   **52.** $\dfrac{3}{14} \div \dfrac{6}{7}$ [3.7b]

53. $20 \div \dfrac{3}{4}$ [3.7b]   **54.** $-\dfrac{5}{36} \div \left(-\dfrac{25}{12}\right)$ [3.7b]

55. $21 \div \dfrac{7}{2a}$ [3.7b]   **56.** $-\dfrac{23}{25} \div \dfrac{23}{25}$ [3.7b]

57. $\dfrac{\dfrac{21}{30}}{\dfrac{14}{15}}$ [3.7b]   **58.** $\dfrac{-\dfrac{2}{3}}{-\dfrac{3}{2}}$ [3.7b]

Solve. [3.8a]

59. $\dfrac{2}{3}x = 160$ **60.** $\dfrac{3}{8} = -\dfrac{5}{4}t$

61. $-\dfrac{1}{7}n = -4$ **62.** $y \cdot \dfrac{1}{2} = \dfrac{1}{3}$

Find the area. [3.6b]

63.

6 m
14 m

64.

10 ft
$\frac{7}{2}$ ft

Solve. [3.8b]

65. A road crew repaves $\frac{1}{12}$ mi of road each day. How long will it take the crew to repave a $\frac{3}{4}$-mi stretch of road?

66. *Level of Education and Median Income.* The median yearly income of individuals with an associate's degree is approximately $\frac{3}{4}$ of the median income of individuals with a bachelor's degree. If the median income for those with bachelor's degrees is $42,780, what is the median income of those with associate's degrees?

Source: U.S. Census Bureau

67. After driving 600 km, the Buxton family has completed $\frac{3}{5}$ of their vacation. How long is the total trip?

68. Molly is making a pepper steak recipe that calls for $\frac{2}{3}$ cup of green bell peppers. How much would be needed to make $\frac{1}{2}$ recipe?

69. The Winchester swim team has 4 swimmers in a $\frac{2}{3}$-mi relay race. If each swims the same distance, how far will each person swim?

70. A book bag requires $\frac{4}{5}$ yd of fabric. How many bags can be made from 48 yd?

71. Solve: $\frac{2}{13} \cdot x = \frac{1}{2}$. [3.8a]

A. $\frac{1}{13}$ **B.** 13   **C.** $\frac{4}{13}$ **D.** $\frac{13}{4}$

72. Multiply and simplify: $\frac{15}{26} \cdot \frac{13}{90}$. [3.6a]

A. $\frac{195}{234}$ **B.** $\frac{1}{12}$ **C.** $\frac{3}{36}$ **D.** $\frac{13}{156}$

Synthesis

73. Simplify: $\frac{15x}{14z} \cdot \frac{17yz}{35xy} \div \left(-\frac{3}{7}\right)^2$. [3.6a], [3.7b]

74. What digit(s) could be inserted in the ones place to make 574 _____ divisible by 6? [3.1b]

75. In the division below, find a and b. [3.7b]

$$\frac{19}{24} \div \frac{a}{b} = \frac{187,853}{268,224}$$

76. A prime number that remains a prime number when its digits are reversed is called a **palindrome prime**. For example, 17 is a palindrome prime because both 17 and 71 are primes. Which of the following numbers are palindrome primes? [3.2b]

13, 91, 16, 11, 15, 24, 29, 101, 201, 37

Understanding Through Discussion and Writing

1. A student incorrectly insists that $\frac{2}{5} \div \frac{3}{4}$ is $\frac{15}{8}$. What mistake is he probably making? [3.7b]

2. Use the number 9432 to explain why the test for divisibility by 9 works. [3.1b]

3. A student claims that "taking $\frac{1}{2}$ of a number is the same as dividing by $\frac{1}{2}$." Explain the error in this reasoning. [3.7b]

4. On p. 176 we explained, using words and pictures, why $\frac{2}{5} \cdot \frac{3}{4}$ equals $\frac{6}{20}$. Present a similar explanation of why $\frac{2}{3} \cdot \frac{4}{7}$ equals $\frac{8}{21}$. [3.4b]

5. Without performing the division, explain why $5 \div \frac{1}{7}$ is a greater number than $5 \div \frac{2}{3}$. [3.5c], [3.7b]

6. If a fraction's numerator and denominator have no factors (other than 1) in common, can the fraction be simplified? Why or why not? [3.5b]

CHAPTER

3   Test

For Extra Help

For step-by-step test solutions, access the Chapter Test Prep Videos in MyMathLab® or on YouTube (search "BittingerPrealgebra" and click on Channels).

1. Determine whether 5682 is divisible by 3. Do not use long division.

2. Determine whether 7018 is divisible by 5. Do not use long division.

3. Find all the factors of 90.

4. Determine whether 93 is prime, composite, or neither.

Find the prime factorization of each number.

5. 36

6. 60

7. Identify the numerator and the denominator of $\frac{4}{9}$.

8. What part is shaded?

9. What part of the set is shaded?

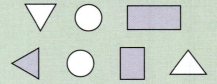

10. *Cholesterol.*   Morrison's cholesterol test showed that his total cholesterol level was 180 mg/dL, his HDL level was 47 mg/dL, and his LDL level was 93 mg/dL.
   a) What was the ratio of total cholesterol to HDL cholesterol?
   b) What was the ratio of HDL cholesterol to LDL cholesterol?

Simplify, if possible. Assume that all variables are nonzero.

11. $\dfrac{32}{1}$

12. $\dfrac{-12}{-12}$

13. $\dfrac{0}{16}$

14. $\dfrac{-8}{24}$

15. $\dfrac{42}{7}$

16. $\dfrac{9x}{45x}$

17. $\dfrac{-62}{0}$

18. $\dfrac{72}{108}$

Use $=$ or $\neq$ for ☐ to write a true sentence.

19. $\dfrac{3}{4}$ ☐ $\dfrac{6}{8}$

20. $\dfrac{5}{4}$ ☐ $\dfrac{9}{7}$

21. Find an equivalent expression for $\dfrac{3}{8}$ with a denominator of 40.

Find the reciprocal.

22. $\dfrac{a}{42}$

23. -9

Perform the indicated operation. Simplify, if possible.

24. $\dfrac{2}{3} \cdot \dfrac{15}{4}$

25. $\dfrac{2}{11} \div \dfrac{3}{4}$

26. $3 \cdot \dfrac{x}{8}$

27. $\dfrac{\dfrac{4}{7}}{-\dfrac{8}{3}}$

28. $12 \div \dfrac{2}{3}$

29. $\dfrac{22c}{15} \cdot \dfrac{5}{33c}$

Solve.

30. A $\frac{3}{4}$-lb slab of cheese is shared equally by 5 people. How much does each person receive?

31. Monroe weighs $\frac{5}{7}$ of his dad's weight. If his dad weighs 175 lb, how much does Monroe weigh?

32. $\dfrac{7}{8} \cdot x = 56$

33. $\dfrac{7}{10} = \dfrac{-2}{5} \cdot t$

34. Find the area.

35. In which figure does the shaded part represent $\frac{7}{6}$ of the figure?

Synthesis

36. Grandma Shelby left $\frac{2}{3}$ of her $\frac{7}{8}$-acre apple farm to Karl. Karl gave $\frac{1}{4}$ of his share to his oldest daughter, Shannon. How much land did Shannon receive?

37. Simplify: $\left(-\dfrac{3}{8}\right)^2 \div \dfrac{6}{7} \cdot \dfrac{2}{9} \div (-5)$.

1. Write a word name: 2,056,783.

Add.

2.
$$\begin{array}{r} 2\ 7\ 4\ 3 \\ +\ 8\ 2\ 3\ 9 \\ \hline \end{array}$$

3. $-29 + (-14)$

4. $-45 + 12$

Subtract.

5.
$$\begin{array}{r} 6\ 3\ 2\ 4 \\ -\ 4\ 1\ 9\ 5 \\ \hline \end{array}$$

6. $27 - 50$

7. $-12 - (-4)$

Multiply and, if possible, simplify.

8.
$$\begin{array}{r} 7\ 3\ 5 \\ \times\quad 2\ 3 \\ \hline \end{array}$$

9. $-52 \cdot 6$

10. $\dfrac{6}{7} \cdot (-35x)$

11. $\dfrac{2}{9} \cdot \dfrac{21}{10}$

Divide and, if possible, simplify.

12. $1\ 3\ \overline{)\ 3\ 0\ 5\ 8}$

13. $-85 \div 5$

14. $-16 \div \dfrac{4}{7}$

15. $\dfrac{3}{7} \div \dfrac{9}{14}$

16. Round 4509 to the nearest ten.

17. Estimate the product by rounding to the nearest hundred. Show your work.
$$\begin{array}{r} 9\ 2\ 1 \\ \times\ 4\ 5\ 3 \\ \hline \end{array}$$

18. Find the absolute value: $|-479|$.

19. Simplify: $10^2 \div 5(-2) - 8(2 - 8)$.

20. Determine whether 98 is prime, composite, or neither.

21. Evaluate $a - b^2$ for $a = -5$ and $b = 4$.

Solve.

22. $a + 24 = 49$

23. $7x = 49$

24. $\dfrac{2}{9} \cdot a = -10$

25. $48 = -4t$

26. $2x + 13 = 3$

27. $-x = -10$

Combine like terms.

28. $8 - 4x - 13 + 9x$

29. $-12x + 7y + 15x$

Simplify, if possible.

30. $\dfrac{97}{97}$

31. $\dfrac{0}{81}$

32. $\dfrac{63x}{1}$

33. $\dfrac{-10}{54}$

Find the reciprocal.

34. $\dfrac{2}{5}$

35. 57

36. Find an equivalent expression for $\frac{3}{10}$ with a denominator of 70. Use multiplying by 1.

37. A 48-oz coffee pot is emptied into 6 mugs. How much will each mug hold if the coffee is poured out evenly?

38. A truck that gets 17 miles per gallon is traded in toward a van that gets 25 miles per gallon. How many more miles per gallon does the van get?

39. There are 7000 students at La Poloma College, and $\frac{5}{8}$ of them live in dorms. How many live in dorms?

40. A thermos of iced tea contained 3 qt of tea when it was $\frac{3}{5}$ full. How much tea could it hold when full?

41. Tony has jogged $\frac{2}{3}$ of a course that is $\frac{9}{10}$ of a mile long. How far has Tony gone?

Synthesis

42. Evaluate $\dfrac{ab}{c}$ for $a = -\dfrac{2}{5}$, $b = \dfrac{10}{13}$, and $c = \dfrac{26}{27}$.

43. Evaluate $-|xy|^2$ for $x = -\dfrac{3}{5}$ and $y = \dfrac{1}{2}$.

44. Wayne and Patty each earn $85 a day, while Janet earns $90 a day. They decide to pool their earnings from three days and spend $\frac{2}{5}$ of that on entertainment and save the rest. How much will Wayne, Patty, and Janet end up saving?

CHAPTER
4

Fraction Notation: Addition, Subtraction, and Mixed Numerals

STUDYING FOR SUCCESS   Preparing for a Test

☐ Make up your own test questions as you study.
☐ Do an overall review of the chapter focusing on the objectives and the examples.
☐ Do the exercises in the Mid-Chapter Review and in the end-of-chapter Review.
☐ Take the chapter test at the end of the chapter.

4.1 Least Common Multiples

OBJECTIVE

a Find the least common multiple, or LCM, of two or more numbers.

SKILL TO REVIEW

Objective 3.1a: Find some multiples of a number.

Multiply by 1, 2, 3, and so on, to find six multiples of each number.

1. 8   **2.** 25

1. Find the LCM of 9 and 15 by examining lists of multiples.

2. Find the LCM of 8 and 14 by examining lists of multiples.

In order to add or subtract fractions, all fractions must share a common denominator. Before discussing common denominators, we look at how to find the **least common multiple** of two or more numbers.

a FINDING LEAST COMMON MULTIPLES

LEAST COMMON MULTIPLE, LCM

The **least common multiple**, or **LCM**, of two natural numbers is the smallest number that is a multiple of both numbers.

EXAMPLE 1 Find the LCM of 20 and 30.

First, we list some multiples of 20 by multiplying 20 by 1, 2, 3, and so on:

20, 40, 60, 80, 100, 120, 140, 160, 180, 200, 220, 240, ...

Then we list some multiples of 30 by multiplying 30 by 1, 2, 3, and so on:

30, 60, 90, 120, 150, 180, 210, 240, ...

Now we determine the smallest number *common* to both lists. The LCM of 20 and 30 is 60.

◀ **Do Margin Exercises 1 and 2.**

Next we develop two more efficient methods for finding LCMs. You may choose to learn only one method. (Consult your instructor.) However, if you intend to study algebra, you should definitely learn method 2.

Method 1: Finding LCMs Using One List of Multiples

The first method for finding LCMs works especially well when the numbers are relatively small.

Method 1. To find the LCM of a set of numbers using a list of multiples:

1. Determine whether the largest number is a multiple of the others. If it is, it is the LCM. That is, if the largest number has the others as factors, the LCM is that number.

2. If not, check multiples of the largest number until you get one that is a multiple of each of the others.

Answers

Skill to Review:
1. 8, 16, 24, 32, 40, 48
2. 25, 50, 75, 100, 125, 150

Margin Exercises:
1. 45   **2.** 56

EXAMPLE 2 Find the LCM of 8 and 10.

1. 10 is the larger number, but it is not a multiple of 8.

2. Check multiples of 10:

$2 \cdot 10 = 20,$ Not a multiple of 8
$3 \cdot 10 = 30,$ Not a multiple of 8
$4 \cdot 10 = 40.$ A multiple of both 8 and 10

The LCM $= 40$.

EXAMPLE 3 Find the LCM of 4 and 14.

1. 14 is the larger number, but it is not a multiple of 4.

2. Check multiples of 14:

$2 \cdot 14 = 28.$ A multiple of 4

The LCM $= 28$.

EXAMPLE 4 Find the LCM of 8 and 32.

1. 32 is the larger number and 32 is a multiple of 8, so it is the LCM.

The LCM $= 32$.

Do Exercises 3–6. ▶

Find each LCM using one list of multiples.

3. 6, 9   **4.** 12, 15

5. 9, 36   **6.** 3, 20

To find the least common multiple of three numbers, we find the LCM of two of the numbers and then find the LCM of that number and the third number.

EXAMPLE 5 Find the LCM of 4, 10, and 15. We can start by finding the LCM of any two of the numbers. Let's use 10 and 15:

1. 15 is the larger number, but it is not a multiple of 10.

2. Check multiples of 15:

$2 \cdot 15 = 30.$ A multiple of 10. The LCM of 10 and 15 is 30.

Note now that any multiple of 30 will automatically be a multiple of 10 and 15. Thus, to find the LCM of 10, 15, and 4, we need only find the LCM of 30 and 4:

1. 30 is not a multiple of 4.

2. Check multiples of 30:

$2 \cdot 30 = 60.$ Since it is a multiple of 4, we know that 60 is the LCM of 30 and 4.

The LCM of 4, 10, and 15 is 60.

Do Exercises 7 and 8. ▶

Find each LCM using lists of multiples.

7. 30, 40, 50   **8.** 10, 20, 40

Method 2: Finding LCMs Using Prime Factorizations

A second method for finding LCMs uses prime factorizations and is usually the best method when the numbers are large. Consider again 20 and 30. Their prime factorizations are $20 = 2 \cdot 2 \cdot 5$ and $30 = 2 \cdot 3 \cdot 5$. Let's look at these prime factorizations in order to find the LCM. Any multiple of 20 will have to have *two* 2's as factors and *one* 5 as a factor. Any multiple of 30 will need to have *one* 2, *one* 3, and *one* 5 as factors.

Answers

3. 18   **4.** 60   **5.** 36   **6.** 60   **7.** 600
8. 40

The smallest number satisfying these conditions is

——— Two 2's, one 5; $2 \cdot 2 \cdot 3 \cdot 5$ is a multiple of 20.

$2 \cdot 2 \cdot 3 \cdot 5$.

——— One 2, one 3, one 5; $2 \cdot 2 \cdot 3 \cdot 5$ is a multiple of 30.

Thus, the LCM of 20 and 30 is $2 \cdot 2 \cdot 3 \cdot 5$, or 60. It has all the factors of 20 and all the factors of 30, but the factors are not repeated when they are common to both numbers.

Note that each prime factor is used the greatest number of times that it occurs in either of the individual factorizations.

Method 2. To find the LCM of two numbers using prime factorizations:

1. Write the prime factorization of each number.
2. Select one of the factorizations and see whether it contains the other.
   a) If it does, it is the LCM.
   b) If it does not, multiply that factorization by those prime factors of the other number that it lacks. The final product is the LCM.
3. As a check, make sure that the LCM includes each factor the greatest number of times that it occurs in either factorization.

EXAMPLE 6 Find the LCM of 18 and 21.

1. We begin by writing the prime factorization of each number:

   $18 = 2 \cdot 3 \cdot 3$ and $21 = 3 \cdot 7$.

2. We select the factorization of 18: $2 \cdot 3 \cdot 3$.
   a) We note that $2 \cdot 3 \cdot 3$ does not contain the other factorization, $3 \cdot 7$.
   b) To find the LCM of 18 and 21, we multiply $2 \cdot 3 \cdot 3$ by the factor of 21 that it lacks, 7:

   ——— 18 is a factor.

   $\text{LCM} = 2 \cdot 3 \cdot 3 \cdot 7$.

   ——— 21 is a factor.

3. The greatest number of times that 2 occurs as a factor of 18 or 21 is **one** time; the greatest number of times that 3 occurs as a factor of 18 or 21 is **two** times; and the greatest number of times that 7 occurs as a factor of 18 or 21 is **one** time. To check, note that the LCM has exactly **one** 2, **two** 3's, and **one** 7. The LCM is $2 \cdot 3 \cdot 3 \cdot 7$, or 126.

EXAMPLE 7 Find the LCM of 7 and 21.

1. Because 7 is prime, we think of $7 = 7$ as a "factorization":

   $7 = 7$ and $21 = 3 \cdot 7$.

2. One factorization, $3 \cdot 7$, contains the other. Thus, the LCM is $3 \cdot 7$, or 21.

EXAMPLE 8 Find the LCM of 24 and 36.

1. We write the prime factorization of each number:

   $24 = 2 \cdot 2 \cdot 2 \cdot 3$ and $36 = 2 \cdot 2 \cdot 3 \cdot 3$.

2. We select the factorization of 24.

   a) The factorization of 24 does not contain the factorization of 36.

   b) To find the LCM of 24 and 36, we multiply the factorization of 24 by any prime factors of 36 that it lacks. We need another factor of 3.

   24 is a factor.

   $LCM = 2 \cdot 2 \cdot 2 \cdot 3 \cdot 3$.

   36 is a factor.

3. Note that the LCM includes 2 and 3 the greatest number of times that each appears as a factor of either 24 or 36. The LCM is $2 \cdot 2 \cdot 2 \cdot 3 \cdot 3$, or 72.

Do Exercises 9–12. ▶

Exponential notation is often helpful when writing least common multiples. Let's reconsider Example 8 using exponents. The largest exponents indicate the greatest number of times that 2 and 3 occur as factors.

$24 = 2 \cdot 2 \cdot 2 \cdot 3 = 2^3 \cdot 3^1$ 2^3 is the greatest power of 2

$36 = 2 \cdot 2 \cdot 3 \cdot 3 = 2^2 \cdot 3^2$ 3^2 is the greatest power of 3

$LCM = 2 \cdot 2 \cdot 2 \cdot 3 \cdot 3 = 2^3 \cdot 3^2$, or 72.

Note that the greatest power of each factor is used to construct the LCM.

Lining up the different prime numbers in the factorizations can help us construct the LCM. This method also works well when finding the LCM of more than two numbers.

EXAMPLE 9 Find the LCM of 27, 90, and 84.

We find the prime factorization of each number and write the factorizations in exponential notation.

$27 = 3 \cdot 3 \cdot 3 = 3^3$

$90 = 2 \cdot 3 \cdot 3 \cdot 5 = 2 \cdot 3^2 \cdot 5$

$84 = 2 \cdot 2 \cdot 3 \cdot 7 = 2^2 \cdot 3 \cdot 7$

No one factorization contains the others. The prime numbers 2, 3, 5, and 7 appear as factors.

We write the factorizations, lining up all the powers of 2, the powers of 3, and so on.

$27 = 3^3$

$90 = 2 \cdot 3^2 \cdot 5$

$84 = 2^2 \cdot 3 \cdot 7$

The LCM is formed by choosing the greatest power of each factor:

$2^2 \cdot 3^3 \cdot 5 \cdot 7 = 3780.$

The LCM of 27, 90, and 84 is 3780.

Do Exercises 13 and 14. ▶

Use prime factorizations to find the LCM.

9. 8, 10   **10.** 5, 30

11. 12, 48

 12. 18, 40

   1. $18 = 2 \cdot 3 \cdot$ ☐

   $40 = 2 \cdot 2 \cdot 2 \cdot$ ☐

   2. Select the factorization of 40:

   $2 \cdot 2 \cdot 2 \cdot 5.$

   This is not a multiple of 18. We need two factors of ☐.

   3. $LCM = 2 \cdot 2 \cdot 2 \cdot 5 \cdot$ ☐ $\cdot$ ☐

   $= $ ☐

Find the LCM.

13. 8, 18, 30   **14.** 10, 20, 25

EXAMPLE 10 Find the LCM of 8 and 25.

We write the prime factorization of each number in exponential notation.

$$8 = 2 \cdot 2 \cdot 2 = 2^3$$
$$25 = 5 \cdot 5 = 5^2$$

The prime numbers 2 and 5 appear as factors. We write the factorizations as

$$8 = 2^3$$
$$25 = 5^2.$$

Note that the two numbers, 8 and 25, have no common prime factor. When this is the case, the LCM is just the product of the two numbers. Thus, the LCM is $2^3 \cdot 5^2 = 8 \cdot 25 = 200$.

Find the LCM.

15. 4, 9   **16.** 5, 6, 7

◀ **Do Exercises 15 and 16.**

The same method works perfectly with variables.

EXAMPLE 11 Find the LCM of $7a^2b$ and ab^3.

We have the following factorizations:

$$7a^2b = 7 \cdot a \cdot a \cdot b \quad \text{and} \quad ab^3 = a \cdot b \cdot b \cdot b.$$

No one factorization contains the other.

Consider the factorization of $7a^2b$, which is $7 \cdot a \cdot a \cdot b$. Since ab^3 contains two more factors of b, we multiply the factorization of $7a^2b$ by $b \cdot b$.

$$7 \cdot a \cdot a \cdot b \cdot b \cdot b$$

$7a^2b$ is a factor.

ab^3 is a factor.

Find the LCM.

17. xy, yz

As a second approach, we find the greatest power of each factor using exponential notation.

$$7a^2b = 7 \cdot a^2 \cdot b$$
$$ab^3 = a \cdot b^3$$

The LCM is $7 \cdot a \cdot a \cdot b \cdot b \cdot b$, or $7a^2b^3$.

18. $5a^2, a^3b$ **GS**
$$5a^2 = 5 \cdot \boxed{}$$
$$a^3b = a^3 \cdot \boxed{}$$
$$\text{LCM} = 5 \cdot a^3 \cdot \boxed{}, \text{ or } \boxed{}$$

◀ **Do Exercises 17 and 18.**

EXAMPLE 12 Find the LCM of $12x^2y^3z$ and $18x^4z^3$.

We write the factorizations using exponential notation:

$$12x^2y^3z = 2^2 \cdot 3 \cdot x^2 \cdot y^3 \cdot z \quad \text{and} \quad 18x^4z^3 = 2 \cdot 3^2 \cdot x^4 \cdot z^3.$$

No one factorization contains the other.

We form the LCM using the greatest power of each factor.

$$12x^2y^3z = 2^2 \cdot 3 \cdot x^2 \cdot y^3 \cdot z$$
$$18x^4z^3 = 2 \cdot 3^2 \cdot x^4 \cdot z^3$$

The LCM is $2^2 \cdot 3^2 \cdot x^4 \cdot y^3 \cdot z^3$, or $36x^4y^3z^3$.

19. Find the LCM of $8a^3b^2$ and $10a^2c^4$.

◀ **Do Exercise 19.**

Answers

15. 36   **16.** 210   **17.** xyz **18.** $5a^3b$

19. $40a^3b^2c^4$

Guided Solution:

18. $a^2, b; b, 5a^3b$

✓ Reading Check

Determine whether each statement is true or false.

_____ **RC1.** Any two numbers have more than one common multiple.

_____ **RC2.** If one number is a multiple of a second number, the larger number is the LCM of the two numbers.

_____ **RC3.** If two numbers have no common prime factor, then the LCM of the numbers is their product.

_____ **RC4.** LCMs cannot be found using prime factorizations.

a Find the LCM of each set of numbers or expressions.

1. 2, 4   **2.** 3, 15   **3.** 10, 25   **4.** 10, 15   **5.** 20, 40

6. 8, 12   **7.** 18, 27   **8.** 9, 11   **9.** 30, 50   **10.** 8, 36

11. 30, 40   **12.** 21, 27   **13.** 18, 24   **14.** 12, 18   **15.** 60, 70

16. 35, 45   **17.** 16, 36   **18.** 24, 32   **19.** 18, 20   **20.** 36, 48

21. 2, 3, 7   **22.** 2, 5, 9   **23.** 3, 6, 15   **24.** 6, 12, 18   **25.** 24, 36, 12

26. 8, 16, 22   **27.** 5, 12, 15   **28.** 12, 18, 40   **29.** 9, 12, 6   **30.** 8, 16, 12

31. 180, 100, 450 **32.** 18, 30, 50, 48 **33.** 8, 48   **34.** 16, 32   **35.** 10, 21

36. 14, 15   **37.** 75, 100   **38.** 81, 90   **39.** 12, 15, 60   **40.** 24, 36, 72

41. ab, bc **42.** $7x, xy$ **43.** $3x, 9x^2$ **44.** $10x^4, 5x^3$ **45.** $4x^3, x^2y$

46. $6ab^2, 9a^3b$ **47.** $6r^3st^4, 8rs^2t$ **48.** $3m^2n^4p^5, 9mn^2p^4$ **49.** a^3b, b^2c, ac^2 **50.** x^2z^3, x^3y, y^2z

Applications of LCMs: Planet Orbits.   Jupiter, Saturn, and Uranus all revolve around the sun. Jupiter takes 12 yr, Saturn 30 yr, and Uranus 84 yr to make a complete revolution. On a certain night, you look at Jupiter, Saturn, and Uranus and wonder how many years it will take before they have the same position again. (*Hint*: The number of years is the LCM of 12, 30, and 84.)

Source: *The Handy Science Answer Book*

51. How often will Jupiter and Saturn appear in the same direction in the night sky as seen from the earth?

52. How often will Jupiter and Uranus appear in the same direction in the night sky as seen from the earth?

53. How often will Saturn and Uranus appear in the same direction in the night sky as seen from the earth?

54. How often will Jupiter, Saturn, and Uranus appear in the same direction in the night sky as seen from the earth?

Skill Maintenance

Perform the indicated operation and, if possible, simplify.

55. $-38 + 52$ [2.2a]

56. $-18 \div \left(\dfrac{2}{3} \right)$ [3.7b]

57. $23 \cdot 345$ [1.4a]

58. $\dfrac{4}{5} \cdot \dfrac{10}{12}$ [3.6a]

59. $\dfrac{4}{5} \div \left(-\dfrac{7}{10} \right)$ [3.7b]

60. $382 - 549$ [2.3a]

Synthesis

Use a calculator and the multiples method to find the LCM of each pair of numbers.

61. 288; 324   **62.** 2700; 7800   **63.** 7719; 18,011   **64.** 17,385; 24,339

65. The tables at a flea market are either 6 ft long or 8 ft long. Each row consists entirely of 6-ft tables or entirely of 8-ft tables, and all rows are the same length. What is the shortest possible length of the rows at the flea market?

African Artistry. In southern Africa, the design of every woven handbag, or *gipatsi* (plural *sipatsi*), is created by repeating two or more geometric patterns. Each pattern encircles the bag, sharing the strands of fabric with any pattern above or below. The length, or period, of each pattern is the number of strands required to construct the pattern. For a gipatsi to be considered beautiful, each individual pattern must fit a whole number of times around the bag.

Source: Gerdes, Paulus. *Women, Art and Geometry in Southern Africa.* Asmara, Eritrea: Africa World Press, Inc., p. 5.

66. A weaver is using two patterns to create a gipatsi. Pattern A is 10 strands long, and pattern B is 3 strands long. What is the smallest number of strands that can be used to complete the gipatsi?

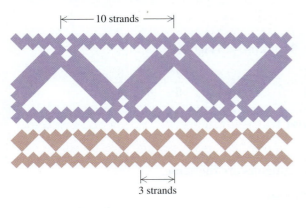

← 10 strands →

3 strands

67. A weaver is using a four-strand pattern, a six-strand pattern, and an eight-strand pattern. What is the smallest number of strands that can be used to complete the gipatsi?

68. *Prescriptions.* Prescriptions for a 30-day supply of simvastatin and a 14-day supply of pain medication are filled at a pharmacy. Assuming the prescriptions are refilled regularly, how long will it be until they are both refilled on the same day?

69. Consider a^3b^2 and a^2b^5. Determine whether each of the following is the LCM of a^3b^2 and a^2b^5. Tell why or why not.
  a) a^3b^3
  b) a^2b^5
  c) a^3b^5

70. Use Example 9 to help find the LCM of 27, 90, 84, 210, 108, and 50.

71. Use Examples 6 and 8 to help find the LCM of 18, 21, 24, 36, 63, 56, and 20.

72. Find three different pairs of numbers for which 56 is the LCM. Do not use 56 itself in any of the pairs.

73. Find three different pairs of numbers for which 54 is the LCM. Do not use 54 itself in any of the pairs.

OBJECTIVES

a   Add using fraction notation when denominators are the same.

b   Add using fraction notation when denominators are different.

c   Use < or > to form a true statement with fraction notation.

d   Solve problems involving addition with fraction notation.

SKILL TO REVIEW

Objective 3.2c: Find the prime factorization of a composite number.

Find the prime factorization of each number.

1. 96   **2.** 1400

1. Find $\dfrac{1}{5} + \dfrac{3}{5}$.

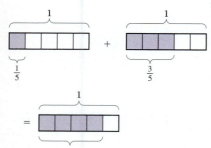

Add and, if possible, simplify.

2. $\dfrac{5}{13} + \dfrac{9}{13}$ **3.** $\dfrac{1}{3} + \dfrac{2}{3}$

4. $\dfrac{-5}{12} + \left(\dfrac{-1}{12}\right)$ **5.** $\dfrac{3}{x} + \dfrac{-7}{x}$

Answers

See next page.

a   LIKE DENOMINATORS

Addition using fraction notation corresponds to combining or putting like things together, just as when we combined like terms. For example,

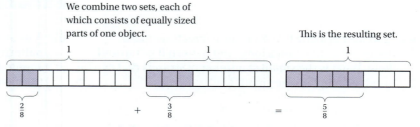

We combine two sets, each of which consists of equally sized parts of one object.

This is the resulting set.

2 eighths + 3 eighths = 5 eighths,

or $2 \cdot \dfrac{1}{8} + 3 \cdot \dfrac{1}{8} = 5 \cdot \dfrac{1}{8},$ or $\dfrac{2}{8} + \dfrac{3}{8} = \dfrac{5}{8}.$

◀ **Do Margin Exercise 1.**

To add when denominators are the same,

a) add the numerators,

b) keep the denominator, and

$$\dfrac{2}{6} + \dfrac{5}{6} = \dfrac{2+5}{6} = \dfrac{7}{6}$$

c) simplify, if possible.

EXAMPLES   Add and, if possible, simplify.

1. $\dfrac{2}{4} + \dfrac{1}{4} = \dfrac{2+1}{4} = \dfrac{3}{4}$ No simplifying is possible.

2. $\dfrac{3}{12} + \dfrac{5}{12} = \dfrac{3+5}{12} = \dfrac{8}{12}$ Adding numerators; the denominator remains unchanged.

$= \dfrac{4}{4} \cdot \dfrac{2}{3} = \dfrac{2}{3}$ Simplifying by removing a factor equal to 1: $\frac{4}{4} = 1$

3. $\dfrac{-11}{6} + \dfrac{3}{6} = \dfrac{-11+3}{6} = \dfrac{-8}{6}$

$= \dfrac{2}{2} \cdot \dfrac{-4}{3} = \dfrac{-4}{3}, \text{ or } -\dfrac{4}{3}$ Removing a factor equal to 1: $\frac{2}{2} = 1$

4. $-\dfrac{2}{a} + \left(-\dfrac{3}{a}\right) = \dfrac{-2}{a} + \dfrac{-3}{a}$ Recall that $-\dfrac{m}{n} = \dfrac{-m}{n}$.

$= \dfrac{-2+(-3)}{a} = \dfrac{-5}{a}, \text{ or } -\dfrac{5}{a}$

◀ **Do Exercises 2–5.**

We may need to add fractions when combining like terms.

EXAMPLE 5 Simplify by combining like terms: $\frac{2}{7}x + \frac{3}{7}x$.

$$\frac{2}{7}x + \frac{3}{7}x = \left(\frac{2}{7} + \frac{3}{7}\right)x \qquad \text{Try to do this step mentally.}$$

$$= \frac{5}{7}x$$

Do Exercises 6 and 7. ▶

Simplify by combining like terms.

6. $\frac{3}{10}a + \frac{1}{10}a$ 7. $-\frac{3}{4}x + \frac{1}{4}x$

b DIFFERENT DENOMINATORS

We cannot add $\frac{1}{2} + \frac{1}{3}$ by simply adding numerators. However, by rewriting $\frac{1}{2}$ as $\frac{1}{2} \cdot \frac{3}{3} = \frac{3}{6}$ and $\frac{1}{3}$ as $\frac{1}{3} \cdot \frac{2}{2} = \frac{2}{6}$, we can determine the sum.

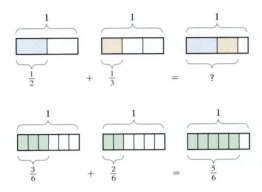

When denominators are different, we can find a common denominator by multiplying by 1. Consider the addition $\frac{3}{4} + \frac{1}{6}$ using two different denominators.

A. We use 24 as a common denominator:

$$\frac{3}{4} + \frac{1}{6} = \frac{3}{4} \cdot \frac{6}{6} + \frac{1}{6} \cdot \frac{4}{4}$$

$$= \frac{18}{24} + \frac{4}{24} = \frac{22}{24} = \frac{11}{12}.$$

B. We use 12 as a common denominator:

$$\frac{3}{4} + \frac{1}{6} = \frac{3}{4} \cdot \frac{3}{3} + \frac{1}{6} \cdot \frac{2}{2}$$

$$= \frac{9}{12} + \frac{2}{12} = \frac{11}{12}.$$

We had to simplify at the end of (A), but not in (B). In (B), we used the *least* common multiple of the denominators, 12, as the common denominator. That number is called the **least common denominator**, or **LCD**. We may still need to simplify when using the LCD, but it is usually easier than when we use a larger denominator.

To add when denominators are different:

a) Find the least common multiple of the denominators. That number is the least common denominator, LCD.

b) Multiply by 1, writing 1 in the form of n/n, to express each fraction in an equivalent form that contains the LCD.

c) Add the numerators, keeping the same denominator.

d) Simplify, if possible.

Answers

Skill to Review:
1. $2 \cdot 2 \cdot 2 \cdot 2 \cdot 2 \cdot 3$
2. $2 \cdot 2 \cdot 2 \cdot 5 \cdot 5 \cdot 7$

Margin Exercises:

1. $\frac{4}{5}$ 2. $\frac{14}{13}$ 3. 1   4. $-\frac{1}{2}$ 5. $-\frac{4}{x}$

6. $\frac{2}{5}a$ 7. $-\frac{1}{2}x$

EXAMPLE 6 Add: $\dfrac{1}{8} + \dfrac{3}{4}$.

a) Since 4 is a factor of 8, the LCM of 4 and 8 is 8. Thus, the LCD is 8.

b) We need to find a fraction equivalent to $\dfrac{3}{4}$ with a denominator of 8:

$$\frac{1}{8} + \frac{3}{4} = \frac{1}{8} + \frac{3}{4} \cdot \frac{2}{2}. \qquad \textit{Think}: 4 \times \square = 8. \text{ The answer is 2,}$$
$$\text{so we multiply by 1, using } \tfrac{2}{2}.$$

c) We add: $\dfrac{1}{8} + \dfrac{6}{8} = \dfrac{7}{8}$. Adding numerators

d) No simplifying is possible. The sum is $\dfrac{7}{8}$.

In Examples 7–10, we follow the same steps without labeling them.

EXAMPLE 7 Add: $\dfrac{5}{6} + \dfrac{1}{9}$.

The LCD is 18. $6 = 2 \cdot 3$ and $9 = 3 \cdot 3$,
so the LCM of 6 and 9 is $2 \cdot 3 \cdot 3$, or 18.

$$\frac{5}{6} + \frac{1}{9} = \frac{5}{6} \cdot \frac{3}{3} + \frac{1}{9} \cdot \frac{2}{2} \qquad \textit{Think}: 9 \times \square = 18. \text{ The answer is 2, so}$$
$$\text{we multiply by 1, using } \tfrac{2}{2}.$$

$$\textit{Think}: 6 \times \square = 18. \text{ The answer is 3, so}$$
$$\text{we multiply by 1, using } \tfrac{3}{3}.$$

$$= \frac{15}{18} + \frac{2}{18} = \frac{17}{18}$$

◀ **Do Exercises 8 and 9.**

EXAMPLE 8 Add: $\dfrac{3}{-5} + \dfrac{11}{10}$.

$$\frac{3}{-5} + \frac{11}{10} = \frac{-3}{5} + \frac{11}{10}$$

Recall that $\dfrac{m}{-n} = \dfrac{-m}{n}$. We generally avoid negative signs in the denominator. The LCD is 10.

$$= \frac{-3}{5} \cdot \frac{2}{2} + \frac{11}{10}$$

$$= \frac{-6}{10} + \frac{11}{10}$$

$$= \frac{5}{10} = \frac{1}{2}$$

> We may still have to simplify, but simplifying is almost always easier if the LCD has been used.

◀ **Do Exercise 10.**

EXAMPLE 9 Add: $\dfrac{5}{8} + 2$.

$$\frac{5}{8} + 2 = \frac{5}{8} + \frac{2}{1} \qquad \text{Rewriting 2 in fraction notation}$$

$$= \frac{5}{8} + \frac{2}{1} \cdot \frac{8}{8} \qquad \text{The LCD is 8.}$$

$$= \frac{5}{8} + \frac{16}{8} = \frac{21}{8}$$

◀ **Do Exercise 11.**

Add using the least common denominator.

8. $\dfrac{2}{3} + \dfrac{1}{6}$

9. $\dfrac{3}{8} + \dfrac{5}{6}$ **GS**

The LCD is [ ].

$$\frac{3}{8} + \frac{5}{6} = \frac{3}{8} \cdot 1 + \frac{5}{6} \cdot 1$$

$$= \frac{3}{8} \cdot \frac{3}{3} + \frac{5}{6} \cdot \frac{}{}$$

$$= \frac{}{24} + \frac{}{24}$$

$$= \frac{}{24}$$

10. Add: $\dfrac{1}{-6} + \dfrac{7}{18}$.

11. Add: $7 + \dfrac{3}{5}$.

Answers

8. $\dfrac{5}{6}$ **9.** $\dfrac{29}{24}$ **10.** $\dfrac{2}{9}$ **11.** $\dfrac{38}{5}$

Guided Solution:
9. $24; \dfrac{4}{4}, 9, 20, 29$

232 CHAPTER 4 Fraction Notation: Addition, Subtraction, and Mixed Numerals

EXAMPLE 10 Add: $\dfrac{9}{70} + \dfrac{11}{21} + \dfrac{-4}{15}$.

We need to determine the LCM of 70, 21, and 15:

$$70 = 2 \cdot 5 \cdot 7, \qquad 21 = 3 \cdot 7, \qquad 15 = 3 \cdot 5.$$

The LCM is $2 \cdot 3 \cdot 5 \cdot 7$, or 210.

$$\frac{9}{70} + \frac{11}{21} + \frac{-4}{15} = \frac{9}{70} \cdot \frac{3}{3} + \frac{11}{21} \cdot \frac{2 \cdot 5}{2 \cdot 5} + \frac{-4}{15} \cdot \frac{7 \cdot 2}{7 \cdot 2}$$

In each case, we multiply by 1 to obtain the LCD. To form 1, look at the prime factorization of the LCD and use the factor(s) missing from each denominator.

$$= \frac{9 \cdot 3}{70 \cdot 3} + \frac{11 \cdot 10}{21 \cdot 10} + \frac{-4 \cdot 14}{15 \cdot 14}$$

$$= \frac{27}{210} + \frac{110}{210} + \frac{-56}{210}$$

$$= \frac{137 + (-56)}{210} = \frac{81}{210}$$

$$= \frac{3 \cdot 3 \cdot 3 \cdot 3}{2 \cdot 3 \cdot 5 \cdot 7} = \frac{3}{3} \cdot \frac{3 \cdot 3 \cdot 3}{2 \cdot 5 \cdot 7} = \frac{27}{70}$$

Do Exercises 12 and 13. ▶

Add.

12. $\dfrac{4}{10} + \dfrac{1}{100} + \dfrac{3}{1000}$

13. $\dfrac{7}{10} + \dfrac{-2}{21} + \dfrac{1}{7}$

c ORDER

When two fractions share a common denominator, the larger number can be found by comparing numerators. For example, 4 is greater than 3, so $\frac{4}{5}$ is greater than $\frac{3}{5}$.

$$\frac{4}{5} > \frac{3}{5}$$

Similarly, because -6 is less than -2, we have

$$\frac{-6}{7} < \frac{-2}{7}, \quad \text{or} \quad -\frac{6}{7} < -\frac{2}{7}.$$

Do Exercises 14–16. ▶

Use $<$ or $>$ for ☐ to form a true sentence.

14. $\dfrac{3}{8}$ ☐ $\dfrac{5}{8}$

15. $\dfrac{7}{10}$ ☐ $\dfrac{6}{10}$ **16.** $\dfrac{-2}{9}$ ☐ $\dfrac{-5}{9}$

EXAMPLE 11 Use $<$ or $>$ for ☐ to form a true sentence:

$$\frac{5}{8} \,\square\, \frac{2}{3}.$$

You can confirm that the LCD is 24. We multiply by 1 to find two fractions equivalent to $\frac{5}{8}$ and $\frac{2}{3}$ with denominators the same:

$$\frac{5}{8} \cdot \frac{3}{3} = \frac{15}{24}; \quad \frac{2}{3} \cdot \frac{8}{8} = \frac{16}{24}. \qquad \frac{5}{8} \,\square\, \frac{2}{3} \text{ is equivalent to } \frac{15}{24} \,\square\, \frac{16}{24}.$$

Since $15 < 16$, it follows that $\frac{15}{24} < \frac{16}{24}$. Thus,

$$\frac{5}{8} < \frac{2}{3}.$$

Use < or > for ☐ to form a true sentence.

17. $\dfrac{2}{3}$ ☐ $\dfrac{3}{4}$ **18.** $\dfrac{-3}{4}$ ☐ $\dfrac{-8}{12}$

19. $\dfrac{7}{8}$ ☐ $\dfrac{5}{6}$

EXAMPLE 12 Use $<$ or $>$ for ☐ to form a true sentence:

$$-\dfrac{89}{100} \;\square\; -\dfrac{9}{10}.$$

We rewrite $-\dfrac{9}{10}$ with a denominator of 100: $-\dfrac{9}{10} \cdot \dfrac{10}{10} = -\dfrac{90}{100}$. We then have

$$\dfrac{-89}{100} \;\square\; \dfrac{-90}{100}. \qquad \text{Recall that } -\dfrac{m}{n} = \dfrac{-m}{n}.$$

Since $-89 > -90$, it follows that $-\dfrac{89}{100} > -\dfrac{90}{100}$, so

$$-\dfrac{89}{100} > -\dfrac{9}{10}.$$

◀ **Do Exercises 17–19.**

d **APPLICATIONS AND PROBLEM SOLVING**

EXAMPLE 13 *Construction.* A contractor uses two layers of subflooring under a ceramic tile floor. First, she installs a $\frac{3}{4}$-in. layer of oriented strand board (OSB). Then a $\frac{1}{2}$-in. sheet of cement board is mortared to the OSB. The mortar is $\frac{1}{8}$-in. thick. What is the total thickness of the two installed subfloors?

1. **Familiarize.** We first make a drawing. We let $T =$ the total thickness of the subfloors.

2. **Translate.** The problem can be translated to an equation as follows.

OSB	plus	mortar	plus	cement board	is	total thickness
$\dfrac{3}{4}$	$+$	$\dfrac{1}{8}$	$+$	$\dfrac{1}{2}$	$=$	T

3. **Solve.** To solve the equation, we carry out the addition.

$$\dfrac{3}{4} + \dfrac{1}{8} + \dfrac{1}{2} = T \qquad \text{The LCM of the denominators is 8.}$$

$$\dfrac{3}{4} \cdot \dfrac{2}{2} + \dfrac{1}{8} + \dfrac{1}{2} \cdot \dfrac{4}{4} = T \qquad \text{Multiplying by 1 to obtain the LCD}$$

$$\dfrac{6}{8} + \dfrac{1}{8} + \dfrac{4}{8} = T$$

$$\dfrac{11}{8} = T$$

4. **Check.** We check by repeating the calculation.

5. **State.** The total thickness of the installed subfloors is $\frac{11}{8}$ in.

◀ **Do Exercise 20.**

---

20. *Catering.* A caterer prepares a mixed berry salad with $\frac{7}{8}$ qt of strawberries, $\frac{3}{4}$ qt of raspberries, and $\frac{5}{16}$ qt of blueberries. How many quarts of berries are in the salad? **GS**

1. **Familiarize.** Let $T =$ the total amount of berries in the salad.
2. **Translate.** To find the total amount, we add.

$$\dfrac{7}{\square} + \dfrac{3}{\square} + \dfrac{5}{\square} = T$$

3. **Solve.** The LCD is ☐.

$$\dfrac{7}{8} \cdot \dfrac{2}{2} + \dfrac{3}{4} \cdot \dfrac{\square}{\square} + \dfrac{5}{16} = T$$

$$\dfrac{\square}{16} + \dfrac{\square}{16} + \dfrac{5}{16} = T$$

$$\dfrac{\square}{16} = T$$

4. **Check.** The answer is reasonable because it is larger than any of the individual amounts.
5. **State.** There are ☐ qt of berries in the salad.

Answers

17. $<$ **18.** $<$ **19.** $>$ **20.** $\dfrac{31}{16}$ qt

Guided Solution:

20. 8, 4, 16; 16; $\dfrac{4}{4}$, 14, 12, 31; $\dfrac{31}{16}$

☑ Reading Check

Determine whether each statement is true or false.

_____ **RC1.** Before we can add two fractions, they must have the same denominator.

_____ **RC2.** To add fractions, we add numerators and add denominators.

_____ **RC3.** If we use the LCD to add fractions, we never need to simplify the result.

_____ **RC4.** Adding fractions with different denominators involves multiplying at least one fraction by 1.

a, **b**   Add and, if possible, simplify.

1. $\dfrac{4}{9} + \dfrac{1}{9}$

2. $\dfrac{3}{11} + \dfrac{5}{11}$

3. $\dfrac{4}{7} + \dfrac{3}{7}$

4. $\dfrac{7}{8} + \dfrac{1}{8}$

5. $\dfrac{7}{10} + \dfrac{3}{-10}$

6. $\dfrac{1}{-6} + \dfrac{5}{6}$

7. $\dfrac{9}{a} + \dfrac{4}{a}$

8. $\dfrac{2}{t} + \dfrac{3}{t}$

9. $\dfrac{-1}{4} + \dfrac{-1}{4}$

10. $\dfrac{7}{12} + \dfrac{-5}{12}$

11. $\dfrac{2}{9}x + \dfrac{5}{9}x$

12. $\dfrac{3}{11}a + \dfrac{2}{11}a$

13. $\dfrac{3}{32}t + \dfrac{13}{32}t$

14. $\dfrac{3}{25}x + \dfrac{12}{25}x$

15. $-\dfrac{2}{x} + \left(-\dfrac{7}{x}\right)$

16. $-\dfrac{7}{a} + \dfrac{5}{a}$

17. $\dfrac{1}{8} + \dfrac{1}{6}$

18. $\dfrac{1}{9} + \dfrac{1}{6}$

19. $\dfrac{-4}{5} + \dfrac{7}{10}$

20. $\dfrac{-3}{4} + \dfrac{-1}{12}$

21. $\dfrac{7}{12} + \dfrac{3}{8}$

22. $\dfrac{7}{8} + \dfrac{1}{16}$

23. $\dfrac{3}{20} + 4$

24. $\dfrac{2}{15} + 3$

25. $\dfrac{5}{-8} + \dfrac{5}{6}$

26. $\dfrac{5}{-6} + \dfrac{7}{9}$

27. $\dfrac{3}{10}x + \dfrac{7}{100}x$

28. $\dfrac{9}{20}a + \dfrac{3}{40}a$

29. $\dfrac{5}{12} + \dfrac{8}{15}$

30. $\dfrac{3}{16} + \dfrac{1}{12}$

31. $\dfrac{7}{8} + \dfrac{0}{1}$

32. $\dfrac{0}{6} + \dfrac{5}{3}$

33. $\dfrac{-7}{10} + \dfrac{-29}{100}$

34. $\dfrac{-3}{10} + \dfrac{-27}{100}$

35. $-\dfrac{1}{10}x + \dfrac{1}{15}x$

36. $-\dfrac{1}{6}x + \dfrac{1}{4}x$

37. $-5t + \dfrac{2}{7}t$

38. $-4x + \dfrac{3}{5}x$

39. $-\dfrac{5}{12} + \dfrac{7}{-24}$

40. $-\dfrac{1}{18} + \dfrac{5}{-12}$

41. $\dfrac{3}{16} + \dfrac{5}{16} + \dfrac{4}{16}$

42. $\dfrac{3}{8} + \dfrac{1}{8} + \dfrac{2}{8}$

43. $\dfrac{4}{10} + \dfrac{3}{100} + \dfrac{7}{1000}$

44. $\dfrac{7}{10} + \dfrac{2}{100} + \dfrac{9}{1000}$

45. $\dfrac{3}{10} + \dfrac{5}{12} + \dfrac{8}{15}$

46. $\dfrac{1}{2} + \dfrac{3}{8} + \dfrac{1}{4}$

47. $\dfrac{5}{6} + \dfrac{25}{52} + \dfrac{7}{4}$

48. $\dfrac{15}{24} + \dfrac{7}{36} + \dfrac{91}{48}$

49. $\dfrac{2}{9} + \dfrac{7}{10} + \dfrac{-4}{15}$

50. $\dfrac{5}{12} + \dfrac{-3}{8} + \dfrac{1}{10}$

51. $-\dfrac{3}{4} + \dfrac{1}{5} + \dfrac{-7}{10}$

52. $\dfrac{1}{3} + \dfrac{-7}{9} + \dfrac{-1}{2}$

c Use < or > for ☐ to form a true sentence.

53. $\dfrac{3}{8} \ \square \ \dfrac{2}{8}$

54. $\dfrac{7}{9} \ \square \ \dfrac{5}{9}$

55. $\dfrac{2}{3} \ \square \ \dfrac{5}{6}$

56. $\dfrac{11}{18} \ \square \ \dfrac{5}{9}$

57. $\dfrac{-2}{7} \ \square \ \dfrac{-5}{7}$

58. $\dfrac{-4}{5} \ \square \ \dfrac{-3}{5}$

59. $\dfrac{9}{15} \ \square \ \dfrac{7}{10}$

60. $\dfrac{5}{14} \ \square \ \dfrac{8}{21}$

61. $\dfrac{3}{4} \ \square \ -\dfrac{1}{5}$

62. $\dfrac{3}{8} \ \square \ -\dfrac{13}{16}$

63. $\dfrac{-7}{20} \ \square \ \dfrac{-6}{15}$

64. $\dfrac{-7}{12} \ \square \ \dfrac{-9}{16}$

Arrange each group of fractions from smallest to largest.

65. $\dfrac{3}{10}, \dfrac{5}{12}, \dfrac{4}{15}$

66. $\dfrac{5}{6}, \dfrac{19}{21}, \dfrac{11}{14}$

d Solve.

67. *Riding a Segway.*® Tate rode a Segway® Personal Transporter $\frac{5}{6}$ mi to the library, then $\frac{3}{4}$ mi to class, and then $\frac{3}{2}$ mi to his part-time job. How far did he ride his Segway®?

68. *Volunteering.* For a community project, an earth science class volunteered one hour per day for three days to join the state highway beautification project. The students collected trash along a $\frac{4}{5}$-mi stretch of highway the first day, a $\frac{5}{8}$-mi stretch the second day, and a $\frac{1}{2}$-mi stretch the third day. How many miles along the highway did they clean?

69. *Caffeine.* To cut back on caffeine intake, Michelle and Gerry mix caffeinated and decaffeinated coffee beans before grinding for a customized mix. They mix $\frac{3}{16}$ lb of decaffeinated beans with $\frac{5}{8}$ lb of caffeinated beans. What is the total amount of coffee beans in the mixture?

70. *Purchasing Tea.* Alyse bought $\frac{1}{3}$ lb of orange pekoe tea and $\frac{1}{2}$ lb of English cinnamon tea. How many pounds of tea did she buy?

71. *Culinary Arts.* The campus culinary arts department is preparing brownies for the international student reception. Students in the catering program iced the $\frac{11}{16}$-in. $\left(\frac{11''}{16}\right)$ brownies with a $\frac{5}{32}$-in. $\left(\frac{5''}{32}\right)$ layer of butterscotch icing. What is the thickness of the iced brownies?

72. *Carpentry.* A carpenter glues two kinds of plywood together. He glues a $\frac{1}{4}$-in. $\left(\frac{1''}{4}\right)$ piece of cherry plywood to a $\frac{3}{8}$-in. $\left(\frac{3''}{8}\right)$ piece of less expensive plywood. What is the total thickness of these pieces?

73. *Baking.* A baker used $\frac{1}{2}$ lb of flour for rolls, $\frac{1}{4}$ lb for donuts, and $\frac{1}{3}$ lb for cookies. How much flour was used?

74. *Baking.* A recipe for muffins calls for $\frac{1}{2}$ qt (quart) of buttermilk, $\frac{1}{3}$ qt of skim milk, and $\frac{1}{16}$ qt of oil. How many quarts of liquid ingredients does the recipe call for?

75. *Meteorology.* On April 15, it rained $\frac{1}{2}$ in. in the morning and $\frac{3}{8}$ in. in the afternoon. How much did it rain altogether?

76. *Medication.* Janine took $\frac{1}{5}$ g of ibuprofen before lunch and $\frac{1}{2}$ g after lunch. How much did she take altogether?

77. A park naturalist hikes $\frac{3}{5}$ mi to a lookout, another $\frac{3}{10}$ mi to an osprey's nest, and finally $\frac{3}{4}$ mi to a campsite. How far does the naturalist hike?

78. A triathlete runs $\frac{7}{8}$ mi, canoes $\frac{1}{3}$ mi, and swims $\frac{1}{6}$ mi. How many miles does the triathlete cover?

79. Punch Recipe. A recipe for strawberry punch calls for $\frac{1}{5}$ qt of ginger ale and $\frac{3}{5}$ qt of strawberry soda. How much liquid is needed? If the recipe is doubled, how much liquid is needed? If the recipe is halved, how much liquid is needed?

80. Concrete Mix. A cubic meter of concrete mix contains 420 kg (kilograms) of cement, 150 kg of stone, and 120 kg of sand. What is the total weight of a cubic meter of the mix? What fractional part is cement? stone? sand? Add these fractional amounts. What is the result?

Skill Maintenance

Subtract. [2.3a]

81. $-7 - 6$

82. $-5 - (-9)$

83. $9 - 17$

84. $-8 - 23$

Evaluate. [2.6a]

85. $\dfrac{x - y}{3}$, for $x = 7$ and $y = -3$

86. $3(x + y)$ and $3x + 3y$, for $x = 5$ and $y = 9$

Solve.

87. $48 \cdot t = 1680$ [1.7b]

88. $10{,}000 = m + 3593$ [1.7b]

89. $3x - 8 = 25$ [2.8d]

90. $5x + 9 = 24$ [2.8d]

91. $\dfrac{2}{3}x = \dfrac{6}{7}$ [3.8a]

92. $-\dfrac{5}{8}t = \dfrac{1}{4}$ [3.8a]

Synthesis

Add and, if possible, simplify.

93. $\dfrac{3}{10}t + \dfrac{2}{7} + \dfrac{2}{15}t + \dfrac{3}{5}$

94. $\dfrac{2}{9} + \dfrac{4}{21}x + \dfrac{4}{15} + \dfrac{3}{14}x$

95. $5t^2 + \dfrac{6}{a}t + 2t^2 + \dfrac{3}{a}t$

Use $<$, $>$, or $=$ for $\square$ to form a true sentence.

96. ▦ $\dfrac{10}{97} + \dfrac{67}{137} \ \square \ \dfrac{8123}{13{,}289}$

97. ▦ $\dfrac{12}{169} + \dfrac{53}{103} \ \square \ \dfrac{10{,}192}{17{,}407}$

98. ▦ $\dfrac{37}{157} + \dfrac{20}{107} \ \square \ \dfrac{6942}{16{,}799}$

99. A guitarist's band is booked for Friday and Saturday nights at a local club. The guitarist's group is a trio on Friday and expands to a quintet on Saturday. Thus, the guitarist is paid one-third of one-half the weekend's pay for Friday and one-fifth of one-half the weekend's pay for Saturday. What fractional part of the total pay did the guitarist receive for the weekend's work? If the band was paid $1200, how much did the guitarist receive?

100. ▦ Consider only the numbers 2, 3, 4, and 5. Assume each is placed in a blank in the following.

$$\dfrac{\square}{\square} + \dfrac{\square}{\square} = \ ?$$

What placement of the numbers in the blanks yields the largest sum?

101. ▦ In the sum below, a and b are digits (so $1b$ is a two-digit number and $35a$ is a three-digit number). Find a and b. (*Hint:* $a < 4$ and $b > 6$.)

$$\dfrac{a}{17} + \dfrac{1b}{23} = \dfrac{35a}{391}$$

102. ▦ Use a standard calculator. Arrange the following in order from smallest to largest.

$$\dfrac{3}{4}, \ \dfrac{17}{21}, \ \dfrac{13}{15}, \ \dfrac{7}{9}, \ \dfrac{15}{17}, \ \dfrac{13}{12}, \ \dfrac{19}{22}$$

Subtraction, Equations, and Applications

a SUBTRACTION

Like Denominators

Let's consider the difference $\frac{4}{8} - \frac{3}{8}$.

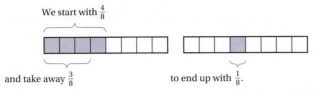

We start with $\frac{4}{8}$

and take away $\frac{3}{8}$

to end up with $\frac{1}{8}$.

We start with 4 eighths and take away 3 eighths:

4 eighths − 3 eighths = 1 eighth,

or $\quad 4 \cdot \dfrac{1}{8} - 3 \cdot \dfrac{1}{8} = \dfrac{1}{8},$ or $\quad \dfrac{4}{8} - \dfrac{3}{8} = \dfrac{1}{8}.$

> To subtract when denominators are the same,
>
> **a)** subtract the numerators,
>
> **b)** keep the denominator, and
>
> **c)** simplify, if possible.
>
> $$\frac{7}{10} - \frac{4}{10} = \frac{7-4}{10} = \frac{3}{10}$$

EXAMPLES Subtract and, if possible, simplify.

1. $\dfrac{8}{13} - \dfrac{3}{13} = \dfrac{8-3}{13} = \dfrac{5}{13}$

2. $\dfrac{3}{35} - \dfrac{13}{35} = \dfrac{3-13}{35} = \dfrac{-10}{35} = \dfrac{5}{5} \cdot \dfrac{-2}{7} = \dfrac{-2}{7},$ or $-\dfrac{2}{7}$ Removing a factor equal to 1: $\frac{5}{5} = 1$

3. $\dfrac{13}{2a} - \dfrac{5}{2a} = \dfrac{13-5}{2a} = \dfrac{8}{2a} = \dfrac{2}{2} \cdot \dfrac{4}{a} = \dfrac{4}{a}$ Removing a factor equal to 1: $\frac{2}{2} = 1$

4. $-\dfrac{7}{t} - \dfrac{2}{t} = \dfrac{-7-2}{t} = \dfrac{-9}{t},$ or $-\dfrac{9}{t}$

Do Margin Exercises 1–4. ▶

Different Denominators

> To subtract when denominators are different:
>
> **a)** Find the least common multiple of the denominators. That number is the least common denominator, LCD.
>
> **b)** Multiply by 1, writing 1 in the form n/n, to express each fraction in an equivalent form that contains the LCD.
>
> **c)** Subtract the numerators, keeping the same denominator.
>
> **d)** Simplify, if possible.

Subtract and, if possible, simplify.

1. $\dfrac{7}{8} - \dfrac{3}{8}$ **2.** $\dfrac{5}{9a} - \dfrac{1}{9a}$

3. $\dfrac{7}{10} - \dfrac{13}{10}$ **4.** $-\dfrac{2}{x} - \dfrac{4}{x}$

Answers

Skill to Review:
1. -6 **2.** 0

Margin Exercises:
1. $\dfrac{1}{2}$ **2.** $\dfrac{4}{9a}$ **3.** $-\dfrac{3}{5}$ **4.** $-\dfrac{6}{x}$

EXAMPLE 5 Subtract: $\frac{2}{5} - \frac{3}{8}$.

a) The LCM of 5 and 8 is 40, so the LCD is 40.

b) We need to find numbers equivalent to $\frac{2}{5}$ and $\frac{3}{8}$ with denominators of 40:

$$\frac{2}{5} - \frac{3}{8} = \frac{2}{5} \cdot \frac{8}{8} - \frac{3}{8} \cdot \frac{5}{5}. \longleftarrow$$

Think: $8 \times \square = 40$. The answer is 5, so we multiply by 1, using $\frac{5}{5}$.

Think: $5 \times \square = 40$. The answer is 8, so we multiply by 1, using $\frac{8}{8}$.

c) We subtract: $\dfrac{16}{40} - \dfrac{15}{40} = \dfrac{16 - 15}{40} = \dfrac{1}{40}$.

d) Since $\frac{1}{40}$ cannot be simplified, we are finished. The answer is $\frac{1}{40}$.

◀ **Do Exercise 5.**

5. Subtract: $\dfrac{3}{4} - \dfrac{2}{3}$.

EXAMPLE 6 Subtract: $\frac{7}{12} - \frac{5}{6}$.

Since 12 is a multiple of 6, the LCM of 6 and 12 is 12. The LCD is 12.

$$\frac{7}{12} - \frac{5}{6} = \frac{7}{12} - \frac{5}{6} \cdot \frac{2}{2}$$

Think: $6 \times \square = 12$. The answer is 2, so we multiply by 1, using $\frac{2}{2}$.

$$= \frac{7}{12} - \frac{10}{12}$$

$$= \frac{7 - 10}{12} = \frac{-3}{12} \qquad 7 - 10 = 7 + (-10) = -3$$

$$= \frac{3}{3} \cdot \frac{-1}{4} = \frac{-1}{4}, \text{ or } -\frac{1}{4} \qquad \begin{array}{l}\text{Simplifying by removing a}\\\text{factor equal to 1: } \frac{3}{3} = 1\end{array}$$

Subtract.

6. $\dfrac{5}{6} - \dfrac{1}{9}$ **GS**

The LCD is $\boxed{}$.

$$\frac{5}{6} - \frac{1}{9} = \frac{5}{6} \cdot \frac{3}{3} - \frac{1}{9} \cdot \frac{\boxed{}}{\boxed{}}$$

$$= \frac{\boxed{}}{18} - \frac{\boxed{}}{18}$$

$$= \frac{\boxed{}}{18}$$

EXAMPLE 7 Subtract: $\frac{17}{24} - \frac{4}{15}$.

We need to find the LCM of 24 and 15:

$$\left.\begin{array}{l} 24 = 2 \cdot 2 \cdot 2 \cdot 3, \\ 15 = 3 \cdot 5. \end{array}\right\} \quad \text{The LCM is } 2 \cdot 2 \cdot 2 \cdot 3 \cdot 5, \text{ or } 120.$$

$$\frac{17}{24} - \frac{4}{15} = \frac{17}{24} \cdot \frac{5}{5} - \frac{4}{15} \cdot \frac{8}{8}$$

Multiplying by 1 to obtain the LCD. To form 1, use the factors of the LCM that each denominator lacks. Note that $2 \cdot 2 \cdot 2 = 8$.

$$= \frac{85}{120} - \frac{32}{120} = \frac{85 - 32}{120} = \frac{53}{120}$$

7. $\dfrac{2}{5} - \dfrac{7}{10}$ **8.** $\dfrac{2}{3} - \dfrac{5}{6}$

9. $\dfrac{11}{28} - \dfrac{5}{16}$

◀ **Do Exercises 6–9.**

10. Simplify: $\dfrac{9}{10}x - \dfrac{3}{5}x$.

EXAMPLE 8 Simplify by combining like terms: $\frac{7}{8}x - \frac{3}{4}x$.

$$\frac{7}{8}x - \frac{3}{4}x = \left(\frac{7}{8} - \frac{3}{4}\right)x \qquad \text{Try to do this step mentally.}$$

$$= \left(\frac{7}{8} - \frac{6}{8}\right)x = \frac{1}{8}x \qquad \text{Multiplying } \frac{3}{4} \text{ by } \frac{2}{2} \text{ and subtracting}$$

◀ **Do Exercise 10.**

Answers

5. $\frac{1}{12}$ **6.** $\frac{13}{18}$ **7.** $-\frac{3}{10}$ **8.** $-\frac{1}{6}$

9. $\frac{9}{112}$ **10.** $\frac{3}{10}x$

Guided Solution:

6. $18; \frac{2}{2}, 15, 2, 13$

240 CHAPTER 4 Fraction Notation: Addition, Subtraction, and Mixed Numerals

b SOLVING EQUATIONS

We can use the addition principle to solve equations containing fractions.

EXAMPLE 9 Solve: $x - \dfrac{1}{3} = -\dfrac{2}{7}$.

$$x - \frac{1}{3} = -\frac{2}{7}$$

$$x - \frac{1}{3} + \frac{1}{3} = -\frac{2}{7} + \frac{1}{3}$$ Using the addition principle: adding $\frac{1}{3}$ to both sides

$$x + 0 = -\frac{2}{7} + \frac{1}{3}$$ Adding $\frac{1}{3}$ "undid" the subtraction of $\frac{1}{3}$ on the left-hand side of the equation.

$$x = -\frac{2}{7} \cdot \frac{3}{3} + \frac{1}{3} \cdot \frac{7}{7}$$ Multiplying by 1 to obtain the LCD, 21

$$x = -\frac{6}{21} + \frac{7}{21} = \frac{1}{21}$$ The solution appears to be $\frac{1}{21}$.

Check:

$$\begin{array}{c|c} x - \dfrac{1}{3} = -\dfrac{2}{7} \\[2mm] \hline \dfrac{1}{21} - \dfrac{1}{3} \quad ? \quad -\dfrac{2}{7} \\[3mm] \dfrac{1}{21} - \dfrac{1}{3} \cdot \dfrac{7}{7} \\[3mm] \dfrac{1}{21} - \dfrac{7}{21} \\[3mm] \dfrac{-6}{21} \\[3mm] -\dfrac{2}{7} \cdot \dfrac{3}{3} \\[3mm] -\dfrac{2}{7} \end{array}$$ TRUE

Our answer checks. The solution is $\frac{1}{21}$.

Recall that we can also create equivalent equations if we subtract the same number on both sides of an equation.

EXAMPLE 10 Solve: $x + \dfrac{1}{4} = \dfrac{3}{5}$.

$$x + \frac{1}{4} - \frac{1}{4} = \frac{3}{5} - \frac{1}{4}$$ Using the addition principle: adding $-\frac{1}{4}$ to, or subtracting $\frac{1}{4}$ from, both sides

$$x + 0 = \frac{3}{5} \cdot \frac{4}{4} - \frac{1}{4} \cdot \frac{5}{5}$$ The LCD is 20. We multiply by 1 to get the LCD.

$$x = \frac{12}{20} - \frac{5}{20} = \frac{7}{20}$$

The solution is $\frac{7}{20}$. We leave the check to the student.

Do Exercises 11–13. ▶

Solve.

11. $x - \dfrac{2}{5} = \dfrac{1}{5}$

12. $x + \dfrac{2}{3} = \dfrac{5}{6}$

GS **13.** $\dfrac{3}{5} + t = -\dfrac{7}{8}$

$$\frac{3}{5} + t - \frac{\square}{\square} = -\frac{7}{8} - \frac{\square}{\square}$$

$$t + 0 = -\frac{7}{8} \cdot \frac{5}{\square} - \frac{3}{5} \cdot \frac{8}{\square}$$

$$t = -\frac{\square}{40} - \frac{\square}{40}$$

$$t = \frac{-35 - \square}{40}$$

$$t = \frac{-35 + (\square)}{40}$$

$$t = \frac{\square}{40} = -\frac{\square}{\square}$$

11	12	13	14	15	16	17
$\frac{1}{5}$	$\frac{1}{4}$	$\frac{9}{25}$	First quarter $\frac{1}{2}$	$\frac{3}{5}$	$\frac{7}{10}$	$\frac{4}{5}$
18	19	20	21	22	23	24
$\frac{17}{20}$	$\frac{19}{20}$	$\frac{99}{100}$	Full moon 1	$\frac{49}{50}$	$\frac{19}{20}$	$\frac{17}{20}$
25	26	27	28	29	30	31
$\frac{4}{5}$	$\frac{7}{10}$	$\frac{3}{5}$	Last quarter $\frac{1}{2}$	$\frac{2}{5}$	$\frac{1}{3}$	$\frac{6}{25}$

SOURCE: Astronomical Applications Department, U.S. Naval Observatory, Washington, DC 20392-5420

c APPLICATIONS AND PROBLEM SOLVING

EXAMPLE 11 *Phases of the Moon.* The moon rotates in such a way that the same side always faces the earth. Throughout a lunar cycle, the portion of the moon that appears illuminated increases from nearly none (new moon) to nearly all (full moon), then decreases back to nearly none. These *phases* of the moon can be described by fractions between 0 and 1, indicating the portion of the moon illuminated. The partial calendar from August 2013 shows the fraction of the moon illuminated at midnight, Eastern Standard Time, for each day.

How much more of the moon appeared illuminated on August 18, 2013, than on August 15, 2013?

1. **Familiarize.** From the calendar, we see that $\frac{3}{5}$ of the moon was illuminated on August 15 and that $\frac{17}{20}$ of the moon was illuminated on August 18. We let $m =$ the additional part of the moon that appeared illuminated on August 18.

2. **Translate.** We translate to an equation.

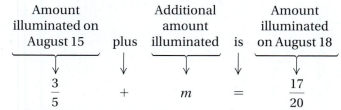

$$\frac{3}{5} + m = \frac{17}{20}$$

3. **Solve.** To solve the equation, we subtract $\frac{3}{5}$ on both sides:

$$\frac{3}{5} + m = \frac{17}{20}$$

$$\frac{3}{5} + m - \frac{3}{5} = \frac{17}{20} - \frac{3}{5} \qquad \text{Subtracting } \frac{3}{5} \text{ on both sides}$$

$$m + 0 = \frac{17}{20} - \frac{3}{5} \cdot \frac{4}{4} \qquad \text{The LCD is 20. We multiply by 1 to obtain the LCD.}$$

$$m = \frac{17}{20} - \frac{12}{20} = \frac{5}{20} = \frac{5 \cdot 1}{5 \cdot 4} = \frac{5}{5} \cdot \frac{1}{4} = \frac{1}{4}.$$

4. **Check.** To check, we add:

$$\frac{3}{5} + \frac{1}{4} = \frac{3}{5} \cdot \frac{4}{4} + \frac{1}{4} \cdot \frac{5}{5} = \frac{12}{20} + \frac{5}{20} = \frac{17}{20}.$$

This is the amount of the moon illuminated on August 18, so the answer checks.

5. **State.** On August 18, $\frac{1}{4}$ more of the moon was illuminated than on August 15.

◀ **Do Exercise 14.**

14. *Phases of the Moon.* Use the calendar in Example 11 to find how much less of the moon appeared illuminated on August 31, 2013, than on August 23, 2013.

Answer

14. $\frac{71}{100}$

Translating for Success

1. **Bubble Wrap.** One-Stop Postal Center orders bubble wrap in 64-yd rolls. On average, $\frac{3}{4}$ yd is used per small package. How many small packages can be prepared with 2 rolls of bubble wrap?

2. **Distance from College.** The post office is $\frac{7}{9}$ mi from the community college. The medical clinic is $\frac{2}{5}$ as far from the college as the post office is. How far is the clinic from the college?

3. **Swimming.** Andrew swims $\frac{7}{9}$ mi every day. One day, he swims $\frac{2}{5}$ mi by 11:00 A.M. How much farther must Andrew swim to reach his daily goal?

4. **Tuition.** The average tuition at Waterside University is $12,000. If a loan is obtained for $\frac{1}{3}$ of the tuition, how much is the loan?

5. **Thermos Bottle Capacity.** A thermos bottle holds $\frac{11}{12}$ gal. How much is in the bottle when it is $\frac{4}{7}$ full?

The goal of these matching questions is to practice step (2), Translate, of the five-step problem-solving process. Translate each word problem to an equation and select a correct translation from equations A–O.

A. $\frac{3}{4} \cdot 64 = x$

B. $\frac{1}{3} \cdot 12{,}000 = x$

C. $\frac{1}{3} + \frac{2}{5} = x$

D. $\frac{2}{5} + x = \frac{7}{9}$

E. $\frac{2}{5} \cdot \frac{7}{9} = x$

F. $\frac{3}{4} \cdot x = 64$

G. $\frac{4}{7} = x + \frac{11}{12}$

H. $\frac{2}{5} = x + \frac{7}{9}$

I. $\frac{4}{7} \cdot \frac{11}{12} = x$

J. $\frac{3}{4} \cdot x = 2 \cdot 64$

K. $\frac{1}{3} \cdot x = 12{,}000$

L. $\frac{1}{3} + \frac{2}{5} + x = 1$

M. $\frac{2}{5} = \frac{7}{9}x$

N. $\frac{4}{7} + x = \frac{11}{12}$

O. $\frac{1}{3} + x = \frac{2}{5}$

Answers on page A-7

6. **Cutting Rope.** A piece of rope $\frac{11}{12}$ yd long is cut into two pieces. One piece is $\frac{4}{7}$ yd long. How long is the other piece?

7. **Planting Corn.** Each year, Prairie State Farm plants 64 acres of corn. With good weather, $\frac{3}{4}$ of the planting can be completed by April 20. How many acres can be planted by April 20 with good weather?

8. **Painting Trim.** A painter used $\frac{1}{3}$ gal of white paint for the trim in the library and $\frac{2}{5}$ gal for the trim in the family room. How much paint was used for the trim in the two rooms?

9. **Lottery Winnings.** Sally won $12,000 in a state lottery and decided to give the net amount after taxes to three charities. One received $\frac{1}{3}$ of the money, and a second received $\frac{2}{5}$. What fractional part did the third charity receive?

10. **Reading Assignment.** When Lowell had read 64 pages of his political science assignment, he had completed $\frac{3}{4}$ of his required reading. How many total pages were assigned?

For Extra Help

MyMathLab

MathXL®
PRACTICE

WATCH

READ

REVIEW

✓ Reading Check

Complete each statement with the appropriate word or words from the following list. A word may be used more than once or not at all.

    denominator   numerator
    denominators   numerators

RC1. To subtract fractions with like denominators, we subtract the ———————— and keep the ————————.

RC2. Before we can subtract fractions, the ———————— must be the same.

RC3. To subtract fractions when denominators are different, we find the LCM of the ————————.

RC4. To subtract fractions when denominators are different, we multiply one or both fractions by 1 to make the ———————— the same.

a   Subtract and, if possible, simplify.

1. $\dfrac{5}{6} - \dfrac{1}{6}$

2. $\dfrac{7}{5} - \dfrac{2}{5}$

3. $\dfrac{9}{16} - \dfrac{13}{16}$

4. $\dfrac{5}{12} - \dfrac{7}{12}$

5. $\dfrac{8}{a} - \dfrac{6}{a}$

6. $\dfrac{4}{t} - \dfrac{9}{t}$

7. $-\dfrac{3}{8} - \dfrac{1}{8}$

8. $-\dfrac{3}{10} - \dfrac{1}{10}$

9. $\dfrac{3}{5a} - \dfrac{7}{5a}$

10. $\dfrac{2}{7t} - \dfrac{10}{7t}$

11. $\dfrac{10}{3t} - \dfrac{4}{3t}$

12. $\dfrac{9}{2a} - \dfrac{5}{2a}$

13. $\dfrac{7}{8} - \dfrac{1}{16}$

14. $\dfrac{4}{3} - \dfrac{5}{6}$

15. $\dfrac{7}{15} - \dfrac{4}{5}$

16. $\dfrac{3}{28} - \dfrac{3}{4}$

17. $\dfrac{3}{4} - \dfrac{1}{20}$

18. $\dfrac{3}{4} - \dfrac{4}{16}$

19. $\dfrac{2}{15} - \dfrac{5}{12}$

20. $\dfrac{11}{16} - \dfrac{9}{10}$

21. $\dfrac{7}{10} - \dfrac{23}{100}$

22. $\dfrac{9}{10} - \dfrac{3}{100}$

23. $\dfrac{7}{15} - \dfrac{3}{25}$

24. $\dfrac{18}{25} - \dfrac{4}{35}$

25. $\dfrac{-41}{100} - \dfrac{3}{10}$

26. $\dfrac{-13}{100} - \dfrac{7}{20}$

27. $\dfrac{2}{3} - \dfrac{1}{8}$

28. $\dfrac{3}{4} - \dfrac{1}{2}$

29. $-\dfrac{3}{10} - \dfrac{7}{25}$

30. $-\dfrac{5}{18} - \dfrac{2}{27}$

31. $\dfrac{3}{8} - \dfrac{5}{12}$

32. $\dfrac{2}{9} - \dfrac{7}{12}$

33. $\dfrac{-5}{18} - \dfrac{7}{24}$

34. $\dfrac{-7}{25} - \dfrac{2}{15}$

35. $\dfrac{13}{90} - \dfrac{17}{120}$

36. $\dfrac{8}{25} - \dfrac{29}{150}$ **37.** $\dfrac{2}{3}x - \dfrac{4}{9}x$ **38.** $\dfrac{7}{4}x - \dfrac{5}{12}x$ **39.** $\dfrac{2}{5}a - \dfrac{3}{4}a$ **40.** $\dfrac{4}{7}a - \dfrac{1}{3}a$

b   Solve.

41. $x - \dfrac{4}{9} = \dfrac{3}{9}$ **42.** $x - \dfrac{3}{11} = \dfrac{7}{11}$ **43.** $a + \dfrac{2}{11} = \dfrac{6}{11}$ **44.** $a + \dfrac{4}{15} = \dfrac{13}{15}$

45. $y + \dfrac{1}{30} = \dfrac{1}{10}$ **46.** $y + \dfrac{1}{3} = \dfrac{5}{6}$ **47.** $a - \dfrac{3}{8} = \dfrac{3}{4}$ **48.** $x - \dfrac{3}{10} = \dfrac{2}{5}$

49. $\dfrac{2}{3} + x = \dfrac{4}{5}$ **50.** $\dfrac{4}{5} + x = \dfrac{6}{7}$ **51.** $\dfrac{3}{8} + a = \dfrac{1}{12}$ **52.** $\dfrac{5}{6} + a = \dfrac{2}{9}$

53. $n - \dfrac{3}{10} = -\dfrac{1}{6}$ **54.** $n - \dfrac{3}{4} = -\dfrac{5}{12}$ **55.** $x + \dfrac{3}{4} = -\dfrac{1}{2}$ **56.** $x + \dfrac{5}{6} = -\dfrac{11}{12}$

c   Solve.

57. For a research paper, Kaitlyn spent $\frac{3}{4}$ hr searching the Internet on google.com and $\frac{1}{3}$ hr on yahoo.com. How much longer did she spend on google.com than on yahoo.com?

58. As part of a fitness program, Deb swims $\frac{1}{2}$ mi every day. One day she had already swum $\frac{1}{5}$ mi. How much farther did Deb need to swim?

59. The tread depth of an IRL Indy Car Series tire is $\frac{3}{32}$ in. Tires for a normal car have a tread depth of $\frac{5}{16}$ in. when new and are considered bald at $\frac{1}{16}$ in. How much deeper is the tread depth of an Indy Car tire than that of a bald tire for a normal car?

60. Gerry uses $\frac{1}{3}$ lb of fresh mozzarella cheese and $\frac{1}{4}$ lb of grated Parmesan cheese on a homemade margherita pizza. How much more mozzarella cheese does he use than Parmesan cheese?

$\frac{3}{32}$ in.

Sources: Indy500.com; *Consumer Reports*

61. From a $\frac{4}{5}$-lb wheel of cheese, a $\frac{1}{4}$-lb piece was served. How much cheese remained on the wheel?

62. A baker has a dispenser containing $\frac{15}{16}$ cup of icing and puts $\frac{1}{12}$ cup on a cinnamon roll. How much icing remains in the dispenser?

63. Jorge's $\frac{3}{4}$-hr drive to a job was part city and part country driving. If $\frac{2}{5}$ hr was city driving, how much time was spent on country driving?

64. Keri exercises $\frac{5}{6}$ of an hour every day. She jogs for $\frac{7}{12}$ of an hour and spends the remaining time warming up and cooling down. How much time does she spend warming up and cooling down?

65. *Woodworking.* Natalie is replacing a $\frac{3}{4}$-in.-thick shelf in her bookcase. If her replacement board is $\frac{15}{16}$ in. thick, how much must it be planed down before the repair can be completed?

66. *Furniture Cleaner.* A $\frac{2}{3}$-cup mixture of lemon juice and olive oil makes an excellent cleaner for wood furniture. If the mixture contains $\frac{1}{4}$ cup of lemon juice, how much olive oil is in the cleaner?

$\frac{1}{4}$ cup $+$? $=$ $\frac{2}{3}$ cup

$\frac{15}{16}$ in.

r in.

$\frac{3}{4}$ in.

67. Blake used $\frac{1}{3}$ cup of maple syrup in preparing the batter for a batch of maple oatbran muffins. Sheila pointed out that the recipe actually calls for $\frac{5}{8}$ cup of syrup. How much more syrup should Blake add to the batter?

68. Amber added $\frac{1}{3}$ qt of two-cycle oil to a fuel mixture for her lawn mower. She then noticed that the owner's manual indicates $\frac{1}{2}$ qt should have been added. How much more two-cycle oil should Amber add to the mixture?

Skill Maintenance

Divide, if possible. If not possible, write "Not defined." [1.5a], [3.3b]

69. $\dfrac{38}{38}$

70. $\dfrac{38}{0}$

71. $\dfrac{124}{0}$

72. $\dfrac{124}{31}$

Divide and simplify. [3.7b]

73. $\dfrac{3}{7} \div \dfrac{9}{4}$

74. $\dfrac{9}{10} \div \dfrac{3}{5}$

75. $7 \div \dfrac{1}{3}$

76. $\dfrac{1}{4} \div 8$

Multiply and simplify. [3.6a]

77. $18 \cdot \dfrac{2}{3}$

78. $\dfrac{5}{12} \cdot 6$

79. $-\dfrac{17}{25} \cdot \dfrac{15}{34}$

80. $\dfrac{7}{20} \cdot \left(-\dfrac{45}{49}\right)$

Synthesis

Simplify.

81. $\dfrac{7}{8} - \dfrac{3}{4} - \dfrac{1}{16}$

82. $\dfrac{9}{10} - \dfrac{1}{2} - \dfrac{2}{15}$

83. $\dfrac{2}{5} - \dfrac{1}{6}(-3)^2$

84. $\dfrac{7}{8} - \dfrac{1}{10}\left(-\dfrac{5}{6}\right)^2$

85. $-4 \cdot \dfrac{3}{7} - \dfrac{1}{7} \cdot \dfrac{4}{5}$

86. $\left(\dfrac{5}{6}\right)^2 - \left(\dfrac{3}{4}\right)^2$

87. $\left(-\dfrac{2}{5}\right)^3 - \left(-\dfrac{3}{10}\right)^3$

88. $\dfrac{3}{17} - \dfrac{2}{19} - \left(\dfrac{3}{17} - \dfrac{2}{19}\right)$

89. At a party, three friends, Ashley, Cole, and Lauren, shared a big tub of popcorn. Within 30 min, the tub was empty. Ashley ate $\frac{7}{12}$ of the tub while Lauren ate only $\frac{1}{6}$ of the tub. How much did Cole eat?

90. A small community garden was divided among four local residents. Based on the time they could spend on their garden sections and their individual crop plans, each resident received a different-size plot to tend. One received $\frac{1}{4}$ of the garden, the second $\frac{1}{16}$, and the third $\frac{3}{8}$ of the garden. How much did the fourth gardener receive?

91. The circle graph below shows how long shoppers stay when visiting a mall. What portion of shoppers stay for 0–2 hr?

Less than
1 hour
$\frac{26}{50}$

2–3 hours
$\frac{5}{50}$

1–2 hours

More than
3 hours
$\frac{2}{50}$

92. Four business partners plan to upgrade their computer system at a cost of $12,600. Celia will pay $\frac{1}{2}$ of the cost, Reba will pay $\frac{1}{4}$ of the cost, Jon will pay $\frac{1}{6}$ of the cost, and Karl will pay the rest.

  a) How much will Karl pay?
  b) What fractional part will Karl pay?

93. Mark Romano owns $\frac{7}{12}$ of Romano-Chrenka Chevrolet and Lisa Romano owns $\frac{1}{6}$. If Paul and Ella Chrenka own the remaining share of the dealership equally, what fractional piece does Paul own?

94. As part of a rehabilitation program, an athlete must swim and then walk a total of $\frac{9}{10}$ km each day. If one lap in the swimming pool is $\frac{3}{80}$ km, how far must the athlete walk after swimming 10 laps?

95. 🖩 Solve: $1x + \dfrac{16}{323} = \dfrac{10}{187}$.

96. 🖩 Determine what whole number a must be in order for the following to be true:

$$\frac{10 + a}{23} = \frac{330}{391} - \frac{a}{17}.$$

97. *Microsoft Interview.* The following is a question taken from an employment interview with Microsoft. "Given a gold bar that can be cut exactly twice and a contractor who must be paid one-seventh of a gold bar every day for seven days, how should the bar be cut?"

Source: *Fortune Magazine,* January 22, 2001

OBJECTIVES

a Solve equations that involve fractions and require use of both the addition principle and the multiplication principle.

b Solve equations by using the multiplication principle to clear fractions.

SKILL TO REVIEW

Objective 3.8a: Use the multiplication principle to solve equations.

Solve.

1. $\frac{2}{3}x = \frac{5}{6}$

2. $18 = -\frac{2}{3}t$

We have used the multiplication principle to solve equations like

$$\frac{2}{3}x = \frac{5}{6} \quad \text{and} \quad 7 = \frac{5}{4}t,$$

and we have used the addition principle to solve equations like

$$\frac{4}{5} + x = \frac{1}{2} \quad \text{and} \quad \frac{7}{3} = t - \frac{2}{9}.$$

We are now ready to solve equations in which both principles are required.

a USING THE PRINCIPLES TOGETHER

Recall that we use the addition and multiplication principles to write equivalent equations. In the following steps, all five equations are equivalent:

$5x - 2 = 43$	We first isolate $5x$.
$5x - 2 + 2 = 43 + 2$	Using the addition principle
$5x = 45$	We now isolate x.
$\frac{1}{5} \cdot 5x = \frac{1}{5} \cdot 45$	Using the multiplication principle
$x = 9.$	The solution of $x = 9$ is the solution of $5x - 2 = 43$.

As a check, note that $5 \cdot 9 - 2 = 45 - 2 = 43$, as desired. The solution is 9.

EXAMPLE 1 Solve: $\frac{3}{4}x - \frac{1}{8} = \frac{1}{2}$.

We first isolate $\frac{3}{4}x$ by adding $\frac{1}{8}$ to both sides:

$$\frac{3}{4}x - \frac{1}{8} = \frac{1}{2}$$

$$\frac{3}{4}x - \frac{1}{8} + \frac{1}{8} = \frac{1}{2} + \frac{1}{8} \qquad \text{Using the addition principle}$$

$$\frac{3}{4}x + 0 = \frac{4}{8} + \frac{1}{8} \qquad \text{Writing with a common denominator}$$

$$\frac{3}{4}x = \frac{5}{8}.$$

Next, we isolate x by multiplying both sides by $\frac{4}{3}$:

$$\frac{3}{4}x = \frac{5}{8} \qquad \text{Note that the reciprocal of } \frac{3}{4} \text{ is } \frac{4}{3}.$$

$$\frac{4}{3} \cdot \frac{3}{4}x = \frac{4}{3} \cdot \frac{5}{8} \qquad \text{Using the multiplication principle}$$

$$1x = \frac{20}{24}, \text{ or } \frac{5}{6}. \qquad \text{Simplifying; the solution appears to be } \frac{5}{6}.$$

Answers

Skill to Review:

1. $\frac{5}{4}$ **2.** -27

Check: $\dfrac{3}{4}x - \dfrac{1}{8} = \dfrac{1}{2}$

$$\dfrac{3}{4} \cdot \dfrac{5}{6} - \dfrac{1}{8} \;\overset{?}{\vert}\; \dfrac{1}{2}$$

$$\dfrac{3 \cdot 5}{4 \cdot 2 \cdot 3} - \dfrac{1}{8}$$ Removing a factor equal to 1: $\frac{3}{3} = 1$

$$\dfrac{5}{8} - \dfrac{1}{8}$$

$$\dfrac{1}{2} \qquad \text{TRUE}$$

The solution is $\frac{5}{6}$.

Do Exercises 1 and 2. ▶

EXAMPLE 2 Solve: $5 + \dfrac{9}{2}t = -\dfrac{7}{2}$.

We first isolate $\frac{9}{2}t$ by subtracting 5 from both sides:

$$5 + \dfrac{9}{2}t = -\dfrac{7}{2}$$

$$5 + \dfrac{9}{2}t - 5 = -\dfrac{7}{2} - 5 \qquad \text{Subtracting 5 from both sides}$$

$$\dfrac{9}{2}t = -\dfrac{7}{2} - \dfrac{10}{2} \qquad \text{Writing 5 as } \tfrac{10}{2} \text{ to use the LCD}$$

$$\dfrac{9}{2}t = -\dfrac{17}{2} \qquad \text{Note that the reciprocal of } \tfrac{9}{2} \text{ is } \tfrac{2}{9}.$$

$$\dfrac{2}{9} \cdot \dfrac{9}{2}t = \dfrac{2}{9} \cdot \left(-\dfrac{17}{2}\right) \qquad \text{Multiplying both sides by } \tfrac{2}{9}$$

$$1t = -\dfrac{2 \cdot 17}{9 \cdot 2} \qquad \text{Removing a factor equal to 1: } \tfrac{2}{2} = 1$$

$$t = -\dfrac{17}{9}.$$

Check: $5 + \dfrac{9}{2}t = -\dfrac{7}{2}$

$$5 + \dfrac{9}{2}\left(-\dfrac{17}{9}\right) \;\overset{?}{\vert}\; -\dfrac{7}{2} \qquad \text{Removing a factor equal to 1: } \tfrac{9}{9} = 1$$

$$5 + \left(-\dfrac{17}{2}\right)$$

$$\dfrac{10}{2} + \left(\dfrac{-17}{2}\right)$$

$$\dfrac{10 - 17}{2}$$

$$\dfrac{-7}{2} \qquad \text{TRUE}$$

The solution is $-\frac{17}{9}$.

Do Exercises 3 and 4. ▶

Sometimes the variable appears on the right side of the equation. The strategy for solving the equation remains the same.

Solve.

1. $\dfrac{3}{5}t - \dfrac{8}{15} = \dfrac{2}{15}$

GS 2. $\dfrac{1}{2}x - \dfrac{1}{5} = \dfrac{7}{10}$

$$\dfrac{1}{2}x - \dfrac{1}{5} + \dfrac{\square}{\square} = \dfrac{7}{10} + \dfrac{1}{5}$$

$$\dfrac{1}{2}x = \dfrac{7}{10} + \dfrac{\square}{10}$$

$$\dfrac{1}{2}x = \dfrac{\square}{10}$$

$$\square \cdot \dfrac{1}{2}x = 2 \cdot \dfrac{9}{10}$$

$$1x = \dfrac{2 \cdot 3 \cdot 3}{2 \cdot \square}$$

$$x = \dfrac{\square}{\square}$$

Solve.

3. $3 + \dfrac{14}{5}t = -\dfrac{21}{5}$

4. $2x + 4 = \dfrac{1}{2}$

Answers

1. $\dfrac{10}{9}$ 2. $\dfrac{9}{5}$ 3. $-\dfrac{18}{7}$ 4. $-\dfrac{7}{4}$

Guided Solution:

2. $\dfrac{1}{5}$, 2, 9, 2, 5, $\dfrac{9}{5}$

EXAMPLE 3 Solve: $20 = 6 - \frac{2}{3}x$.

Our plan is to first use the addition principle to isolate $-\frac{2}{3}x$ and then use the multiplication principle to isolate x.

$$20 = 6 - \frac{2}{3}x$$

$$20 - 6 = 6 - \frac{2}{3}x - 6 \qquad \text{Subtracting 6 (or adding } -6\text{) on both sides}$$

$$14 = -\frac{2}{3}x$$

$$\left(-\frac{3}{2}\right)14 = \left(-\frac{3}{2}\right)\left(-\frac{2}{3}x\right) \qquad \text{Multiplying both sides by } -\frac{3}{2}$$

$$-\frac{3 \cdot 14}{2} = 1x$$

$$-\frac{3 \cdot 7 \cdot 2}{2} = 1x \qquad \text{Removing a factor equal to 1: } \frac{2}{2} = 1$$

$$-21 = x$$

Check:
$$20 = 6 - \frac{2}{3}x$$
$$\overline{}$$
$$20 \ ? \ 6 - \frac{2}{3}(-21)$$
$$6 + \frac{42}{3}$$
$$6 + 14$$
$$20 \qquad \textbf{TRUE}$$

The solution is -21.

5. Solve: $9 - \frac{3}{4}x = 21$.

◀ **Do Exercise 5.**

b **CLEARING FRACTIONS**

We now look at an alternative approach for solving Examples 1–3. Key to this approach is using the multiplication principle in the *first* step to produce an equivalent equation that is "cleared of fractions."

To "clear fractions," we identify the LCM of the denominators and use the multiplication principle. Because the LCM is a common multiple of the denominators, when both sides of the equation are multiplied by the LCM, the resulting terms can all be simplified. An equivalent equation can then be written without using fractions. We demonstrate this approach by solving Examples 1 and 2 by clearing fractions.

···················· **Caution!** ····················

We can "clear fractions" in equations, not in expressions. Do not multiply to clear fractions when simplifying an expression.

··

Either of the methods discussed in this section can be used to solve equations that contain fractions, but it is important for students planning to continue in algebra to thoroughly understand *both* methods.

Answer

5. -16

EXAMPLE 4 Solve Example 1 by clearing fractions:

$$\frac{3}{4}x - \frac{1}{8} = \frac{1}{2}.$$

The LCM of the denominators is 8, so we begin by multiplying both sides of the equation by 8:

$$\frac{3}{4}x - \frac{1}{8} = \frac{1}{2}$$

$$8\left(\frac{3}{4}x - \frac{1}{8}\right) = 8 \cdot \frac{1}{2}$$ Multiplying both sides by 8. We use parentheses when we are multiplying more than one term.

.... **Caution!** ....

$$\frac{8 \cdot 3}{4}x - 8 \cdot \frac{1}{8} = \frac{8}{2}$$ Use the distributive law carefully! Here we multiply every term inside the parentheses by 8.

....

$$\frac{4 \cdot 2 \cdot 3}{4}x - 1 = 4$$ Factoring and simplifying

$$6x - 1 = 4$$ The equation is now cleared of fractions. This is the advantage of using this method.

$$6x - 1 + 1 = 4 + 1$$ Adding 1 to both sides

$$6x = 5$$

$$\frac{6x}{6} = \frac{5}{6}$$ Dividing both sides by 6 (or multiplying both sides by $\frac{1}{6}$)

$$x = \frac{5}{6}.$$ Simplifying

Since $\frac{5}{6}$ was the solution in Example 1, we have a check. The solution is $\frac{5}{6}$.

EXAMPLE 5 Solve Example 2 by clearing fractions:

$$5 + \frac{9}{2}t = -\frac{7}{2}.$$

The LCM of the denominators is 2, so we begin by multiplying both sides of the equation by 2:

$$2\left(5 + \frac{9}{2}t\right) = 2\left(-\frac{7}{2}\right)$$ Using the multiplication principle

$$2 \cdot 5 + \frac{2 \cdot 9}{2}t = -\frac{2 \cdot 7}{2}$$ Using the distributive law; multiplying every term by 2

$$10 + 9t = -7$$ Simplifying and removing a factor equal to 1: $\frac{2}{2} = 1$. The equation is now cleared of fractions.

$$10 + 9t - 10 = -7 - 10$$ Subtracting 10 from both sides

$$9t = -17$$ Simplifying

$$\frac{9t}{9} = -\frac{17}{9}$$ Dividing both sides by 9 (or multiplying both sides by $\frac{1}{9}$)

$$t = -\frac{17}{9}.$$ Simplifying

Since the solution in Example 2 is also $-\frac{17}{9}$, we have a check. The solution is $-\frac{17}{9}$.

Do Exercises 6 and 7. ▶

GS 6. Solve Example 3 by clearing fractions:

$$20 = 6 - \frac{2}{3}x.$$

$$\boxed{}(20) = 3\left(6 - \frac{2}{3}x\right)$$

$$\boxed{} = 3 \cdot \boxed{} - \frac{3 \cdot 2}{3}x$$

$$60 = \boxed{} - \boxed{}$$

$$60 - \boxed{} = 18 - 2x - 18$$

$$\boxed{} = -2x$$

$$\frac{42}{\boxed{}} = \frac{-2x}{-2}$$

$$\boxed{} = x$$

7. Solve Margin Exercise 1 by clearing fractions:

$$\frac{3}{5}t - \frac{8}{15} = \frac{2}{15}.$$

Answers

6. −21   **7.** $\frac{10}{9}$

Guided Solution:
6. 3, 60, 6, 18, 2x, 18, 42, −2, −21

SECTION 4.4 Solving Equations: Using the Principles Together   **251**

✓ Reading Check

Determine whether each statement is true or false.

_____ **RC1.** We cannot use both the additon principle and the multiplication principle in the solution of an equation.

_____ **RC2.** To solve for a variable, it must appear on the left side of an equation.

_____ **RC3.** The first step in the solution of an equation always involves the addition principle.

_____ **RC4.** We clear fractions by multiplying both sides of an equation by the LCM of the denominators.

a   Solve using the addition principle and/or the multiplication principle. Don't forget to check!

1. $6x - 3 = 15$

2. $7x - 6 = 22$

3. $5x + 7 = 10$

4. $19 = 2x + 4$

5. $8 = 3x + 11$

6. $2a + 9 = -7$

7. $\frac{2}{3}y - 8 = 1$

8. $\frac{3}{10}y - 7 = 3$

9. $\frac{3}{2}t - \frac{1}{4} = \frac{1}{2}$

10. $\frac{1}{4}t + \frac{1}{8} = \frac{1}{2}$

11. $\frac{1}{5}x + \frac{3}{10} = \frac{3}{5}$

12. $\frac{4}{3}x - \frac{2}{15} = \frac{2}{15}$

13. $5 - \frac{3}{4}x = 6$

14. $3 - \frac{2}{5}x = 6$

15. $-1 + \frac{2}{5}t = -\frac{4}{5}$

16. $-2 + \frac{1}{6}t = -\frac{7}{4}$

17. $12 = 8 + \frac{7}{2}t$

18. $7 = 5 + \frac{3}{2}t$

19. $-11 = \frac{2}{3}x - 7$

20. $-10 = \frac{2}{5}x - 4$

21. $7 = a + \frac{14}{5}$

22. $9 = a + \frac{47}{10}$

23. $\frac{2}{5}t - 1 = \frac{7}{5}$

24. $-\frac{53}{4} = \frac{3}{2}a + 2$

25. $\frac{39}{8} = \frac{11}{4} - \frac{1}{2}x$

26. $\frac{7}{2} = \frac{13}{2} - \frac{1}{7}y$

27. $-\frac{13}{3}s + \frac{11}{2} = \frac{35}{4}$

28. $-\frac{11}{5}t + \frac{36}{5} = \frac{7}{2}$

b   Solve by using the multiplication principle to clear fractions.

29. $\dfrac{1}{2}x - \dfrac{1}{4} = \dfrac{1}{2}$

30. $\dfrac{1}{3}x - \dfrac{1}{6} = \dfrac{2}{3}$

31. $7 = \dfrac{4}{9}t + 5$

32. $5 = \dfrac{4}{7}t + 3$

33. $-3 = \dfrac{3}{4}t - \dfrac{1}{2}$

34. $-2 = \dfrac{4}{3}t - \dfrac{5}{6}$

35. $\dfrac{4}{3} - \dfrac{5}{6}x = \dfrac{3}{2}$

36. $\dfrac{3}{2} - \dfrac{5}{3}x = \dfrac{5}{6}$

37. $-\dfrac{3}{4} = -\dfrac{5}{6} - \dfrac{1}{2}x$

38. $-\dfrac{1}{4} = -\dfrac{2}{3} - \dfrac{1}{6}x$

39. $\dfrac{4}{3} - \dfrac{1}{5}t = \dfrac{3}{4}$

40. $\dfrac{2}{5} - \dfrac{3}{4}t = \dfrac{4}{3}$

Skill Maintenance

Divide.   [2.5a]

41. $39 \div (-3)$

42. $56 \div (-7)$

43. $(-72) \div (-4)$

44. $(-81) \div (-3)$

Solve.   [2.3b]

45. Jeremy withdraws $200 from his bank account, makes a $90 deposit, and then withdraws another $40. How much has Jeremy's account balance changed?

46. Animal Instinct, a pet supply store, makes a profit of $850 on Friday and a profit of $375 on Saturday, but suffers a loss of $45 on Sunday. Find the total profit or loss for the three days.

Divide and simplify.   [3.7b]

47. $\dfrac{10}{7} \div (2m)$

48. $45n \div \dfrac{9}{4}$

Synthesis

Solve.

49. 🖩 $\dfrac{553}{2451}a - \dfrac{13}{57} = \dfrac{29}{43}$

50. 🖩 $\dfrac{1081}{3599}x - \dfrac{17}{61} = \dfrac{19}{59}$

51. 🖩 $\dfrac{11}{17} = \dfrac{13}{41} - \dfrac{23}{29}t$

52. $-\dfrac{a}{5} + \dfrac{31}{4} = \dfrac{16}{3}$

53. $\dfrac{47}{5} - \dfrac{a}{4} = \dfrac{44}{7}$

54. $\dfrac{49}{8} + \dfrac{2x}{9} = 4$

55. The perimeter of the figure shown is 15 cm. Solve for x.

56. The perimeter of the figure is 15 cm. Solve for n.

SKILL TO REVIEW

Objective 1.5a: Divide whole numbers.

Divide.

1. $735 \div 16$

2. $23\overline{)6023}$

Convert to a mixed numeral.

1. $1 + \dfrac{2}{3} = \square\dfrac{\square}{\square}$

$\underbrace{\dfrac{1}{3} \quad \dfrac{1}{3} \quad \dfrac{1}{3}}_{1} \qquad \underbrace{\dfrac{1}{3} \quad \dfrac{1}{3}}_{\frac{2}{3}}$

2. $2 + \dfrac{3}{4} = \square\dfrac{\square}{\square}$

3. $12 + \dfrac{2}{7}$

a   MIXED NUMERALS

The following figure illustrates the use of a **mixed numeral**. The bolt shown is $2\frac{3}{8}$ in. long. The length is given as a whole-number part, 2, and a fractional part less than 1, $\frac{3}{8}$. We can represent the measurement as $\frac{19}{8}$, but $2\frac{3}{8}$ makes the length easier to visualize and is thus more descriptive.

A mixed numeral $2\frac{3}{8}$ represents a sum:

$$2\dfrac{3}{8} \quad \text{means} \quad 2 + \dfrac{3}{8}.$$

This is a whole number.   This is a fraction less than 1.

EXAMPLES   Convert to a mixed numeral.

1. $7 + \dfrac{2}{5} = 7\dfrac{2}{5}$ 　　　　　　　**2.** $4 + \dfrac{3}{10} = 4\dfrac{3}{10}$

◀ **Do Margin Exercises 1–3.**

The notation $2\frac{3}{4}$ has a plus sign left out. To aid in understanding, we sometimes write the missing plus sign: $2 + \frac{3}{4}$. Similarly, the notation $-5\frac{2}{3}$ has a minus sign left out, since $-5\frac{2}{3} = -(5 + \frac{2}{3}) = -5 - \frac{2}{3}$.

Mixed numerals can be displayed on the number line, as shown here.

EXAMPLES Convert to fraction notation.

3. $2\frac{3}{4} = 2 + \frac{3}{4}$ Inserting the missing plus sign

$= \frac{2}{1} + \frac{3}{4}$ $2 = \frac{2}{1}$

$= \frac{2}{1} \cdot \frac{4}{4} + \frac{3}{4}$ Finding a common denominator

$= \frac{8}{4} + \frac{3}{4}$

$= \frac{11}{4}$ Adding

4. $4\frac{3}{10} = 4 + \frac{3}{10} = \frac{4}{1} + \frac{3}{10} = \frac{4}{1} \cdot \frac{10}{10} + \frac{3}{10} = \frac{40}{10} + \frac{3}{10} = \frac{43}{10}$

Do Exercises 4 and 5. ▶

Convert to fraction notation.

4. $4\frac{2}{5}$ **5.** $6\frac{1}{10}$

Using Example 4, we can develop a faster way to convert.

To convert from a mixed numeral like $4\frac{3}{10}$ to fraction notation:

ⓐ Multiply the whole number by the denominator: $4 \cdot 10 = 40$.

ⓑ Add the result to the numerator: $40 + 3 = 43$.

ⓒ Keep the denominator.

$\underset{\times}{\overset{+}{4\frac{3}{10}}} = \frac{43}{10}$ ⟵ ⓒ

Convert to fraction notation. Use the faster method.

EXAMPLES Convert to fraction notation.

5. $6\frac{2}{3} = \frac{20}{3}$ $6 \cdot 3 = 18, 18 + 2 = 20$

6. $8\frac{2}{9} = \frac{74}{9}$ $8 \cdot 9 = 72, 72 + 2 = 74$

7. $10\frac{7}{8} = \frac{87}{8}$ $10 \cdot 8 = 80, 80 + 7 = 87$

Do Exercises 6–9. ▶

GS **6.** $4\frac{5}{6}$

$4 \cdot 6 = \boxed{}$

$24 + \boxed{} = 29$

$4\frac{5}{6} = \frac{\boxed{}}{6}$

7. $9\frac{1}{4}$ **8.** $20\frac{2}{3}$

9. $1\frac{9}{13}$

To find the opposite of the number in Example 5, we can write either $-6\frac{2}{3}$ or $-\frac{20}{3}$. Thus, to convert a negative mixed numeral to fraction notation, we remove the negative sign for purposes of computation and then include it in the answer.

Convert to fraction notation.

10. $-6\frac{2}{5}$ **11.** $-7\frac{2}{9}$

EXAMPLES Convert to fraction notation.

8. $-5\frac{1}{3} = -\left(5 + \frac{1}{3}\right) = -\frac{16}{3}$ $5 \cdot 3 = 15; 15 + 1 = 16;$ include the negative sign

9. $-7\frac{5}{6} = -\left(7 + \frac{5}{6}\right) = -\frac{47}{6}$ $7 \cdot 6 = 42; 42 + 5 = 47$

Do Exercises 10 and 11. ▶

Answers

4. $\frac{22}{5}$ **5.** $\frac{61}{10}$ **6.** $\frac{29}{6}$ **7.** $\frac{37}{4}$
8. $\frac{62}{3}$ **9.** $\frac{22}{13}$ **10.** $-\frac{32}{5}$ **11.** $-\frac{65}{9}$

Guided Solution:
6. 24, 5, 29

Writing Mixed Numerals

We can find a mixed numeral for $\frac{5}{3}$ as follows:

$$\frac{5}{3} = \frac{3}{3} + \frac{2}{3} = 1 + \frac{2}{3} = 1\frac{2}{3}.$$

In terms of objects, we can think of $\frac{5}{3}$ as $\frac{3}{3}$, or 1, plus $\frac{2}{3}$, as shown below.

$$\frac{5}{3} = \qquad \frac{3}{3}, \text{ or } 1 \qquad + \qquad \frac{2}{3}$$

Fraction symbols like $\frac{5}{3}$ also indicate division; $\frac{5}{3}$ means $5 \div 3$. Let's divide the numerator by the denominator.

$$\begin{array}{r} 1 \\ 3\overline{)5} \\ \underline{3} \\ 2 \end{array} \leftarrow 2 \div 3 = \tfrac{2}{3}$$

Thus, $\frac{5}{3} = 1\frac{2}{3}$.

To convert from fraction notation to a mixed numeral, divide.

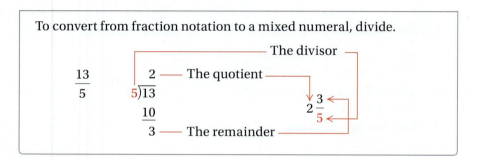

EXAMPLES Convert to a mixed numeral.

10. $\dfrac{69}{10}$

$$\begin{array}{r} 6 \\ 10\overline{)69} \\ \underline{60} \\ 9 \end{array} \qquad \frac{69}{10} = 6\frac{9}{10}$$

11. $\dfrac{122}{8}$

$$\begin{array}{r} 15 \\ 8\overline{)122} \\ \underline{8} \\ 42 \\ \underline{40} \\ 2 \end{array} \qquad \frac{122}{8} = 15\frac{2}{8} = 15\frac{1}{4}$$

> Simplify the fraction part of a mixed numeral, if possible.

Convert to a mixed numeral.

12. $\dfrac{7}{3}$

13. $\dfrac{11}{10}$

14. $\dfrac{110}{6}$

15. $\dfrac{231}{18}$

◀ **Do Exercises 12–15.**

A fraction larger than 1, such as $\frac{27}{8}$, is sometimes referred to as an "improper" fraction. However, the use of notation such as $\frac{27}{8}, \frac{11}{9}$, and $\frac{89}{10}$ is quite "proper" and very common in algebra.

Answers

12. $2\frac{1}{3}$ 13. $1\frac{1}{10}$ 14. $18\frac{1}{3}$ 15. $12\frac{5}{6}$

The same procedure also works with negative numbers. Of course, the result will be a negative mixed numeral.

EXAMPLE 12 Convert $\dfrac{-9}{4}$ to a mixed numeral.

Since $\quad 4\overline{)9},\quad$ we have $\quad \dfrac{9}{4} = 2\dfrac{1}{4}.\quad$ Thus, $\quad \dfrac{-9}{4} = -2\dfrac{1}{4}.$

$$\begin{array}{r} 2 \\ 4\overline{)9} \\ 8 \\ \hline 1 \end{array}$$

Do Exercises 16 and 17. ▶

b FINDING QUOTIENTS AND AVERAGES

It is quite common when performing long division to express the quotient as a mixed numeral. As in Examples 10–12, the remainder becomes the numerator of the fraction part of the mixed numeral.

EXAMPLE 13 Divide. Write a mixed numeral for the quotient.

$$7\,\overline{)\,6\ 3\ 4\ 1}$$

We first divide as usual.

$$\begin{array}{r} 9\ 0\ 5 \\ 7\,\overline{)\,6\ 3\ 4\ 1} \\ 6\ 3\ 0\ 0 \\ \hline 4\ 1 \\ 3\ 5 \\ \hline 6 \end{array}$$

$$\dfrac{6341}{7} = 905\dfrac{6}{7}$$

The answer is 905 R 6, or $905\dfrac{6}{7}$. Using fraction notation, we write $\dfrac{6341}{7} = 905\dfrac{6}{7}$.

Do Exercises 18 and 19. ▶

EXAMPLE 14 *Dietary Fiber.* Each of the following five fruits is a good source of dietary fiber. This list gives the amount of fiber, in grams, contained in each serving of fruit. How much fiber is contained, on average, in these fruits?

Source: www.NationalFiberCouncil.org

Raspberries (1 cup)	8 g
Pear (1 medium)	6 g
Blueberries (1 cup)	4 g
Apple (1 medium)	3 g
Banana (1 medium)	3 g

To find the *average* of a set of values, we add the values and divide the sum by the number of values being added.

$$\text{Average grams of fiber} = \dfrac{8\,\text{g} + 6\,\text{g} + 4\,\text{g} + 3\,\text{g} + 3\,\text{g}}{5} = \dfrac{24\,\text{g}}{5} = 4\dfrac{4}{5}\,\text{g}$$

On average, these fruits contain $4\dfrac{4}{5}$ g of fiber per serving.

Do Exercise 20. ▶

Convert to a mixed numeral.

GS **16.** $\dfrac{-17}{5}$

17. $-\dfrac{134}{12}$

Divide. Write a mixed numeral for the answer.

18. $6\,\overline{)\,4\ 8\ 4\ 6}$

19. $4\ 5\,\overline{)\,6\ 0\ 5\ 3}$

20. *Fitness.* Shelby recently added jogging to her exercise routine. The following list gives the total amount of time she jogged for each of the first four weeks of her new routine. Find the average number of minutes she jogged per week.

Week 1: 24 min
Week 2: 30 min
Week 3: 27 min
Week 4: 32 min

Answers

16. $-3\dfrac{2}{5}$ **17.** $-11\dfrac{1}{6}$ **18.** $807\dfrac{2}{3}$

19. $134\dfrac{23}{45}$ **20.** $28\dfrac{1}{4}$ min

Guided Solution:

16. 5, 17, $3\dfrac{2}{5}$, $3\dfrac{2}{5}$

✓ Reading Check

Determine whether each statement is true or false.

_____ **RC1.** A mixed numeral consists of a whole-number part and a fraction less than 1.

_____ **RC2.** The mixed numeral $5\frac{1}{4}$ represents $5 + \frac{1}{4}$.

_____ **RC3.** It is never appropriate to use fraction notation such as $\frac{33}{25}$.

_____ **RC4.** When a quotient is written as a mixed numeral, the divisor is the denominator of the fraction part (assuming that the fraction has not been simplified).

a Convert to fraction notation.

1. $7\frac{2}{3}$

2. $6\frac{2}{5}$

3. $6\frac{1}{4}$

4. $8\frac{1}{2}$

5. $-20\frac{1}{8}$

6. $-10\frac{1}{3}$

7. $5\frac{1}{10}$

8. $8\frac{1}{10}$

9. $20\frac{3}{5}$

10. $30\frac{4}{5}$

11. $-33\frac{1}{3}$

12. $-66\frac{2}{3}$

13. $1\frac{5}{8}$

14. $1\frac{3}{5}$

15. $-12\frac{3}{4}$

16. $-15\frac{2}{3}$

17. $5\frac{7}{10}$

18. $7\frac{3}{100}$

19. $-5\frac{7}{100}$

20. $-6\frac{4}{15}$

Convert to a mixed numeral.

21. $\frac{16}{3}$

22. $\frac{19}{8}$

23. $\frac{45}{6}$

24. $\frac{30}{9}$

25. $\frac{57}{10}$

26. $\frac{-89}{10}$

27. $\frac{65}{9}$

28. $\frac{65}{8}$

29. $\frac{-33}{6}$

30. $\frac{-50}{8}$

31. $\frac{46}{4}$

32. $\frac{39}{9}$

33. $\frac{-12}{8}$

34. $-\frac{57}{6}$

35. $\frac{307}{5}$

36. $\frac{227}{4}$

37. $-\frac{413}{50}$

38. $\frac{467}{100}$

b Divide. Write a mixed numeral for the answer.

39. $8\overline{)869}$ **40.** $3\overline{)2126}$ **41.** $7\overline{)6345}$ **42.** $9\overline{)9110}$ **43.** $21\overline{)852}$

44. $85\overline{)7670}$ **45.** $-302 \div 15$ **46.** $-475 \div 13$ **47.** $471 \div (-21)$ **48.** $545 \div (-25)$

Education.   Since 2001, the non-profit organization School on Wheels has provided one-on-one tutoring and educational advocacy for school-aged children experiencing homelessness in Indianapolis. The table below shows the number of children who received services each year, as well as other related information.

Sources: www.indysow.org; School on Wheels annual reports

49. What is the average number of backpacks with school supplies given to children for the school years 2007–2008 through 2012–2013?

50. What is the average number of homeless children served for the school years 2007–2008 through 2012–2013?

51. What is the average number of volunteer tutor hours for the school years 2007–2008 through 2012–2013?

52. What is the average number of volunteer tutor hours for the school years 2001–2002 through 2006–2007?

SCHOOL YEAR	NUMBER OF HOMELESS CHILDREN SERVED	NUMBER OF VOLUNTEER TUTOR HOURS	NUMBER OF BACKPACKS WITH SCHOOL SUPPLIES GIVEN TO CHILDREN
2012–2013	359	5511	278
2011–2012	407	6018	310
2010–2011	415	6708	320
2009–2010	373	5965	242
2008–2009	365	4796	303
2007–2008	322	4027	201
2006–2007	363	4083	101
2005–2006	388	3385	105
2004–2005	375	3131	87
2003–2004	240	1298	92
2002–2003	256	338	46
2001–2002	50	158	63

Skill Maintenance

Use the distributive law to write an equivalent expression.   [2.6b]

53. $6(3x + y - 4)$

54. $-3(a - 2b + 4c)$

Synthesis

Write a mixed numeral for each number or sum listed.

55. ▦ $\dfrac{128,236}{541}$ **56.** ▦ $\dfrac{103,676}{349}$ **57.** $\dfrac{56}{7} + \dfrac{2}{3}$ **58.** $\dfrac{72}{12} + \dfrac{5}{6}$ **59.** $\dfrac{12}{5} + \dfrac{19}{15}$

60. There are $\frac{366}{7}$ weeks in a leap year.

61. There are $\frac{365}{7}$ weeks in a year.

62. *Athletics.*   At a track and field meet, the hammer that is thrown has a wire length ranging from 3 ft $10\frac{1}{4}$ in. to 3 ft $11\frac{3}{4}$ in., a $4\frac{1}{8}$-in. grip, and a 16-lb ball with a diameter of $4\frac{3}{8}$ in. to $5\frac{1}{8}$ in. Give specifications for the wire length and diameter of an "average" hammer.

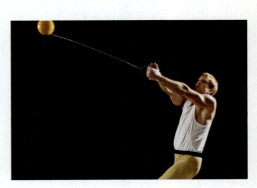

Concept Reinforcement

Determine whether each statement is true or false.

_____ **1.** If $\dfrac{a}{b} > \dfrac{c}{b}, b \neq 0$, then $a > c$. [4.2c]

_____ **2.** The mixed numeral $1\frac{1}{2}$ and the fraction $\frac{6}{4}$ represent the same number. [4.5a]

_____ **3.** The least common multiple of two natural numbers is the smallest number that is a factor of both. [4.1a]

_____ **4.** To add fractions when denominators are the same, we keep the numerator and add the denominators. [4.2a]

Guided Solutions

 Fill in each blank with the number that creates a correct solution.

5. Subtract: $\dfrac{11}{42} - \dfrac{3}{35}$. [4.3a]

$$\dfrac{11}{42} - \dfrac{3}{35} = \dfrac{11}{2 \cdot \boxed{} \cdot 7} - \dfrac{3}{\boxed{} \cdot 7}$$ Factoring the denominators

$$= \dfrac{11}{2 \cdot 3 \cdot 7} \cdot \left(\dfrac{\boxed{}}{\boxed{}} \right) - \dfrac{3}{5 \cdot 7} \cdot \left(\dfrac{2 \cdot 3}{2 \cdot 3} \right)$$ Multiplying by 1 to get the LCD

$$= \dfrac{11 \cdot \boxed{}}{2 \cdot 3 \cdot 7 \cdot \boxed{}} - \dfrac{3 \cdot 2 \cdot 3}{5 \cdot 7 \cdot 2 \cdot 3}$$ Multiplying

$$= \dfrac{\boxed{}}{2 \cdot 3 \cdot 5 \cdot 7} - \dfrac{\boxed{}}{2 \cdot 3 \cdot 5 \cdot 7}$$ Simplifying

$$= \dfrac{\boxed{} - \boxed{}}{2 \cdot 3 \cdot 5 \cdot 7} = \dfrac{\boxed{}}{\boxed{}}$$ Subtracting and simplifying

6. Solve: $x + \dfrac{1}{8} = \dfrac{2}{3}$. [4.3b]

$$x + \dfrac{1}{8} = \dfrac{2}{3}$$

$$x + \dfrac{1}{8} - \boxed{} = \dfrac{2}{3} - \boxed{}$$ Subtracting on both sides

$$x + 0 = \dfrac{2}{3} \cdot \dfrac{\boxed{}}{\boxed{}} - \dfrac{1}{8} \cdot \dfrac{\boxed{}}{\boxed{}}$$ Multiplying by 1 to get the LCD

$$x = \dfrac{\boxed{}}{24} - \dfrac{\boxed{}}{24}$$ Simplifying and multiplying

$$x = \dfrac{\boxed{}}{\boxed{}}$$ Subtracting

The solution is $\dfrac{\boxed{}}{\boxed{}}$.

Mixed Review

7. Match each set of numbers in the first column with its least common multiple in the second column by drawing connecting lines. [4.1a]

45 and 50

50 and 80   120

30 and 24   720

18, 24, and 80   400

30, 45, and 50   450

Calculate and simplify. [4.2b], [4.3a]

8. $\dfrac{1}{5} + \dfrac{7}{45}$

9. $\dfrac{5}{6} + \dfrac{2}{3} + \dfrac{7}{12}$

10. $\dfrac{2}{9} - \dfrac{1}{6}$

11. $\dfrac{1}{15} - \dfrac{5}{18}$

12. $\dfrac{19}{48} - \dfrac{11}{30}$

13. $-\dfrac{3}{8}x + \dfrac{1}{12}x$

14. $\dfrac{-3}{40} + \dfrac{-5}{24}$

15. $\dfrac{8}{65} - \dfrac{2}{35}$

Solve.

16. Miguel jogs for $\frac{4}{5}$ mi, rests, and then jogs for another $\frac{2}{3}$ mi. How far does he jog in all? [4.2d]

17. One weekend, Kirby spent $\frac{39}{5}$ hr playing two iPod games—Brain Challenge: Cerebral Burn and Scrabble: Go for a Triple Word Score. She spent $\frac{11}{4}$ hr playing Scrabble. How many hours did she spend playing Brain Challenge? [4.3c]

18. Arrange in order from smallest to largest:

$\dfrac{4}{9}, \dfrac{3}{10}, \dfrac{2}{7}$, and $\dfrac{1}{5}$. [4.2c]

19. Solve: $\dfrac{2}{5} + x = \dfrac{9}{16}$. [4.3b]

20. Solve: $\dfrac{3}{4}x + 1 = \dfrac{1}{3}$. [4.4a], [4.4b]

21. Divide: $15\overline{)263}$. Write a mixed numeral for the answer. [4.5b]

22. Fraction notation for $9\frac{3}{8}$ is which of the following? [4.5a]

A. $\dfrac{27}{8}$ **B.** $\dfrac{93}{8}$ **C.** $\dfrac{75}{8}$ **D.** $\dfrac{80}{3}$

23. Mixed numeral notation for $-\frac{39}{4}$ is which of the following? [4.5a]

A. $-35\dfrac{1}{4}$ **B.** $-\dfrac{4}{39}$ **C.** $-9\dfrac{3}{4}$ **D.** $-36\dfrac{3}{4}$

Understanding Through Discussion and Writing

24. Is the LCM of two numbers always larger than either number? Why or why not? [4.1a]

25. Explain the role of multiplication when adding using fraction notation with different denominators. [4.2b]

26. A student made the following error:

$$\dfrac{8}{5} - \dfrac{8}{2} = \dfrac{8}{3}.$$

Find at least two ways to convince him of the mistake. [4.3a]

27. Are the numbers $2\frac{1}{3}$ and $2 \cdot \frac{1}{3}$ equal? Why or why not? [4.5a]

4.6   Addition and Subtraction Using Mixed Numerals; Applications

OBJECTIVES

a   Add using mixed numerals.

b   Subtract using mixed numerals.

c   Solve applied problems involving addition and subtraction with mixed numerals.

d   Add and subtract using negative mixed numerals.

SKILL TO REVIEW

Objective 3.5b: Simplify fraction notation.

Simplify.

1. $\dfrac{18}{32}$ **2.** $\dfrac{78}{117}$

Add.

1. $\begin{aligned} &2\tfrac{3}{10}\\ +\;&5\tfrac{1}{10}\end{aligned}$ **2.** $\begin{aligned} &8\tfrac{2}{5}\\ +\;&3\tfrac{7}{10}\end{aligned}$

Answers

Skill to Review:

1. $\dfrac{9}{16}$ 2. $\dfrac{2}{3}$

Margin Exercises:

1. $7\dfrac{2}{5}$ 2. $12\dfrac{1}{10}$

---

a   ADDITION USING MIXED NUMERALS

To add mixed numerals, we first add the fractions. Then we add the whole numbers.

EXAMPLE 1   Add: $1\tfrac{5}{8} + 3\tfrac{1}{8}$. Write a mixed numeral for the answer.

$$
\begin{array}{r}
1\dfrac{5}{8}\\[4pt]
+\;3\dfrac{1}{8}\\[2pt]
\hline
\dfrac{6}{8}
\end{array}
\qquad = \qquad
\begin{array}{r}
1\dfrac{5}{8}\\[4pt]
+\;3\dfrac{1}{8}\\[2pt]
\hline
4\dfrac{6}{8} = 4\dfrac{3}{4}
\end{array}
$$

Simplifying: $\dfrac{6}{8} = \dfrac{3}{4}$

↑ Add the fractions.   ↑ Add the whole numbers.

Sometimes we must write the fractional parts with a common denominator before we can add.

EXAMPLE 2   Add: $5\tfrac{2}{3} + 3\tfrac{5}{6}$. Write a mixed numeral for the answer.

The LCD is 6.

$$
\begin{array}{r}
5\dfrac{2}{3}\cdot\dfrac{2}{2}\\[4pt]
+\;3\dfrac{5}{6}\\[2pt]
\hline
\end{array}
\qquad
\begin{aligned}
&= \quad 5\dfrac{4}{6}\\[4pt]
&= +\;3\dfrac{5}{6}\\[2pt]
\hline
&\;\;\;8\dfrac{9}{6} = 8 + \dfrac{9}{6}
\end{aligned}
$$

$$
= 8 + 1\tfrac{1}{2} \qquad \text{Writing } \tfrac{9}{6} \text{ as a mixed numeral; } \tfrac{9}{6} = 1\tfrac{3}{6} = 1\tfrac{1}{2}
$$

$$
= 9\tfrac{1}{2}
$$

◀ **Do Margin Exercises 1 and 2.**

The fractional part of a mixed numeral should always be less than 1.

EXAMPLE 3 Add: $10\frac{5}{6} + 7\frac{3}{8}$.

The LCD is 24.

$$
\begin{array}{r}
10\ \dfrac{5}{6}\cdot\dfrac{4}{4} = 10\dfrac{20}{24} \\[2mm]
+\ 7\ \dfrac{3}{8}\cdot\dfrac{3}{3} = +\ 7\dfrac{9}{24} \\[2mm]
\hline
\end{array}
$$

$$17\frac{29}{24} = 17 + \frac{29}{24}$$

$$= 17 + 1\frac{5}{24} \qquad \text{Writing } \tfrac{29}{24} \text{ as a mixed numeral, } 1\tfrac{5}{24}$$

$$= 18\frac{5}{24}$$

Do Exercise 3. ▶

3. Add.

$$
\begin{array}{r}
9\dfrac{3}{4} \\[2mm]
+\ 3\dfrac{5}{6} \\[2mm]
\hline
\end{array}
$$

b SUBTRACTION USING MIXED NUMERALS

EXAMPLE 4 Subtract: $7\frac{3}{4} - 2\frac{1}{4}$.

$$
\begin{array}{r}
7\ \dfrac{3}{4} = \\[2mm]
-\ 2\ \dfrac{1}{4} = \\[2mm]
\hline
\dfrac{2}{4}
\end{array}
\qquad
\begin{array}{r}
7\ \dfrac{3}{4} \\[2mm]
-\ 2\ \dfrac{1}{4} \\[2mm]
\hline
5\ \dfrac{2}{4} = 5\dfrac{1}{2}
\end{array}
\ \leftarrow \text{Simplifying: } \frac{2}{4} = \frac{1}{2}
$$

↑ Subtract the fractions.   ↑ Subtract the whole numbers.

EXAMPLE 5 Subtract: $9\frac{4}{5} - 3\frac{1}{2}$.

The LCD is 10.

$$
\begin{array}{r}
9\ \dfrac{4}{5}\cdot\dfrac{2}{2} = 9\dfrac{8}{10} \\[2mm]
-\ 3\ \dfrac{1}{2}\cdot\dfrac{5}{5} = -\ 3\dfrac{5}{10} \\[2mm]
\hline
6\dfrac{3}{10}
\end{array}
$$

Do Exercises 4 and 5. ▶

EXAMPLE 6 Subtract: $12 - 9\frac{3}{8}$.

$$
\begin{array}{r}
12\ \ = \ 11\dfrac{8}{8} \qquad 12 = 11 + 1 = 11 + \dfrac{8}{8} = 11\dfrac{8}{8} \\[2mm]
-\ 9\dfrac{3}{8} = -\ 9\dfrac{3}{8} \\[2mm]
\hline
2\dfrac{5}{8}
\end{array}
$$

Do Exercise 6. ▶

Subtract.

4.
$$
\begin{array}{r}
10\dfrac{7}{8} \\[2mm]
-\ 9\dfrac{3}{8} \\[2mm]
\hline
\end{array}
$$

GS **5.**
$$
\begin{array}{r}
8\dfrac{2}{3} = \ 8\dfrac{\square}{6} \\[2mm]
-\ 5\dfrac{1}{2} = -\ 5\dfrac{\square}{6} \\[2mm]
\hline
3\dfrac{\square}{6}
\end{array}
$$

GS **6.** Subtract: $5 - 1\frac{1}{3}$.

$$
\begin{array}{r}
5\ \ = \ 4\dfrac{\square}{\square} \\[2mm]
-\ 1\dfrac{1}{3} = -\ 1\dfrac{1}{3} \\[2mm]
\hline
3\dfrac{\square}{3}
\end{array}
$$

Answers

3. $13\frac{7}{12}$ **4.** $1\frac{1}{2}$ **5.** $3\frac{1}{6}$ **6.** $3\frac{2}{3}$

Guided Solutions:

5. 4, 3, 1 **6.** $\frac{3}{3}$, 2

SECTION 4.6 Addition and Subtraction Using Mixed Numerals; Applications   **263**

EXAMPLE 7 Subtract: $7\frac{1}{6} - 2\frac{1}{4}$.

$$\left. \begin{array}{r} 7\dfrac{1}{6} \cdot \dfrac{2}{2} = \quad 7\dfrac{2}{12} \\[2mm] -\ 2\dfrac{1}{4} \cdot \dfrac{3}{3} = -\ 2\dfrac{3}{12} \end{array} \right\} \longleftarrow \text{The LCD is 12.}$$

$$\begin{array}{r} 7\dfrac{2}{12} = \quad 6\dfrac{14}{12} \\[2mm] -\ 2\dfrac{3}{12} = -\ 2\dfrac{3}{12} \\[2mm] \hline 4\dfrac{11}{12} \end{array}$$

We cannot subtract $\frac{3}{12}$ from $\frac{2}{12}$. We borrow 1, or $\frac{12}{12}$, from 7:
$7\frac{2}{12} = 6 + 1 + \frac{2}{12} = 6 + \frac{12}{12} + \frac{2}{12} = 6\frac{14}{12}.$

◀ **Do Exercise 7.**

7. Subtract.

$$\begin{array}{r} 8\dfrac{1}{9} \\[2mm] -4\dfrac{5}{6} \\[2mm] \hline \end{array}$$

Combine like terms.

8. $7\frac{1}{12}t + 1\frac{5}{12}t$

9. $7\frac{11}{12}x - 5\frac{2}{3}x$

10. $5\frac{11}{15}x + 8\frac{3}{10}x$

EXAMPLE 8 Combine like terms: $9\frac{3}{4}x + 4\frac{1}{2}x$.

$$\begin{aligned} 9\frac{3}{4}x + 4\frac{1}{2}x &= \left(9\frac{3}{4} + 4\frac{1}{2}\right)x && \text{Using the distributive law;} \\ && &\text{this is often done mentally.} \\ &= \left(9\frac{3}{4} + 4\frac{2}{4}\right)x && \text{The LCD is 4.} \\ &= 13\frac{5}{4}x = 14\frac{1}{4}x \end{aligned}$$

◀ **Do Exercises 8–10.**

C APPLICATIONS AND PROBLEM SOLVING

EXAMPLE 9 *Men's Long-Jump World Records.* On October 18, 1968, Bob Beamon set a world record of $29\frac{3}{16}$ ft for the long jump, a record that was not broken for nearly 23 years. This record-setting jump was significantly longer than the previous one of $27\frac{19}{48}$ ft, accomplished on May 29, 1965, by Ralph Boston. How much longer was Beamon's jump than Boston's?
Source: "Long jump world record progression," en.wikipedia.org

1. **Familiarize.** The phrase "how much longer" indicates subtraction. We let $w =$ the difference in the world records.

2. **Translate.** We translate as follows:

Beamon's jump	−	Boston's jump	=	Difference in length
↓	↓	↓	↓	↓
$29\frac{3}{16}$	−	$27\frac{19}{48}$	=	w.

3. **Solve.** To solve the equation, we subtract. The LCD is 48.

$$\begin{array}{r} 29\dfrac{3}{16} = \quad 29\dfrac{3}{16} \cdot \dfrac{3}{3} = \quad 29\dfrac{9}{48} = \quad 28\dfrac{57}{48} \\[2mm] -\ 27\dfrac{19}{48} = -\ 27\dfrac{19}{48} \quad = -\ 27\dfrac{19}{48} = -\ 27\dfrac{19}{48} \\[2mm] \hline 1\dfrac{38}{48} = 1\dfrac{19}{24} \end{array}$$

Thus, $w = 1\frac{19}{24}$.

Answers

7. $3\frac{5}{18}$ **8.** $8\frac{1}{2}t$ **9.** $2\frac{1}{4}x$ **10.** $14\frac{1}{30}x$

4. Check. To check, we add the difference, $1\frac{19}{24}$, to Boston's jump.

Since $27\frac{19}{48} + 1\frac{19}{24} = 29\frac{3}{16}$, the answer checks.

5. State. Beamon's jump was $1\frac{19}{24}$ ft longer than Boston's.

Do Exercise 11. ▶

EXAMPLE 10 *Locks of Love.* Locks of Love, Inc., is a non-profit organization that provides hairpieces for children who have lost their hair as a result of medical conditions. The minimum length for hair that is donated is 10 in. Karissa and Cayla allowed their hair to grow in order to donate it. The length cut from Karissa's hair was $15\frac{1}{4}$ in., and the length cut from Cayla's hair was $14\frac{1}{2}$ in. After the hair was cut, it was discovered that the ends of each lock had highlighting that needed to be trimmed. Because of the highlighting, $1\frac{1}{2}$ in. was cut from Karissa's lock of hair and $2\frac{3}{4}$ in. was cut from Cayla's. In all, what was the total usable length of hair that Karissa and Cayla donated?

1. Familiarize. We let $l = $ the total usable length of hair that Karissa and Cayla donated.

2. Translate. The length l is the sum of the lengths that were cut, minus the sum of the lengths that were trimmed from the locks. Thus we have

$$l = \left(15\frac{1}{4} + 14\frac{1}{2}\right) - \left(1\frac{1}{2} + 2\frac{3}{4}\right).$$

3. Solve. This is a three-step problem.

a) We first add the two lengths $15\frac{1}{4}$ and $14\frac{1}{2}$.

$$\begin{array}{r} 15\frac{1}{4} = 15\frac{1}{4} \\ + \, 14\frac{1}{2} = +\,14\frac{2}{4} \\ \hline 29\frac{3}{4} \end{array}$$

b) Next, we add the two lengths $1\frac{1}{2}$ and $2\frac{3}{4}$.

$$\begin{array}{r} 1\frac{1}{2} = 1\frac{2}{4} \\ + \, 2\frac{3}{4} = +\,2\frac{3}{4} \\ \hline 3\frac{5}{4} = 4\frac{1}{4} \end{array}$$

c) Finally, we subtract $4\frac{1}{4}$ from $29\frac{3}{4}$.

$$\begin{array}{r} 29\frac{3}{4} \\ - 4\frac{1}{4} \\ \hline 25\frac{2}{4} = 25\frac{1}{2} \end{array}$$

Thus, $l = 25\frac{1}{2}$.

4. Check. We can check by doing the problem a different way. We can subtract the trimmed length from each lock, then add the adjusted lengths together.

Karissa's lock	Cayla's lock	Sum of lengths
$15\frac{1}{4} = 14\frac{5}{4}$	$14\frac{1}{2} = 13\frac{6}{4}$	$13\frac{3}{4}$
$- 1\frac{1}{2} = -\,1\frac{2}{4}$	$- 2\frac{3}{4} = -\,2\frac{3}{4}$	$+ \, 11\frac{3}{4}$
$13\frac{3}{4}$	$11\frac{3}{4}$	$24\frac{6}{4} = 25\frac{1}{2}$

We obtained the same answer, so our answer checks.

5. State. The sum of the usable lengths of hair donated was $25\frac{1}{2}$ in.

Do Exercise 12. ▶

11. *Travel Distance.* On a two-day business trip, Paul drove $213\frac{7}{10}$ mi the first day and $107\frac{5}{8}$ mi the second day. What was the total distance that Paul drove?

$15\frac{1}{4}$ in. $14\frac{1}{2}$ in.

$\Big\} 1\frac{1}{2}$ in. $\Big\} 2\frac{3}{4}$ in.

12. *Liquid Fertilizer.* There is $283\frac{5}{8}$ gal of liquid fertilizer in a fertilizer application tank. After applying $178\frac{2}{3}$ gal to a soybean field, the farmer requests that Braden's Farm Supply deliver an additional 250 gal to the tank. How many gallons of fertilizer are in the tank after the delivery?

Answers

11. $321\frac{13}{40}$ mi **12.** $354\frac{23}{24}$ gal

d NEGATIVE MIXED NUMERALS

Consider the numbers $5\frac{3}{4}$ and $-5\frac{3}{4}$ on the number line.

Note that just as $5\frac{3}{4}$ means $5 + \frac{3}{4}$, we can regard $-5\frac{3}{4}$ as $-5 - \frac{3}{4}$.

Consider the subtraction $4 - 4\frac{1}{2}$. We know that if we have \$4 and make a \$$4\frac{1}{2}$ purchase, we will owe half a dollar. Thus,

$$4 - 4\frac{1}{2} = -\frac{1}{2}.$$

We can subtract by rewriting the subtraction as addition:

$$4 - 4\frac{1}{2} = 4 + \left(-4\frac{1}{2}\right).$$

Because $-4\frac{1}{2}$ has the greater absolute value, the answer will be negative. The difference in absolute value is $4\frac{1}{2} - 4 = \frac{1}{2}$, so

$$4 - 4\frac{1}{2} = -\frac{1}{2}.$$

Another way to see this is to convert to fraction notation and subtract:

$$4 - 4\frac{1}{2} = \frac{8}{2} - \frac{9}{2} = \frac{8}{2} + \left(-\frac{9}{2}\right) = -\frac{1}{2}.$$

EXAMPLE 11 Subtract: $3\frac{2}{7} - 4\frac{2}{5}$.

Since $4\frac{2}{5}$ is greater than $3\frac{2}{7}$, the answer will be negative. We can also see this by rewriting the subtraction as $3\frac{2}{7} + \left(-4\frac{2}{5}\right)$. The difference in absolute values is

$$
\begin{array}{rcccc}
4\dfrac{2}{5} = & 4\,\dfrac{2\cdot 7}{5\cdot 7} = & 4\dfrac{14}{35} \\[2mm]
-\;3\dfrac{2}{7} = & -\,3\,\dfrac{2\cdot 5}{7\cdot 5} = & -\,3\dfrac{10}{35} \\[2mm]
\hline
& & 1\dfrac{4}{35}.
\end{array}
$$

Because $-4\frac{2}{5}$ has the larger absolute value, we make the answer negative.

Thus, $3\frac{2}{7} - 4\frac{2}{5} = -1\frac{4}{35}$. ◄

◄ **Do Exercises 13–15.**

Subtract.

13. $7 - 7\frac{3}{4}$

14. $5\frac{1}{2} - 9\frac{3}{4}$

15. $4\frac{2}{3} - 7\frac{1}{6}$

EXAMPLE 12 Subtract: $-6\frac{4}{5} - \left(-9\frac{3}{10}\right)$.

We write the subtraction as addition:

$$-6\frac{4}{5} - \left(-9\frac{3}{10}\right) = -6\frac{4}{5} + 9\frac{3}{10}. \qquad \text{\color{red}Instead of subtracting,}$$
$$\text{\color{red}we add the opposite.}$$

Since $9\frac{3}{10}$ has the greater absolute value, the answer will be positive. The difference in absolute values is

$$
\begin{array}{rcccccl}
9\dfrac{3}{10} &=& 9\ \dfrac{3}{10} &=& 9\dfrac{3}{10} &=& 8\dfrac{13}{10} \\[2mm]
-6\dfrac{4}{5} &=& -6\ \dfrac{4}{5}\cdot\dfrac{2}{2} &=& -6\dfrac{8}{10} &=& -6\dfrac{8}{10} \\[2mm]
\hline
& & & & & & 2\dfrac{5}{10} = 2\dfrac{1}{2}.
\end{array}
$$

Thus, $-6\frac{4}{5} - (-9\frac{3}{10}) = 2\frac{1}{2}$.

We can check by converting to fraction notation and redoing the calculation:

$$
\begin{aligned}
-6\frac{4}{5} - \left(-9\frac{3}{10}\right) &= -\frac{34}{5} - \left(-\frac{93}{10}\right) \\[2mm]
&= -\frac{34}{5} + \frac{93}{10} &&\text{\color{red}Adding the opposite} \\[2mm]
&= -\frac{68}{10} + \frac{93}{10} &&\text{\color{red}Writing with a common} \\
&&&\text{\color{red}denominator} \\[2mm]
&= \frac{25}{10} = \frac{5}{2} = 2\frac{1}{2}. &&\text{\color{red}$-68 + 93 = 25$}
\end{aligned}
$$

Do Exercises 16 and 17. ▶

Subtract.

16. $-7\frac{1}{3} - \left(-5\frac{1}{2}\right)$

17. $-4\frac{1}{10} - \left(-7\frac{2}{5}\right)$

To add two negative numbers we add absolute values and make the answer negative.

EXAMPLE 13 Add: $-4\frac{1}{6} + \left(-5\frac{2}{9}\right)$.

We add the absolute values and make the answer negative.

$$
\begin{array}{rclcl}
4\dfrac{1}{6} &=& 4\dfrac{1}{6}\cdot\dfrac{3}{3} &=& 4\dfrac{3}{18} \qquad \text{\color{red}Adding absolute values} \\[2mm]
+\,5\dfrac{2}{9} &=& +\,5\dfrac{2}{9}\cdot\dfrac{2}{2} &=& +\,5\dfrac{4}{18} \\[2mm]
\hline
& & & & 9\dfrac{7}{18}
\end{array}
$$

Thus, $-4\frac{1}{6} + (-5\frac{2}{9}) = -9\frac{7}{18}$.

Do Exercise 18. ▶

18. Add: $-7\frac{1}{10} + \left(-6\frac{2}{15}\right)$.

Answers

16. $-1\frac{5}{6}$ **17.** $3\frac{3}{10}$ **18.** $-13\frac{7}{30}$

✓ Reading Check

Match each addition or subtraction with the correct first step from the following list.

a) Add the fractions.

b) Write the fractional parts with a common denominator.

c) Rename 5 as $4\frac{9}{9}$.

d) Borrow 1 from 5 and add it to $\frac{1}{9}$.

_____ **RC1.** $\quad 5\frac{1}{9}$ $\qquad\qquad$ _____ **RC2.** $\quad 5\frac{4}{9}$ $\qquad\qquad$ _____ **RC3.** $\quad 5\frac{4}{9}$ $\qquad\qquad$ _____ **RC4.** $\quad 5$

$\qquad\quad -3\frac{4}{9}$ $\qquad\qquad\qquad\quad +3\frac{1}{9}$ $\qquad\qquad\qquad\quad +3\frac{1}{18}$ $\qquad\qquad\qquad\quad -3\frac{1}{9}$

a , **b** Perform the indicated operation. Write a mixed numeral for each answer.

1. $\quad 6$
$\quad +5\frac{2}{5}$

2. $\quad 3$
$\quad +6\frac{5}{7}$

3. $\quad 2\frac{7}{8}$
$\quad +6\frac{5}{8}$

4. $\quad 2\frac{5}{6}$
$\quad +5\frac{5}{6}$

5. $\quad 4\frac{1}{4}$
$\quad +1\frac{1}{12}$

6. $\quad 4\frac{1}{12}$
$\quad +5\frac{1}{6}$

7. $\quad 7\frac{3}{4}$
$\quad +5\frac{5}{6}$

8. $\quad 4\frac{3}{8}$
$\quad +6\frac{5}{12}$

9. $\quad 3\frac{2}{5}$
$\quad +8\frac{7}{10}$

10. $\quad 5\frac{1}{2}$
$\quad +3\frac{7}{10}$

11. $\quad 6\frac{3}{8}$
$\quad +10\frac{5}{6}$

12. $\quad \frac{5}{8}$
$\quad +1\frac{5}{6}$

13. $\quad 18\frac{4}{5}$
$\quad +\ 2\frac{7}{10}$

14. $\quad 15\frac{5}{8}$
$\quad +11\frac{3}{4}$

15. $\quad 14\frac{5}{8}$
$\quad +13\frac{1}{4}$

16. $\quad 16\frac{1}{4}$
$\quad +15\frac{7}{8}$

17. $\quad 8\frac{9}{10}$
$\quad -1\frac{7}{10}$

18. $\quad 6\frac{7}{8}$
$\quad -5\frac{3}{8}$

19. $\quad 9\frac{3}{5}$
$\quad -3\frac{1}{2}$

20. $\quad 8\frac{2}{3}$
$\quad -7\frac{1}{2}$

21. $\quad 4\frac{1}{5}$
$\quad -2\frac{3}{5}$

22. $\quad 5\frac{1}{8}$
$\quad -2\frac{3}{8}$

23. $\quad 19$
$\quad -5\frac{3}{4}$

24. $\quad 17$
$\quad -3\frac{7}{8}$

25. $\quad 34$
$\quad -18\frac{5}{8}$

26.
$$23$$
$$-\ 19\frac{3}{4}$$

27.
$$21\frac{1}{6}$$
$$-\ 13\frac{3}{4}$$

28.
$$42\frac{1}{10}$$
$$-\ 23\frac{7}{12}$$

29.
$$25\frac{1}{9}$$
$$-\ 13\frac{5}{6}$$

30.
$$23\frac{5}{16}$$
$$-\ 14\frac{7}{12}$$

Combine like terms.

31. $1\frac{1}{8}t + 7\frac{5}{8}t$

32. $6\frac{1}{4}x + 8\frac{3}{4}x$

33. $9\frac{1}{2}x - 7\frac{1}{2}x$

34. $7\frac{3}{4}x - 2\frac{1}{4}x$

35. $5\frac{9}{10}y + 2\frac{2}{5}y$

36. $9\frac{3}{4}t + 2\frac{3}{8}t$

37. $37\frac{5}{9}t - 25\frac{2}{3}t$

38. $23\frac{1}{6}t - 19\frac{2}{3}t$

39. $2\frac{5}{6}x + 7\frac{3}{8}x$

40. $7\frac{3}{20}t + 1\frac{2}{15}t$

41. $11a - 8\frac{2}{3}a$

42. $6a - 3\frac{7}{10}a$

 Solve.

43. *Widening a Driveway.* Sherry and Woody are widening their existing $17\frac{1}{4}$-ft driveway by adding $5\frac{9}{10}$ ft on one side. What is the new width of the driveway?

44. *Plumbing.* A plumber uses two pipes, each of length $51\frac{5}{16}$ in., and one pipe of length $34\frac{3}{4}$ in. when installing a shower. How much pipe is used in all?

45. *Upholstery Fabric.* Executive Car Care sells 45-in. upholstery fabric for car restoration. Art bought $9\frac{1}{4}$ yd and $10\frac{5}{6}$ yd for two car projects. How many yards did Art buy?

46. *Painting.* A painter used $1\frac{3}{4}$ gal of paint for the Garcias' living room and $1\frac{1}{3}$ gal for their family room. How much paint was used in all?

47. *Height.* Casey's beagle is $14\frac{1}{4}$ in. from shoulder to floor, and her basset hound is $13\frac{5}{16}$ in. from shoulder to floor. How much shorter is her basset hound?

48. *Winterizing a Swimming Pool.* To winterize their swimming pool, the Jablonskis are draining the water into a nearby field. The distance to the field is $103\frac{1}{2}$ ft. Because their only hose measures $62\frac{3}{4}$ ft, they need to buy an additional hose. How long must the new hose be?

Find the perimeter of (distance around) each figure.

49.

$36\frac{5}{8}$ in.

$30\frac{1}{2}$ in.

50.

$8\frac{1}{2}$ in.

$9\frac{3}{4}$ in.

51.

$5\frac{3}{4}$ yd $5\frac{3}{4}$ yd

$5\frac{3}{4}$ yd $5\frac{3}{4}$ yd

$5\frac{3}{4}$ yd

52.

$3\frac{7}{16}$ ft

$3\frac{7}{16}$ ft

$6\frac{7}{8}$ ft

$6\frac{7}{8}$ ft

53.

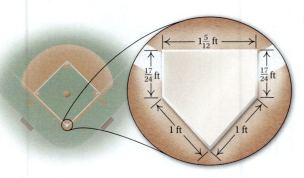

$1\frac{5}{12}$ ft

$\frac{17}{24}$ ft $\frac{17}{24}$ ft

1 ft   1 ft

54.

$44\frac{1}{2}$ ft

$30\frac{1}{2}$ ft

$12\frac{1}{3}$ ft

14 ft   14 ft

Find the length *d* in each figure.

55.

$2\frac{3}{4}$ ft $2\frac{3}{4}$ ft

d

$12\frac{7}{8}$ ft

56.

$2\frac{1}{5}$ in. $2\frac{1}{5}$ in.

d

$10\frac{1}{2}$ in.

57. Stone Bench. Baytown Village Stone Creations is making a custom stone bench as shown below. The recommended height for the bench is 18 in. The depth of the stone bench is $3\frac{3}{8}$ in. Each of the two supporting legs is made up of three stacked stones. Two of the stones measure $3\frac{1}{2}$ in. and $5\frac{1}{4}$ in. How much must the third stone measure?

58. Window Dimensions. The Sanchez family is replacing a window in their home. The original window measures $4\frac{5}{6}$ ft $\times$ $8\frac{1}{4}$ ft. The new window is $2\frac{1}{3}$ ft wider. What are the dimensions of the new window?

59. Carpentry. When cutting wood with a saw, a carpenter must take into account the thickness of the saw blade. Suppose that from a piece of wood 36 in. long, a carpenter cuts a $15\frac{3}{4}$-in. length with a saw blade that is $\frac{1}{8}$ in. thick. How long is the piece that remains?

60. Cutco Cutlery. The Essentials 5-piece set sold by Cutco contains three knives with different blade lengths: $7\frac{5}{8}''$ Petite Chef, $6\frac{3}{4}''$ Petite Carver, and $2\frac{3}{4}''$ Paring Knife. How much larger is the blade of the Petite Chef than that of the Petite Carver? than that of the Paring Knife?

Source: Cutco Cutlery Corporation

61. Interior Design. Eric worked $10\frac{1}{2}$ hr over a three-day period on an interior design project. If he worked $2\frac{1}{2}$ hr on the first day and $4\frac{1}{5}$ hr on the second, how many hours did Eric work on the third day?

62. Painting. Geri had $3\frac{1}{2}$ gal of paint. It took $2\frac{3}{4}$ gal to paint the family room. She estimated that it would take $2\frac{1}{4}$ gal to paint the living room. How much more paint did Geri need?

63. Fly Fishing. Bryan is putting together a fly fishing line and uses $58\frac{5}{8}$ ft of slow-sinking fly line and $8\frac{3}{4}$ ft of leader line. He uses $\frac{3}{8}$ ft of the slow-sinking fly line to connect the two lines. The knot used to connect the fly to the leader line uses $\frac{1}{6}$ ft of the leader line. How long is the finished fly fishing line?

64. Find the smallest length of a bolt that will pass through a piece of tubing with an outside diameter of $\frac{1}{2}$ in., a washer $\frac{1}{16}$ in. thick, a piece of tubing with a $\frac{3}{4}$-in. outside diameter, another washer, and a nut $\frac{3}{16}$ in. thick.

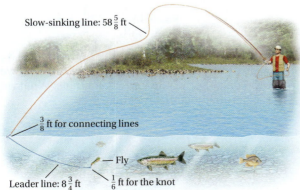

Slow-sinking line: $58\frac{5}{8}$ ft

$\frac{3}{8}$ ft for connecting lines

Fly

Leader line: $8\frac{3}{4}$ ft $\frac{1}{6}$ ft for the knot

d Perform the indicated operation. Write a mixed numeral for each answer.

65. $9 - 9\frac{2}{5}$

66. $8 - 8\frac{3}{7}$

67. $3\frac{1}{2} - 6\frac{3}{4}$

68. $5\frac{1}{2} - 7\frac{3}{4}$

69. $3\frac{4}{5} + \left(-7\frac{2}{3}\right)$

70. $2\frac{3}{7} + \left(-5\frac{1}{2}\right)$

71. $-3\frac{1}{5} - 4\frac{2}{5}$

72. $-5\frac{3}{8} - 4\frac{1}{8}$

73. $-4\frac{1}{12} + 6\frac{2}{3}$

74. $-2\frac{3}{4} + 5\frac{3}{8}$

75. $-6\frac{1}{9} - \left(-4\frac{2}{9}\right)$

76. $-2\frac{3}{5} - \left(-1\frac{1}{5}\right)$

Skill Maintenance

77. Write expanded notation for 38,125. [1.1b]

78. Write a word name for 2,005,689. [1.1c]

79. Write exponential notation for $9 \cdot 9 \cdot 9 \cdot 9$. [1.9a]

80. Evaluate: 3^4. [1.9b]

Determine whether the first number is divisible by the second. [3.1a], [3.1b]

81. 9993 by 3

82. 9993 by 9

83. 2345 by 9

84. 2345 by 5

85. 2335 by 10

86. 7764 by 6

87. 18,888 by 2

88. 18,888 by 6

89. Multiply and simplify: $\frac{15}{9} \cdot \frac{18}{39}$. [3.6a]

90. Divide and simplify: $\frac{12}{25} \div \frac{24}{5}$. [3.7b]

Synthesis

Calculate each of the following. Write the result as a mixed numeral.

91. ▦ $3289\frac{1047}{1189} + 5278\frac{32}{41}$

92. ▦ $4230\frac{19}{73} - 5848\frac{17}{29}$

Solve.

93. $35\frac{2}{3} + n = 46\frac{1}{4}$

94. $42\frac{7}{9} = x - 13\frac{2}{5}$

95. $-15\frac{7}{8} = 12\frac{1}{2} + t$

96. A post for a pier is 29 ft long. Half of the post extends above the water's surface and $8\frac{3}{4}$ ft of the post is buried in mud. How deep is the water at that location?

97. An algebra text is $1\frac{1}{8}$ in. thick, $9\frac{3}{4}$ in. long, and $8\frac{1}{2}$ in. wide. If the front, back, and spine of the book were unfolded, they would form a rectangle. What would the perimeter of that rectangle be?

Multiplication and Division of Mixed Numerals; Applications

4.7

Carrying out addition and subtraction with mixed numerals is usually easier if the numbers are left as mixed numerals. With multiplication and division, however, it is easier to convert the numbers to fraction notation first.

a MULTIPLICATION

> **MULTIPLICATION USING MIXED NUMERALS**
>
> To multiply using mixed numerals, first convert to fraction notation and multiply. Then convert the answer to a mixed numeral, if appropriate.

EXAMPLE 1 Multiply: $6 \cdot 2\frac{1}{2}$.

$$6 \cdot 2\frac{1}{2} = \frac{6}{1} \cdot \frac{5}{2} = \frac{6 \cdot 5}{1 \cdot 2} = \frac{2 \cdot 3 \cdot 5}{2 \cdot 1} = 15 \qquad \text{Removing a factor equal to 1: } \frac{2}{2} = 1$$

Here we write fraction notation.

EXAMPLE 2 Multiply: $3\frac{1}{2} \cdot \frac{3}{4}$.

$$3\frac{1}{2} \cdot \frac{3}{4} = \frac{7}{2} \cdot \frac{3}{4} = \frac{21}{8} = 2\frac{5}{8}$$

> Recall that common denominators are *not* required when multiplying fractions.

Do Margin Exercises 1 and 2.

EXAMPLE 3 Multiply: $-10 \cdot 5\frac{2}{3}$.

$$-10 \cdot 5\frac{2}{3} = -\frac{10}{1} \cdot \frac{17}{3} = -\frac{170}{3} = -56\frac{2}{3}$$

EXAMPLE 4 Multiply: $2\frac{1}{4} \cdot 5\frac{2}{3}$.

$$2\frac{1}{4} \cdot 5\frac{2}{3} = \frac{9}{4} \cdot \frac{17}{3} = \frac{9 \cdot 17}{4 \cdot 3} = \frac{3 \cdot 3 \cdot 17}{2 \cdot 2 \cdot 3} = \frac{51}{4} = 12\frac{3}{4}$$

Caution!

$2\frac{1}{4} \cdot 5\frac{2}{3} \neq 10\frac{2}{12}$. A common error is to multiply the whole numbers and then the fractions. The correct answer, $12\frac{3}{4}$, is found only after converting to fraction notation.

Do Exercises 3 and 4.

OBJECTIVES

a Multiply using mixed numerals.

b Divide using mixed numerals.

c Evaluate expressions using mixed numerals.

d Solve applied problems involving multiplication and division with mixed numerals.

SKILL TO REVIEW

Objective 3.7b: Divide and simplify using fraction notation.

Divide and simplify.

1. $85 \div \frac{17}{5}$ **2.** $\frac{7}{65} \div \frac{21}{25}$

Multiply.

1. $8 \cdot 3\frac{1}{2}$ **2.** $5\frac{1}{2} \cdot \frac{3}{7}$

Multiply.

GS 3. $-2 \cdot 6\frac{2}{5} = -\frac{2}{1} \cdot \frac{\square}{5}$

$$= -\frac{\square}{5} = -\square\frac{\square}{\square}$$

4. $3\frac{1}{3} \cdot 2\frac{1}{2}$

Answers

Skill to Review:

1. 25 **2.** $\frac{5}{39}$

Margin Exercises:

1. 28 **2.** $2\frac{5}{14}$ **3.** $-12\frac{4}{5}$ **4.** $8\frac{1}{3}$

Guided Solution:

3. 32, 64, $12\frac{4}{5}$

b DIVISION

The division $1\frac{1}{2} \div \frac{1}{6}$ is shown here. This division means "How many $\frac{1}{6}$'s are in $1\frac{1}{2}$?" We see that the answer is 9.

$\frac{1}{6}$ goes into $1\frac{1}{2}$ nine times.

When we divide using mixed numerals, we convert to fraction notation first.

$$1\frac{1}{2} \div \frac{1}{6} = \frac{3}{2} \div \frac{1}{6} \qquad \text{We convert } 1\frac{1}{2} \text{ to fraction notation.}$$

$$= \frac{3}{2} \cdot \frac{6}{1} \qquad \text{To divide by } \frac{1}{6}, \text{ multiply by the reciprocal, } \frac{6}{1}.$$

$$= \frac{3 \cdot 6}{2 \cdot 1} = \frac{3 \cdot 3 \cdot 2}{2 \cdot 1} = \frac{3 \cdot 3}{1} \cdot \frac{2}{2} = \frac{3 \cdot 3}{1} \cdot 1 = 9$$

DIVISION USING MIXED NUMERALS

To divide using mixed numerals, first write fraction notation and divide. Then convert the answer to a mixed numeral, if appropriate.

---

5. Divide: $63 \div 5\frac{1}{4}$.

EXAMPLE 5 Divide: $32 \div 3\frac{1}{5}$.

$$32 \div 3\frac{1}{5} = \frac{32}{1} \div \frac{16}{5} \qquad \text{Converting to fraction notation}$$

$$= \frac{32}{1} \cdot \frac{5}{16} = \frac{32 \cdot 5}{1 \cdot 16} = \frac{2 \cdot 16 \cdot 5}{1 \cdot 16} = 10 \qquad \begin{array}{l}\text{Removing a factor}\\\text{equal to 1: } \frac{16}{16} = 1\end{array}$$

⬑ Remember to multiply by the reciprocal of the divisor.

Divide.

6. $2\frac{1}{4} \div 1\frac{1}{5}$

$$= \frac{\square}{4} \div \frac{\square}{5}$$

$$= \frac{9}{4} \cdot \frac{5}{\square}$$

$$= \frac{3 \cdot 3 \cdot 5}{2 \cdot 2 \cdot 2 \cdot \square}$$

$$= \frac{\square}{\square} \cdot \frac{3 \cdot 5}{2 \cdot 2 \cdot 2}$$

$$= \frac{15}{\square}$$

$$= \square\frac{\square}{\square}$$

GS

·········· **Caution!** ··········

The reciprocal of $3\frac{1}{5}$ is neither $5\frac{1}{3}$ nor $3\frac{5}{1}$!

◀ Do Exercise 5.

EXAMPLE 6 Divide: $2\frac{1}{3} \div 1\frac{3}{4}$.

$$2\frac{1}{3} \div 1\frac{3}{4} = \frac{7}{3} \div \frac{7}{4} = \frac{7}{3} \cdot \frac{4}{7} = \frac{7 \cdot 4}{7 \cdot 3} = \frac{4}{3} = 1\frac{1}{3} \qquad \begin{array}{l}\text{Removing a factor}\\\text{equal to 1: } \frac{7}{7} = 1\end{array}$$

7. $1\frac{3}{4} \div \left(-2\frac{1}{2}\right)$

EXAMPLE 7 Divide: $-1\frac{3}{5} \div \left(-3\frac{1}{3}\right)$.

$$-1\frac{3}{5} \div \left(-3\frac{1}{3}\right) = -\frac{8}{5} \div \left(-\frac{10}{3}\right) = \frac{8}{5} \cdot \frac{3}{10} \qquad \begin{array}{l}\text{The product or}\\\text{quotient of two}\\\text{negatives is positive.}\end{array}$$

$$= \frac{2 \cdot 4 \cdot 3}{5 \cdot 2 \cdot 5} = \frac{12}{25} \qquad \begin{array}{l}\text{Removing a factor}\\\text{equal to 1: } \frac{2}{2} = 1\end{array}$$

Answers

5. 12   **6.** $1\frac{7}{8}$ **7.** $-\frac{7}{10}$

Guided Solution:

6. $9, 6, 6, 3, \frac{3}{3}, 8, 1\frac{7}{8}$

◀ Do Exercises 6 and 7.

c EVALUATING EXPRESSIONS

Mixed numerals can appear in algebraic expressions.

EXAMPLE 8 A train traveling r miles per hour for t hours travels a total of rt miles. (*Remember*: Distance $=$ Rate $\cdot$ Time.)

a) Find the distance traveled by a train moving at 60 mph in $2\frac{3}{4}$ hr.

b) Find the distance traveled if the speed of the train is $26\frac{1}{2}$ mph and the time is $2\frac{2}{3}$ hr.

a) We evaluate rt for $r = 60$ and $t = 2\frac{3}{4}$:

$$rt = 60 \cdot 2\frac{3}{4}$$
$$= \frac{60}{1} \cdot \frac{11}{4}$$
$$= \frac{15 \cdot 4 \cdot 11}{1 \cdot 4} = 165. \qquad \text{Removing a factor equal to 1: } \tfrac{4}{4} = 1$$

In $2\frac{3}{4}$ hr, a train moving at 60 mph travels 165 mi.

b) We evaluate rt for $r = 26\frac{1}{2}$ and $t = 2\frac{2}{3}$:

$$rt = 26\frac{1}{2} \cdot 2\frac{2}{3}$$
$$= \frac{53}{2} \cdot \frac{8}{3} = \frac{53 \cdot 2 \cdot 4}{2 \cdot 3} \qquad \text{Removing a factor equal to 1: } \tfrac{2}{2} = 1$$
$$= \frac{212}{3} = 70\frac{2}{3}.$$

In $2\frac{2}{3}$ hr, a train moving at $26\frac{1}{2}$ mph travels $70\frac{2}{3}$ mi.

EXAMPLE 9 Evaluate $x + yz$ for $x = 7\frac{1}{3}$, $y = \frac{1}{3}$, and $z = 5$.

We substitute and follow the rules for order of operations:

$$x + yz = 7\frac{1}{3} + \frac{1}{3} \cdot 5$$
$$= 7\frac{1}{3} + \frac{1}{3} \cdot \frac{5}{1} \qquad \text{Multiply first; then add.}$$
$$= 7\frac{1}{3} + \frac{5}{3}$$
$$\left. \begin{array}{l} = 7\frac{1}{3} + 1\frac{2}{3} \\[2mm] = 8\frac{3}{3} = 9. \end{array} \right\} \quad \text{Adding mixed numerals}$$

Do Exercises 8–10. ▶

Evaluate.

8. rt, for $r = 78$ and $t = 2\frac{1}{4}$

9. $7xy$, for $x = 9\frac{2}{5}$ and $y = 2\frac{3}{7}$

10. $x - y \div z$, for $x = 5\frac{7}{8}, y = \frac{1}{4}$, and $z = 2$

Answers

8. $175\frac{1}{2}$ 9. $159\frac{4}{5}$ 10. $5\frac{3}{4}$

d APPLICATIONS AND PROBLEM SOLVING

EXAMPLE 10 *Training Regimens.* Fitness trainers suggest training regimens for athletes who are preparing to run marathons and mini-marathons. One suggested twelve-week regimen combines days of short, easy running with other days of cross-training, rest, and long-distance running. During week nine, this regimen calls for a long-distance run of 10 mi, which is $2\frac{1}{2}$ times the length of the long-distance run recommended for week one. What is the length of the long-distance run recommended for week one?

Source: shape.com

1. **Familiarize.** We ask the question "10 is $2\frac{1}{2}$ times what number?" We let $r =$ the length of the long-distance run recommended for week one.

2. **Translate.** The problem can be translated to an equation.

$$10 \quad = \quad 2\frac{1}{2} \quad \cdot \quad r$$

3. **Solve.** To solve the equation, we multiply on both sides.

$$10 = 2\frac{1}{2} \cdot r$$

$$10 = \frac{5}{2} \cdot r \qquad \text{Converting } 2\frac{1}{2} \text{ to fraction notation}$$

$$\frac{2}{5} \cdot 10 = \frac{2}{5} \cdot \frac{5}{2}r \qquad \text{Using the multiplication principle}$$

$$\frac{2 \cdot 10}{5} = 1 \cdot r \qquad \text{Multiplying}$$

$$4 = r \qquad \text{Simplifying: } \frac{2 \cdot 10}{5} = \frac{20}{5} = 4$$

4. **Check.** If the length of the long-distance run recommended for week one is 4 mi, we find the length of the run recommended for week nine by multiplying 4 by $2\frac{1}{2}$.

$$2\frac{1}{2} \cdot 4 = \frac{5}{2} \cdot 4 = \frac{20}{2} = 10$$

The answer checks.

5. **State.** The regimen recommends a long-distance run of 4 mi for week one.

◀ **Do Exercises 11 and 12.**

EXAMPLE 11 *Flooring.* Ann and Tony plan to redo the floor of their living room. Since part of the room will be covered by a rug, they want to lay a hardwood floor only on the part of the room not under the rug. If the room is $22\frac{1}{2}$ ft by $15\frac{1}{2}$ ft and the rug is 9 ft by 12 ft, how much hardwood flooring do they need? How much hardwood flooring would it take to cover the entire floor of the room?

Solve.

11. Kyle's pickup truck travels on an interstate highway at 65 mph for $3\frac{1}{2}$ hr. How far does it travel?

12. Holly's minivan traveled 302 mi on $15\frac{1}{10}$ gal of gas. How many miles per gallon did it get?

Answers

11. $227\frac{1}{2}$ mi   12. 20 mpg

1. **Familiarize.** We draw a diagram and let $B = $ the area of the room, $R = $ the area of the rug, and $H = $ the area to be covered by hardwood flooring.

2. **Translate.** This is a multistep problem. We first find the area of the room, B, and the area of the rug, R. Then $H = B - R$. We find each area using the formula for the area of a rectangle: $A = l \times w$.

3. **Solve.** We carry out the calculations.

$$B = \text{length} \times \text{width}$$
$$= 22\frac{1}{2} \cdot 15\frac{1}{2}$$
$$= \frac{45}{2} \cdot \frac{31}{2}$$
$$= \frac{1395}{4} = 348\frac{3}{4} \text{ sq ft}$$

$$R = \text{length} \times \text{width}$$
$$= 12 \cdot 9$$
$$= 108 \text{ sq ft}$$

Then $H = B - R$

$$= 348\frac{3}{4} \text{ sq ft} - 108 \text{ sq ft}$$

$$= 240\frac{3}{4} \text{ sq ft}$$

4. **Check.** We can perform a check by repeating the calculations.

5. **State.** Ann and Tony will need $240\frac{3}{4}$ sq ft of hardwood flooring. It would take $348\frac{3}{4}$ sq ft of hardwood flooring to cover the entire floor of the room.

Do Exercise 13. ▶

13. *Koi Pond.* Colleen designed a koi fish pond for her backyard. Using the dimensions shown in the diagram below, determine the area of Colleen's backyard remaining after the pond was completed.

Sources: en.wikipedia.org; pondliner.com

Answer

13. 945 sq yd

🖩 **CALCULATOR CORNER**

Operations on Fractions and Mixed Numerals Fraction calculators can add, subtract, multiply, and divide fractions and mixed numerals. The ⌊$a^b/_c$⌋ key is used to enter fractions and mixed numerals. The fraction $\frac{3}{4}$ is entered by pressing ⌊3⌋ ⌊$a^b/_c$⌋ ⌊4⌋ , and it appears on the display as 3 ⌋ 4 . The mixed numeral $1\frac{5}{16}$ is entered by pressing ⌊1⌋ ⌊$a^b/_c$⌋ ⌊5⌋ ⌊$a^b/_c$⌋ ⌊1⌋ ⌊6⌋ , and it is displayed as

1 ⌋ 5 ⌋ 16 . Fraction results that are greater than 1 are always displayed as mixed numerals. To express the result for $1\frac{5}{16}$ as a fraction, we press ⌊SHIFT⌋ ⌊d/c⌋. We get 21 ⌋ 16 , or $\frac{21}{16}$. Some calculators display fractions and mixed numerals in the way in which we write them.

EXERCISES Perform each calculation. Give the answer in fraction notation.

1. $\frac{1}{3} + \frac{1}{4}$

2. $\frac{7}{5} - \frac{3}{10}$

3. $\frac{15}{4} \cdot \frac{7}{12}$

4. $-\frac{4}{5} \div \frac{8}{3}$

Perform each calculation. Give the answer as a mixed numeral.

5. $4\frac{1}{3} + 5\frac{4}{5}$

6. $9\frac{2}{7} - 8\frac{1}{4}$

7. $-2\frac{1}{3} \cdot \left(-4\frac{3}{5}\right)$

8. $10\frac{7}{10} \div 3\frac{5}{6}$

Translating for Success

1. **Raffle Tickets.** At the Happy Hollow Camp Fall Festival, Rico and Becca, together, spent $270 on raffle tickets that sell for $\$\frac{9}{20}$ each. How many tickets did they buy?

2. **Irrigation Pipe.** Jed uses two pipes, one of which measures $5\frac{1}{3}$ ft, to repair the irrigation system in the Aguilars' lawn. The total length of the two pipes is $8\frac{7}{12}$ ft. How long is the other pipe?

3. **Vacation Days.** Together, Oscar and Claire have 36 vacation days a year. Oscar has 22 vacation days per year. How many does Claire have?

4. **Enrollment in Japanese Classes.** Last year at Lakeside Community College, 225 students enrolled in basic mathematics. This number is $4\frac{1}{2}$ times as many as the number who enrolled in Japanese. How many enrolled in Japanese?

5. **Bicycling.** Cole rode his bicycle $5\frac{1}{3}$ mi on Saturday and $8\frac{7}{12}$ mi on Sunday. How far did he ride on the weekend?

The goal of these matching questions is to practice step (2), Translate, of the five-step problem-solving process. Translate each word problem to an equation and select a correct translation from equations A–O.

A. $13\frac{11}{12} = x + 5\frac{1}{3}$

B. $\frac{3}{4} \cdot x = 1\frac{2}{3}$

C. $270 - \frac{20}{9} = x$

D. $225 = 4\frac{1}{2} \cdot x$

E. $98 \div 2\frac{1}{3} = x$

F. $22 + x = 36$

G. $x = 4\frac{1}{2} \cdot 225$

H. $x = 5\frac{1}{3} + 8\frac{7}{12}$

I. $22 \cdot x = 36$

J. $x = \frac{3}{4} \cdot 1\frac{2}{3}$

K. $5\frac{1}{3} + x = 8\frac{7}{12}$

L. $\frac{9}{20} \cdot 270 = x$

M. $1\frac{2}{3} + \frac{3}{4} = x$

N. $98 - 2\frac{1}{3} = x$

O. $\frac{9}{20} \cdot x = 270$

Answers on page A-8

6. **Deli Order.** For a promotional open house for contractors last year, the Bayside Builders Association ordered 225 turkey sandwiches. Due to increased registrations this year, $4\frac{1}{2}$ times as many sandwiches are needed. How many sandwiches should be ordered?

7. **Dog Ownership.** In Sam's community, $\frac{9}{20}$ of the households own at least one dog. There are 270 households. How many own dogs?

8. **Magic Tricks.** Samantha has 98 ft of rope and needs to cut it into $2\frac{1}{3}$-ft pieces to be used in a magic trick. How many pieces can be cut from the rope?

9. **Painting.** Laura needs $1\frac{2}{3}$ gal of paint to paint the ceiling of the exercise room and $\frac{3}{4}$ gal of the same paint for the bathroom. How much paint does Laura need?

10. **Chocolate Fudge Bars.** A recipe for chocolate fudge bars that serves 16 includes $1\frac{2}{3}$ cups of sugar. How much sugar is needed for $\frac{3}{4}$ of this recipe?

☑ **Reading Check**

Determine whether each statement is true or false.

_____ **RC1.** To multiply using mixed numerals, we first convert to fraction notation.

_____ **RC2.** To divide using mixed numerals, we first convert to fraction notation.

_____ **RC3.** The product of mixed numerals is generally written as a mixed numeral, unless it is a whole number or less than 1.

_____ **RC4.** To divide fractions, we multiply by the reciprocal of the divisor.

a   Multiply. Write a mixed numeral for each answer.

1. $8 \cdot 2\frac{5}{6}$

2. $10 \cdot 3\frac{3}{4}$

3. $6\frac{2}{3} \cdot \frac{1}{4}$

4. $-\frac{1}{3} \cdot 5\frac{2}{5}$

5. $20\left(-2\frac{5}{6}\right)$

6. $6\frac{3}{8} \cdot 4\frac{1}{3}$

7. $3\frac{1}{2} \cdot 4\frac{2}{3}$

8. $4\frac{1}{5} \cdot 5\frac{1}{4}$

9. $-2\frac{3}{10} \cdot 4\frac{2}{5}$

10. $4\frac{7}{10} \cdot 5\frac{3}{10}$

11. $\left(-6\frac{3}{10}\right)\left(-5\frac{7}{10}\right)$

12. $-20\frac{1}{2} \cdot \left(-10\frac{1}{5}\right)$

b   Divide. Write a mixed numeral for each answer whenever possible.

13. $20 \div 3\frac{1}{5}$

14. $18 \div 2\frac{1}{4}$

15. $8\frac{2}{5} \div 7$

16. $3\frac{3}{8} \div 3$

17. $6\frac{1}{4} \div 3\frac{3}{4}$

18. $5\frac{4}{5} \div 2\frac{1}{2}$

19. $-1\frac{7}{8} \div 1\frac{2}{3}$

20. $-4\frac{3}{8} \div 2\frac{5}{6}$

21. $5\frac{1}{10} \div 4\frac{3}{10}$

22. $4\frac{1}{10} \div 2\frac{1}{10}$

23. $-20\frac{1}{4} \div (-90)$

24. $-12\frac{1}{2} \div (-50)$

c Evaluate.

25. lw, for $l = 2\frac{3}{5}$ and $w = 9$

26. mv, for $m = 7$ and $v = 3\frac{2}{5}$

27. rs, for $r = 5$ and $s = 3\frac{1}{7}$

28. rt, for $r = 5\frac{2}{3}$ and $t = 4\frac{1}{5}$

29. mt, for $m = 6\frac{2}{9}$ and $t = -4\frac{3}{8}$

30. $M \div NP$, for $M = 2\frac{1}{4}$, $N = -5$, and $P = 2\frac{1}{3}$

31. $R \cdot S \div T$, for $R = 4\frac{2}{3}$, $S = 1\frac{3}{7}$, and $T = -5$

32. $a - bc$, for $a = 18$, $b = 2\frac{1}{5}$, and $c = 3\frac{3}{4}$

33. $r + ps$, for $r = 5\frac{1}{2}$, $p = 3$, and $s = 2\frac{1}{4}$

34. $s + rt$, for $s = 3\frac{1}{2}$, $r = 5\frac{1}{2}$, and $t = 7\frac{1}{2}$

35. $m + n \div p$, for $m = 7\frac{2}{5}$, $n = 4\frac{1}{2}$, and $p = 6$

36. $x - y \div z$, for $x = 9$, $y = 2\frac{1}{2}$, and $z = 3\frac{3}{4}$

d Solve.

37. *Art Prices.* A 1966 Andy Warhol portrait of Marlon Brando on his motorcycle sold for $5 million at Christie's in 2003. On November 14, 2012, the work was sold again by Christie's for about $4\frac{4}{5}$ times the amount paid in 2003. How much was paid for the portrait in 2012?

Source: Businessweek.com

38. *Spreading Grass Seed.* Emily seeds lawns for Sam's Superior Lawn Care. When she walks at a rapid pace, the wheel on the broadcast spreader completes $150\frac{2}{3}$ revolutions per minute. How many revolutions does the wheel complete in 15 min?

39. *Population.* The population of Alabama is $6\frac{2}{3}$ times that of Alaska. The population of Alaska is approximately 720,000. What is the population of Alabama?

Source: U.S. Census Bureau

40. *Population.* The population of New York is $32\frac{1}{2}$ times the population of Wyoming. The population of Wyoming is approximately 600,000. What is the population of New York?

Source: U.S. Census Bureau

41. Apple Net Income.   Apple, Inc., reported net income of about $8,000,000,000 in fiscal year 2009. In fiscal year 2012, Apple's net income was about $5\frac{1}{4}$ times that amount. What was Apple's net income for fiscal year 2012?

42. Median Income.   Median household income in the United States was about $12,000 in 1975. By 2011, median household income was $4\frac{1}{6}$ times that amount. What was the median household income in 2011?

Source: U.S. Census Bureau

43. Average Speed in Indianapolis 500.   Tony Kanaan won the Indianapolis 500 in 2013 with the highest average speed—about $187\frac{1}{2}$ mph. This record speed is about $2\frac{1}{2}$ times the average speed of the first winner, Ray Harroun, in 1911. What was the average speed in the first Indianapolis 500?

Source: Indianapolis Motor Speedway

44. Population.   The population of Cleveland is about $1\frac{1}{3}$ times the population of Cincinnati. In 2013, the population of Cincinnati was approximately 294,750. What was the population of Cleveland in 2013?

Source: U.S. Census Bureau

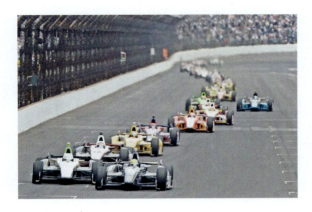

45. Sidewalk.   A sidewalk alongside a garden at the conservatory is to be $14\frac{2}{5}$ yd long. Rectangular stone tiles that are each $1\frac{1}{8}$ yd long are used to form the sidewalk. How many tiles are used?

46. Aeronautics.   Most space shuttles orbit the earth once every $1\frac{1}{2}$ hr. How many orbits are made every 24 hr?

47. Doubling a Recipe.   The chef of a five-star hotel is doubling a recipe for chocolate cake. The original recipe requires $2\frac{3}{4}$ cups of flour and $1\frac{1}{3}$ cups of sugar. How much flour and sugar will she need?

48. Half of a Recipe.   A caterer is following a salad dressing recipe that calls for $1\frac{7}{8}$ cups of mayonnaise and $1\frac{1}{6}$ cups of sugar. How much mayonnaise and sugar will he need if he prepares $\frac{1}{2}$ of the amount of salad dressing?

49. Mileage.   A car traveled 213 mi on $14\frac{2}{10}$ gal of gas. How many miles per gallon did it get?

50. Mileage.   A car traveled 385 mi on $15\frac{4}{10}$ gal of gas. How many miles per gallon did it get?

51. Mural.   A student artist painted a mural on the wall under a bridge. The dimensions of the mural are $6\frac{2}{3}$ ft by $9\frac{3}{8}$ ft. What is the area of the mural?

52. Weight of Water.   The weight of water is $62\frac{1}{2}$ lb per cubic foot. What is the weight of $2\frac{1}{4}$ cubic feet of water?

53. *Weight of Water.* The weight of water is $62\frac{1}{2}$ lb per cubic foot. How many cubic feet would be occupied by 25,000 lb of water?

54. *Weight of Water.* The weight of water is $8\frac{1}{3}$ lb per gallon. Harry rolls his lawn with an 800-lb capacity roller. Express the water capacity of the roller in gallons.

55. *Servings of Salmon.* A serving of filleted fish is generally considered to be about $\frac{1}{3}$ lb. How many servings can be prepared from $5\frac{1}{2}$ lb of salmon fillet?

56. *Servings of Tuna.* A serving of fish steak (cross section) is generally $\frac{1}{2}$ lb. How many servings can be prepared from a cleaned $18\frac{3}{4}$-lb tuna?

57. *Landscaping.* The previous owners of Ashley's new home had a large L-shaped vegetable garden consisting of a rectangle that was $15\frac{1}{2}$ ft by 20 ft adjacent to one that was $10\frac{1}{2}$ ft by $12\frac{1}{2}$ ft. Ashley wants to cover the garden with sod. What is the total area of the sod she must purchase?

58. *Home Furnishings.* An L-shaped sunroom consists of a rectangle that is $9\frac{1}{2}$ ft by 12 ft adjacent to one that is $9\frac{1}{2}$ ft by 8 ft. What is the total area of a carpet that covers the floor?

Find the area of each shaded region.

59.

60.

61.

62.

Copyright © 2016 Pearson Education, Inc.

63. Temperature. Fahrenheit temperature can be obtained from Celsius (Centigrade) temperature by multiplying by $1\frac{4}{5}$ and adding 32°. What Fahrenheit temperature corresponds to a Celsius temperature of 20°?

64. Temperature. Fahrenheit temperature can be obtained from Celsius (Centigrade) temperature by multiplying by $1\frac{4}{5}$ and adding 32°. What Fahrenheit temperature corresponds to the Celsius temperature of boiling water, 100°?

65. Building a Ziggurat. The dimensions of all of the square bricks that King Nebuchadnezzar used over 2500 years ago to build ziggurats were $13\frac{1}{4}$ in. $\times$ $13\frac{1}{4}$ in. $\times$ $3\frac{1}{4}$ in. What are the perimeter and the area of the $13\frac{1}{4}$ in. $\times$ $13\frac{1}{4}$ in. side? of the $13\frac{1}{4}$ in. $\times$ $3\frac{1}{4}$ in. side?

Source: www.eartharchitecture.org

66. Word Processing. For David's design report, he needs to create a table containing two columns, each $1\frac{1}{2}$ in. wide, and five columns, each $\frac{3}{4}$ in. wide. Will this table fit on a piece of standard paper that is $8\frac{1}{2}$ in. wide? If so, how wide will each side margin be if the margins on each side are to be of equal width?

Skill Maintenance

67. Find the perimeter of the figure. [1.2b]

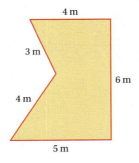

68. Find the area of the region. [1.4b]

Synthesis

Simplify. Write each answer as a mixed numeral whenever possible.

69. $-8 \div \frac{1}{2} + \frac{3}{4} + \left(-5 - \frac{5}{8}\right)^2$

70. $\left(\frac{5}{9} - \frac{1}{4}\right)(-12) + \left(-4 - \frac{3}{4}\right)^2$

71. $\frac{1}{3} \div \left(\frac{1}{2} - \frac{1}{5}\right) \times \frac{1}{4} + \frac{1}{6}$

72. $\frac{7}{8} - 1\frac{1}{8} \times \frac{2}{3} + \frac{9}{10} \div \frac{3}{5}$

73. Find r if

$$\frac{1}{r} = \frac{1}{40} + \frac{1}{60} + \frac{1}{80}.$$

74. Heights. Find the average height of the following NBA players:

LeBron James	6 ft 8 in.
Dwayne Wade	6 ft 4 in.
Dwight Howard	6 ft 11 in.
Dirk Nowitzki	7 ft 0 in.
Al Jefferson	6 ft 10 in.

75. Water Consumption. According to the U.S. Department of Energy, washing one load of clothes uses $2\frac{2}{3}$ times the amount of hot water required for the average shower. If the average shower uses 12 gal of hot water, how much hot water will two showers and two loads of wash require?

4.8 Order of Operations and Complex Fractions

OBJECTIVES

a Simplify expressions containing fraction notation using the rules for order of operations.

b Simplify complex fractions.

SKILL TO REVIEW

Objective 1.9c: Simplify expressions using the rules for order of operations.

Simplify.

1. $22 - 3 \cdot 4$

2. $6(4 + 1)^2 - 3^4 \div 3$

Simplify.

1. $\dfrac{2}{5} \cdot \dfrac{5}{8} + \dfrac{1}{4}$

2. $\dfrac{1}{3} \cdot \dfrac{3}{4} \div \dfrac{5}{8} - \dfrac{1}{10}$ **GS**

$= \dfrac{\square}{12} \div \dfrac{5}{8} - \dfrac{1}{10} = \dfrac{3}{12} \cdot \dfrac{\square}{5} - \dfrac{1}{10}$

$= \dfrac{3 \cdot 2 \cdot 2 \cdot \square}{3 \cdot 2 \cdot 2 \cdot 5} - \dfrac{1}{10} = \dfrac{\square}{5} - \dfrac{1}{10}$

$= \dfrac{\square}{10} - \dfrac{1}{10} = \dfrac{\square}{10}$

a ORDER OF OPERATIONS

Like expressions containing integers, expressions containing fraction notation follow the rules for order of operations.

> ### RULES FOR ORDER OF OPERATIONS
>
> 1. Do all calculations within parentheses ( ), brackets [ ], braces { }, absolute-value symbols, numerators, or denominators.
> 2. Evaluate all exponential expressions.
> 3. Do all multiplications and divisions in order from left to right.
> 4. Do all additions and subtractions in order from left to right.

EXAMPLE 1 Simplify: $\dfrac{1}{6} + \dfrac{2}{3} \div \dfrac{1}{2} \cdot \dfrac{5}{8}$.

$\dfrac{1}{6} + \dfrac{2}{3} \div \dfrac{1}{2} \cdot \dfrac{5}{8} = \dfrac{1}{6} + \dfrac{2}{3} \cdot \dfrac{2}{1} \cdot \dfrac{5}{8}$ Doing the division first by multiplying by the reciprocal of $\frac{1}{2}$

$= \dfrac{1}{6} + \dfrac{2 \cdot 2 \cdot 5}{3 \cdot 1 \cdot 8}$ Doing the multiplications in order from left to right

$= \dfrac{1}{6} + \dfrac{2 \cdot 2 \cdot 5}{3 \cdot 1 \cdot 2 \cdot 2 \cdot 2}$ Factoring

$= \dfrac{1}{6} + \dfrac{5}{6}$ Removing a factor equal to 1: $\dfrac{2 \cdot 2}{2 \cdot 2} = 1$; simplifying

$= \dfrac{6}{6}$, or 1   Doing the addition

◀ Do Margin Exercises 1 and 2.

EXAMPLE 2 Simplify: $\dfrac{2}{3} \cdot 24 - 11\dfrac{1}{2}$.

$\dfrac{2}{3} \cdot 24 - 11\dfrac{1}{2} = \dfrac{2 \cdot 24}{3 \cdot 1} - 11\dfrac{1}{2}$ Doing the multiplication first

$= \dfrac{2 \cdot 3 \cdot 8}{3 \cdot 1} - 11\dfrac{1}{2}$ Factoring

$= 2 \cdot 8 - 11\dfrac{1}{2}$ Removing a factor equal to 1: $\dfrac{3}{3} = 1$

$= 16 - 11\dfrac{1}{2}$ Multiplying

$= 15\dfrac{2}{2} - 11\dfrac{1}{2}$

$= 4\dfrac{1}{2}$, or $\dfrac{9}{2}$ Subtracting

Answers

Skill to Review:
1. 10   2. 123

Margin Exercises:
1. $\dfrac{1}{2}$ 2. $\dfrac{3}{10}$

Guided Solution:
2. 3, 8, 2, 2, 4, 3

Do Exercise 3. ▶

3. Simplify: $\dfrac{3}{4} \cdot 16 + 8\dfrac{2}{3}$.

EXAMPLE 3 Simplify.

a) $-\dfrac{1}{2} - \dfrac{2}{3}\left(-\dfrac{1}{2}\right)^2$

b) $1 - \dfrac{2}{3}\left(\dfrac{1}{3} - \dfrac{1}{2}\right)$

a) The parentheses here are not grouping symbols, so we begin by evaluating the exponential expression.

$$-\dfrac{1}{2} - \dfrac{2}{3}\left(-\dfrac{1}{2}\right)^2 = -\dfrac{1}{2} - \dfrac{2}{3} \cdot \dfrac{1}{4} \qquad \left(-\dfrac{1}{2}\right)^2 = \left(-\dfrac{1}{2}\right) \cdot \left(-\dfrac{1}{2}\right) = \dfrac{1}{4}$$

$$= -\dfrac{1}{2} - \dfrac{2 \cdot 1}{3 \cdot 2 \cdot 2} \qquad \text{Multiplying and factoring}$$

$$= -\dfrac{1}{2} - \dfrac{1}{6} \qquad \text{Simplifying}$$

$$= -\dfrac{3}{6} - \dfrac{1}{6} \qquad \text{Writing with the LCD, 6, in order to subtract}$$

$$= -\dfrac{4}{6} = -\dfrac{2}{3} \qquad \text{Subtracting and simplifying}$$

b) We first subtract within the parentheses.

$$1 - \dfrac{2}{3}\left(\dfrac{1}{3} - \dfrac{1}{2}\right) = 1 - \dfrac{2}{3}\left(\dfrac{2}{6} - \dfrac{3}{6}\right) \qquad \text{Writing with the LCD, 6, in order to subtract}$$

$$= 1 - \dfrac{2}{3}\left(-\dfrac{1}{6}\right) \qquad \begin{array}{l}\text{Subtracting within the parentheses:} \\ \dfrac{2}{6} - \dfrac{3}{6} = \dfrac{2}{6} + \left(-\dfrac{3}{6}\right) = -\dfrac{1}{6}\end{array}$$

$$= 1 - \left(-\dfrac{2 \cdot 1}{3 \cdot 2 \cdot 3}\right) \qquad \begin{array}{l}\text{There are no exponential expressions, so we multiply and factor.}\end{array}$$

$$= 1 - \left(-\dfrac{1}{9}\right) \qquad \text{Simplifying}$$

$$= \dfrac{9}{9} + \dfrac{1}{9} = \dfrac{10}{9} \qquad \begin{array}{l}\text{Writing with the LCD, 9, and} \\ \text{adding the opposite of } -\dfrac{1}{9}\end{array}$$

Do Exercises 4–6. ▶

Simplify.

4. $\left(\dfrac{3}{4}\right)^2 - \dfrac{1}{2} \div \left(-\dfrac{4}{5}\right)$

5. $1 - \left(\dfrac{2}{3} - \dfrac{3}{4}\right)^2$

6. $\left(\dfrac{2}{3} + \dfrac{3}{4}\right) \div 2\dfrac{1}{3} - \left(\dfrac{1}{2}\right)^3$

b COMPLEX FRACTIONS

A **complex fraction** is a fraction in which the numerator and/or denominator contain one or more fractions. The following are some examples of complex fractions.

$$\dfrac{\frac{2}{3}}{7} \;\leftarrow\; \begin{array}{l}\text{The numerator} \\ \text{contains a fraction.}\end{array}$$

$$\dfrac{-\frac{1}{5}}{-\frac{9}{10}} \begin{array}{l}\leftarrow \text{The numerator} \\ \quad\text{contains a fraction.} \\ \leftarrow \text{The denominator} \\ \quad\text{contains a fraction.}\end{array}$$

Since a fraction bar represents division, complex fractions can be rewritten using the division symbol ÷ .

Answers

3. $20\dfrac{2}{3}$, or $\dfrac{62}{3}$ **4.** $\dfrac{19}{16}$, or $1\dfrac{3}{16}$

5. $\dfrac{143}{144}$ **6.** $\dfrac{27}{56}$

EXAMPLE 4 Simplify.

a) $\dfrac{\frac{2}{3}}{7}$
b) $\dfrac{-\frac{1}{5}}{-\frac{9}{10}}$

a) $\dfrac{\frac{2}{3}}{7} = \dfrac{2}{3} \div 7$ Rewriting using a division symbol

$= \dfrac{2}{3} \cdot \left(\dfrac{1}{7}\right)$ Multiplying by the reciprocal of the divisor

$= \dfrac{2 \cdot 1}{3 \cdot 7}$ Multiplying numerators and multiplying denominators

$= \dfrac{2}{21}$ This expression cannot be simplified.

b) $\dfrac{-\frac{1}{5}}{-\frac{9}{10}} = -\dfrac{1}{5} \div \left(-\dfrac{9}{10}\right)$ Rewriting using a division symbol

$= -\dfrac{1}{5} \cdot \left(-\dfrac{10}{9}\right)$ Multiplying by the reciprocal of the divisor; the product will be positive.

$= \dfrac{1 \cdot 2 \cdot \cancel{5}}{\cancel{5} \cdot 3 \cdot 3}$ Multiplying numerators and multiplying denominators; factoring

$= \dfrac{2}{9}$ Removing a factor equal to 1 and simplifying

◀ **Do Exercises 7 and 8.**

Complex fractions may contain variables.

EXAMPLE 5 Simplify: $\dfrac{\frac{x}{24}}{\frac{15}{28}}$.

$\dfrac{\frac{x}{24}}{\frac{15}{28}} = \dfrac{x}{24} \div \dfrac{15}{28}$ Rewriting using a division symbol

$= \dfrac{x}{24} \cdot \dfrac{28}{15}$ Multiplying by the reciprocal of the divisor

$= \dfrac{x \cdot 2 \cdot 2 \cdot 7}{2 \cdot 2 \cdot 2 \cdot 3 \cdot 3 \cdot 5}$ Multiplying numerators and multiplying denominators; factoring

$= \dfrac{x \cdot 7}{2 \cdot 3 \cdot 3 \cdot 5}$ Removing a factor equal to 1: $\dfrac{2 \cdot 2}{2 \cdot 2} = 1$

$= \dfrac{7x}{90}$ Multiplying

◀ **Do Exercises 9 and 10.**

Simplify.

7. $\dfrac{10}{\frac{5}{8}} = 10 \div \dfrac{\square}{\square}$

$= 10 \cdot \dfrac{\square}{\square}$

$= \dfrac{10 \cdot \square}{5}$

$= \dfrac{\square \cdot \cancel{5} \cdot 8}{\cancel{5} \cdot 1}$

$= \square$

8. $\dfrac{\frac{7}{5}}{-\frac{10}{7}}$

Simplify.

9. $\dfrac{\frac{x}{20}}{\frac{3}{10}}$

10. $\dfrac{-\frac{7}{12}}{-\frac{x}{18}}$

Answers

7. 16   **8.** $-\dfrac{49}{50}$ **9.** $\dfrac{x}{6}$ **10.** $\dfrac{21}{2x}$

Guided Solution:

7. $\dfrac{5}{8}, \dfrac{8}{5}$, 8, 2, 16

CHAPTER 4   Fraction Notation: Addition, Subtraction, and Mixed Numerals

When the numerator or denominator of a complex fraction consists of more than one term, first simplify the numerator and/or denominator separately.

EXAMPLE 6 Simplify: $\dfrac{\frac{1}{2}-\frac{2}{3}}{1\frac{7}{8}}$.

$\dfrac{\frac{1}{2}-\frac{2}{3}}{1\frac{7}{8}}=\dfrac{\frac{3}{6}-\frac{4}{6}}{\frac{15}{8}}$ Writing the fractions in the numerator with a common denominator

Writing the mixed numeral in the denominator as a fraction

$=\dfrac{-\frac{1}{6}}{\frac{15}{8}}$ Subtracting in the numerator of the complex fraction

$=-\dfrac{1}{6}\div\dfrac{15}{8}$ Rewriting using a division symbol

$=-\dfrac{1}{6}\cdot\dfrac{8}{15}$ Multiplying by the reciprocal of the divisor

$=-\dfrac{1\cdot2\cdot2\cdot2}{2\cdot3\cdot3\cdot5}$ Multiplying numerators and multiplying denominators; factoring

$=-\dfrac{4}{45}$ Removing a factor equal to 1: $\dfrac{2}{2}=1$

Do Exercises 11 and 12. ▶

Simplify.

11. $\dfrac{\frac{7}{12}+\frac{5}{6}}{\frac{4}{9}}$

12. $\dfrac{-\frac{3}{5}}{\frac{2}{3}-\frac{7}{10}}$

EXAMPLE 7 *Harvesting Walnut Trees.* A woodland owner decided to harvest five walnut trees in order to improve the growing conditions of the remaining trees. The logs she sold measured $7\frac{5}{8}$ ft, $8\frac{1}{4}$ ft, $8\frac{3}{4}$ ft, $9\frac{1}{8}$ ft, and $10\frac{1}{2}$ ft. What is the average length of the logs?

Recall that to compute an average, we add the numbers and then divide the sum by the number of addends. We have

$$\dfrac{7\frac{5}{8}+8\frac{1}{4}+8\frac{3}{4}+9\frac{1}{8}+10\frac{1}{2}}{5}=\dfrac{7\frac{5}{8}+8\frac{2}{8}+8\frac{6}{8}+9\frac{1}{8}+10\frac{4}{8}}{5}$$

$=\dfrac{42\frac{18}{8}}{5}=\dfrac{42\frac{9}{4}}{5}=\dfrac{\frac{177}{4}}{5}$ Adding, simplifying, and converting to fraction notation

$=\dfrac{177}{4}\div5$ Rewriting using a division symbol

$=\dfrac{177}{4}\cdot\dfrac{1}{5}$ Multiplying by the reciprocal of 5

$=\dfrac{177}{20}=8\dfrac{17}{20}.$ Converting to a mixed numeral

The average length of the logs is $8\frac{17}{20}$ ft.

Do Exercises 13–15. ▶

13. Rachel has triplets. Their birth weights are $3\frac{1}{2}$ lb, $2\frac{3}{4}$ lb, and $3\frac{1}{8}$ lb. What is the average weight of her babies?

14. Find the average of $\dfrac{1}{2},\dfrac{1}{3}$, and $\dfrac{5}{6}$.

15. Find the average of $\dfrac{3}{4}$ and $\dfrac{4}{5}$.

Answers

11. $3\frac{3}{16}$ **12.** 18   **13.** $3\frac{1}{8}$ lb   **14.** $\frac{5}{9}$

15. $\frac{31}{40}$

✓ Reading Check

Match the beginning of each statement with the correct ending from the list at the right so that the rules for order of operations are listed in the correct order.

RC1. Do all _____.

a) multiplications and divisions in order from left to right.

RC2. Evaluate all _____.

b) additions and subtractions in order from left to right.

RC3. Do all _____.

c) calculations within parentheses, brackets, braces, and so forth.

RC4. Do all _____.

d) exponential expressions.

a Simplify.

1. $\dfrac{1}{8} + \dfrac{1}{4} \cdot \dfrac{2}{3}$

2. $\dfrac{2}{5} - \dfrac{4}{5} \div \dfrac{2}{3}$

3. $-\dfrac{1}{6} - 3\left(-\dfrac{5}{9}\right)$

4. $1 + \dfrac{2}{3}\left(-\dfrac{6}{25}\right)$

5. $\dfrac{9}{10} - \left(\dfrac{2}{5} - \dfrac{3}{8}\right)$

6. $-\dfrac{5}{9} + \left(\dfrac{1}{3} - \dfrac{5}{6}\right)$

7. $\dfrac{5}{8} \div \dfrac{1}{4} - \dfrac{2}{3} \cdot \dfrac{4}{5}$

8. $\dfrac{4}{7} \cdot \dfrac{7}{15} + \dfrac{2}{3} \div 8$

9. $\dfrac{7}{8} \div \dfrac{1}{2} \cdot \dfrac{1}{4}$

10. $\dfrac{7}{10} \cdot \dfrac{4}{5} \div \dfrac{2}{3}$

11. $\dfrac{3}{4} - \dfrac{2}{3} \cdot \left(\dfrac{1}{2} + \dfrac{2}{5}\right)$

12. $\dfrac{3}{4} \div \dfrac{1}{2} \cdot \left(\dfrac{8}{9} - \dfrac{2}{3}\right)$

13. $\dfrac{4}{5} \div \left(\dfrac{2}{9} \cdot \dfrac{1}{2}\right) \cdot \left(-\dfrac{5}{6}\right)$

14. $-\dfrac{4}{9} \cdot \left(\dfrac{3}{8} \div \dfrac{1}{2}\right) \div \left(-\dfrac{2}{3}\right)$

15. $\left(\dfrac{2}{3}\right)^2 - \dfrac{1}{3} \cdot 1\dfrac{1}{4}$

16. $-1\dfrac{3}{5} - \dfrac{9}{10} + \left(-\dfrac{1}{2}\right)^2$

17. $-\dfrac{12}{25}\left(\dfrac{3}{4} - \dfrac{1}{2}\right)^2$

18. $\dfrac{2}{3}\left(\dfrac{1}{3} - \dfrac{1}{5}\right)^2$

19. $-\dfrac{3}{4} \div \left(\dfrac{2}{3} - \dfrac{1}{6}\right) + \dfrac{1}{2}$

20. $-6 \div \left(\dfrac{1}{2} - \dfrac{2}{3}\right) + \dfrac{1}{3}$

21. $\left(-\dfrac{3}{2}\right)^2 - 2\left(\dfrac{1}{4} - \dfrac{3}{2}\right)$

22. $\left(-\dfrac{1}{2}\right)^3 - \dfrac{3}{4}\left(\dfrac{1}{3} - \dfrac{1}{6}\right)$

23. $\dfrac{1}{2} - \left(\dfrac{1}{2}\right)^2 + \left(\dfrac{1}{2}\right)^3$

24. $1 + \dfrac{1}{4} + \left(\dfrac{1}{4}\right)^2 - \left(\dfrac{1}{4}\right)^3$

25. $\left(\dfrac{3}{5} - \dfrac{1}{2}\right) \div \left(\dfrac{3}{4} - \dfrac{3}{10}\right)$

26. $\left(\dfrac{2}{3} + \dfrac{3}{4}\right) \div \left(\dfrac{5}{6} - \dfrac{1}{3}\right)$

b   Simplify.

27. $\dfrac{\dfrac{3}{8}}{\dfrac{11}{8}}$

28. $\dfrac{-\dfrac{1}{8}}{\dfrac{3}{4}}$

29. $\dfrac{-4}{\dfrac{6}{7}}$

30. $\dfrac{-\dfrac{3}{8}}{-12}$

31. $\dfrac{\dfrac{1}{40}}{-\dfrac{1}{50}}$

32. $\dfrac{\dfrac{7}{9}}{\dfrac{3}{9}}$

33. $\dfrac{-\dfrac{1}{10}}{-10}$

34. $\dfrac{28}{-\dfrac{7}{4}}$

35. $\dfrac{\dfrac{5}{18}}{-1\dfrac{2}{3}}$

36. $\dfrac{-2\dfrac{1}{5}}{\dfrac{7}{10}}$

37. $\dfrac{\dfrac{x}{28}}{\dfrac{5}{8}}$

38. $\dfrac{\dfrac{x}{15}}{\dfrac{3}{4}}$

39. $\dfrac{\dfrac{n}{14}}{\dfrac{2}{3}}$

40. $\dfrac{-\dfrac{t}{10}}{\dfrac{4}{45}}$

41. $\dfrac{-\dfrac{3}{35}}{-\dfrac{x}{10}}$

42. $\dfrac{\dfrac{7}{40}}{-\dfrac{6}{x}}$

43. $\dfrac{-\dfrac{5}{8}}{\left(\dfrac{3}{2}\right)^2}$

44. $\dfrac{\left(-\dfrac{2}{3}\right)^2}{\left(\dfrac{9}{10}\right)^2}$

45. $\dfrac{\dfrac{1}{6}-\dfrac{5}{9}}{\dfrac{2}{3}}$

46. $\dfrac{\dfrac{7}{12}}{\dfrac{1}{4}-\dfrac{5}{8}}$

47. $\dfrac{\dfrac{1}{4}-\dfrac{3}{8}}{\dfrac{1}{2}-\dfrac{7}{8}}$

48. $\dfrac{\dfrac{1}{2}-\dfrac{3}{5}}{\dfrac{2}{5}-\dfrac{1}{2}}$

49. Find the average of $\dfrac{2}{3}$ and $\dfrac{7}{8}$.

50. Find the average of $\dfrac{1}{4}$ and $\dfrac{1}{5}$.

51. Find the average of $\dfrac{1}{6}, \dfrac{1}{8},$ and $\dfrac{3}{4}$.

52. Find the average of $\dfrac{4}{5}, \dfrac{1}{2},$ and $\dfrac{1}{10}$.

53. Find the average of $3\dfrac{1}{2}$ and $9\dfrac{3}{8}$.

54. Find the average of $10\dfrac{2}{3}$ and $24\dfrac{5}{6}$.

55. *Hiking the Appalachian Trail.*   Ellen camped and hiked for three consecutive days along a section of the Appalachian Trail. The distances she hiked on the three days were $15\frac{5}{32}$ mi, $20\frac{3}{16}$ mi, and $12\frac{7}{8}$ mi. Find the average of these distances.

56. *Vertical Leaps.*   Eight-year-old Zachary registered vertical leaps of $12\frac{3}{4}$ in., $13\frac{3}{4}$ in., $13\frac{1}{2}$ in., and 14 in. Find his average vertical leap.

57. Black Bear Cubs. Black bears typically have two cubs. In January 2007 in northern New Hampshire, a black bear sow gave birth to a litter of 5 cubs. This is so rare that Tom Sears, a wildlife photographer, spent 28 hr per week for six weeks watching for the perfect opportunity to photograph this family of six. At the time of this photo, an observer estimated that the cubs weighed $7\frac{1}{2}$ lb, 8 lb, $9\frac{1}{2}$ lb, $10\frac{5}{8}$ lb, and $11\frac{3}{4}$ lb. What was the average weight of the cubs?

Source: Andrew Timmins, New Hampshire Fish and Game Department, *Northcountry News*, Warren, NH; Tom Sears, photographer

58. Acceleration. The results of a *Road & Track* road acceleration test for five cars are given in the graph below. The test measures the time in seconds required to go from 0 mph to 60 mph. What was the average time?

Acceleration: 0 mph to 60 mph

SOURCE: *Road & Track*, October 2008, pp.156–157

Skill Maintenance

Simplify.

59. $12 + 30 \div 3 - 2$ [1.9c]

60. $5 \cdot 2^2 \div 10$ [1.9c]

61. $10^2 - [3 \cdot 2^4 \div (10 - 2) + 5 \cdot 2]$ [1.9d]

62. $(10 + 3 \cdot 4 \div 6)^2 - 11 \cdot 2^2$ [1.9c]

63. List all the factors of 42. [3.2a]

64. Determine whether 114 is divisible by 7. [3.1a]

65. Classify the given numbers as prime, composite, or neither. [3.2b]

   1, 5, 7, 9, 14, 23, 43

66. Find the prime factorization of 150. [3.2c]

Synthesis

Simplify.

67. $\left(1\frac{1}{2} - 1\frac{1}{3}\right)^2 \cdot 144 - \frac{9}{10} \div 4\frac{1}{5}$

68. ▦ $\left(3\frac{1}{2} - 2\frac{1}{3}\right)^3 - 30 \cdot 2\frac{1}{2} \div (-2)^5$

69. $\dfrac{\frac{2}{3}x - \frac{1}{2}x}{\left(1\frac{1}{4} + \frac{1}{2}\right)^2}$

70. $\dfrac{\frac{x}{24}}{\frac{x}{48}}$

Estimate each of the following as 0, $\frac{1}{2}$, or 1.

71. $\frac{2}{99}$

72. $\frac{19}{20}$

73. $\frac{13}{27}$

74. $\frac{101}{100}$

75. $\frac{215}{429}$

76. $\frac{1}{1000}$

Vocabulary Reinforcement

Complete each statement with the correct term from the given list. Some of the choices may not be used and some may be used more than once.

1. The _____ of two numbers is the smallest number that is a multiple of both numbers. [4.1a]

2. A _____ represents a sum of a whole number and a fraction less than 1. [4.5a]

3. To multiply using mixed numerals, we first convert to _____ notation. [4.7a]

4. A _____ contains a fraction in its numerator and/or denominator. [4.8b]

5. To add fractions, the _____ of the fractions must be the same. [4.2a]

6. The least common denominator of two fractions is the _____ of the denominators of the fractions. [4.2b]

7. When finding the LCM of a set of numbers using prime factorizations, we use each prime number the _____ number of times that it appears in any one factorization. [4.1a]

8. To compare two fractions with a common denominator, we compare their _____. [4.2c]

greatest

least

numerators

denominators

fraction

decimal

mixed numeral

complex fraction

least common multiple

greatest common factor

Concept Reinforcement

Determine whether each statement is true or false.

_____ 1. The mixed numeral $5\frac{2}{3}$ can be represented by the sum $5 \cdot \frac{3}{3} + \frac{2}{3}$. [4.5a]

_____ 2. The least common multiple of two numbers is always larger than or equal to the larger number. [4.1a]

_____ 3. To clear fractions in an equation, multiply both sides by the LCM of all denominators in the equation. [4.4b]

_____ 4. The product of any two mixed numerals is greater than 1. [4.7a]

Study Guide

Objective 4.1a Find the least common multiple, or LCM, of two or more numbers.

Example Find the LCM of 105 and 90.
$$105 = 3 \cdot 5 \cdot 7,$$
$$90 = 2 \cdot 3 \cdot 3 \cdot 5;$$
$$LCM = 2 \cdot 3 \cdot 3 \cdot 5 \cdot 7 = 630$$

Practice Exercise
1. Find the LCM of 52 and 78.

Objective 4.2b Add using fraction notation when denominators are different.

Example Add: $\dfrac{5}{24} + \dfrac{7}{45}$.

$$\dfrac{5}{24} + \dfrac{7}{45} = \dfrac{5}{2 \cdot 2 \cdot 2 \cdot 3} \cdot \dfrac{3 \cdot 5}{3 \cdot 5} + \dfrac{7}{3 \cdot 3 \cdot 5} \cdot \dfrac{2 \cdot 2 \cdot 2}{2 \cdot 2 \cdot 2}$$

$$= \dfrac{75}{360} + \dfrac{56}{360} = \dfrac{131}{360}$$

Practice Exercise

2. Add: $\dfrac{19}{60} + \dfrac{11}{36}$.

Objective 4.2c Use $<$ or $>$ to form a true statement with fraction notation.

Example Use $<$ or $>$ for $\square$ to write a true sentence:

$$\dfrac{5}{12} \,\square\, \dfrac{9}{16}.$$

The LCD is 48. Thus, we have

$$\dfrac{5}{12} \cdot \dfrac{4}{4} \,\square\, \dfrac{9}{16} \cdot \dfrac{3}{3}, \quad \text{or} \quad \dfrac{20}{48} \,\square\, \dfrac{27}{48}.$$

Since $20 < 27$, $\dfrac{20}{48} < \dfrac{27}{48}$ and thus $\dfrac{5}{12} < \dfrac{9}{16}$.

Practice Exercise

3. Use $<$ or $>$ for $\square$ to write a true sentence:

$$\dfrac{3}{13} \,\square\, \dfrac{5}{12}.$$

Objective 4.3a Subtract using fraction notation.

Example Subtract: $\dfrac{7}{12} - \dfrac{11}{60}$.

$$\dfrac{7}{12} - \dfrac{11}{60} = \dfrac{7}{12} \cdot \dfrac{5}{5} - \dfrac{11}{60} = \dfrac{35}{60} - \dfrac{11}{60}$$

$$= \dfrac{35 - 11}{60} = \dfrac{24}{60} = \dfrac{2 \cdot \cancel{12}}{5 \cdot \cancel{12}} = \dfrac{2}{5}$$

Practice Exercise

4. Subtract: $\dfrac{29}{35} - \dfrac{5}{7}$.

Objective 4.4a Solve equations that involve fractions and require use of both the addition principle and the multiplication principle.

Example Solve: $\dfrac{1}{2}x + \dfrac{1}{6} = \dfrac{5}{8}$.

$$\dfrac{1}{2}x + \dfrac{1}{6} = \dfrac{5}{8}$$

$$\dfrac{1}{2}x + \dfrac{1}{6} - \dfrac{1}{6} = \dfrac{5}{8} - \dfrac{1}{6}$$

$$\dfrac{1}{2}x = \dfrac{5}{8} \cdot \dfrac{3}{3} - \dfrac{1}{6} \cdot \dfrac{4}{4}$$

$$\dfrac{1}{2}x = \dfrac{15}{24} - \dfrac{4}{24}$$

$$\dfrac{1}{2}x = \dfrac{11}{24}$$

$$\dfrac{2}{1}\left(\dfrac{1}{2}x\right) = \dfrac{2}{1}\left(\dfrac{11}{24}\right)$$

$$x = \dfrac{\cancel{2} \cdot 11}{1 \cdot \cancel{2} \cdot 12} = \dfrac{11}{12}$$

The solution is $\dfrac{11}{12}$.

Practice Exercise

5. Solve: $\dfrac{2}{9} + \dfrac{2}{3}x = \dfrac{1}{6}$.

Objective 4.5a Convert between mixed numerals and fraction notation.

Example Convert $2\frac{5}{13}$ to fraction notation: $2\frac{5}{13} = \frac{31}{13}$.

Example Convert $\frac{40}{9}$ to a mixed numeral: $\frac{40}{9} = 4\frac{4}{9}$.

Practice Exercises

6. Convert $8\frac{2}{3}$ to fraction notation.

7. Convert $\frac{47}{6}$ to a mixed numeral.

Objective 4.6b Subtract using mixed numerals.

Example Subtract: $3\frac{3}{8} - 1\frac{4}{5}$.

$$3\frac{3}{8} = 3\frac{15}{40} = 2\frac{55}{40}$$
$$-1\frac{4}{5} = -1\frac{32}{40} = -1\frac{32}{40}$$
$$\overline{\phantom{-1\frac{4}{5} = -1\frac{32}{40} =}\; 1\frac{23}{40}}$$

Practice Exercise

8. Subtract: $10\frac{5}{7} - 2\frac{3}{4}$.

Objective 4.7a Multiply using mixed numerals.

Example Multiply: $7\frac{1}{4} \cdot 5\frac{3}{10}$. Write a mixed numeral for the answer.

$$7\frac{1}{4} \cdot 5\frac{3}{10} = \frac{29}{4} \cdot \frac{53}{10}$$
$$= \frac{1537}{40} = 38\frac{17}{40}$$

Practice Exercise

9. Multiply: $4\frac{1}{5} \cdot 3\frac{7}{15}$.

Objective 4.7d Solve applied problems involving multiplication and division with mixed numerals.

Example The population of New York is $3\frac{1}{3}$ times that of Missouri. The population of New York is approximately 19,000,000. What is the population of Missouri?

Translate:

$$\underbrace{\text{Population of New York}}_{19,000,000} \underset{=}{\text{ is }} \underset{3\frac{1}{3}}{3\frac{1}{3}} \underset{\cdot}{\text{ times }} \underbrace{\text{population of Missouri}}_{x.}$$

Solve:

$$19,000,000 = \frac{10}{3} \cdot x$$

$$\frac{19,000,000}{\frac{10}{3}} = \frac{\frac{10}{3} \cdot x}{\frac{10}{3}}$$

$$19,000,000 \cdot \frac{3}{10} = x$$

$$5,700,000 = x.$$

The population of Missouri is about 5,700,000.

Practice Exercise

10. The population of Louisiana is $2\frac{1}{2}$ times the population of West Virginia. The population of West Virginia is approximately 1,800,000. What is the population of Louisiana?

Objective 4.8a Simplify expressions containing fraction notation using the rules for order of operations.

Example Simplify: $\left(\dfrac{4}{5}\right)^2 - \dfrac{1}{5} \cdot 2\dfrac{1}{8}$.

$$\left(\dfrac{4}{5}\right)^2 - \dfrac{1}{5} \cdot 2\dfrac{1}{8} = \dfrac{16}{25} - \dfrac{1}{5} \cdot \dfrac{17}{8}$$

$$= \dfrac{16}{25} - \dfrac{17}{40} = \dfrac{16}{25} \cdot \dfrac{8}{8} - \dfrac{17}{40} \cdot \dfrac{5}{5}$$

$$= \dfrac{128}{200} - \dfrac{85}{200} = \dfrac{43}{200}$$

Practice Exercise

11. Simplify: $\dfrac{3}{2} \cdot 1\dfrac{1}{3} \div \left(\dfrac{2}{3}\right)^2$.

Review Exercises

Find the LCM. [4.1a]

1. 12 and 18

2. 18 and 45

3. 3, 6, and 30

4. 26, 36, and 54

Perform the indicated operation and, if possible, simplify. [4.2a, b], [4.3a]

5. $\dfrac{2}{9} + \dfrac{5}{9}$

6. $\dfrac{7}{x} + \dfrac{2}{x}$

7. $-\dfrac{6}{5} + \dfrac{11}{15}$

8. $\dfrac{5}{16} + \dfrac{3}{24}$

9. $\dfrac{7}{9} - \dfrac{5}{9}$

10. $\dfrac{1}{4} - \dfrac{3}{8}$

11. $\dfrac{10}{27} - \dfrac{2}{9}$

12. $\dfrac{5}{6} - \dfrac{7}{9}$

Use $<$ or $>$ for $\square$ to form a true sentence. [4.2c]

13. $\dfrac{4}{7} \,\square\, \dfrac{5}{9}$

14. $-\dfrac{8}{9} \,\square\, -\dfrac{11}{13}$

Solve. [4.3b], [4.4a]

15. $x + \dfrac{2}{5} = \dfrac{7}{8}$

16. $\dfrac{1}{2}a - 3 = \dfrac{5}{2}$

17. $5 + \dfrac{16}{3}x = \dfrac{5}{9}$

18. $\dfrac{22}{5} = \dfrac{16}{5} + \dfrac{5}{2}x$

Solve by using the multiplication principle to clear fractions. [4.4b]

19. $\dfrac{5}{3}x + \dfrac{5}{6} = \dfrac{3}{2}$

Convert to fraction notation. [4.5a]

20. $7\dfrac{1}{2}$

21. $8\dfrac{3}{8}$

22. $4\dfrac{1}{3}$

23. $-1\dfrac{5}{7}$

Convert to a mixed numeral. [4.5a]

24. $\dfrac{7}{3}$

25. $\dfrac{-27}{4}$

26. $\dfrac{63}{5}$

27. $\dfrac{7}{2}$

28. Divide. Write a mixed numeral for the answer.
$7896 \div (-9)$ [4.5b]

29. Gina's golf scores were 80, 82, and 85. What was her average score? [4.5b]

Perform the indicated operation. Write a mixed numeral for each answer. [4.6a, b, d]

30. $\begin{array}{r} 7\dfrac{3}{5} \\ + 2\dfrac{4}{5} \\ \hline \end{array}$

31. $\begin{array}{r} 6\dfrac{1}{3} \\ + 5\dfrac{2}{5} \\ \hline \end{array}$

32. $-3\dfrac{5}{6} + \left(-5\dfrac{1}{6}\right)$

33. $-2\dfrac{3}{4} + 4\dfrac{1}{2}$

34. $\begin{array}{r} 14 \\ - 6\dfrac{2}{9} \\ \hline \end{array}$

35. $\begin{array}{r} 9\dfrac{3}{5} \\ - 4\dfrac{13}{15} \\ \hline \end{array}$

36. $4\dfrac{5}{8} - 9\dfrac{3}{4}$

37. $-7\dfrac{1}{2} - 6\dfrac{3}{4}$

Combine like terms.   [4.2b], [4.6b]

38. $\dfrac{4}{9}x + \dfrac{1}{3}x$

39. $8\dfrac{3}{10}a - 5\dfrac{1}{8}a$

Perform the indicated operation. Write a mixed numeral or integer for each answer, unless the answer is less than 1.   [4.7a, b]

40. $6 \cdot 2\dfrac{2}{3}$

41. $-5\dfrac{1}{4} \cdot \dfrac{2}{3}$

42. $2\dfrac{1}{5} \cdot 1\dfrac{1}{10}$

43. $2\dfrac{2}{5} \cdot 2\dfrac{1}{2}$

44. $-54 \div 2\dfrac{1}{4}$

45. $2\dfrac{2}{5} \div \left(-1\dfrac{7}{10}\right)$

46. $3\dfrac{1}{4} \div 26$

47. $4\dfrac{1}{5} \div 4\dfrac{2}{3}$

Evaluate.   [4.7c]

48. $5x - y$, for $x = 3\dfrac{1}{5}$ and $y = 2\dfrac{2}{7}$

49. $2a \div b$, for $a = 5\dfrac{2}{11}$ and $b = 3\dfrac{4}{5}$

Solve.   [4.2d], [4.3c], [4.6c], [4.7d]

50. *Sewing.* Kim wants to make slacks and a jacket. She needs $1\dfrac{5}{8}$ yd of 60-in. fabric for the slacks and $2\dfrac{5}{8}$ yd for the jacket. How many yards in all does Kim need to make the outfit?

51. The San Diaz drama club had $\dfrac{3}{8}$ of a vegetarian pizza, $1\dfrac{1}{2}$ cheese pizzas, and $1\dfrac{1}{4}$ pepperoni pizzas remaining after a cast party. How many pizzas remained altogether?

52. *Turkey Servings.* Turkey contains $1\dfrac{1}{3}$ servings per pound. How many pounds are needed for 32 servings?

53. *Humane Society Pie Sale.* Green River's Humane Society recently hosted its annual dessert social. Each of the 83 pies donated was cut into 6 pieces. At the end of the evening, 382 pieces of pie had been sold. How many pies were sold? How many were left over? Express your answers in mixed numerals.

54. *Running.* Janelle has mapped a $1\dfrac{1}{2}$-mi running route in her neighborhood. One Saturday, she ran this route $2\dfrac{1}{2}$ times. How many miles did she run?

55. What is the sum of the areas in the figure below?

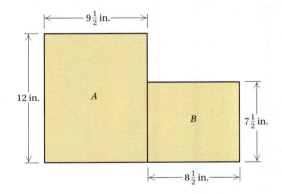

56. In the figure above, how much larger is the area of rectangle *A* than the area of rectangle *B*?

57. *Painting a Border.* Katie hired an artist to paint a decorative border around the top of her son's bedroom. The artist charges $20 per foot. The room measures $11\dfrac{3}{4}$ ft $\times$ $9\dfrac{1}{2}$ ft. What is Katie's cost for the project?

58. *NCAA Football Goalposts.* In college football, the distance between goalposts was reduced from $23\frac{1}{3}$ ft to $18\frac{1}{2}$ ft. By how much was it reduced?

Source: NCAA

$23\frac{1}{3}$ ft

$18\frac{1}{2}$ ft

Simplify each expression using the rules for order of operations. [4.8a]

59. $\dfrac{1}{2} + \dfrac{1}{8} \div \dfrac{1}{4}$

60. $\dfrac{4}{5} - \dfrac{1}{2} \cdot \left(\dfrac{1}{4} - \dfrac{3}{5} \right)$

61. $20\dfrac{3}{4} - 1\dfrac{1}{2} \times 12 + \left(\dfrac{1}{2} \right)^2$

62. Find the average of $\dfrac{1}{2}, \dfrac{1}{4}, \dfrac{1}{3},$ and $\dfrac{1}{5}.$ [4.4a]

Simplify. [4.8b]

63. $\dfrac{\dfrac{1}{3}}{\dfrac{5}{8} - 1}$

64. $\dfrac{-\dfrac{x}{7}}{\dfrac{3}{14}}$

65. Simplify: $\dfrac{1}{4} + \dfrac{2}{5} \div 5^2.$ [4.8a]

    A. $\dfrac{133}{500}$ **B.** $\dfrac{3}{500}$

    C. $\dfrac{117}{500}$ **D.** $\dfrac{5}{2}$

66. Solve: $x + \dfrac{2}{3} = 5.$ [4.3b]

    A. $\dfrac{15}{2}$ **B.** $5\dfrac{2}{3}$

    C. $\dfrac{10}{3}$ **D.** $4\dfrac{1}{3}$

Synthesis

67. Find r if

$$\dfrac{1}{r} = \dfrac{1}{100} + \dfrac{1}{150} + \dfrac{1}{200}.$$ [4.2b], [4.4b]

68. Place the numbers 3, 4, 5, and 6 in the boxes in order to make a true equation: [4.5a]

$$\dfrac{\square}{\square} + \dfrac{\square}{\square} = 3\dfrac{1}{4}.$$

69. Find the largest integer for which each fraction is greater than 1. [4.2c]

    a) $\dfrac{7}{\square}$ **b)** $\dfrac{11}{\square}$

    c) $\dfrac{\square}{-27}$ **d)** $\dfrac{\square}{-\frac{1}{2}}$

Understanding Through Discussion and Writing

1. Is the sum of two mixed numerals always a mixed numeral? Why or why not? [4.6a]

2. Write a problem for a classmate to solve. Design the problem so that its solution is found by performing the multiplication $4\frac{1}{2} \cdot 33\frac{1}{3}.$ [4.7d]

3. A student insists that $3\frac{2}{5} \cdot 1\frac{3}{7} = 3\frac{6}{35}.$ What mistake is he making and how should he have proceeded? [4.7a]

4. Discuss the role of least common multiples in adding and subtracting with fraction notation. [4.2b], [4.3a]

5. Find a real-world situation that fits this equation:

$$2 \cdot 15\dfrac{3}{4} + 2 \cdot 28\dfrac{5}{8} = 88\dfrac{3}{4}.$$ [4.6c], [4.7d]

6. A student insists that $5 \cdot 3\frac{2}{7} = (5 \cdot 3) \cdot (5 \cdot \frac{2}{7}).$ What mistake is she making and how should she have proceeded? [4.7a]

CHAPTER

4 **Test**

For Extra Help For step-by-step test solutions, access the Chapter Test Prep Videos in MyMathLab® or on YouTube (search "BittingerPrealgebra" and click on Channels).

1. Find the LCM of 12 and 16.

Perform the indicated operation and, if possible, simplify.

2. $\dfrac{1}{2} + \dfrac{5}{2}$

3. $-\dfrac{7}{8} + \dfrac{2}{3}$

4. $\dfrac{5}{t} - \dfrac{3}{t}$

5. $\dfrac{5}{6} - \dfrac{3}{4}$

6. $\dfrac{5}{8} - \dfrac{17}{24}$

Solve.

7. $x + \dfrac{2}{3} = \dfrac{11}{12}$

8. $-5x - 3 = 9$

9. $\dfrac{3}{4} = \dfrac{1}{2} + \dfrac{5}{3}x$

10. Use $<$ or $>$ for $\square$ to form a true sentence.

$\dfrac{6}{7} \square \dfrac{21}{25}$

Convert to fraction notation.

11. $3\dfrac{1}{2}$

12. $-9\dfrac{3}{8}$

13. Convert to a mixed numeral:

$-\dfrac{74}{9}.$

14. Divide. Write a mixed numeral for the answer.

$1\,1\,)\overline{1\,7\,8\,9}$

Perform the indicated operation. Write a mixed numeral for each answer.

15. $\quad 6\dfrac{2}{5}$
$+\ 7\dfrac{4}{5}$
___

16. $\quad 3\dfrac{1}{4}$
$+\ 9\dfrac{1}{6}$
___

17. $\quad 10\dfrac{1}{6}$
$-\ 5\dfrac{7}{8}$
___

18. $14 + \left(-5\dfrac{3}{7}\right)$

19. $3\dfrac{4}{5} - 9\dfrac{1}{2}$

Combine like terms.

20. $\dfrac{3}{8}x - \dfrac{1}{2}x$

21. $5\dfrac{2}{11}a - 3\dfrac{1}{5}a$

Perform the indicated operation.

22. $9 \cdot 4\frac{1}{3}$

23. $6\frac{3}{4} \cdot \left(-2\frac{2}{3}\right)$

24. $33 \div 5\frac{1}{2}$

25. $2\frac{1}{3} \div 1\frac{1}{6}$

Evaluate.

26. $\frac{2}{3}ab$, for $a = 7$ and $b = 4\frac{1}{5}$

27. $4 + mn$, for $m = 7\frac{2}{5}$ and $n = 3\frac{1}{4}$

Solve.

28. *Weightlifting.* In 2002, Hossein Rezazadeh of Iran completed a clean and jerk of 263 kg. This amount was about $2\frac{1}{2}$ times his body weight. How much did Rezazadeh weigh?

Source: *The Guinness Book of Records,* 2005

29. *Book Order.* An order of books for a math course weighs 220 lb. Each book weighs $2\frac{3}{4}$ lb. How many books are in the order?

30. *Carpentry.* The following diagram shows a middle drawer support guide for a cabinet drawer. Find each of the following.

   a) The short length a across the top
   b) The length b across the bottom

31. *Carpentry.* In carpentry, some pieces of plywood that are called "$\frac{3}{4}$-inch" plywood are actually $\frac{11}{16}$ in. thick. How much thinner is such a piece than its name indicates?

32. *Women's Dunks.* The first three women in the history of college basketball able to dunk a basketball are listed below. Their names, heights, and universities are

   Michelle Snow, $6\frac{5}{12}$ ft, Tennessee;
   Charlotte Smith, $5\frac{11}{12}$ ft, North Carolina;
   Georgeann Wells, $6\frac{7}{12}$ ft, West Virginia.

Find the average height of these women.

Source: *USA Today,* 11/30/00, p. 3C

Simplify.

33. $\frac{2}{3} + 1\frac{1}{3} \cdot 2\frac{1}{8}$

34. $-1\frac{1}{2} - \frac{1}{2}\left(\frac{1}{2} \div \frac{1}{4}\right) + \left(\frac{1}{2}\right)^2$

35. $\dfrac{\frac{1}{3} - \frac{7}{9}}{\frac{1}{2} + \frac{1}{6}}$

36. Find the LCM of 12, 36, and 60.
   A. 6   **B.** 12
   C. 60   **D.** 180

Synthesis

37. The students in a math class can be organized into study groups of 8 each so that no students are left out. The same class of students can also be organized into groups of 6 so that no students are left out.

   a) Find some class sizes for which this will work.
   b) Find the smallest such class size.

38. Rebecca walks 17 laps at her health club. Trent walks 17 laps at his health club. If the track at Rebecca's health club is $\frac{1}{7}$ mi long and the track at Trent's is $\frac{1}{8}$ mi long, who walks farther? How much farther?

Solve.

1. There are $20\frac{1}{3}$ gal of water in a rainbarrel; $5\frac{3}{4}$ gal are poured out and $8\frac{2}{3}$ gal are returned after a heavy rainfall. How many gallons of water are then in the barrel?

2. How many people can receive equal $16 shares from a total of $496?

3. A recipe calls for $\frac{4}{5}$ tsp of salt. How much salt should be used for $\frac{1}{2}$ recipe? for 5 recipes?

4. How many pieces, each $2\frac{3}{8}$ ft long, can be cut from a piece of wire 38 ft long?

5. An emergency food pantry fund contains $423. From this fund, $148 and $167 are withdrawn for expenses. How much is left in the fund?

6. In a walkathon, Jermaine walked $\frac{9}{10}$ mi and Oleta walked $\frac{3}{4}$ mi. What was the total distance they walked?

What part is shaded?

7.

8.

Calculate and simplify.

9. $\begin{array}{r} 3704 \\ +5278 \end{array}$

10. $\begin{array}{r} 7605 \\ -3087 \end{array}$

11. $\begin{array}{r} 278 \\ \times\ 18 \end{array}$

12. $29(-5)$

13. $\frac{3}{8}+\frac{1}{24}$

14. $\begin{array}{r} 2\frac{3}{4} \\ +\ 5\frac{1}{2} \end{array}$

15. $\frac{4}{t}-\frac{9}{t}$

16. $\begin{array}{r} 2\frac{1}{3} \\ -\ 1\frac{1}{6} \end{array}$

17. $\frac{9}{10}\cdot\frac{5}{3}$

18. $18\left(-\frac{5}{6}\right)$

19. $-9-(-25)$

20. $2\frac{1}{5}\div\frac{3}{10}$

Divide. If there is a remainder, write the answer in the form 34 R 7.

21. $6 \overline{)4290}$ **22.** $45 \overline{)2531}$

23. Write a mixed numeral for the answer in Exercise 22.

24. In the number 2753, what digit names tens?

25. *Room Carpeting.* The Chandlers are carpeting an L-shaped family room consisting of one rectangle that is $8\frac{1}{2}$ ft by 11 ft and another rectangle that is $6\frac{1}{2}$ ft by $7\frac{1}{2}$ ft.
   a) Find the area of the carpet.
   b) Find the perimeter of the carpet.

26. Round 38,478 to the nearest hundred.

27. Find the LCM of 18 and 24.

28. Simplify:
$$\left(\frac{1}{2} + \frac{2}{5}\right)^2 \div 3 + 6 \times \left(2 + \frac{1}{4}\right).$$

Use $<$, $>$, or $=$ for ☐ to write a true sentence.

29. $\frac{4}{5} \,\square\, \frac{4}{6}$ **30.** $\frac{3}{13} \,\square\, \frac{9}{39}$ **31.** $-\frac{5}{12} \,\square\, -\frac{3}{7}$

32. Evaluate $\dfrac{t + p}{3}$ for $t = -4$ and $p = 16$.

Simplify.

33. $\dfrac{36}{45}$ **34.** $\dfrac{0}{27}$ **35.** $\dfrac{-320}{10}$

36. Convert to fraction notation: $4\frac{5}{8}$.

37. Convert to a mixed numeral: $-\dfrac{17}{3}$.

Solve.

38. $x + 24 = 117$ **39.** $x + \dfrac{7}{9} = \dfrac{4}{3}$

40. $\dfrac{7}{9} \cdot t = -\dfrac{4}{3}$ **41.** $\dfrac{5}{7} = \dfrac{1}{3} + 4a$

42. *Matching.* Match each item in the first column with the appropriate item in the second column by drawing connecting lines. There can be more than one correct correspondence for an item.

Factors of 68	12, 54, 72, 300
Factorization of 68	2, 3, 17, 19, 23, 31, 47, 101
Prime factorization of 68	2 · 2 · 17
Numbers divisible by 6	2 · 34
Numbers divisible by 8	8, 16, 24, 32, 40, 48, 64, 864
Numbers divisible by 5	1, 2, 4, 17, 34, 68
Prime numbers	70, 95, 215

Synthesis

43. Find the smallest prime number that is larger than 2000.

44. Solve: $7x - \dfrac{2}{3}(x - 6) = 6\dfrac{5}{7}$.

CHAPTER
5

Decimal Notation

5.1 Decimal Notation, Order, and Rounding

OBJECTIVES

a Given decimal notation, write a word name.

b Convert between decimal notation and fraction notation.

c Given a pair of numbers in decimal notation, tell which is larger.

d Round decimal notation to the nearest thousandth, hundredth, tenth, one, ten, hundred, or thousand.

SKILL TO REVIEW

Objective 1.6a: Round to the nearest ten, hundred, or thousand.

Round 4735 to the nearest

1. Ten.

2. Hundred.

The set of **rational numbers** consists of the **integers** ..., -3, -2, -1, 0, 1, 2, 3, ..., and fractions like $\frac{1}{2}$, $\frac{2}{3}$, $\frac{-7}{8}$, $\frac{17}{-10}$, and so on. In this chapter, we will use *decimal notation* to represent rational numbers. Using decimal notation, we can write 0.875 for $\frac{7}{8}$, for example, or 48.97 for $48\frac{97}{100}$. A number written in decimal notation is often simply referred to as a *decimal*.

The word *decimal* comes from the Latin word *decima*, meaning a *tenth part*. Since our usual counting system is based on tens, decimal notation is a natural extension of an already familiar system.

a DECIMAL NOTATION AND WORD NAMES

One model of the Magellan GPS navigation system sells for $249.98. The dot in $249.98 is called a **decimal point**. Since $0.98, or 98¢, is $\frac{98}{100}$ of a dollar, it follows that

$$\$249.98 = \$249 + \$0.98.$$

Also, since $0.98, or 98¢, has the same value as 9 dimes + 8 cents and 1 dime is $\frac{1}{10}$ of a dollar and 1 cent is $\frac{1}{100}$ of a dollar, we can write

$$249.98 = 2 \cdot 100 + 4 \cdot 10 + 9 \cdot 1 + 9 \cdot \frac{1}{10} + 8 \cdot \frac{1}{100}.$$

This is an extension of expanded notation for whole numbers. The place values are 100, 10, 1, $\frac{1}{10}$, $\frac{1}{100}$, and so on. We can see this on a **place-value chart**. The value of each place is $\frac{1}{10}$ as large as that of the one to its left.

Let's consider decimal notation using a place-value chart to represent 2.0677 min, the women's 200-meter backstroke world record held by Missy Franklin from the United States.

PLACE-VALUE CHART							
Hundreds	Tens	Ones	Tenth*s*	Hundredth*s*	Thousandth*s*	Ten-Thousandth*s*	Hundred-Thousandth*s*
100	10	1	$\frac{1}{10}$	$\frac{1}{100}$	$\frac{1}{1000}$	$\frac{1}{10,000}$	$\frac{1}{100,000}$
2 .	0	6	7	7			

Answers

Skill to Review:
1. 4740   2. 4700

The decimal notation 2.0677 means

2 ones + 0 tenths + 6 hundredths + 7 thousandths + 7 ten-thousandths,

or

$$2 + \frac{0}{10} + \frac{6}{100} + \frac{7}{1000} + \frac{7}{10,000}, \quad \text{or} \quad 2\frac{677}{10,000}.$$

We read both 2.0677 and $2\frac{677}{10,000}$ as

"Two *and* six hundred seventy-seven ten-thousandths."

We read the decimal point as "and." Note that the place values to the right of the decimal point always end in *th*. We can also read 2.0677 as "Two *point* zero six seven seven."

To write a word name from decimal notation,

a) write a word name for the whole number (the number named to the left of the decimal point),

397.685

       → Three hundred ninety-seven

b) write the word "and" for the decimal point, and

397.685

       Three hundred ninety-seven and

c) write a word name for the number named to the right of the decimal point, followed by the place value of the last digit.

397.685

       Three hundred ninety-seven and six hundred eighty-five *thousandths*

EXAMPLE 1 *Median Age.* The median age of residents in Maine is 42.7 years. The median age of residents in Utah is 29.2 years. Write word names for 42.7 and 29.2.

Source: U.S. Census Bureau

Forty-two and seven tenths

Twenty-nine and two tenths

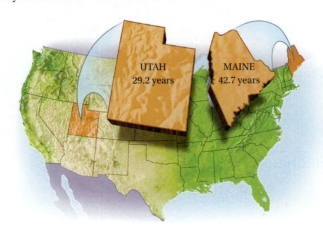

UTAH
29.2 years

MAINE
42.7 years

1. Life Expectancy. The life expectancy at birth in Kenya in 2011 was 61.3 years for males and 64.2 years for females. Write word names for 61.3 and 64.2.

Source: World Health Organization

2. 10,000-Meter Record. Wang Junxia of China holds the women's world record for the 10,000-meter run: 29.5297 min. Write a word name for 29.5297.

Write a word name for each number.

3. 245.89

4. 34.0064

5. 31,079.756

EXAMPLE 2 Write a word name for 410.87.

Four hundred ten and eighty-seven hundredths

EXAMPLE 3 *Record Price.* Edvard Munch's painting *The Scream* sold for $119.9225 million at Sotheby's in New York City on May 2, 2012. This price set a world record as the highest paid for a work of art sold at auction. Write a word name for 119.9225.

Source: Sotheby's

One hundred nineteen and nine thousand two hundred twenty-five ten-thousandths

EXAMPLE 4 Write a word name for 1788.045.

One thousand, seven hundred eighty-eight and forty-five thousandths

◀ Do Exercises 1–5.

b CONVERTING BETWEEN DECIMAL NOTATION AND FRACTION NOTATION

Given decimal notation, we can convert to fraction notation as follows:

$$9.875 = 9 + \frac{8}{10} + \frac{7}{100} + \frac{5}{1000}$$

$$= 9 \cdot \frac{1000}{1000} + \frac{8}{10} \cdot \frac{100}{100} + \frac{7}{100} \cdot \frac{10}{10} + \frac{5}{1000}$$

$$= \frac{9000}{1000} + \frac{800}{1000} + \frac{70}{1000} + \frac{5}{1000} = \frac{9875}{1000}.$$

Decimal notation   Fraction notation

9.875 $\frac{9875}{1000}$

3 decimal places   3 zeros

To convert from decimal notation to fraction notation,

a) count the number of decimal places,   4.98

                                          2 places

b) move the decimal point that many   4.98.   Move
places to the right, and   2 places.

c) write the answer over a denominator $\dfrac{498}{100}$ 2 zeros
of 1 followed by that number of zeros.

For a number like 0.876, we write a 0 to call attention to the presence of the decimal point.

EXAMPLE 5 Write fraction notation for 0.876. Do not simplify.

     0.876   0.876. $0.876 = \dfrac{876}{1000}$

                    3 places   3 zeros

Decimals greater than 1 or less than −1 can be written either as fractions or as mixed numerals.

EXAMPLE 6 Write 56.23 as a fraction and as a mixed numeral.

     56.23   56.23. $56.23 = \dfrac{5623}{100}$,   and $56.23 = 56\dfrac{23}{100}$

                  2 places   2 zeros

As a check, note that both 56.23 and $56\frac{23}{100}$ are read as "fifty-six and twenty-three hundredths."

EXAMPLE 7 Write −2.6073 as a fraction and as a mixed numeral.

    $-2.6073. = -\dfrac{26{,}073}{10{,}000}$ and $-2.6073 = -2\dfrac{6073}{10{,}000}$

     4 places   4 zeros

Do Exercises 6–9. ▶

To write $\frac{5328}{10}$ as a decimal, we can first divide to find an equivalent mixed numeral.

$$\frac{5328}{10} = 532\frac{8}{10}$$

Next note that

$$532\frac{8}{10} = 532 + \frac{8}{10}$$
$$= 532.8.$$

```
      5 3 2
10 ) 5 3 2 8
      5 0
      3 2
      3 0
        2 8
        2 0
          8
```

Write fraction notation. Do not simplify.

6. 0.5491

GS **7.** 0.896

Write as a fraction and as a mixed numeral.

8. 75.069   **9.** −312.9

Answers

6. $\dfrac{5491}{10{,}000}$ **7.** $\dfrac{896}{1000}$ **8.** $\dfrac{75{,}069}{1000}; 75\dfrac{69}{1000}$

9. $-\dfrac{3129}{10}; -312\dfrac{9}{10}$

Guided Solution:

7. 3; 000

Thus, if fraction notation has a denominator that is a power of ten, such as 10, 100, 1000, and so on, we convert to decimal notation by reversing the procedure that we just used in Examples 5–7.

To convert from fraction notation to decimal notation when the denominator is 10, 100, 1000, and so on,

a) count the number of zeros and $\dfrac{8679}{1000}$

                                   —— 3 zeros

b) move the decimal point that number of places to the left. Leave off the denominator.   8.679.   Move 3 places.

                                 $\dfrac{8679}{1000} = 8.679$

EXAMPLE 8   Write decimal notation for $\dfrac{47}{10}$.

$\dfrac{47}{10}$ 4.7. $\dfrac{47}{10} = 4.7$ The decimal point is moved to the left.

     1 zero   1 place

EXAMPLE 9   Write decimal notation for $\dfrac{123{,}067}{10{,}000}$.

$\dfrac{123{,}067}{10{,}000}$ 12.3067. $\dfrac{123{,}067}{10{,}000} = 12.3067$

     4 zeros   4 places

To move the decimal point to the left, we may need to write extra 0's.

EXAMPLE 10   Write decimal notation for $-\dfrac{9}{100}$.

$-\dfrac{9}{100}$ −0.09. $-\dfrac{9}{100} = -0.09$

     2 zeros   2 places

◀ **Do Exercises 10–13.**

To convert fractions with denominators other than 10, 100, and so on to decimal notation, we will usually perform long division.

If a mixed numeral has a fraction part with a denominator that is a power of ten, such as 10, 100, or 1000, and so on, we first write the mixed numeral as a sum of a whole number and a fraction. Then we convert to decimal notation.

EXAMPLE 11   Write decimal notation for $23\dfrac{59}{100}$.

$$23\frac{59}{100} = 23 + \frac{59}{100} = 23 \text{ and } \frac{59}{100} = 23.59$$

◀ **Do Exercises 14 and 15.**

Write decimal notation for each number.

10. $\dfrac{743}{100}$

     $\dfrac{743}{100}$ 7.43.

     [ ] zeros   [ ] places

     $\dfrac{743}{100} = 7.$[ ]

11. $-\dfrac{73}{1000}$ **12.** $\dfrac{67{,}089}{10{,}000}$

13. $-\dfrac{9}{10}$

Write decimal notation for each number.

14. $-7\dfrac{3}{100}$ **15.** $23\dfrac{47}{1000}$

Answers

10. 7.43   11. −0.073   12. 6.7089
13. −0.9   14. −7.03   15. 23.047

Guided Solution:
10. 2, 2; 43

c ORDER

To understand how to compare numbers in decimal notation, consider 0.85 and 0.9. First note that $0.9 = 0.90$ because $\frac{9}{10} = \frac{90}{100}$. Since $0.85 = \frac{85}{100}$, it follows that $\frac{85}{100} < \frac{90}{100}$ and $0.85 < 0.9$. This leads us to a quick way to compare two numbers in decimal notation.

> To compare two positive numbers in decimal notation, start at the left and compare corresponding digits, moving from left to right. When two digits differ, the number with the larger digit is the larger of the two numbers. Extra zeros can be written to the right of the last decimal place.

EXAMPLE 12 Which is larger: 2.109 or 2.1?

2.109	2.109	2.109	2.109
↑ The same	↑ The same	↑ The same	↑ Different;
↓	↓	↓	↓ 9 is larger
2.1	2.1	2.10	2.100 than 0.

Thus, 2.109 is larger than 2.1. In symbols, $2.109 > 2.1$.

EXAMPLE 13 Which is larger: 0.09 or 0.108?

0.09	0.09
↑ The same	↑ Different; 1 is larger than 0.
↓	↓
0.108	0.108

Thus, 0.108 is larger than 0.09. In symbols, $0.108 > 0.09$.

Do Exercises 16–21. ▶

We can use the number line to visualize order. We illustrate Examples 12 and 13 below. Larger numbers are always to the right.

Note from the number line that $-2 < -1$. Similarly, $-1.57 < -1.52$.

> To compare two negative numbers in decimal notation, start at the left and compare corresponding digits, moving from left to right. When two digits differ, the number with the smaller digit is the larger of the two numbers.

EXAMPLE 14 Which is larger: -3.8 or -3.82?

Thus, -3.8 is larger than -3.82. In symbols, $-3.8 > -3.82$. (See the number line above.)

Do Exercises 22 and 23. ▶

Which number is larger?

16. 2.04, 2.039

17. 0.06, 0.008

18. 0.5, 0.58

19. 1, 0.9999

20. 0.8989, 0.09898

21. 21.006, 21.05

Which number is larger?

22. -34.01, -34.008

23. -9.12, -8.98

Answers

16. 2.04 **17.** 0.06 **18.** 0.58 **19.** 1
20. 0.8989 **21.** 21.05 **22.** -34.008
23. -8.98

d ROUNDING

We round decimals in much the same way that we round whole numbers. To see how, we use the number line.

EXAMPLE 15 Round 0.37 to the nearest tenth.

Here is part of the number line, magnified.

We see that 0.37 is closer to 0.40 than to 0.30. Thus, 0.37 rounded to the nearest tenth is 0.4.

To round to a certain place:

a) Locate the digit in that place.

b) Consider the next digit to the right.

c) If the digit to the right is 5 or greater, add 1 to the original digit. If the digit to the right is 4 or less, the original digit does not change. In either case, drop all numbers to the right of the original digit.

EXAMPLE 16 Round 72.3846 to the nearest hundredth.

a) Locate the digit in the hundredths place, 8.

$$7\,2\,.\,3\,8\,4\,6$$

b) Consider the next digit to the right, 4.

c) Since that digit, 4, is less than 5, the original digit does not change. The rounded number is 72.38.

.... **Caution!** ....

72.39 is not a correct answer to Example 16. It is *incorrect* to round sequentially from right to left as follows: 72.3846, 72.385, 72.39.

72.3846 is closer to 72.38 than to 72.39.

....

EXAMPLE 17 Round −0.064 to the nearest tenth.

a) Locate the digit in the tenths place, 0.

$$-0\,.\,0\,6\,4$$

b) Consider the next digit to the right, 6.

c) Since that digit, 6, is greater than 5, round from −0.064 to −0.1.

The answer is −0.1. Since −0.1 < −0.064, we actually rounded *down*.

◀ **Do Exercises 24–42.**

Round to the nearest tenth.

24. 2.76   **25.** 13.85

26. −234.448   **27.** 7.009

Round to the nearest hundredth.

28. 0.6362   **29.** −7.8348

30. 34.67514   **31.** −0.02521

Round to the nearest thousandth.

32. 0.94347   **33.** −8.00382

34. −43.111943   **35.** 37.400526

Round 7459.3548 to the nearest

36. Thousandth.

37. Hundredth.

38. Tenth.   **39.** One.

40. Ten. (*Caution:* "Tens" are not "tenths.")

41. Hundred.   **42.** Thousand.

Answers

24. 2.8   **25.** 13.9   **26.** −234.4   **27.** 7.0
28. 0.64   **29.** −7.83   **30.** 34.68
31. −0.03   **32.** 0.943   **33.** −8.004
34. −43.112   **35.** 37.401   **36.** 7459.355
37. 7459.35   **38.** 7459.4   **39.** 7459
40. 7460   **41.** 7500   **42.** 7000

✓ **Reading Check**

Name the digit that represents each place value in the number 436.81205.

RC1. Hundred-thousandths _____

RC2. Thousandths _____

RC3. Tens _____

RC4. Ten-thousandths _____

RC5. Tenths _____

RC6. Hundreds _____

RC7. Hundredths _____

RC8. Ones _____

a

1. *Currency Conversion.* One Japanese yen was worth about $0.0119 in U.S. currency recently. Write a word name for 0.0119.

   Source: finance.yahoo.com

2. *Currency Conversion.* One U.S. dollar was worth 0.9949 Canadian dollars recently. Write a word name for 0.9949.

   Source: finance.yahoo.com

3. *Birth Rate.* There were 137.6 triplet births per 100,000 live births in a recent year in the United States. Write a word name for 137.6.

   Source: U.S. Centers for Disease Control and Prevention

4. *Coffee Consumption.* Annual per capita coffee consumption in Finland is 26.7 lb. In Spain, annual per capita coffee consumption is only 9.4 lb. Write word names for 26.7 and 9.4.

   Source: International Coffee Organization

5. *Stock Price.* Apple, Inc.'s stock price was recently $519.22 per share. Write a word name for 519.22.

   Source: NASDAQ

6. *Indianapolis 500.* Ryan Hunter-Reay won the 2014 Indianapolis 500 race with a time of 2 hr 40 min 48.2305 sec. Write a word name for 48.2305.

   Source: Indianapolis Motor Speedway

Write a word name for the number in each sentence.

7. One gallon of paint is equal to 3.785 L of paint.

8. *Water Weight.* One gallon of water weighs 8.35 lb.

b Write each number as a fraction and, if possible, as a mixed numeral. Do not simplify.

9. 7.3

10. 4.9

11. 21.67

12. −57.32

13. −2.703

14. 0.079

15. 0.0109

16. 1.0008

17. −4.0003

18. −9.012

19. −0.0207

20. −0.00104

21. 70.00105

22. 60.0403

Write decimal notation for each number.

23. $\dfrac{3}{10}$

24. $\dfrac{73}{10}$

25. $-\dfrac{59}{100}$

26. $-\dfrac{67}{100}$

27. $\dfrac{3798}{1000}$

28. $\dfrac{780}{1000}$

29. $\dfrac{78}{10,000}$

30. $\dfrac{56,788}{100,000}$

31. $\dfrac{-18}{100,000}$

32. $\dfrac{-2347}{100}$

33. $\dfrac{486,197}{1,000,000}$

34. $\dfrac{8,953,074}{1,000,000}$

35. $7\dfrac{13}{1000}$

36. $4\dfrac{909}{1000}$

37. $-8\dfrac{431}{1000}$

38. $-49\dfrac{32}{1000}$

39. $2\dfrac{1739}{10,000}$

40. $9243\dfrac{1}{10}$

41. $8\dfrac{953,073}{1,000,000}$

42. $2256\dfrac{3059}{10,000}$

c Which number is larger?

43. 0.06, 0.58

44. 0.008, 0.8

45. 0.403, 0.410

46. 42.06, 42.1

47. −5.046, −5.043

48. −324.19, −325.19

49. 234.07, 235.07

50. 0.99999, 1

51. 0.007, $\dfrac{7}{100}$

52. $\dfrac{73}{10}$, 0.73

53. −0.872, −0.873

54. −0.8437, −0.84384

d Round to the nearest tenth.

55. 0.23

56. 0.85

57. −0.372

58. −0.261

59. 2.951

60. 7.532

61. −327.2347

62. −8.7493

Round to the nearest hundredth.

63. 0.893

64. 0.675

65. −0.6666

66. −7.5252

67. 0.9952

68. 207.9976

69. −0.03488

70. −9.27481

Round to the nearest thousandth.

71. 0.5724

72. 0.6666

73. 17.0015

74. 123.4562

75. −20.20202

76. −0.10346

77. 9.98487

78. 67.100602

Round 809.47321 to the nearest

79. Tenth.

80. Thousandth.

81. Hundredth.

82. One.

Skill Maintenance

Add or subtract, as indicated.

83.
$$\begin{array}{r} 6\ 8\ 1 \\ +\ 1\ 4\ 9 \\ \hline \end{array}$$ [1.2a]

84. $\dfrac{681}{1000} + \dfrac{149}{1000}$ [4.2a]

85.
$$\begin{array}{r} 2\ 6\ 7 \\ -\ \ \ 8\ 5 \\ \hline \end{array}$$ [1.3a]

86. $\dfrac{267}{100} - \dfrac{85}{100}$ [4.3a]

87. $\dfrac{37}{55} - \dfrac{49}{55}$ [4.3a]

88. $-\dfrac{29}{34} + \dfrac{14}{34}$ [4.2a]

89.
$$\begin{array}{r} 3\ 4,9\ 0\ 3 \\ -\ \ \ 1,9\ 4\ 5 \\ \hline \end{array}$$ [1.3a]

90.
$$\begin{array}{r} 4\ 9\ 3\ 7 \\ +\ 5\ 7\ 8\ 9 \\ \hline \end{array}$$ [1.2a]

Synthesis

91. Arrange the following numbers in order from smallest to largest.

−0.989, −0.898, −1.009, −1.09, −0.098

92. Arrange the following numbers in order from smallest to largest.

−2.018, −2.1, −2.109, −2.0119, −2.108

Truncating. There are other methods of rounding decimal notation. A computer often uses a method called **truncating**. To truncate we drop off decimal places right of the rounding place, which is the same as changing all digits to the right of the rounding place to zeros. For example, rounding 6.78093456285102 to the ninth decimal place, using truncating, gives us 6.780934562. Use truncating to round each of the following to the fifth decimal place, that is, the hundred-thousandth place.

93. 6.78346123

94. 6.783461902

95. 99.999999999

96. 0.030303030303

Temperature Change. The graph below is based on the average global temperature from 1951 through 1980. Each bar indicates, in Celsius degrees, how much above or below average the temperature was for the year. Use the graph for Exercises 97–100.

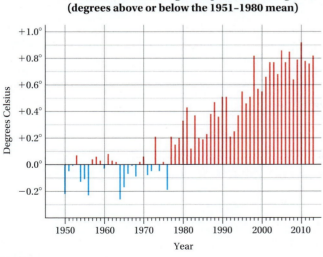

Annual Global Average Temperature Change
(degrees above or below the 1951–1980 mean)

SOURCE: Goddard Institute for Space Studies

97. For what year(s) was the yearly temperature more than 0.8 degree above average?

98. What was the last year in which the yearly temperature was more than 0.2 degree below average?

99. What was the last year in which the yearly temperature was below average?

100. For what year(s) was the yearly temperature more than 0.9 degree above average?

5.2 Addition and Subtraction of Decimals

Add.

1.
```
     0.8 4 7
 + 1 0.0 7
```

2.
```
     2.1
     0.7 3
 + 3 1.3 6 8
```

Add.

3. 0.02 + 4.3 + 0.649

4. 0.12 + 3.006 + 0.4357

5. 0.4591 + 0.2374 + 8.70894

Answers

Skills to Review:
1. 641 **2.** 2356

Margin Exercises:
1. 10.917 **2.** 34.198 **3.** 4.969
4. 3.5617 **5.** 9.40544

a ADDITION

Adding with decimal notation is similar to adding whole numbers. First, we line up the decimal points so that we can add corresponding place-value digits. Then we add digits from the right. For example, we add the thousandths, then the hundredths, and so on, carrying if necessary. If desired, we can write extra zeros to the right of the decimal point so that the number of places is the same in all of the addends.

EXAMPLE 1 Add: 56.314 + 17.78.

```
   5 6 . 3 1 4      Lining up the decimal points in order to add
 + 1 7 . 7 8 0      Writing an extra zero to the right of the
                    decimal point
```

```
   5 6 . 3 1 4
 + 1 7 . 7 8 0
               4    Adding thousandths
```

```
   5 6 . 3 1 4
 + 1 7 . 7 8 0
             9 4    Adding hundredths
```

```
       1
   5 6 . 3 1 4      Adding tenths
 + 1 7 . 7 8 0      We get 10 tenths = 1 one + 0 tenths,
       . 0 9 4      so we carry the 1 to the ones column.
                    Writing a decimal point in the answer
```

```
   1   1
   5 6 . 3 1 4      Adding ones
 + 1 7 . 7 8 0      We get 14 ones = 1 ten + 4 ones, so we
     4 . 0 9 4      carry the 1 to the tens column.
```

```
 1 1   1
   5 6 . 3 1 4
 + 1 7 . 7 8 0
 7 4 . 0 9 4       Adding tens
```

◀ **Do Margin Exercises 1–2.**

EXAMPLE 2 Add: 3.42 + 0.237 + 14.1.

```
     3 . 4 2 0     Lining up the decimal points
     0 . 2 3 7     and writing extra zeros
 + 1 4 . 1 0 0
   1 7 . 7 5 7     Adding
```

◀ **Do Exercises 3–5.**

Now we consider the addition 3456 + 19.347. Keep in mind that any whole number has an "unwritten" decimal point at the right that can be followed by zeros. For example, 3456 can also be written 3456.000. When adding, we can always write in the decimal point and extra zeros if desired.

EXAMPLE 3 Add: 3456 + 19.347.

$$
\begin{array}{r}
\overset{1}{} \\
3\,4\,5\,6\,.\,0\,0\,0 \\
+\quad 1\,9\,.\,3\,4\,7 \\
\hline
3\,4\,7\,5\,.\,3\,4\,7
\end{array}
$$

  Writing in the decimal point and extra zeros
  Lining up the decimal points
  Adding

Do Exercises 6 and 7. ▶

b SUBTRACTION

Subtracting with decimal notation is similar to subtracting whole numbers. First, we line up the decimal points so that we can subtract corresponding place-value digits. Then we subtract digits from the right. For example, we subtract the thousandths, then the hundredths, the tenths, and so on, borrowing if necessary.

EXAMPLE 4 Subtract: 56.314 − 17.78.

$$
\begin{array}{r}
5\,6\,.\,3\,1\,4 \\
-\,1\,7\,.\,7\,8\,\underline{0}
\end{array}
$$
  Lining up the decimal points in order to subtract
  Writing an extra 0

$$
\begin{array}{r}
5\,6\,.\,3\,1\,4 \\
-\,1\,7\,.\,7\,8\,0 \\
\hline
4
\end{array}
$$
  Subtracting thousandths

$$
\begin{array}{r}
\overset{2\ 11}{} \\
5\,6\,.\,3\,\cancel{1}\,4 \\
-\,1\,7\,.\,7\,8\,0 \\
\hline
3\ 4
\end{array}
$$
  Borrowing tenths to subtract hundredths
  Subtracting hundredths

$$
\begin{array}{r}
\overset{\ \ 12}{5\ \overset{2}{\ }\,11} \\
5\,6\,.\,3\,\cancel{1}\,4 \\
-\,1\,7\,.\,7\,8\,0 \\
\hline
.\,5\ 3\ 4
\end{array}
$$
  Borrowing ones to subtract tenths
  Subtracting tenths; writing a decimal point

$$
\begin{array}{r}
\overset{15\ \ 12}{4\ \overset{5}{\ }\,\overset{2}{\ }\,11} \\
\cancel{5}\,6\,.\,3\,\cancel{1}\,4 \\
-\,1\,7\,.\,7\,8\,0 \\
\hline
8\,.\,5\ 3\ 4
\end{array}
$$
  Borrowing tens to subtract ones
  Subtracting ones

$$
\begin{array}{r}
\overset{15\ \ 12}{4\ \overset{5}{\ }\,\overset{2}{\ }\,11} \\
\cancel{5}\,6\,.\,3\,\cancel{1}\,4 \\
-\,1\,7\,.\,7\,8\,0 \\
\hline
3\,8\,.\,5\,3\,4
\end{array}
$$
  Subtracting tens

Check by adding:
$$
\begin{array}{r}
\overset{1\ \ 1\ \ 1}{3\,8\,.\,5\,3\,4} \\
+\,1\,7\,.\,7\,8\,0 \\
\hline
5\,6\,.\,3\,1\,4
\end{array}
$$
← | The answer checks because this is the top number in the subtraction. |

Do Exercises 8 and 9. ▶

Add.

6. 789 + 123.67

GS **7.** 45.78 + 2467 + 1.993

$$
\begin{array}{r}
\square\ \overset{1}{\ }\ \square\ \ \overset{1}{\ } \\
4\,5\,.\,7\,8\,0 \\
2\,4\,6\,7\,.\,0\,0\,0 \\
+\quad\quad 1\,.\,9\,9\,3 \\
\hline
2\ \square\ 1\,4\,.\,\square\,7\,3
\end{array}
$$

Subtract.

GS **8.** 37.428 − 26.674

$$
\begin{array}{r}
\square \\
\overset{6}{\ }\ \overset{3}{\ }\ \square \\
3\,7\,.\,4\,2\,8 \\
-\,2\,6\,.\,6\,7\,4 \\
\hline
1\ \square\,.\,7\ \square\,4
\end{array}
$$

9.
$$
\begin{array}{r}
0\,.\,3\,4\,7 \\
-\,0\,.\,0\,0\,8
\end{array}
$$

Answers

6. 912.67   **7.** 2514.773   **8.** 10.754
9. 0.339

Guided Solutions:
7. 1, 1; 5, 7   **8.** 13, 12; 0, 5

Subtract.

10. 2.9 − 0.36

11. 0.43 − 0.18762

12. 5.27 − 0.00008

Subtract.

13. 1277 − 82.78

14. 5 − 0.0089

$$\begin{array}{r} \boxed{}^{10} \\ 5.0000 \\ -\ 0.0089 \\ \hline 4.\quad 9\ 1\boxed{} \end{array}$$

Add.

15. 7.42 + (−9.38)

16. −4.201 + 7.36

17. Add: −7.49 + (−5.8).

Answers

10. 2.54   **11.** 0.24238   **12.** 5.26992
13. 1194.22   **14.** 4.9911   **15.** −1.96
16. 3.159   **17.** −13.29

Guided Solution:
14. 4999; 9, 1

EXAMPLE 5   Subtract: 23.08 − 5.0053.

$$\begin{array}{r} {}^{1\ \ 13\quad\ \ 7\ \ 9\ 10} \\ 2\,3.0\,8\,0\,0 \\ -\ \ \ \ 5.0\,0\,5\,3 \\ \hline 1\,8.0\,7\,4\,7 \end{array}$$ Writing two extra zeros to the right of the last digit

Subtracting

◀ **Do Exercises 10–12.**

When subtraction involves an integer, the "unwritten" decimal point can be written in. Extra zeros can then be written in to the right of the decimal point.

EXAMPLE 6   Subtract: 456 − 2.467.

$$\begin{array}{r} {}^{5\ \ 9\ \ 9\ 10} \\ 4\,5\,6.0\,0\,0 \\ -\ \ \ \ \ \ 2.4\,6\,7 \\ \hline 4\,5\,3.5\,3\,3 \end{array}$$ Writing in the decimal point and extra zeros

Subtracting

◀ **Do Exercises 13 and 14.**

C   ADDING AND SUBTRACTING WITH NEGATIVES

Negative decimals are added or subtracted like negative integers.

> To add a negative number and a positive number:
>
> **a)** Determine the sign of the number with the greater absolute value.
>
> **b)** Subtract the smaller absolute value from the larger one.
>
> **c)** The answer is the difference from part (b) with the sign from part (a).

EXAMPLE 7   Add: −13.82 + 4.69.

a) Since $|-13.82| > |4.69|$, the sign of the number with the greater absolute value is negative.

b)
$$\begin{array}{r} {}^{7\ 12} \\ 1\,3.8\,2 \\ -\ \ \ 4.6\,9 \\ \hline 9.1\,3 \end{array}$$ Finding the difference of the absolute values

c) The answer is negative: −13.82 + 4.69 = −9.13.

◀ **Do Exercises 15 and 16.**

> To add two negative numbers:
>
> **a)** Add the absolute values.
>
> **b)** Make the answer negative.

EXAMPLE 8   Add: −2.306 + (−3.125).

a)
$$\left.\begin{array}{r} {}^{\ \ \ \ \ 1} \\ 2.3\,0\,6 \\ +\ 3.1\,2\,5 \end{array}\right\} \quad |-2.306| = 2.306 \text{ and } |-3.125| = 3.125$$
$$\begin{array}{r} \hline 5.4\,3\,1 \end{array}$$ Adding the absolute values

b) −2.306 + (−3.125) = −5.431   The answer is negative.

◀ **Do Exercise 17.**

To subtract, we add the opposite of the number being subtracted.

EXAMPLE 9 Subtract: $-3.1 - 4.8$.

$$-3.1 - 4.8 = -3.1 + (-4.8) \qquad \text{Adding the opposite of 4.8}$$
$$= -7.9 \qquad \text{The sum of two negative numbers is negative.}$$

EXAMPLE 10 Subtract: $-7.9 - (-8.5)$.

$$-7.9 - (-8.5) = -7.9 + 8.5 \qquad \text{Adding the opposite of } -8.5$$
$$= 0.6 \qquad \text{Subtracting absolute values. The answer is positive since 8.5 has the larger absolute value.}$$

Do Exercises 18–21. ▶

d COMBINING LIKE TERMS

Recall that like, or similar, terms have exactly the same variable factors. To combine like terms, we add or subtract coefficients to form an equivalent expression.

EXAMPLE 11 Combine like terms: $3.2x + 4.6x$.

These are the coefficients.

$$3.2x + 4.6x = (3.2 + 4.6)x \qquad \text{Using the distributive law—try to do this step mentally.}$$
$$= 7.8x \qquad \text{Adding}$$

A similar procedure is used when subtracting like terms.

EXAMPLE 12 Combine like terms: $4.13a - 7.56a$.

$$4.13a - 7.56a = (4.13 - 7.56)a \qquad \text{Using the distributive law}$$
$$= (4.13 + (-7.56))a \qquad \text{Adding the opposite of 7.56}$$
$$= -3.43a \qquad \text{Subtracting absolute values. The coefficient is negative since } |-7.56| > |4.13|.$$

When more than one pair of like terms is present, we can rearrange the terms and then simplify.

EXAMPLE 13 Combine like terms: $5.7x - 3.9y - 2.4x + 4.5y$.

$$5.7x - 3.9y - 2.4x + 4.5y$$
$$= 5.7x + (-3.9y) + (-2.4x) + 4.5y \qquad \text{Rewriting as addition}$$
$$= 5.7x + (-2.4x) + (-3.9y) + 4.5y \qquad \text{Using the commutative law to rearrange}$$
$$= 3.3x + 0.6y \qquad \text{Combining like terms}$$

With practice, you will be able to perform many of the above steps mentally.

Do Exercises 22–24. ▶

Subtract.

18. $9.25 - 13.41$

19. $-5.72 - 4.19$

20. $9.8 - (-2.6)$

21. $-5.9 - (-3.2)$

Combine like terms.

22. $5.8x - 2.1x$

23. $-5.9a + 7.6a$

24. $-4.8y + 7.5 + 2.1y - 2.1$

Answers

18. -4.16 **19.** -9.91 **20.** 12.4
21. -2.7 **22.** $3.7x$ **23.** $1.7a$
24. $-2.7y + 5.4$

✓ Reading Check

Complete each subtraction and its check by selecting a number from the list at the right.

RC1.
```
    2 3 . 7
 −   1 . 8 7 6
  ┌─────────┐
  └─────────┘
```
Check:
```
    2 1 . 8 2 4
 +   1 . 8 7 6
  ┌─────────┐
  └─────────┘
```

RC2.
```
  1 8 7 . 6 2 3
 −    4 0 . 9
  ┌─────────┐
  └─────────┘
```
Check:
```
  1 4 6 . 7 2 3
 + ┌─────────┐
   └─────────┘
  1 8 7 . 6 2 3
```

| |
| 23.7 |
| 187.623 |
| 21.824 |
| 40.9 |
| 1.876 |
| 146.723 |

a   Add.

1.
```
  4 2 6 . 2 5
 +   3 8 . 1 2
```

2.
```
  6 4 1 . 8 2 3
 +    1 4 . 9 1 5
```

3.
```
  6 5 9 . 4 0 3
 + 9 1 6 . 6 1 2
```

4.
```
  8 7 5 . 7 9 5
 + 3 2 4 . 8 6 2
```

5.
```
      9 . 1 0 4
 + 1 2 3 . 4 5 6
```

6.
```
    3 . 4 0 9
 + 8 1 . 0 0 1
```

7. 2.006 + 5.817

8. 0.8096 + 0.7856

9. 20.7 + 30.0124

10. 0.263 + 0.8

11. 1.06 + 9

12. 12 + 18.08

13. 0.34 + 3.5 + 0.127 + 768

14. 2.3 + 0.729 + 23

15. 17 + 3.24 + 0.256 + 0.3689

16.
```
      4 7 . 8
    2 1 9 . 8 5 2
      4 3 . 5 9
 +  6 6 6 . 7 1 3
```

17.
```
        2 . 7 0 3
       7 8 . 3 3
       2 8 . 0 0 0 9
 +  1 1 8 . 4 3 4 1
```

18.
```
       1 3 . 7 2
        9 . 1 1 2
    6 5 4 2 . 7 9 0 8
 +     2 3 . 9 0 1
```

b Subtract.

19.
$$\begin{array}{r} 47.596 \\ -6.215 \\ \hline \end{array}$$

20.
$$\begin{array}{r} 11.345 \\ -2.105 \\ \hline \end{array}$$

21.
$$\begin{array}{r} 51.31 \\ -2.29 \\ \hline \end{array}$$

22.
$$\begin{array}{r} 37.45 \\ -6.32 \\ \hline \end{array}$$

23.
$$\begin{array}{r} 3.6 \\ -0.036 \\ \hline \end{array}$$

24.
$$\begin{array}{r} 28.0 \\ -0.28 \\ \hline \end{array}$$

25.
$$\begin{array}{r} 92.341 \\ -6.42 \\ \hline \end{array}$$

26.
$$\begin{array}{r} 0.346 \\ -0.0346 \\ \hline \end{array}$$

27.
$$\begin{array}{r} 3.0074 \\ -1.3408 \\ \hline \end{array}$$

28.
$$\begin{array}{r} 32.7978 \\ -0.0592 \\ \hline \end{array}$$

29.
$$\begin{array}{r} 6.07 \\ -2.0078 \\ \hline \end{array}$$

30.
$$\begin{array}{r} 1.0 \\ -0.9999 \\ \hline \end{array}$$

31. $30.24 - 0.241$

32. $100.12 - 0.112$

33. $34.07 - 30.7$

34. $36.2 - 16.28$

35. $8.45 - 7.405$

36. $3.801 - 2.81$

37. $6.003 - 2.3$

38. $1 - 0.0098$

39. $2 - 1.0908$

40. $100 - 0.34$

41. $624 - 18.79$

42. $7.48 - 2.6$

43. $57.803 - 4.6$

44. $25.008 - 12.4$

45. $263.7 - 102.08$

46. $19 - 1.198$

47. $45 - 0.999$

48. $10.05 - 0.392$

c Add or subtract, as indicated.

49. $-5.02 + 1.73$

50. $-4.31 + 7.66$

51. $12.9 - 15.4$

52. $27.2 - 31.9$

53. $-2.9 + (-4.3)$

54. $-7.49 - 1.82$

55. $-4.301 + 7.68$

56. $-5.952 + 7.98$

57. $-12.9 - 3.7$

58. $-8.7 - 12.4$

59. $-2.1 - (-4.6)$

60. $-4.3 - (-2.5)$

61. $14.301 + (-17.82)$

62. $13.45 + (-18.701)$

63. $7.201 - (-2.4)$

64. $2.901 - (-5.7)$

65. $96.9 + (-21.4)$

66. $43.2 + (-10.9)$

67. $-3 - (-12.7)$

68. $-4.5 - (-7)$

69. $-4.9 - 5.392$

70. $89.3 - 100$

71. $14.7 - 15$

72. $-7.201 - 1.9$

d Combine like terms.

73. $1.8x + 3.9x$

74. $7.9x + 1.3x$

75. $17.59a - 12.73a$

76. $23.28a - 15.79a$

77. $15.2t + 7.9 + 5.9t$

78. $29.5t - 4.8 + 7.6t$

79. $5.217x - 8.134x$

80. $6.317t - 9.429t$

81. $4.906y - 7.1 + 3.2y$

82. $9.108y + 4.2 + 3.7y$

83. $4.8x + 1.9y - 5.7x + 1.2y$

84. $3.2r - 4.1t - 5.6t + 1.9r$

85. $4.9 - 3.9t - 6 - 4.5t$

86. $5 + 9.7x - 7.2 - 12.8x$

318   CHAPTER 5   Decimal Notation

Copyright © 2016 Pearson Education, Inc.

Skill Maintenance

Multiply. [3.4b]

87. $\dfrac{3}{5} \cdot \dfrac{4}{7}$

88. $\dfrac{2}{9} \cdot \dfrac{7}{5}$

89. $\dfrac{3}{10} \cdot \dfrac{21}{100}$

Evaluate. [2.6a]

90. $8 - 2x^2$, for $x = 3$

91. $5 - 3x^2$, for $x = -2$

92. $7 + 2x^2 \div 3$, for $x = 6$

Synthesis

Combine like terms.

93. ▦ $-3.928 - 4.39a + 7.4b - 8.073 + 2.0001a - 9.931b - 9.8799a + 12.897b$

94. ▦ $79.02x + 0.0093y - 53.14z - 0.02001y - 37.987z - 97.203x - 0.00987y$

95. ▦ $39.123a - 42.458b - 72.457a + 31.462b - 59.491 + 37.927a$

96. Ryan presses the wrong key when using a calculator and adds 235.7 instead of subtracting it. The incorrect answer is 817.2. What is the correct answer?

97. Alicia presses the wrong key when using a calculator and subtracts 349.2 instead of adding it. The incorrect answer is −836.9. What is the correct answer?

98. ▦ Find the errors, if any, in the balances in this checkbook.

20___		RECORD ALL CHARGES OR CREDITS THAT AFFECT YOUR ACCOUNT					
DATE	CHECK NUMBER	TRANSACTION DESCRIPTION	√ T	(−) PAYMENT/ DEBIT	(+ OR −) OTHER	(+) DEPOSIT/ CREDIT	BALANCE FORWARD
							2767 73
8/16	432	Burch Laundry		23 56			2744 16
8/19	433	Rogers TV		20 49			2764 65
8/20		Deposit				85 00	2848 65
8/21	434	Galaxy Records		48 60			2801 05
8/22	435	Electric Works		267 95			2533 09

Find a.

99.
$$\begin{array}{r} 9\,3.a\,4\,3 \\ -\ 8\,7.9\,6\,9 \\ \hline 5.2\,7\,4 \end{array}$$

100.
$$\begin{array}{r} 4\,8\,1.a\,2\,4 \\ -\ \ \ 7\,2.9\,7\,8 \\ \hline 4\,0\,8.3\,4\,6 \end{array}$$

5.3 Multiplication of Decimals

OBJECTIVES

a Multiply using decimal notation.

b Convert from notation like 45.7 million to standard notation and convert between dollars and cents.

c Evaluate algebraic expressions using decimal notation.

SKILL TO REVIEW

Objective 1.4a: Multiply whole numbers.

Multiply.

1.   4 2
   × 6 3

2.   7 1 6
   ×   5 8

a MULTIPLICATION

To develop an understanding of decimal multiplication, consider 2.3×1.12:

$$2.3 \times 1.12 = \frac{23}{10} \times \frac{112}{100} = \frac{2576}{1000} = 2.576.$$

Note that the number of decimal places in the product is the sum of the numbers of decimal places in the factors.

$$\begin{array}{r} 1.1\,2 \quad (\textbf{2}\text{ decimal places}) \\ \times \quad 2.3 \quad (\textbf{1}\text{ decimal place}) \\ \hline 2.5\,7\,6 \quad (\textbf{3}\text{ decimal places}) \end{array}$$

Now consider 0.02×3.412:

$$0.02 \times 3.412 = \frac{2}{100} \times \frac{3412}{1000} = \frac{6824}{100,000} = 0.06824.$$

Again, note that the number of decimal places in the product is the sum of the numbers of decimal places in the factors.

$$\begin{array}{r} 3.4\,1\,2 \quad (\textbf{3}\text{ decimal places}) \\ \times \quad 0.0\,2 \quad (\textbf{2}\text{ decimal places}) \\ \hline 0.0\,6\,8\,2\,4 \quad (\textbf{5}\text{ decimal places}) \end{array}$$

The 0 after the decimal point is necessary.

To multiply using decimal notation: 0.8×0.43

a) Ignore the decimal points, for the moment, and multiply as though both factors were integers.

$$\begin{array}{r} \overset{2}{0.4}\,3 \\ \times \quad 0.8 \\ \hline 3\,4\,4 \end{array}$$
Ignore the decimal points for now.

b) Place the decimal point in the result. The number of decimal places in the product is the sum of the numbers of places in the factors.

$$\begin{array}{r} 0.4\,3 \quad (\textbf{2}\text{ decimal places}) \\ \times \quad 0.8 \quad (\textbf{1}\text{ decimal place}) \\ \hline 0.3\,4\,4 \quad (\textbf{3}\text{ decimal places}) \end{array}$$

Count the number of decimal places by starting at the far right and moving the decimal point to the left.

EXAMPLE 1 Multiply: 8.3×74.6.

a) Ignore the decimal points and multiply as if both factors were integers.

$$\begin{array}{r} \overset{3}{}\overset{4}{} \\ \overset{1}{}\overset{1}{} \\ 7\,4.6 \\ \times \quad 8.3 \\ \hline 2\,2\,3\,8 \\ 5\,9\,6\,8\,0 \\ \hline 6\,1\,9\,1\,8 \end{array}$$
We are not yet finished.

b) Place the decimal point in the result. The number of decimal places in the product is the sum, $1 + 1$, of the numbers of decimal places in the factors.

$$
\begin{array}{r}
7\ 4.6 \\
\times \quad\ 8.3 \\
\hline
2\ 2\ 3\ 8 \\
5\ 9\ 6\ 8\ 0 \\
\hline
6\ 1\ 9.1\ 8
\end{array}
$$

(**1** decimal place)
(**1** decimal place)

(**2** decimal places)

Do Exercise 1. ▶

1. Multiply:

$$
\begin{array}{r}
7\ 6.3 \\
\times\quad 8.2 \\
\hline
\end{array}
$$

EXAMPLE 2 Multiply: 0.0032×2148.

$$
\begin{array}{r}
2\ 1\ 4\ 8 \\
\times\ 0.0\ 0\ 3\ 2 \\
\hline
4\ 2\ 9\ 6 \\
6\ 4\ 4\ 4\ 0 \\
\hline
6.8\ 7\ 3\ 6
\end{array}
$$

(**0** decimal places)
(**4** decimal places)

(**4** decimal places)

EXAMPLE 3 Multiply: -0.104×0.86.

Multiplying the absolute values, we have

$$
\begin{array}{r}
0.8\ 6 \\
\times\ \ 0.1\ 0\ 4 \\
\hline
3\ 4\ 4 \\
8\ 6\ 0\ 0 \\
\hline
0.0\ 8\ 9\ 4\ 4
\end{array}
$$

(**2** decimal places)
(**3** decimal places)

(**5** decimal places)   We write an extra zero.

Since the product of a negative number and a positive number is negative, the answer is -0.08944.

Do Exercises 2–4. ▶

Multiply.

2.
$$
\begin{array}{r}
4\ 2\ 1\ 3 \\
\times\ 0.0\ 0\ 5\ 1 \\
\hline
\end{array}
$$

GS **3.** 42.65×0.804

$$
\begin{array}{r}
4\ 2.6\ 5 \\
\times\ \ 0.8\ 0\ 4 \\
\hline
1\ \ \square\ 0\ 6\ 0 \\
3\ 4\ 1\ 2\ 0\ \ \ 0 \\
\hline
3\ 4.\ \square\ 9\ 0\ \ \ 0
\end{array}
$$

4. $5.2014 \times (-2.41)$

Multiplying by 0.1, 0.01, 0.001, and So On

Now let's consider some special kinds of products. The first involves multiplying by a tenth, hundredth, thousandth, ten-thousandth, and so on. We can see a pattern in the following products.

$$0.1 \times 38 = \frac{1}{10} \times 38 = \frac{38}{10} = 3.8$$

$$0.01 \times 38 = \frac{1}{100} \times 38 = \frac{38}{100} = 0.38$$

$$0.001 \times 38 = \frac{1}{1000} \times 38 = \frac{38}{1000} = 0.038$$

$$0.0001 \times 38 = \frac{1}{10,000} \times 38 = \frac{38}{10,000} = 0.0038$$

Note in each case that the product is *smaller* than 38. That is, the decimal point in each product is farther to the left than the unwritten decimal point in 38. Note also that each product can be obtained from 38 by moving the decimal point.

Answers

1. 625.66   **2.** 21.4863   **3.** 34.2906
4. -12.535374

Guided Solution:
3. 7; 0; 2, 6

To multiply any number by 0.1, 0.01, 0.001, and so on:

a) count the number of decimal places in the tenth, hundredth, or thousandth, and so on, and

$$0.001 \times 34.45678$$
$$\longrightarrow 3 \text{ places}$$

b) move the decimal point in the other number that many places to the left. Use zeros as placeholders when necessary.

$$0.001 \times 34.45678 = 0.034.45678$$

Move 3 places to the left.

$$0.001 \times 34.45678 = 0.03445678$$

EXAMPLES Multiply.

Multiply.

5. 0.1×746

6. 0.001×732.4

7. $(-0.01) \times 6.2$

8. 0.0001×723.6

4. $0.1 \times 45 = 4.5$ Moving the decimal point one place to the left

5. $0.01 \times 243.7 = 2.437$ Moving the decimal point two places to the left

6. $0.001 \times (-8.2) = -0.0082$ Moving the decimal point three places to the left. This requires writing two extra zeros.

7. $0.0001 \times 536.9 = 0.05369$ Moving the decimal point four places to the left. This requires writing one extra zero.

◀ **Do Exercises 5–8.**

Multiplying by 10, 100, 1000, and So On

Next we consider multiplying by 10, 100, 1000, and so on. We see a pattern in the following.

$$
\begin{array}{r}
5.237 \\
\times \quad 10 \\
\hline
0000 \\
5237 \\
\hline
52.370
\end{array}
\qquad
\begin{array}{r}
5.237 \\
\times \quad 100 \\
\hline
0000 \\
0000 \\
5237 \\
\hline
523.700
\end{array}
\qquad
\begin{array}{r}
5.237 \\
\times \quad 1000 \\
\hline
0000 \\
0000 \\
0000 \\
5237 \\
\hline
5237.000
\end{array}
$$

Note in each case that the product is *larger* than 5.237. That is, the decimal point in each product is farther to the right than the decimal point in 5.237. Also, each product can be obtained from 5.237 by moving the decimal point.

CALCULATOR CORNER

Multiplication with Decimal Notation To use a calculator to multiply with decimal notation, we use the $\boxed{\cdot}$, $\boxed{\times}$, and $\boxed{=}$ keys. To find 4.78×0.34, for example, we press $\boxed{4}$ $\boxed{\cdot}$ $\boxed{7}$ $\boxed{8}$ $\boxed{\times}$ $\boxed{\cdot}$ $\boxed{3}$ $\boxed{4}$ $\boxed{=}$. The display reads 1.6252, so $4.78 \times 0.34 = 1.6252$.

EXERCISES Use a calculator to multiply.

1. 17.63
      $\times \quad 8.1$

2. $0.04 \times (-12.69)$

3. 586.4×13.5

4. $-4.003 \times (-5.1)$

To multiply any number by 10, 100, 1000, and so on:

a) count the number of zeros and

$$1000 \times 34.45678$$
$$\longrightarrow 3 \text{ zeros}$$

b) move the decimal point in the other number that many places to the right. Use zeros as placeholders when necessary.

$$1000 \times 34.45678 = 34.456.78$$

Move 3 places to the right.

$$1000 \times 34.45678 = 34{,}456.78$$

Answers

5. 74.6   **6.** 0.7324
7. −0.062   **8.** 0.07236

EXAMPLES Multiply.

8. $10 \times 32.98 = 329.8$ Moving the decimal point one place to the right

9. $100 \times 4.7 = 470$ Moving the decimal point two places to the right. The 0 in 470 is a placeholder.

10. $1000 \times (-2.4167) = -2416.7$ Moving the decimal point three places to the right

11. $10{,}000 \times 7.52 = 75{,}200$ Moving the decimal point four places to the right and using two zeros as placeholders

Do Exercises 9–12. ▶

Multiply.

9. 10×53.917

10. $100 \times (-62.417)$

11. 1000×83.9

12. $10{,}000 \times 57.04$

b   APPLICATIONS USING MULTIPLICATION WITH DECIMAL NOTATION

Naming Large Numbers

We often see notation like the following in newspapers and magazines and on television and the Internet.

- In 2013, wildfires burned over 4.3 million acres in the United States.
  Source: National Interagency Fire Center

- The number of valid U.S. passports in circulation in 2011 exceeded 113.4 million.
  Source: U.S. State Department

- Each day, about 144.8 billion e-mails are sent worldwide.
  Source: Mashable.com

- At one point in 2014, the U.S. national debt was approximately $17.478 trillion. The national debt has increased, on average, by $3.81 billion per day since 2007.
  Source: www.usdebtclock.org

To understand such notation, consider the information in the following table.

NAMING LARGE NUMBERS

1 hundred $= 100 = 10 \cdot 10 = 10^2$
       → 2 zeros

1 thousand $= 1000 = 10 \cdot 10 \cdot 10 = 10^3$
       → 3 zeros

1 million $= 1{,}000{,}000 = 10 \cdot 10 \cdot 10 \cdot 10 \cdot 10 \cdot 10 = 10^6$
       → 6 zeros

1 billion $= 1{,}000{,}000{,}000 = 10^9$
       → 9 zeros

1 trillion $= 1{,}000{,}000{,}000{,}000 = 10^{12}$
       → 12 zeros

Answers

9. 539.17   **10.** −6241.7
11. 83,900   **12.** 570,400

To convert a large number to standard notation, we proceed as follows.

EXAMPLE 12 *Text Messages.* Worldwide, 8.6 trillion text messages are sent each year. Convert 8.6 trillion to standard notation.

Source: CNNTech, Portio Research

$$8.6 \text{ trillion} = 8.6 \times 1 \text{ trillion}$$
$$= 8.6 \times \underline{1,000,000,000,000}$$

↳ 12 zeros

$$= 8,600,000,000,000$$ Moving the decimal point 12 places to the right

◀ **Do Exercises 13 and 14.**

Money Conversion

Converting from dollars to cents is like multiplying by 100. To see why, consider $19.43.

$$\$19.43 = 19.43 \times \$1$$ We think of $19.43 as 19.43 × 1 dollar, or 19.43 × $1.
$$= 19.43 \times 100¢$$ Substituting 100¢ for $1: $1 = 100¢
$$= 1943¢$$ Multiplying

DOLLARS TO CENTS

To convert from dollars to cents, move the decimal point two places to the right and change the $ sign in front to a ¢ sign at the end.

EXAMPLES Convert from dollars to cents.

13. $189.64 = 18,964¢

14. $0.75 = 75¢

◀ **Do Exercises 15 and 16.**

Converting from cents to dollars is like multiplying by 0.01. To see why, consider 65¢.

$$65¢ = 65 \times 1¢$$ We think of 65¢ as 65 × 1 cent, or 65 × 1¢.
$$= 65 \times \$0.01$$ Substituting $0.01 for 1¢: 1¢ = $0.01
$$= \$0.65$$ Multiplying

CENTS TO DOLLARS

To convert from cents to dollars, move the decimal point two places to the left and change the ¢ sign at the end to a $ sign in front.

EXAMPLES Convert from cents to dollars.

15. 395¢ = $3.95

16. 8503¢ = $85.03

◀ **Do Exercises 17 and 18.**

Convert the number in each sentence to standard notation.

13. The largest building in the world is the Pentagon, which has 3.7 million square feet of floor space.

14. *Text Messages.* In the United States, 2.2 trillion text messages are sent each year.
Source: CNNTech, Forrester Research

Convert from dollars to cents.

15. $15.69
$$\$15.69 = 15.69 \times \$1$$
$$= 15.69 \times \boxed{}¢$$
$$= \boxed{}¢$$

16. $0.17

Convert from cents to dollars.

17. 35¢

18. 577¢

Answers

13. 3,700,000   **14.** 2,200,000,000,000
15. 1569¢   **16.** 17¢   **17.** $0.35
18. $5.77

Guided Solution:
15. 100, 1569

c EVALUATING

Algebraic expressions are often evaluated using numbers written in decimal notation.

EXAMPLE 17 Evaluate Prt for $P = 780$, $r = 0.12$, and $t = 0.5$.

This product can be used to determine the interest paid on $780, borrowed at 12 percent simple interest, for half a year. We substitute as follows.

$Prt = 780 \cdot 0.12 \cdot 0.5 = 780 \cdot 0.06 = 46.8$ This would represent $46.80.

Do Exercise 19. ▷

EXAMPLE 18 Find the perimeter of a stamp that is 3.25 cm long and 2.5 cm wide.

Recall that the perimeter, P, of a rectangle of length l and width w is given by the formula

$$P = 2l + 2w.$$

Thus, we evaluate $2l + 2w$ for $l = 3.25$ and $w = 2.5$.

$$2l + 2w = 2 \cdot 3.25 + 2 \cdot 2.5$$
$$= 6.5 + 5.0$$ Remember the rules for order of operations.
$$= 11.5$$

The perimeter is 11.5 cm.

Do Exercise 20. ▷

EXAMPLE 19 *Multiple Births.* The expression $2.67t + 65.02$ can be used to predict the number of twin births, in thousands, in the United States t years after 1980. Predict the number of twin births in 2015.

Source: Based on information from the Centers for Disease Control

2015 is 35 years after 1980, so we evaluate $2.67t + 65.02$ for $t = 35$.

$$2.67t + 65.02 = 2.67 \cdot 35 + 65.02$$
$$= 93.45 + 65.02$$
$$= 158.47$$

In 2015, there will be approximately 158.47 thousand, or 158,470, twin births.

Do Exercise 21. ▷

19. Evaluate lwh for $l = 3.2$, $w = 2.6$, and $h = 0.8$. (This is the formula for the volume of a rectangular box.)

20. Find the area of the stamp in Example 18.

21. Evaluate $6.28rh + 3.14r^2$ for $r = 1.5$ and $h = 5.1$. (This is the formula for the surface area of an open can.)

Answers

19. 6.656   **20.** 8.125 sq cm   **21.** 55.107

✓ Reading Check

Match each expression with an equivalent expression from the list at the right.

RC1. 0.001×38 _____   **a)** 3800¢

RC2. 1000×38 _____   **b)** 380

RC3. 38¢ _____   **c)** 0.038

RC4. \$38 _____   **d)** \$3.80

RC5. 380¢ _____   **e)** 38,000

RC6. 10×38 _____   **f)** \$0.38

a   Multiply.

1.
$$\begin{array}{r} 6.8 \\ \times\ \ \ 7 \\ \hline \end{array}$$

2.
$$\begin{array}{r} 5.7 \\ \times\ 0.9 \\ \hline \end{array}$$

3.
$$\begin{array}{r} 0.8\,4 \\ \times\ \ \ \ \ 8 \\ \hline \end{array}$$

4.
$$\begin{array}{r} 7.3 \\ \times\ 0.6 \\ \hline \end{array}$$

5.
$$\begin{array}{r} 6.3 \\ \times\ 0.0\,4 \\ \hline \end{array}$$

6.
$$\begin{array}{r} 7.8 \\ \times\ 0.0\,9 \\ \hline \end{array}$$

7.
$$\begin{array}{r} 8\,7 \\ \times\ 0.0\,0\,6 \\ \hline \end{array}$$

8.
$$\begin{array}{r} 2\,5.9 \\ \times\ 0.0\,0\,5 \\ \hline \end{array}$$

9. 10×42.63

10. 100×2.8793

11. -1000×783.686852

12. -0.34×1000

13. -7.8×100

14. $0.00238 \times (-10)$

15. 0.1×79.18

16. 0.01×789.235

17. 0.001×97.68

18. 8976.23×0.001

19. $28.7 \times (-0.01)$

20. $0.0325 \times (-0.1)$

21.
$$\begin{array}{r} 2.7\,3 \\ \times\ \ \ 1\,6 \\ \hline \end{array}$$

22.
$$\begin{array}{r} 8.2\,7 \\ \times\ \ \ 5.4 \\ \hline \end{array}$$

23.
$$\begin{array}{r} 0.9\,8\,4 \\ \times\ 0.0\,3\,1 \\ \hline \end{array}$$

24.
$$\begin{array}{r} 7.4\,8\,9 \\ \times\ \ \ \ \ 1.7 \\ \hline \end{array}$$

25. $(-37.4)(-2.4)$

26. $569(-1.05)$

27. $749(-0.43)$

28. $(-0.876)(-0.0204)$

29.
$$\begin{array}{r} 0.8\,7 \\ \times\ \ \ 6\,4 \\ \hline \end{array}$$

30.
$$\begin{array}{r} 7.2\,5 \\ \times\ \ \ 6\,0 \\ \hline \end{array}$$

31.
$$\begin{array}{r} 4\,6.5\,0 \\ \times\ \ \ \ \ 7\,5 \\ \hline \end{array}$$

32.
$$\begin{array}{r} 8.2\,4 \\ \times\ 7\,0\,3 \\ \hline \end{array}$$

33. $(-0.231)(-0.5)$

34. $(-12.3)(-1.08)$

35. $9.42 \times (-1000)$

36. -7.6×1000

37. $-95.3 \times (-0.0001)$

38. $-4.23 \times (-0.001)$

b Convert from dollars to cents.

39. $57.06   **40.** $49.85   **41.** $0.95   **42.** $0.49   **43.** $0.01   **44.** $0.09

Convert from cents to dollars.

45. 72¢   **46.** 52¢   **47.** 2¢   **48.** 5¢   **49.** 6399¢   **50.** 5238¢

51. *Farming Area.* China has 3.48 million hectares of land devoted to farming. This area is approximately 24% of China's total area. Convert 3.48 million to standard notation.

Source: Viking Cruises

52. *Doll Sales.* In the United States, spending on dolls totaled $2.7 billion in 2011. Convert 2.7 billion to standard notation.

Source: The NPD Group

53. *Spending on Pets.* In 2011, Americans spent approximately $50.96 billion on their pets, of which $13.4 billion was for veterinary care. Convert 50.96 billion and 13.41 billion to standard notation.

Source: American Pet Association

54. *Library of Congress.* The Library of Congress is the largest public library in the United States. It holds about 33.5 million books. Convert 33.5 million to standard notation.

Source: American Library Association

55. *Safe Water.* Worldwide, about 2.2 million people die each year because of diseases caused by unsafe water. Convert 2.2 million to standard notation.

Source: United Nations Environment Programme

56. *Overdraft Charges.* Revenue of U.S. banks from checking account overdraft charges rose to $31.5 billion in the fiscal year ending in June 2012. Convert 31.5 billion to standard notation.

Sources: Moebs Services; abcNews, *Consumer Report*, September 27, 2012

c Evaluate.

57. $P + Prt$, for $P = 10{,}000$, $r = 0.04$, and $t = 2.5$
(*A formula for adding interest*)

58. $6.28r(h + r)$, for $r = 10$ and $h = 17.2$
(*Surface area of a cylinder*)

59. $vt + 0.5at^2$, for $v = 10$, $t = 1.5$, and $a = 9.8$
(*A physics formula*)

60. $4lh + 2h^2$, for $l = 3.5$ and $h = 1.2$
(*Surface area of a rectangular prism*)

Find **(a)** the perimeter and **(b)** the area of a rectangular room with the given dimensions.

61. 12.5 ft long, 9.5 ft wide

62. 10.25 ft long, 8 ft wide

63. 8.4 m wide, 10.5 m long

64. 8.2 yd long, 6.4 yd wide

Nursing. The expression $0.0375t + 2.2$ can be used to predict the number of registered nurses, in millions, in the United States t years after 2000. Predict the number of registered nurses in the United States in the year indicated.

Source: Based on information from the Bureau of Labor Statistics, U.S. Dept. of Labor

65. 2015

66. 2020

Skill Maintenance

Divide.

67. $-162 \div 6$ [2.5a]

68. $-216 \div (-6)$ [2.5a]

69. $-1035 \div (-15)$ [2.5a]

70. $-423 \div 3$ [2.5a]

71. $17\overline{)20{,}006}$ [1.5a]

72. $675 \div (-25)$ [2.5a]

Synthesis

Consider the following names for large numbers in addition to those already discussed in this section:

$$1 \text{ quadrillion} = 1{,}000{,}000{,}000{,}000{,}000 = 10^{15};$$
$$1 \text{ quintillion} = 1{,}000{,}000{,}000{,}000{,}000{,}000 = 10^{18};$$
$$1 \text{ sextillion} = 1{,}000{,}000{,}000{,}000{,}000{,}000{,}000 = 10^{21};$$
$$1 \text{ septillion} = 1{,}000{,}000{,}000{,}000{,}000{,}000{,}000{,}000 = 10^{24}.$$

Find each of the following. Express the answer with a name that is a power of 10.

73. (1 trillion) $\cdot$ (1 billion)

74. (1 million) $\cdot$ (1 billion)

75. (1 trillion) $\cdot$ (1 trillion)

76. Is a billion millions the same as a million billions? Explain.

77. In Great Britain, France, and Germany, a billion means a million millions. Write standard notation for the British number 6.6 billion.

78. One light-year (LY) is 9.46×10^{12} km. The star Regulus is 85 LY from the earth. How many billions of kilometers (km) from the earth is Regulus?

Source: *The Cambridge Factfinder*, 4th ed

Evaluate using a calculator.

79. $d + vt + at^2$, for $d = 79.2$, $v = 3.029$, $t = 7.355$, and $a = 4.9$ (*A physics formula for distance traveled*)

80. $0.5(b_1 + b_2)h$, for $b_1 = 9.7$ cm, $b_2 = 13.4$ cm, and $h = 6.32$ cm (*A geometry formula for the area of a trapezoid*)

81. *Electric Bills* Recently, electric bills from the Central Vermont Public Service Corporation consisted of a "customer charge" of $0.374 per day plus an "energy charge" of $0.1174 per kilowatt-hour (kWh) for the first 250 kWh used and $0.09079 per kilowatt-hour for each kilowatt-hour in excess of 250. From April 20 to May 20, the Coy-Bergers used 480 kWh of electricity. What was their bill for the period?

Division of Decimals

a DIVISION

Whole-Number Divisors

Now that we have studied multiplication of decimals, we can develop a procedure for division. The following divisions are justified by the multiplication in each *check*:

This is the dividend. ⟶ $\dfrac{651}{7} = 93$ This is the quotient.
This is the divisor. ⟶

Check: $7 \cdot 93 = 651$

$\dfrac{65.1}{7} = 9.3$ *Check*: $7 \cdot 9.3 = 65.1$

$\dfrac{6.51}{7} = 0.93$ *Check*: $7 \cdot 0.93 = 6.51$

$\dfrac{0.651}{7} = 0.093$ *Check*: $7 \cdot 0.093 = 0.651$

Note that the number of decimal places in each quotient is the same as the number of decimal places in the dividend.

> To perform long division of a decimal number by a whole number:
>
> **a)** place the decimal point directly above the decimal point in the dividend, and
> **b)** divide as though dividing whole numbers.
>
> $$\begin{array}{r} 0.8\,4 \leftarrow \text{Quotient} \\ 7\,\overline{)\,5.8\,8} \leftarrow \text{Dividend} \\ 5\,6\,0 \\ \underline{2\,8} \\ 2\,8 \\ \underline{} \\ 0 \leftarrow \text{Remainder} \end{array}$$
>
> Divisor ↑

EXAMPLE 1 Divide: $82.08 \div 24$.

Place the decimal point.

$$\begin{array}{r} 3.4\,2 \\ 24\,\overline{)\,8\,2.0\,8} \\ \underline{7\,2} \\ 1\,0\,0 \\ \underline{9\,6} \\ 4\,8 \\ \underline{4\,8} \\ 0 \end{array}$$

Divide as though dividing whole numbers.

Since $(3.42)(24) = 82.08$, the answer checks.

Do Margin Exercises 1–3. ▶

OBJECTIVES

a Divide using decimal notation.

b Simplify expressions using the rules for order of operations.

SKILL TO REVIEW

Objective 1.5a: Divide whole numbers.

Divide.

1. $5\,\overline{)\,2\,4\,5}$

2. $2\,3\,\overline{)\,1\,9\,7\,8}$

Divide.

1. $9\,\overline{)\,5.4}$

2. $1\,5\,\overline{)\,2\,2.5}$

3. $8\,2\,\overline{)\,3\,8.5\,4}$

Answers

Skill to Review:
1. 49   **2.** 86

Margin Exercises:
1. 0.6   **2.** 1.5   **3.** 0.47

We can think of a whole-number dividend as having a decimal point at the end with as many 0's as we wish after the decimal point. For example, 12 = 12. = 12.0 = 12.00 = 12.000, and so on. We can also add 0's after the last digit in the decimal portion of a number: 3.6 = 3.60 = 3.600, and so on.

EXAMPLE 2 Divide: 30 ÷ 8.

```
        3.
   8 ) 3 0.
       2 4
         6
```
Place the decimal point and divide to find how many ones.

```
        3.
   8 ) 3 0.0
       2 4
         6 0
```
Write an extra zero to the right of the decimal point. This does not change the number.

```
        3.7
   8 ) 3 0.0
       2 4
         6 0
         5 6
           4
```
Divide to find how many tenths.

```
        3.7
   8 ) 3 0.0 0
       2 4
         6 0
         5 6
           4 0
```
Write another zero.

```
        3.7 5
   8 ) 3 0.0 0
       2 4
         6 0
         5 6
           4 0
           4 0
             0
```
Divide to find how many hundredths.

Check:
```
         6 4
      3.7 5
    ×     8
    3 0.0 0
```

Divide.

4. 2 5) 8

5. −23 ÷ 4

6. 2.15 ÷ 86

```
        0.  2
   8 6 ) 2. 1 5 0
         1 7 2
           4 3
           4 3 0
               0
```

EXAMPLE 3 Divide: −4.5 ÷ 250.

We first consider 4.5 ÷ 250.

```
         0.0 1 8
   2 5 0 ) 4.5 0 0
           2 5 0
           2 0 0 0
           2 0 0 0
               0  ← Since the remainder is 0, we are finished.
```

Since a negative number divided by a positive number is negative, the answer is −0.018. To check, note that (−0.018)(250) = −4.5.

◀ **Do Exercises 4–6.**

Answers

4. 0.32   5. −5.75   6. 0.025

Guided Solution:

6. 0, 5; 0

Divisors That Are Not Whole Numbers

Consider the division

$$0.2\,4\,\overline{)\,8.2\,0\,8}$$

We write the division as $\dfrac{8.208}{0.24}$. Then we multiply by 1 to change to a whole-number divisor:

The division $0.24\overline{)8.208}$ is the same as $24\overline{)820.8}$.

$$\frac{8.208}{0.24} = \frac{8.208}{0.24} \times \frac{100}{100} = \frac{820.8}{24}.$$

The divisor is now a whole number.

To divide when the divisor is not a whole number:

a) move the decimal point (multiply by 10, 100, and so on) to make the divisor a whole number,

$$0.2\,4\,\overline{)\,8.2\,0\,8}$$

Move 2 places to the right.

b) move the decimal point in the dividend the same number of places (multiply the same way), and

$$0.2\,4\,\overline{)\,8.2\,0\,8}$$

Move 2 places to the right.

c) place the decimal point for the answer directly above the new decimal point in the dividend and divide as though dividing whole numbers.

$$
\begin{array}{r}
3\,4.2 \\
0.2\,4\,\overline{)\,8.2\,0_\wedge 8} \\
7\,2 \\
\hline
1\,0\,0 \\
9\,6 \\
\hline
4\,8 \\
4\,8 \\
\hline
0
\end{array}
$$

(The new decimal point in the dividend is indicated by a caret.)

EXAMPLE 4 Divide: $5.848 \div 8.6$.

$$8.6\,\overline{)\,5.8\,4\,8}$$

Multiply the divisor by 10. (Move the decimal point 1 place.) Multiply the same way in the dividend. (Move 1 place.)

$$
\begin{array}{r}
0.6\,8 \\
8.6\,\overline{)\,5.8_\wedge 4\,8} \\
5\,1\,6 \\
\hline
6\,8\,8 \\
6\,8\,8 \\
\hline
0
\end{array}
$$

Place a decimal point above the new decimal point in the dividend and then divide.

Note: $\dfrac{5.848}{8.6} = \dfrac{5.848}{8.6} \cdot \dfrac{10}{10} = \dfrac{58.48}{86}$.

Check: $(0.68)(8.6) = 5.848$

Do Exercises 7–9. ▶

CALCULATOR CORNER

Finding Remainders We can find a whole-number remainder by multiplying the decimal portion of a quotient by the divisor. For example, to find the quotient and the whole-number remainder for $567 \div 13$, we can use a calculator to find that

$$567 \div 13 \approx 43.61538462.$$

To isolate the portion to the right of the decimal point, we can subtract 43. When the decimal part of the quotient is multiplied by the divisor, we have

$$0.61538462 \times 13 = 8.00000006.$$

The rounding error in the result may vary, depending on the calculator used. We see that $567 \div 13 = 43\,\text{R}\,8$.

EXERCISES Find the quotient and the whole-number remainder for each of the following.

1. $478 \div 17$

2. $815 \div 7$

3. $824 \div 11$

4. $7888 \div 19$

Divide.

GS **7.** $0.375 \div 0.25$

$$\frac{0.375}{0.25} = \frac{0.375}{0.25} \times \frac{\boxed{}}{100}$$

$$= \frac{37.5}{\boxed{}}$$

$$
\begin{array}{r}
1.\boxed{} \\
0.2\,5\,\overline{)\,0.3\,7_\wedge 5} \\
2\,5 \\
\hline
1\,2\,\boxed{} \\
1\,2\,5 \\
\hline
0
\end{array}
$$

8. $4.067 \div (-0.83)$

9. $-44.8 \div (-3.5)$

Answers

7. 1.5 **8.** −4.9 **9.** 12.8

Guided Solution:

7. 100, 25; 5, 5

EXAMPLE 5 Divide: $-12 \div (-0.64)$.

Note first that a negative number divided by a negative number is positive. To find the quotient, we consider $12 \div 0.64$.

$$0.64 \overline{)\ 12.}$$

Place a decimal point at the end of the whole number.

$$0.64 \overline{)\ 12.00}$$

Multiply the divisor by 100. (Move the decimal point 2 places.) Multiply the same way in the dividend. (Move 2 places after adding extra zeros.)

$$
\begin{array}{r}
1\ 8\ .\ 7\ 5 \\
0.64\ \overline{)\ 12.00\,00} \\
\underline{6\ 4} \\
5\ 6\ 0 \\
\underline{5\ 1\ 2} \\
4\ 8\ 0 \\
\underline{4\ 4\ 8} \\
3\ 2\ 0 \\
\underline{3\ 2\ 0} \\
0
\end{array}
$$

Place a decimal point above the new decimal point in the dividend and then divide.

We have $-12 \div (-0.64) = 18.75$.

◀ **Do Exercises 10 and 11.**

Divide.

10. $1.6 \overline{)\ 2.5}$

11. $-9 \div 0.03$

Dividing by 10, 100, 1000, and So On

We can divide quickly by a ten, hundred, or thousand, or by a tenth, hundredth, or thousandth. Each procedure we use is based on multiplying by 1. Consider the following example:

$$\frac{23.789}{1000} = \frac{23.789}{1000} \cdot \frac{0.001}{0.001} = \frac{0.023789}{1} = 0.023789.$$

We are dividing by a number greater than 1: The result is *smaller* than 23.789.

To divide by 10, 100, 1000, and so on,

a) count the number of zeros in the divisor, and

$$\frac{713.49}{100}$$

2 zeros ↵

b) write the quotient by moving the decimal point in the dividend that number of places to the left.

$$\frac{713.49}{100} = \frac{713.49}{100.} = \frac{7.1349}{1.00} = 7.1349$$

2 places to the left to change 100 to 1

EXAMPLE 6 Divide: $\dfrac{0.0104}{10}$.

$$\frac{0.0104}{10} = \frac{0.0104}{10.} = \frac{0.00104}{1.0} = 0.00104$$

1 zero   1 place to the left to change 10 to 1

10. 1.5625   **11.** −300

CHAPTER 5   Decimal Notation

EXAMPLE 7 Divide −213.75 by 100.

$$\frac{-213.75}{100} = \frac{-213.75}{100.} = \frac{-2.1375}{1.00} = -2.1375$$

2 zeros   2 places to the left

The answer is −2.1375.

Do Exercises 12 and 13. ▶

Divide.

12. $\dfrac{0.176}{100}$

13. $\dfrac{-98.47}{1000}$

Dividing by 0.1, 0.01, 0.001, and So On

Now consider the following example:

$$\frac{23.789}{0.01} = \frac{23.789}{0.01} \cdot \frac{100}{100} = \frac{2378.9}{1} = 2378.9.$$

We are dividing by a number less than 1: The result is *larger* than 23.789.

To divide by 0.1, 0.01, 0.001, and so on,

a) count the number of decimal places in the divisor, and

$$\frac{713.49}{0.001}$$

3 places ←⤶

b) write the quotient by moving the decimal point in the dividend that number of places to the right.

$$\frac{713.49}{0.001} = \frac{713.49}{0.001} = \frac{713,490}{1} = 713,490$$

3 places to the right to change 0.001 to 1

EXAMPLE 8 Divide: $\dfrac{67.8}{0.1}$.

$$\frac{67.8}{0.1} = \frac{67.8}{0.1} = \frac{678}{1} = 678$$

1 place   1 place to the right to change 0.1 to 1

The answer is 678.

Do Exercises 14 and 15. ▶

Divide.

14. $\dfrac{-6.7832}{0.1}$

15. $\dfrac{12.78}{0.01}$

b ORDER OF OPERATIONS: DECIMAL NOTATION

The rules for order of operations apply when simplifying expressions involving decimal notation.

RULES FOR ORDER OF OPERATIONS

1. Do all calculations within grouping symbols first.
2. Evaluate all exponential expressions.
3. Do all multiplications and divisions in order from left to right.
4. Do all additions and subtractions in order from left to right.

Answers

12. 0.00176   **13.** −0.09847   **14.** −67.832
15. 1278

EXAMPLE 9   Simplify: $2.5 \times 25 \div 25{,}000 \times 250$.

$$2.5 \times 25 \div 25{,}000 \times 250 = 62.5 \div 25{,}000 \times 250$$
$$= 0.0025 \times 250$$
$$= 0.625.$$

Doing all multiplications and divisions in order from left to right

EXAMPLE 10   Simplify: $(5 - 0.06) \div 2 + 3.42 \times 0.1$.

$$(5 - 0.06) \div 2 + 3.42 \times 0.1 = 4.94 \div 2 + 3.42 \times 0.1$$

Working inside the parentheses

$$= 2.47 + 0.342$$

Multiplying and dividing in order from left to right

$$= 2.812$$

Simplify.

16. $625 \div 62.5 \times 25 \div 6250$

$= \boxed{} \times 25 \div 6250$

$= \boxed{} \div 6250$

$= 0.04$

17. $0.25 \cdot (1 + 0.08) - 0.0274$

18. $[(19.7 - 17.2)^2 + 3] \div (-1.25)$

EXAMPLE 11   Simplify: $13 - [5.4(1.3^2 + 0.21) \div 0.6]$.

$$13 - [5.4(1.3^2 + 0.21) \div 0.6]$$
$$= 13 - [5.4(1.69 + 0.21) \div 0.6]$$
$$= 13 - [5.4 \times 1.9 \div 0.6]$$

Working in the innermost parentheses first

$$= 13 - [10.26 \div 0.6]$$

Multiplying

$$= 13 - 17.1$$

Dividing

$$= -4.1$$

◀ **Do Exercises 16–18.**

EXAMPLE 12   *Population Density.*   The table below shows the number of residents per square mile in the six New England states. Find the average number of residents per square mile for this group of states.
Source: 2010 U.S. Census

19. *Population Density.*   The table below shows the number of residents per square mile in five northwestern states. Find the average number of residents per square mile for this group of states.

Source: 2010 U.S. Census

STATE	RESIDENTS PER SQUARE MILE
Washington	101.2
Oregon	39.9
Idaho	19.0
Montana	6.8
Wyoming	5.8

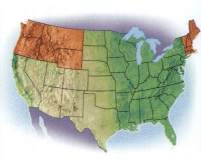

STATE	NUMBER OF RESIDENTS PER SQUARE MILE
Maine	43.1
New Hampshire	147.0
Vermont	67.9
Massachusetts	839.4
Rhode Island	1018.1
Connecticut	738.1

The **average** of a set of numbers is the sum of the numbers divided by the number of addends. We find the sum of the population densities per square mile and divide it by the number of addends, 6:

$$\frac{43.1 + 147 + 67.9 + 839.4 + 1018.1 + 738.1}{6} = \frac{2853.6}{6} = 475.6.$$

Thus the average number of residents per square mile for these six states is 475.6.

◀ **Do Exercise 19.**

Answers

16. 0.04   **17.** 0.2426   **18.** −7.4   **19.** 34.54

Guided Solution:

16. 10, 250

✓ Reading Check

Name the operation that should be performed first in evaluating each expression. Do not calculate.

RC1. $(2 - 0.04) \div 4 + 8.5$ ————————

RC2. $0.02 + 2.06 \div 0.01$ ————————

RC3. $5 \times 2.1 + 0.1 - 8^3$ ————————

RC4. $18.2 - (4.1 + 6.9)$ ————————

RC5. $16 - 9 \div 3 + 7.3$ ————————

RC6. $4(10 - 5) \times 14.2$ ————————

a   Divide.

1. $2 \overline{)5.9\,8}$

2. $5 \overline{)13.5}$

3. $4 \overline{)9\,5.1\,2}$

4. $8 \overline{)2\,5.9\,2}$

5. $1\,2 \overline{)8\,4.9\,6}$

6. $2\,3 \overline{)2\,5.0\,7}$

7. $1\,5 \overline{)1\,8}$

8. $3\,0 \overline{)5\,4}$

9. $5.4 \div (-6)$

10. $3.6 \div (-4)$

11. $-30 \div 0.005$

12. $-100 \div 0.0002$

13. $0.0\,6 \overline{)8.4}$

14. $0.0\,4 \overline{)1.6\,8}$

15. $2.6 \overline{)1\,0\,4}$

16. $3.2 \overline{)1\,9\,2}$

17. $1.8 \div (-12)$

18. $6 \div (-15)$

19. $8.5 \overline{)2\,7.2}$

20. $6.2 \overline{)4\,6.5}$

21. $-31.59 \div 8.1$

22. $-39.06 \div 4.2$

23. $-5 \div (-8)$

24. $-7 \div (-8)$

25. $0.4\,7 \overline{)0.1\,2\,2\,2}$

26. $0.5\,4 \overline{)0.2\,7}$

27. $0.0\,3\,2 \overline{)0.0\,7\,4\,8\,8}$

28. $0.0\,1\,7 \overline{)1.5\,8\,1}$

29. $-24.969 \div 82$

30. $-25.221 \div 42$

31. $\dfrac{213.4567}{100}$ **32.** $\dfrac{769.3265}{1000}$ **33.** $\dfrac{-23.59}{10}$ **34.** $\dfrac{-83.57}{10}$ **35.** $\dfrac{1.0237}{0.001}$ **36.** $\dfrac{3.4029}{0.001}$

37. $\dfrac{-92.36}{0.01}$ **38.** $\dfrac{-56.78}{0.001}$ **39.** $\dfrac{0.8172}{10}$ **40.** $\dfrac{0.5678}{1000}$ **41.** $\dfrac{0.97}{0.1}$ **42.** $\dfrac{0.97}{0.001}$

43. $\dfrac{52.7}{-1000}$ **44.** $\dfrac{8.9}{-100}$ **45.** $\dfrac{75.3}{-0.001}$ **46.** $\dfrac{63.47}{-0.1}$ **47.** $\dfrac{-75.3}{1000}$ **48.** $\dfrac{23,001}{100}$

b   Simplify.

49. $14 \times (82.6 + 67.9)$ **50.** $(26.2 - 14.8) \times 12$ **51.** $0.003 + 3.03 \div (-0.01)$

52. $42 \times (10.6 + 0.024)$ **53.** $(4.9 - 18.6) \times 13$ **54.** $4.2 \times 5.7 + 0.7 \div 3.5$

55. $210.3 - 4.24 \times 1.01$ **56.** $-7.32 + 0.04 \div 0.1^2$ **57.** $0.04 \times 0.1 \div 0.4 \times 50$

58. $30 \div 0.2 \times 0.4 \div 10$ **59.** $12 \div (-0.03) - 12 \times 0.03^2$ **60.** $(5 - 0.04)^2 \div 4 + 8.7 \times 0.4$

61. $(4 - 2.5)^2 \div 100 + 0.1 \times 6.5$ **62.** $4 \div 0.4 - 0.1 \times 5 + 0.1^2$

63. $6 \times 0.9 - 0.1 \div 4 + 0.2^3$ **64.** $5.5^2 \times [(6 - 7.8) \div 0.06 + 0.12]$

65. $12^2 \div (12 + 2.4) - [(2 - 2.4) \div 0.8]$ **66.** $0.01 \times \{[(4 - 0.25) \div 2.5] - (4.5 - 4.025)\}$

67. Mountain Peaks in Colorado. The elevations of four mountain peaks in Colorado are listed in the table below. Find the average elevation of these peaks.

MOUNTAIN PEAK	ELEVATION (in feet)
Mount Elbert	14,440
Mount Evans	14,271
Pikes Peak	14,115
Crested Butte	12,168

68. Driving Costs. The table below shows the cost per mile when specific types of vehicles are driven 15,000 miles in a year. Find the average cost per mile for the listed vehicles.

TYPE OF VEHICLE	COST PER MILE (in cents)
Small sedan	44.9
Medium sedan	58.5
Minivan	63.4
Large sedan	75.5
SUV 4WD	75.7

SOURCE: AAA

69. Camping in National Parks. The graph below shows the numbers of overnight camping stays in Park Service campgrounds in the National Park system from 2009 to 2013. Find the average number of stays per year during this period.

Camping in Park Service Campgrounds

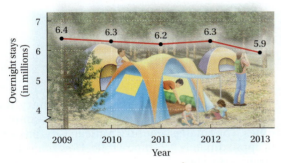

SOURCE: U.S. National Park Service

70. Life Expectancy. The information in the following bar graph shows the six highest life expectancies in the world. Determine the average life expectancy for these countries.

Life Expectancy at Birth as of 2013

SOURCE: *CIA World Factbook*

The following table lists the lengths of the longest railway tunnels in the world. Use the table for Exercises 71 and 72.

TUNNEL	Gotthard Base, Switzerland	Brenner Base, Austria	Sei-Kan, Japan	Eurotunnel, England-France	Lötschberg, Switzerland
LENGTH, IN MILES	35.5	34.4	33.5	31.3	21.5
LENGTH, IN KILOMETERS	57.1	55.4	53.9	50.5	34.6

SOURCE: lotsberg.net/data/rail

71. Find the average length of the tunnels, in miles.

72. Find the average length of the tunnels, in kilometers.

Skill Maintenance

Simplify. [3.5b]

73. $\dfrac{38}{146}$

74. $\dfrac{4r}{20r}$

Find the prime factorization. [3.2c]

75. 684

76. 2005

77. Add: $10\frac{1}{2} + 4\frac{5}{8}$. [4.6a]

78. Subtract: $10\frac{1}{2} - 4\frac{5}{8}$. [4.6b]

Evaluate. [1.9b]

79. 7^3

80. 2^6

Solve. [1.7b]

81. $235 = 5 \cdot z$

82. $q + 31 = 72$

Synthesis

Calculate each of the following.

83. ▦ $7.434 \div (-1.2) \times 9.5 + 1.47^2$

84. ▦ $-9.46 \times 2.1^2 \div 3.5 + 4.36$

85. ▦ $9.0534 - 2.041^2 \times 0.731 \div 1.043^2$

86. ▦ $23.042(7 - 4.037 \times 1.46 - 0.932^2)$

Solve.

87. $439.57 \times 0.01 \div 1000 \cdot x = 4.3957$

88. $5.2738 \div 0.01 \times 1000 \div t = 52.738$

89. $0.0329 \div 0.001 \times 10^4 \div x = 3290$

90. $-4.302 \times 0.1^2 \div 0.001 \cdot t = -430.2$

91. ▦ *Television Ratings.* A television rating point represents 1,150,000 households. The 2014 Super Bowl was viewed in approximately 53.7 million households. How many rating points did the game receive? Round to the nearest tenth.

Source: Nielsen Media Research

92. *Size of Country.* The world's largest country, Russia, has an area of approximately 6.6 million square miles. The smallest country, Vatican City, has an area of approximately 0.2 square mile. How many times larger is Russia?

Source: U.S. Bureau of the Census, International Data Base

▦ *Electric Bills.* Recently, electric bills from the Central Vermont Public Service Corporation consisted of a "customer charge" of $0.374 per day plus an "energy charge" of $0.1174 per kilowatt-hour (kWh) for the first 250 kWh used and $0.09079 per kilowatt-hour for each kilowatt-hour in excess of 250.

93. From August 20 to September 20, the Kaufmans' bill was $59.10. How many kilowatt-hours of electricity did they use? (Round to the nearest kilowatt-hour.)

94. From July 20 to August 20, the McGuires' bill was $70. How many kilowatt-hours of electricity did they use? (Round to the nearest kilowatt-hour.)

Concept Reinforcement

Determine whether each statement is true or false.

_____ **1.** In the number 308.00567, the digit 6 names the tens place. [5.1a]

_____ **2.** When writing a word name for decimal notation, we write the word "and" for the decimal point. [5.1a]

_____ **3.** On the number line, -2.3 is to the left of -2.2. [5.1c]

Guided Solutions

 Fill in each blank with the number that creates a correct statement or solution.

4. Evaluate $P(1 + r)$ for $P = 5000$ and $r = 0.045$. [5.3c]

$P(1 + r) =$ [____] $(1 +$ [____] $)$ Substituting

$= 5000($ [____] $)$ Adding within parentheses

$=$ [____]   Multiplying

5. Simplify: $5.6 + 4.3 \times (6.5 - 0.25)^2$. [5.4b]

$5.6 + 4.3 \times (6.5 - 0.25)^2 = 5.6 + 4.3 \times ($ [____] $)^2$ Carrying out the operation inside parentheses

$= 5.6 + 4.3 \times$ [____]   Evaluating the exponential expression

$= 5.6 +$ [____]   Multiplying

$=$ [____]   Adding

Mixed Review

6. *Mile Run Record.* The difference between the men's record for the mile run, held by Hicham El Guerrouj of Morocco (with a time of 3 min 43.13 sec) and the women's record for the mile run, held by Svetlana Masterkova of Russia (with a time of 4 min 12.56 sec) is 29.43 sec. Write a word name for 29.43. [5.1a]

Source: International Association of Athletics Federations (iaaf.org)

7. *Skin Allergies.* Skin allergies are the most common allergies among children. In 2010, 9.4 million children in the United States suffered from skin allergies. Convert 9.4 million to standard notation. [5.3b]

Source: CDC National Center for Health Statistics

Write each number as a fraction and, if possible, as a mixed numeral. [5.1b]

8. 4.53

9. 0.287

Which number is larger? [5.1c]

10. 0.07, 0.13

11. $-5.2, -5.09$

Write decimal notation. [5.1b]

12. $\dfrac{7}{10}$

13. $\dfrac{639}{100}$

14. $-35\dfrac{67}{100}$

15. $8\dfrac{2}{1000}$

Round 28.4615 to the nearest [5.1d]

16. Thousandth.

17. Hundredth.

18. Tenth.

19. One.

Add. [5.2a], [5.2c]

20.
```
  4 7.6 3 8
+    2.4 5 7
```

21.
```
      1 5.6
    2 3 4.7 2 9
        3.0 8
+   9 6 1.4 5 3
```

22. $-10.5 + 0.27$

23. $16 + 0.34 + 1.9$

Subtract. [5.2b], [5.2c]

24.
```
  3 2 1.5 7
−    4 9.3 8
```

25.
```
    5.6
−   0.0 0 7
```

26. $34.3 - 18.75$

27. $-6.9 - 13$

Multiply. [5.3a]

28.
```
    4.6
×   0.9
```

29.
```
    1 5.3
×     6.0 7
```

30. 100×81.236

31. $0.1 \times (-0.483)$

Divide. [5.4a]

32. $-20.24 \div (-4)$

33. $21.76 \div 6.8$

34. $76.3 \div 0.1$

35. $914.036 \div 1000$

36. Convert \$20.45 to cents.   [5.3b]

37. Convert 147¢ to dollars.   [5.3b]

38. Combine like terms: $3.08x - 7.1 - 4.3x$.   [5.2d]

39. Evaluate $2(l + w)$ for $l = 1.3$ and $w = 0.8$.   [5.3c]

Simplify. [5.4b]

40. $6.594 + 0.5318 \div 0.01$

41. $7.3 \times 4.6 - 0.8 \div 3.2$

Understanding Through Discussion and Writing

42. A classmate rounds 236.448 to the nearest one and gets 237. Explain the possible error.   [5.1d]

43. Explain the error in the following:
Subtract.
$$73.089 - 5.0061 = 2.3028 \quad [5.2b]$$

44. Explain why $10 \div 0.2 = 100 \div 2$.   [5.4a]

45. Kayla made these two computational mistakes:
$$0.247 \div 0.1 = 0.0247; \quad 0.247 \div 10 = 2.47.$$
In each case, how could you convince her that a mistake has been made?   [5.4a]

Using Fraction Notation with Decimal Notation

5.5

a   USING DIVISION TO FIND DECIMAL NOTATION

Recall that $\frac{a}{b}$ means $a \div b$. Thus, using division, we can express *any* fraction as a decimal. This means that any *rational* number (ratio of integers) can be written as a decimal.

EXAMPLE 1   Find decimal notation for $\frac{3}{20}$.

Because $\frac{3}{20}$ means $3 \div 20$, we can perform long division.

$$\begin{array}{r} 0.1\,5 \\ 20\overline{)3.0\,0} \\ \underline{2\;0} \\ 1\;0\;0 \\ \underline{1\;0\;0} \\ 0 \end{array}$$

We are finished when the remainder is 0.

We have $\frac{3}{20} = 0.15$.

EXAMPLE 2   Find decimal notation for $\frac{-7}{8}$.

Since $\frac{-7}{8}$ means $-7 \div 8$ and a negative number divided by a positive number is negative, we know that the decimal will be negative. We divide 7 by 8 and make the result negative.

$$\begin{array}{r} 0.8\,7\,5 \\ 8\overline{)7.0\,0\,0} \\ \underline{6\;4} \\ 6\;0 \\ \underline{5\;6} \\ 4\;0 \\ \underline{4\;0} \\ 0 \end{array}$$

Thus, $\frac{-7}{8} = -0.875$.

Do Margin Exercises 1 and 2. ▶

When division with decimals ends with a remainder of 0, or *terminates*, as in Examples 1 and 2, the result is called a **terminating decimal**. If the division does not terminate, the result will be a **repeating decimal**.

OBJECTIVES

a   Use division to convert fraction notation to decimal notation.

b   Round numbers named by repeating decimals.

c   Convert certain fractions to decimal notation by using equivalent fractions.

d   Simplify expressions that contain both fraction notation and decimal notation.

SKILL TO REVIEW

Objective 5.4a: Divide using decimal notation.

Divide.

1. $3 \div 4$ **2.** $25 \div 8$

Find decimal notation.

1. $\dfrac{2}{5}$ **2.** $\dfrac{-5}{8}$

Answers

Skill to Review:
1. 0.75   **2.** 3.125

Margin Exercises:
1. 0.4   **2.** −0.625

Find decimal notation.

3. $\dfrac{1}{6}$ GS

$$\dfrac{1}{6} = \boxed{} \div 6$$

$$\begin{array}{r} 0.1\ \boxed{}\ 6 \\ \boxed{})\overline{1.0\ 0\ 0} \\ \underline{6} \\ 4\ 0 \\ 3\ 6 \\ \boxed{}\ 0 \\ 3\ 6 \\ \boxed{} \end{array}$$

$$\dfrac{1}{6} = 0.1666\ldots = 0.1\overline{6}$$

4. $\dfrac{2}{3}$

EXAMPLE 3 Find decimal notation for $\frac{5}{6}$.

Since $\frac{5}{6}$ means $5 \div 6$, we have

$$\begin{array}{r} 0.8\ 3\ 3 \\ 6)\overline{5.0\ 0\ 0} \\ \underline{4\ 8} \\ 2\ 0 \\ \underline{1\ 8} \\ 2\ 0 \\ \underline{1\ 8} \\ 2 \end{array}$$

Since 2 keeps reappearing as a remainder, the digits repeat and will continue to do so; therefore,

$$\dfrac{5}{6} = 0.83333\ldots\ .$$ The dots indicate an endless sequence of repeating digits in the quotient.

When there is a repeating pattern, we often use an overbar to indicate the repeating part, in this case, only the 3.

$$\dfrac{5}{6} = 0.8\overline{3}$$

◀ **Do Exercises 3 and 4.**

EXAMPLE 4 Find decimal notation for $-\frac{4}{11}$.

Since $-\frac{4}{11}$ is negative, we divide 4 by 11 and make the result negative.

$$\begin{array}{r} 0.3\ 6\ 3\ 6 \\ 1\ 1)\overline{4.0\ 0\ 0\ 0} \\ \underline{3\ 3} \\ 7\ 0 \\ \underline{6\ 6} \\ 4\ 0 \\ \underline{3\ 3} \\ 7\ 0 \\ \underline{6\ 6} \\ 4 \end{array}$$

Since 7 and 4 keep repeating as remainders, the sequence of digits "36" repeats in the quotient, and

$$\dfrac{4}{11} = 0.363636\ldots,\quad \text{or}\quad 0.\overline{36}.$$

Find decimal notation.

5. $\dfrac{5}{11}$ **6.** $-\dfrac{12}{11}$

Thus, $-\frac{4}{11} = -0.\overline{36}$.

◀ **Do Exercises 5 and 6.**

When a fraction is written in simplified form, we can tell from the denominator whether its decimal notation will repeat or terminate.

For a fraction in simplified form,

- if the denominator has a prime factor other than 2 or 5, the decimal notation repeats.

- if the denominator has no prime factor other than 2 or 5, the decimal notation terminates.

EXAMPLE 5 Find decimal notation for $\frac{3}{7}$.

Because 7 is not a product of 2's and/or 5's, we expect a repeating decimal.

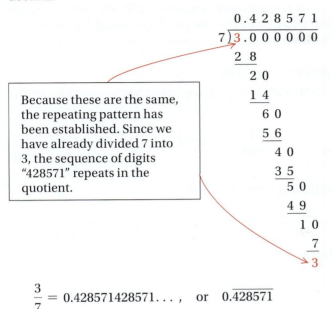

Because these are the same, the repeating pattern has been established. Since we have already divided 7 into 3, the sequence of digits "428571" repeats in the quotient.

$$\frac{3}{7} = 0.428571428571\ldots, \quad \text{or} \quad 0.\overline{428571}$$

Do Exercise 7. ▶

CALCULATOR CORNER

Recognizing Repeating Decimals   Short repeating patterns can usually be recognized on a calculator's display. For example, to convert $\frac{1}{11}$ to decimal notation, press $\boxed{1}\ \boxed{\div}\ \boxed{1}\ \boxed{1}\ \boxed{=}$. The result for a 10-digit display is $\boxed{.0909090909}$, so the decimal notation is $0.\overline{09}$.

For $-\frac{12}{11}$, the same calculator displays $\boxed{-1.090909091}$. Note that the last digit has been rounded from 0 to 1. The decimal notation for $-\frac{12}{11}$ is $-1.\overline{09}$.

EXERCISES   Convert to decimal notation.

1. $-\dfrac{1}{6}$ **2.** $\dfrac{7}{11}$

3. $\dfrac{19}{3}$ **4.** $-\dfrac{514}{9}$

b ROUNDING REPEATING DECIMALS

The repeating part of a decimal can be so long that it will not fit on a calculator display. For example, when $\frac{5}{97}$ is written as a decimal, its repeating part is 96 digits long! Most calculators round repeating decimals to 9 or 10 decimal places.

In applied problems, repeating decimals are rounded to get approximate answers. To round a repeating decimal, we can extend the decimal notation at least one place past the rounding digit and then round as before.

EXAMPLES   Round each to the nearest tenth, hundredth, and thousandth.

	Nearest tenth	*Nearest hundredth*	*Nearest thousandth*
6. $0.8\overline{3} = 0.83333\ldots$	0.8	0.83	0.833
7. $3.\overline{09} = 3.090909\ldots$	3.1	3.09	3.091
8. $-4.1\overline{763} = -4.1763763\ldots$	-4.2	-4.18	-4.176

Do Exercises 8–11. ▶

EXAMPLE 9 *Gas Mileage.*   A car travels 457 mi on 16.4 gal of gasoline. The ratio of number of miles driven to amount of gasoline used is *gas mileage*. Find the gas mileage, and convert the ratio to decimal notation rounded to the nearest tenth.

$$\frac{\text{Miles driven}}{\text{Gasoline used}} = \frac{457}{16.4} \approx 27.86 \qquad \text{Dividing to 2 decimal places}$$

$$\approx 27.9 \qquad \text{Rounding to 1 decimal place}$$

The gas mileage is 27.9 miles per gallon, or 27.9 mpg.

Do Exercise 12. ▶

7. Find decimal notation for $\dfrac{5}{7}$.

Round each to the nearest tenth, hundredth, and thousandth.

8. $0.\overline{6}$ **9.** $0.6\overline{08}$

10. $-7.3\overline{49}$ **11.** $2.6\overline{891}$

12. *Gas Mileage.*   A car travels 380 mi on 15.7 gal of gasoline. Find the gasoline mileage, and convert the ratio to decimal notation rounded to the nearest tenth.

Answers

7. $0.\overline{714285}$ **8.** 0.7; 0.67; 0.667
9. 0.6; 0.61; 0.608   **10.** -7.3; -7.35; -7.349
11. 2.7; 2.69; 2.689   **12.** 24.2 miles per gallon

Multiply by a form of 1 to find decimal notation for each number.

13. $\dfrac{4}{5}$ **14.** $-\dfrac{9}{20}$

15. $\dfrac{7}{200}$ **16.** $\dfrac{33}{25}$

c   MORE WITH CONVERSIONS

Fractions like $\frac{3}{10}$ or $-\frac{71}{1000}$ can be converted to decimal notation without using long division. When a denominator is a factor of 10, 100, and so on, we can convert to decimal notation by finding (perhaps mentally) an equivalent fraction in which the denominator is a power of 10.

EXAMPLE 10 Find decimal notation for $-\frac{7}{500}$.

Since $500 \cdot 2 = 1000$, and 1000 is a power of 10, we use $\frac{2}{2}$ as an expression for 1.

$$-\frac{7}{500} = -\frac{7}{500} \cdot \frac{2}{2} = -\frac{14}{1000} = -0.014 \qquad \textit{Think: } 1000 \div 500 = 2, \text{ and } 7 \cdot 2 = 14.$$

EXAMPLE 11 Find decimal notation for $\frac{9}{25}$.

$$\frac{9}{25} = \frac{9}{25} \cdot \frac{4}{4} = \frac{36}{100} = 0.36 \qquad \textit{Using } \frac{4}{4} \text{ for 1 to get a denominator of 100}$$

As a check, we can divide.

```
      0.3 6
2 5 )9.0 0
      7 5          Note that multiplication by 1 is much faster.
      1 5 0
      1 5 0
          0
```

EXAMPLE 12 Find decimal notation for $\frac{7}{4}$.

$$\frac{7}{4} = \frac{7}{4} \cdot \frac{25}{25} = \frac{175}{100} = 1.75 \qquad \textit{Using } \frac{25}{25} \text{ for 1 to get a denominator of 100. You might also note that 7 quarters is \$1.75.}$$

◄ **Do Exercises 13–16.**

d   CALCULATIONS WITH FRACTION AND DECIMAL NOTATION TOGETHER

Fraction notation and decimal notation can occur together in a calculation. In such cases, there are at least three ways in which we might proceed.

EXAMPLE 13 Calculate: $\frac{2}{3} \times 0.576$.

Method 1: Perhaps the quickest method is to treat 0.576 as $\frac{0.576}{1}$. Then we multiply 0.576 by 2 and divide the result by 3.

$$\frac{2}{3} \times 0.576 = \frac{2}{3} \times \frac{0.576}{1}$$

$$= \frac{2 \times 0.576}{3} = \frac{1.152}{3}$$

$$= 0.384$$

```
      0.3 8 4
3 )1.1 5 2
      9
      2 5
      2 4
        1 2
        1 2
          0
```

Method 2: A second way to do this calculation is to convert the fraction notation to decimal notation so that both numbers are in decimal notation. Since $\frac{2}{3}$ converts to repeating decimal notation, it is first rounded to some chosen decimal place. We choose three decimal places because 0.576 has three decimal places. Then, using decimal notation, we multiply.

$$\frac{2}{3} \times 0.576 = 0.\overline{6} \times 0.576 \approx 0.667 \times 0.576 = 0.384192$$

Method 3: A third method is to convert the decimal notation to fraction notation so that both numbers are in fraction notation. The answer can be left in fraction notation and simplified, or we can convert back to decimal notation and, if appropriate, round.

$$\frac{2}{3} \times 0.576 = \frac{2}{3} \cdot \frac{576}{1000} = \frac{2 \cdot 576}{3 \cdot 1000}$$

$$= \frac{2 \cdot 2 \cdot 2 \cdot 2 \cdot 2 \cdot 2 \cdot 2 \cdot 3 \cdot 3}{2 \cdot 2 \cdot 2 \cdot 3 \cdot 5 \cdot 5 \cdot 5} \quad \color{red}{\text{Factoring}}$$

$$= \frac{2 \cdot 2 \cdot 2 \cdot 3}{2 \cdot 2 \cdot 2 \cdot 3} \cdot \frac{2 \cdot 2 \cdot 2 \cdot 2 \cdot 3}{5 \cdot 5 \cdot 5} \quad \color{red}{\text{Removing a factor equal to 1: } \frac{2 \cdot 2 \cdot 2 \cdot 3}{2 \cdot 2 \cdot 2 \cdot 3} = 1}$$

$$= \frac{2 \cdot 2 \cdot 2 \cdot 2 \cdot 3}{5 \cdot 5 \cdot 5} = \frac{48}{125}, \text{ or } 0.384$$

Note that we get an exact answer with methods 1 and 3, but method 2 gives an approximation since we rounded decimal notation for $\frac{2}{3}$.

Do Exercises 17 and 18. ▶

EXAMPLE 14 *Boating.* A triangular sail on a single-sail day cruiser is 3.4 m wide and 4.2 m tall. Find the area of the sail.

1. **Familiarize.** We make a drawing and recall that the formula for the area, A, of a triangle with base b and height h is $A = \frac{1}{2}bh$.

2. **Translate.** We substitute 3.4 for b and 4.2 for h.

$$A = \frac{1}{2}bh$$

$$= \frac{1}{2}(3.4)(4.2) \quad \color{red}{\text{Substituting}}$$

3. **Solve.** We simplify as follows.

$$A = \frac{1}{2}(3.4)(4.2)$$

$$= \frac{3.4}{2}(4.2) \quad \color{red}{\text{Multiplying } \frac{1}{2} \text{ and } \frac{3.4}{1}}$$

$$= 1.7(4.2) \quad \color{red}{\text{Dividing}}$$

$$= 7.14 \quad \color{red}{\text{Multiplying}}$$

4. **Check.** To check, we repeat the calculations using the commutative law. We also rewrite $\frac{1}{2}$ as 0.5.

$$\frac{1}{2}(3.4)(4.2) = 0.5(4.2)(3.4) = (2.1)(3.4) = 7.14$$

Our answer checks.

5. **State.** The area of the sail is 7.14 m² (square meters).

Do Exercise 19. ▶

Calculate.

 17. $\frac{3}{4} \times 0.62$.

Method 1:

$$\frac{3}{4} \times 0.62 = \frac{3}{4} \times \frac{0.62}{\boxed{}}$$

$$= \frac{\boxed{}}{4} = 0.465$$

Method 2:

$$\frac{3}{4} \times 0.62 = \boxed{} \times 0.62$$

$$= 0.465$$

Method 3:

$$\frac{3}{4} \times 0.62 = \frac{3}{4} \cdot \frac{62}{\boxed{}}$$

$$= \frac{\boxed{}}{400}$$

$$= \frac{\boxed{}}{200} = 0.465$$

18. $\frac{1}{3} \times 0.384 + \frac{5}{8} \times 0.6784$

19. Find the area of a triangular window that is 3.25 ft wide and 2.6 ft tall.

Answers

17. $\frac{93}{200}$, or 0.465   **18.** 0.552   **19.** 4.225 ft²

Guided Solution:
17. 1, 1.86; 0.75; 100, 186, 93

✓ Reading Check

Determine whether the decimal notation for each fraction is terminating or repeating.

RC1. $\dfrac{4}{9}$ _____

RC2. $\dfrac{3}{32}$ _____

RC3. $\dfrac{39}{40}$ _____

RC4. $\dfrac{7}{12}$ _____

RC5. $\dfrac{2}{11}$ _____

RC6. $\dfrac{80}{125}$ _____

a , c   Find decimal notation for each number.

1. $\dfrac{3}{8}$

2. $\dfrac{3}{5}$

3. $\dfrac{-1}{2}$

4. $\dfrac{-1}{4}$

5. $\dfrac{3}{25}$

6. $\dfrac{7}{20}$

7. $\dfrac{9}{40}$

8. $\dfrac{3}{40}$

9. $\dfrac{13}{25}$

10. $\dfrac{17}{25}$

11. $\dfrac{-17}{20}$

12. $\dfrac{-13}{20}$

13. $-\dfrac{9}{16}$

14. $-\dfrac{5}{16}$

15. $\dfrac{7}{5}$

16. $\dfrac{3}{2}$

17. $\dfrac{28}{25}$

18. $\dfrac{31}{20}$

19. $\dfrac{11}{-8}$

20. $\dfrac{17}{-10}$

21. $-\dfrac{39}{40}$

22. $-\dfrac{17}{40}$

23. $\dfrac{121}{200}$

24. $\dfrac{32}{125}$

25. $\dfrac{8}{15}$

26. $\dfrac{7}{9}$

27. $\dfrac{1}{3}$

28. $\dfrac{1}{9}$

29. $\dfrac{-4}{3}$

30. $\dfrac{-8}{9}$

31. $\dfrac{7}{6}$

32. $\dfrac{7}{11}$

33. $-\dfrac{14}{11}$

34. $-\dfrac{7}{11}$

35. $\dfrac{-5}{12}$

36. $\dfrac{-11}{12}$

37. $\dfrac{127}{500}$ **38.** $\dfrac{83}{500}$ **39.** $\dfrac{4}{33}$ **40.** $\dfrac{5}{33}$

41. $\dfrac{-12}{55}$ **42.** $\dfrac{-5}{22}$ **43.** $\dfrac{4}{7}$ **44.** $\dfrac{2}{7}$

b   Round each to the nearest tenth, hundredth, and thousandth.

45. $0.\overline{18}$ **46.** $0.\overline{83}$ **47.** $0.2\overline{7}$ **48.** $3.5\overline{4}$

For Exercises 49–60, round the decimal notation for each number to the nearest tenth, hundredth, and thousandth.

49. $\dfrac{4}{11}$ **50.** $\dfrac{3}{11}$ **51.** $-\dfrac{5}{3}$ **52.** $-\dfrac{19}{16}$

53. $\dfrac{-8}{17}$ **54.** $\dfrac{-7}{13}$ **55.** $\dfrac{7}{12}$ **56.** $\dfrac{2}{15}$

57. $\dfrac{29}{-150}$ **58.** $\dfrac{37}{-150}$ **59.** $\dfrac{7}{-9}$ **60.** $\dfrac{5}{-13}$

61. For this set of people, what is the ratio, in decimal notation rounded to the nearest thousandth, where appropriate, of:

   a) women to the total number of people?
   b) women to men?
   c) men to the total number of people?
   d) men to women?

62. For this set of pennies and quarters, what is the ratio, in decimal notation rounded to the nearest thousandth, where appropriate, of:

   a) pennies to quarters?
   b) quarters to pennies?
   c) pennies to total number of coins?
   d) total number of coins to pennies?

Gas Mileage. In each of Exercises 63–66, find the gas mileage rounded to the nearest tenth.

63. 285 mi; 18 gal

64. 396 mi; 17 gal

65. 324.8 mi; 18.2 gal

66. 264.8 mi; 12.7 gal

d Calculate and write the result as a decimal.

67. $\dfrac{7}{8} \times 12.64$

68. $\dfrac{4}{5} \times 384.8$

69. $6.84 \div 2\dfrac{1}{2}$

70. $8\dfrac{1}{2} \div 2.125$

71. $\dfrac{47}{9}(-79.95)$

72. $\dfrac{7}{11}(-2.7873)$

73. $\dfrac{1}{2} - 0.5$

74. $3\dfrac{1}{8} - 2.75$

75. $\left(\dfrac{1}{6}\right)0.0765 + \left(\dfrac{3}{4}\right)0.1124$

76. $\left(\dfrac{2}{5}\right)6384.1 - \left(\dfrac{5}{8}\right)156.56$

77. $\dfrac{3}{4} \times 2.56 - \dfrac{7}{8} \times 3.94$

78. $\dfrac{2}{5} \times 3.91 - \dfrac{7}{10} \times 4.15$

79. $5.2 \times 1\dfrac{7}{8} \div 0.4$

80. $4\dfrac{3}{4} \times 0.5 \div 0.1$

Solve.

81. Find the area of a triangular shawl that is 1.8 m long and 1.2 m wide.

82. Find the area of a triangular sign that is 1.5 m wide and 1.5 m tall.

83. Find the area of a triangular stamp that is 3.4 cm wide and 3.4 cm tall.

84. Find the area of a triangular reflector that is 7.4 cm wide and 9.1 cm tall.

85. Find the area of the kite shown on the left.

86. Find the area of the kite shown on the right.

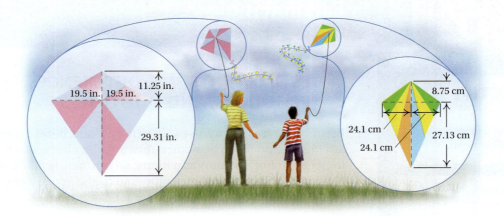

Skill Maintenance

Calculate.

87. $9 \cdot 2\frac{1}{3}$ [4.7a]

88. $16\frac{1}{10} - 14\frac{3}{5}$ [4.6b]

89. $84 \div 8\frac{2}{5}$ [4.7b]

90. $14\frac{3}{5} + 16\frac{1}{10}$ [4.6a]

Solve. [4.6c]

91. A recipe for bread calls for $\frac{2}{3}$ cup of water, $\frac{1}{4}$ cup of milk, and $\frac{1}{8}$ cup of oil. How many cups of liquid ingredients does the recipe call for?

92. A board $\frac{7}{10}$ in. thick is glued to a board $\frac{3}{5}$ in. thick. The glue is $\frac{3}{100}$ in. thick. How thick is the result?

Synthesis

93. 🔲 Find decimal notation for $\frac{1}{7}, \frac{2}{7}, \frac{3}{7}, \frac{4}{7}$, and $\frac{5}{7}$. Observe the pattern and predict the decimal notation for $\frac{6}{7}$. Check your answer on a calculator.

94. 🔲 Find decimal notation for $\frac{1}{9}, \frac{1}{99}$, and $\frac{1}{999}$. Observe the pattern and predict the decimal notation for $\frac{1}{9999}$. Check your answer on a calculator.

The formula $A = \pi r^2$ is used to find the area, A, of a circle with radius r. For Exercises 95 and 96, find the area of a circle with the given radius, using $\frac{22}{7}$ for π. For Exercises 97 and 98, use 3.14 for π or a calculator with a π key.

95. $r = 2.1$ cm

96. $r = 1.4$ cm

97. 🔲 $r = \frac{3}{4}$ ft

98. 🔲 $r = 4\frac{1}{2}$ yd

5.6 Estimating

OBJECTIVE

a Estimate sums, differences, products, and quotients.

SKILL TO REVIEW

Objective 1.6b: Estimate sums, differences, products, and quotients by rounding.

Estimate by first rounding to the nearest ten.

1.
```
  4 6 7
− 2 8 4
```

2.
```
    5 4
×   2 9
```

a ESTIMATING SUMS, DIFFERENCES, PRODUCTS, AND QUOTIENTS

Estimating has many uses. It can be done before a problem is even attempted and it can be done afterward as a check, even when we are using a calculator. In many situations, an estimate is all we need. We usually estimate by rounding the numbers so that there are one or two nonzero digits. Consider the following prices for Examples 1–3.

$289.95

8.2 Megapixel Digital Camera

Four Burner Gas Grill

$139.97

19" Flat-Panel HDTV

$449.99

EXAMPLE 1 Estimate by rounding to the nearest ten the total cost of one grill and one TV.

We are estimating the sum

$289.95 + $449.99 = Total cost.

The estimate found by rounding the addends to the nearest ten is

$290 + $450 = $740.   (Estimated total cost)

◀ Do Margin Exercise 1.

EXAMPLE 2 About how much more does the TV cost than the camera? Estimate by rounding to the nearest ten.

We are estimating the difference

$449.99 − $139.97 = Price difference.

The estimate found by rounding each price to the nearest ten is

$450 − $140 = $310.   (Estimated price difference)

◀ Do Exercise 2.

1. Estimate by rounding to the nearest ten the total cost of one grill and one camera. Which of the following is an appropriate estimate?

   a) $43   b) $400
   c) $410   d) $430

2. About how much more does the TV cost than the grill? Estimate by rounding to the nearest ten. Which of the following is an appropriate estimate?

   a) $100   b) $150
   c) $160   d) $300

Answers

Skill to Review:
1. 190   2. 1500

Margin Exercises:
1. (d)   2. (c)

EXAMPLE 3 Estimate the total cost of 4 cameras.

We are estimating the product

$$4 \times \$139.97 = \text{Total cost}.$$

The estimate is found by rounding 139.97 to the nearest ten:

$$4 \times \$140 = \$560.$$

Do Exercise 3. ▶

EXAMPLE 4 A student government group is planning an event for first-year students. Since the local weather is often rainy, the group decides to purchase umbrellas bearing the university logo for door prizes. About how many umbrellas that cost $39.99 each can be purchased for $356?

We are estimating the quotient

$$\$356 \div \$39.99.$$

Since we want a whole-number estimate, we need to round appropriately. Rounding $39.99 to the nearest one, we get $40. Since $356 is close to $360, which is a multiple of $40, we estimate

$$\$360 \div \$40 = 9.$$

The answer is about 9 umbrellas.

Do Exercise 4. ▶

When estimating, we usually look for numbers that are easy to work with. For example, if multiplying, we might round 0.43 to 0.5 and 8.9 to 10, because 0.5 and 10 are convenient numbers to multiply.

EXAMPLE 5 Estimate: 4.8×52. Do not find the actual product. Which of the following is an appropriate estimate?

a) 25   **b)** 250   **c)** 2500   **d)** 360

We round 4.8 to the nearest one and 52 to the nearest ten:

$$5 \times 50 = 250. \quad \text{(Estimated product)}$$

Thus, an approximate estimate is (b).

Other estimates that we might have used in Example 5 are

$$5 \times 52 = 260 \quad \text{or} \quad 4.8 \times 50 = 240.$$

The estimate in Example 5, $5 \times 50 = 250$, is the easiest to do because the factors have the fewest nonzero digits. You could probably do it mentally. In general, we try to round so that a computation has as few nonzero digits as possible while still keeping the estimated value close to the original value.

Do Exercises 5–10. ▶

EXAMPLE 6 Estimate: $82.08 \div 24$. Which of the following is an appropriate estimate?

a) 400   **b)** 16   **c)** 40   **d)** 4

This is about $80 \div 20$, so the answer is about 4. Thus, an appropriate estimate is (d).

3. Estimate the total cost of 6 TVs. Which of the following is an appropriate estimate?

a) $450   **b)** $2700
c) $4500   **d)** $27,000

4. Refer to the umbrella price in Example 4. About how many umbrellas can be purchased for $675?

Estimate each product. Do not find the actual product. Which of the following is an appropriate estimate?

5. 2.4×8

a) 16   **b)** 34
c) 125   **d)** 5

6. 24×0.6

a) 200   **b)** 5
c) 110   **d)** 20

7. 0.86×0.432

a) 0.04   **b)** 0.4
c) 1.1   **d)** 4

8. 0.82×0.1

a) 800   **b)** 8
c) 0.08   **d)** 80

9. 0.12×18.248

a) 180   **b)** 1.8
c) 0.018   **d)** 18

10. 24.234×5.2

a) 200   **b)** 120
c) 12.5   **d)** 234

Answers

3. (b)   **4.** 17 umbrellas   **5.** (a)   **6.** (d)
7. (b)   **8.** (c)   **9.** (b)   **10.** (b)

Estimate each quotient. Which of the following is an appropriate estimate?

11. 59.78 ÷ 29.1

  a) 200   **b)** 20

  c) 2   **d)** 0.2

12. 82.08 ÷ 2.4

  a) 40   **b)** 4.0

  c) 400   **d)** 0.4

13. 0.1768 ÷ 0.08

  a) 8   **b)** 10

  c) 2   **d)** 20

EXAMPLE 7 Estimate: 94.18 ÷ 3.2. Which of the following is an appropriate estimate?

a) 30   **b)** 300   **c)** 3   **d)** 60

    This is about 90 ÷ 3, so the answer is about 30. Thus, an appropriate estimate is (a).

EXAMPLE 8 Estimate: 0.0156 ÷ 1.3. Which of the following is an appropriate estimate?

a) 0.2   **b)** 0.002   **c)** 0.02   **d)** 20

    This is about 0.02 ÷ 1, so the answer is about 0.02. Thus, an appropriate estimate is (c).

◀ **Do Exercises 11–13.**

    In some cases, it is easier to estimate a quotient by checking products rather than by rounding the divisor and the dividend.

EXAMPLE 9 Estimate: 0.0074 ÷ 0.23. Which of the following is an appropriate estimate?

a) 0.3   **b)** 0.03   **c)** 300   **d)** 3

    We estimate 3 for a quotient. We check by multiplying.

$$0.23 \times 3 = 0.69$$

We make the estimate smaller. We estimate 0.3 and check by multiplying.

$$0.23 \times 0.3 = 0.069$$

We make the estimate smaller. We estimate 0.03 and check by multiplying.

$$0.23 \times 0.03 = 0.0069$$

This is about 0.0074, so the quotient is about 0.03. Thus, an appropriate estimate is (b).

14. Estimate: 0.0069 ÷ 0.15. Which of the following is an appropriate estimate?

  a) 0.5   **b)** 50

  c) 0.05   **d)** 0.004

◀ **Do Exercise 14.**

Answers

11. (c)   **12.** (a)   **13.** (c)   **14.** (c)

5.6   Exercise Set

For Extra Help

MyMathLab®

MathXL®

PRACTICE   WATCH   READ   REVIEW

☑ Reading Check

Match each calculation with the most appropriate estimate from the list at the right.

RC1. 0.1003 × 0.8 _____   **a)** 0.8

RC2. 38.41 + 41.777 _____   **b)** 0.08

RC3. 0.00152 × 4025 _____   **c)** 8

RC4. 1632 ÷ 1.9 _____   **d)** 80

RC5. 9.054 − 8.3111 _____   **e)** 800

RC6. 162,105 × 0.0496 _____   **f)** 8000

For Exercises 1–8, use the prices shown below to estimate each sum, difference, product, or quotient. Indicate which of the choices is an appropriate estimate.

2 Quart Ice Cream Maker $79.⁹⁹

Satellite Radio Receiver $149.⁹⁹

Black & White Multifunction Printer/Copier/Scanner $279.⁸⁹

1. Estimate the total cost of one printer and one satellite radio.
   a) $43   b) $4300   c) $360   d) $430

2. Estimate the total cost of one satellite radio and one ice cream maker.
   a) $230   b) $23   c) $2300   d) $400

3. About how much more does the printer cost than the satellite radio?
   a) $1300   b) $200   c) $130   d) $13

4. About how much more does the satellite radio cost than the ice cream maker?
   a) $7000   b) $70   c) $130   d) $700

5. Estimate the total cost of 6 ice cream makers.
   a) $480   b) $48   c) $240   d) $4800

6. Estimate the total cost of 4 printers.
   a) $1200   b) $1120   c) $11,200   d) $600

7. About how many ice cream makers can be purchased for $830?
   a) 120   b) 100   c) 10   d) 1000

8. About how many printers can be purchased for $5627?
   a) 200   b) 20   c) 1800   d) 2000

Estimate by rounding as directed.

9. 0.02 + 1.31 + 0.34;
   nearest tenth

10. 0.88 + 2.07 + 1.54;
    nearest one

11. 6.03 + 0.007 + 0.214;
    nearest one

12. 1.11 + 8.888 + 99.94;
    nearest one

13. 52.367 + 1.307 + 7.324;
    nearest one

14. 12.9882 + 1.2115;
    nearest tenth

15. 2.678 − 0.445;
nearest tenth

16. 12.9882 − 1.0115;
nearest one

17. 198.67432 − 24.5007;
nearest ten

Estimate. Choose a rounding digit that gives one or two nonzero digits. Indicate which of the choices is an appropriate estimate.

18. 234.12321 − 200.3223
  a) 600   **b)** 60
  c) 300   **d)** 30

19. 49 × 7.89
  a) 400   **b)** 40
  c) 4   **d)** 0.4

20. 7.4 × 8.9
  a) 95   **b)** 63
  c) 124   **d)** 6

21. 98.4 × 0.083
  a) 80   **b)** 12
  c) 8   **d)** 0.8

22. 78 × 5.3
  a) 400   **b)** 800
  c) 40   **d)** 8

23. 3.6 ÷ 4
  a) 10   **b)** 1
  c) 0.1   **d)** 0.01

24. 0.0713 ÷ 1.94
  a) 3.5   **b)** 0.35
  c) 0.035   **d)** 35

25. 74.68 ÷ 24.7
  a) 9   **b)** 3
  c) 12   **d)** 120

26. 914 ÷ 0.921
  a) 10   **b)** 100
  c) 1000   **d)** 1

27. *Fence Posts.* Melanie needs to build a fence around a new horse pasture. The perimeter of the area to be fenced is 1760 ft. Estimate the number of wooden fence posts needed if the posts are placed 8.625 ft apart.

28. *Ticketmaster.* Recently, Ticketmaster stock sold for $22.25 per share. Estimate how many shares can be purchased for $4400.

29. *Day-Care Supplies.* Helen wants to buy 12 boxes of crayons at $1.89 per box for the day care center that she runs. Estimate the total cost of the crayons.

30. *Batteries.* Oscar buys 6 packages of AAA batteries at $5.29 per package. Estimate the total cost of the purchase.

Skill Maintenance

31. *Honey Production.* In 2011, 176,462,000 lb of honey was produced in the United States. The two states with the greatest honey production were North Dakota and California. North Dakota produced 46,410,000 lb of honey, and California produced 27,470,000 lb. How many more pounds of honey were produced in North Dakota than in California? [1.8a]

Source: U.S. Department of Agriculture

32. *Jewelry.* Christie has designed a pair of kite-shaped earrings, as shown below. Determine the surface area of the front of one earring. [3.6b]

33. About $\frac{9}{25}$ of all pizzas that Americans order have pepperoni as a topping. If Americans eat 350 slices of pizza every second, how many of those slices are topped with pepperoni? [3.6b]

Source: inventors.about.com

34. A batch of fudge requires $\frac{3}{4}$ cup of sugar. How much sugar is needed to make 12 batches? [3.6b]

35. After her company was restructured, Meghan's pay was $\frac{9}{10}$ of what it had been. If she is now making $32,850 a year, what was she making before the reorganization? [3.8b]

36. Rick's Market sells Swiss cheese in $\frac{3}{4}$-lb packages. How many packages can be made from a 12-lb slab of cheese? [3.8b]

Synthesis

The following were done on a calculator. Estimate to determine whether the decimal point was placed correctly.

37. $178.9462 \times 61.78 = 11,055.29624$

38. $14,973.35 \div 298.75 = 501.2$

39. $19.7236 - 1.4738 \times 4.1097 = 1.366672414$

40. $28.46901 \div 4.9187 - 2.5081 = 3.279813473$

41. 🖩 Use one of $+, -, \times,$ and $\div$ in each blank to make a true sentence.

    a) $(0.37 \boxed{} 18.78) \boxed{} 2^{13} = 156,876.8$
    b) $2.56 \boxed{} 6.4 \boxed{} 51.2 \boxed{} 17.4 = 312.84$

5.7 Solving Equations

OBJECTIVES

a Solve equations containing decimals and one variable term.

b Solve equations containing decimals and two or more variable terms.

SKILL TO REVIEW

Objective 2.8d: Solve equations that require use of both the addition principle and the division principle.

Solve.

1. $5x + 7 = -3$
2. $3x - 5 - x = 4 + x$

We now use the addition and division principles to solve equations involving decimals.

a EQUATIONS WITH ONE VARIABLE TERM

Recall that equations like $5x + 7 = -3$ are normally solved by first "undoing" the addition and then "undoing" the multiplication. This reverses the order of operations in which we add last and multiply first.

EXAMPLE 1 Solve: $0.5x + 5 = 8$.

$$0.5x + 5 = 8$$
$$0.5x + 5 - 5 = 8 - 5 \qquad \text{Subtracting 5 from both sides}$$
$$0.5x = 3 \qquad \text{Simplifying}$$
$$\frac{0.5x}{0.5} = \frac{3}{0.5} \qquad \text{Dividing both sides by 0.5}$$
$$x = 6 \qquad \text{Simplifying}$$

Check:
$$\begin{array}{c|c} 0.5x + 5 = 8 \\ \hline 0.5(6) + 5 \ ? \ 8 \\ 3 + 5 \\ 8 & 8 \quad \text{TRUE} \end{array}$$

The solution is 6.

◀ **Do Margin Exercises 1 and 2.**

EXAMPLE 2 Solve: $4.2x + 3.7 = -26.12$.

$$4.2x + 3.7 = -26.12$$
$$4.2x + 3.7 - 3.7 = -26.12 - 3.7 \qquad \text{Subtracting 3.7 from both sides}$$
$$4.2x = -29.82 \qquad \text{Simplifying}$$
$$\frac{4.2x}{4.2} = \frac{-29.82}{4.2} \qquad \text{Dividing both sides by 4.2}$$
$$x = -7.1 \qquad \text{Simplifying}$$

Check:
$$\begin{array}{c|c} 4.2x + 3.7 = -26.12 \\ \hline 4.2(-7.1) + 3.7 \ ? \ -26.12 \\ -29.82 + 3.7 \\ -26.12 & -26.12 \quad \text{TRUE} \end{array}$$

The solution is -7.1.

◀ **Do Exercises 3 and 4.**

Solve.

1. $6x + 7.4 = 11$

2. $0.2 - 0.1x = 1.4$

Solve.

3. $7.4t + 1.25 = 27.89$

4. $-5.7 + 4.8x = -14.82$

Answers

Skill to Review:
1. -2 2. 9

Margin Exercises:
1. 0.6   2. -12 3. 3.6   4. -1.9

b EQUATIONS WITH TWO OR MORE VARIABLE TERMS

Some equations have variable terms on both sides. To solve such an equation, we use the addition principle to get all variable terms on one side of the equation and all constant terms on the other side.

EXAMPLE 3 Solve: $10x - 7 = 2x + 13$.

We begin by subtracting $2x$ from (or adding $-2x$ to) each side. This will group all variable terms on one side of the equation.

$$10x - 7 - 2x = 2x + 13 - 2x \qquad \text{Adding } -2x \text{ to both sides}$$
$$8x - 7 = 13 \qquad\qquad\qquad \text{Combining like terms}$$

We use the addition principle to isolate all constant terms on one side.

$$8x - 7 = 13$$
$$8x - 7 + 7 = 13 + 7 \qquad \text{Adding 7 to both sides}$$
$$8x = 20 \qquad\qquad\quad \text{Simplifying (combining like terms)}$$
$$\frac{8x}{8} = \frac{20}{8} \qquad\qquad\quad \text{Dividing both sides by 8}$$
$$x = 2.5$$

Check:
$$\begin{array}{c|c} \multicolumn{2}{c}{10x - 7 = 2x + 13} \\ \hline 10(2.5) - 7 \ ? & 2(2.5) + 13 \\ 25 - 7 & 5 + 13 \\ 18 & 18 \qquad \text{TRUE} \end{array}$$

The solution is 2.5.

Sometimes it may be easier to combine all variable terms on the right side and all constant terms on the left side.

EXAMPLE 4 Solve: $11 - 3t = 7t + 8$.

We can combine all variable terms on the right side by adding $3t$ to both sides.

$$11 - 3t = 7t + 8$$
$$11 - 3t + 3t = 7t + 8 + 3t \qquad \text{Adding } 3t \text{ to both sides}$$
$$11 = 10t + 8 \qquad\qquad\quad \text{Combining like terms}$$
$$11 - 8 = 10t + 8 - 8 \qquad \text{Subtracting 8 from both sides}$$
$$3 = 10t$$
$$\frac{3}{10} = \frac{10t}{10} \qquad\qquad\quad \text{Dividing both sides by 10}$$
$$0.3 = t$$

Check:
$$\begin{array}{c|c} \multicolumn{2}{c}{11 - 3t = 7t + 8} \\ \hline 11 - 3(0.3) \ ? & 7(0.3) + 8 \\ 11 - 0.9 & 2.1 + 8 \\ 10.1 & 10.1 \qquad \text{TRUE} \end{array}$$

The solution is 0.3.

Solve.

5. $10t - 3 = 4t + 18$

6. $8 + 4x = 9x - 3$

$8 + 4x - 4x = 9x - 3 - \boxed{}$

$8 = \boxed{} - 3$

$8 + 3 = 5x - 3 + \boxed{}$

$\boxed{} = 5x$

$\dfrac{11}{5} = \dfrac{5x}{\boxed{}}$

$\boxed{} = x$

7. $2.1x - 45.3 = 17.3x + 23.1$

8. Solve: $3(x + 5) = 20 - x.$

$3(x + 5) = 20 - x$

$3x + \boxed{} = 20 - x$

$3x + 15 + x = 20 - x + \boxed{}$

$\boxed{} + 15 = 20$

$4x + 15 - \boxed{} = 20 - 15$

$4x = \boxed{}$

$\dfrac{4x}{4} = \dfrac{5}{\boxed{}}$

$x = \boxed{}$

9. Solve: $8(x - 2) - 15 = 4x + 2.$

Answers

5. 3.5   **6.** 2.2   **7.** −4.5

8. 1.25   **9.** 8.25

Guided Solutions:

6. $4x, 5x, 3, 11, 5, 2.2$

8. $15, x, 4x, 15, 5, 4, 1.25$

Note that in Example 4 the variable appears on the right side of the last equation. It does not matter whether the variable is isolated on the right or left side. What is important is that you have a clear direction to your work as you proceed from step to step.

◀ **Do Exercises 5–7.**

EXAMPLE 5   Solve: $5(x + 1) = 7x + 12.$

$$5(x + 1) = 7x + 12$$

$$5 \cdot x + 5 \cdot 1 = 7x + 12 \qquad \text{Using the distributive law to remove parentheses}$$

$$5x + 5 = 7x + 12 \qquad \text{Simplifying}$$

$$5x + 5 - 7x = 7x + 12 - 7x \qquad \text{Subtracting } 7x \text{ from both sides}$$

$$-2x + 5 = 12 \qquad \text{Simplifying}$$

$$-2x + 5 - 5 = 12 - 5 \qquad \text{Subtracting 5 from both sides}$$

$$-2x = 7$$

$$\frac{-2x}{-2} = \frac{7}{-2} \qquad \text{Dividing both sides by } -2$$

$$x = -3.5$$

Check:
$$\begin{array}{c|c} \multicolumn{2}{c}{5(x + 1) = 7x + 12} \\ \hline 5(-3.5 + 1) & 7(-3.5) + 12 \\ 5(-2.5) & -24.5 + 12 \\ -12.5 & -12.5 \end{array} \quad \text{TRUE}$$

Check in the original equation.

The solution is −3.5.

◀ **Do Exercise 8.**

EXAMPLE 6   Solve: $9(x - 3) + 7 = 5x - 47.$

We use the distributive law and combine like terms before using the addition and division principles.

$$9(x - 3) + 7 = 5x - 47$$

$$9x - 27 + 7 = 5x - 47 \qquad \text{Using the distributive law}$$

$$9x - 20 = 5x - 47 \qquad \text{Simplifying}$$

$$9x - 20 - 5x = 5x - 47 - 5x \qquad \text{Subtracting } 5x \text{ from both sides}$$

$$4x - 20 = -47 \qquad \text{Simplifying}$$

$$4x - 20 + 20 = -47 + 20 \qquad \text{Adding 20 to both sides}$$

$$4x = -27 \qquad \text{Simplifying}$$

$$\frac{4x}{4} = -\frac{27}{4} \qquad \text{Dividing both sides by 4}$$

$$x = -6.75$$

Check:
$$\begin{array}{c|c} \multicolumn{2}{c}{9(x - 3) + 7 = 5x - 47} \\ \hline 9(-6.75 - 3) + 7 & 5(-6.75) - 47 \\ 9(-9.75) + 7 & -33.75 - 47 \\ -87.75 + 7 & -80.75 \\ -80.75 & -80.75 \end{array} \quad \text{TRUE}$$

The solution is −6.75.

◀ **Do Exercise 9.**

✓ Reading Check

Determine whether each statement is true or false.

_____ **RC1.** Equations involving decimals never have integer solutions.

_____ **RC2.** An equation that does not involve decimals may have a decimal solution.

_____ **RC3.** In the final step of a solution of an equation, the variable may appear on the right side or on the left side of the equation.

_____ **RC4.** You should be able to give a reason for each step in your solution.

a Solve. Remember to check.

1. $5x = 27$

2. $4x = 75$

3. $16 \cdot y = 3.2$

4. $36 \cdot y = 14.76$

5. $-1.5t = 9.36$

6. $-1.2t = -11.4$

7. $x + 15.7 = 3.1$

8. $x + 13.9 = 4.2$

9. $1.25 = 3.8 + x$

10. $-1.3 = 4.05 + x$

11. $x - 0.37 = 1.6$

12. $x - 0.4 = 12$

13. $y - 9.8 = -1.42$

14. $y - 6 = -0.2$

15. $-9 = t - 3.1$

16. $-23 = t - 0.01$

17. $5x - 8 = 22$

18. $4x - 7 = 13$

19. $6.9x - 8.4 = 4.02$

20. $7.1x - 9.3 = 8.45$

21. $21.6 + 4.1t = 6.43$

22. $12.4 + 3.7t = 2.04$

23. $-26.25 = 7.5x + 9$

24. $-43.72 = 8.7x + 5$

25. $-4.2x + 3.04 = -4.1$

26. $-2.9x - 2.24 = -17.9$

27. $-3.05 = 7.24 - 3.5t$

28. $-4.62 = 5.68 - 2.5t$

29. $3 - 1.2y = -2.4$

30. $6 - 3.5y = 12.3$

b Solve. Remember to check.

31. $9x - 2 = 5x + 34$

32. $8x - 5 = 6x + 9$

33. $2x + 6 = 7x - 10$

34. $3x + 4 = 11x - 6$

35. $5y - 3 = 4 + 9y$

36. $6y - 5 = 8 + 10y$

37. $5.9x + 67 = 7.6x + 16$

38. $2.1x + 42 = 5.2x - 20$

39. $7.8a + 2 = 2.4a + 19.28$

40. $7.5a - 5.16 = 3.1a + 12$

41. $6(x + 2) = 4x + 30$

42. $5(x + 3) = 3x + 23$

43. $5(x + 3) = 15x - 6$

44. $2(x + 3) = 4x - 11$

45. $7a - 9 = 15(a - 3)$

46. $2a - 7 = 12(a - 3)$

47. $1.5(y - 6) = 1.3 - y$

48. $2.3(5 - y) = 0.2y + 1.7$

49. $2.9(x + 8.1) = 7.8x - 3.95$

50. $2(x + 7.3) = 6x - 0.83$

51. $-6.21 - 4.3t = 9.8(t + 2.1)$

52. $-7.37 - 3.2t = 4.9(t + 6.1)$

53. $4(x - 2) - 9 = 2x + 9$

54. $9(x - 4) + 13 = 4x - 23$

55. $2(4y - 1.8) + 0.4 = 8(2y - 0.4)$

56. $3(1.2y + 5) = 2(2.5y - 7) + 1$

57. $43(7 - 2x) + 34 = 50(x - 4.1) + 744$

58. $34(5 - 3.5x) = 12(3x - 8) + 653.5$

Skill Maintenance

Find the area of each figure.   [3.6b]

59.

4 m

7 m

60.

6 cm

9 cm

61.

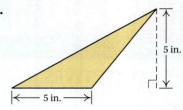

5 in.

5 in.

62.

8 ft

6 ft

63.

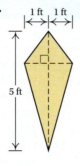

1 ft   1 ft

5 ft

64.

6 m

2 m

2 m

65. Subtract: $\dfrac{3}{25} - \dfrac{7}{10}$.   [4.3a]

66. Simplify: $\dfrac{0}{-18}$.   [3.3b]

67. Add: $-17 + 24 + (-9)$.   [2.2a]

68. Solve: $3x - 10 = 14$.   [2.8d]

Synthesis

Solve.

69. ⊞ $7.035(4.91x - 8.21) + 17.401 = 23.902x - 7.372815$

70. ⊞ $8.701(3.4 - 5.1x) - 89.321 = 5.401x + 74.65787$

71. $5(x - 4.2) + 3[2x - 5(x + 7)] = 39 + 2(7.5 - 6x) + 3x$

72. $14(2.5x - 3) + 9x + 5 = 4(3.25 - x) + 2[5x - 3(x + 1)]$

73. ⊞ $3.5(4.8x - 2.9) + 4.5 = 9.4x - 3.4(x - 1.9)$

74. ⊞ $4.19 - 1.8(4.5x - 6.4) = 3.1(9.8 + x)$

5.8

Applications and Problem Solving

OBJECTIVES

a Translate key phrases to algebraic expressions.

b Solve applied problems involving decimals.

SKILL TO REVIEW

Objective 1.8a: Solve applied problems involving addition, subtraction, multiplication, or division of whole numbers.

Solve.

1. A piece of pecan pie has 502 calories. This is 186 calories more than a piece of pumpkin pie contains. How many calories does a piece of pumpkin pie have?

a TRANSLATING TO ALGEBRAIC EXPRESSIONS

To translate problems to equations, we need to be able to translate phrases to algebraic expressions. Certain key words in phrases help direct the translation.

KEY WORDS	SAMPLE PHRASE OR SENTENCE	TRANSLATION
Addition (+)		
added to	350 lb was added to the car's weight.	$w + 350$
sum of	The sum of a number and 10	$n + 10$
plus	13 plus some number	$13 + x$
more than	176 more than last year's enrollment	$r + 176$
increased by	The original estimate, increased by 50	$y + 50$
Subtraction (−)		
subtracted from	2 oz was subtracted from the bag's weight.	$w - 2$
difference of	The difference of two prices	$p - q$
minus	A construction crew of size c, minus 4 workers	$c - 4$
less than	18 less than the number of volunteers last month	$v - 18$
decreased by	An essay's grade, decreased by 5 points	$g - 5$
Multiplication (·)		
multiplied by	The number of mints in a box, multiplied by 5	$5 \cdot m$
product of	The product of two numbers	$a \cdot b$
times	10 times Jon's age	$10j$
twice	Twice the wholesale price	$2w$
of	$\frac{1}{2}$ of the number of pages assigned	$\frac{1}{2}p$
Division (÷)		
divided by	A 16-oz bag of almonds, divided by 5	$16 \div 5$
quotient of	The quotient of 60 and 3	$60 \div 3$
divided into	6 divided into the cost of the meal	$c \div 6$
ratio of	The ratio of 456 to the number of gallons of gasoline	$456/g$

It is helpful to choose a descriptive variable to represent the unknown. For example, w suggests weight and g suggests the number of gallons of gasoline.

Answer

Skill to Review:
1. 316 calories

The following tips are helpful in translating phrases to algebraic expressions.

> **TIPS FOR TRANSLATING PHRASES TO ALGEBRAIC EXPRESSIONS**
>
> - Replace the unknown in the statement with a specific number before translating using a variable.
> - Write down what each variable represents.
> - Check the translation using another number to see if it matches the phrase.
> - Be especially careful with order when subtracting and dividing.

EXAMPLE 1 Translate each phrase to an algebraic expression.

a) A number added to 5

b) A number subtracted from 5

a) Let n represent the number. Then the phrase "a number added to 5" can be translated $5 + n$, or $n + 5$.

b) Let n represent the number. Then the phrase "a number subtracted from 5" can be translated $5 - n$.

Do Exercises 1 and 2. ▶

······················· **Caution!** ·······················

Note that in Example 1, there are two correct translations for part (a) and only one correct translation for part (b). Recall that numbers can be added in either order, so $5 + n$ and $n + 5$ represent the same value. However, order is important in subtraction, so $5 - n$ and $n - 5$ do NOT represent the same value.

EXAMPLE 2 Translate each phrase to an algebraic expression.

a) Than's age increased by six

b) Half of some number

c) Twice the cost

d) Seven more than twice the weight

e) Fifteen divided by a number

f) A number divided by fifteen

g) Six less than the product of two numbers

h) Nine times the sum of two numbers

Phrase	*Variable(s)*	*Algebraic Expression*
a) Than's age increased by six	Let a = Than's age.	$a + 6$, or $6 + a$
b) Half of some number	Let n = the number.	$\frac{1}{2}n$, or $\frac{n}{2}$, or $n \div 2$
c) Twice the cost	Let c = the cost.	$2c$
d) Seven more than twice the weight	Let w = the weight.	$2w + 7$, or $7 + 2w$

Translate to an algebraic expression.

1. Twenty more than a number

2. Twenty less than a number

Answers

1. Let n = the number; $n + 20$, or $20 + n$
2. Let n = the number; $n - 20$

Translate to an algebraic expression.

3. The height of a tree, decreased by 50

4. Four less than ten times a number

5. Fourteen more than the product of the hourly rate and the number of hours worked

e)	Fifteen divided by a number	Let x = the number.	$15 \div x$, or $\dfrac{15}{x}$
f)	A number divided by fifteen	Let x = the number.	$x \div 15$, or $\dfrac{x}{15}$
g)	Six less than the product of two numbers	Let m and n = the numbers.	$mn - 6$
h)	Nine times the sum of two numbers	Let a and b = the numbers.	$9(a + b)$

◀ **Do Exercises 3–5.**

b SOLVING APPLIED PROBLEMS

EXAMPLE 3 *Canals.* The Panama Canal in Panama is 50.7 mi long. The Suez Canal in Egypt is 119.9 mi long. How much longer is the Suez Canal?

Panama Canal   Suez Canal

6. *Debit-Card Transactions.* U.S. debit-card transactions in 2009 totaled 38.5 billion. The number of transactions in 2012 was 52.6 billion. How many more debit-card transactions were there in 2012 than in 2009?

Source: *The Nilson Report*

1. **Familiarize.** We let l = the distance in miles that the length of the longer canal differs from the length of the shorter canal.

2. **Translate.** We translate as follows, using the given information:

Length of Panama Canal, the shorter canal	plus	Additional length	is	Length of Suez Canal, the longer canal
50.7 mi	+	l	=	119.9 mi.

3. **Solve.** We solve the equation by subtracting 50.7 mi on both sides:

$$50.7 + l = 119.9$$
$$50.7 + l - 50.7 = 119.9 - 50.7$$
$$l = 69.2.$$

4. **Check.** We can check by adding.

```
    5 0.7
+   6 9.2
  1 1 9.9
```

The answer checks.

5. **State.** The Suez Canal is 69.2 mi longer than the Panama Canal.

◀ **Do Exercise 6.**

Answers

3. Let h = the height of a tree; $h - 50$
4. Let x = the number; $10x - 4$
5. Let r = the hourly rate and w = the number of hours worked; $rw + 14$, or $14 + rw$
6. 14.1 billion transactions

EXAMPLE 4 *iPad Purchase.* A large architectural firm spent $10,399.74 on 26 iPads for its architects. How much did each iPad cost?

1. **Familiarize.** We let $c =$ the cost of each iPad.

2. **Translate.** We translate as follows:

Number of iPads purchased	times	Cost of each iPad	is	Total cost of purchase
26	·	c	=	$10,399.74.

3. **Solve.** We solve the equation by dividing by 26 on both sides.

$$\frac{26 \cdot c}{26} = \frac{10{,}399.74}{26}$$

$$c = \frac{10{,}399.74}{26}$$

$$c = 399.99$$

4. **Check.** We check by estimating:

$$10{,}399.74 \div 26 \approx 10{,}000 \div 25 = 400.$$

Since 400 is close to 399.99, the answer is probably correct.

5. **State.** The cost of each iPad was $399.99.

Do Exercise 7. ▶

7. *Mileage Rates.* For a recent year, the Internal Revenue Service allowed a tax deduction of 23.5¢ per mile driven for medical or moving purposes. What deduction, in dollars, would be allowed for driving 1862 mi during a move?

Source: Internal Revenue Service

Multistep Problems

EXAMPLE 5 *Tracking a Bank Balance.* Revenue of U.S. banks from checking account overdraft charges was $31.5 billion in a recent fiscal year. (**Source:** Moebs Services) To avoid overdrafts and to track her spending, Maggie keeps a running account of her banking transactions. She checks her balance online, but because of pending amounts that post later, she keeps a separate record. Maggie had $2432.27 in her account. She used her debit card to pay her rent of $835 and make purchases of $14.13, $38.60, and $205.98. She then deposited her weekly pay of $748.35. What was her balance after these transactions?

1. **Familiarize.** We first find the total of the debits. Then we find how much is left in the account after the debits are deducted. Finally, we add to this amount the deposit to find the balance in the account after all the transactions.

2, 3. **Translate and Solve.** We let $d =$ the total amount of the debits. We are combining amounts: $835 + $14.13 + $38.60 + $205.98 = d$. To solve the equation, we add.

835.00	First debit
14.13	Second debit
38.60	Third debit
+ 205.98	Fourth debit
1093.71	Total debits

Thus, $d = 1093.71$.

Now let a = the amount in the account after the debits are deducted. We subtract: $2432.27 - $1093.71 = a$.

$$\begin{array}{r} 2432.27 \\ -\ 1093.71 \\ \hline 1338.56 \end{array} \begin{array}{l} \text{Original amount} \\ \text{Total debits} \\ \text{New amount} \end{array}$$

Thus, $a = 1338.56$.

Finally, we let f = the amount in the account after the paycheck is deposited.

Amount after debits	plus	Amount of deposit	is	Final amount
↓	↓	↓	↓	↓
1338.56	+	748.35	=	f

To solve the equation, we add.

$$\begin{array}{r} 1338.56 \\ +\ 748.35 \\ \hline 2086.91 \end{array} \begin{array}{l} \text{Balance after debits} \\ \text{Paycheck deposit} \\ \text{Final amount} \end{array}$$

4. Check. We repeat the computations.

5. State. Maggie had $2086.91 in her account after all the transactions.

◄ **Do Exercise 8.**

8. *Bank Balance.* Stephen had $915.22 in his checking account. He used his debit card to pay a charge card minimum payment of $36 and to make purchases of $67.50, $178.23, and $429.05. He then deposited his weekly pay of $570.91. How much was in his account after these transactions?

EXAMPLE 6 *Gas Mileage.* Ava filled her gas tank and noted that the odometer read 67,507.8. After the next fill-up, the odometer read 68,006.1. It took 16.5 gal to fill the tank. How many miles per gallon did Ava's car get?

1. Familiarize. We first make a drawing.

n miles, 16.5 gallons

This is a two-step problem. First, we find the number of miles that have been driven between fill-ups. We let n = the number of miles driven and m = the number of miles per gallon.

2, 3. Translate and Solve. We translate and solve as follows:

First reading	plus	Number of miles driven	is	Second reading
↓	↓	↓	↓	↓
67,507.8	+	n	=	68,006.1.

To solve the equation, we subtract 67,507.8 on both sides:

$$\begin{aligned} n &= 68{,}006.1 - 67{,}507.8 \\ &= 498.3. \end{aligned}$$

$$\begin{array}{r} 6\,8{,}0\,0\,6.1 \\ -\ 6\,7{,}5\,0\,7.8 \\ \hline 4\,9\,8.3 \end{array}$$

Next, we divide the total number of miles driven by the number of gallons. This gives us m, that is, the gas mileage. The division that corresponds to the situation is $498.3 \div 16.5 = m$.

Thus, $m = 30.2$.

$$\begin{array}{r} 3\ 0.2 \\ 16.5\overline{)4\,9\,8.3\,0} \\ \underline{4\,9\,5} \\ 3\ 3\ 0 \\ \underline{3\ 3\ 0} \\ 0 \end{array}$$

4. Check. To check, we first multiply the number of miles per gallon by the number of gallons to find the number of miles driven:

$$16.5 \times 30.2 = 498.3.$$

Then we add 498.3 to 67,507.8 to find the new odometer reading:

$$67,507.8 + 498.3 = 68,006.1.$$

The gas mileage of 30.2 checks.

5. State. Ava's car got 30.2 miles per gallon.

Do Exercise 9. ▶

9. *Gas Mileage.* John filled his gas tank and noted that the odometer read 38,320.8. After the next fill-up, the odometer read 38,735.5. It took 14.5 gal to fill the tank. How many miles per gallon did John get?

Example 7 involves a formula giving the area of a circle.

In any circle, a **diameter** is a segment that passes through the center of the circle with endpoints on the circle. A **radius** is a segment with one endpoint on the center and the other endpoint on the circle. The area, A, of a circle with radius of length r is given by

$$A = \pi \cdot r^2,$$

where $\pi \approx 3.14$.
   The length r of a radius of a circle is half the length d of a diameter:

$$r = \frac{1}{2}d.$$

EXAMPLE 7 The Northfield Tap and Die Company stamps 6-cm-wide discs out of metal squares that are 6 cm by 6 cm. How much metal remains after the disc has been punched out?

1. Familiarize. We make, and label, a drawing. We let a = the amount of metal remaining, in square centimeters, and list the relevant formulas.

   For a square with sides of length s, $Area = s^2$.

   For a circle with radius of length r, $Area = \pi \cdot r^2$, where $\pi \approx 3.14$.

   The radius r of a circle is half of its diameter: $r = \frac{1}{2}d$.

We also list the known information about the square and the disc:

   Square: The side s = 6 cm.
   Disc: The diameter d = 6 cm.

   The radius $r = \frac{1}{2}(6 \text{ cm}) = 3 \text{ cm}.$

2. Translate. To find the amount left over, we subtract the area of the disc from the area of the square.

The shaded area is the amount of metal remaining.

Answer
9. 28.6 mpg

3. Solve. We simplify as follows.

$$6^2 - 3.14(3)^2 = a$$
$$36 - 3.14 \cdot 9 = a$$
$$36 - 28.26 = a$$
$$7.74 = a$$

4. Check. We can repeat our calculation as a check. Note that 7.74 is less than the area of the disc, which in turn is less than the area of the square. This agrees with the impression given by our drawing.

5. State. The amount of material left over is 7.74 cm².

◀ **Do Exercise 10.**

10. Suppose that an 8-in.-wide disc is punched out of an 8-in. by 8-in. sheet of metal. How much material is left over?

EXAMPLE 8 *Photo Books.* Users of Apple's iPhoto application can create their own photo books. The price of a small 20-page book is $24.99. Additional pages are 79 cents per page. Marta has $35 to spend on a book. What is the greatest number of pages she can put in the book?
Source: www.apple.com

1. Familiarize. Suppose that Marta put 30 pages in the book. She would have to pay an additional price per page for $30 - 20$, or 10, pages. The price would be the base price plus the charge for extra pages, or

$$\$24.99 + \$0.79 \cdot 10 = \$24.99 + \$7.90 = \$32.89.$$

This familiarizes us with the way in which price is calculated. Note that we convert 79 cents to $0.79 so that only one unit, dollars, is used. Note also that Marta can put more than 30 pages in the book.

We let $p =$ the number of extra pages Marta adds to the 20-page book. Note that the total number of pages in the book is then $20 + p$.

2. Translate. The problem can be rephrased and translated as follows.

Base price of book	plus	Cost per page	times	Number of extra pages	is	Cost of book
↓	↓	↓	↓	↓	↓	↓
$24.99	+	$0.79	·	p	=	$35

3. Solve. We solve the equation.

$$24.99 + 0.79p = 35$$
$$0.79p = 10.01 \qquad \text{Subtracting 24.99 from both sides}$$
$$p = \frac{10.01}{0.79} \qquad \text{Dividing both sides by 0.79}$$
$$p \approx 12.7 \qquad \text{Rounding}$$

4. Check. We check in the original problem. Since Marta cannot pay for parts of a page, we must round the answer *down* to 12 in order to keep the cost under $35. This makes the total length of the book $20 + 12$, or 32, pages. If Marta adds 12 pages to the book, the additional cost will be 12 times $0.79, or $9.48. If we add $9.48 to the base price of $24.99, we get $34.47, which is just under the $35 Marta has to spend.

5. State. With $35, Marta can make a 32-page book.

◀ **Do Exercise 11.**

11. Bike Rentals. Mike's Bikes rents mountain bikes. The shop charges $4.00 insurance for each rental plus $6.00 per hour. For how many hours can a person rent a bike with $25.00?

Answers

10. 13.76 in²   **11.** 3.5 hr

Problems with More Than One Unknown

EXAMPLE 9 *Multi-sport Recreation.* Around the Bend Expeditions offers a two-day canyon trip combining biking and canoeing. Those participating will ride mountain bikes for 12 miles longer than the distance paddled. The total length of the trip is 43 miles. How long is the biking portion of the trip, and how long is the canoeing portion of the trip?

1. **Familiarize.** We first list the quantities we are asked to find:

    Length of biking portion, in miles

    Length of canoeing portion, in miles

    We will want to represent both quantities using only one variable. To do so, we use the second sentence in the problem: *Those participating will ride mountain bikes for 12 miles longer than the distance paddled.* Since the biking portion of the trip is described in terms of the canoeing portion, we let $x =$ the length of the canoeing portion, in miles. Then the length of the biking portion, also in miles, is $x + 12$.

    Length of biking portion, in miles: $x + 12$

    Length of canoeing portion, in miles: x

2. **Translate.** We use the total length of the trip to translate to an equation. Note that all lengths are in miles.

Length of biking portion	plus	Length of canoeing portion	is	Total length of trip
$(x + 12)$	$+$	x	$=$	43

3. **Solve.** We solve the equation.

$$(x + 12) + x = 43$$
$$x + 12 + x = 43 \qquad \text{Removing parentheses}$$
$$2x + 12 = 43 \qquad \text{Combining like terms}$$
$$2x = 31 \qquad \text{Subtracting 12 from both sides}$$
$$x = 15.5 \qquad \text{Dividing both sides by 2}$$

Recall that we are looking for two quantities. We will use the value of x to find both of them. We return to the list of unknowns in the *Familiarize* step.

Length of biking portion, in miles: $x + 12 = 15.5 + 12 = 27.5$

Length of canoeing portion, in miles: $x = 15.5$

4. **Check.** There are two statements in the problem to verify. First, since $27.5 - 15.5 = 12$, the biking portion is 12 miles longer than the canoeing portion. Second, since $27.5 + 15.5 = 43$, the total length of the trip is 43 miles.

5. **State.** The biking portion of the trip is 27.5 miles, and the canoeing portion is 15.5 miles.

Do Exercise 12. ▶

> **Caution!**
>
> Do not skip the *Familiarize* step, particularly in problems involving more than one unknown. Describing the unknowns using one variable is often the most challenging part of the solution process.

12. Holly and Lorenzo worked together on a project for a psychology class. Holly spent twice as much time on the project as Lorenzo did. They spent a total of 10.5 hours on the project. How much time did each person work on the project?

Answer

12. Holly: 7 hr; Lorenzo: 3.5 hr

Translating for Success

1. Gas Mileage. Art filled his SUV's gas tank and noted that the odometer read 38,271.8. At the next fill-up, the odometer read 38,677.9. It took 28.4 gal to fill the tank. How many miles per gallon did the SUV get?

2. Dimensions of a Parking Lot. A store's parking lot is a rectangle that measures 85.2 ft by 52.3 ft. What is the area of the parking lot?

3. Game Snacks. Three students pay $18.40 for snacks at a football game. What is each student's share of the cost?

4. Electrical Wiring. An electrician needs 1314 ft of wiring cut into $2\frac{1}{2}$-ft pieces. How many pieces will she have?

5. College Tuition. Wayne needs $4638 for the fall semester's tuition. On the day of registration, he has only $3092. How much does he need to borrow?

The goal of these matching questions is to practice step (2), Translate, of the five-step problem-solving process. Translate each word problem to an equation and select a correct translation from equations A–O.

A. $2\frac{1}{2} \cdot n = 1314$

B. $18.4 \times 3.87 = n$

C. $n = 85.2 \times 52.3$

D. $19 - (-4) = n$

E. $3 \times 18.40 = n$

F. $2\frac{1}{2} \cdot 1314 = n$

G. $3092 + n = 4638$

H. $18.4 \cdot n = 3.87$

I. $\dfrac{406.1}{28.4} = n$

J. $52.3 \cdot n = 85.2$

K. $n = 19 + (-4)$

L. $52.3 + n = 85.2$

M. $3092 + 4638 = n$

N. $3 \cdot n = 18.40$

O. $85.2 + 52.3 = n$

Answers on page A-10

6. Cost of Gasoline. What is the cost, in dollars, of 18.4 gal of gasoline at $3.87 per gallon?

7. Temperature. At noon, the temperature in Pierre was 19°F. At midnight, the temperature had fallen to −4°F. By how many degrees had the temperature fallen?

8. Acres Planted. This season Sam planted 85.2 acres of corn and 52.3 acres of soybeans. Find the total number of acres that he planted.

9. Amount Inherited. Tara inherited $2\frac{1}{2}$ times as much as her cousin. Her cousin received $1314. How much did Tara receive?

10. Travel Funds. The athletic department needs travel funds of $4638 for the tennis team and $3092 for the golf team. What is the total amount needed for travel?

✓ Reading Check

Complete each step in the five-step problem-solving strategy with the correct word from the following list.

Solve   Familiarize   State   Translate   Check

Five-Step Problem-Solving Strategy

RC1. _____ yourself with the problem situation.

RC2. _____ the problem to an equation.

RC3. _____ the equation.

RC4. _____ the solution.

RC5. _____ the answer using a complete sentence.

a   Translate to an algebraic expression. Choice of variables used may vary.

1. Five more than Ron's age

2. The product of four and t

3. 6 more than b

4. 7 more than Jen's weight

5. 9 less than c

6. 4 less than d

7. A number decreased by 16

8. A number increased by 20

9. 8 times Nate's speed

10. The ratio of a number and 100

11. x divided by 17

12. 100 divided by the hourly rate

13. 20 added to half of a number

14. 18 subtracted from twice a number

15. 20 less than 4 times a number

16. 35 more than 10 times a number

17. The sum of the box's length and width

18. The Cessna's speed minus the wind speed

19. 10 more than the product of the rate and the time

20. 50 less than the product of the length and width

21. The sum of 10 times a number and the number

22. The difference of a number and twice the number

23. 5 times the difference of two numbers

24. One fourth of the sum of two numbers

Hurricane Damage. The amounts of damage caused by the five most costly Atlantic hurricanes in the United States are shown in the table below. Use this table to do Exercises 25 and 26.

Most Costly Hurricanes

RANK	HURRICANE	YEAR	COST IN 2010 DOLLARS (in billions)
1	Katrina	2005	$105.8
2	Andrew	1992	45.6
3	Ike	2008	27.8
4	Wilma	2005	20.6
5	Ivan	2004	19.8

SOURCE: National Hurricane Center

25. How much more costly was Hurricane Katrina than Hurricane Andrew?

26. What was the total cost of the two hurricanes that occurred in 2005?

27. *Gasoline Cost.* What is the cost, in dollars, of 20.4 gal of gasoline at 324.9 cents per gallon? (324.9 cents = $3.249) Round the answer to the nearest cent.

28. *Gasoline Cost.* What is the cost, in dollars, of 15.3 gal of gasoline at 389.9 cents per gallon? (389.9 cents = $3.899) Round the answer to the nearest cent.

29. *Cost of Bottled Water.* The cost of a year's supply of a popular brand of bottled water, based on consumption of 64 oz per day, at the supermarket price of $3.99 for a six-pack of half-liter bottles is $918.82. This is $918.31 more than the cost of drinking the same amount of tap water for a year. What is the cost of drinking tap water for a year?

Source: American Water Works Association

30. *Body Temperature.* Normal body temperature is 98.6°F. During an illness, a patient's temperature rose 4.2°. What was the new temperature?

31. *Lottery Winnings.* The largest lottery jackpot in the United States totaled $656,000,000 and was shared equally by 3 winners. How much was each winner's share? Round to the nearest cent.

Source: Mega Millions

32. *Lunch Costs.* A group of 4 students pays $47.84 for lunch and splits the cost equally. What is each person's share?

33. *100-Meter Record.* The fastest speed clocked for a cheetah running a distance of 100 m is 5.95 sec. The men's world record for the 100-m dash is held by Jamaican Usain Bolt. His time was 3.63 sec more than the cheetah's. What is the men's 100-m record held by Usain Bolt?

Source: "Cheetahs on the Edge," by Roff Smith, *National Geographic*, November 2012.

34. *Record Movie Openings.* The movie *Marvel's The Avengers* took in $207.4 million on its first weekend. This topped the previous record for opening-weekend revenue, set by *Harry Potter and the Deathly Hallows: Part II*, by $38.2 million. How much did *Harry Potter and the Deathly Hallows: Part II* take in on its opening weekend?

Source: Nash Information Services

35. Odometer Reading. The Binford family's odometer reads 22,456.8 at the beginning of a trip. The family's online driving directions tell them that they will be driving 234.7 mi. What will the odometer read at the end of the trip?

36. Miles Driven. Petra bought gasoline when the odometer read 14,296.3. At the next gasoline purchase, the odometer read 14,515.8. How many miles had been driven?

37. Gas Mileage. Peggy filled her van's gas tank and noted that the odometer read 26,342.8. After the next filling, the odometer read 26,736.7. It took 19.5 gal to fill the tank. How many miles per gallon did the van get?

38. Gas Mileage. Henry filled his Honda's gas tank and noted that the odometer read 18,943.2. After the next filling, the odometer read 19,306.2. It took 13.2 gal to fill the tank. How many miles per gallon did the car get?

39. Andrew bought a DVD of the movie *Horton Hears a Who* for his nephew for $23.99 plus $1.68 sales tax. He paid for it with a $50 bill. How much change did he receive?

40. Claire bought a copy of the book *Make Way for Ducklings* for her daughter for $16.95 plus $0.85 sales tax. She paid for it with a $20 bill. How much change did she receive?

41. Health Care. Phil injects 38 units of insulin each day for a week. Each unit is 0.01 cc (cubic centimeter). How many cc's of insulin does he use in a week?

42. Chemistry. The water in a filled tank weighs 748.45 lb. One cubic foot of water weighs 62.5 lb. How many cubic feet of water does the tank hold?

43. Egg Costs. A restaurant owner bought 20 dozen eggs for $25.80. Find the cost of each egg to the nearest tenth of a cent (thousandth of a dollar).

44. Weight Loss. A person weighing 170 lb burns 8.6 calories per minute while mowing a lawn. One must burn about 3500 calories in order to lose 1 lb. How many pounds would be lost by mowing for 2 hr? Round to the nearest tenth.

45. Loose Change. In 2012, passengers at Miami International Airport left $39,613 in loose change at airport checkpoints. This was $12,712.79 more than passengers left that year at Las Vegas McCarran International Airport. How much change was left at the Las Vegas airport?

Source: Transportation Security Administration

46. Online Ad Spending. It is projected that $132.1 billion will be spent for online advertising in 2015. This is $63.7 billion more than in 2010. What was spent on online advertising in 2010?

Source: eMarketer, June 2011

47. Pole Vault Pit. Find the area and the perimeter of the landing area of the pole vault pit shown here.

Landing area

48. Stamp. Find the area and the perimeter of the stamp shown here.

49. *Study Cards.* An instructor allows her students to bring one 7.6-cm by 12.7-cm index card to the final exam, with notes of any sort written on the card. If both sides of the card are used, how much area is available for notes?

7.6 cm

12.7 cm

50. *Stamps.* Find the total area of the stamps shown.

2.5 cm

2.25 cm

Find the length d in each figure.

51.

0.8 cm d 0.8 cm

3.91 cm

52.

0.9 cm d 0.9 cm

4.52 cm

53. *Carpentry.* A round, 6-ft-wide hot tub is being built into a 12-ft by 30-ft rectangular deck. How much decking is needed for the surface of the deck?

12 ft   6 ft   30 ft

54. *Travel Poster.* Sam is decorating his dorm room with a travel poster. The dimensions of the poster are as shown. How much area is not devoted to the painting?

15.7 in.

22.2 in.

27.4 in.

19.3 in.

THE GALAPAGOS ISLANDS

55. *Web Space.* Penn State charges $8 a month for Internet use plus 7 cents for each gigabyte over 10 gigabytes of Web space used. Nikki's bill for one month was $16.75. How many additional gigabytes did she pay for?

Source: Penn State Information Technology Services

56. *Service Calls.* JoJo's Service Center charges $30 for a house call plus $37.50 for each hour the job takes. For how long has a repairperson worked on a house call if the bill comes to $123.75?

57. Lot A measures 250.1 ft by 302.7 ft. Lot B measures 389.4 ft by 566.2 ft. What is the total area of the two lots?

58. A 4-ft by 4-ft tablecloth is cut from a round tablecloth that is 6 ft wide. Find the area of the cloth left over.

59. Frank has been sent to the store with $40 to purchase 6 lb of cheese at $4.79 a pound and as many bottles of seltzer, at $0.64 a bottle, as possible. How many bottles of seltzer should Frank buy?

60. Janice has been sent to the store with $30 to purchase 5 pt of strawberries at $2.49 a pint and as many bags of chips, at $1.39 a bag, as possible. How many bags of chips should Janice buy?

61. *Endangered Species.* Part of the giant panda's habitat is protected by the Chinese government. There are 0.4 million more acres unprotected than there are protected acres. The total size of the giant panda's habitat is 5.4 million acres. How many acres are protected and how many are not protected?

Source: www.worldwildlife.org

62. *Hours of Daylight.* On December 22, Fairbanks, Alaska, has 16.6 more hours of darkness than it has daylight. How many hours of daylight and how many hours of darkness are there in Fairbanks on that day? [*Hint:* There are 24 hours in a day.]

Source: "Alaska's Winter Daylight," by Kimi Ross, at www.bellaonline.com

63. *Vacation Spending.* Emily spent three times as much on lodging as she did for food on a recent vacation. She spent a total of $261.20 for food and lodging. How much did she spend for each?

64. *Homework.* Ian spends twice as long on his science homework as he does on his history homework. One week, he spent a total of 13.5 hours on science and history homework. How much time did he spend on each subject?

65. *Construction Pay.* A construction worker is paid $18.50 per hour for the first 40 hr of work, and time and a half, or $27.75 per hour, for any overtime exceeding 40 hr per week. One week she works 46 hr. How much is her pay?

66. *Summer Work.* Zachary worked 53 hr during a week one summer. He earned $7.50 per hour for the first 40 hr and $11.25 per hour for overtime (hours exceeding 40). How much did Zachary earn during the week?

67. Loan Payment. In order to make money on loans, financial institutions are paid back more money than they loan. Suppose you borrow $120,000 to buy a house and agree to make monthly payments of $880.52 for 30 years. How much do you pay back altogether? How much more do you pay back than the amount of the loan?

68. Loan Payment. In order to make money on loans, financial institutions are paid back more money than they loan. Suppose you borrow $270,000 to buy a house and agree to make monthly payments of $1105.73 for 30 years. How much do you pay back altogether? How much more do you pay back than the amount of the loan?

Skill Maintenance

Simplify, if possible.

69. $\dfrac{0}{-13}$ [3.3b]

70. $\dfrac{12}{0}$ [3.3b]

71. $\dfrac{-76}{-76}$ [3.3b]

72. Add: $-\dfrac{4}{5} + \dfrac{7}{10}$.   [4.2b]

73. Subtract: $\dfrac{8}{11} - \dfrac{4}{3}$.   [4.3a]

74. Add: $4\dfrac{1}{3} + 2\dfrac{1}{2}$.   [4.6a]

Synthesis

75. A poster that is 61.8 cm by 73.2 cm includes a 2-cm border. What is the area of the poster inside the border?

61.8 cm

73.2 cm

76. A French press coffeepot requires no filters, but costs $34.95. Kenny could buy a plastic drip cone for $4.49, but the cone requires filters which cost $0.04 per pot dripped. How many pots of coffee must Kenny make for the French press pot to be the more economical purchase?

77. If the daily rental for a car is $18.90 plus a certain price per mile, and Lindsey must drive 190 mi in one day and still stay within a $55.00 budget, what is the highest price per mile that Lindsey can afford?

78. A 25-ft by 30-ft yard contains an 8-ft-wide, round fountain. How many 1-lb bags of grass seed should be purchased to seed the lawn if 1 lb of seed covers 300 ft²?

79. Find the shaded area. What assumptions must you make?

10 cm

80. You can drive from home to work using either of two routes:

Route A: Via interstate highway, 7.6 mi, with a speed limit of 65 mph.

Route B: Via a country road, 5.6 mi, with a speed limit of 50 mph.

Assuming you drive at the posted speed limit, how much time can you save by taking the faster route?

Vocabulary Reinforcement

Complete each statement with the correct word from the list at the right.

1. A _____ decimal occurs when we convert a fraction to a decimal and the denominator of the fraction has at least one factor other than 2 or 5.   [5.5a]

2. A _____ decimal occurs when we convert a fraction to a decimal and the denominator of the fraction has only 2's or 5's, or both, as factors.   [5.5a]

3. One _____ = 1,000,000,000.   [5.3b]

4. One _____ = 1,000,000.   [5.3b]

5. One _____ = 1,000,000,000,000.   [5.3b]

6. The _____ consist of the integers, . . . , −3, −2, −1, 0, 1, 2, 3, . . . , and fractions like $\frac{-1}{2}, \frac{4}{5}$, and $\frac{31}{25}$.   [5.1a]

trillion

million

billion

rational numbers

repeating

terminating

Concept Reinforcement

Determine whether each statement is true or false.

_____ 1. One thousand billion is one trillion.   [5.3b]

_____ 2. The number of decimal places in the product of two numbers is the product of the numbers of places in the factors.   [5.3a]

_____ 3. When we divide a positive number by 0.1, 0.01, 0.001, and so on, the quotient is larger than the dividend.   [5.4a]

_____ 4. For a fraction with a factor other than 2 or 5 in its denominator, decimal notation terminates.   [5.5a]

_____ 5. An estimate found by rounding to the nearest ten is usually more accurate than one found by rounding to the nearest hundred.   [5.6a]

Study Guide

Objective 5.1b   Convert between decimal notation and fraction notation.

Example   Write fraction notation for 5.347.

$$5.347 \qquad 5.347. \qquad \frac{5347}{1000}$$

3 decimal places   Move 3 places to the right.   3 zeros

Practice Exercise

1. Write fraction notation for 50.93.

Example   Write decimal notation for $\frac{29}{1000}$.

$$\frac{29}{1000} \qquad 0.029. \qquad \frac{29}{1000} = 0.029$$

3 zeros   Move 3 places to the left.

Practice Exercise

2. Write decimal notation for $\frac{817}{10}$.

Example Write decimal notation for $4\dfrac{63}{100}$.

$$4\dfrac{63}{100} = 4 + \dfrac{63}{100} = 4 \text{ and } \dfrac{63}{100} = 4.63$$

Practice Exercise

3. Write decimal notation for $42\dfrac{159}{1000}$.

Objective 5.1d Round decimal notation to the nearest thousandth, hundredth, tenth, one, ten, hundred, or thousand.

Example Round 19.7625 to the nearest hundredth.

Locate the digit in the hundredths place, 6. Consider the next digit to the right, 2. Since that digit, 2, is 4 or lower, the original digit does not change.

19.7625
  ↓
19.76

Practice Exercise

4. Round 153.346 to the nearest hundredth.

Objective 5.2a Add using decimal notation.

Example Add: 14.26 + 63.589.

$$\begin{array}{r} \overset{1}{} \\ 1\,4.2\,6\,0 \\ +\,6\,3.5\,8\,9 \\ \hline 7\,7.8\,4\,9 \end{array}$$ Writing an extra zero

Practice Exercise

5. Add: 5.54 + 33.071.

Objective 5.2b Subtract using decimal notation.

Example Subtract: 67.345 − 24.28.

$$\begin{array}{r} \overset{2}{}\overset{14}{} \\ 6\,7.\cancel{3}\,\cancel{4}\,5 \\ -\,2\,4.2\,8\,0 \\ \hline 4\,3.0\,6\,5 \end{array}$$ Writing an extra zero

Practice Exercise

6. Subtract: 221.04 − 13.192.

Objective 5.3a Multiply using decimal notation.

Example Multiply: 1.8 × 0.04.

$$\begin{array}{r} 1.8 \quad (\text{1 decimal place}) \\ \times\,0.0\,4 \quad (\text{2 decimal places}) \\ \hline 0.0\,7\,2 \quad (\text{3 decimal places}) \end{array}$$

Practice Exercise

7. Multiply: 5.46 × 3.5.

Example Multiply: 0.001 × 87.1.

0.001 × 87.1   0.087.1

3 decimal places   Move 3 places to the left. We write an extra zero.

0.001 × 87.1 = 0.0871

Practice Exercise

8. Multiply: 17.6 × 0.01.

Example Multiply: 63.4 × 100.

63.4 × 100   63.40.

2 zeros   Move 2 places to the right. We write an extra zero.

63.4 × 100 = 6340

Practice Exercise

9. Multiply: 1000 × 60.437.

Objective 5.4a Divide using decimal notation.

Example Divide: $21.35 \div 6.1$.

$$
\begin{array}{r}
3.5 \\
6.1 \,)\overline{2\,1.3\,5} \\
1\,8\,3 \\
\hline
3\,0\,5 \\
3\,0\,5 \\
\hline
0
\end{array}
$$

Practice Exercise

10. Divide: $26.64 \div 3.6$.

Example Divide: $\dfrac{16.7}{1000}$.

$$\frac{16.7}{1000} = \frac{016.7}{1\,000.} = \frac{0.0167}{1.0} = 0.0167$$

3 zeros   Move 3 places to the left to change 1000 to 1.

$$\frac{16.7}{1000} = 0.0167$$

Practice Exercise

11. Divide: $\dfrac{4.7}{100}$.

Example Divide: $\dfrac{42.93}{0.001}$.

$$\frac{42.93}{0.001} = \frac{42.930}{0.001} = \frac{42,930}{1.} = 42,930$$

3 decimal places   Move 3 places to the right to change 0.001 to 1.

$$\frac{42.93}{0.001} = 42,930$$

Practice Exercise

12. Divide: $\dfrac{156.9}{0.01}$.

Review Exercises

Convert the number in each sentence to standard notation.   [5.3b]

1. Russia has the largest total area of any country in the world, at 6.59 million square miles.

2. Americans eat more than 3.1 billion lb of chocolate each year.

   Source: Chocolate Manufacturers' Association

Write a word name.   [5.1a]

3. 3.47   4. 0.031

5. 27.0001   6. 0.9

Write in fraction notation and, if possible, as a mixed numeral.   [5.1b]

7. 0.09   8. −4.561

9. −0.089   10. 3.0227

Write in decimal notation.   [5.1b]

11. $\dfrac{34}{1000}$ 12. $\dfrac{42,603}{10,000}$

13. $27\dfrac{91}{100}$ 14. $-867\dfrac{6}{1000}$

Which number is larger?   [5.1c]

15. 0.034, 0.0185   16. −0.91, −0.19

17. 0.741, 0.6943   18. 1.038, 1.041

Round 17.4287 to the nearest   [5.1d]

19. Tenth.   **20.** Hundredth.

21. Thousandth.   **22.** One.

Perform the indicated operation.

23. $\begin{array}{r} 2\,3\,6\,.\,2\,3\,1 \\ 2\,6\,3\,.\,4 \\ +\quad\ 0\,.\,1\,9\,8 \\ \hline \end{array}$ [5.2a]

24. $\begin{array}{r} 3\,7\,.\,6\,4\,5 \\ -\quad\ 8\,.\,4\,9\,7 \\ \hline \end{array}$ [5.2b]

25. $219.3 + 2.8 + 7$ [5.2a]

26. $745.0109 - 59.959$ [5.2b]

27. $-37.8 + (-19.5)$ [5.2c]

28. $-7.52 - (-9.89)$ [5.2c]

29. $\begin{array}{r} 4\,8 \\ \times\ 0\,.\,2\,7 \\ \hline \end{array}$ [5.3a]

30. $-3.7\,(0.29)$ [5.3a]

31. $\begin{array}{r} 2\,4\,.\,6\,8 \\ \times\ 1\,0\,0\,0 \\ \hline \end{array}$ [5.3a]

32. $2\,5\,\overline{)8\,0}$ [5.4a]

33. $11.52 \div (-7.2)$ [5.4a]

34. $\dfrac{276.3}{1000}$ [5.4a]

35. 3.056×0.001 [5.3a]

36. $\dfrac{-4.38}{0.001}$ [5.4a]

Combine like terms.   [5.2d]

37. $3.7x - 5.2y - 1.5x - 3.9y$

38. $7.94 - 3.89a + 4.63 + 1.05a$

39. Evaluate: $P - Prt$ for $P = 1000$, $r = 0.05$, and $t = 1.5$. (*A formula for depreciation*)   [5.3c]

40. Simplify: $9 - 3.2(-1.5) + 5.2^2$.   [5.4b]

41. Convert 1549 cents to dollars.   [5.3b]

42. Round $248.\overline{27}$ to the nearest hundredth.   [5.5b]

Multiply by a form of 1 to find decimal notation for each number.   [5.5c]

43. $\dfrac{13}{5}$

44. $\dfrac{32}{25}$

Use division to find decimal notation for each number.   [5.5a]

45. $\dfrac{13}{4}$

46. $-\dfrac{7}{6}$

47. Calculate: $\dfrac{4}{15} \times 79.05$.   [5.5d]

Solve. Remember to check.

48. $t - 4.3 = -7.5$ [5.7a]

49. $4.1x + 5.6 = -6.7$ [5.7a]

50. $6x - 11 = 8x + 4$ [5.7b]

51. $3(x + 2) = 5x - 7$ [5.7b]

Solve.   [5.8b]

52. *Record Movie Revenue.*   *Avatar* took in $2.78 billion over its entire run in movie theaters. This is $0.59 billion more than *Titanic* took in. How much did *Titanic* take in during its entire run?

Source: boxofficemojo.com/alltime/world

53. Stacia, a coronary intensive care nurse, earned $620.74 during a recent 40-hr week. What was her hourly wage? Round to the nearest cent.

54. *Landscaping.*   A rectangular yard is 20 ft by 15 ft. The yard is covered with grass except for a circular flower garden with an 8-ft diameter. Find the area of grass in the yard.

55. Derek had $1034.46 in his bank account. He used his debit card to buy a Wii system for $249.99. How much was left in his account?

56. *Credit Card Processing.*   Pay Right charges $150 for software, $8.95 a month for service, and 21¢ per online transaction. Timeless Treasures paid $178.90 for its first month of service. How many transactions did the store process that month?

57. *Recycling.*   Lisa volunteers at a local recycling center. One Saturday, she processed 130 more pounds of newspaper than of glass. She processed a total of 261.4 pounds of newspaper and glass. How many pounds of each did she process?

58. *Gas Mileage.* Inge wants to estimate gas mileage per gallon. With an odometer reading 36,057.1, she fills up. At 36,217.6 mi, the tank is refilled with 11.1 gal. Find the mileage per gallon. Round to the nearest tenth.

59. *Books in Libraries.* The table below lists the numbers of books, in millions, held in the five largest public libraries in the United States. Find the average number of books per library. Round to the nearest tenth.

LIBRARY	NUMBER OF BOOKS (in millions)
Library of Congress	33.5
Boston Public Library	24.1
New York Public Library	16.6
Harvard University	16.6
University of Illinois–Urbana	12.8

SOURCE: American Library Association

60. *Scanning Posters.* A high school club needs to scan posters designed by students and load them onto a flash drive. The copy center charges $12.99 for the flash drive and $1.09 per square foot for scanning. If the club needs to scan 13 posters at 3 sq ft per poster, what will the total cost be?

61. One pound of lean boneless ham contains 4.5 servings. It costs $5.99 per pound. What is the cost per serving? Round to the nearest cent.

62. *Construction.* A rectangular room measures 14.5 ft by 16.25 ft. How many feet of crown molding are needed to go around the top of the room? How many square feet of bamboo tiles are needed for the floor of the room?

63. Estimate the quotient $82.304 \div 17.287$ by rounding to the nearest ten. [5.6a]

A. 0.4   **B.** 4
C. 40   **D.** 400

64. Translate to an algebraic expression: 15 less than twice the price. [5.8a]

A. $2p - 15$ **B.** $15 - 2p$
C. $2 + p - 15$ **D.** $2(p - 15)$

Synthesis

65. 🖩 In each of the following, use $+$, $-$, $\times$, or $\div$ in each blank to make a true sentence. [5.4b]
a) $2.56 - 6.4 \;\square\; 51.2 - 17.4 + 89.7 = 119.66$
b) $(11.12 \;\square\; 0.29)3^4 = 877.23$

66. Arrange from smallest to largest:
$$-\frac{2}{3}, \; -\frac{15}{19}, \; -\frac{11}{13}, \; \frac{-5}{7}, \; \frac{-13}{15}, \; \frac{-17}{20}.$$

[5.1c], [5.5a]

67. *Automobile Leases.* Quentin leases a Ford Fusion Energi for $279 a month. He must pay an additional 20 cents per mile for all miles over 10,000 in one year. In 2014, his total bill for leasing the car was $6600. How many miles did he drive the car in 2014? [5.8b]

68. Use the fact that $\frac{1}{3} = 0.\overline{3}$ to find repeating decimal notation for 1. Explain how you got your answer. [5.5a]

69. 🖩 Sal's sells Sicilian pizza as a 17-in. by 20-in. pie for $15 or as an 18-in.-diameter round pie for $14. Which is a better buy and why? [5.8b]

Understanding Through Discussion and Writing

1. Describe in your own words a procedure for converting from decimal notation to fraction notation. [5.1b]

2. A student insists that $346.708 \times 0.1 = 3467.08$. How could you convince him that a mistake had been made without checking on a calculator? [5.3a]

3. When is long division *not* the fastest way to convert from fraction notation to decimal notation? [5.5a]

4. Consider finding decimal notation for $\frac{44}{125}$. Discuss as many ways as you can for finding such notation and give the answer. [5.5a]

CHAPTER

5   **Test**

For Extra Help

For step-by-step test solutions, access the Chapter Test Prep Videos in MyMathLab® or on YouTube (search "BittingerPrealgebra" and click on Channels).

1. In a recent year, 2.6 billion ducks were killed for food worldwide. Convert 2.6 billion to standard notation.

   Sources: FAO; *National Geographic*, May 2011, Nigel Holmes

2. Write a word name for 123.0047.

Write in fraction notation and, if possible, as a mixed numeral.

3. −0.91

4. 2.769

Write in decimal notation.

5. $\dfrac{74}{1000}$

6. $-\dfrac{37,047}{10,000}$

7. $756\dfrac{9}{100}$

8. $91\dfrac{703}{1000}$

Which number is larger?

9. 0.07, 0.162

10. 8.049, 8.0094

11. −0.09, −0.9

Round 5.6783 to the nearest

12. One.

13. Hundredth.

14. Thousandth.

15. Tenth.

Perform the indicated operation.

16.
```
   4 0 2.3
     2.8 1
+    0.1 0 9
```

17.
```
    0.1 2 5
×   0.2 4
```

18.
```
   2 1 3.4 5
×    0.0 0 1
```

19.
```
   5 2.0 9 1
−     7.3 4 5
```

20. 342.9 + 8.1 + 5.37

21. −9.5 + 7.3

22. 2 − 0.0054

23. 1000 × 73.962

24. $4\overline{)1\,9}$

25. $3.3\overline{)1\,0\,0.3\,2}$

26. $\dfrac{-346.82}{1000}$

27. $\dfrac{346.82}{0.01}$

28. Convert $179.82 to cents.

29. Combine like terms:
$$4.1x + 5.2 - 3.9y + 5.7x - 9.8.$$

30. Evaluate: $2l + 4w + 2h$ for $l = 2.4$, $w = 1.3$, and $h = 0.8$. (*The total girth of a postal package*)

31. Simplify: $20 \div 5(-2)^2 - 8.4$.

Multiply by a form of 1 to find decimal notation for each number.

32. $\dfrac{8}{5}$

33. $\dfrac{21}{4}$

Use division to find decimal notation for each number.

34. $-\dfrac{7}{16}$

35. $\dfrac{14}{9}$

36. Round $1.\overline{5}$ to the nearest hundredth.

Calculate.

37. $3 \div (-0.3) \cdot 2 - 1.5^2$

38. $(8 - 1.23) \div 4 + 5.6 \times 0.02$

39. $\dfrac{3}{8} \times 45.6 - \dfrac{1}{5} \times 36.9$

Solve. Remember to check.

40. $17y - 3.12 = -58.2$

41. $9t - 4 = 6t + 26$

42. $4 + 2(x - 3) = 7x - 9$

43. *Scanning Blueprints.* A building contractor needs to scan and load blueprints onto a flash drive. The copy center charges $10.99 for the flash drive and $1.19 per square foot for scanning. If the contractor needs to scan 5 blueprints at 6 sq ft per print, what will the total cost be?

44. *Gas Mileage.* Tina wants to estimate the gas mileage in her economy car. At 76,843 mi, she fills the tank with 14.3 gal of gasoline. At 77,310 mi, she fills the tank with 16.5 gal of gasoline. Find the mileage per gallon. Round to the nearest tenth.

45. *Checking Account Balance.* Nicholas had a balance of $820 in his checking account before making purchases of $123.89, $56.68, and $46.98 with his debit card. What was the balance after the purchases had been made?

46. The office manager for the Drake, Smith, and Hartner law firm buys 7 cases of copy paper at $41.99 per case. What is the total cost?

47. Life Expectancy. Life expectancies at birth for seven Asian countries are listed in the table below. Find the average life expectancy for this group of countries. Round to the nearest tenth.

COUNTRY	LIFE EXPECTANCY (in years)
Japan	82.25
South Korea	79.05
People's Republic of China	73.47
India	69.9
Russia	66.03
North Korea	63.81
Afghanistan	44.64

SOURCE: *CIA Factbook* 2011

48. About how many gallons of gasoline, at $2.749 per gallon, can be bought with $20?

A. 1 gallon   **B.** 3 gallons

C. 7 gallons   **D.** 12 gallons

Synthesis

49. Use one of the words *sometimes*, *never*, or *always* to complete each of the following.

a) The product of two numbers greater than 0 and less than 1 is _____ less than 1.

b) The product of two numbers greater than 1 is _____ less than 1.

c) The product of a number greater than 1 and a number less than 1 is _____ equal to 1.

d) The product of a number greater than 1 and a number less than 1 is _____ equal to 0.

50. Silver's Health Club charges a $79 membership fee and $42.50 a month. Allise has a coupon that will allow her to join the club for $299 for six months. How much will Allise save if she uses the coupon?

51. Travel Costs. Roundtrip airfare between Burlington, VT, and Newark, NJ, often costs $359. One estimate of the true cost of driving is $1.35 per mile. Is it more economical to fly or drive the 600 mi for **(a)** an individual; **(b)** a couple; **(c)** a family of 4?

Source: commutesolutions.org

Convert to fraction notation.

1. $2\frac{2}{9}$

2. 3.051

Find decimal notation.

3. $-\frac{7}{5}$

4. $\frac{6}{11}$

5. Determine whether 43 is prime, composite, or neither.

6. Determine whether 2,053,752 is divisible by 5.

Calculate.

7. $48 + 12 \div 4 - 10 \times 2 + 6892 \div 4$

8. $4.7 - \{0.1[1.2(3.95 - 1.65) + 1.5 \div 2.5]\}$

Round to the nearest hundredth.

9. 584.973

10. $218.\overline{5}$

11. Estimate the product 16.392×9.715 by rounding to the nearest one.

12. Estimate by rounding to the nearest tenth:
$2.714 + 4.562 - 3.31 - 0.0023$.

13. Evaluate $a \div 3 \cdot b$ for $a = 18$ and $b = 2$.

14. Combine like terms:
$-4p + 9 + 11p - 17$.

Add and simplify.

15. $2\frac{1}{4} + 3\frac{4}{5}$

16.
$$\begin{array}{r} 34,921 \\ 93,092 \\ +\ 11,103 \\ \hline \end{array}$$

17. $\frac{1}{6} + \frac{2}{3} + \frac{8}{9}$

18. $143.9 + 2.053$

Subtract and simplify.

19. $723,041 - 12,904$

20. $5.903 - 19$

21. $5\frac{1}{7} - 4\frac{3}{7}$

22. $\frac{9}{10} - \frac{10}{11}$

Multiply and simplify.

23. $-\frac{3}{8} \cdot \frac{4}{9}$

24.
$$\begin{array}{r} 2532 \\ \times\ 2100 \\ \hline \end{array}$$

25.
$$\begin{array}{r} 23.9 \\ \times\ 0.2 \\ \hline \end{array}$$

26.
$$\begin{array}{r} 27.9431 \\ \times\ 0.001 \\ \hline \end{array}$$

Divide and simplify.

27. $16.5\overline{)35.013}$

28. $26\overline{)47,918}$

29. $13.8621 \div 0.001$

30. $-\dfrac{4}{9} \div \left(-\dfrac{8}{15} \right)$

Solve.

31. $3(x - 7) = 5x + 2$

32. $-75 \cdot x = 2100$

33. $2.4x - 7.1 = 2.05$

34. $1062 + y = 368{,}313$

35. $t + \dfrac{5}{6} = \dfrac{8}{9}$

36. $-\dfrac{7}{8} \cdot t = \dfrac{7}{16}$

37. *Dominant Languages.* The table below lists the numbers of first-language speakers (people who acquired the language as infants) for the top five languages spoken worldwide. How many people in the world are first-language speakers of Spanish, Arabic, or Hindi?

LANGUAGE	FIRST-LANGUAGE SPEAKERS (in millions)
Chinese	1213
Spanish	329
English	328
Arabic	221
Hindi	182

SOURCES: Living Tongues Institute for Endangered Languages; UNESCO; SIL International

38. Refer to the table in Exercise 37. How many more people in the world are first-language speakers of Chinese than first-language speakers of English?

39. After Lauren made a $450 down payment on a sofa, $\frac{3}{10}$ of the total cost was paid. How much did the sofa cost?

40. Joshua's tuition was $3600. He obtained a loan for $\frac{2}{3}$ of the tuition. How much was the loan?

41. The balance in Elliott's checking account is $314.79. After a check is written for $56.02, what is the balance in the account?

42. A clerk in Leah's Delicatessen sold $1\frac{1}{2}$ lb of ham, $2\frac{3}{4}$ lb of turkey, and $2\frac{1}{4}$ lb of roast beef. How many pounds of meat were sold altogether?

43. A triangular sail has a height of 16 ft and a base of 11 ft. Find its area.

44. A 4-in. by 5-in. rectangle is punched from a round piece of steel that is 9 in. wide. How much steel will be left over?

Simplify.

45. $\left(\dfrac{3}{4} \right)^2 - \dfrac{1}{8} \cdot \left(3 - 1\dfrac{1}{2} \right)^2$

46. $1.2 \times 12.2 \div 0.1 \times 3.6$

Synthesis

47. Using a manufacturer's coupon, Lucy bought 2 cartons of orange juice and received a third carton free. The price of each carton was $3.59. What was the cost per carton with the coupon? Round to the nearest cent.

48. A carton of gelatin mix packages weighs $15\frac{3}{4}$ lb. Each package weighs $1\frac{3}{4}$ oz. How many packages are in the carton? (1 lb = 16 oz)

Introduction to Graphing and Statistics

6.1   Interpreting Data from Tables and Graphs

OBJECTIVES

a Extract and interpret data from tables.

b Extract and interpret data from graphs.

c Extract and interpret data from histograms.

SKILL TO REVIEW

Objective 4.7a: Multiply using mixed numerals.

Multiply.

1. $3\frac{1}{2} \times 800$

2. $1\frac{3}{4} \times 3020$

We use tables and graphs to display data and to communicate information about the data. For example, the following table and graphs display data on the resting heart rate for several mammals.

Table

	MOUSE	GIRAFFE	CAT	HUMAN	HORSE	ELEPHANT
Average Resting Heart Rate (in beats per minute)	500	170	130	70	35	28

SOURCES: elephantnaturepark.org; vetmedicine.about.com; giraffeconservation.org; learningabouthorses.com

Pictograph

Resting Heart Rate

Mouse / Giraffe / Cat / Human / Horse / Elephant

♥ = 50 beats per minute

Bar Graph

Resting heart rate (in number of beats per minute) vs. Mammal: Mouse, Giraffe, Cat, Human, Horse, Elephant

Answers

Skill to Review:
1. 2800   2. 5285

Comparing the table and the graphs reveals that the exact data values are most easily read in a table. The fastest and slowest heart rates can be determined easily from the graphs. The graph below communicates additional information about the data. The size of each mammal in this graph indicates the heart rate, not the actual size of the animal. The unexpected relative sizes of the mammals in the graph emphasize the fact that many smaller animals have faster heart rates than larger animals.

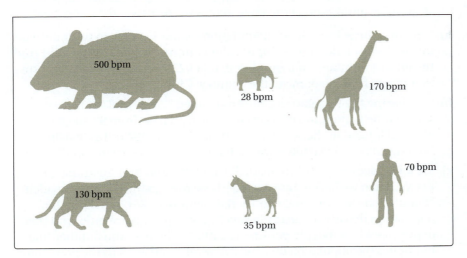

a READING AND INTERPRETING TABLES

A **table** is often used to present data in rows and columns.

EXAMPLE 1 *Population Density.*   The following table lists populations and land areas of 10 countries.

COUNTRY	LAND AREA (in square miles)	POPULATION		POPULATION DENSITY (per square mile)	
		2008	2012	2008	2012
Australia	2,941,299	20,434,176	22,015,576	7	7
Brazil	3,265,077	190,010,647	199,321,413	58	61
China	3,600,947	1,321,851,888	1,343,239,923	367	373
Finland	117,558	5,238,460	5,262,930	45	45
Germany	134,836	82,400,996	81,305,856	611	603
India	1,147,955	1,129,866,154	1,205,073,612	984	1050
Japan	144,689	127,433,494	127,368,088	881	880
Kenya	219,789	36,913,721	43,013,341	168	196
Mexico	742,490	108,700,891	114,975,406	146	155
United States	3,537,439	301,139,947	313,847,465	85	89

SOURCES: *World Almanac 2008*; U.S. Census Bureau

a) Which country had the largest population in 2012?

b) In which country or countries did the population decrease from 2008 to 2012?

c) In terms of population density in 2012, how many times more people per square mile were there in Kenya than in Australia?

d) Find the average land area of the four largest countries in the table.

Careful examination of the table allows us to answer the questions.

a) Note that the column head "Population" actually refers to two table columns. We look down the Population column headed "2012" and find the largest number. That number is 1,343,239,923. Then we look across that row to find the name of the country: China.

b) Comparing the Population columns headed "2008" and "2012," we see that the population decreased in the fifth row (from 82,400,996 to 81,305,856) and in the seventh row (from 127,433,494 to 127,368,088). Looking across these rows, we find the countries: Germany and Japan.

c) We look down the column headed "Country" and find Australia and Kenya. Then we look across each of those rows to the columns headed "Population Density" and choose the column for 2012. In 2012, the population density of Australia was 7 people per square mile, and that of Kenya was 196 people per square mile. We let $n =$ the number that multiplies the population density of Australia and gives us the population density of Kenya. Then we reword the question and translate.

Rewording: 7   times   what number   is   196?

Translating: 7 $\cdot$ n $=$ 196

To solve, we use the division principle.

$$7 \cdot n = 196$$
$$\frac{7 \cdot n}{7} = \frac{196}{7} \qquad \text{Dividing both sides by 7}$$
$$n = 28$$

In 2012, Kenya had 28 times more people per square mile than Australia did.

d) By looking down the column headed "Land Area," we determine that the four largest countries in the table are Australia, Brazil, China, and the United States. We find the average land area of these countries:

$$\frac{2{,}941{,}299 + 3{,}265{,}077 + 3{,}600{,}947 + 3{,}537{,}439}{4} = \frac{13{,}344{,}762}{4}$$
$$= 3{,}336{,}190.5 \text{ sq mi.}$$

◀ Do Exercises 1–5.

b READING AND INTERPRETING GRAPHS

Pictographs (or *picture graphs*) are another way to show information. Instead of actually listing the amounts to be considered, a **pictograph** uses symbols to represent the amounts. A pictograph includes a *key* that tells what each symbol represents.

Use the table in Example 1 to answer Margin Exercises 1–5.

1. Which country has the smallest land area?

2. What was the population of Mexico in 2012?

3. In terms of population density in 2008, how many times more people per square mile were there in Kenya than in Australia?   **GS**

The population density of Australia in 2008 was ☐ people per square mile. The population density of Kenya in 2008 was ☐ people per square mile.

$$7 \cdot n = \boxed{}$$
$$\frac{7 \cdot n}{7} = \frac{\boxed{}}{7}$$
$$n = \boxed{}$$

In 2008, Kenya had ☐ times more people per square mile than Australia did.

4. Which country had the greatest increase in population from 2008 to 2012?

5. Find the average population density in 2012 of the four most crowded countries in the table.

Answers

1. Finland   2. 114,975,406
3. 24 times the number of people per square mile
4. India, with an increase of 75,207,458
5. 726.5 people per square mile

Guided Solution:
3. 7, 168; 168, 168, 24; 24

EXAMPLE 2 *Roller Coasters.* The following pictograph shows the number of roller coasters listed in the Roller Coaster Data Base for six continents. Below the graph is a key that tells you that each symbol represents 100 roller coasters.

Roller Coasters of the World

Africa	🎢
Asia	🎢🎢🎢🎢🎢🎢🎢🎢🎢🎢🎢🎢🎢
Australia	🎢
Europe	🎢🎢🎢🎢🎢🎢🎢🎢
North America	🎢🎢🎢🎢🎢🎢🎢
South America	🎢🎢

 = 100 roller coasters

a) Which continent has the greatest number of roller coasters?

b) About how many roller coasters are there in Australia?

c) How many more roller coasters are there in Europe than in North America?

We can determine the answers by reading the pictograph.

a) The continent with the most symbols is Asia, so Asia has the greatest number of roller coasters.

b) The pictograph shows about $\frac{1}{4}$ symbol for Australia. Since each symbol represents 100 roller coasters, there are about $\frac{1}{4} \times 100$, or 25, roller coasters in Australia.

c) From the graph, we see that there are 8×100, or 800, roller coasters in Europe and about $7\frac{1}{2} \times 100$, or 750, roller coasters in North America. Thus there are $800 - 750$, or 50, more roller coasters in Europe than in North America. We could also estimate this difference by noting that Europe has $\frac{1}{2}$ of a symbol more than North America does, and $\frac{1}{2} \times 100 = 50$.

Do Exercises 6–8. ▶

When representing data with graphs, we must be sure that the areas of regions of the graph are proportional to the numbers that the regions represent. For example, in pictographs, each symbol is the same size, and the number of symbols is proportional to the actual data values. Thus the total area of the symbols is proportional to the data values. This area principle is illustrated in the following example.

Use the pictograph in Example 2 to answer Margin Exercises 6–8.

6. Which continent has the smallest number of roller coasters?

7. About how many roller coasters are there in Asia?

GS **8.** How many more roller coasters are there in South America than in Africa?

The graph shows $1\frac{1}{2}$ symbols for South America.

This represents ____ roller coasters.

The graph shows $\frac{1}{2}$ symbol for Africa.

This represents ____ roller coasters.

There are ____ more roller coasters in South America than in Africa.

EXAMPLE 3 *Electricity Generation.* The following graph illustrates the different methods used in the United States to generate electricity. Each yellow circle represents the amount of electricity generated, in terawatt-hours per year (TWh). Some methods of electricity generation are more efficient than others. The larger circle surrounding each yellow circle represents the amount of electricity that could have been generated if the method were 100% efficient.

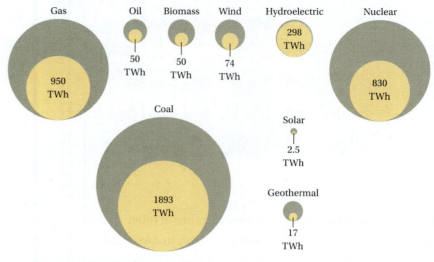

U.S. Electricity Generation

SOURCES: International Energy Agency; Eurelectric

a) How much electricity is generated annually with wind?

b) Which method generates the least electricity?

c) Which method of electricity generation is the most efficient?

d) Is generation of electricity from oil or from biomass more efficient?

We use the information in the graph to answer the questions. Note that the area of the larger circle that is not covered by the yellow circle represents the amount of energy lost or wasted during the generation process.

a) From the yellow circle labeled "Wind," we see that 74 TWh of electricity is generated annually with wind.

b) The smallest yellow circle is labeled "Solar," so the least electricity is generated by solar methods.

c) The yellow circle that most nearly fills its outer circle is labeled "Hydroelectric," so hydroelectric generation is the most efficient.

d) The same amount of electricity, 50 TWh, is generated using oil and biomass. Since the outer circle corresponding to oil is smaller than that corresponding to biomass, generation of electricity from oil is more efficient.

◀ **Do Exercises 9–12.**

Use the graph in Example 3 to answer Margin Exercises 9–12.

9. How much electricity is generated annually from gas?

10. Which method generates the most electricity?

11. Solar, geothermal, hydroelectric, and wind are all considered renewable energy sources. How much electricity is generated annually from renewable sources?

12. Is nuclear generation of electricity more or less efficient than generation of electricity from gas?

Answers

9. 950 TWh   **10.** Coal   **11.** 391.5 TWh
12. Nuclear generation of electricity is less efficient than generation from gas.

C HISTOGRAMS

A **histogram** is a special kind of graph that shows how often certain numbers appear in a set of data. Such *frequencies* are usually considered in terms of ranges of values.

EXAMPLE 4 *Fuel Economy.* Listed below are the fuel economy ratings, in miles per gallon, for combined city and highway driving for all 2013 midsize car models sold in the United States.

23, 20, 21, 21, 28, 24, 21, 22, 20, 20, 21, 19, 28, 26, 24, 23, 24, 20, 17, 26,

19, 17, 16, 14, 13, 29, 29, 21, 21, 22, 24, 22, 22, 20, 23, 23, 22, 16, 14, 21,

21, 19, 22, 21, 29, 31, 30, 33, 30, 29, 27, 33, 31, 30, 28, 26, 24, 30, 28, 22,

23, 24, 32, 29, 27, 22, 17, 19, 21, 18, 17, 23, 24, 31, 32, 27, 13, 13, 47, 29,

28, 26, 26, 25, 28, 43, 43, 29, 28, 25, 25, 22, 30, 32, 32, 32, 31, 32, 30, 30,

20, 19, 18, 29, 21, 22, 20, 20, 21, 19, 18, 23, 26, 28, 28, 29, 29, 30, 26, 26,

23, 21, 31, 24, 40, 19, 18, 19, 18, 20, 26, 25, 21, 22, 45, 21, 25, 24, 31, 32,

21, 22, 23, 20, 19, 23, 22, 23, 22, 26, 25, 31, 25, 22, 30, 34, 33, 34, 24, 27,

20, 50, 50, 24, 28, 40, 41, 40, 25, 25, 34, 23, 35, 26, 25, 21, 23

It is difficult to make sense of the 177 numbers in this data set, so the data are displayed below in a histogram.

Gas Mileages

Gas mileage
(in miles per gallons)

SOURCE: www.fueleconomy.gov

a) In which range of gas mileages did the greatest number of midsize models fall?

b) About how many midsize models had gas mileages that were less than 16 mpg?

c) About how many more midsize models had gas mileages between 26 mpg and 30 mpg than between 31 mpg and 35 mpg?

We use the histogram to answer the questions.

a) The tallest rectangle in the histogram is above the range 21–25, so the range 21 mpg to 25 mpg included the greatest number of midsize models.

b) The rectangle corresponding to 11–15 is 5 units high, so 5 midsize models had gas mileages that were less than 16 mpg.

c) From the histogram, we estimate that about 44 midsize models had gas mileages in the 26–30 range and about 21 models had gas mileages in the 31–35 range. Thus about 44 − 21, or 23, more midsize models had gas mileages between 26 mpg and 30 mpg than between 31 mpg and 35 mpg.

Do Exercises 13–15. ▶

The following histogram illustrates test grades for a class of 100 students. Use the histogram for Margin Exercises 13–15.

Test grades

13. Which range of grades included the greatest number of students?

14. About how many students received a test grade between 90 and 99?

15. About how many more students received a grade between 90 and 99 than a grade between 50 and 59?

Answers

13. 80–89 **14.** 25 students **15.** 20 students

✓ Reading Check

Determine whether each statement is true or false.

RC1. _____ There is only one correct way to represent a set of data.

RC2. _____ It is usually easy to read exact amounts from a pictograph.

RC3. _____ Histograms show frequencies.

RC4. _____ If the same data were displayed in a table and in a pictograph, we would have to use the pictograph to determine a maximum or a minimum.

Heat Index. In warm weather, a person can feel hot because of reduced heat loss from the skin caused by higher humidity. The **temperature–humidity index**, or **apparent temperature**, is what the temperature would have to be with no humidity in order to give the same heat effect. The following table lists the apparent temperatures for various actual temperatures and relative humidities. Use this table for Exercises 1–12.

ACTUAL TEMPERATURE (°F)	RELATIVE HUMIDITY									
	10%	20%	30%	40%	50%	60%	70%	80%	90%	100%
	APPARENT TEMPERATURE (°F)									
75°	75	77	79	80	82	84	86	88	90	92
80°	80	82	85	87	90	92	94	97	99	102
85°	85	88	91	94	97	100	103	106	108	111
90°	90	93	97	100	104	107	111	114	118	121
95°	95	99	103	107	111	115	119	123	127	131
100°	100	105	109	114	118	123	127	132	137	141
105°	105	110	115	120	125	131	136	141	146	151

In Exercises 1–4, find the apparent temperature for the given actual temperature and humidity combinations.

1. 80°, 60%

2. 90°, 70%

3. 85°, 90%

4. 95°, 80%

5. Which temperature–humidity combinations give an apparent temperature of 100°?

6. Which temperature–humidity combinations give an apparent temperature of 111°?

7. At a relative humidity of 50%, what actual temperatures give an apparent temperature above 100°?

8. At a relative humidity of 90%, what actual temperatures give an apparent temperature above 100°?

9. At an actual temperature of 95°, what relative humidities give an apparent temperature above 100°?

10. At an actual temperature of 85°, what relative humidities give an apparent temperature above 100°?

11. At an actual temperature of 85°, what is the difference in humidities required to raise the apparent temperature from 94° to 108°?

12. At an actual temperature of 80°, what is the difference in humidities required to raise the apparent temperature from 87° to 102°?

Planets. Use the following table, which lists information about the planets, for Exercises 13–20.

PLANET	AVERAGE DISTANCE FROM SUN (in miles)	DIAMETER (in miles)	LENGTH OF PLANET'S DAY IN EARTH TIME (in days)	TIME OF REVOLUTION IN EARTH TIME (in years)
Mercury	35,983,000	3,031	58.82	0.24
Venus	67,237,700	7,520	224.59	0.62
Earth	92,955,900	7,926	1.00	1.00
Mars	141,634,800	4,221	1.03	1.88
Jupiter	483,612,200	88,846	0.41	11.86
Saturn	888,184,000	74,898	0.43	29.46
Uranus	1,782,000,000	31,763	0.45	84.01
Neptune	2,794,000,000	31,329	0.66	164.78

SOURCE: *The Handy Science Answer Book*, Gale Research, Inc.

13. Find the average distance from the sun to Jupiter.

14. How long is a day on Venus?

15. Which planet has a time of revolution of 164.78 years?

16. Which planet has a diameter of 4221 mi?

17. About how many Earth diameters would it take to equal one Jupiter diameter?

18. How much longer is the longest time of revolution than the shortest?

19. What is the average of the diameters of the planets?

20. What is the average of the average distances from the sun of the planets?

Nutrition Facts. Most foods are required by law to provide factual information regarding nutrition, like that in the following table of nutrition facts from a box of Frosted Flakes cereal. Use the nutrition data for Exercises 21–26 on the next page.

Nutrition Facts

Serving Size 3/4 Cup (30g/1.1oz)
Servings Per Container About 16

Amount Per Serving	Cereal	Cereal with 1/2 Cup Vitamins A&D Fat Free Milk
Calories	110	150
Calories from Fat	0	0

	% Daily Value**	
Total Fat 0g*	0%	0%
Saturated Fat 0g	0%	0%
Trans Fat 0g		
Cholesterol 0mg	0%	0%
Sodium 140mg	6%	9%
Potassium 20mg	1%	6%
Total Carbohydrate 27g	9%	11%
Dietary Fiber 1g	3%	3%
Sugars 11g		
Other Carbohydrate 15g		
Protein 1g		

Vitamin A	10%	15%
Vitamin C	10%	10%
Calcium	0%	15%
Iron	25%	25%
Vitamin D	10%	25%
Thiamin	25%	30%
Riboflavin	25%	35%
Niacin	25%	25%
Vitamin B₆	25%	25%
Folic Acid	25%	25%
Vitamin B₁₂	25%	35%

*Amount in cereal. One half cup of fat free milk contributes an additional 40 calories, 65mg sodium, 6g total carbohydrates (6g sugars), and 4g protein.

**Percent Daily Values are based on a 2,000 calorie diet. Your daily values may be higher or lower depending on your calorie needs.

SOURCE: © 2013 Kellogg North America Company

21. Suppose your morning bowl of cereal consists of $1\frac{1}{2}$ cups of Frosted Flakes with 1 cup of fat-free milk. How many calories do you consume?

22. Suppose your morning bowl of cereal consists of $1\frac{1}{2}$ cups of Frosted Flakes with 1 cup of fat-free milk. What percent of the daily value of dietary fiber do you consume?

23. A nutritionist recommends that you look for foods that provide 10% or more of the daily value of vitamin C. Do you get that with 1 serving of Frosted Flakes and $\frac{1}{2}$ cup of fat-free milk?

24. Suppose you are trying to limit your daily caloric intake to 2000 calories. How many servings of cereal alone would it take to exceed 2000 calories?

25. Suppose your morning bowl of cereal consists of $1\frac{1}{2}$ cups of Frosted Flakes with 1 cup of fat-free milk. How much sodium do you consume? (*Hint*: Use the data listed in the first footnote below the table of nutrition facts.)

26. Suppose your morning bowl of cereal consists of $1\frac{1}{2}$ cups of Frosted Flakes with 1 cup of fat-free milk. How much protein do you consume? (*Hint*: Use the data listed in the first footnote below the table of nutrition facts.)

Rhino Population. The rhinoceros is considered one of the world's most endangered animals. The worldwide total number of rhinoceroses is approximately 20,700. The following pictograph shows the populations of the five remaining rhino species. Located in the graph is a key that tells you that each symbol represents 300 rhinos. Use the pictograph for Exercises 27–32.

SOURCE: World Wildlife Fund, 2008

27. Which species has the greatest number of rhinos?

28. Which species has the least number of rhinos?

29. How many more black rhinos are there than Indian rhinos?

30. How many more white rhinos are there than black rhinos?

31. What is the average number of rhinos for the five species?

32. How does the white rhino population compare with the Indian rhino population?

Personal Consumption Expenditures. The following graph shows the amounts of personal consumption expenditures, in dollars per person per year, in the United States, for four years. The graph also shows the amounts spent on food and on financial services and insurance for those years, labeled as percents of the personal consumption expenditures. Use the graph for Exercises 33–40.

Personal Consumption Expenditures

SOURCES: *TIME*, October 10, 2011, p. 32; infoplease.com; U.S. Census Bureau

33. How much were personal consumption expenditures per person in 1950?

34. How much were personal consumption expenditures per person in 2010?

35. For which of the years shown was more spent on food than on financial services and insurance?

36. For which of the years shown was more spent on financial services and insurance than on food?

37. How much per person was spent on food in 1990?

38. How much per person was spent on financial services and insurance in 1970?

39. a) How much less, as a percent of personal consumption expenditures, was spent on food in 2010 than in 1950?

b) How much more, in dollars, was spent on food in 2010 than in 1950?

40. a) How much more, as a percent of personal consumption expenditures, was spent on financial services and insurance in 2010 than in 1950?

b) How much more, in dollars, was spent on financial services and insurance in 2010 than in 1950?

Basketball. The following histogram illustrates the number of points scored per game by the Los Angeles Lakers during the 2012–2013 regular basketball season. Use the graph for Exercises 41–44.

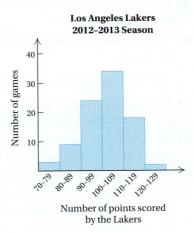

41. In how many games did the Lakers score 90–99 points?

42. In what point range did the highest number of Laker scores lie?

43. In what point range did the lowest number of Laker scores lie?

44. In how many more games did the Lakers score 100–109 points than 90–99 points?

6.2 Interpreting and Drawing Bar Graphs and Line Graphs

OBJECTIVES

a Extract and interpret data from bar graphs.

b Draw bar graphs.

c Extract and interpret data from line graphs.

d Draw line graphs.

SKILL TO REVIEW

Objective 5.1c: Given a pair of numbers in decimal notation, tell which is larger.

Which number is larger?

1. 0.078, 0.1
2. 36.4, 9.875

a READING AND INTERPRETING BAR GRAPHS

A **bar graph** is convenient for showing comparisons because you can tell at a glance which quantity is the largest or smallest. A *scale* is usually included with a bar graph so that estimates of values can be made with some accuracy. Bar graphs may be drawn horizontally or vertically, and sometimes a double bar graph is used to make comparisons.

EXAMPLE 1 *Coffee and Tea Consumption.* The following horizontal bar graph is a double bar graph, showing per capita consumption, in pounds per person per year, of both coffee and tea for several countries.

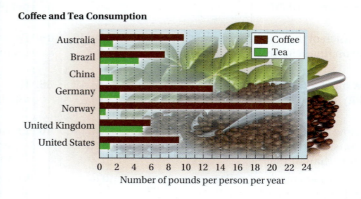

a) Which country has the highest per capita coffee consumption?

b) What is the per capita tea consumption in Brazil?

c) In which country do people consume about the same number of pounds of coffee and pounds of tea per year?

d) In which two countries do people consume about the same amount of tea?

e) In which countries is per capita coffee consumption greater than 10 pounds per year?

We use the graph to answer the questions.

a) The longest brown bar is for Norway. Thus Norway has the highest coffee consumption per capita.

b) We look to the right along the green bar associated with Brazil. Since it ends halfway between 4 and 5, we estimate Brazil's per capita tea consumption to be 4.5 pounds per year.

c) The brown and green bars representing data for the United Kingdom are about the same length, so people in the United Kingdom consume about the same number of pounds of coffee and pounds of tea per year.

Answers

Skill to Review:
1. 0.1 2. 36.4

d) The green bars are the same length for Australia and China; both countries have a per capita consumption of 1.5 pounds of tea. Thus people in Australia and China, on average, consume about the same amount of tea per year.

e) We move across the horizontal scale to 10. From there we move up, noting any brown bars that are longer than 10 units. We see that per capita coffee consumption is greater than 10 pounds per year in Germany and Norway.

Do Exercises 1–3. ▶

b DRAWING BAR GRAPHS

EXAMPLE 2 *Population by Age.* Listed below are U.S. population data for selected age groups. Make a vertical bar graph of the data.

AGE GROUP	PERCENT OF POPULATION
5–17 years	17%
18 years and older	76
10–49 years	54
16–64 years	66
55 years and older	25
65 years and older	13
85 years and older	2

SOURCE: U.S. Census Bureau

First, we indicate the age groups in seven equally spaced intervals on the horizontal scale and give the horizontal scale the title "Age category." (See Figure 1 below.)

Next, we scale the vertical axis. To do so, we look over the data and note that it ranges from 2% to 76%. We start the vertical scaling at 0, labeling the marks by 10's from 0 to 80. We give the vertical scale the title "Percent of population" and the graph the overall title "U.S. Population by Age."

Finally, we draw vertical bars to show the various percents, as shown in Figure 2.

FIGURE 1

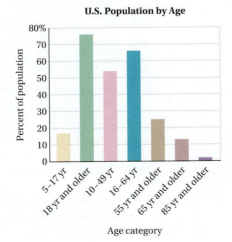

FIGURE 2

Do Exercise 4. ▶

Use the bar graph in Example 1 to answer Margin Exercises 1–3.

1. What is the per capita coffee consumption in the United Kingdom?

2. In which countries is per capita tea consumption less than 2 pounds per year?

3. How many more pounds of coffee are consumed per person in the United States than pounds of tea?

4. *Planetary Moons.* Make a horizontal bar graph to show the numbers of moons orbiting the various planets.

PLANET	MOONS
Earth	1
Mars	2
Jupiter	63
Saturn	60
Uranus	27
Neptune	13

SOURCE: National Aeronautics and Space Administration

Answers

1. About 5.9 pounds per year
2. Australia, China, Norway, and the United States
3. About 8 pounds per person
4.

C READING AND INTERPRETING LINE GRAPHS

Line graphs are often used to show a change over time as well as to indicate patterns or trends.

EXAMPLE 3 *Gold.* The following line graph shows the average price of gold, in dollars per ounce, for various years from 1970 to 2012.

Historic Gold Prices

SOURCE: kitco.com

a) For which year before 2000 was the average price of gold the highest?

b) Between which years did the average price of gold decrease?

c) For which year was the average price of gold about $450 per ounce?

d) By how much did the average price of gold increase from 2010 to 2012?

We look at the graph to answer the questions.

a) Before 2000, the highest point on the graph corresponds to 1980. Thus, before 2000, the average price of gold was highest in 1980.

b) Reading the graph from left to right, we see that the average price of gold decreased from 1980 to 1985 and from 1995 to 2000.

c) We look from left to right along a line at $450 per ounce. We see that the average price of gold was about $450 per ounce in 2005.

Historic Gold Prices

SOURCE: kitco.com

d) The graph shows that the average price of gold was about $1225 per ounce in 2010 and about $1675 per ounce in 2012. Thus the average price of gold increased by $1675 − $1225 = $450 per ounce.

◀ **Do Exercises 5–7.**

Use the line graph in Example 3 to answer Margin Exercises 5–7.

5. For which year after 1980 was the average price of gold the lowest?

6. Between which years did the average price of gold increase by about $800 per ounce?

7. For which years was the average price of gold less than $400 per ounce?

We look from left to right along a line at $ ⬚ per ounce. The points on the graph that are below this line correspond to the years 1970, 1975, 1985, 1990, 1995, and ⬚ .

Answers

5. 2000   **6.** Between 2005 and 2010
7. 1970, 1975, 1985, 1990, 1995, 2000

Guided Solution:
7. 400; 2000

d DRAWING LINE GRAPHS

EXAMPLE 4 *Temperature in Enclosed Vehicle.* The temperature inside an enclosed vehicle increases rapidly with time. Listed in the table below are the inside temperatures of an enclosed vehicle for specified elapsed times when the outside temperature is 80°F. Make a line graph of the data.

ELAPSED TIME	TEMPERATURE IN ENCLOSED VEHICLE WITH OUTSIDE TEMPERATURE 80°F
10 min	99°
20 min	109°
30 min	114°
40 min	118°
50 min	120°
60 min	123°

SOURCES: General Motors; Jan Null, Golden Gate Weather Services

First, we indicate the 10-min elapsed time intervals on the horizontal scale and give the horizontal scale the title "Elapsed time (in minutes)." See the figure on the left below. Next, we scale the vertical axis by 10's beginning with 80 to show the number of degrees and give the vertical scale the title "Temperature (in degrees)." The jagged line at the base of the vertical scale indicates that an unused portion of the scale has been omitted. We also give the graph the overall title "Temperature in Enclosed Vehicle with Outside Temperature 80°F."

Next, we mark the temperature at the appropriate level above each elapsed time. (See the figure on the right above.) Then we draw line segments connecting the points. The rapid change in temperature can be observed easily from the graph.

Do Exercise 8. ▶

8. *Military Technologies.* Listed below are the numbers of bachelor's degrees in military technologies earned in the United States for the years 2003–2009. Make a line graph of the data.

YEAR	NUMBER OF DEGREES EARNED
2003	6
2004	10
2005	40
2006	33
2007	168
2008	39
2009	55

SOURCE: U.S. Census Bureau

Answer

8.

☑ **Reading Check**

Determine whether each statement is true or false.

RC1. _____ Bar graphs may be drawn horizontally or vertically.

RC2. _____ A double bar graph indicates two amounts for each category.

RC3. _____ A line graph is always used to show trends over time.

RC4. _____ Some data could be illustrated using either a line graph or a bar graph.

a

Bearded Irises.   A gardener planted six varieties of bearded iris in a new garden on campus. Students from the horticulture department were assigned to record data on the range of heights for each variety. The vertical bar graph below shows their results. The length of the light green shaded portion of each bar (stem) along with the blossom illustrates the range of heights for a variety. For example, the range of heights for the miniature dwarf bearded iris is 2 in. to 9 in.

1. Which variety of iris has a minimum height of 17 in.?

2. Which variety of iris has a maximum height of 28 in.?

3. What is the range of heights for the border bearded iris?

4. What is the range of heights for the standard dwarf bearded iris?

5. Which variety of iris has the smallest range in heights?

6. Which irises have a maximum height less than 16 in.?

7. What is the difference between the maximum heights of the tallest iris and the shortest iris?

8. Which irises have a range in heights less than 10 in.?

Bearded Irises

Height (in inches)

Variety of bearded iris

SOURCE: www.irises.org/classification.htm

Chocolate Desserts. The following horizontal bar graph shows the average caloric content of various kinds of chocolate desserts. Use the bar graph for Exercises 9–16.

9. Estimate how many calories there are in 1 cup of hot cocoa with skim milk.

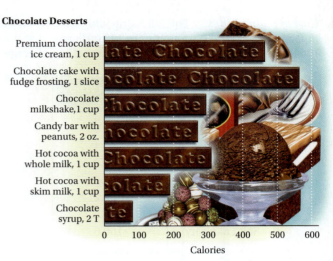

Chocolate Desserts

Premium chocolate ice cream, 1 cup
Chocolate cake with fudge frosting, 1 slice
Chocolate milkshake, 1 cup
Candy bar with peanuts, 2 oz.
Hot cocoa with whole milk, 1 cup
Hot cocoa with skim milk, 1 cup
Chocolate syrup, 2 T

0 100 200 300 400 500 600

Calories

10. Estimate how many calories there are in a 2-oz candy bar with peanuts.

11. Which dessert has the highest caloric content?

12. Which dessert has the lowest caloric content?

13. Which dessert contains about 460 calories?

14. Which desserts contain about 300 calories?

15. How many more calories are there in 1 cup of hot cocoa made with whole milk than in 1 cup of hot cocoa made with skim milk?

16. If Emily drinks a 4-cup chocolate milkshake, how many calories does she consume?

Bachelor's Degrees. The graph at right provides data on the numbers of bachelor's degrees conferred on men and on women in selected years. Use the bar graph for Exercises 17–20.

17. In which years were more bachelor's degrees conferred on men than on women?

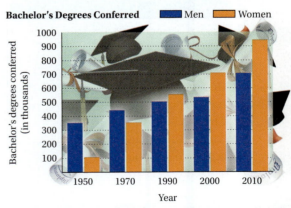

Bachelor's Degrees Conferred   ■ Men ■ Women

Bachelor's degrees conferred (in thousands)

1000 900 800 700 600 500 400 300 200 100

1950 1970 1990 2000 2010

Year

SOURCE: National Center for Education Statistics, U.S. Department of Education

18. How many more bachelor's degrees were conferred on women in 2010 than in 1970?

19. How many more bachelor's degrees were conferred on women than on men in 2000?

20. In which years were the numbers of bachelor's degrees conferred on men and on women each greater than 500,000?

b

21. *Cost of Living Index.* The following table lists the cost of living index for several cities. The national average of this index is 100. An index greater than 100 indicates that the cost of living is higher than average, and an index less than 100 indicates that the cost of living is lower than average. Make a horizontal bar graph to illustrate the data.

CITY	COST OF LIVING INDEX
Chicago	116.9
Denver	99.4
New York City	185.8
Juneau	136.5
Indianapolis	87.2
San Diego	132.3
Salt Lake City	100.6

SOURCE: U.S. Census Bureau

Use the data and the bar graph in Exercise 21 to do Exercises 22–25.

22. Which city has the highest cost of living index?

23. In which cities is the cost of living index less than 100?

24. In which cities is the cost of living approximately the national average?

25. How much higher is the cost of living index in New York City than in Chicago?

26. *Commuting Time.* The following table lists the average commuting time in six metropolitan areas with more than 1 million people. Make a vertical bar graph to illustrate the data.

CITY	COMMUTING TIME (in minutes)
New York City	38.3
Los Angeles	29.0
Phoenix	24.5
Houston	25.8
Indianapolis	21.7
Chicago	33.2

SOURCE: U.S. Census Bureau

Use the data and the bar graph in Exercise 26 to do Exercises 27–30.

27. Which city has the longest commuting time?

28. Which city has the shortest commuting time?

29. How much longer is the commuting time in New York City than that in Indianapolis?

30. Which cities have average commuting times of more than a half-hour?

c

Facebook Stock.   The line graph below shows the price per share of Facebook stock when it was first offered in May 2012 and at the beginning of each month for the remainder of that year. Use the graph for Exercises 31–34.

31. Estimate the opening price per share of Facebook stock in May 2012.

Stock Performance of Facebook

SOURCE: finance.yahoo.com

32. How much higher was the opening price of Facebook stock than its price at the beginning of September?

33. Between which months did the price of Facebook stock increase?

34. Between which months was the decrease in the price of Facebook stock the greatest?

Monthly Loan Payment.   Suppose you borrow $110,000 at an interest rate of $5\frac{1}{2}\%$ to buy a condominium. The following graph shows the monthly payment required to pay off the loan, depending on the length of the loan. Use the graph for Exercises 35–42.

35. Estimate the monthly payment for a loan of 15 years.

$110,000 Loan Repayment

36. Estimate the monthly payment for a loan of 25 years.

37. What time period corresponds to a monthly payment of about $760?

38. What time period corresponds to a monthly payment of about $625?

39. By how much does the monthly payment decrease when the loan period is increased from 10 years to 20 years?

40. By how much does the monthly payment decrease when the loan period is increased from 5 years to 20 years?

41. For a 10-year loan, there are 120 monthly payments. In all, how much will you pay back for a 10-year loan?

42. For a 20-year loan, there are 240 monthly payments. In all, how much will you pay back for a 20-year loan?

d

43. *Longevity Beyond Age 65.* The data in the table below indicate how many years beyond age 65 a male who is 65 in the given year could expect to live. Draw a line graph using the horizontal axis to scale "Year."

YEAR	AVERAGE NUMBER OF YEARS MEN ARE ESTIMATED TO LIVE BEYOND AGE 65
1980	14
1990	15
2000	15.9
2010	16.4
2020	16.9
2030	17.5

SOURCE: 2000 Social Security Report

44. What was the increase in longevity (years beyond 65) between 1980 and 2000?

45. What is the expected increase in longevity between 1980 and 2030?

46. What is the expected increase in longevity between 2020 and 2030?

47. What is the expected increase in longevity between 2000 and 2030?

Skill Maintenance

Solve.

48. $32 + n = 115$ [1.7b]

49. $x \cdot \dfrac{2}{3} = \dfrac{8}{9}$ [3.8b]

50. $y + \dfrac{5}{8} = \dfrac{11}{12}$ [4.3b]

51. $5 \cdot x = 11.3$ [5.7a]

52. $t + 4.752 = 11.1$ [5.7a]

53. $6x - 1 = 4x + 15$ [5.7b]

Calculate.

54. $3 \times [11 + (18 - 10) \div 2^3 - 5]$ [1.9d]

55. $2.56 \div (4 - 3.84) + 6.3 \times 0.2$ [5.4b]

56. $\dfrac{9}{10} \div \dfrac{1}{2} \cdot \dfrac{1}{3} - \left(\dfrac{1}{4} - \dfrac{1}{6}\right)$ [4.8a]

57. $6.25 \times 7\dfrac{1}{5}$ [5.5d]

Ordered Pairs and Equations in Two Variables

By using two perpendicular number lines as **axes** (pronounced ăk′ sēz; singular: **axis**), we can use points to represent solutions of certain equations. First we look at graphing points using such axes.

a   POINTS AND ORDERED PAIRS

When two number lines are used as axes, a grid can be formed. Just as a location in a city might be given as the intersection of an avenue and a side street, a point on a plane might be described as the intersection of a vertical line and a horizontal line on a grid. For the point shown in the figure below, these lines pass through 3 on the horizontal axis and 4 on the vertical axis. Thus, the **first coordinate** of this point is 3 and the **second coordinate** is 4. **Ordered pair** notation, (3, 4), provides a quick way of stating this.

Cault!!

When writing an ordered pair, you should *always* list the coordinate from the horizontal axis first.

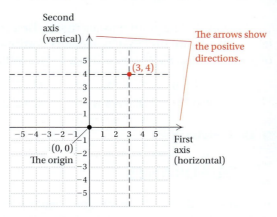

The point (0, 0), where the axes cross each other, is called the **origin**. To graph, or *plot*, the point (3, 4), we can begin at the origin and move horizontally (along the first axis) to the number 3. From there, we move up 4 units vertically and make a "dot."

Do Margin Exercises 1 and 2. ▶

EXAMPLE 1 Plot the points $(-5, 2)$ and $(2, -5)$.

To plot $(-5, 2)$, we locate -5 on the first, or horizontal, axis. From there we go up 2 units and make a dot.

To plot $(2, -5)$, we locate 2 on the first, or horizontal, axis. Then we go down 5 units and make a dot. Note that the order of the numbers within a pair is important: $(2, -5) \neq (-5, 2)$.

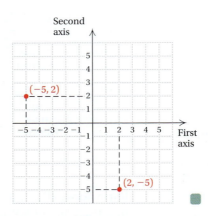

OBJECTIVES

a Plot a point, given its coordinates. Find coordinates, given a point.

b Determine the quadrant in which a point lies.

c Determine whether an ordered pair is a solution of an equation with two variables.

SKILL TO REVIEW

Objective 2.6a: Evaluate an algebraic expression by substitution.

1. Evaluate $2x + 3y$ for $x = 5$ and $y = -3$.
2. Evaluate $x - y$ for $x = 2$ and $y = -10$.

Plot these points on the graph below.

1. $(2, 5)$ **2.** $(4, 1)$

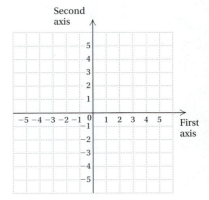

Answers

Skill to Review:
1. 1   **2.** 12

Margin Exercises:
1 and 2.

Plot these points on the graph below.

3. $(-2, 5)$ **4.** $(-3, -4)$

5. $(5, -3)$ **6.** $(-2, -1)$

7. $(0, -3)$ **8.** $(2\frac{1}{2}, 0)$

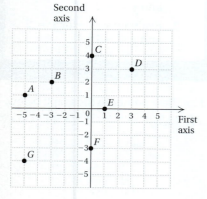

9. Determine the coordinates of points A, B, C, D, E, F, and G on the graph below.

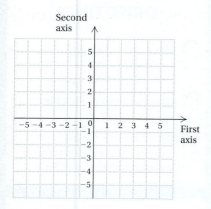

◀ **Do Exercises 3–8.**

To determine the coordinates of a given point, we first look directly above or below the point to find the point's horizontal coordinate. Then we look to the left or right of the point to identify the vertical coordinate.

EXAMPLE 2   Determine the coordinates of points A, B, C, D, E, F, and G.

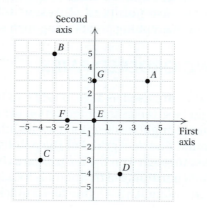

We look below point A to see that its first coordinate is 4. Looking to the left of point A, we find that its second coordinate is 3. Thus, the coordinates of point A are $(4, 3)$. The coordinates of the other points are

B: $(-3, 5)$; C: $(-4, -3)$; D: $(2, -4)$;

E: $(0, 0)$; F: $(-2, 0)$; G: $(0, 3)$.

◀ **Do Exercise 9.**

b   QUADRANTS

The axes divide the plane into four regions, or **quadrants**. For any point in region I (the *first quadrant*), both coordinates are positive. For any point in region II (the *second quadrant*), the first coordinate is negative and the second coordinate is positive. In region III (the *third quadrant*), both coordinates are negative. In region IV (the *fourth quadrant*), the first coordinate is positive and the second coordinate is negative.

As the figure above illustrates, the point $(2, 4)$ is in the first quadrant, $(-4.5, 5)$ is in the second quadrant, $(-2, -5)$ is in the third quadrant, and $(5, -5)$ is in the fourth quadrant.

Answers

3–8.

9. A: $(-5, 1)$; B: $(-3, 2)$; C: $(0, 4)$; D: $(3, 3)$; E: $(1, 0)$; F: $(0, -3)$; G: $(-5, -4)$

Do Exercises 10–15.

C SOLUTIONS OF EQUATIONS

The coordinate system we have just introduced is called the **Cartesian** coordinate system, in honor of the great mathematician and philosopher René Descartes (1596–1650). Legend has it that Descartes hit upon the idea of the coordinate system after watching a fly stop several times on the ceiling over his bed. Descartes used this coordinate system as a method of presenting solutions of equations containing two variables. Equations like $3x + 2y = 8$ have ordered pairs as solutions. In Section 6.4, we will find solutions and graph them. Here we simply practice checking to see if an ordered pair is a solution.

To determine whether an ordered pair is a solution of an equation, we normally substitute the first coordinate for the variable that comes first alphabetically and the second coordinate for the variable that is last alphabetically. The letters x and y are used most often.

EXAMPLE 3 Determine whether the ordered pair $(2, 1)$ is a solution of the equation $3x + 2y = 8$.

$$\begin{array}{c|c} 3x + 2y = 8 & \\ \hline 3 \cdot 2 + 2 \cdot 1 \;?\; 8 & \\ 6 + 2 & \\ 8 & 8 \quad \text{TRUE} \end{array}$$

Substituting 2 for x and 1 for y (alphabetical order of variables)

Since the equation becomes true, $(2, 1)$ is a solution.

In a similar manner, we can show that $(0, 4)$ and $(4, -2)$ are also solutions of $3x + 2y = 8$. In fact, there are an infinite number of solutions of $3x + 2y = 8$.

EXAMPLE 4 Determine whether the ordered pair $(-2, 3)$ is a solution of the equation $2t = 4s - 8$.

We substitute:

$$\begin{array}{c|c} 2t = 4s - 8 & \\ \hline 2 \cdot 3 \;?\; 4(-2) - 8 & \\ 6 & -8 - 8 \\ 6 & -16 \quad \text{FALSE} \end{array}$$

Using alphabetical order, substituting -2 for s and 3 for t

Since the equation becomes false, $(-2, 3)$ is not a solution.

Unless it is stated otherwise, the coordinates of an ordered pair correspond alphabetically to the variables.

Do Exercises 16 and 17.

10. What can you say about the coordinates of a point in the third quadrant?

11. What can you say about the coordinates of a point in the fourth quadrant?

In which quadrant is each point located?

GS 12. $(5, 3)$
To plot the point $(5, 3)$, we locate ☐ on the horizontal axis and go up ☐ units. We are now in the ☐ quadrant, or quadrant I.

13. $(-6, -4)$ 14. $(10, -14)$

15. $\left(-13, 9\tfrac{1}{2}\right)$

GS 16. Determine whether $(5, 1)$ is a solution of $y = 2x + 3$.
We substitute ☐ for x and ☐ for y.

$$\begin{array}{c|c} y = 2x + 3 & \\ \hline 1 \;?\; 2 \cdot 5 + 3 & \\ 1 & \;\square \end{array}$$

Since $1 = 13$ is ☐, <br> true/false
 $(5, 1)$ ☐ a solution.
 is/is not

17. Determine whether $(-13.6, 25.4)$ is a solution of $3x + 2y = 10$.

Answers

10. Both are negative numbers.
11. The first, or horizontal, coordinate is positive; the second, or vertical, coordinate is negative.
12. I   13. III   14. IV   15. II   16. No
17. Yes

Guided Solutions:
12. 5, 3; first
16. 5, 1; 13; false, is not

☑ Reading Check

Match the letter from the graph with the most appropriate term.
Not all letters will be used.

RC1. _____ Third quadrant

RC2. _____ Origin

RC3. _____ Second coordinate

RC4. _____ Vertical axis

RC5. _____ Ordered pair

a   Plot each group of points on the given graph below.

1. $(4, 4)$ $(-2, 4)$ $(5, -3)$ $(-5, -5)$ $(0, 4)$ $(0, -4)$
$(3, 0)$ $(-4, 0)$

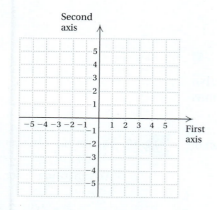

2. $(2, 5)$ $(-1, 3)$ $(3, -2)$ $(-2, -4)$ $(0, 4)$ $(0, -5)$
$(5, 0)$ $(-5, 0)$

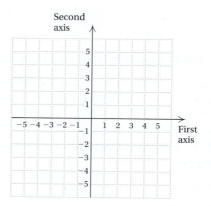

3. $(-2, -4)$ $(5, -4)$ $(0, 3\frac{1}{2})$ $(4, 3\frac{1}{2})$ $(-1, -3)$ $(-1, 5)$
$(4, -1)$ $(-2, 0)$

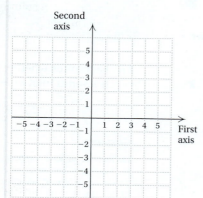

4. $(-3, -1)$ $(5, 1)$ $(-1, -5)$ $(0, 0)$ $(0, 1)$ $(-4, 0)$
$(2, 3\frac{1}{2})$ $(4\frac{1}{2}, -2)$

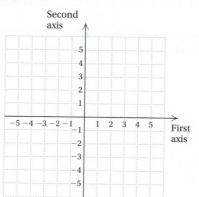

Determine the coordinates of points *A*, *B*, *C*, *D*, *E*, and *F*.

5.

6.

7.

8.

b In which quadrant is each point located?

9. $(-5, 3)$

10. $(-12, 1)$

11. $(100, -1)$

12. $\left(35\frac{1}{2}, -2\frac{1}{2}\right)$

13. $(-6.5, -1.9)$

14. $(-3.4, -5.9)$

15. $\left(3\frac{7}{10}, 9\frac{1}{11}\right)$

16. $(1895, 1492)$

Complete each sentence using the words *positive* and *negative* and/or the numeral I, II, III, and IV.

17. In quadrant IV, first coordinates are always _____ and second coordinates are always _____.

18. In quadrant III, first coordinates are always _____ and second coordinates are always _____.

19. In quadrant _____, both coordinates are always negative.

20. In quadrant _____, both coordinates are always positive.

21. In quadrants I and _____, the first coordinate is always _____.

22. In quadrants II and _____, the second coordinate is always _____.

C Determine whether each ordered pair is a solution of the given equation.

23. $(4, 3)$; $y = 2x - 5$

24. $(1, 7)$; $y = 2x + 5$

25. $(2, -3)$; $3x - y = 4$

26. $(-1, 4)$; $2x + y = 6$

27. $(-2, -1)$; $3x + 2y = -8$

28. $(0, -4)$; $4x + 2y = -9$

29. $(5, -4)$; $3x + y = 19$

30. $(-1, 7)$; $x - y = -8$

31. $\left(2\frac{1}{3}, 6\right)$; $2y - 3x = 5$

32. $\left(1\frac{1}{4}, 3\right)$; $2y - 4x = 1$

33. $(2.4, 0.7)$; $y = 5x - 11.3$

34. $(1.8, 7.4)$; $y = 3x + 2$

Skill Maintenance

Solve.

35. $3x - 4 = 17$ [2.8d]

36. $5(x - 2) = 3x - 4$ [5.7b]

37. $-\frac{1}{9}t = \frac{2}{3}t$ [3.8b]

38. Simplify: $\frac{90}{51}$. [3.5b]

39. Combine like terms: [4.6b]
$$7\frac{2}{11}a - 5\frac{1}{3}a.$$

40. Simplify: [2.7a]
$$3(x - 5) + 4x - 9.$$

Synthesis

Determine whether each ordered pair is a solution of the given equation.

41. ▦ $(-2.37, 1.23)$; $5.2x + 6.1y = -4.821$

42. ▦ $(4.16, -9.35)$; $6.5x - 7.2y = -94.35$

In Exercises 43–46, determine the quadrant(s) in which the point could be located.

43. The first coordinate is positive.

44. The second coordinate is negative.

45. The first and second coordinates are equal.

46. The first coordinate is the opposite of the second coordinate.

47. The points $(-1, 1)$, $(4, 1)$, and $(4, -5)$ are three vertices of a rectangle. Find the coordinates of the fourth vertex.

48. A parallelogram is a four-sided polygon with two pairs of parallel sides (two examples are shown below). Three parallelograms share the vertices $(-2, -3)$, $(-1, 2)$, and $(4, -3)$. Find the fourth vertex of each parallelogram.

Graphing Linear Equations

Once we can find a few ordered pairs that make an equation true, we can graph the equation.

a FINDING SOLUTIONS

To find a solution of an equation with two variables, we first replace one variable with some number choice and then solve the resulting equation for the other variable.

EXAMPLE 1   Find a solution of $x + y = 7$. Let $x = 5$.

If x is 5, then the solution will be an ordered pair $(5, y)$. We find y by substituting 5 for x in $x + y = 7$:

$$5 + y = 7.$$

We solve as follows:

$$
\begin{aligned}
5 + y &= 7 \\
5 + y - 5 &= 7 - 5 \qquad \text{Subtracting 5 from both sides} \\
y &= 2.
\end{aligned}
$$

Since $5 + 2 = 7$, the ordered pair $(5, 2)$ is a solution of $x + y = 7$.

Do Margin Exercise 1. ▶

EXAMPLE 2   Complete these solutions of $2x + 3y = 8$: $(\square, 2)$; $(-2, \square)$.

Recall that each solution will be of the form (x, y). Thus, to complete the pair $(\square, 2)$, we replace y with 2 and solve for x:

$$
\begin{aligned}
2x + 3y &= 8 \\
2x + 3 \cdot 2 &= 8 \qquad \text{Substituting 2 for } y \\
2x + 6 &= 8 \\
2x + 6 - 6 &= 8 - 6 \qquad \text{Subtracting 6 from both sides} \\
2x &= 2 \\
\tfrac{1}{2} \cdot 2x &= \tfrac{1}{2} \cdot 2 \qquad \text{Multiplying both sides by } \tfrac{1}{2} \\
x &= 1.
\end{aligned}
$$

Thus, $(1, 2)$ is a solution of $2x + 3y = 8$. Since $2(1) + 3(2) = 8$, the solution checks.

To complete the pair $(-2, \square)$, we replace x with -2 and solve for y:

$$
\begin{aligned}
2x + 3y &= 8 \\
2(-2) + 3y &= 8 \qquad \text{Substituting } -2 \text{ for } x \\
-4 + 3y &= 8 \\
3y &= 12 \qquad \text{Adding 4 to both sides} \\
y &= 4. \qquad \text{Dividing both sides by 3}
\end{aligned}
$$

Thus, $(-2, 4)$ is also a solution of $2x + 3y = 8$. Since $2(-2) + 3(4) = 8$, the solution checks.

Do Exercise 2. ▶

SKILL TO REVIEW

Objective 2.8d: Solve equations that require use of both the addition principle and the division principle.

Solve.

1. $3x + 2 \cdot 4 = 23$

2. $4(-1) - y = 7$

GS **1.** Find a solution of $x - y = 3$. Let $y = 5$.

Substitute 5 for y and solve for x.

$$
\begin{aligned}
x - y &= 3 \\
x - \blacksquare &= 3 \\
x - 5 + 5 &= 3 + \blacksquare \\
x &= \blacksquare
\end{aligned}
$$

One solution of $x - y = 3$ is $(\square, 5)$.

2. Complete these solutions of $5x + y = 10$: $(1, \square)$; $(\square, -5)$.

Answers

Skill to Review:
1. 5   2. −11

Margin Exercises:
1. $(8, 5)$ 2. $(1, 5)$; $(3, -5)$

Guided Solution:
1. 5, 5, 8; 8

3. Find three solutions of $x + 2y = 5$. Answers may vary.

4. Find three solutions of $y = -2x + 1$. Answers may vary.

Answers

3. $(1, 2), (5, 0), (3, 1)$
4. $(0, 1), (2, -3), (-2, 5)$

EXAMPLE 3 Find three solutions of $2x - y = 5$.

We are free to use *any* number as a replacement for either *x* or *y*. To find one solution, we select 1 as a replacement for *x*. We then solve for *y*:

$$2x - y = 5 \qquad \text{We are looking for an ordered pair } (1, \square).$$
$$2 \cdot 1 - y = 5 \qquad \text{Substituting 1 for } x$$
$$2 - y = 5$$
$$-y = 3 \qquad \text{Subtracting 2 from both sides}$$
$$-1y = 3 \qquad \text{Recall that } -a = -1 \cdot a.$$
$$y = -3. \qquad \text{Dividing both sides by } -1$$

Thus, $(1, -3)$ is one solution of $2x - y = 5$.

To find a second solution, we choose to replace *y* with 0 and solve for *x*:

$$2x - y = 5 \qquad \text{We are looking for an ordered pair } (\square, 0).$$
$$2x - 0 = 5 \qquad \text{Substituting 0 for } y$$
$$2x = 5 \qquad \text{Simplifying}$$
$$x = 2.5. \qquad \text{Dividing both sides by 2}$$

Thus, $(2.5, 0)$ is a second solution of $2x - y = 5$.

To find a third solution, we can replace *x* with 0 and solve for *y*:

$$2x - y = 5 \qquad \text{We are looking for an ordered pair } (0, \square).$$
$$2 \cdot 0 - y = 5 \qquad \text{Substituting 0 for } x$$
$$0 - y = 5$$
$$-y = 5$$
$$-1y = 5 \qquad \text{Try to do this step mentally.}$$
$$y = -5. \qquad \text{Dividing both sides by } -1$$

The pair $(0, -5)$ is a third solution of $2x - y = 5$.

Note that three different choices for *x* or *y* would have given three different solutions. There are an infinite number of ordered pairs that are solutions, so it is unlikely for two students to have solutions that match entirely.

To find a solution of an equation with two variables:

1. Choose a replacement for one variable.

2. Solve for the other variable.

3. Write the solution as an ordered pair that reflects the alphabetical order of the variables.

◄ **Do Exercises 3 and 4.**

b GRAPHING EQUATIONS

Equations like those considered in Examples 1–3 are in the form $Ax + By = C$. All equations that can be written this way are said to be **linear** because the solutions of such an equation, when graphed, form a straight line. An equation $Ax + By = C$ is called the **standard form** of a linear equation. When the line representing the solutions is drawn, we say that we have *graphed* the equation.

Since solutions of $Ax + By = C$ are written in the form (x, y), we label the horizontal axis as the *x*-axis and the vertical axis as the *y*-axis.

EXAMPLE 4 Graph: $2x - y = 5$.

We first need to calculate several solutions of $2x - y = 5$. In Example 3, we found that $(1, -3)$, $(2.5, 0)$, and $(0, -5)$ are solutions of the equation.

Next, we plot the points. As expected, the points describe a straight line. We draw the line using a straightedge and label it with the equation, as shown on the right below.

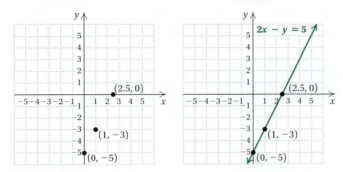

Note that two points are enough to determine a line, but we generally calculate and graph at least three ordered pairs before drawing each line. If the points do not all line up, we know that a mistake has been made.

Do Exercise 5. ▶

Linear equations are not always written in standard form. Equations like $y = 2x$ or $y = x + 2$ are also linear. To find solutions of equations like $y = 2x$, we usually choose values for x and then calculate y.

EXAMPLE 5 Graph: $y = 2x$.

First, we find some ordered pairs that are solutions. To find three ordered pairs, we can choose any three values for x and then calculate the corresponding values for y. One good choice is 0; we also choose -2 and 3.

If x is 0, then $y = 2x = 2 \cdot 0 = 0$.   Thus, $(0, 0)$ is a solution.
If x is -2, then $y = 2x = 2(-2) = -4$.   Thus, $(-2, -4)$ is a solution.
If x is 3, then $y = 2x = 2 \cdot 3 = 6$.   Thus, $(3, 6)$ is a solution.

We can compute additional pairs if we wish and list the ordered pairs that are solutions in a table.

x	y $y = 2x$	(x, y)
0	0	$(0, 0)$
-2	-4	$(-2, -4)$
3	6	$(3, 6)$
1	2	$(1, 2)$

↑ Substitute for x.
↑ Compute the value of y.
↑ Form the ordered pair (x, y).
Plot the points.
Draw and label the graph.

Next, we plot these points. We draw the line, or graph, with a ruler and label it $y = 2x$.

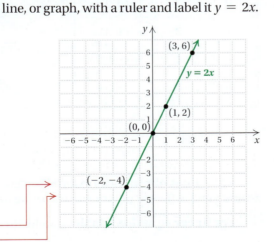

Do Exercises 6 and 7. ▶

---

5. Graph $x + 2y = 5$. Use the results from Margin Exercise 3.

Graph.

6. $y = 3x$

7. $y = \frac{1}{2}x$

Answers

5.

$x + 2y = 5$

6.

$y = 3x$

7.

$y = \frac{1}{2}x$

Graph.

8. $y = -x$ (or $y = -1 \cdot x$)

9. $y = -2x$

EXAMPLE 6 Graph: $y = -3x$.

We make a table of solutions. Then we plot the points, draw the line, and label the line $y = -3x$.

If x is 0, then $y = -3 \cdot 0 = 0$.
If x is 1, then $y = -3 \cdot 1 = -3$.
If x is -2, then $y = -3(-2) = 6$.
If x is 2, then $y = -3 \cdot 2 = -6$.

x	y $y = -3x$	(x, y)
0	0	$(0, 0)$
1	-3	$(1, -3)$
-2	6	$(-2, 6)$
2	-6	$(2, -6)$

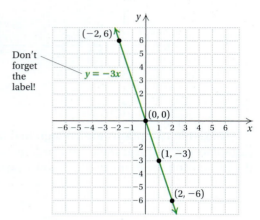

Don't forget the label!

◄ **Do Exercises 8 and 9.**

EXAMPLE 7 Graph: $y = x + 2$.

We make a table of solutions. Then we plot the points, draw the line, and label it.

If x is 0, then $y = 0 + 2 = 2$.
If x is 1, then $y = 1 + 2 = 3$.
If x is -1, then $y = -1 + 2 = 1$.
If x is 3, then $y = 3 + 2 = 5$.

x	y $y = x + 2$	(x, y)
0	2	$(0, 2)$
1	3	$(1, 3)$
-1	1	$(-1, 1)$
3	5	$(3, 5)$

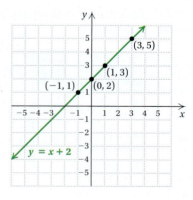

The values of x in these examples were *chosen*. Different choices for x would yield different points, but the same line.

For linear equations, tables can be formed using any numbers for x.

Answers

8.
$y = -x$

9.
$y = -2x$

EXAMPLE 8 Graph: $y = \frac{2}{3}x + 1$.

We make a table of solutions, plot the points, and draw and label the line. For this table, we selected multiples of 3 as x-values to avoid fraction values for y.

If x is 6, then $y = \frac{2}{3} \cdot 6 + 1 = 4 + 1 = 5$.
If x is 3, then $y = \frac{2}{3} \cdot 3 + 1 = 2 + 1 = 3$.
If x is 0, then $y = \frac{2}{3} \cdot 0 + 1 = 0 + 1 = 1$.
If x is -3, then $y = \frac{2}{3} \cdot (-3) + 1 = -2 + 1 = -1$.

x	$y = \frac{2}{3}x + 1$	(x, y)
6	5	$(6, 5)$
3	3	$(3, 3)$
0	1	$(0, 1)$
-3	-1	$(-3, -1)$

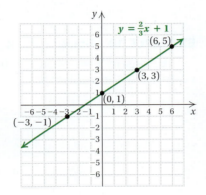

Graph.

10. $y = x + 1$

Do Exercises 10–12. ▶

C GRAPHING HORIZONTAL OR VERTICAL LINES

Any equation in the form $Ax + By = C$ is linear, provided A and B are not both zero. If A is 0 and B is nonzero, there is no x-term and the graph is a horizontal line. If B is 0 and A is nonzero, there is no y-term and the graph is a vertical line.

EXAMPLE 9 Graph: $y = 3$.

We regard $y = 3$ as $0 \cdot x + y = 3$. No matter what number we choose for x, we find that y must be 3 if the equation is to be true.

Choose any number for x. →

x	$y = 3$	(x, y)
-2	3	$(-2, 3)$
0	3	$(0, 3)$
4	3	$(4, 3)$

All pairs have 3 as the y-coordinate.

y must be 3.

When we plot $(-2, 3)$, $(0, 3)$, and $(4, 3)$ and connect the points, we obtain a horizontal line. Any ordered pair of the form $(x, 3)$ is a solution, so the line is 3 units above the x-axis, as shown in the graph at the top of the next page.

11. $y = -2x + 1$

12. $y = \frac{3}{5}x$

Answers

10.

$y = x + 1$

11.

$y = -2x + 1$

12.

$y = \frac{3}{5}x$

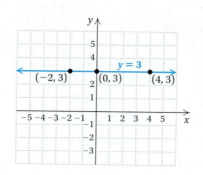

Graph.

13. $y = 4$

14. $x = 5$

Answers

13.

$y = 4$

14.

$x = 5$

EXAMPLE 10 Graph: $x = -4$.

We regard $x = -4$ as $x + 0 \cdot y = -4$ and make a table with -4 in every row in the x-column.

x must be $-4. \rightarrow$

x $x = -4$	y	(x, y)
-4	-5	$(-4, -5)$
-4	1	$(-4, 1)$
-4	3	$(-4, 3)$

All pairs have -4 as the x-coordinate.

Choose any number for y.

When we plot $(-4, -5)$, $(-4, 1)$, and $(-4, 3)$ and connect them, we obtain a vertical line. Any ordered pair of the form $(-4, y)$ is a solution, so the line is 4 units left of the y-axis.

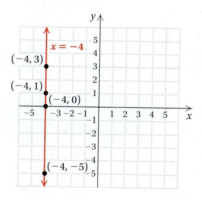

◀ **Do Exercises 13 and 14.**

HORIZONTAL AND VERTICAL LINES

The graph of $y = b$ is a horizontal line.
The graph of $x = a$ is a vertical line.

✓ Reading Check

Determine whether each statement is true or false.

RC1. _____ A linear equation in two variables has at most one solution.

RC2. _____ Every solution of $y = 3x - 7$ is an ordered pair.

RC3. _____ The graph of $y = 3x - 7$ represents all solutions of the equation.

RC4. _____ If a point is on the graph of $y = 3x - 7$, the corresponding ordered pair is a solution of the equation.

RC5. _____ To find a solution of $y = 3x - 7$, we can choose any value for x and calculate the corresponding y-value.

RC6. _____ The graph of every equation is a straight line.

a   For each equation, use the indicated value to find an ordered pair that is a solution.

1. $x + y = 8$; let $x = 5$

2. $x + y = 5$; let $x = 4$

3. $2x + y = 7$; let $x = 3$

4. $x + 2y = 9$; let $y = 4$

5. $y = 3x - 1$; let $x = 5$

6. $y = 2x + 7$; let $x = 3$

7. $x + 3y = 1$; let $x = 10$

8. $5x + y = 7$; let $y = -8$

9. $2x + 5y = 15$; let $x = 0$

10. $5x + 2y = 18$; let $x = 0$

11. $3x - 2y = 8$; let $y = -1$

12. $2x - 5y = 12$; let $y = -2$

For each equation, complete the given ordered pairs.

13. $x + y = 4$; $(\square, 3)$; $(-1, \square)$

14. $x - y = 6$; $(\square, 2)$; $(9, \square)$

15. $x - y = 4$; $(\square, 3)$; $(10, \square)$

16. $x + y = 10$; $(\square, 8)$; $(3, \square)$

17. $2x + 3y = 30$; $(0, \square)$; $(\square, 0)$

18. $3x + 2y = 24$; $(0, \square)$; $(\square, 0)$

19. $3x + 5y = 14$; $(3, \square)$; $(\square, 4)$

20. $4x + 3y = 11$; $(5, \square)$; $(\square, 2)$

21. $y = 4x$; $(\square, 4)$; $(-2, \square)$

22. $y = 6x$; $(\square, 6)$; $(-2, \square)$

23. $2x + 5y = 3$; $(0, \square)$; $(\square, 0)$

24. $5x + 7y = 9$; $(0, \square)$; $(\square, 0)$

For each equation, find three solutions. Answers may vary.

25. $x + y = 9$　　　　**26.** $x + y = 19$　　　　**27.** $y = 4x$　　　　**28.** $y = 5x$

29. $3x + y = 13$　　　　**30.** $x + 5y = 12$　　　　**31.** $y = 3x - 1$　　　　**32.** $y = 2x + 5$

33. $y = -7x$　　　　**34.** $y = -4x$　　　　**35.** $4 + y = x$　　　　**36.** $3 + y = x$

37. $3x + 2y = 12$　　　　**38.** $2x + 3y = 18$　　　　**39.** $y = \frac{1}{3}x + 2$　　　　**40.** $y = \frac{1}{2}x + 5$

b　Graph each equation.

41. $x + y = 6$

42. $x + y = 4$

43. $x - 1 = y$

44. $x - 2 = y$

45. $y = x - 4$

46. $y = x - 5$

47. $y = \dfrac{1}{3}x$

48. $y = -\dfrac{1}{3}x$

49. $y = x$

50. $y = x - 3$

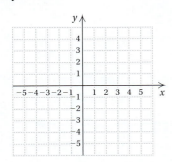

51. $y = 2x - 1$

52. $y = 2x - 3$

53. $y = 2x + 1$

54. $y = 3x + 1$

55. $y = \dfrac{2}{5}x$

56. $y = \dfrac{3}{4}x$

57. $y = -x + 4$

58. $y = -x + 5$

59. $y = \dfrac{2}{3}x + 1$

60. $y = \dfrac{2}{5}x - 1$

61. $y = 2$

62. $y = 1$

63. $x = 2$

64. $x = 3$

65. $x = -3$

66. $x = -1$

67. $y = -4$

68. $y = -2$

Skill Maintenance

69. The tunes on Miles Davis's classic *Kind of Blue* album are approximately 9 min, $9\frac{1}{2}$ min, $5\frac{1}{2}$ min, $11\frac{1}{2}$ min, and $9\frac{1}{2}$ min long. Find the average length of a tune on that album. [4.6c], [4.7d]

70. The books on Sherry's nightstand are 243, 410, 352, and 274 pages long. What is the average length of a book on the nightstand? [5.8b]

71. A recipe for a batch of chili calls for $\frac{3}{4}$ cup of red wine vinegar. How much vinegar is needed to make $2\frac{1}{2}$ batches of chili? [4.7d]

Simplify.

72. $-\dfrac{49}{77}$ [3.5b]

73. $-8 - 5^2 \cdot 2(3 - 4)$ [2.5b]

74. $\dfrac{3}{10}\left(-\dfrac{25}{12}\right)$ [3.4b]

Synthesis

Find three solutions of each equation. Then graph the equation.

75. ▦ $21x - 70y = -14$

76. ▦ $25x + 80y = 100$

77. ▦ $50x + 75y = 180$

78. Use the graph in Example 4 to find three solutions of $2x - y = 5$. Do not use the ordered pairs already listed.

79. List all solutions of $x + y = 6$ that use only whole numbers.

80. Graph three solutions of $y = |x|$ in the second quadrant and another three solutions in the first quadrant.

To the Student and the Instructor: Exercises marked with a ◣◥ symbol are meant to be solved using a graphing calculator.

81. ◣◥ Use a graphing calculator to graph each of the following.

    (a) $y = -0.63x + 2.8$ **(b)** $y = 2.3x - 4.1$

Mid-Chapter Review

Concept Reinforcement

Determine whether each statement is true or false.

_____ **1.** The key on a pictograph indicates what each symbol represents.   [6.1b]

_____ **2.** The solution of an equation in two variables is written as an ordered pair.   [6.3c]

_____ **3.** The graph of the equation $x = 5$ is a horizontal line.   [6.4c]

Guided Solutions

 Fill in each blank with the number that creates a correct solution.

4. Determine whether the ordered pair $(2, -1)$ is a solution of the equation $3x + y = 5$.   [6.3c]

$$3x + y = 5$$
$$3 \cdot \boxed{} + (\boxed{}) \; ? \; 5$$
$$\boxed{} + (\boxed{})$$
$$\boxed{} \bigm| 5 \quad \text{TRUE}$$

Since $3x + y = 5$ becomes true, $(2, -1)$ is a solution.

5. Find a solution of $x - y = 6$. Let $x = 1$.   [6.4a]

$$x - y = 6$$
$$\boxed{} - y = 6$$
$$-y = \boxed{}$$
$$y = \boxed{}$$

Thus, $(1, \boxed{})$ is a solution of $x - y = 6$.

Mixed Review

Downsizing. Companies sometimes downsize their products. That is, they charge the same price for a package that contains less product. The following table lists products that have been downsized. Use this table for Exercises 6–10.   [6.1a]

PRODUCT	SIZE (in ounces) OLD	SIZE (in ounces) NEW	PERCENT SMALLER
Breyer's ice cream	56	48	14
Hellmann's mayonnaise	32	30	6
Hershey's Special Dark chocolate bar	8	6.8	15
Iams cat food	6	5.5	8
Nabisco Chips Ahoy cookies	16	15.25	5
Skippy creamy peanut butter	18	16.3	9
Tropicana orange juice	96	89	7

SOURCE: *Consumer Reports*

6. How much less ice cream is in the new Breyer's ice cream package than in the old package?

7. By what percent has the size of a jar of Hellmann's mayonnaise changed in the downsizing process?

8. Which product in the table showed the greatest percent decrease?

9. How much less orange juice is in the new Tropicana orange juice package than in the old package?

10. Which product in the table showed the smallest percent decrease?

In which quadrant is each point located?   [6.3b]

11. $(-2, -12)$

12. $(-5, 6)$

13. $\left(\dfrac{1}{2}, -8\right)$

14. Determine whether the ordered pair $(0, -5)$ is a solution of $5x + y = 5$.   [6.3c]

15. Find a solution of $y = x + 3$. Let $x = -10$.   [6.4a]

16. Complete this solution of $3x - y = 7$: $(-1, \square)$.   [6.4a]

17. Find three solutions of $2x + y = 5$. Answers may vary.   [6.4a]

Graph each equation.   [6.4b], [6.4c]

18. $x + y = 3$

19. $y = x - 2$

20. $y = 4x$

21. $y = 2x - 4$

22. $y = -3$

23. $x = 1$

Understanding Through Discussion and Writing

24. Under what conditions will the points (a, b) and (b, a) be in the same quadrant?   [6.3b]

25. In which quadrant, if any, is the point $(5, 0)$? Why?   [6.3b]

26. To graph a linear equation, a student plots three points and discovers that the points do not line up with each other. What should the student do next?   [6.4b]

27. What is the greatest number of quadrants that a line can pass through? Why?   [6.3b], [6.4b]

6.5   Means, Medians, and Modes

OBJECTIVES

a Find the mean of a set of numbers and solve applied problems involving means.

b Find the median of a set of numbers and solve applied problems involving medians.

c Find the mode of a set of numbers and solve applied problems involving modes.

SKILLS TO REVIEW

Objectives 1.9c and 5.4b: Simplify expressions using the rules for order of operations.

1. Find the average of 282, 137, 5280, and 193.

2. Find the average of $23.40, $89.15, and $148.17 to the nearest cent.

A **statistic** is a number that is derived from a set of data and used to describe the data. One type of statistic is a *center point* or *measure of central tendency.* Three commonly used center points are the mean, the median, and the mode.

a   MEANS

The most commonly used center point is the *average* of the set of numbers. Although the word "average" is often used in everyday speech, in math we generally use *mean* instead.

MEAN

To find the **mean** of a set of numbers, add the numbers and then divide the sum by the number of items of data.

EXAMPLE 1   On 4 days of a trip, a car was driven the following number of miles each day: 240, 302, 280, 320. What was the mean number of miles per day?

$$\frac{240 + 302 + 280 + 320}{4} = \frac{1142}{4}, \quad \text{or} \quad 285.5$$

The car was driven an average of 285.5 mi per day. Had the car been driven exactly 285.5 mi each day, the same total distance (1142 mi) would have been traveled.

◀ **Do Margin Exercises 1–3.**

Find the mean.

1. 12, 15, 27   2. 10.5, 9.5, 8.2, 7.2

3. In a five-game series, Antonio scored 26, 21, 13, 14, and 23 points. Find the mean number of points scored per game.

EXAMPLE 2   *Scoring Average.*   Michael Jordan is the third highest scorer in the history of the National Basketball Association. He scored 32,292 points in 1072 games. What was the average number of points scored per game? Round to the nearest tenth.

Source: National Basketball Association

We know the total number of points, 32,292, and the number of games, 1072. We divide and round to the nearest tenth:

$$\frac{32{,}292}{1072} = 30.123134 \ldots \approx 30.1.$$

Michael Jordan's average score was about 30.1 points per game.

Answers

Skills to Review:
1. 1473   2. $86.91

Margin Exercises:
1. 18   2. 8.85   3. 19.4 points per game

Do Exercise 4. ▶

In a *weighted average*, more importance, or *weight*, is assigned to some values than to others. For example, a course syllabus may include the following description:

COURSE COMPONENT	WEIGHT FOR GRADE
Quizzes	20
Homework	30
Tests	50

If Allison has scored 70% on quizzes, 100% on homework, and 92% on tests, she cannot calculate her course grade by averaging 70, 100, and 92, because each category is weighted differently. Instead, she must multiply each percentage by its weight, add the results, and divide by the total of the weights:

$$\text{Course grade} = \frac{70 \cdot 20 + 100 \cdot 30 + 92 \cdot 50}{20 + 30 + 50}$$

$$= \frac{9000}{100} = 90.$$

Allison's course grade is 90%.

A grade point average is another example of a weighted average.

EXAMPLE 3 *Grade Point Average.* In many schools, students are assigned grade point values for grades obtained. The **grade point average**, or **GPA**, is the average of the grade point values for each credit hour taken. At Meg's college, grade point values are assigned as follows:

A: 4.0   B: 3.0   C: 2.0   D: 1.0   F: 0.0.

Meg earned the following grades for one semester. What was her grade point average?

COURSE	GRADE	NUMBER OF CREDIT HOURS IN COURSE
Colonial History	B	3
Basic Mathematics	A	4
English Literature	A	3
French	C	4
Time Management	D	1

To find the GPA, we first multiply the grade point value for each grade by the number of credit hours in the course to determine the number of *quality points,* and then add. Here each grade is weighted by the number of credit hours in the course.

Colonial History	$3.0 \cdot 3 = 9$
Basic Mathematics	$4.0 \cdot 4 = 16$
English Literature	$4.0 \cdot 3 = 12$
French	$2.0 \cdot 4 = 8$
Time Management	$1.0 \cdot 1 = \underline{1}$
	46 (Total)

4. *Home-Run Batting Average.* Babe Ruth hit 714 home runs in 22 seasons in the major leagues. What was his average number of home runs per season? Round to the nearest tenth.

Source: Major League Baseball

Answer

4. 32.5 home runs per season

5. Soha's sociology professor included the following in the course syllabus:

COURSE COMPONENT	WEIGHT FOR GRADE
Participation	15
Book reports	25
Research paper	40

Soha received 88% on her research paper and 92% on her book reports, and she anticipates a score of 100% for participation. What is her course grade?

Course grade

$$= \frac{100 \cdot 15 + 92 \cdot \boxed{} + 88 \cdot \boxed{}}{15 + 25 + 40}$$

$$= \frac{7320}{\boxed{}} = \boxed{}$$

Soha's course grade is ⬜%.

6. *Grade Point Average.* Alex earned the following grades one semester.

GRADE	NUMBER OF CREDIT HOURS IN COURSE
B	3
C	4
C	4
A	2

What was Alex's grade point average? Assume that the grade point values are 4.0 for an A, 3.0 for a B, and so on. Round to the nearest tenth.

7. *Grading.* To get an A in math, Rosa must score a mean of 90 on four tests. On the first three tests, her scores were 80, 100, and 86. What is the lowest score that Rosa can get on the last test and still get an A?

The total number of credit hours taken is $3 + 4 + 3 + 4 + 1$, or 15. We divide the number of quality points, 46, by the number of credit hours, 15, and round to the nearest tenth:

$$\text{GPA} = \frac{46}{15} \approx 3.1.$$

Meg's grade point average was 3.1.

◄ **Do Exercises 5 and 6.**

EXAMPLE 4 *Grading.* To get a B in math, Geraldo must score a mean of 80 on five tests. On the first four tests, his scores were 79, 88, 64, and 78. What is the lowest score that Geraldo can get on the last test and still get a B?
We can find the total of the five scores needed as follows:

$$80 + 80 + 80 + 80 + 80 = 5 \cdot 80, \quad \text{or} \quad 400.$$

The total of the scores on the first four tests is

$$79 + 88 + 64 + 78 = 309.$$

Thus Geraldo needs to get at least

$$400 - 309, \quad \text{or} \quad 91$$

in order to get a B. We can check this as follows:

$$\frac{79 + 88 + 64 + 78 + 91}{5} = \frac{400}{5}, \quad \text{or} \quad 80.$$

◄ **Do Exercise 7.**

b MEDIANS

Another type of center-point statistic is the *median*. Medians are useful when we wish to de-emphasize unusually extreme numbers. For example, suppose a small class scored as follows on an exam.

Jae:	78	Pat:	56
Jill:	81	Carmen:	84
Matt:	82		

Let's first list the scores in order from smallest to largest:

$$56, \quad 78, \quad 81, \quad 82, \quad 84.$$

↑
Middle score

The middle score—in this case, 81—is called the **median**. Note that because of the extremely low score of 56, the mean of the scores is 76.2. In this example, the median may be a more appropriate center-point statistic.

Answers

5. 91.5% **6.** 2.5 **7.** 94

Guided Solution:
5. 25, 40, 80, 91.5; 91.5

EXAMPLE 5 What is the median of this set of numbers?

99, 870, 91, 98, 106, 90, 98

We first rearrange the numbers in order from smallest to largest. Then we locate the middle number, 98.

90, 91, 98, 98, 99, 106, 870

↑
Middle number

The median is 98.

Do Exercises 8–10. ▷

Do Exercises 8–10.

Find the median.

8. 17, 13, 18, 14, 19

9. 20, 14, 13, 19, 16, 18, 17

10. 78, 81, 83, 91, 103, 102, 122, 119, 88

MEDIAN

Once the data in a set are listed in order, from smallest to largest, the **median** is the middle number if there is an odd number of data items. If there is an even number of items, the median is the number that is the average of the two middle numbers.

EXAMPLE 6 What is the median of this set of numbers?

69, 80, 61, 63, 62, 65

We first rearrange the numbers in order from smallest to largest. There is an even number of numbers. We look for the middle two, which are 63 and 65. The median is halfway between 63 and 65, the number 64.

61, 62, 63, 65, 69, 80

The average of the middle numbers is

$$\frac{63 + 65}{2} = \frac{128}{2}, \text{ or } 64.$$

└── The median is 64.

EXAMPLE 7 *Salaries.* The following are the salaries of the four highest-paid players in the National Hockey League. What is the median salary?

PLAYER	SALARY
Sidney Crosby	$8,700,000
Alexander Ovechkin	9,500,000
Evgeni Malkin	8,700,000
Eric Staal	8,250,000

SOURCE: Forbes.com

We rearrange the numbers in order from smallest to largest:

$8,250,000, $8,700,000, $8,700,000, $9,500,000.

The two middle numbers are $8,700,000 and $8,700,000. Since they are the same number, their average is $8,700,000. Thus the median salary is $8,700,000.

Do Exercises 11 and 12. ▷

Do Exercises 11 and 12.

Find the median.

11. *Salaries of Part-Time Typists.* $3300, $4000, $3900, $3600, $3800, $3400

GS **12.** 68, 34, 67, 69, 34, 70

Rearrange the numbers in order from smallest to largest:

34, 34, ___, 68, ___, 70.

The middle numbers are ___ and 68.

The average of 67 and 68 is ___.

The median is ___.

Answers

8. 17 **9.** 17 **10.** 91 **11.** $3700

12. 67.5

Guided Solution:

12. 67, 69; 67, 67.5, 67.5

C MODES

The final type of center-point statistic we will consider is the *mode*.

MODE

The **mode** of a set of data is the number or numbers that occur most often. If each number occurs the same number of times, there is *no* mode.

EXAMPLE 8 Find the mode of these data.

17, 13, 18, 17, 14, 19

To find the mode, it is helpful to first rearrange the numbers in order from smallest to largest.

13, 14, 17, 17, 18, 19

The number that occurs most often is 17. Thus the mode is 17.

A set of data has just one mean and just one median, but it can have no mode, one mode, or more than one mode.

EXAMPLE 9 Find the mode of these data.

5, 5, 11, 11, 13, 13

The numbers in this set of data are 5, 11, and 13. Each occurs twice, so all the numbers are equally represented. There is *no mode*.

EXAMPLE 10 Find the modes of these data.

33, 34, 34, 34, 35, 36, 37, 37, 37, 38, 39, 40

There are two numbers that occur most often, 34 and 37. Thus the modes are 34 and 37.

◀ Do Exercises 13–16.

Which statistic is best for a particular situation? If someone is bowling, the *mean* score from several games is a good indicator of that person's ability. If someone is applying for a job, the *median* salary at that business is often most indicative of what people are earning there because although executives tend to make a lot more money, there are fewer of them. For similar reasons, the selling price of homes is usually reported as a *median* price. Finally, if someone is reordering stock for a clothing store, the *mode* of the sizes sold is probably the most important statistic.

Find the modes of these data.

13. 23, 45, 45, 45, 78

14. 34, 34, 67, 67, 68, 70

15. 24, 89, 13, 28, 67, 27
   Rearrange the numbers in order from smallest to largest.

   13, 24, ⬚, 28, 67, ⬚.

   Each number occurs ⬚ time. There is no mode.

16. In a lab, Gina determined the mass, in grams, of each of five eggs:

   15 g, 19 g, 19 g, 14 g, 18 g.

   a) What is the mean?
   b) What is the median?
   c) What is the mode?

Answers

13. 45 **14.** 34, 67 **15.** No mode exists.

16. **(a)** 17 g; **(b)** 18 g; **(c)** 19 g

Guided Solution:

15. 27, 89; one

Translating for Success

1. Perimeter. A rectangular garden is 35 ft long and 27 ft wide. How many feet of fencing are needed to surround the garden?

2. Craft Sale. Marissa sold crocheted mittens at a craft sale. She sold 35 pairs on Friday and 27 pairs on Saturday. What was the mean number of pairs of mittens sold each day?

3. Jogging. David must jog 2 mi a day for training for basketball. If the alley behind his house is $\frac{1}{5}$ mi long, how many times must he run the length of the alley?

4. Taxi Fare. A taxi driver charges $2.25 plus 95 cents a mile. How far can Tonya travel on $11.75?

5. Fudge. Amber made $3\frac{1}{2}$ lb of fudge and divided it evenly among 3 packages for gifts. How much fudge was in each package?

The goal of these matching questions is to practice step (2), Translate, of the five-step problem-solving process. Translate each word problem to an equation and select a correct translation from equations A–O.

A. $x = 2 \div \frac{1}{5}$

B. $x = 4 + (-8)$

C. $x = 3 \cdot 3\frac{1}{2}$

D. $x = 4 - (-8)$

E. $35 + 27 = x$

F. $x = 2 \cdot 35 + 2 \cdot 27$

G. $x = 3\frac{1}{2} \div 3$

H. $x = 248 \div 15\frac{5}{10}$

I. $x = 3 + 3\frac{1}{2}$

J. $x = 2 \cdot \frac{1}{5}$

K. $35 \cdot 27 = x$

L. $2.25 + 0.95 + x = 11.75$

M. $248 \div x = 15\frac{5}{10}$

N. $x = \dfrac{35 + 27}{2}$

O. $2.25 + 0.95x = 11.75$

Answers on page A-13

6. Temperature. The weather forecast for the low temperature in Muncy on December 14 was 4°F. The actual low was −8°F. How many degrees lower than the forecast was the actual temperature?

7. Pizza Recipe. A recipe for pizza crust calls for $3\frac{1}{2}$ cups of flour. How many cups of flour are needed for 3 pizza crusts?

8. Total Purchase. Oliver spent $11.75 at the Country Store. He purchased candy for 95 cents, nuts for $2.25, and coffee. How much did the coffee cost?

9. Mileage. A car traveled 248 mi on $15\frac{5}{10}$ gal of gas. How many miles per gallon did it get?

10. Cell Phone Minutes. On Friday, Lathan made phone calls of lengths 35 min and 27 min. How many minutes did he use?

✓ Reading Check

Complete each sentence with the appropriate word from the list on the right. Not all choices will be used.

RC1. A(n) _____ is a number describing a set of data.

RC2. To find the _____ of a set of numbers, add the numbers and then divide by the number of items of data.

RC3. To find the weighted average of a set of numbers, multiply each number by its _____ , add the results, and divide by the total of the weights.

RC4. The _____ of a set of numbers is the number or numbers that occur most often.

mean

median

mode

statistic

weight

 For each set of numbers, find the mean, the median, and any modes that exist.

1. *Great Smoky Mountains National Park.* More people visit Great Smoky Mountains National Park each year than any other U.S. national park. The following bar graph shows the numbers of visitors the park had for 2005 to 2012. What is the mean number of visitors for the 8 years? the median? the mode?

Visitors to Great Smoky Mountains National Park

2. *Congestion.* The following bar graph shows the annual number of hours of traffic delay per auto commuter for 8 U.S. cities. What is the mean delay time? the median? the mode?

Traffic Delays

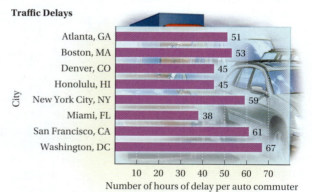

Number of hours of delay per auto commuter

SOURCE: *2012 Annual Urban Mobility Report*, Texas A&M Transportation Institute

3. 17, 19, 29, 18, 14, 29

4. 72, 83, 85, 88, 92

5. 5, 37, 20, 20, 35, 5, 25

6. 13, 32, 25, 27, 13

7. 4.3, 7.4, 1.2, 5.7, 8.3

8. 13.4, 13.4, 12.6, 42.9

9. 234, 228, 234, 229, 234, 278

10. $29.95, $28.79, $30.62, $28.79, $29.95

11. *Gas Mileage.* The 2013 Kia Rio does 396 mi of highway driving on 11 gal of gasoline. What is the mean number of miles expected per gallon—that is, what is the gas mileage?

Source: fueleconomy.gov

12. *Gas Mileage.* When using gas only, the 2013 Chevrolet Volt does 315 mi of city driving on 9 gal of gasoline. What is the mean number of miles expected per gallon—that is, what is the gas mileage?

Source: fueleconomy.gov

Grade Point Average. The tables in Exercises 13 and 14 show the grades of a student for one semester. In each case, find the grade point average. Assume that the grade point values are 4.0 for an A, 3.0 for a B, and so on. Round to the nearest tenth.

13.

GRADE	NUMBER OF CREDIT HOURS IN COURSE
B	4
A	5
D	3
C	4

14.

GRADE	NUMBER OF CREDIT HOURS IN COURSE
A	5
C	4
F	3
B	5

15. *Brussels Sprouts.* The following prices per stalk of Brussels sprouts were found at five farmers' markets:

$3.99, $4.49, $4.99, $3.99, $3.49.

What was the mean price per stalk? the median price? the mode?

16. *Mangoes.* The most popular fruit in the world is the mango, which is grown in over 2000 varieties. The following prices per pound of mangoes were found at five supermarkets:

$2.49, $1.59, $2.29, $2.49, $2.29.

What was the mean price per pound? the median price? the mode?

17. *Grading.* To get a B in math, Rich must score an average of 80 on five tests. His scores on the first four tests were 80, 74, 81, and 75. What is the lowest score that Rich can get on the last test and still receive a B?

18. *Grading.* To get an A in math, Cybil must score an average of 90 on five tests. Her scores on the first four tests were 90, 91, 81, and 92. What is the lowest score that Cybil can get on the last test and still receive an A?

19. *Length of Pregnancy.* Marta was pregnant 270 days, 259 days, and 272 days for her first three pregnancies. In order for the mean length of Marta's pregnancies to equal the worldwide mean length of 266 days, how long must her fourth pregnancy last?

Source: Vardaan Hospital, Dr. Rekha Khandelwal, M.S.

20. *Male Height.* Jason's brothers are 174 cm, 180 cm, 179 cm, and 172 cm tall. The average male is 176.5 cm tall. How tall is Jason if he and his brothers have a mean height of 176.5 cm?

21. *Median Home Prices.* The following table lists the selling prices of homes in two counties during one month.
a) Find the median home price for each county.
b) Which county had the lower median home price?

JEFFERSON COUNTY	HAMILTON COUNTY
$122,587	$387,262
138,291	146,989
121,103	262,105
768,407	253,289
532,194	112,681
129,683	127,092
278,104	131,612
110,329	

22. *Median Salaries.* The following table lists salaries for two small companies.
a) Find the median salary for each company.
b) Which company has the higher median salary?

VALUE SERVICES	DEPENDABLE CARE
$ 48,267	$18,242
32,193	21,607
189,607	98,322
56,189	87,212
28,394	56,812
152,693	42,394
42,681	50,112
	52,987

23. Movie Ticket Sales. The following table lists the number of movie tickets sold annually, in billions, from 2002 to 2012.

a) Find the mean number of tickets sold for the 8 years from 2002 to 2009.

b) Find the mean number of tickets sold for the 8 years from 2005 to 2012.

c) On average, were more tickets sold per year from 2002 to 2009 or from 2005 to 2012?

YEAR	NUMBER OF MOVIE TICKETS SOLD (in billions)	YEAR	NUMBER OF MOVIE TICKETS SOLD (in billions)
2002	1.58	2008	1.39
2003	1.55	2009	1.42
2004	1.49	2010	1.33
2005	1.40	2011	1.30
2006	1.41	2012	1.37
2007	1.40		

SOURCE: the-numbers.com

24. Movies Released. The following table lists the number of movies released to U.S. theaters annually from 2005 to 2012.

a) Find the mean number of movies released for the 5 years from 2005 to 2009.

b) Find the mean number of movies released for the 5 years from 2008 to 2012.

c) On average, were more movies released from 2005 to 2009 or from 2008 to 2012?

YEAR	NUMBER OF MOVIES RELEASED	YEAR	NUMBER OF MOVIES RELEASED
2005	507	2009	557
2006	594	2010	563
2007	611	2011	609
2008	638	2012	677

SOURCE: mpaa.org

Skill Maintenance

Multiply.

25. 12.86×17.5 [5.3a]

26. 222×0.5678 [5.3a]

27. $\dfrac{4}{5} \cdot \dfrac{3}{28}$ [3.6a]

28. $\dfrac{28}{45} \cdot \dfrac{3}{2}$ [3.6a]

Synthesis

Bowling Averages. Bowling averages are always computed by rounding down to the nearest integer. For example, suppose a bowler gets a total of 599 for 3 games. To find the average, we divide 599 by 3 and drop the amount to the right of the decimal point.

$$\frac{599}{3} \approx 199.67 \qquad \text{The bowler's average is 199.}$$

29. If Frances bowls 4176 in 23 games, what is her average?

30. If Eric bowls 4621 in 27 games, what is his average?

31. Hank Aaron. Hank Aaron averaged $34\frac{7}{22}$ home runs per year over a 22-yr career. After 21 yr, Aaron had averaged $35\frac{10}{21}$ home runs per year. How many home runs did Aaron hit in his final year?

32. Because of a poor grade on the fifth and final test, Chris's mean test grade fell from 90.5 to 84.0. What did Chris score on the fifth test? Assume that all tests are equally important.

33. Price Negotiations. Amy offers $6400 for a used Ford Taurus advertised at $8000. The first offer from Jim, the car's owner, is to "split the difference" and price the car at $(6400 + 8000) \div 2$, or $7200. Amy's second offer is to split the difference between Jim's offer and her first offer. Jim's second offer is to split the difference between Amy's second offer and his first offer. If this pattern continues and Amy accepts Jim's third (and final) offer, how much will she pay for the car?

34. The ordered set of data 18, 21, 24, a, 36, 37, b has a median of 30 and a mean of 32. Find a and b.

Predictions and Probability

a MAKING PREDICTIONS

OBJECTIVES

a Make predictions from a set of data using interpolation or extrapolation.

b Calculate the probability of an event occurring.

Sometimes we use data to make predictions or estimates of missing data points. One process for doing so is called **interpolation**. Interpolation enables us to estimate missing "in-between values" on the basis of known information.

EXAMPLE 1 *Monthly Mortgage Payments.* When money is borrowed and then repaid in monthly installments, the payment amount increases as the total number of payments decreases. The table below lists the size of a monthly payment when $110,000 is borrowed (at 9% interest) for various lengths of time. Use interpolation to estimate the monthly payment on a 35-yr loan.

YEAR	MONTHLY PAYMENT
5	$2283.42
10	1393.43
15	1115.69
20	989.70
25	923.12
30	885.08
35	?
40	848.50

SKILL TO REVIEW

Objective 5.5a: Use division to convert fraction notation to decimal notation.

Convert to decimal notation.

1. $\dfrac{4}{5}$ **2.** $\dfrac{2}{25}$

To use interpolation, we first plot the points and look for a trend. It seems reasonable to draw a line between the points corresponding to 30 and 40. We can "zoom in" to better visualize the situation. To estimate the second coordinate that is paired with 35, we trace a vertical line up from 35 to the graph and then left to the vertical axis. Thus we estimate the value to be $867. We can also estimate this value by averaging $885.08 and $848.50.

$$\frac{\$885.08 + \$848.50}{2} = \$866.79$$

$110,000 Loan Repayment

1. *Study Time and Test Scores.* A professor gathered the following data comparing study time and test scores. Use interpolation to estimate the test score received when studying for 21 hr.

STUDY TIME (in hours)	TEST SCORE (in percent)
19	83
20	85
21	?
22	91
23	93

When we estimate in this manner to find an in-between value, we are *interpolating*. Real-world information about the data might tell us that an estimate found in this way is unreliable. For example, data from the stock market might be too erratic for interpolation.

◀ **Do Exercise 1.**

We often analyze data with the intention of going "beyond" the data. One process for doing so is called **extrapolation**.

EXAMPLE 2 *Media.* The data in the following table and graphs show the mobile media share for U.S. adults in three years. Use extrapolation to estimate the mobile media share in 2015.

YEAR	MOBILE MEDIA SHARE
2012	13.4%
2013	19.2%
2014	23.3%
2015	?

SOURCES: eMarketer; statista.com

Mobile Media Share

 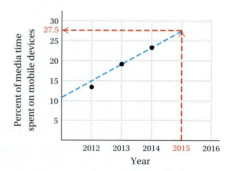

First, we analyze the data and note that they increase from 2012 through 2014. Then we draw a "representative" line through the data and beyond. We choose to draw the line through the two most recent points. To estimate a value for 2015, we draw a vertical line up from 2015 until it hits the representative line. We go to the left and read off a value—about 27.5. Thus we estimate the mobile media share for U.S. adults to be 27.5% in 2015. When we estimate in this way to find a "go-beyond value," we are *extrapolating*. Estimates found with this method vary depending on the "representative" line chosen.

◀ **Do Exercise 2.**

2. *Computer Media Share.* Use extrapolation to estimate the online media share for desktop and laptop computers in 2015. Use the information in the following table.

YEAR	DESKTOP AND LAPTOP COMPUTER ONLINE MEDIA SHARE
2012	20.7%
2013	19.2%
2014	18.0%
2015	?

SOURCES: eMarketer; statista.com

In calculus and statistics, other methods of interpolation and extrapolation are developed. The two basic concepts remain unchanged, but more complicated methods of determining what line "best fits" the given data are used. These methods are often computer-based.

Answers

1. 88%   **2.** About 16.8%

b PROBABILITY

Using *probability*, we can attach a numerical value to the likelihood that a specific event will occur.

Suppose we flip a fair coin. Because the coin is just as likely to land heads as it is to land tails, we say that the *probability* of it landing heads is $\frac{1}{2}$. Similarly, if we roll a fair die (plural: dice), we are as likely to roll a ⚃ as we are to roll a ⚀, ⚁, ⚂, ⚃, or ⚄. Because of this, we say that the probability of rolling a ⚃ is $\frac{1}{6}$.

EXAMPLE 3 A die is about to be rolled. Find the probability that a number greater than 4 will be rolled.

Since ⚀, ⚁, ⚂, ⚃, ⚄, and ⚅ are all equally likely to be rolled, and since two of these possibilities involve numbers greater than 4, we have

$$\begin{array}{l}\text{The probability of rolling} \\ \text{a number greater than 4}\end{array} = \frac{2}{6} \begin{array}{l}\leftarrow \text{Number of ways to roll a 5 or 6} \\ \leftarrow \text{Number of (equally likely) possible} \\ \text{outcomes}\end{array}$$

$$= \frac{1}{3}.$$

A coin landing heads and rolling a ⚃ are examples of "events." To find the probability, or likelihood, of an event occurring, we use the following principle.

> ### THE PRIMARY PRINCIPLE OF PROBABILITY
>
> If an event E can occur m ways out of n equally likely possible outcomes, then
>
> $$\text{The probability of } E \text{ occurring} = \frac{m}{n}.$$

EXAMPLE 4 A spinner such as the one shown below is used by flicking the arrow so that it rotates quickly. Find the probability that the arrow will come to rest on a green space. (Assume that the arrow will not rest on a line.)

There are 12 sections of equal size, so there are 12 equally likely possible outcomes. Since 4 of the possibilities are green, we have

$$\begin{array}{l}\text{Probability of} \\ \text{landing on green}\end{array} = \frac{\text{Number of ways to land on green}}{\text{Number of ways to land on a space}}$$

$$= \frac{4}{12} = \frac{1}{3}.$$

Do Exercise 3. ▶

GS **3.** Find the probability of the arrow in Example 4 landing on red or blue.

Probability of landing on red or blue

$$= \frac{\begin{array}{l}\text{Number of ways to} \\ \text{land on red or blue}\end{array}}{\begin{array}{l}\text{Number of ways to} \\ \text{land on a space}\end{array}}$$

$$= \frac{\boxed{}}{12}$$

$$= \frac{\boxed{}}{3}$$

EXAMPLE 5 A cloth bag contains 20 equal-size marbles: 5 are red, 7 are blue, and 8 are yellow. A marble is randomly selected. Find the probability that **(a)** a red marble is selected; **(b)** a blue marble is selected; **(c)** a yellow marble is selected.

4. A school play is attended by 500 people: 250 children, 100 seniors, and 150 (nonsenior) adults. After everyone has been seated, one audience member is selected at random. Find the probability of each of the following.

a) A child is selected.

b) A senior is selected.

c) A (nonsenior) adult is selected.

a) Since all 20 marbles are equally likely to be selected, we have

$$\text{The probability of selecting a red marble} = \frac{\text{Number of ways to select a red marble}}{\text{Number of ways to select any marble}}$$

$$= \frac{5}{20} = \frac{1}{4}, \text{ or } 0.25$$

b)
$$\text{The probability of selecting a blue marble} = \frac{\text{Number of ways to select a blue marble}}{\text{Number of ways to select any marble}}$$

$$= \frac{7}{20}, \text{ or } 0.35$$

c)
$$\text{The probability of selecting a yellow marble} = \frac{\text{Number of ways to select a yellow marble}}{\text{Number of ways to select any marble}}$$

$$= \frac{8}{20} = \frac{2}{5}, \text{ or } 0.4$$

◀ **Do Exercise 4.**

Many probability problems involve a standard deck of 52 playing cards. Such a deck is made up as shown below.

A deck of 52 cards

EXAMPLE 6 A card is randomly selected from a well-shuffled (mixed) deck of cards. Find the probability that **(a)** the card is a jack; **(b)** the card is a club.

5. A card is randomly selected from a well-shuffled deck of cards. Find the probability of each of the following.

a) The card is a diamond.

b) The card is a king or queen.

a)
$$\text{The probability of selecting a jack} = \frac{\text{Number of ways to select a jack}}{\text{Number of ways to select any card}}$$

$$= \frac{4}{52} = \frac{1}{13}$$

b)
$$\text{The probability of selecting a club} = \frac{\text{Number of ways to select a club}}{\text{Number of ways to select any card}}$$

$$= \frac{13}{52} = \frac{1}{4}$$

Answers

4. **(a)** $\frac{1}{2}$, or 0.5; **(b)** $\frac{1}{5}$, or 0.2;

(c) $\frac{3}{10}$, or 0.3 5. **(a)** $\frac{1}{4}$, or 0.25; **(b)** $\frac{2}{13}$

◀ **Do Exercise 5.**

✓ Reading Check

Complete each sentence with the appropriate word from the following list.

extrapolation   interpolation   probability

RC1. We use _____ to estimate values between known values.

RC2. We use _____ to estimate values beyond known values.

RC3. The _____ of an event occurring is a numerical value indicating its likelihood of occurrence.

a   Use interpolation or extrapolation to find the missing data values.

1. *Study Time and Grades.*   A math instructor asked her students to keep track of how much time each spent studying the chapter on decimal notation. They collected the information together with test scores from that chapter's test. The data are given in the following table. Estimate the missing value.

STUDY TIME (in hours)	TEST GRADE
13	80
15	85
17	80
19	?
21	86
23	91

2. *Maximum Heart Rate.*   A person's maximum heart rate depends on his or her gender, age, and resting heart rate. The following table relates resting heart rate and maximum heart rate for a 30-yr-old woman. Estimate the missing value.

RESTING HEART RATE (in beats per minute)	MAXIMUM HEART RATE (in beats per minute)
58	173
65	178
70	?
78	185
85	188

SOURCE: American Heart Association

3. *Digital Music Revenue.*

YEAR	DIGITAL MUSIC REVENUE (in billions of dollars)
2008	4.0
2009	4.4
2010	4.6
2011	5.1
2012	5.6
2013	5.9
2014	?

SOURCES: International Federation of the Phonographic Industry; statista.com

4. *Veterans Benefits.*

YEAR	COST OF MEDICAL CARE FOR U.S. VETERANS (in billions of dollars)
2007	$31.9
2008	35.3
2009	40.1
2010	42.4
2011	43.1
2012	45.5
2013	?

SOURCE: U.S. Department of Veterans Affairs

5. *International Students.*

YEAR	NEW INTERNATIONAL STUDENT ENROLLMENT IN U.S. INSTITUTIONS
2008–2009	200,000
2009–2010	203,000
2010–2011	215,000
2011–2012	?
2012–2013	251,000

SOURCE: Institute of International Education

6. *Baseball Ticket Price.*

YEAR	AVERAGE MAJOR LEAGUE BASEBALL TICKET PRICE
2010	$26.74
2011	26.91
2012	?
2013	27.73
2014	27.93

SOURCES: *Team Marketing Report*, annual reports

7. *Home Electricity Use.*

YEAR	AVERAGE U.S. HOUSEHOLD ELECTRICITY USE (in thousands of kWh)
2010	11.5
2011	11.3
2012	10.8
2013	?

SOURCE: Energy Information Administration, cited in Anne R. Carey and Paul Trap, *USA Today*, January 19, 2014

8. *Print Media Share.*

YEAR	SHARE OF MEDIA TIME SPENT WITH PRINT MEDIA
2012	5.4%
2013	4.4%
2014	3.5%
2015	?

SOURCES: eMarketer; statista.com

b Find each of the following probabilities.

Rolling a Die. In Exercises 9–12, assume that a die is about to be rolled.

9. Find the probability that a ⚁ is rolled.

10. Find the probability that a ⚃ is rolled.

11. Find the probability that an odd number is rolled.

12. Find the probability that a number greater than 2 is rolled.

Playing Cards. In Exercises 13–18, assume that one card is randomly selected from a well-shuffled deck (see p. 438).

13. Find the probability that the card is the jack of spades.

14. Find the probability that the card is a picture card (jack, queen, or king).

15. Find the probability that an 8 or a 6 is selected.

16. Find the probability that a black 5 is selected.

17. Find the probability that a red picture card (jack, queen, or king) is selected.

18. Find the probability that a 10 is selected.

Candy Colors. A box of candy was found to contain the following numbers of gumdrops.

Strawberry	7
Lemon	8
Orange	9
Cherry	4
Lime	5
Grape	6

In Exercises 19–22, assume that one of the gumdrops is randomly chosen from the box.

19. Find the probability that a cherry gumdrop is selected.

20. Find the probability that an orange gumdrop is selected.

21. Find the probability that the gumdrop is *not* lime.

22. Find the probability that the gumdrop is *not* lemon.

Skill Maintenance

23. *Population of Africa.* The population of South America is approximately $\frac{9}{25}$ of the population of Africa. In 2014, the population of South America was about 387,000,000. What was the population of Africa in 2014? [3.8b]

24. *Community Garden.* The Bingham community garden is to be split into 16 equally sized plots. If the garden occupies $\frac{3}{4}$ acre of land, how large will each plot be? [3.8b]

Synthesis

25. A coin is flipped twice. What is the probability that two heads will occur?

26. A coin is flipped twice. What is the probability that one head and one tail will occur?

27. A die is rolled twice. What is the probability that a ⚅ is rolled twice?

28. A day is chosen randomly during a leap year. What is the probability that the day is in July?

29. A die is rolled. What is the probability that a 7 is rolled?

30. A die is rolled. What is the probability that a ⚀, ⚁, ⚂, ⚃, ⚄, or ⚅ is rolled?

Vocabulary Reinforcement

Choose the term from the list on the right that best completes each sentence. Not every term will be used.

1. A(n) _____ presents data in rows and columns. [6.1a]

2. The _____ is the point where the axes cross. [6.3a]

3. A(n) _____ uses symbols to represent amounts. [6.1b]

4. The _____ of a set of data is the number or numbers that occur most often. [6.5c]

5. The _____ of a set of data is the sum of the numbers in the set divided by the number of items of data. [6.5a]

mean

median

mode

table

pictograph

histogram

origin

quadrant

Concept Reinforcement

Determine whether each statement is true or false.

_____ 1. To find the mean of a set of numbers, add the numbers and then multiply by the number of items of data. [6.5a]

_____ 2. If each number in a set of data occurs the same number of times, there is no mode. [6.5c]

_____ 3. If there is an odd number of items in a set of data, the middle number is the median. [6.5b]

Study Guide

Objective 6.1a Extract and interpret data from tables.

Example Which oatmeal listed below has the greatest number of calories?

PRODUCT	COST	CALORIES	FAT (g)	FIBER (g)	SUGARS (g)
	PER PACKET (instant) OR SERVING (longer-cooking)				
Quaker Quick-1 Minute	0.19	150	3.0	4	1
Kashi Heart to Heart Golden Brown Maple	0.44	160	2.0	5	12
Quaker Organic Maple & Brown Sugar	0.54	150	2.0	3	12
Nature's Path Organic Maple Nut	0.47	200	4.0	4	12

SOURCE: *Consumer Reports*, November 2008

We look down the column headed "Calories" and find the largest number, 200. The name of the oatmeal in that row is Nature's Path Organic Maple Nut.

Practice Exercises

1. Which oatmeal has the greatest cost per serving? What is that cost?

2. How many grams of sugar are in the Kashi oatmeal?

Objective 6.2a   Extract and interpret data from bar graphs.

Example   The horizontal bar graph below shows the building costs of selected stadiums. Estimate how much more Lucas Oil Stadium cost than Invesco Field cost.

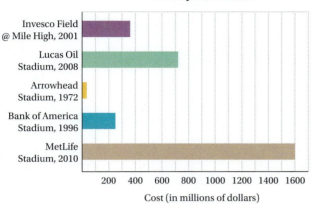

Cost of Sports Stadiums

SOURCE: National Football League

We estimate each cost.

Lucas Oil Stadium: $720 million; *Invesco Field:*
$360 million

Difference: $720 million $-$ $360 million $=$
$360 million

Practice Exercises

3. Which stadium cost less than $100 million?

4. Estimate how much more Invesco Field cost than Bank of America Stadium did.

Objective 6.3c   Determine whether an ordered pair is a solution of an equation with two variables.

Example   Determine whether the ordered pair $(-2, 1)$ is a solution of the equation $x + 5y = 3$.

$$x + 5y = 3$$
$$\overline{-2 + 5 \cdot 1} \ ? \ 3$$
$$-2 + 5 \ \Big|$$
$$3 \ \Big| \ 3 \quad \text{TRUE}$$

Since the equation becomes true for $x = -2$ and $y = 1$, $(-2, 1)$ is a solution.

Practice Exercise

5. Determine whether the ordered pair $(2, -4)$ is a solution of the equation $3x + y = 10$.

Objective 6.4b   Graph linear equations in two variables.

Example   Graph: $y = 2x - 3$.

We make a table of solutions. Then we plot the points, draw the line, and label it.

x	y	(x, y)
0	-3	$(0, -3)$
1	-1	$(1, -1)$
2	1	$(2, 1)$
3	3	$(3, 3)$

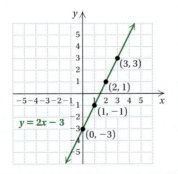

Practice Exercise

6. Graph: $y = x - 4$.

Objectives 6.5a, b, c Find the mean, the median, and the mode of a set of numbers.

Example Find the mean, the median, and the mode of this set of numbers:

2.6, 3.5, 61.8, 10.4, 3.5, 21.6, 10.4, 3.5.

Mean: We add the numbers and divide by the number of data items:

$$\frac{2.6 + 3.5 + 61.8 + 10.4 + 3.5 + 21.6 + 10.4 + 3.5}{8}$$

$$= \frac{117.3}{8} = 14.6625.$$

Median: We first rearrange the numbers from smallest to largest:

2.6, 3.5, 3.5, 3.5, 10.4, 10.4, 21.6, 61.8.

The median is halfway between the middle two numbers, which are 3.5 and 10.4. The average of these middle numbers is 6.95.

Mode: The number that occurs most often is 3.5, so it is the mode.

Practice Exercise

7. Find the mean, the median, and the mode of this set of numbers:

8, 13, 1, 4, 8, 7, 15.

Objective 6.6b Calculate the probability of an event occurring.

Example A company randomly selects a month of the year for an annual party. What is the probability that they choose a month whose name begins with J?

There are 12 months in a year, so there are 12 equally likely possible outcomes. There are 3 months whose names begin with J: January, June, and July. We have

The probability that the name of the month begins with J $= \dfrac{\text{Number of months whose names begin with J}}{\text{Number of months in the year}}$

$$= \frac{3}{12} = \frac{1}{4}.$$

Practice Exercise

8. A month of the year is randomly selected. What is the probability that the name of the month begins with A?

Review Exercises

Smartphone and Tablet Ownership. The table below lists the percents of the populations of several countries who own a smartphone and who own a tablet. Use this table for Exercises 1–3. [6.1a]

COUNTRY	PERCENT OWNING SMARTPHONE	PERCENT OWNING TABLET
Singapore	56%	18%
China	52	30
United States	35	17
Mexico	15	2
Germany	14	3
Philippines	8	5

SOURCE: Based on information from mastercard.com

1. What percent of China's population owns a smartphone?

2. In what countries does less than 10% of the population own a tablet?

3. In which country do approximately twice as many people own a smartphone as own a tablet?

Major League World Series. Except for four years, the World Series of Major League Baseball has been a best-of-seven series. In 1903, 1919, 1920, and 1921, the championship was a best-of-nine series. The championships have all been decided in 4, 5, 6, 7, or 8 games. The following pictograph shows the number of times the series has extended to each number of games. Use this graph for Exercises 4–6. [6.1b]

Number of Games Needed to Decide the World Series

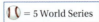

= 5 World Series

4. How many World Series were decided in 4 games?

5. In what number of games were the most World Series decided?

6. How many more World Series were decided in 7 games than were decided in 4 games?

Governors' Salaries. The histogram below shows the numbers of state governors in the United States who receive annual salaries in the given ranges. Use the graph for Exercises 7–9. [6.1c]

State Governors' Salaries

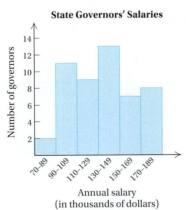

Annual salary
(in thousands of dollars)

SOURCE: knowledgecenter.csg.org

7. Which salary range has the smallest number of governors?

8. How many more governors make between $130,000 and $149,000 than make between $90,000 and $109,000?

9. How many governors make less than $130,000?

Tornadoes. The bar graph below shows the total number of tornadoes that occurred in the United States from 2010 through 2012, by month. Use the graph for Exercises 10–12. [6.2a]

Number of Tornadoes in 2010–2012

Month

10. Which month had the greatest number of tornadoes?

11. How many more tornadoes occurred in May than in June?

12. Do more tornadoes occur in the winter or in the spring?

Homelessness. The line graph below shows the average number of homeless children in New York City's shelter system each night for various years. Use the graph for Exercises 13–17. [6.2c]

New York City's Homeless Children

SOURCE: NYC Department of Homeless Services and Human Resources Administration and NYC Stat, shelter census reports

13. During which year after 1990 were there the fewest children in the shelter system?

14. How many children were in the shelter system each night in 2001?

15. In which years were there about 17,000 children each night in the shelter system?

16. Between which years did the number of children in the shelter system decrease?

17. By how much did the number of children in the shelter system each night increase from 2010 to 2013?

First-Class Postage. The table below lists the cost of first-class postage in various years. Use the table for Exercises 18 and 19.

YEAR	FIRST-CLASS POSTAGE
2001	34¢
2002	37
2006	39
2007	41
2008	42
2009	44
2012	45
2013	46

SOURCE: U.S. Postal Service

18. Make a vertical bar graph of the data. [6.2b]

19. Make a line graph of the data. [6.2d]

Determine the coordinates for each point. [6.3a]

20. *A*

21. *B*

22. *C*

23. *D*

Plot each point on the graph below. [6.3a]

24. (2, 5) **25.** (0, −3) **26.** (−4, −2)

In which quadrant is each point located? [6.3b]

27. (3, −8) **28.** (−20, −14) **29.** $\left(4\frac{9}{10}, 1\frac{3}{10}\right)$

30. Complete these solutions of $2x + 4y = 10$:
$(1, \square); (\square, -2)$. [6.4a]

Graph on a plane. [6.4b, c]

31. $y = 2x - 5$

32. $y = -\dfrac{3}{4}x$

33. $x + y = 4$

34. $x = -5$

35. $y = 6$

Find the mean. [6.5a]

36. 26, 34, 43, 51

37. 11, 14, 17, 18, 7

38. 0.2, 1.7, 1.9, 2.4

39. 700, 2700, 3000, 900, 1900

40. $2, $14, $17, $17, $21, $29

41. 20, 190, 280, 470, 470, 500

42. To get an A in math, Naomi must average at least 90 on four tests. Her scores on the first three tests were 94, 78, and 92. What is the lowest score she can receive on the last test and still get an A? [6.5a]

43. *Gas Mileage.* A 2012 Mazda Miata does 336 mi of highway driving on 12 gal of gasoline. What is the gas mileage? [6.5a]

44. *Grade Point Average.* Find the grade point average for one semester given the following grades. Assume the grade point values are 4.0 for A, 3.0 for B, and so on. Round to the nearest tenth. [6.5a]

COURSE	GRADE	NUMBER OF CREDIT HOURS IN COURSE
Math	A	5
English	B	3
Computer Science	C	4
Spanish	B	3
College Skills	B	1

Find the median. [6.5b]

45. 26, 34, 43, 51

46. 7, 11, 14, 17, 18

47. 0.2, 1.7, 1.9, 2.4

48. 700, 900, 1900, 2700, 3000

49. $2, $17, $21, $29, $14, $17

50. 470, 20, 190, 280, 470, 500

51. One summer, a student worked part time as a veterinary assistant. She earned the following weekly amounts over a six-week period: $360, $192, $240, $216, $420, and $132. What was the mean amount earned per week? the median? [6.5a, b]

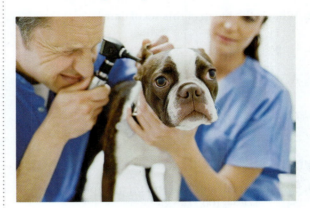

Find the mode or modes, if any exist. [6.5c]

52. 26, 34, 43, 26, 51

53. 17, 7, 11, 11, 14, 17, 18

54. 0.2, 0.2, 1.7, 1.9, 2.4, 0.2

55. 700, 700, 800, 2700, 800

56. $14, $17, $21, $29, $17, $2

57. 20, 20, 20, 20, 20, 500

58. Use extrapolation and the graph for Exercises 13–17 to estimate the number of children in New York City's shelter system each night in 2014. [6.6a]

A deck of 52 playing cards is thoroughly shuffled and a card is randomly selected. [6.6b]

59. Find the probability that the 5 of clubs is selected.

60. Find the probability that a red card is selected.

61. Find the mode(s), if any exist, of this set of data.

6, 9, 6, 8, 8, 5, 10, 5, 9, 10 [6.5c]

A. 8 **B.** 5, 6, 8, 9, 10
C. 9 **D.** No mode exists.

62. What is the mean of this set of data?

$\dfrac{1}{2}, \dfrac{1}{3}, \dfrac{1}{4}, \dfrac{1}{5}$ [6.5a]

A. $\dfrac{77}{240}$ **B.** $\dfrac{1}{3}$

C. $\dfrac{7}{24}$ **D.** $\dfrac{1}{4}$

Synthesis

63. 🔲 Find three solutions and then graph $34x + 47y = 100$. [6.4a, b]

64. A typing pool consists of four senior typists who earn $12.35 per hour and nine other typists who earn $11.15 per hour. Find the mean hourly wage. [6.5a]

65. The ordered set of data 298, 301, 305, a, 323, b, 390 has a median of 316 and a mean of 326. Find a and b. [6.5a, b]

Graph on a plane. [6.4b]

66. $1\dfrac{2}{3}x + \dfrac{3}{4}y = 2$

67. Find the perimeter of a rectangle with vertices at $(5, 3)$, $(5, -2)$, $(-3, -2)$, and $(-3, 3)$. [6.3a]

68. Find the area of a rectangle with vertices at $(0, 9)$, $(0, -4)$, $(5, -4)$, and $(5, 9)$. [6.3a]

Understanding Through Discussion and Writing

1. Find a real-world situation that fits this equation:

$T = \dfrac{20,500 + 22,800 + 23,400 + 26,000}{4}$. [6.5a]

2. Can bar graphs always, sometimes, or never be converted to line graphs? Why? [6.2b, d]

3. Is it possible for the mean of a set of numbers to be greater than all but one of the numbers in the set? Why or why not? [6.5a]

4. Is it possible for the median of a set of four numbers to be one of the numbers in the set? Why or why not? [6.5b]

5. One way the U.S. Census Bureau reports income is by giving a 3-yr mean of median incomes. Why would this statistic be chosen? [6.5a, b]

6. Would a company considering expansion be more interested in interpolation or extrapolation? Why? [6.6a]

CHAPTER

6   **Test**

For Extra Help   For step-by-step test solutions, access the Chapter Test Prep Videos in MyMathLab® or on You Tube (search "BittingerPrealgebra" and click on Channels).

This table lists the number of calories burned during various walking activities. Use it for Exercises 1 and 2.

WALKING ACTIVITY	CALORIES BURNED IN 30 MIN		
	110 LB	132 LB	154 LB
Walking			
Fitness (5 mph)	183	213	246
Mildly energetic (3.5 mph)	111	132	159
Strolling (2 mph)	69	84	99
Hiking			
3 mph with 20-lb load	210	249	285
3 mph with 10-lb load	195	228	264
3 mph with no load	183	213	246

1. Which activity provides the greatest benefit in burned calories for a person who weighs 132 lb?

2. What activities would allow a person weighing 154 lb to burn at least 250 calories every 30 min?

Waste Generated. The number of pounds of waste generated per person per year varies greatly among countries around the world. In the pictograph at right, each symbol represents approximately 100 lb of waste. Use the pictograph for Exercises 3–6.

3. In which country does each person generate 1300 lb of waste per year?

4. In which countries does each person generate more than 1500 lb of waste per year?

5. How many pounds of waste per person per year are generated in Canada?

6. How many more pounds of waste per person per year are generated in the United States than in Mexico?

Amount of Waste Generated (per person per year)

SOURCE: OECD, Key Environmental Indicators   = 100 pounds

7. *Animal Speeds.*   The following table lists maximum speeds of movement for various animals, in miles per hour, compared to the speed of the fastest human. Make a vertical bar graph of the data.

ANIMAL	SPEED (in miles per hour)
Human	28
Chicken	9
Elephant	25
Cheetah	70
Zebra	40
Lion	50
Pronghorn antelope	61
Elk	45

SOURCE: *Natural History Magazine,* © American Museum of Natural History

Refer to the table in Exercise 7 for Exercises 8–11.

8. By how much does the fastest speed exceed the slowest speed?

9. How many times faster can a cheetah run than the fastest human?

10. Find the mean of all the speeds.

11. Find the median of all the speeds.

Hurricanes.   The following line graph shows the numbers of Atlantic hurricanes for the years 2000–2012. Use the graph for Exercises 12–17.

Atlantic Hurricanes

12. What year had the greatest number of Atlantic hurricanes?

13. In what year were there 3 Atlantic hurricanes?

14. How many hurricanes were there in 2012?

15. How many more hurricanes were there in 2005 than in 2006?

16. Find the mean number of hurricanes per year for the years 2008–2012.

17. Use extrapolation to estimate the number of Atlantic hurricanes in 2013.

In which quadrant is each point located?

18. $\left(-\frac{1}{2}, 7\right)$ **19.** $(-5, -6)$

Determine the coordinates of each point.

20. A

21. B

22. C

23. Complete the following solution of the equation $y - 3x = -10$: (, 2).

Graph.

24. $y = 2x - 2$

25. $y = -\dfrac{3}{2}x$

26. $x = -2$

Find the mean.

27. 45, 49, 52, 54   **28.** 1, 2, 3, 4, 5   **29.** 3, 17, 17, 18, 18, 20

Find the median and any modes that exist.

30. 45, 47, 54, 54   **31.** 1, 2, 3, 4, 5   **32.** 20, 17, 17, 18, 3, 18

33. To get a C in chemistry, Kegan must score an average of 70 on four tests. Scores on his first three tests were 68, 71, and 65. What is the lowest score that Kegan can receive on the last test and still get a C?

34. *Grade Point Average.*   Find the grade point average for one semester given the following grades. Assume the grade point values are 4.0 for A, 3.0 for B, and so on. Round to the nearest tenth.

COURSE	GRADE	NUMBER OF CREDIT HOURS IN COURSE
Introductory Algebra	B	3
English	A	3
Business	C	4
Spanish	B	3
Typing	B	2

35. Tomas has 15 cherry jellybeans, 22 vanilla jellybeans, and 18 lime jellybeans. What is the probability that a randomly chosen jellybean will be vanilla?

Synthesis

Graph.

36. $\frac{1}{4}x + 3\frac{1}{2}y = 1$

37. $\frac{5}{6}x - 2\frac{1}{3}y = 1$

38. Find the area of a rectangle whose vertices are $(-3, 1)$, $(5, 1)$, $(5, 8)$, and $(-3, 8)$.

1. *Net Worth.* In 2013, the net worth of Warren Buffett of the United States was $53.5 billion. Write standard notation for 53.5 billion.

2. *Gas Mileage.* A 2013 Subaru Outback does 312 mi of city driving on 13 gal of gasoline. What is the gas mileage?

3. In 402,513, what does the digit 5 mean?

4. Evaluate: $3 + 5^3$.

5. Find all the factors of 60.

6. Round 52.045 to the nearest tenth.

7. Convert to fraction notation: $3\dfrac{3}{10}$.

8. Convert from cents to dollars: 210¢.

9. Find $-x$ when $x = -9$

10. Evaluate $2x - y$ for $x = 3$ and $y = 8$.

Perform the indicated operation and, if possible, simplify.

11. $\dfrac{-2}{15} + \dfrac{3}{10}$

12. $-18 + (-21)$

13. $\dfrac{14}{15} - \dfrac{3}{5}$

14. $350 - 24.57$

15. $3\dfrac{3}{7} \cdot 4\dfrac{3}{8}$

16. $(17.4)(-2.43)$

17. $\dfrac{13}{15} \div \dfrac{26}{27}$

18. $-1334.183 \div (-21.4)$

Solve.

19. $x + \dfrac{2}{3} = -\dfrac{1}{5}$

20. $\dfrac{2}{5} \cdot y = \dfrac{3}{10}$

21. $\dfrac{3}{8}x + 2 = 11$

22. $3(x - 5) = 7x + 2$

Solve.

23. *Energy Consumption.* In a recent year, American utility companies generated 1464 billion kilowatt-hours (kWh) of electricity using coal, 455 billion using nuclear power, 273 billion using natural gas, 250 billion using hydroelectric plants, 118 billion using petroleum, and 12 billion using geothermal technology and other methods. How many kilowatt-hours of electricity were produced that year?

24. A piece of fabric $1\frac{3}{4}$ yd long is cut into 7 equal strips. What is the length of each strip?

25. A recipe calls for $\frac{3}{4}$ cup of sugar. How much sugar should be used for $\frac{1}{2}$ of the recipe?

26. *Peanut Products.* In any given year, the average American eats 2.7 lb of peanut butter, 1.5 lb of salted peanuts, 1.2 lb of peanut candy, 0.7 lb of in-shell peanuts, and 0.1 lb of peanuts in other forms. How many pounds of peanuts and products containing peanuts does the average American eat in a year?

27. A landscaper bought 22 evergreen trees for $210. What was the cost of each tree? Round to the nearest cent.

28. Find the perimeter and the area of the rectangle.

29. In which quadrant is the point $(-4, 9)$ located?

30. Graph: $y = \dfrac{1}{2}x - 4$.

31. Combine like terms: $6x + 4y - 8x - 3y$.

32. Multiply: $5(2a - 3b + 1)$.

FedEx. The following table lists the costs of delivering a package by FedEx Priority Overnight shipping from zip code 46143 to zip code 80403. Use the table for Exercises 33–35.

WEIGHT (in pounds)	COST
1	$52.55
2	58.23
3	64.47
4	70.48
5	77.12
6	83.08
7	89.32
8	95.00
9	101.13
10	104.19

SOURCE: Federal Express Corporation

33. Find the mean and the median of these costs.

34. Make a vertical bar graph of the data.

35. Make a line graph of the data.

Synthesis

36. Simplify:

$$\left(\frac{3}{4}\right)^2 - \frac{1}{8} \cdot \left(3 - 1\frac{1}{2}\right)^2.$$

37. Add and write the answer as a mixed numeral:

$$-5\frac{42}{100} + \frac{355}{100} + \frac{89}{10} + \frac{17}{1000}.$$

38. A square with sides parallel to the axes has the point $(2, 3)$ at its center. Find the coordinates of the square's vertices if each side is 8 units long.

Ratio and Proportion

7.1   Introduction to Ratios

OBJECTIVES

a Find fraction notation for ratios.

b Simplify ratios.

SKILL TO REVIEW

Objective 3.5b: Simplify fraction notation.

Simplify.

1. $\dfrac{16}{64}$ **2.** $\dfrac{40}{24}$

a RATIOS

RATIO

A **ratio** is the quotient of two quantities.

The average wind speed in Chicago is 10.4 mph. The average wind speed in Boston is 12.5 mph. The *ratio* of average wind speed in Chicago to average wind speed in Boston is written using colon notation,

$$\text{Chicago wind speed} \rightarrow 10.4 : 12.5, \leftarrow \text{Boston wind speed}$$

or fraction notation,

$$\dfrac{10.4}{12.5}. \begin{array}{l}\leftarrow \text{Chicago wind speed} \\ \leftarrow \text{Boston wind speed}\end{array}$$

We read both forms of notation as "the ratio of 10.4 to 12.5."

RATIO NOTATION

The **ratio** of a to b is expressed by the fraction notation $\dfrac{a}{b}$, where a is the numerator and b is the denominator, or by the colon notation $a : b$.

1. Find the ratio of 5 to 11.

2. Find the ratio of 57.3 to 86.1.

3. Find the ratio of $6\frac{3}{4}$ to $7\frac{2}{5}$.

EXAMPLE 1   Find the ratio of 7 to 8.

The ratio is $\dfrac{7}{8}$,   or $7 : 8$.

EXAMPLE 2   Find the ratio of 31.4 to 100.

The ratio is $\dfrac{31.4}{100}$,   or $31.4 : 100$.

EXAMPLE 3   Find the ratio of $4\frac{2}{3}$ to $5\frac{7}{8}$. You need not simplify.

The ratio is $\dfrac{4\frac{2}{3}}{5\frac{7}{8}}$,   or $4\frac{2}{3} : 5\frac{7}{8}$.

◀ **Do Margin Exercises 1–3.**

Answers

Skill to Review:

1. $\frac{1}{4}$ **2.** $\frac{5}{3}$

Margin Exercises:

1. $\frac{5}{11}$, or 5 : 11   **2.** $\frac{57.3}{86.1}$, or 57.3 : 86.1

3. $\frac{6\frac{3}{4}}{7\frac{2}{5}}$, or $6\frac{3}{4} : 7\frac{2}{5}$

In most of our work, we will use fraction notation for ratios.

EXAMPLE 4 *Media Usage.* In 2012, Americans spent an average of 6 hr/month on Facebook and 147 hr/month watching TV. Find the ratio of average time spent on Facebook to average time spent watching TV.

Sources: comScore and Nielsen

The ratio is $\dfrac{6}{147}$.

EXAMPLE 5 *Record Rainfall.* The greatest rainfall ever recorded in the United States during a 12-month period was 739 in. in Kukui, Maui, Hawaii, from December 1981 to December 1982. What is the ratio of amount of rainfall, in inches, to time, in months? of time, in months, to amount of rainfall, in inches?

Source: *Time Almanac*

The ratio of amount of rainfall, in inches, to time, in months, is

$\dfrac{739}{12}$. ← Rainfall
 ← Time

The ratio of time, in months, to amount of rainfall, in inches, is

$\dfrac{12}{739}$. ← Time
 ← Rainfall

EXAMPLE 6 Refer to the triangle below.

a) What is the ratio of the length of the longest side to the length of the shortest side?

$\dfrac{5}{3}$ ← Longest side
 ← Shortest side

b) What is the ratio of the length of the shortest side to the length of the longest side?

$\dfrac{3}{5}$ ← Shortest side
 ← Longest side

Do Exercises 4–6. ▶

4. *Record Snowfall.* The greatest snowfall recorded in North America during a 24-hr period was 76 in. in Silver Lake, Colorado, on April 14–15, 1921. What is the ratio of amount of snowfall, in inches, to time, in hours?

Source: U.S. Army Corps of Engineers

5. *Frozen Fruit Drinks.* A Berries & Kreme Chiller from Krispy Kreme contains 960 calories, while Smoothie King's MangoFest drink contains 258 calories. What is the ratio of the number of calories in the Krispy Kreme drink to the number of calories in the Smoothie King drink? of the number of calories in the Smoothie King drink to the number of calories in the Krispy Kreme drink?

Source: Physicians Committee for Responsible Medicine

GS **6.** In the triangle below, what is the ratio of the length of the shortest side to the length of the longest side?

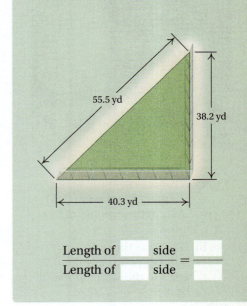

Length of ▢ side   ▢
─────────────── = ──
Length of ▢ side   ▢

Answers

4. $\dfrac{76}{24}$ **5.** $\dfrac{960}{258}, \dfrac{258}{960}$ **6.** $\dfrac{38.2}{55.5}$

Guided Solution:

6. $\dfrac{\text{shortest}}{\text{longest}}, \dfrac{38.2}{55.5}$

EXAMPLE 7 *Grammy Awards.* From 1959 to 2011, a total of 798 musical artists won Grammy awards for Record of the Year, Album of the Year, and Best New Artist. The bar graph below shows the numbers of award winners from four music genres.

Grammy Award Winners

SOURCES: *LA Times*; cbsnews.com

a) What is the ratio of the number of country winners to the number of rock winners?

b) What is the ratio of the number of pop, R&B, and country winners to the number of rock winners?

c) What is the ratio of the number of Grammy winners from genres other than rock, pop, R&B, and country to the total number of Grammy winners?

a) The ratio of the number of country winners to the number of rock winners is

$$\frac{57}{286}. \quad \leftarrow \text{Country winners} \\ \leftarrow \text{Rock winners}$$

b) The total number of pop, R&B, and country winners is

$$151 + 135 + 57 = 343.$$

The ratio of the number of these winners to the number of rock winners is

$$\frac{343}{286}. \quad \leftarrow \text{Pop, R\&B, and country winners} \\ \leftarrow \text{Rock winners}$$

c) The total number of rock, pop, R&B, and country Grammy winners is

$$286 + 151 + 135 + 57 = 629.$$

Thus the number of Grammy winners from other genres is

$$798 - 629 = 169.$$

The ratio of the number of Grammy winners from other genres to the total number of Grammy winners is

$$\frac{169}{798}. \quad \leftarrow \text{Winners from other genres} \\ \leftarrow \text{Total number of winners}$$

◀ **Do Exercise 7.**

7. *NBA Playoffs.* During one game of the 2014 NBA Eastern Conference finals, the Indiana Pacers made a total of 64 baskets. Of these, 29 were free throws, 27 were two-point field goals, and the remainder were three-point field goals.

Source: nba.com

a) What was the ratio of the number of two-point field goals to the number of free throws?

b) What was the ratio of the number of three-point field goals to the total number of baskets?

Answers

7. (a) $\frac{27}{29}$; (b) $\frac{8}{64}$

b SIMPLIFYING NOTATION FOR RATIOS

Sometimes a ratio can be simplified. Simplifying provides a means of finding other numbers with the same ratio.

EXAMPLE 8 Find the ratio of 6 to 8. Then simplify to find two other numbers in the same ratio.

We write the ratio in fraction notation and then simplify:

$$\frac{6}{8} = \frac{2 \cdot 3}{2 \cdot 4} = \frac{2}{2} \cdot \frac{3}{4} = 1 \cdot \frac{3}{4} = \frac{3}{4}.$$

Thus, 3 and 4 have the same ratio as 6 and 8. We can express this by saying "6 is to 8 as 3 is to 4."

Do Exercise 8. ▶

EXAMPLE 9 Find the ratio of 2.4 to 10. Then simplify to find two other numbers in the same ratio.

We first write the ratio in fraction notation. Next, we multiply by 1 to clear the decimal from the numerator. Then we simplify.

$$\frac{2.4}{10} = \frac{2.4}{10} \cdot \frac{10}{10} = \frac{24}{100} = \frac{4 \cdot 6}{4 \cdot 25} = \frac{4}{4} \cdot \frac{6}{25} = \frac{6}{25}$$

Thus, 2.4 is to 10 as 6 is to 25.

Do Exercises 9 and 10. ▶

EXAMPLE 10 An HDTV screen that measures approximately 46 in. diagonally has a width of 40 in. and a height of $22\frac{1}{2}$ in. Find the ratio of width to height and simplify.

The ratio is $\dfrac{40}{22\frac{1}{2}} = \dfrac{40}{22.5} = \dfrac{40}{22.5} \cdot \dfrac{10}{10} = \dfrac{400}{225}$

$$= \frac{25 \cdot 16}{25 \cdot 9} = \frac{25}{25} \cdot \frac{16}{9}$$

$$= \frac{16}{9}.$$

Thus we can say that the ratio of width to height is 16 to 9, which can also be expressed as 16 : 9.

Do Exercise 11. ▶

---

8. Find the ratio of 18 to 27. Then simplify to find two other numbers in the same ratio.

GS **9.** Find the ratio of 3.6 to 12. Then simplify to find two other numbers in the same ratio.

Ratio of 3.6 to 12: $\dfrac{\boxed{}}{\boxed{}}$

Simplifying:

$$\frac{3.6}{12} \cdot \frac{10}{\boxed{}} = \frac{\boxed{}}{120} = \frac{\boxed{} \cdot 3}{12 \cdot 10}$$

$$= \frac{\boxed{}}{12} \cdot \frac{3}{10} = \frac{\boxed{}}{10}$$

10. Find the ratio of 1.2 to 1.5. Then simplify to find two other numbers in the same ratio.

11. An HDTV screen that measures 44 in. diagonally has a width of 38.4 in. and a height of 21.6 in. Find the ratio of height to width and simplify.

Answers

8. 18 is to 27 as 2 is to 3.
9. 3.6 is to 12 as 3 is to 10.

10. 1.2 is to 1.5 as 4 is to 5.   **11.** $\frac{9}{16}$

Guided Solution:

9. $\frac{3.6}{12}$; 10, 36, 12, 12, 3

✔ Reading Check

Determine whether each statement is true or false.

_____ **RC1.** A ratio is a quotient.

_____ **RC2.** If there are 2 teachers and 27 students in a classroom, the ratio of students to teachers is 27 : 2.

_____ **RC3.** The ratio 6 : 7 can also be written $\frac{7}{6}$.

_____ **RC4.** The numbers 2 and 3 are in the same ratio as the numbers 4 and 9.

a   Find fraction notation for each ratio. You need not simplify.

1. 4 to 5

2. 3 to 2

3. 178 to 572

4. 329 to 967

5. 0.4 to 12

6. 2.3 to 22

7. 3.8 to 7.4

8. 0.6 to 0.7

9. 56.78 to 98.35

10. 456.2 to 333.1

11. $8\frac{3}{4}$ to $9\frac{5}{6}$

12. $10\frac{1}{2}$ to $43\frac{1}{4}$

Shark Attacks.   Of the 75 unprovoked shark attacks recorded worldwide in 2011, 29 occurred in U.S. waters. The bar graph below shows the breakdown by state. Use the graph for Exercises 13–16.

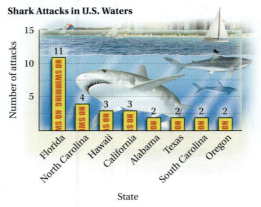

Shark Attacks in U.S. Waters

Number of attacks

States: Florida (11), North Carolina (4), Hawaii (3), California (3), Alabama (2), Texas (2), South Carolina (2), Oregon (2)

State

SOURCE: University of Florida

13. What is the ratio of the number of shark attacks in U.S. waters to the number of shark attacks worldwide?

14. What is the ratio of the number of shark attacks in North Carolina to the number of shark attacks in California?

15. What is the ratio of the number of shark attacks in Florida to the total number of shark attacks in the other seven states?

16. What is the ratio of the total number of shark attacks in states bordering the Pacific Ocean (Hawaii, California, and Oregon) to the total number of shark attacks in the other five states?

17. *Tax Freedom Day.* Of the 366 days in 2012 (a leap year), the average American worked 107 days to pay his or her federal, state, and local taxes. Find the ratio of the number of days worked to pay taxes in 2012 to the number of days in the year.

Source: Tax Foundation

18. *Careers in Medicine.* The number of jobs for nurses is expected to increase by 711,900 between 2010 and 2020. During the same decade, the number of jobs for physicians is expected to increase by 168,300. Find the ratio of the increase in jobs for physicians to the increase in jobs for nurses.

Source: Bureau of Labor Statistics

19. *Music Album Sales.* In 2011, 104.8 million albums were downloaded digitally and 3100.7 million albums were sold on CDs. What is the ratio of downloads to CD sales? of CD sales to downloads?

Source: Recording Industry of America

20. *Silicon in the Earth's Crust.* Every 100 tons of the earth's crust contains about 28 tons of silicon. What is the ratio of the weight of silicon to the weight of crust? of the weight of crust to the weight of silicon?

Source: *The Handy Science Answer Book*

21. *Field Hockey.* A diagram of the playing area for field hockey is shown below. What is the ratio of width to length? of length to width?

Source: *Sports: The Complete Visual Reference*

22. *The Leaning Tower of Pisa.* The Leaning Tower of Pisa was reopened to the public in 2001 following a 10-yr stabilization project. The 184.5-ft tower now leans about 13 ft from its base. What is the ratio of the distance that it leans to its height? of its height to the distance that it leans?

Source: CNN

b Find the ratio of the first number to the second and simplify.

23. 4 to 6

24. 6 to 10

25. 18 to 24

26. 28 to 36

27. 4.8 to 10

28. 5.6 to 10

29. 2.8 to 3.6

30. 4.8 to 6.4

31. 20 to 30

32. 40 to 60

33. 56 to 100

34. 42 to 100

35. 128 to 256

36. 232 to 116

37. 0.48 to 0.64

38. 0.32 to 0.96

39. For this rectangle, find the ratios of length to width and of width to length.

478 ft

213 ft

40. For this right triangle, find the ratios of the length of the shortest side to the length of the longest side and of longest length to shortest length.

107.3 m

47.5 m

96.2 m

Skill Maintenance

Add. Simplify if possible.

41. $18,468 + 390,082$ [1.2a]

42. $24 + 3.006$ [5.2a]

43. $4.2 + 28.07 + 365$ [5.2a]

44. $\dfrac{1}{3} + \dfrac{3}{10}$ [4.2b]

45. $\dfrac{3}{8} + \dfrac{7}{12}$ [4.2b]

46. $4\dfrac{5}{6} + 9\dfrac{1}{2}$ [4.6a]

Subtract. Simplify if possible.

47. $982,001 - 39,782$ [1.3a]

48. $12.5 - 9.9$ [5.2b]

49. $96 - 1.625$ [5.2b]

50. $\dfrac{9}{16} - \dfrac{1}{2}$ [4.3a]

51. $8\dfrac{2}{3} - 6\dfrac{4}{5}$ [4.6b]

52. $11 - 6\dfrac{1}{3}$ [4.6b]

Synthesis

53. Find the ratio of $3\dfrac{3}{4}$ to $5\dfrac{7}{8}$ and simplify.

Fertilizer.   Exercises 54 and 55 refer to a common lawn fertilizer known as "5, 10, 15." This mixture contains 5 parts of potassium for every 10 parts of phosphorus and 15 parts of nitrogen. (This is often denoted 5 : 10 : 15.)

54. Find and simplify the ratio of potassium to nitrogen and of nitrogen to phosphorus.

55. Simplify the ratio 5 : 10 : 15.

Rates and Unit Prices

a RATES

A 2013 Honda Odyssey can travel 504 mi on 18 gal of gasoline. Let's consider the ratio of miles to gallons:

Source: www.fueleconomy.gov.

$$\frac{504 \text{ mi}}{18 \text{ gal}} = \frac{504 \text{ miles}}{18 \text{ gallon}} = \frac{28 \text{ miles}}{1 \text{ gallon}}$$

$$= 28 \text{ miles per gallon} = 28 \text{ mpg}.$$

"per" means "division," or "for each."

The ratio

$$\frac{504 \text{ mi}}{18 \text{ gal}}, \quad \text{or} \quad \frac{504 \text{ mi}}{18 \text{ gal}}, \quad \text{or} \quad 28 \text{ mpg},$$

is called a **rate**.

OBJECTIVES

a Give the ratio of two different measures as a rate.

b Find unit prices and use them to compare purchases.

SKILL TO REVIEW

Objective 5.4a: Divide using decimal notation.

Divide.

1. $15\overline{)634.5}$ **2.** $2.4\overline{)6}$

RATE

When a ratio is used to compare two different kinds of measure, we call it a **rate**.

Suppose David's car travels 475.4 mi on 16.8 gal of gasoline. Is the gas mileage (mpg) of his car better than that of the Honda Odyssey above? To determine this, it helps to convert the ratio to decimal notation and perhaps round. Thus we have

$$\frac{475.4 \text{ miles}}{16.8 \text{ gallons}} = \frac{475.4}{16.8} \text{ mpg} \approx 28.298 \text{ mpg}.$$

Since $28.298 > 28$, David's car gets better gas mileage than the Honda Odyssey does.

EXAMPLE 1 It takes 60 oz of grass seed to seed 3000 sq ft of lawn. What is the rate in ounces per square foot?

$$\frac{60 \text{ oz}}{3000 \text{ sq ft}} = \frac{1}{50} \frac{\text{oz}}{\text{sq ft}}, \quad \text{or} \quad 0.02 \frac{\text{oz}}{\text{sq ft}}$$

EXAMPLE 2 Martina bought 5 lb of organic russet potatoes for $4.99. What was the rate in cents per pound?

$$\frac{\$4.99}{5 \text{ lb}} = \frac{499 \text{ cents}}{5 \text{ lb}} = 99.8\text{¢}/\text{lb}$$

Answers

Skill to Review:
1. 42.3   **2.** 2.5

EXAMPLE 3 *Hourly Wage.*

In 2011, Walmart employees earned, on average, $470 for a 40-hr work week. What was the rate of pay per hour?

Source: "Living Wage Policies and Big-Box Retail," UC Berkeley Center for Labor Research and Education, Ken Jacobs, Dave Graham-Squire, and Stephanie Luce, at http://laborcenter.berkeley.edu/retail/bigbox_livingwage_policies11.pdf

The rate of pay is the ratio of money earned to length of time worked, or

$$\frac{\$470}{40\,\text{hr}} = \frac{470}{40}\frac{\text{dollars}}{\text{hr}} = 11.75\frac{\text{dollars}}{\text{hr}}, \quad \text{or} \quad \$11.75\,\text{per hr.}$$

EXAMPLE 4 *Ratio of Strikeouts to Home Runs.*

In the 2012 baseball season, Miguel Cabrera of the Detroit Tigers had 98 strikeouts and 44 home runs. What was his strikeout to home-run rate?

Source: Major League Baseball

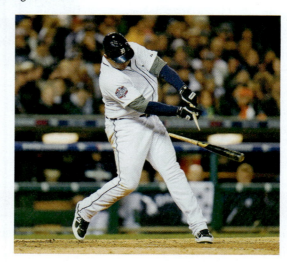

$$\frac{98\,\text{strikeouts}}{44\,\text{home runs}} = \frac{98}{44}\frac{\text{strikeouts}}{\text{home runs}} = \frac{98}{44}\,\text{strikeouts per home run}$$

$$\approx 2.227\,\text{strikeouts per home run}$$

◀ Do Exercises 1–5.

b UNIT PRICING

> ### UNIT PRICE
>
> A **unit price**, or **unit rate**, is the ratio of price to the number of units.

EXAMPLE 5 *Unit Price of Pears.*

Ruth bought a 15.2-oz can of pears for $1.30. What is the unit price in cents per ounce?

$$\text{Unit price} = \frac{\text{Price}}{\text{Number of units}}$$

$$= \frac{\$1.30}{15.2\,\text{oz}} = \frac{130\,\text{cents}}{15.2\,\text{oz}} = \frac{130}{15.2}\frac{\text{cents}}{\text{oz}}$$

$$\approx 8.553\,\text{cents per ounce, or } 8.553¢/\text{oz}$$

◀ Do Exercise 6.

A ratio of distance traveled to time is called a *speed.* What is the rate, or speed, in miles per hour?

1. 45 mi, 9 hr

2. 120 mi, 10 hr

What is the rate, or speed, in feet per second?

3. 2200 ft, 2 sec

4. 52 ft, 13 sec **GS**

5. *Babe Ruth.* In his baseball career, Babe Ruth had 1330 strikeouts and 714 home runs. What was his home-run to strikeout rate?

Source: Major League Baseball

6. **Unit Price of Pasta Sauce.** **GS** Gregory bought a 26-oz jar of pasta sauce for $2.79. What is the unit price in cents per ounce?

$$\text{Unit price} = \frac{\text{Price}}{\text{Number of units}}$$

$$= \frac{\$\boxed{}}{26\,\text{oz}} = \frac{\boxed{}\,\text{cents}}{26\,\text{oz}}$$

$$= \frac{\boxed{}}{26}\frac{\text{cents}}{\text{oz}} \approx \boxed{}\,¢/\text{oz}$$

Answers

1. 5 mi/hr, or 5 mph   **2.** 12 mi/hr, or 12 mph
3. 1100 ft/sec   **4.** 4 ft/sec
5. $\frac{714}{1330}$ home run per strikeout ≈ 0.537 home run per strikeout   **6.** 10.731¢/oz

Guided Solutions:
4. $\frac{52\,\text{ft}}{13\,\text{sec}} = 4\,\text{ft/sec}$ **6.** 2.79, 279, 279, 10.731

Unit prices enable us to do comparison shopping and determine the best buy for a product on the basis of price. It is often helpful to change all prices to cents so that we can compare unit prices more easily.

EXAMPLE 6 *Unit Price of Salad Dressing.* At the request of his customers, Angelo started bottling and selling the basil and lemon salad dressing that he serves in his café. The dressing is sold in four sizes of containers as listed in the table below. Compute the unit price of each size of container and determine which size is the best buy on the basis of unit price alone.

Size	Price	Unit Price
10 oz	$2.49	
16 oz	$3.59	
20 oz	$4.09	
32 oz	$6.79	

We compute the unit price of each size and fill in the chart:

10 oz: $\dfrac{\$2.49}{10 \text{ oz}} = \dfrac{249 \text{ cents}}{10 \text{ oz}} = \dfrac{249}{10} \dfrac{\text{cents}}{\text{oz}} = 24.900\text{¢/oz};$

16 oz: $\dfrac{\$3.59}{16 \text{ oz}} = \dfrac{359 \text{ cents}}{16 \text{ oz}} = \dfrac{359}{16} \dfrac{\text{cents}}{\text{oz}} \approx 22.438\text{¢/oz};$

20 oz: $\dfrac{\$4.09}{20 \text{ oz}} = \dfrac{409 \text{ cents}}{20 \text{ oz}} = \dfrac{409}{20} \dfrac{\text{cents}}{\text{oz}} = 20.450\text{¢/oz};$

32 oz: $\dfrac{\$6.79}{32 \text{ oz}} = \dfrac{679 \text{ cents}}{32 \text{ oz}} = \dfrac{679}{32} \dfrac{\text{cents}}{\text{oz}} \approx 21.219\text{¢/oz}.$

Size	Price	Unit Price
10 oz	$2.49	24.900¢/oz
16 oz	$3.59	22.438¢/oz
20 oz	$4.09	20.450¢/oz Lowest unit price
32 oz	$6.79	21.219¢/oz

On the basis of unit price alone, we see that the 20-oz container is the best buy.

Do Exercise 7. ▶

Although we often think that "bigger is cheaper," this is not always the case, as we see in Example 6. In addition, even when a larger package has a lower unit price than a smaller package, it still might not be the best buy for you. For example, some of the food in a large package could spoil before it is used, or you might not have room to store a large package.

7. *Cost of Mayonnaise.*
Complete the following table for Hellmann's mayonnaise sold on an online shopping site. Which size has the lowest unit price?

Source: Peapod.com

Size	Price	Unit Price
8 oz	$2.79	
10 oz	$3.69	
30 oz	$5.39	

✓ Reading Check

Complete each sentence with the correct quantity from the list at the right, given the following information.

Jason wrote an 8-page essay in 9 hr.

Keri graded 8 essays in 3 hr.

RC1. Jason wrote at a rate of _____ pages/hr.   **a)** $\frac{8}{3}$

RC2. Jason wrote at a rate of _____ hr/page.   **b)** $\frac{8}{9}$

RC3. Keri graded at a rate of _____ hr/essay.   **c)** $\frac{9}{8}$

RC4. Keri graded at a rate of _____ essays/hr.   **d)** $\frac{3}{8}$

a In Exercises 1–4, find each rate, or speed, as a ratio of distance to time. Round to the nearest hundredth where appropriate.

1. 120 km, 3 hr   **2.** 18 mi, 9 hr   **3.** 217 mi, 29 sec   **4.** 443 m, 48 sec

5. *Chevrolet Malibu LS—City Driving.* A 2013 Chevrolet Malibu LS will travel 312.5 mi on 12.5 gal of gasoline in city driving. What is the rate in miles per gallon?

Source: Chevrolet

6. *Mazda3—City Driving.* A 2013 Mazda3 i SV will travel 348 mi on 14.5 gal of gasoline in city driving. What is the rate in miles per gallon?

Source: mazdausa.com

7. *Mazda3—Highway Driving.* A 2013 Mazda3 i SV will travel 643.5 mi on 19.5 gal of gasoline in highway driving. What is the rate in miles per gallon?

Source: mazdausa.com

8. *Chevrolet Malibu LS—Highway Driving.* A 2013 Chevrolet Malibu LS will travel 499.5 mi on 13.5 gal of gasoline in highway driving. What is the rate in miles per gallon?

Source: Chevrolet

9. *Population Density of Monaco.* Monaco is a tiny country on the Mediterranean coast of France. It has an area of 0.75 sq mi and a population of 35,427 people. What is the rate of number of people per square mile? The rate of people per square mile is called the *population density.* Monaco has the highest population density of any country in the world.

Source: World Bank

10. *Population Density of Australia.* The continent of Australia, with the island state of Tasmania, has an area of 2,967,893 sq mi and a population of 22,620,600 people. What is the rate of number of people per square mile? The rate of people per square mile is called the *population density.* Australia has one of the lowest population densities in the world.

Source: World Bank

11. A car is driven 500 mi in 20 hr. What is the rate in miles per hour? in hours per mile?

12. A student eats 3 hamburgers in 15 min. What is the rate in hamburgers per minute? in minutes per hamburger?

13. *Broadway Musicals.* In the 17 years from 1987 through 2003, the musical *Les Misérables* was performed on Broadway 6680 times. What was the average rate of performances per year?

Source: broadwaymusicalhome.com

14. *Employment Growth.* In the 10 years from 2010 to 2020, the number of jobs for interpreters and translators is expected to grow by 24,600. What is the expected average rate of growth in jobs per year?

Source: U.S. Bureau of Labor Statistics

15. *Speed of Light.* Light travels 186,000 mi in 1 sec. What is its rate, or speed, in miles per second?

Source: *The Handy Science Answer Book*

16. *Speed of Sound.* Sound travels 1100 ft in 1 sec. What is its rate, or speed, in feet per second?

Source: *The Handy Science Answer Book*

17. Impulses in nerve fibers travel 310 km in 2.5 hr. What is the rate, or speed, in kilometers per hour?

18. A black racer snake can travel 4.6 km in 2 hr. What is its rate, or speed, in kilometers per hour?

19. *Lawn Watering.* Watering a lawn adequately requires 623 gal of water for every 1000 ft². What is the rate in gallons per square foot?

20. A car is driven 200 km on 40 L of gasoline. What is the rate in kilometers per liter?

21. *Elephant Heart Rate.* The heart of an elephant, at rest, beats an average of 1680 beats in 60 min. What is the rate in beats per minute?

Source: *The Handy Science Answer Book*

22. *Human Heart Rate.* The heart of a human, at rest, beats an average of 4200 beats in 60 min. What is the rate in beats per minute?

Source: *The Handy Science Answer Book*

b Find each unit price in Exercises 23–30. Then, in each exercise, determine which size is the better buy based on unit price alone.

23. *Hidden Valley Ranch Dressing.*

Size	Price	Unit Price
16 oz	$4.19	
20 oz	$5.29	

24. *Miracle Whip.*

Size	Price	Unit Price
32 oz	$5.29	
48 oz	$8.29	

25. *Cascade Powder Detergent.*

Size	Price	Unit Price
45 oz	$5.90	
75 oz	$8.59	

26. *Bush's Homestyle Baked Beans.*

Size	Price	Unit Price
16 oz	$2.99	
28 oz	$4.51	

27. *Jif Creamy Peanut Butter.*

Size	Price	Unit Price
18 oz	$3.28	
28 oz	$4.89	

28. *Hills Brothers Coffee.*

Size	Price	Unit Price
26 oz	$9.99	
39 oz	$14.99	

29. *Campbell's Condensed Tomato Soup.*

Size	Price	Unit Price
10.7 oz	$1.09	
26 oz	$2.69	

30. *Nabisco Saltines.*

Size	Price	Unit Price
16 oz	$3.59	
32 oz	$5.19	

Find the unit price of each brand in Exercises 31–34. Then, in each exercise, determine which brand is the better buy based on unit price alone.

31. *Vanilla Ice Cream.*

Brand	Size	Price
B	32 oz	$5.99
E	48 oz	$6.99

32. *Orange Juice.*

Brand	Size	Price
M	54 oz	$4.79
T	59 oz	$5.99

33. *Tomato Ketchup.*

Brand	Size	Price
A	24 oz	$2.49
B	36 oz	$3.29
H	46 oz	$3.69

34. *Yellow Mustard.*

Brand	Size	Price
F	14 oz	$1.29
G	19 oz	$1.99
P	20 oz	$2.49

Skill Maintenance

Multiply. Simplify, if possible.

35. 25×462 [1.4a]

36. 8.4×80.892 [5.3a]

37. 0.01×274.568 [5.3a]

38. $\dfrac{50}{9} \cdot \dfrac{6}{5}$ [3.6a]

39. $3\dfrac{4}{5} \cdot 2\dfrac{1}{4}$ [4.7a]

40. $4\dfrac{1}{10} \cdot 3\dfrac{1}{3}$ [4.7a]

Divide. Simplify, if possible.

41. $4000 \div 32$ [1.5a]

42. $95 \div 10$ [5.4a]

43. $80.892 \div 8.4$ [5.4a]

44. $\dfrac{50}{9} \div \dfrac{6}{5}$ [3.7b]

45. $200 \div 1\dfrac{1}{3}$ [4.7b]

46. $4\dfrac{6}{7} \div \dfrac{1}{4}$ [4.7b]

Synthesis

47. Manufacturers sometimes change the sizes of their containers in such a way that the consumer thinks the price of a product has been lowered when, in reality, a higher unit price is being charged.

Some aluminum juice cans are now concave (curved in) on the bottom. Suppose the volume of the can in the figure has been reduced from 6 fl oz to 5.5 fl oz, and the price of each can has been reduced from 65¢ to 60¢. Find the unit price of each container in cents per ounce.

SKILL TO REVIEW

Objective 3.5c: Test to determine whether two fractions are equivalent.

Use = or ≠ for ☐ to write a true sentence.

1. $\frac{18}{12}$ ☐ $\frac{9}{6}$ **2.** $\frac{4}{7}$ ☐ $\frac{5}{8}$

Determine whether the two pairs of numbers are proportional.

1. 3, 4 and 6, 8

2. 1, 4 and 10, 39

3. 1, 2 and 20, 39   **GS**

We compare cross products.

1 · ☐ = ☐ $\frac{1}{2}$? $\frac{20}{39}$

2 · ☐ = ☐

Since 39 ≠ 40, the numbers ☐ proportional.
are/are not

Determine whether the two pairs of numbers are proportional.

4. 6.4, 12.8 and 5.3, 10.6

5. 6.8, 7.4 and 3.4, 4.2

Answers

Skill to Review:
1. =   **2.** ≠

Margin Exercises:
1. Yes   **2.** No   **3.** No   **4.** Yes   **5.** No

Guided Solution:
3. 39, 39, 20, 40; are not

a PROPORTIONS

When two pairs of numbers, such as 3, 2 and 6, 4, have the same ratio, we say that they are **proportional**. The equation

$$\frac{3}{2} = \frac{6}{4}$$

states that the pairs 3, 2 and 6, 4 are proportional. Such an equation is called a **proportion**. We sometimes read $\frac{3}{2} = \frac{6}{4}$ as "3 is to 2 as 6 is to 4."

Since ratios can be written using fraction notation, we can use the test for equality of fractions to determine whether two ratios are the same.

A TEST FOR EQUALITY OF FRACTIONS

Two fractions are equal if their cross products are equal.

EXAMPLE 1 Determine whether 1, 2 and 3, 6 are proportional.

We can use cross products.

$$1 \cdot 6 = 6 \qquad \frac{1}{2} \overset{?}{=} \frac{3}{6} \qquad 2 \cdot 3 = 6$$

Since the cross products are the same, $6 = 6$, we know that $\frac{1}{2} = \frac{3}{6}$, so the numbers are proportional.

EXAMPLE 2 Determine whether 2, 5 and 4, 7 are proportional.

We can use cross products.

$$2 \cdot 7 = 14 \qquad \frac{2}{5} \overset{?}{=} \frac{4}{7} \qquad 5 \cdot 4 = 20$$

Since the cross products are not the same, $14 \neq 20$, we know that $\frac{2}{5} \neq \frac{4}{7}$, so the numbers are not proportional.

◀ **Do Margin Exercises 1–3.**

EXAMPLE 3 Determine whether 3.2, 4.8 and 0.16, 0.24 are proportional.

We can use cross products.

$$3.2 \times 0.24 = 0.768 \qquad \frac{3.2}{4.8} \overset{?}{=} \frac{0.16}{0.24} \qquad 4.8 \times 0.16 = 0.768$$

Since the cross products are the same, $0.768 = 0.768$, we know that $\frac{3.2}{4.8} = \frac{0.16}{0.24}$, so the numbers are proportional.

◀ **Do Exercises 4 and 5.**

EXAMPLE 4 Determine whether $4\frac{2}{3}, 5\frac{1}{2}$ and $8\frac{7}{8}, 16\frac{1}{3}$ are proportional.

We can use cross products:

$$4\frac{2}{3} \cdot 16\frac{1}{3} = \frac{14}{3} \cdot \frac{49}{3} \qquad \frac{4\frac{2}{3}}{5\frac{1}{2}} \overset{?}{=} \frac{8\frac{7}{8}}{16\frac{1}{3}} \qquad 5\frac{1}{2} \cdot 8\frac{7}{8} = \frac{11}{2} \cdot \frac{71}{8}$$

$$= \frac{686}{9} \qquad\qquad\qquad\qquad = \frac{781}{16}$$

$$= 76\frac{2}{9}; \qquad\qquad\qquad\qquad = 48\frac{13}{16}.$$

Since the cross products are not the same, $76\frac{2}{9} \neq 48\frac{13}{16}$, we know that the numbers are not proportional.

Do Exercise 6. ▶

6. Determine whether $4\frac{2}{3}, 5\frac{1}{2}$ and $14, 16\frac{1}{2}$ are proportional.

b SOLVING PROPORTIONS

One way to solve a proportion is to use cross products. Then we can divide on both sides to get the variable alone.

EXAMPLE 5 Solve the proportion $\dfrac{x}{3} = \dfrac{4}{6}$.

$$\frac{x}{3} = \frac{4}{6}$$

$x \cdot 6 = 3 \cdot 4$ Equating cross products (finding cross products and setting them equal)

$$\frac{x \cdot 6}{6} = \frac{3 \cdot 4}{6}$$ Dividing both sides by 6

$$x = \frac{3 \cdot 4}{6} = \frac{12}{6} = 2$$

We can check that 2 is the solution by replacing x with 2 and finding cross products:

$$2 \cdot 6 = 12 \qquad \frac{2}{3} \overset{?}{=} \frac{4}{6} \qquad 3 \cdot 4 = 12.$$

Since the cross products are the same, it follows that $\frac{2}{3} = \frac{4}{6}$. Thus the pairs of numbers 2, 3 and 4, 6 are proportional, and 2 is the solution of the proportion.

SOLVING PROPORTIONS

To solve $\dfrac{x}{a} = \dfrac{c}{d}$ for x, equate *cross products* and divide on both sides to get x alone.

Do Exercise 7. ▶

7. Solve: $\dfrac{x}{63} = \dfrac{2}{9}$.

EXAMPLE 6 Solve: $\dfrac{x}{7} = \dfrac{5}{3}$. Write a mixed numeral for the answer.

We have

$$\frac{x}{7} = \frac{5}{3}$$

$$x \cdot 3 = 7 \cdot 5 \qquad \text{Equating cross products}$$

$$\frac{x \cdot 3}{3} = \frac{7 \cdot 5}{3} \qquad \text{Dividing both sides by 3}$$

$$x = \frac{7 \cdot 5}{3} = \frac{35}{3}, \text{ or } 11\frac{2}{3}.$$

The solution is $11\frac{2}{3}$.

◀ **Do Exercise 8.**

EXAMPLE 7 Solve: $\dfrac{7.7}{15.4} = \dfrac{y}{2.2}$.

We have

$$\frac{7.7}{15.4} = \frac{y}{2.2}$$

$$7.7 \times 2.2 = 15.4 \times y \qquad \text{Equating cross products}$$

$$\frac{7.7 \times 2.2}{15.4} = \frac{15.4 \times y}{15.4} \qquad \text{Dividing both sides by 15.4}$$

$$\frac{7.7 \times 2.2}{15.4} = y$$

$$\frac{16.94}{15.4} = y \qquad \text{Multiplying}$$

$$1.1 = y. \qquad \text{Dividing: } 15.4\overline{\smash)16.9{\scriptstyle\wedge}4}$$
$$ \begin{array}{r} 1.1 \\ \hline 154 \\ \hline 154 \\ 154 \\ \hline 0 \end{array}$$

The solution is 1.1.

EXAMPLE 8 Solve: $\dfrac{8}{x} = \dfrac{5}{3}$. Write decimal notation for the answer.

We have

$$\frac{8}{x} = \frac{5}{3}$$

$$8 \cdot 3 = x \cdot 5 \qquad \text{Equating cross products}$$

$$\frac{8 \cdot 3}{5} = x \qquad \text{Dividing by 5}$$

$$\frac{24}{5} = x \qquad \text{Multiplying}$$

$$4.8 = x. \qquad \text{Simplifying}$$

The solution is 4.8.

◀ **Do Exercises 9 and 10.**

9. Solve: $\dfrac{21}{5} = \dfrac{n}{2.5}$.

10. Solve: $\dfrac{6}{x} = \dfrac{25}{11}$. Write decimal notation for the answer.

Answers

8. $11\frac{1}{4}$ **9.** 10.5   **10.** 2.64

Guided Solution:
8. 5, 4, 4, 45, 11, 1

EXAMPLE 9 Solve: $\dfrac{3.4}{4.93} = \dfrac{10}{n}$.

We have

$$\dfrac{3.4}{4.93} = \dfrac{10}{n}$$

$3.4 \times n = 4.93 \times 10$ Equating cross products

$\dfrac{3.4 \times n}{3.4} = \dfrac{4.93 \times 10}{3.4}$ Dividing both sides by 3.4

$n = \dfrac{4.93 \times 10}{3.4}$

$n = \dfrac{49.3}{3.4}$ Multiplying

$n = 14.5.$ Dividing

The solution is 14.5.

Do Exercise 11. ▶

11. Solve: $\dfrac{0.4}{0.9} = \dfrac{4.8}{t}$.

EXAMPLE 10 Solve: $\dfrac{4\frac{2}{3}}{5\frac{1}{2}} = \dfrac{14}{x}$.

We have

$$\dfrac{4\frac{2}{3}}{5\frac{1}{2}} = \dfrac{14}{x}$$

$4\dfrac{2}{3} \cdot x = 5\dfrac{1}{2} \cdot 14$ Equating cross products

$\dfrac{14}{3} \cdot x = \dfrac{11}{2} \cdot 14$ Converting to fraction notation

$\dfrac{\dfrac{14}{3} \cdot x}{\dfrac{14}{3}} = \dfrac{\dfrac{11}{2} \cdot 14}{\dfrac{14}{3}}$ Dividing both sides by $\dfrac{14}{3}$; we could instead multiply both sides by $\dfrac{3}{14}$

$x = \dfrac{11}{2} \cdot 14 \div \dfrac{14}{3}$

$x = \dfrac{11}{2} \cdot 14 \cdot \dfrac{3}{14}$ Multiplying by the reciprocal of the divisor

$x = \dfrac{11 \cdot 3}{2}$ Simplifying by removing a factor of 1: $\dfrac{14}{14} = 1$

$x = \dfrac{33}{2}$, or $16\dfrac{1}{2}$.

The solution is $\frac{33}{2}$, or $16\frac{1}{2}$.

Do Exercise 12. ▶

12. Solve: $\dfrac{8\frac{1}{3}}{x} = \dfrac{10\frac{1}{2}}{3\frac{3}{4}}$.

Answers

11. 10.8 **12.** $\dfrac{125}{42}$, or $2\dfrac{41}{42}$

For Extra Help

MyMathLab® MathXL®
PRACTICE   WATCH   READ   REVIEW

✓ Reading Check

Complete each statement with the appropriate word from the following list.

cross products   proportion   proportional   ratio

RC1. The quotient of two quantities is their _____.

RC2. Two pairs of numbers that have the same ratio are _____.

RC3. A _____ states that two pairs of numbers have the same ratio.

RC4. For the equation $\frac{2}{x} = \frac{3}{y}$, the _____ are $2y$ and $3x$.

---

a Determine whether the two pairs of numbers are proportional.

1. $5, 6$ and $7, 9$

2. $7, 5$ and $6, 4$

3. $1, 2$ and $10, 20$

4. $7, 3$ and $21, 9$

5. $2.4, 3.6$ and $1.8, 2.7$

6. $4.5, 3.8$ and $6.7, 5.2$

7. $5\frac{1}{3}, 8\frac{1}{4}$ and $2\frac{1}{5}, 9\frac{1}{2}$

8. $2\frac{1}{3}, 3\frac{1}{2}$ and $14, 21$

b Solve.

9. $\frac{18}{4} = \frac{x}{10}$

10. $\frac{x}{45} = \frac{20}{25}$

11. $\frac{x}{8} = \frac{9}{6}$

12. $\frac{8}{10} = \frac{n}{5}$

13. $\frac{t}{12} = \frac{5}{6}$

14. $\frac{12}{4} = \frac{x}{3}$

15. $\frac{2}{5} = \frac{8}{n}$

16. $\frac{10}{6} = \frac{5}{x}$

17. $\frac{n}{15} = \frac{10}{30}$

18. $\frac{2}{24} = \frac{x}{36}$

19. $\frac{16}{12} = \frac{24}{x}$

20. $\frac{8}{12} = \frac{20}{x}$

21. $\frac{6}{11} = \frac{12}{x}$

22. $\frac{8}{9} = \frac{32}{n}$

23. $\frac{20}{7} = \frac{80}{x}$

24. $\frac{36}{x} = \frac{9}{5}$

25. $\dfrac{12}{9} = \dfrac{x}{7}$ **26.** $\dfrac{x}{20} = \dfrac{16}{15}$ **27.** $\dfrac{x}{13} = \dfrac{2}{9}$ **28.** $\dfrac{8}{11} = \dfrac{x}{5}$

29. $\dfrac{100}{25} = \dfrac{20}{n}$ **30.** $\dfrac{35}{125} = \dfrac{7}{m}$ **31.** $\dfrac{6}{y} = \dfrac{18}{15}$ **32.** $\dfrac{15}{y} = \dfrac{3}{4}$

33. $\dfrac{x}{3} = \dfrac{0}{9}$ **34.** $\dfrac{x}{6} = \dfrac{1}{6}$ **35.** $\dfrac{1}{2} = \dfrac{7}{x}$ **36.** $\dfrac{2}{5} = \dfrac{12}{x}$

37. $\dfrac{1.2}{4} = \dfrac{x}{9}$ **38.** $\dfrac{x}{11} = \dfrac{7.1}{2}$ **39.** $\dfrac{8}{2.4} = \dfrac{6}{y}$ **40.** $\dfrac{3}{y} = \dfrac{5}{4.5}$

41. $\dfrac{t}{0.16} = \dfrac{0.15}{0.40}$ **42.** $\dfrac{0.12}{0.04} = \dfrac{t}{0.32}$ **43.** $\dfrac{0.5}{n} = \dfrac{2.5}{3.5}$ **44.** $\dfrac{6.3}{0.9} = \dfrac{0.7}{n}$

45. $\dfrac{1.28}{3.76} = \dfrac{4.28}{y}$ **46.** $\dfrac{10.4}{12.4} = \dfrac{6.76}{t}$ **47.** $\dfrac{7}{\frac{1}{4}} = \dfrac{28}{x}$ **48.** $\dfrac{5}{\frac{1}{3}} = \dfrac{3}{x}$

49. $\dfrac{\frac{1}{5}}{\frac{1}{10}} = \dfrac{\frac{1}{10}}{x}$ **50.** $\dfrac{\frac{1}{4}}{\frac{1}{2}} = \dfrac{\frac{1}{2}}{x}$ **51.** $\dfrac{y}{\frac{3}{5}} = \dfrac{\frac{7}{12}}{\frac{14}{15}}$ **52.** $\dfrac{\frac{5}{8}}{\frac{5}{4}} = \dfrac{y}{\frac{3}{2}}$

53. $\dfrac{x}{1\frac{3}{5}} = \dfrac{2}{15}$ **54.** $\dfrac{1}{7} = \dfrac{x}{4\frac{1}{2}}$ **55.** $\dfrac{2\frac{1}{2}}{3\frac{1}{3}} = \dfrac{x}{4\frac{1}{4}}$ **56.** $\dfrac{3\frac{1}{2}}{y} = \dfrac{6\frac{1}{2}}{4\frac{2}{3}}$

57. $\dfrac{5\frac{1}{5}}{6\frac{1}{6}} = \dfrac{y}{3\frac{1}{2}}$

58. $\dfrac{10\frac{3}{8}}{12\frac{2}{3}} = \dfrac{5\frac{3}{4}}{y}$

Skill Maintenance

59. *Drive-in Movie Theaters.* The number of drive-in movie theaters in the United States has declined steadily since the 1950s. In 2012, there were 365 drive-in theaters. This is 3698 fewer than in 1958. How many drive-in movie theaters were there in 1958? [1.8a]

Source: Drive-Ins.com

60. *Bird Feeders.* After Raena poured a 4-lb bag of birdseed into her new bird feeder, the feeder was $\frac{2}{3}$ full. How much seed does the feeder hold when it is full? [3.8b]

61. Mariah bought $\frac{3}{4}$ lb of cheese at the city market and gave $\frac{1}{2}$ lb of it to Lindsay. How much cheese did Mariah have left? [4.3c]

62. David bought $\frac{3}{4}$ lb of fudge and gave $\frac{1}{2}$ of it to Chris. How much fudge did he give to Chris? [3.6b]

63. Rocky is $187\frac{1}{10}$ cm tall and his daughter is $180\frac{3}{4}$ cm tall. How much taller is Rocky? [4.6c]

64. A serving of fish steak (cross section) is generally $\frac{1}{2}$ lb. How many servings can be prepared from a cleaned $18\frac{3}{4}$-lb tuna? [4.7d]

65. *Expense Needs.* Aaron has $34.97 to spend on a book that costs $49.95, a cap that costs $14.88, and a sweatshirt that costs $29.95. How much more money does Aaron need to make these purchases? [5.8b]

66. *Gas Mileage.* Joanna filled her van's gas tank and noted that the odometer read 42,598.2. After the next fill-up, the odometer read 42,912.1. It took 14.6 gal to fill the tank. How many miles per gallon did the van get? [5.8b]

Synthesis

▦ Solve.

67. $\dfrac{1728}{5643} = \dfrac{836.4}{x}$

68. $\dfrac{x}{4} = \dfrac{x-1}{6}$

69. *Strikeouts per Home Run.* Baseball Hall-of-Famer Babe Ruth had 1330 strikeouts and 714 home runs in his career. Hall-of-Famer Mike Schmidt had 1883 strikeouts and 548 home runs in his career. Find the rate of strikeouts per home run for each player. (These rates were considered among the highest in the history of the game and yet each player was voted into the Hall of Fame.)

Mid-Chapter Review

Concept Reinforcement

Determine whether each statement is true or false.

_____ 1. A ratio can be written in fraction notation or in colon notation. [7.1a]

_____ 2. A rate is a ratio. [7.2a]

_____ 3. The largest size package of an item always has the lowest unit price. [7.2b]

_____ 4. If $\dfrac{x}{t} = \dfrac{y}{s}$, then $xy = ts$. [7.3a]

Guided Solutions

5. What is the rate, or speed, in miles per hour? [7.2a]

    120 mi, 2 hr

$$\frac{120 \quad \boxed{}}{\boxed{} \text{ hr}} = \frac{120 \quad \boxed{}}{\boxed{} \text{ hr}} = \boxed{} \text{ mi/hr}$$

6. Solve: $\dfrac{x}{4} = \dfrac{3}{6}$. [7.3b]

$$\frac{x}{4} = \frac{3}{6}$$

$x \cdot \boxed{} = \boxed{} \cdot 3$ Equating cross products

$\dfrac{x \cdot 6}{\boxed{}} = \dfrac{4 \cdot 3}{\boxed{}}$ Dividing on both sides

$x = \boxed{}$ Simplifying

Mixed Review

Find fraction notation for each ratio. [7.1a]

7. 4 to 7

8. 313 to 199

9. 35 to 17

10. 59 to 101

Find the ratio of the first number to the second and simplify. [7.1b]

11. 8 to 12

12. 25 to 75

13. 32 to 28

14. 100 to 76

15. 112 to 56

16. 15 to 3

17. 2.4 to 8.4

18. 0.27 to 0.45

Find each rate, or speed, as a ratio of distance to time. Round to the nearest hundredth where appropriate. [7.2a]

19. 243 mi, 4 hr   **20.** 146 km, 3 hr   **21.** 65 m, 5 sec   **22.** 97 ft, 6 sec

23. *Record Snowfall.* The greatest recorded snowfall in a single storm occurred in the Mt. Shasta Ski Bowl in California, when 189 in. fell during a seven-day storm in 1959. What is the rate in inches per day? [7.2a]

Source: U.S. Army Corps of Engineers

24. *Free Throws.* During the 2011–2012 basketball season, Kevin Durant of the Oklahoma City Thunder attempted 501 free throws and made 431 of them. What is the rate in number of free throws made to number of free throws attempted? Round to the nearest thousandth. [7.2a]

Source: National Basketball Association

25. Jerome bought an 18-oz jar of grape jelly for $2.09. What is the unit price in cents per ounce? [7.2b]

26. Martha bought 12 oz of deli honey ham for $5.99. What is the unit price in cents per ounce? [7.2b]

Determine whether the two pairs of numbers are proportional. [7.3a]

27. 3, 7 and 15, 35   **28.** 9, 7 and 7, 5   **29.** 2.4, 1.5 and 3.2, 2.1   **30.** $1\frac{3}{4}, 1\frac{1}{3}$ and $8\frac{3}{4}, 6\frac{2}{3}$

Solve. [7.3b]

31. $\dfrac{9}{15} = \dfrac{x}{20}$

32. $\dfrac{x}{24} = \dfrac{30}{18}$

33. $\dfrac{12}{y} = \dfrac{20}{15}$

34. $\dfrac{2}{7} = \dfrac{10}{y}$

35. $\dfrac{y}{1.2} = \dfrac{1.1}{0.6}$

36. $\dfrac{0.24}{0.02} = \dfrac{y}{0.36}$

37. $\dfrac{\frac{1}{4}}{x} = \dfrac{\frac{1}{8}}{\frac{1}{4}}$

38. $\dfrac{1\frac{1}{2}}{3\frac{1}{4}} = \dfrac{7\frac{1}{2}}{x}$

Understanding Through Discussion and Writing

39. Can every ratio be written as the ratio of some number to 1? Why or why not? [7.1a]

40. What can be concluded about a rectangle's width if the ratio of length to perimeter is 1 to 3? Make some sketches and explain your reasoning. [7.1a, b]

41. Instead of equating cross products, a student solves $\dfrac{x}{7} = \dfrac{5}{3}$ by multiplying on both sides by the least common denominator, 21. Is his approach a good one? Why or why not? [7.3b]

42. An instructor predicts that a student's test grade will be proportional to the amount of time the student spends studying. What is meant by this? Write an example of a proportion that represents the grades of two students and their study times. [7.3b]

Applications of Proportions

a   APPLICATIONS AND PROBLEM SOLVING

Proportions have applications in such diverse fields as business, chemistry, health sciences, and home economics, as well as in many areas of daily life. Proportions are useful in making predictions.

EXAMPLE 1 *Predicting Total Distance.*   Donna drives her delivery van 800 mi in 3 days. At this rate, how far will she drive in 15 days?

1. **Familiarize.**   We let d = the distance traveled in 15 days.

2. **Translate.**   We translate to a proportion. We make each side the ratio of distance to time, with distance in the numerator and time in the denominator.

$$\text{Distance in 15 days} \rightarrow \frac{d}{15} = \frac{800}{3} \leftarrow \text{Distance in 3 days}$$
$$\text{Time} \rightarrow \qquad\qquad\qquad \leftarrow \text{Time}$$

It may help to verbalize the proportion above as "the unknown distance d is to 15 days as the known distance 800 mi is to 3 days."

3. **Solve.**   Next, we solve the proportion:

$$d \cdot 3 = 15 \cdot 800 \qquad \text{Equating cross products}$$

$$\frac{d \cdot 3}{3} = \frac{15 \cdot 800}{3} \qquad \text{Dividing both sides by 3}$$

$$d = \frac{15 \cdot 800}{3}$$

$$d = 4000. \qquad \text{Multiplying and dividing}$$

4. **Check.**   We substitute into the proportion and check cross products:

$$\frac{4000}{15} = \frac{800}{3};$$

$$4000 \cdot 3 = 12{,}000; \qquad 15 \cdot 800 = 12{,}000.$$

The cross products are the same, so the answer checks.

5. **State.**   Donna will drive 4000 mi in 15 days.

Do Margin Exercise 1. ▶

Problems involving proportions can be translated in more than one way. For Example 1, any one of the following is also a correct translation:

$$\frac{15}{d} = \frac{3}{800}, \qquad \frac{15}{3} = \frac{d}{800}, \qquad \frac{800}{d} = \frac{3}{15}.$$

Equating the cross products in each proportion gives us the equation $d \cdot 3 = 15 \cdot 800$, which is the equation we obtained in Example 1.

OBJECTIVE

a   Solve applied problems involving proportions.

SKILL TO REVIEW

Objective 5.7a: Solve equations containing decimals and one variable term.

Solve.

1. $120 \cdot 25 = 1.6n$

2. $0.3w = 1.68 \cdot 0.4$

1. *Burning Calories.*   The readout on Mary's treadmill indicates that she burns 108 calories when she walks for 24 min. How many calories will she burn if she walks at the same rate for 30 min?

EXAMPLE 2 *Recommended Dosage.* To control a fever, a doctor suggests that a child who weighs 28 kg be given 320 mg of a liquid pain reliever. If the dosage is proportional to the child's weight, how much of the medication is recommended for a child who weighs 35 kg?

1. Familiarize. We let t = the number of milligrams of the liquid pain reliever recommended for a child who weighs 35 kg.

2. Translate. We translate to a proportion, keeping the amount of medication in the numerators.

$$\text{Medication suggested} \rightarrow \frac{320}{28} = \frac{t}{35} \leftarrow \text{Medication suggested}$$
$$\text{Child's weight} \rightarrow \qquad\qquad \leftarrow \text{Child's weight}$$

3. Solve. Next, we solve the proportion:

$$320 \cdot 35 = 28 \cdot t \qquad \text{Equating cross products}$$
$$\frac{320 \cdot 35}{28} = \frac{28 \cdot t}{28} \qquad \text{Dividing both sides by 28}$$
$$\frac{320 \cdot 35}{28} = t$$
$$400 = t. \qquad \text{Multiplying and dividing}$$

4. Check. We substitute into the proportion and check cross products:

$$\frac{320}{28} = \frac{400}{35};$$
$$320 \cdot 35 = 11{,}200; \qquad 28 \cdot 400 = 11{,}200.$$

The cross products are the same.

5. State. The dosage for a child who weighs 35 kg is 400 mg.

◀ **Do Exercise 2.**

EXAMPLE 3 *Purchasing Tickets.* Carey bought 8 tickets to an international festival for $52. How many tickets could she purchase with $90?

1. Familiarize. We let n = the number of tickets that can be purchased with $90.

2. Translate. We translate to a proportion, keeping the number of tickets in the numerators.

$$\text{Tickets} \rightarrow \frac{8}{52} = \frac{n}{90} \leftarrow \text{Tickets}$$
$$\text{Cost} \rightarrow \qquad\qquad \leftarrow \text{Cost}$$

3. Solve. Next, we solve the proportion:

$$52 \cdot n = 8 \cdot 90 \qquad \text{Equating cross products}$$

$$\frac{52 \cdot n}{52} = \frac{8 \cdot 90}{52} \qquad \text{Dividing both sides by 52}$$

$$n = \frac{8 \cdot 90}{52}$$

$$n \approx 13.8. \qquad \text{Multiplying and dividing}$$

Because it is impossible to buy a fractional part of a ticket, we must round our answer *down* to 13.

4. Check. As a check, we use a different approach: We find the cost per ticket and then divide $90 by that price. Since $52 \div 8 = 6.50$ and $90 \div 6.50 \approx 13.8$, we have a check.

5. State. Carey could purchase 13 tickets with $90.

Do Exercise 3. ▶

3. *Purchasing Shirts.* If 2 shirts can be bought for $47, how many shirts can be bought with $200?

EXAMPLE 4 *Waist-to-Hip Ratio.* To reduce the risk of heart disease, it is recommended that a man's waist-to-hip ratio be 0.9 or lower. Mac's hip measurement is 40 in. To meet the recommendation, what should his waist measurement be?
Source: Mayo Clinic

Waist measurement is the smallest measurement below the ribs but above the navel.

Hip measurement is the largest measurement around the widest part of the buttocks.

1. Familiarize. Note that $0.9 = \dfrac{9}{10}$. We let $w =$ Mac's waist measurement.

2. Translate. We translate to a proportion as follows.

$$\begin{array}{c} \text{Waist measurement} \to \\ \text{Hip measurement} \to \end{array} \frac{w}{40} = \frac{9}{10} \begin{array}{c} \leftarrow \text{Recommended} \\ \swarrow \text{waist-to-hip ratio} \end{array}$$

3. Solve. Next, we solve the proportion:

$$w \cdot 10 = 40 \cdot 9 \qquad \text{Equating cross products}$$

$$\frac{w \cdot 10}{10} = \frac{40 \cdot 9}{10} \qquad \text{Dividing both sides by 10}$$

$$w = \frac{40 \cdot 9}{10}$$

$$w = 36. \qquad \text{Multiplying and dividing}$$

Answer

3. 8 shirts

4. _Waist-to-Hip Ratio._ It is recommended that a woman's waist-to-hip ratio be 0.85 or lower. Martina's hip measurement is 40 in. To meet the recommendation, what should her waist measurement be?

Source: Mayo Clinic

4. Check. As a check, we divide 36 by 40: $36 \div 40 = 0.9$. This is the desired ratio.

5. State. Mac's recommended waist measurement is 36 in. or less.

◀ **Do Exercise 4.**

EXAMPLE 5 _Construction Plans._ Architects make blueprints for construction projects. These are scale drawings in which lengths are in proportion to actual sizes. The Hennesseys are adding a rectangular deck to their house. The architect's blueprint is rendered such that $\frac{3}{4}$ in. on the drawing is actually 2.25 ft on the deck. The width of the deck on the drawing is 4.3 in. How wide is the deck in reality?

1. Familiarize. We let w = the width of the deck.

2. Translate. Then we translate to a proportion, using 0.75 for $\frac{3}{4}$ in.

$$\text{Measure on drawing} \rightarrow \frac{0.75}{2.25} = \frac{4.3}{w} \begin{matrix} \leftarrow \text{Width on drawing} \\ \leftarrow \text{Width on deck} \end{matrix}$$
$$\text{Measure on deck} \rightarrow$$

3. Solve. Next, we solve the proportion:

$$0.75 \times w = 2.25 \times 4.3 \qquad \text{Equating cross products}$$

$$\frac{0.75 \times w}{0.75} = \frac{2.25 \times 4.3}{0.75} \qquad \text{Dividing both sides by 0.75}$$

$$w = \frac{2.25 \times 4.3}{0.75}$$

$$w = 12.9.$$

4. Check. We substitute into the proportion and check cross products:

$$\frac{0.75}{2.25} = \frac{4.3}{12.9};$$

$$0.75 \times 12.9 = 9.675; \qquad 2.25 \times 4.3 = 9.675.$$

The cross products are the same, so the answer checks.

5. State. The width of the deck is 12.9 ft.

◀ **Do Exercise 5.**

5. _Construction Plans._ In Example 5, the length of the actual deck is 28.5 ft. What is the length of the deck on the blueprint?

ers

less   **5.** 9.5 in.

EXAMPLE 6 *Estimating a Wildlife Population.* Scientists often use proportions to estimate the size of a wildlife population. They begin by collecting and marking, or tagging, a portion of the population. This tagged sample is released and mingles with the entire population. At a later date, the scientists collect a second sample from the population. The proportion of tagged individuals in the second sample is estimated to be the same as the proportion of tagged individuals in the entire population.

The marking can be done by using actual tags or by identifying individuals in other ways. For example, marine biologists can identify an individual whale by the patterns on its tail. Recently, scientists have begun using DNA to identify individuals in populations. For example, to identify individual bears in the grizzly bear population of the Northern Continental Divide ecosystem in Montana, geneticists use DNA from fur samples left on branches near the bears' feeding areas.

In one recent large-scale study in this ecosystem, biologists identified 545 individual grizzly bears. If later a sample of 30 bears contained 25 of the previously identified individuals, estimate the total number of bears in the ecosystem.

Source: Based on information from the Northern Divide Grizzly Bear Project

1. **Familiarize.** We let B = the total number of bears in the ecosystem. We assume that the ratio of the number of identified bears to the total number of bears in the ecosystem is the same as the ratio of the number of identified bears in the later sample to the total number of bears in the later sample.

2. **Translate.** We translate to a proportion as follows.

$$\text{Identified bears} \rightarrow \frac{545}{B} = \frac{25}{30} \leftarrow \text{Identified bears in sample}$$
$$\text{Total number of bears} \rightarrow \qquad\quad \leftarrow \text{Number of bears in sample}$$

3. **Solve.** Next, we solve the proportion:

$$545 \cdot 30 = B \cdot 25 \qquad \text{Equating cross products}$$

$$\frac{545 \cdot 30}{25} = \frac{B \cdot 25}{25} \qquad \text{Dividing both sides by 25}$$

$$\frac{545 \cdot 30}{25} = B$$

$$654 = B. \qquad \text{Multiplying and dividing}$$

4. **Check.** We substitute into the proportion and check cross products:

$$\frac{545}{654} = \frac{25}{30};$$

$$545 \cdot 30 = 16{,}350; \qquad 654 \cdot 25 = 16{,}350.$$

The cross products are the same.

5. **State.** We estimate that there are 654 bears in the ecosystem.

Do Exercise 6. ▶

GS **6. Estimating a Deer Population.** To determine the number of deer in a forest, a conservationist catches 153 deer, tags them, and releases them. Later, 62 deer are caught, and it is found that 18 of them are tagged. Estimate how many deer are in the forest.

1. **Familiarize.** Let D = the number of deer in the forest.

2. **Translate.**

$$\frac{153}{D} = \frac{\boxed{}}{\boxed{}}$$

3. **Solve.**

$$153 \cdot 62 = D \cdot \boxed{}$$

$$\boxed{} = D$$

4. **Check.** The cross products are the same.

5. **State.** There are about $\boxed{}$ deer in the forest.

Answer

6. 527 deer

Guided Solution:

6. $\dfrac{18}{62}$; 18, 527; 527

Translating for Success

1. **Calories in Cereal.** There are 140 calories in a $1\frac{1}{2}$-cup serving of Brand A cereal. How many calories are there in 6 cups of the cereal?

2. **Calories in Cereal.** There are 140 calories in 6 cups of Brand B cereal. How many calories are there in a $1\frac{1}{2}$-cup serving of the cereal?

3. **Gallons of Gasoline.** Jared's SUV traveled 310 mi on 15.5 gal of gasoline. At this rate, how many gallons would be needed to travel 465 mi?

4. **Gallons of Gasoline.** Elizabeth's fuel-efficient car traveled 465 mi on 15.5 gal of gasoline. At this rate, how many gallons will be needed to travel 310 mi?

5. **Perimeter.** What is the perimeter of a rectangular field that measures 83.7 m by 62.4 m?

The goal of these matching questions is to practice step (2), Translate, of the five-step problem-solving process. Translate each word problem to an equation and select a correct translation from equations A–O.

A. $\dfrac{310}{15.5} = \dfrac{465}{x}$

B. $180 = 1\frac{1}{2} \cdot x$

C. $x = 71\frac{1}{8} - 76\frac{1}{2}$

D. $71\frac{1}{8} \cdot x = 74$

E. $74 \cdot 71\frac{1}{8} = x$

F. $x = 83.7 + 62.4$

G. $71\frac{1}{8} + x = 76\frac{1}{2}$

H. $x = 1\frac{2}{3} \cdot 180$

I. $\dfrac{140}{6} = \dfrac{x}{1\frac{1}{2}}$

J. $x = 2(83.7 + 62.4)$

K. $\dfrac{465}{15.5} = \dfrac{310}{x}$

L. $x = 83.7 \cdot 62.4$

M. $x = 180 \div 1\frac{2}{3}$

N. $\dfrac{140}{1\frac{1}{2}} = \dfrac{x}{6}$

O. $x = 1\frac{2}{3} \div 180$

Answers on page A-16

6. **Electric Bill.** Last month Todd's electric bills for his two rental properties were $83.70 and $62.40. What was the total electric bill for the two properties?

7. **Package Tape.** A postal service center uses rolls of package tape that each contain 180 ft of tape. If it takes an average of $1\frac{2}{3}$ ft of tape per package, how many packages can be taped with one roll?

8. **Online Price.** Jane spent $180 for an area rug in a department store. Later, she saw the same rug for sale online and realized she had paid $1\frac{1}{2}$ times the online price. What was the online price?

9. **Heights of Sons.** Henry's three sons play basketball on three different college teams. Jeff's, Jason's, and Jared's heights are 74 in., $71\frac{1}{8}$ in., and $76\frac{1}{2}$ in., respectively. How much taller is Jared than Jason?

10. **Area of a Lot.** Bradley bought a lot that measured 74 yd by $71\frac{1}{8}$ yd. What was the area of the lot?

✓ Reading Check

Complete each proportion to form a correct translation of the following problem.

Christy ran 4 marathons in the first 6 months of the year. At this rate, how many marathons will she run in the first 8 months of the year? Let $m =$ the number of marathons she will run in 8 months.

RC1. $\dfrac{4}{6} = \dfrac{\square}{\square}$

RC2. $\dfrac{6}{4} = \dfrac{\square}{\square}$

RC3. $\dfrac{4}{m} = \dfrac{\square}{\square}$

RC4. $\dfrac{m}{4} = \dfrac{\square}{\square}$

a Solve.

1. *Study Time and Test Grades.* An English instructor asserted that students' test grades are directly proportional to the amount of time spent studying. Lisa studies for 9 hr for a particular test and gets a score of 75. At this rate, for how many hours would she have had to study to get a score of 92?

2. *Study Time and Test Grades.* A mathematics instructor asserted that students' test grades are directly proportional to the amount of time spent studying. Brent studies for 15 hr for a final exam and gets a score of 75. At this rate, what score would he have received if he had studied for 18 hr?

3. *Movies.* If *The Hobbit: An Unexpected Journey* is played at the rate preferred by director Peter Jackson, a moviegoer sees 600 frames in $12\frac{1}{2}$ sec. How many frames does a moviegoer see in 160 sec?

4. *Sugaring.* When 20 gal of maple sap are boiled down, the result is $\frac{1}{2}$ gal of maple syrup. How much sap is needed to produce 9 gal of syrup?

Source: University of Maine

5. *Gas Mileage.* A 2013 Ford Mustang GT will travel 403 mi on 15.5 gal of gasoline in highway driving.

   a) How many gallons of gasoline will it take to drive 2690 mi from Boston to Phoenix?
   b) How far can the car be driven on 140 gal of gasoline?

   Source: Ford Motor Company

6. *Gas Mileage.* A 2013 Volkswagen Beetle will travel 495 mi on 16.5 gal of premium gasoline in highway driving.

   a) How many gallons of gasoline will it take to drive 1650 mi from Pittsburgh to Albuquerque?
   b) How far can the car be driven on 130 gal of gasoline?

   Source: Volkswagen of America, Inc.

7. Overweight Americans.   A recent study determined that of every 100 American adults, 69 are overweight or obese. It is estimated that the U.S. population will be about 337 million in 2020. At the given rate, how many Americans will be considered overweight or obese in 2020?

Source: U.S. Centers for Disease Control and Prevention

8. Prevalence of Diabetes.   A recent study determined that of every 1000 Americans age 65 or older, 269 have been diagnosed with diabetes. It is estimated that there will be about 55 million Americans in this age group in 2020. At the given rate, how many in this age group will have been diagnosed with diabetes in 2020?

Source: American Diabetes Association

9. Quality Control.   A quality-control inspector examined 100 lightbulbs and found 7 of them to be defective. At this rate, how many defective bulbs will there be in a lot of 2500?

10. Grading.   A high school math teacher must grade 32 reports on famous mathematicians. She can grade 4 reports in 90 min. At this rate, how long will it take her to grade all 32 reports?

11. Recommended Dosage.   To control an infection, Dr. Okeke prescribes a dosage of 200 mg of Rocephin every 8 hr for an infant weighing 15.4 lb. At this rate, what would the dosage be for an infant weighing 20.2 lb?

12. Metallurgy.   In Ethan's white gold ring, the ratio of nickel to gold is 3 to 13. If the ring contains 4.16 oz of gold, how much nickel does it contain?

13. Painting.   Fred uses 3 gal of paint to cover 1275 ft² of siding. How much siding can Fred paint with 7 gal of paint?

14. Waterproofing.   Bonnie can waterproof 450 ft² of decking with 2 gal of sealant. How many gallons of the sealant should Bonnie buy for a 1200-ft² deck?

15. Publishing.   Every 6 pages of a manuscript that an author submits correspond to 5 published pages. How many published pages will a 540-page manuscript become?

16. Turkey Servings.   An 8-lb turkey breast contains 36 servings of meat. How many pounds of turkey breast would be needed for 54 servings?

17. *Mileage.* Jean bought a new car. In the first 8 months, it was driven 9000 mi. At this rate, how many miles will the car be driven in 1 year?

18. *Coffee Production.* Coffee beans from 18 trees are required to produce enough coffee each year for a person who drinks 2 cups of coffee per day. Jared brews 15 cups of coffee each day for himself and his coworkers. How many coffee trees are required for this?

19. *Cap'n Crunch's Peanut Butter Crunch® Cereal.* The nutritional chart on the side of a box of Quaker Cap'n Crunch's Peanut Butter Crunch® cereal states that there are 110 calories in a $\frac{3}{4}$-cup serving. How many calories are there in 6 cups of the cereal?

20. *Rice Krispies® Cereal.* The nutritional chart on the side of a box of Kellogg's Rice Krispies® cereal states that there are 130 calories in a $1\frac{1}{4}$-cup serving. How many calories are there in 5 cups of the cereal?

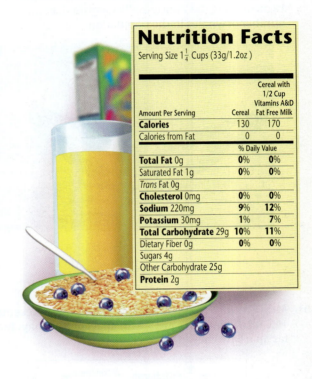

21. *Lefties.* In a class of 40 students, on average, 6 will be left-handed. If a class includes 9 "lefties," how many students would you estimate are in the class?

22. *Class Size.* A college advertises that its student-to-faculty ratio is 27 to 2. If 81 students register for Introductory Spanish, how many sections of the course would you expect to see offered?

23. *Painting.* Helen can paint 950 ft² with 2 gal of paint. How many 1-gal cans does she need in order to paint a 30,000-ft² wall?

24. *Snow to Water.* Under typical conditions, $1\frac{1}{2}$ ft of snow will melt to 2 in. of water. How many inches of water will result when $5\frac{1}{2}$ ft of snow melts?

25. *Gasoline Mileage.* Nancy's van traveled 84 mi on 6.5 gal of gasoline. At this rate, how many gallons would be needed to travel 126 mi?

26. *Bicycling.* Roy bicycled 234 mi in 14 days. At this rate, how far would Roy bicycle in 42 days?

27. Grass-Seed Coverage. It takes 60 oz of grass seed to seed 3000 ft² of lawn. At this rate, how much would be needed for 5000 ft² of lawn?

28. Grass-Seed Coverage. In Exercise 27, how much seed would be needed for 7000 ft² of lawn?

29. Estimating a Whale Population. To determine the number of humpback whales in a population, a marine biologist, using tail markings, identifies 27 individual whales. Several weeks later, 40 whales from the population are randomly sighted. Of the 40 sighted, 12 are among the 27 originally identified. Estimate the number of whales in the population.

30. Estimating a Trout Population. To determine the number of trout in a lake, a conservationist catches 112 trout, tags them, and throws them back into the lake. Later, 82 trout are caught, and it is found that 32 of them are tagged. Estimate how many trout there are in the lake.

31. Map Scaling. On a road atlas map, 1 in. represents 16.6 mi. If two cities are 3.5 in. apart on the map, how far apart are they in reality?

32. Map Scaling. On a map, $\frac{1}{4}$ in. represents 50 mi. If two cities are $3\frac{1}{4}$ in. apart on the map, how far apart are they in reality?

33. Currency Exchange. On January 7, 2013, 1 U.S. dollar was worth about 0.76 euro.
a) How much were 50 U.S. dollars worth in euros on that day?
b) How much would a car that costs 8640 euros cost in U.S. dollars?

34. Currency Exchange. On January 7, 2013, 1 U.S. dollar was worth about 12.74 Mexican pesos.
a) How much were 150 U.S. dollars worth in Mexican pesos on that day?
b) While traveling in Mexico at that time, Jake bought a watch that cost 3600 Mexican pesos. How much did it cost in U.S. dollars?

35. *Basketball.* After playing 32 games in the 2012–2013 National Basketball Association season, Kobe Bryant of the Los Angeles Lakers had scored 977 points.

   a) At this rate, how many games would it take him to score 2000 points?

   b) At this rate, how many points would Bryant score in the whole 82-game season?

Source: National Basketball Association

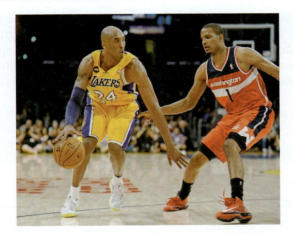

36. *Home Runs.* After playing 118 games in the 2012 Major League Baseball season, Andrew McCutchen of the Pittsburgh Pirates had 24 home runs.

   a) At this rate, how many games would it take him to hit 30 home runs?

   b) At this rate, how many home runs would McCutchen hit in the entire 162-game season?

Source: Major League Baseball

Skill Maintenance

Solve.

37. $12 \cdot x = 1944$ [1.7b]

38. $6807 = m + 2793$ [1.7b]

39. $t + 4.25 = 8.7$ [5.7a]

40. $112.5 \cdot p = 45$ [5.7a]

41. $3.7 + y = 18$ [5.7a]

42. $0.078 = 0.3 \cdot t$ [5.7a]

43. $c + \dfrac{4}{5} = \dfrac{9}{10}$ [4.3b]

44. $\dfrac{5}{6} = \dfrac{2}{3} \cdot x$ [3.8a]

Synthesis

45. Carney College is expanding from 850 to 1050 students. To avoid any rise in the student-to-faculty ratio, the faculty of 69 professors must also increase. How many new faculty positions should be created?

46. In recognition of her outstanding work, Sheri's salary has been increased from $26,000 to $29,380. Tim is earning $23,000 and is requesting a proportional raise. How much more should he ask for?

47. *Baseball Statistics.* Cy Young, one of the greatest baseball pitchers of all time, gave up an average of 2.63 earned runs every 9 innings. Young pitched 7356 innings, more than anyone else in the history of baseball. How many earned runs did he give up?

48. *Real-Estate Values.* According to Coldwell Banker Real Estate Corporation, a home selling for $189,000 in Austin, Texas, would sell for $437,850 in Denver, Colorado. How much would a $350,000 home in Denver sell for in Austin? Round to the nearest $1000.

Source: Coldwell Banker Real Estate Corporation

49. The ratio 1 : 3 : 2 is used to estimate the relative costs of a CD player, receiver, and speakers when shopping for a sound system. That is, the receiver should cost three times the amount spent on the CD player and the speakers should cost twice the amount spent on the CD player. If you had $900 to spend, how would you allocate the money, using this ratio?

7.5 Geometric Applications

OBJECTIVES

a Find lengths of sides of similar triangles using proportions.

b Use proportions to find lengths in pairs of figures that differ only in size.

SKILL TO REVIEW

Objective 7.3b: Solve proportions.

Solve.

1. $\dfrac{7}{x} = \dfrac{8}{3}$ **2.** $\dfrac{2}{1\frac{1}{2}} = \dfrac{p}{\frac{1}{4}}$

---

1. This pair of triangles is similar. **GS** Find the missing length x.

The ratio of x to 20 is the same as the ratio of 9 to [ ].

$$\frac{x}{20} = \frac{9}{\boxed{}}$$

$$x \cdot \boxed{} = 20 \cdot 9$$

$$\frac{x \cdot 12}{\boxed{}} = \frac{20 \cdot 9}{\boxed{}}$$

$$x = \frac{\boxed{}}{12} = \boxed{}$$

Answers

Skill to Review:

1. 2.625 **2.** $\dfrac{1}{3}$

Margin Exercise:

1. 15

Guided Solution:

1. 12; 12, 12, 12, 12, 180, 15

a PROPORTIONS AND SIMILAR TRIANGLES

Look at the pair of triangles below. Note that they appear to have the same shape, but their sizes are different. These are examples of **similar triangles**. You can imagine using a magnifying glass to enlarge the smaller triangle so that it looks the same as the larger one. This process works because the corresponding sides of each triangle have the same ratio. That is, the following proportion is true.

$$\frac{a}{d} = \frac{b}{e} = \frac{c}{f}$$

SIMILAR TRIANGLES

Similar triangles have the same shape. The lengths of their corresponding sides have the same ratio—that is, they are proportional.

EXAMPLE 1 The triangles below are similar triangles. Find the missing length x.

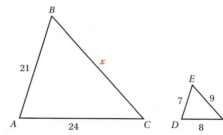

The ratio of x to 9 is the same as the ratio of 24 to 8 or 21 to 7. We get the proportions

$$\frac{x}{9} = \frac{24}{8} \quad \text{and} \quad \frac{x}{9} = \frac{21}{7}.$$

For both proportions, the sides of the larger triangle are in the numerators, and the sides of the smaller triangle are in the denominators.

We can solve either one of these proportions. We use the first one:

$$\frac{x}{9} = \frac{24}{8}$$

$$x \cdot 8 = 9 \cdot 24 \qquad \text{Equating cross products}$$

$$\frac{x \cdot 8}{8} = \frac{9 \cdot 24}{8} \qquad \text{Dividing both sides by 8}$$

$$x = 27. \qquad \text{Simplifying}$$

The missing length x is 27. Other proportions could also be used.

◀ **Do Margin Exercise 1.**

Similar triangles and proportions can often be used to find lengths that would ordinarily be difficult to measure. For example, we could find the height of a flagpole without climbing it or the distance across a river without crossing it.

EXAMPLE 2   How tall is a flagpole that casts a 56-ft shadow at the same time that a 6-ft man casts a 5-ft shadow?

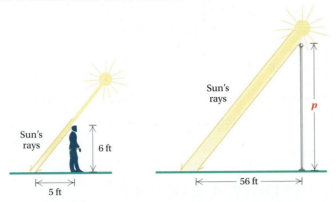

If we use the sun's rays to represent the third side of the triangle in our drawing of the situation, we see that we have similar triangles. Let $p =$ the height of the flagpole. The ratio of 6 to p is the same as the ratio of 5 to 56. Thus we have the proportion

$$\text{Height of man} \rightarrow \frac{6}{p} = \frac{5}{56} \leftarrow \text{Length of shadow of man} \atop \leftarrow \text{Length of shadow of pole}$$

Solve: $6 \cdot 56 = p \cdot 5$ Equating cross products

$$\frac{6 \cdot 56}{5} = \frac{p \cdot 5}{5}$$ Dividing both sides by 5

$$\frac{6 \cdot 56}{5} = p$$ Simplifying

$$67.2 = p.$$

The height of the flagpole is 67.2 ft.

Do Exercise 2. ▶

2. How tall is a flagpole that casts a 45-ft shadow at the same time that a 5.5-ft woman casts a 10-ft shadow?

EXAMPLE 3   *Rafters of a House.*   Carpenters use similar triangles to determine the length of rafters for a house. They first choose the pitch of the roof, or the ratio of the rise over the run. Then using a triangle with that ratio, they calculate the length of the rafters needed for the house. Loren is making rafters for a roof with a 6/12 pitch on a house that is 30 ft wide. Using a rafter guide, Loren knows that the rafter length corresponding to the 6/12 pitch is 13.4. Find the length x of the rafters for this house to the nearest tenth of a foot.

We have the proportion

Length of rafter in 6/12 triangle $\searrow$ Run in 6/12 $\swarrow$ triangle

Length of rafter $\nearrow$ on the house $\dfrac{13.4}{x} = \dfrac{12}{15}.$ Run in similar $\nwarrow$ triangle on the house

Solve: $13.4 \cdot 15 = x \cdot 12$ Equating cross products

$\dfrac{13.4 \cdot 15}{12} = \dfrac{x \cdot 12}{12}$ Dividing both sides by 12

$\dfrac{13.4 \cdot 15}{12} = x$

$16.8 \text{ ft} \approx x.$ Rounding to the nearest tenth of a foot

3. *Rafters of a House.* Referring to Example 3, find the length y of the rise of the rafters of the house to the nearest tenth of a foot.

The length x of the rafters for the house is about 16.8 ft.

◀ **Do Exercise 3.**

b   PROPORTIONS AND OTHER GEOMETRIC SHAPES

When one geometric figure is a magnification of another, the figures are similar. Thus the corresponding lengths are proportional.

EXAMPLE 4 The sides in the photographs below are proportional. Find the width of the larger photograph.

2.5 cm
← 3.5 cm →

x
← 10.5 cm →

4. The sides in the photographs below are proportional. Find the width of the larger photograph.

6 cm
← 10 cm →

x
← 35 cm →

We let $x =$ the width of the photograph. Then we translate to a proportion.

Larger width $\rightarrow$ $\dfrac{x}{2.5} = \dfrac{10.5}{3.5}$ $\leftarrow$ Larger length
Smaller width $\rightarrow$ $\leftarrow$ Smaller length

Solve: $x \cdot 3.5 = 2.5 \cdot 10.5$ Equating cross products

$\dfrac{x \cdot 3.5}{3.5} = \dfrac{2.5 \cdot 10.5}{3.5}$ Dividing both sides by 3.5

$x = \dfrac{2.5 \cdot 10.5}{3.5}$ Simplifying

$x = 7.5.$

Thus the width of the larger photograph is 7.5 cm.

◀ **Do Exercise 4.**

EXAMPLE 5 A scale model of an addition to an athletic facility is 12 cm wide at the base and rises to a height of 15 cm. If the actual base is to be 52 ft, what will the actual height of the addition be?

We let h = the height of the addition. Then we translate to a proportion.

$$\text{Width in model} \rightarrow \frac{12}{52} = \frac{15}{h} \leftarrow \text{Height in model} \atop \leftarrow \text{Actual height}$$

Width in model → ; Actual width → ; Height in model ← ; Actual height ←

Solve: $12 \cdot h = 52 \cdot 15$ Equating cross products

$$\frac{12 \cdot h}{12} = \frac{52 \cdot 15}{12}$$ Dividing both sides by 12

$$h = \frac{52 \cdot 15}{12} = 65.$$

Thus the height of the addition will be 65 ft.

Do Exercise 5. ▶

GS **5.** Refer to the figure in Example 5. If a skylight on the model is 3 cm wide, how wide will an actual skylight be?

Let w = the width of an actual skylight.

$$\frac{12}{52} = \frac{\square}{w}$$

$$12 \cdot w = 52 \cdot \square$$

$$w = \square$$

The width of an actual skylight will be 13 $\square$.

EXAMPLE 6 *Bicycle Design.* Two important dimensions to consider when buying a bicycle are *stack* and *reach*, as illustrated in the diagram at the right. The Country Racer bicycle comes in six different frame sizes, each proportional to the others. In the smallest frame size, the stack is 50 cm and the reach is 37.5 cm. In the largest frame size, the stack is 60 cm. What is the reach in the largest frame size of the Country Racer?

We let r = the reach in the largest frame size. Then we translate to a proportion.

$$\text{Stack in smaller frame} \rightarrow \frac{50}{37.5} = \frac{60}{r} \leftarrow \text{Stack in larger frame} \atop \leftarrow \text{Reach in larger frame}$$

Solve:

$$50 \cdot r = 37.5 \cdot 60$$ Equating cross products

$$\frac{50 \cdot r}{50} = \frac{37.5 \cdot 60}{50}$$ Dividing both sides by 50

$$r = \frac{37.5 \cdot 60}{50} = 45.$$

Thus the reach in the largest frame size is 45 cm.

Do Exercise 6. ▶

6. Refer to Example 6. In another frame size of the Country Racer, the reach is 42 cm. What is the stack?

Answers

5. 13 ft   **6.** 56 cm

Guided Solution:
5. 3, 3, 13; ft

✓ Reading Check

Complete each proportion based on the following similar triangles.

RC1. $\dfrac{a}{3} = \dfrac{c}{\square}$

RC2. $\dfrac{\square}{4} = \dfrac{c}{5}$

RC3. $\dfrac{3}{\square} = \dfrac{5}{c}$

RC4. $\dfrac{8}{4} = \dfrac{a}{\square}$

a   The triangles in each exercise are similar. Find the missing lengths.

1.

2.

3.

4.

5.

6.

7.

8.

9. When a tree with a height of 8 m casts a shadow 5 m long, how long a shadow is cast by a person 2 m tall?

10. How tall is a flagpole that casts a 42-ft shadow at the same time that a $5\frac{1}{2}$-ft woman casts a 7-ft shadow?

11. How tall is a tree that casts a 27-ft shadow at the same time that a 4-ft fence post casts a 3-ft shadow?

12. How tall is a tree that casts a 32-ft shadow at the same time that an 8-ft light pole casts a 9-ft shadow?

13. Find the height h of the wall.

6 ft

h

6 ft

7 ft

14. Find the length L of the lake. Assume that the ratio of L to 120 yd is the same as the ratio of 720 yd to 30 yd.

120 yd

30 yd

720 yd

L

15. Find the distance across the river. Assume that the ratio of d to 25 ft is the same as the ratio of 40 ft to 10 ft.

10 ft

d

25 ft

40 ft

16. To measure the height of a hill, a string is stretched from level ground to the top of the hill. A 3-ft stick is placed under the string, touching it at point P, a distance of 5 ft from point G, where the string touches the ground. The string is then detached and found to be 120 ft long. How high is the hill?

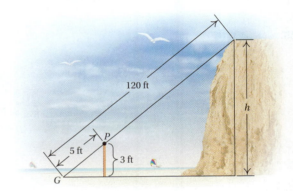

120 ft

h

P

5 ft

3 ft

G

---

b   In each of Exercises 17–26, the sides in each pair of figures are proportional. Find the missing lengths.

17.

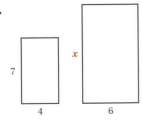

6

x

9

6

18.

x

5

7

14

19.

7

x

4

6

20.

11

x

4

3

21.

22.

23.

24.

25.

26.

27. A scale model of an addition to a medical clinic is 15 cm wide at the base and rises to a height of 19 cm. If the actual base is to be 120 ft, what will the actual height of the addition be?

28. Refer to the figure in Exercise 27. If a large window on the model is 3 cm wide, how wide will the actual window be?

Skill Maintenance

Determine whether each number is prime, composite, or neither.   [3.2b]

29. 83

30. 28

Find the prime factorization of each number.   [3.2c]

31. 808

32. 93

Use = or ≠ for ☐ to write a true sentence.   [3.5c]

33. $\dfrac{12}{8}$ ☐ $\dfrac{6}{4}$

34. $-\dfrac{4}{7}$ ☐ $-\dfrac{5}{9}$

Use < or > for ☐ to write a true sentence.   [4.2c]

35. $\dfrac{7}{12}$ ☐ $\dfrac{11}{15}$

36. $-\dfrac{1}{6}$ ☐ $-\dfrac{2}{11}$

Simplify.

37. $\left(\dfrac{1}{2}\right)^2 + \dfrac{2}{3} \cdot 4\dfrac{1}{2}$ [4.8a]

38. $18.3 + 2.5 \times 4.2 - (2.6 - 0.3^2)$ [5.4b]

39. $9 \times 15 - [2^3 \cdot 6 - (2 \cdot 5 + 3 \cdot 10)]$ [1.9d]

40. $2600 \div 13 - 5^3$ [1.9c]

Synthesis

Hockey Goals.   An official hockey goal is 6 ft wide. To make scoring more difficult, goalies often position themselves far in front of the goal to "cut down the angle." In Exercises 41 and 42, suppose that a slapshot is attempted from point *A* and that the goalie is 2.7 ft wide. Determine how far from the goal the goalie should be located if point *A* is the given distance from the goal. (*Hint*: First find how far the goalie should be from point *A*.)

41. ▦ 25 ft

42. ▦ 35 ft

43. A miniature air conditioning unit is built for the model referred to in Exercise 27. An actual unit is 10 ft high. How high should the model unit be?

▦ Solve. Round the answer to the nearest thousandth.

44. $\dfrac{8664.3}{10{,}344.8} = \dfrac{x}{9776.2}$

45. $\dfrac{12.0078}{56.0115} = \dfrac{789.23}{y}$

▦ The triangles in each exercise are similar triangles. Find the lengths not given.

46.

47.

Vocabulary Reinforcement

Complete each statement with the correct term from the list at the right. Some of the choices may not be used.

1. A ratio is the _____ of two quantities.   [7.1a]

2. Similar triangles have the same _____ .   [7.5a]

3. To solve $\dfrac{x}{a} = \dfrac{c}{d}$ for x, equate the _____ and divide on both sides to get x alone on one side.   [7.3b]

4. A(n) _____ is a ratio used to compare two different kinds of measure.   [7.2a]

5. A(n) _____ states that two pairs of numbers have the same ratio.   [7.3a]

6. A unit price is the ratio of _____ to the number of units.   [7.2b]

cross products

price

product

proportion

quantity

quotient

rate

shape

Concept Reinforcement

Determine whether each statement is true or false.

_____ 1. When we simplify a ratio like $\frac{8}{12}$, we find two other numbers in the same ratio.   [7.1b]

_____ 2. The proportion $\dfrac{a}{b} = \dfrac{c}{d}$ can also be written as $\dfrac{c}{a} = \dfrac{d}{b}$.   [7.3a]

_____ 3. Similar triangles must be the same size.   [7.5a]

_____ 4. Lengths of corresponding sides of similar figures have the same ratio.   [7.5b]

Study Guide

Objective 7.1a   Find fraction notation for ratios.

Example   Find the ratio of 7 to 18.

    Write a fraction with a numerator of 7 and a denominator of 18: $\dfrac{7}{18}$.

Practice Exercise

1. Find the ratio of 17 to 3.

Objective 7.1b   Simplify ratios.

Example   Simplify the ratio 8 to 2.5.

$$\frac{8}{2.5} = \frac{8}{2.5} \cdot \frac{10}{10} = \frac{80}{25} = \frac{5 \cdot 16}{5 \cdot 5}$$

$$= \frac{5}{5} \cdot \frac{16}{5} = \frac{16}{5}$$

Practice Exercise

2. Simplify the ratio 3.2 to 2.8.

Objective 7.2a   Give the ratio of two different measures as a rate.

Example   A driver travels 156 mi on 6.5 gal of gas. What is the rate in miles per gallon?

$$\frac{156 \text{ mi}}{6.5 \text{ gal}} = \frac{156}{6.5}\frac{\text{mi}}{\text{gal}} = 24\frac{\text{mi}}{\text{gal}}, \text{ or } 24 \text{ mpg}$$

Practice Exercise

3. A student earned $120 for working 16 hr. What was the rate of pay per hour?

Objective 7.2b   Find unit prices and use them to compare purchases.

Example   A 16-oz can of tomatoes costs $1.00. A 20-oz can costs $1.23. Which has the lower unit price?

$$16 \text{ oz}: \frac{\$1.00}{16 \text{ oz}} = \frac{100 \text{ cents}}{16 \text{ oz}} = 6.25¢/\text{oz}$$

$$20 \text{ oz}: \frac{\$1.23}{20 \text{ oz}} = \frac{123 \text{ cents}}{20 \text{ oz}} = 6.15¢/\text{oz}$$

Thus the 20-oz can has the lower unit price.

Practice Exercise

4. A 28-oz jar of Brand A spaghetti sauce costs $2.79. A 32-oz jar of Brand B spaghetti sauce costs $3.29. Find the unit price of each brand and determine which is a better buy based on unit price alone.

Objective 7.3a   Determine whether two pairs of numbers are proportional.

Example   Determine whether 3, 4 and 7, 9 are proportional.

$$3 \cdot 9 = 27 \qquad \frac{3}{4} \overset{?}{=} \frac{7}{9} \qquad 4 \cdot 7 = 28$$

Since the cross products are not the same ($27 \neq 28$), $\frac{3}{4} \neq \frac{7}{9}$ and the numbers are not proportional.

Practice Exercise

5. Determine whether 7, 9 and 21, 27 are proportional.

Objective 7.3b   Solve proportions.

Example   Solve: $\frac{3}{4} = \frac{y}{7}$.

$$3 \cdot 7 = 4 \cdot y \qquad \text{Equating cross products}$$

$$\frac{3 \cdot 7}{4} = \frac{4 \cdot y}{4} \qquad \text{Dividing by 4 on both sides}$$

$$\frac{21}{4} = y$$

The solution is $\frac{21}{4}$.

Practice Exercise

6. Solve: $\frac{9}{x} = \frac{8}{3}$.

Objective 7.4a   Solve applied problems involving proportions.

Example   Martina bought 3 tickets to a campus theater production for $16.50. How much would 8 tickets cost?

We translate to a proportion.

$$\text{Tickets} \rightarrow \frac{3}{16.50} = \frac{8}{c} \leftarrow \text{Tickets}$$
$$\text{Cost} \rightarrow \qquad \qquad \leftarrow \text{Cost}$$

$$3 \cdot c = 16.50 \cdot 8 \qquad \text{Equating cross products}$$

$$c = \frac{16.50 \cdot 8}{3}$$

$$c = 44$$

Eight tickets would cost $44.

Practice Exercise

7. On a map, $\frac{1}{2}$ in. represents 50 mi. If two cities are $1\frac{3}{4}$ in. apart on the map, how far apart are they in reality?

Objective 7.5a Find lengths of sides of similar triangles using proportions.

Example The triangles below are similar. Find the missing length x.

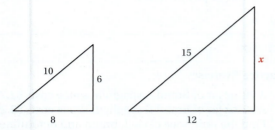

The ratio of 6 to x is the same as the ratio of 8 to 12 (and also as the ratio of 10 to 15). We write and solve a proportion:

$$\frac{6}{x} = \frac{8}{12}$$

$$6 \cdot 12 = x \cdot 8$$

$$\frac{6 \cdot 12}{8} = \frac{x \cdot 8}{8}$$

$$9 = x.$$

The missing length is 9. (We could also have used other proportions, including $\frac{6}{x} = \frac{10}{15}$, to find x.)

Practice Exercise

8. The triangles below are similar. Find the missing length y.

Review Exercises

Write fraction notation for each ratio. Do not simplify. [7.1a]

1. 47 to 84

2. 46 to 1.27

3. 83 to 100

4. 0.72 to 197

5. At Preston Seafood Market, 12,480 lb of tuna and 16,640 lb of salmon were sold one year. [7.1a, b]
   a) Write fraction notation for the ratio of tuna sold to salmon sold.
   b) Write fraction notation for the ratio of salmon sold to the total number of pounds of both kinds of fish sold.

Find the ratio of the first number to the second number and simplify. [7.1b]

6. 9 to 12

7. 3.6 to 6.4

8. *Gas Mileage.* The Chrysler 200 will travel 406 mi on 14.5 gal of gasoline in highway driving. What is the rate in miles per gallon? [7.2a]

   Source: Chrysler Motor Corporation

9. *Flywheel Revolutions.* A certain flywheel makes 472,500 revolutions in 75 min. What is the rate of spin in revolutions per minute? [7.2a]

10. A lawn requires 319 gal of water for every 500 ft². What is the rate in gallons per square foot? [7.2a]

11. *Calcium Supplement.* The price for a particular calcium supplement is $18.99 for 300 tablets. Find the unit price in cents per tablet. [7.2b]

12. Raquel bought a 24-oz loaf of 12-grain bread for $4.69. Find the unit price in cents per ounce. [7.2b]

13. *Vegetable Oil.* Find the unit price. Then determine which size has the lowest unit price. [7.2b]

Package	Price	Unit Price
32 oz	$4.79	
48 oz	$5.99	
64 oz	$9.99	

Determine whether the two pairs of numbers are proportional. [7.3a]

14. $9, 15$ and $36, 60$

15. $24, 37$ and $40, 46.25$

Solve. [7.3b]

16. $\dfrac{8}{9} = \dfrac{x}{36}$

17. $\dfrac{6}{x} = \dfrac{48}{56}$

18. $\dfrac{120}{\frac{3}{7}} = \dfrac{7}{x}$

19. $\dfrac{4.5}{120} = \dfrac{0.9}{x}$

Solve. [7.4a]

20. *Quality Control.* A factory manufacturing computer circuits found 3 defective circuits in a lot of 65 circuits. At this rate, how many defective circuits can be expected in a lot of 585 circuits?

21. *Exchanging Money.* On January 7, 2013, 1 U.S. dollar was worth about 0.99 Canadian dollar.
  a) How much were 250 U.S. dollars worth in Canada on that day?
  b) While traveling in Canada that day, Jamal saw a sweatshirt that cost 50 Canadian dollars. How much would it cost in U.S. dollars?

22. A train travels 448 mi in 7 hr. At this rate, how far will it travel in 13 hr?

23. *Gasoline Consumption.* The United States consumed about 89 billion gallons of gasoline in the first 8 months of 2011. At this rate, how much was consumed in 12 months?

24. *Trash Production.* A study shows that 5 people generate 23 lb of trash each day. The population of Austin, Texas, is 820,611. How many pounds of trash are produced in Austin in one day?

Sources: U.S. Environmental Protection Agency; U.S. Census Bureau

25. *Calories Burned.* Kevin burned 200 calories while playing ultimate frisbee for $\frac{3}{4}$ hr. How many calories would he burn if he played for $1\frac{1}{5}$ hr?

26. *Lawyers in Chicago.* In Illinois, there are about 4.6 lawyers for every 1000 people. The population of Chicago is 2,707,120. How many lawyers would you expect there to be in Chicago?

Sources: American Bar Association; U.S. Census Bureau

Each pair of triangles in Exercises 27 and 28 is similar. Find the missing length(s). [7.5a]

27.

28.

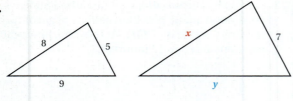

29. How tall is a billboard that casts a 25-ft shadow at the same time that an 8-ft sapling casts a 5-ft shadow? [7.5a]

30. The lengths in the figures below are proportional. Find the missing lengths. [7.5b]

31. *Turkey Servings.* A 25-lb turkey serves 18 people. Find the rate in servings per pound. [7.2a]

A. 0.36 serving/lb
B. 0.72 serving/lb
C. 0.98 serving/lb
D. 1.39 servings/lb

32. If 3 dozen eggs cost $5.04, how much will 5 dozen eggs cost? [7.4a]

A. $6.72
B. $6.96
C. $8.40
D. $10.08

Synthesis

33. *Paper Towels.* Find the unit price and determine which item would be the best buy based on unit price alone. [7.2b]

PACKAGE	PRICE	UNIT PRICE PER SHEET
8 rolls, 60 (2 ply) sheets per roll	$8.67	
15 rolls, 60 (2 ply) sheets per roll	$14.92	
6 big rolls, 165 (2 ply) sheets per roll	$13.95	

34. ▦ The following triangles are similar. Find the missing lengths. [7.5a]

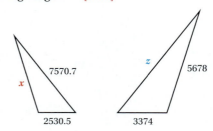

35. Shine-and-Glo Painters uses 2 gal of finishing paint for every 3 gal of primer. Each gallon of finishing paint covers 450 ft^2. If a surface of 4950 ft^2 needs both primer and finishing paint, how many gallons of each should be purchased? [7.4a]

Understanding Through Discussion and Writing

1. If you were a college president, which would you prefer: a low or high faculty-to-student ratio? Why? [7.1a]

2. Can unit prices be used to solve proportions that involve money? Explain why or why not. [7.2b], [7.4a]

3. Write a proportion problem for a classmate to solve. Design the problem so that the solution is "Leslie would need 16 gal of gasoline in order to travel 368 mi." [7.4a]

4. Is it possible for two triangles to have two pairs of sides that are proportional without the triangles being similar? Why or why not? [7.5a]

CHAPTER

7   **Test**

For Extra Help   For step-by-step test solutions, access the Chapter Test Prep Videos in MyMathLab® or on YouTube (search "BittingerPrealgebra" and click on "Channels").

Write fraction notation for each ratio. Do not simplify.

1. 85 to 97

2. 0.34 to 124

Find the ratio of the first number to the second number and simplify.

3. 18 to 20

4. 0.75 to 0.96

5. What is the rate in feet per second?

10 feet, 16 seconds

6. *Ham Servings.*   A 12-lb shankless ham contains 16 servings. What is the rate in servings per pound?

7. *Gas Mileage.*   Jeff's convertible will travel 464 mi on 14.5 gal of gasoline in highway driving. What is the rate in miles per gallon?

8. *Bagged Salad Greens.*   Ron bought a 16-oz bag of salad greens for $2.49. Find the unit price in cents per ounce.

9. The table below lists prices for concentrated liquid laundry detergent. Find the unit price of each size in cents per ounce. Then determine which has the lowest unit price.

Size	Price	Unit Price
40 oz	$6.59	
50 oz	$6.99	
100 oz	$11.49	
150 oz	$24.99	

Determine whether the two pairs of numbers are proportional.

10. 7, 8 and 63, 72

11. 1.3, 3.4 and 5.6, 15.2

Solve.

12. $\dfrac{9}{4} = \dfrac{27}{x}$

13. $\dfrac{150}{2.5} = \dfrac{x}{6}$

14. $\dfrac{x}{100} = \dfrac{27}{64}$

15. $\dfrac{68}{y} = \dfrac{17}{25}$

Solve.

16. *Distance Traveled.*   An ocean liner traveled 432 km in 12 hr. At this rate, how far would the boat travel in 42 hr?

17. *Time Loss.*   A watch loses 2 min in 10 hr. At this rate, how much will it lose in 24 hr?

18. *Map Scaling.* On a map, 3 in. represents 225 mi. If two cities are 7 in. apart on the map, how far apart are they in reality?

19. *Tower Height.* A birdhouse on a pole that is 3 m high casts a shadow 5 m long. At the same time, the shadow of a tower is 110 m long. How tall is the tower?

20. *Charity Work.* Kayla is crocheting hats for a charity. She can make 8 hats from 12 packages of yarn.
a) How many hats can she make from 20 packages of yarn?
b) How many packages of yarn does she need to make 20 hats?

21. *Thanksgiving Dinner.* A traditional turkey dinner for 8 people cost about $33.81 in a recent year. How much would it cost to serve a turkey dinner for 14 people?

Source: American Farm Bureau Federation

The sides in each pair of figures are proportional. Find the missing lengths.

22.

23.

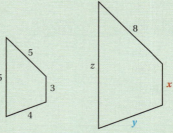

24. Lucita walks $4\frac{1}{2}$ mi in $1\frac{1}{2}$ hr. What is her rate in miles per hour?
A. $\frac{1}{3}$ mph
B. $1\frac{1}{2}$ mph
C. 3 mph
D. $4\frac{1}{2}$ mph

Synthesis

Solve.

25. $\dfrac{x + 3}{4} = \dfrac{5x + 2}{8}$

26. Nancy wants to win a gift card from the campus bookstore by guessing the number of marbles in an 8-gal jar. She knows that there are 128 oz in a gallon. She goes home and fills an 8-oz jar with 46 marbles. How many marbles should she guess are in the 8-gal jar?

Calculate and simplify.

1.
$$\begin{array}{r} 2\,7.6\,8 \\ 3.0\,1\,9 \\ +\ 4\,8\,3.2\,9\,7 \\ \hline \end{array}$$

2.
$$\begin{array}{r} 2\frac{1}{3} \\ +\ 4\frac{5}{12} \\ \hline \end{array}$$

3. $\dfrac{6}{35} + \dfrac{5}{28}$

4.
$$\begin{array}{r} 4\,0.2 \\ -\ \ \ 9.7\,0\,9 \\ \hline \end{array}$$

5. $-32 - (-15)$

6. $\dfrac{3}{20} - \dfrac{4}{15}$

7.
$$\begin{array}{r} 3\,7.6\,4 \\ \times\ \ \ \ \ \ 5.9 \\ \hline \end{array}$$

8. $-43\,(15)$

9. $2\dfrac{1}{3} \cdot 1\dfrac{2}{7}$

10. $2.3\,\overline{)\,9\,8.9}$

11. $-306 \div (-6)$

12. $\dfrac{7}{11} \div \dfrac{14}{33}$

13. Write expanded notation: 30,074.

14. Write a word name for 120.07.

Which number is larger?

15. 0.7, 0.698

16. $-0.799,\ -0.8$

17. Find the prime factorization of 144.

18. Find the LCM of 18 and 30.

19. What part is shaded?

20. Simplify: $\dfrac{90}{144}$.

Calculate.

21. $\dfrac{3}{5} \times 9.53$

22. $7.2 \div 0.4(-1.5) + (1.2)^2$

23. Evaluate $\dfrac{t-7}{w}$ for $t = -3$ and $w = -2$.

24. Write fraction notation for the ratio 0.3 to 15. Do not simplify.

25. Find the mean: 23, 49, 52, 71.

26. Determine whether the pairs 3, 9 and 25, 75 are proportional.

27. What is the rate in meters per second?

        660 meters, 12 seconds

28. *Unit Prices.* An 8-oz can of pineapple chunks costs $0.99. A 24.5-oz jar of pineapple chunks costs $3.29. Which has the lower unit price?

Solve.

29. $\dfrac{14}{25} = \dfrac{x}{54}$

30. $-423 = 16 \cdot t$

31. $\dfrac{2}{3} \cdot y = \dfrac{16}{27}$

32. $9x - 7 = -43$

33. $34.56 + n = 67.9$

34. $2(x - 3) + 9 = 5x - 6$

35. Ramona's recipe for fettuccini alfredo has 520 calories in 1 cup. How many calories are there in $\frac{3}{4}$ cup?

36. A machine can stamp out 925 washers in 5 min. An order is placed for 1295 washers. How long will it take to stamp them out?

37. A 46-oz juice can contains $5\frac{3}{4}$ cups of juice. A recipe calls for $3\frac{1}{2}$ cups of juice. How many cups are left over?

38. It takes a carpenter $\frac{2}{3}$ hr to hang a door. How many doors can the carpenter hang in 8 hr?

39. *Car Travel.* A car travels 337.62 mi in 8 hr. How far does it travel in 1 hr?

40. *Shuttle Orbits.* A space shuttle made 16 orbits a day during an 8.25-day mission. How many orbits were made during the entire mission?

41. How many even prime numbers are there?
    A. 5   **B.** 3   **C.** 2
    D. 1   **E.** None

42. The gas mileage of a car is 28.16 mpg. How many gallons per mile is this?
    A. $\dfrac{704}{25}$ **B.** $\dfrac{25}{704}$ **C.** $\dfrac{2816}{100}$
    D. $\dfrac{250}{704}$ **E.** None

Synthesis

43. A soccer goalie wishing to block an opponent's shot moves toward the shooter to reduce the shooter's view of the goal. If the goalie can only defend a region 10 ft wide, how far in front of the goal should the goalie be? (See the figure at right.)

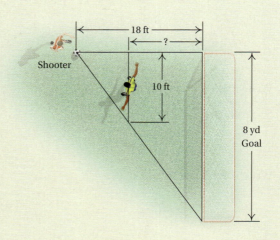

CHAPTER

8

Percent Notation

8.1   Percent Notation

OBJECTIVES

a Write three kinds of notation for a percent.

b Convert between percent notation and decimal notation.

c Convert between fraction notation and percent notation.

SKILL TO REVIEW

Objective 5.3a: Multiply using decimal notation.

Multiply.

1. 68.3×0.01

2. 3013×2.4

Where Coffee Drinkers Get Their Coffee

Home 29%
Starbucks 25%
Dunkin' Donuts 11%
Other 19%
McDonald's 16%

SOURCE: "Starbucks' Big Mug," *TIME*, June 25, 2012

Answers

Skill to Review:
1. 0.683   2. 7231.2

a   UNDERSTANDING PERCENT NOTATION

Today almost half of the world's population lives in cities. It is estimated that by 2050, 70% of the population will live in cities. What does 70% mean? It means that of every 100 people on Earth in 2050, 70 will live in cities.

Thus, 70% is a ratio of 70 to 100, or $\dfrac{70}{100}$.

Sources: DuPont, 2011; *National Geographic*, December 2011

70 of 100 squares are shaded.

70% or $\dfrac{70}{100}$ or 0.70 of the large square is shaded.

We encounter percent notation frequently. Here are some examples.

- The 1964–2013 Washington quarter is 91.67% copper and 8.33% nickel. The 1946–2013 Jefferson nickel is 75% copper and 25% nickel.
  Source: Coinflation.com

- A blood alcohol level of 0.08% is the standard used by most states as the legal limit for drunk driving.
  Source: The National Highway Safety Administration

- Educational loans of $20,000 or more per year are obtained by 16% of college students.
  Source: TRU Survey

- Of senior citizens aged 76 and older, 31% own a laptop.
  Source: Internet & American Life Project

Percent notation is often represented using a circle graph, or pie chart, to show how the parts of a quantity are related. For example, the circle graph at left illustrates the percentages of coffee drinkers who get their coffee at selected locations.

PERCENT NOTATION
The notation **n%** means "*n* per hundred."

This definition leads us to the following equivalent ways of defining percent notation.

NOTATION FOR n%

Percent notation, n%, can be expressed using

a ratio → $n\% = $ the ratio of n to $100 = \dfrac{n}{100}$,

fraction notation → $n\% = n \times \dfrac{1}{100}$, or

decimal notation → $n\% = n \times 0.01$.

EXAMPLE 1 Write three kinds of notation for 35%.

Using a ratio: $35\% = \dfrac{35}{100}$ A ratio of 35 to 100

Using fraction notation: $35\% = 35 \times \dfrac{1}{100}$ Replacing % with $\times \dfrac{1}{100}$

Using decimal notation: $35\% = 35 \times 0.01$ Replacing % with $\times 0.01$

During 2011, 42.2% of electricity in the United States was generated using coal.
Source: U.S. Energy Information Administration

EXAMPLE 2 Write three kinds of notation for 67.8%.

Using a ratio: $67.8\% = \dfrac{67.8}{100}$ A ratio of 67.8 to 100

Using fraction notation: $67.8\% = 67.8 \times \dfrac{1}{100}$ Replacing % with $\times \dfrac{1}{100}$

Using decimal notation: $67.8\% = 67.8 \times 0.01$ Replacing % with $\times 0.01$

Do Exercises 1–4. ▶

Write three kinds of notation for each percent, as in Example 1.

1. 70%   2. 23.4%

3. 100%   4. 0.6%

b   CONVERTING BETWEEN PERCENT NOTATION AND DECIMAL NOTATION

To write decimal notation for a number like 78%, we can think of percent notation as a ratio and write

$78\% = \dfrac{78}{100}$ Using the definition of percent as a ratio

$= 0.78.$ Dividing

Similarly,

$4.9\% = \dfrac{4.9}{100}$ Using the definition of percent as a ratio

$= 0.049.$ Dividing

We can also convert 78% to decimal notation by replacing "%" with "$\times 0.01$" and writing

$78\% = 78 \times 0.01$ Replacing % with $\times 0.01$

$= 0.78.$ Multiplying

It is thought that the Roman emperor Augustus began percent notation by taxing goods sold at a rate of $\frac{1}{100}$. In time, the symbol "%" evolved by interchanging the parts of the symbol "100" to "0/0" and then to "%."

Answers

1. $\dfrac{70}{100}$; $70 \times \dfrac{1}{100}$; 70×0.01

2. $\dfrac{23.4}{100}$; $23.4 \times \dfrac{1}{100}$; 23.4×0.01

3. $\dfrac{100}{100}$; $100 \times \dfrac{1}{100}$; 100×0.01

4. $\dfrac{0.6}{100}$; $0.6 \times \dfrac{1}{100}$; 0.6×0.01

Find decimal notation.

5. 34%   6. 78.9%

7. $3\frac{1}{2}\%$

Find decimal notation for the percent notation(s) in each sentence.

8. *Energy Use.* The United States consumes 19% of the world's energy. Russia consumes only 6% of the world's energy.

   Source: U.S. Energy Information Administration

9. *Blood Alcohol Level.* A blood alcohol level of 0.08% is the standard used by most states as the legal limit for drunk driving.

Similarly,

$$4.9\% = 4.9 \times 0.01 \qquad \text{Replacing \% with } \times 0.01$$
$$= 0.049. \qquad \text{Multiplying}$$

Dividing by 100 amounts to moving the decimal point two places to the left, which is the same as multiplying by 0.01. This leads us to a quick way to convert from percent notation to decimal notation: We drop the percent symbol and move the decimal point two places to the left.

To convert from percent notation to decimal notation,	36.5%
a) replace the percent symbol % with × 0.01, and	36.5 × 0.01
b) multiply by 0.01, which means move the decimal point two places to the left.	0.36.5 Move 2 places to the left. 36.5% = 0.365

EXAMPLE 3 Find decimal notation for 99.44%.

a) Replace the percent symbol with × 0.01.   99.44 × 0.01

b) Move the decimal point two places to the left.   0.99.44

Thus, 99.44% = 0.9944.

EXAMPLE 4 The interest rate on a $2\frac{1}{2}$-year certificate of deposit is $3\frac{3}{4}\%$. Find decimal notation for $3\frac{3}{4}\%$.

a) Convert $3\frac{3}{4}$ to decimal notation and replace the percent symbol with × 0.01. $3\frac{3}{4}\%$ 3.75 × 0.01

b) Move the decimal point two places to the left.   0.03.75

Thus, $3\frac{3}{4}\% = 0.0375$.

◀ **Do Exercises 5–9.**

The procedure used in Examples 3 and 4 can be reversed to write a decimal, like 0.38, as an equivalent percent. To see why, note that

$$0.38 = 0.38 \times 100\% = (0.38 \times 100)\% = 38\%.$$

To convert from decimal notation to percent notation, multiply by 100%. That is,	0.675 = 0.675 × 100%
a) multiply by 100, which means move the decimal point two places to the right, and	0.67.5 Move 2 places to the right.
b) write a % symbol.	67.5% 0.675 = 67.5%

Answers

5. 0.34   6. 0.789   7. 0.035
8. 0.19; 0.06   9. 0.0008

EXAMPLE 5 Of the time that people declare as sick leave, 0.21 is actually used for family issues. Find percent notation for 0.21.

Source: CCH Inc.

a) Move the decimal point two places to the right.

0.21.

b) Write a % symbol.

21%

Thus, 0.21 = 21%.

EXAMPLE 6 Find percent notation for 5.6.

a) Move the decimal point two places to the right, adding an extra zero.

5.60.

b) Write a % symbol.

560%

Thus, 5.6 = 560%.

EXAMPLE 7 Of those who play golf, 0.149 play 8–24 rounds per year. Find percent notation for 0.149.

Source: U.S. Golf Association

a) Move the decimal point two places to the right.

0.14.9

b) Write a % symbol.

14.9%

Thus, 0.149 = 14.9%.

EXAMPLE 8 Find percent notation for 0.00325.

a) Move the decimal point two places to the right.

0.00.325

b) Write a % symbol.

0.325%

Thus, 0.00325 = 0.325%.

Do Exercises 10–15. ▶

Find percent notation.

10. 0.24   **11.** 3.47

12. 1   **13.** 0.005

Find percent notation for the decimal notation(s) in each sentence.

14. *Women in Congress.* In 2012, 0.19 of the members of the U.S. Congress were women.

Source: *Wall Street Journal*

15. *Soccer.* Of Americans in the 18–24 age group, 0.311 have played soccer; of the 12–17 age group, 0.396 have played.

Source: ESPN Sports Poll, a service of TNS Sport

c CONVERTING BETWEEN FRACTION NOTATION AND PERCENT NOTATION

To convert from fraction notation to percent notation,

$$\frac{3}{5}$$ Fraction notation

a) find decimal notation by division, and

$$\begin{array}{r} 0.6 \\ 5\overline{)3.0} \\ \underline{3\ 0} \\ 0 \end{array}$$

b) convert the decimal notation to percent notation.

0.6 = 0.60 = 60%   Percent notation

$$\frac{3}{5} = 60\%$$

EXAMPLE 9 *First Language.* The first language of approximately $\frac{3}{16}$ of the world's population is Chinese. Find percent notation for $\frac{3}{16}$.

Sources: *National Geographic*, "Languages at Risk," Virginia W. Mason, July 2012; U.S. Census Bureau; *The CIA World Factbook 2012*

a) Find decimal notation by division.

$$
\begin{array}{r}
0.1\ 8\ 7\ 5 \\
16\overline{\smash{)}\ 3.0\ 0\ 0\ 0} \\
\underline{1\ 6} \\
1\ 4\ 0 \\
\underline{1\ 2\ 8} \\
1\ 2\ 0 \\
\underline{1\ 1\ 2} \\
8\ 0 \\
\underline{8\ 0} \\
0
\end{array}
\qquad \frac{3}{16} = 0.1875
$$

b) Convert the answer to percent notation.

0.18.75 ↗

$$\frac{3}{16} = 18.75\%, \text{ or } 18\frac{3}{4}\% \qquad 0.75 = \frac{3}{4}$$

Don't forget the % symbol.

Find percent notation.

16. $\frac{1}{4}$ **17.** $\frac{5}{8}$

◀ **Do Exercises 16 and 17.**

Fractions named by repeating decimals also can be converted to percent notation.

EXAMPLE 10 Find percent notation for $\frac{1}{6}$.

a) Find decimal notation by division.

$$
\begin{array}{r}
0.1\ 6\ 6 \\
6\overline{\smash{)}\ 1.0\ 0\ 0} \\
\underline{6} \\
4\ 0 \\
\underline{3\ 6} \\
4\ 0 \\
\underline{3\ 6} \\
4
\end{array}
$$

We get a repeating decimal: $0.16\overline{6}$.

b) Convert the answer to percent notation.

0.16.6̄ ↗

$$\frac{1}{6} = 16.\overline{6}\%, \text{ or } 16\frac{2}{3}\% \qquad 0.\overline{6} = \frac{2}{3}$$

◀ **Do Exercises 18 and 19.**

18. Water is the single most abundant chemical in the body. The human body is about $\frac{2}{3}$ water. Find percent notation for $\frac{2}{3}$.

19. Find percent notation: $\frac{5}{6}$.

Answers

16. 25%   **17.** 62.5%, or $62\frac{1}{2}\%$

18. $66.\overline{6}\%$, or $66\frac{2}{3}\%$ **19.** $83.\overline{3}\%$, or $83\frac{1}{3}\%$

In some cases, division is not the fastest way to convert a fraction to percent notation. The following are some optional ways in which this conversion might be done.

EXAMPLE 11   Find percent notation for $\frac{69}{100}$.

We use the definition of percent as a ratio.

$$\frac{69}{100} = 69\%$$

EXAMPLE 12   Find percent notation for $\frac{17}{20}$.

We want to multiply by 1 to get 100 in the denominator. We think of what we must multiply 20 by in order to get 100. That number is 5, so we multiply by 1 using $\frac{5}{5}$.

$$\frac{17}{20} \cdot \frac{5}{5} = \frac{85}{100} = 85\%$$

Note that this shortcut works only when the denominator of a simplified fraction is a factor of 100.

EXAMPLE 13   Find percent notation for $\frac{18}{25}$.

$$\frac{18}{25} = \frac{18}{25} \cdot \frac{4}{4} = \frac{72}{100} = 72\%$$

Do Exercises 20–23. ▶

The method used in Example 11 is reversed when we convert from percent notation to fraction notation.

To convert from percent notation to fraction notation,

   30%   Percent notation

a) use the definition of percent as a ratio, and

   $\frac{30}{100}$

b) simplify, if possible.

   $\frac{3}{10}$ Fraction notation

   $30\% = \frac{3}{10}$

EXAMPLE 14   Find fraction notation for 75%.

$$75\% = \frac{75}{100} \qquad \text{Using the definition of percent}$$

$$= \frac{3 \cdot 25}{4 \cdot 25}$$

$$= \frac{3}{4} \cdot \frac{25}{25} \qquad \text{Simplifying by removing a factor equal to 1: } \frac{25}{25} = 1$$

$$= \frac{3}{4}$$

Find percent notation.

20. $\frac{57}{100}$

GS **21.** $\dfrac{19}{25} = \dfrac{19}{25} \cdot \dfrac{4}{\boxed{}}$

$\qquad = \dfrac{76}{\boxed{}} = \boxed{}\%$

22. $\frac{7}{10}$ **23.** $\frac{1}{4}$

CALCULATOR CORNER

Converting from Fraction Notation to Percent Notation   A calculator can be used to convert from fraction notation to percent notation. We simply perform the division on the calculator and then use the percent key. To convert $\frac{17}{40}$ to percent notation, for example, we press

| 1 | 7 | ÷ | 4 | 0 | 2nd | % |, or

| 1 | 7 | ÷ | 4 | 0 | SHIFT | % |.

The display reads $\boxed{42.5}$, so $\frac{17}{40} = 42.5\%$.

EXERCISES   Use a calculator to find percent notation. Round to the nearest hundredth of a percent.

1. $\frac{13}{25}$ **2.** $\frac{5}{13}$

3. $\frac{43}{39}$ **4.** $\frac{12}{7}$

5. $\frac{217}{364}$ **6.** $\frac{2378}{8401}$

Answers

20. 57%   **21.** 76%   **22.** 70%
23. 25%

Guided Solution:
21. 4, 100, 76

EXAMPLE 15 Find fraction notation for 62.5%.

$$62.5\% = \frac{62.5}{100} \qquad \text{Using the definition of percent}$$

$$= \frac{62.5}{100} \times \frac{10}{10} \qquad \begin{array}{l}\text{Multiplying by 1 to eliminate the}\\ \text{decimal point in the numerator}\end{array}$$

$$= \frac{625}{1000}$$

$$\left.\begin{array}{l} = \dfrac{5 \cdot 125}{8 \cdot 125} = \dfrac{5}{8} \cdot \dfrac{125}{125} \\[2mm] = \dfrac{5}{8} \end{array}\right\} \quad \text{Simplifying}$$

Find fraction notation.

24. 60%

25. 3.25%

$$= \frac{3.25}{\boxed{}} = \frac{3.25}{100} \times \frac{\boxed{}}{100}$$

$$= \frac{325}{\boxed{}} = \frac{13 \times \boxed{}}{400 \times 25}$$

$$= \frac{13}{400} \times \frac{25}{25} = \frac{\boxed{}}{400}$$

26. $66\frac{2}{3}\%$ **27.** $12\frac{1}{2}\%$

EXAMPLE 16 Find fraction notation for $16\frac{2}{3}\%$.

$$16\frac{2}{3}\% = \frac{50}{3}\% \qquad \begin{array}{l}\text{Converting from the mixed}\\ \text{numeral to fraction notation}\end{array}$$

$$= \frac{50}{3} \times \frac{1}{100} \qquad \text{Using the definition of percent}$$

$$\left.\begin{array}{l} = \dfrac{50 \cdot 1}{3 \cdot 50 \cdot 2} = \dfrac{1}{3 \cdot 2} \cdot \dfrac{50}{50} \\[2mm] = \dfrac{1}{6} \end{array}\right\} \quad \text{Simplifying}$$

◀ **Do Exercises 24–27.**

The table below lists fraction, decimal, and percent equivalents that are used so often it would speed up your work if you memorized them. For example, $\frac{1}{3} = 0.\overline{3}$, so we say that the **decimal equivalent** of $\frac{1}{3}$ is $0.\overline{3}$, or that $0.\overline{3}$ has the **fraction equivalent** $\frac{1}{3}$. This table also appears on the inside back cover of the book.

FRACTION, DECIMAL, AND PERCENT EQUIVALENTS

FRACTION NOTATION	$\frac{1}{10}$	$\frac{1}{8}$	$\frac{1}{6}$	$\frac{1}{5}$	$\frac{1}{4}$	$\frac{3}{10}$	$\frac{1}{3}$	$\frac{3}{8}$	$\frac{2}{5}$	$\frac{1}{2}$	$\frac{3}{5}$	$\frac{5}{8}$	$\frac{2}{3}$	$\frac{7}{10}$	$\frac{3}{4}$	$\frac{4}{5}$	$\frac{5}{6}$	$\frac{7}{8}$	$\frac{9}{10}$	$\frac{1}{1}$
DECIMAL NOTATION	0.1	0.125	$0.16\overline{6}$	0.2	0.25	0.3	$0.33\overline{3}$	0.375	0.4	0.5	0.6	0.625	$0.66\overline{6}$	0.7	0.75	0.8	$0.83\overline{3}$	0.875	0.9	1
PERCENT NOTATION	10%	12.5%, or $12\frac{1}{2}\%$	$16.\overline{6}\%$, or $16\frac{2}{3}\%$	20%	25%	30%	$33.\overline{3}\%$, or $33\frac{1}{3}\%$	37.5%, or $37\frac{1}{2}\%$	40%	50%	60%	62.5%, or $62\frac{1}{2}\%$	$66.\overline{6}\%$, or $66\frac{2}{3}\%$	70%	75%	80%	$83.\overline{3}\%$, or $83\frac{1}{3}\%$	87.5%, or $87\frac{1}{2}\%$	90%	100%

Find fraction notation.

28. $33.\overline{3}\%$ **29.** $83.\overline{3}\%$

EXAMPLE 17 Find fraction notation for $16.\overline{6}\%$.

We can use the table above or recall that $16.\overline{6}\% = 16\frac{2}{3}\% = \frac{1}{6}$. We can also recall from our work with repeating decimals in Chapter 5 that $0.\overline{6} = \frac{2}{3}$. Then we have $16.\overline{6}\% = 16\frac{2}{3}\%$ and can proceed as in Example 16.

◀ **Do Exercises 28 and 29.**

Answers

24. $\frac{3}{5}$ **25.** $\frac{13}{400}$ **26.** $\frac{2}{3}$

27. $\frac{1}{8}$ **28.** $\frac{1}{3}$ **29.** $\frac{5}{6}$

Guided Solution:
25. 100, 100, 10,000, 25, 13

✓ Reading Check

Find percent notation for each shaded area.

RC1.   RC2.   RC3.   RC4.

a Write three kinds of notation, as in Examples 1 and 2 on p. 509.

1. 90%   **2.** 58.7%   **3.** 12.5%   **4.** 130%

b Find decimal notation.

5. 67%   **6.** 17%   **7.** 45.6%   **8.** 76.3%

9. 59.01%   **10.** 30.02%   **11.** 10%   **12.** 80%

13. 1%   **14.** 100%   **15.** 200%   **16.** 300%

17. 0.1%   **18.** 0.4%   **19.** 0.09%   **20.** 0.12%

21. 0.18%   **22.** 5.5%   **23.** 23.19%   **24.** 87.99%

25. $56\frac{1}{2}\%$ **26.** $61\frac{3}{4}\%$ **27.** $14\frac{7}{8}\%$ **28.** $93\frac{1}{8}\%$

Find decimal notation for the percent notation(s) in each sentence.

29. *Daily Calories.* In 2006, American adults got about 13% of their daily calories from fast food. This percentage decreased to 11% in 2010.

Source: *Indianapolis Star,* February 21, 2013, Nanci Hellmich

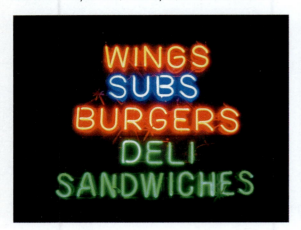

30. *Bachelor's Degrees.* In 1970, 1% of taxi drivers had a bachelor's degree. By 2010, 15% of taxi drivers had a bachelor's degree.

Source: *USA Today,* January, 28, 2013, Mary Beth Marklein

31. *Fuel Efficiency.* Speeding up by only 5 mph on the highway cuts fuel efficiency by approximately 7% to 8%.

Source: *Wall Street Journal,* "Pain Relief," by A. J. Miranda, September 15, 2008

32. *Foreign-Born Population.* In 2010, the U.S. foreign-born population was 12.9%, the highest since 1920.

Source: U.S. Census Bureau

33. *Credit-Card Debt.* In 2012, approximately 13.9% of American households had credit-card debt that exceeded 40% of their income.

Source: statisticbrain.com

34. *Eating Out.* On a given day, 58% of all Americans eat meals and snacks away from home.

Source: U.S. Department of Agriculture

Find percent notation.

35. 0.47   **36.** 0.87   **37.** 0.03   **38.** 0.01   **39.** 8.7

40. 4   **41.** 0.334   **42.** 0.889   **43.** 12   **44.** 16.8

45. 0.4   **46.** 0.5   **47.** 0.006   **48.** 0.008   **49.** 0.017

50. 0.024   **51.** 0.2718   **52.** 0.8911   **53.** 0.0239   **54.** 0.00073

Find percent notation for the decimal notation(s) in each sentence.

55. *Dining Together.* In 2012, 0.34 of American families dined together four or five times per week.

Source: unitedfamiliesinternational.wordpress.com

56. *Age 65 and Older.* In Alaska, 0.057 of the residents are age 65 or older. In Florida, 0.176 are age 65 or older.

Source: U.S. Census Bureau

57. Recycling Aluminum Cans. Over 0.651 of all aluminum cans are recycled.

Source: earth911.com

58. Wasting Food. Americans waste an estimated 0.27 of the food available for consumption. The waste occurs in restaurants, supermarkets, cafeterias, and household kitchens.

Source: *New York Times*, "One Country's Table Scraps, Another Country's Meal," by Andrew Martin, May 18, 2008

59. Residents Aged 15 or Younger. In Haiti, 0.359 of the residents are age 15 or younger. In the United States, 0.2 of the residents are age 15 or younger.

Source: *The World Almanac 2012*

60. Graduation Rates. In 2010, the high school graduation rate in the United States was 0.742. The dropout rate was 0.034.

Source: National Center for Educational Statistics

c Find percent notation.

61. $\dfrac{41}{100}$ **62.** $\dfrac{36}{100}$ **63.** $\dfrac{5}{100}$ **64.** $\dfrac{1}{100}$ **65.** $\dfrac{2}{10}$ **66.** $\dfrac{7}{10}$

67. $\dfrac{3}{10}$ **68.** $\dfrac{9}{10}$ **69.** $\dfrac{1}{2}$ **70.** $\dfrac{3}{4}$ **71.** $\dfrac{7}{8}$ **72.** $\dfrac{1}{8}$

73. $\dfrac{4}{5}$ **74.** $\dfrac{2}{5}$ **75.** $\dfrac{2}{3}$ **76.** $\dfrac{1}{3}$ **77.** $\dfrac{1}{6}$ **78.** $\dfrac{5}{6}$

79. $\dfrac{3}{16}$ **80.** $\dfrac{11}{16}$ **81.** $\dfrac{13}{16}$ **82.** $\dfrac{7}{16}$ **83.** $\dfrac{4}{25}$ **84.** $\dfrac{17}{25}$

85. $\dfrac{1}{20}$ **86.** $\dfrac{31}{50}$ **87.** $\dfrac{17}{50}$ **88.** $\dfrac{11}{20}$

Find percent notation for the fraction notations in each sentence.

89. College Graduation Rates. In 4-yr public colleges in Delaware, $\dfrac{11}{20}$ of the students graduate in 4 yr, and $\dfrac{7}{10}$ of the students graduate in 6 yr.

Source: *The Chronicle of Higher Education: College Completion*

90. College Graduation Rates. In 4-yr public colleges in Alaska, $\dfrac{2}{25}$ of the students graduate in 4 yr, and $\dfrac{13}{50}$ of the students graduate in 6 yr.

Source: *The Chronicle of Higher Education: College Completion*

In Exercises 91–96, write percent notation for the fractions given in the pie chart below.

How Food Dollars Are Spent

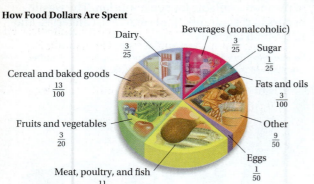

Dairy $\frac{3}{25}$

Beverages (nonalcoholic) $\frac{3}{25}$

Sugar $\frac{1}{25}$

Cereal and baked goods $\frac{13}{100}$

Fats and oils $\frac{3}{100}$

Fruits and vegetables $\frac{3}{20}$

Other $\frac{9}{50}$

Meat, poultry, and fish $\frac{11}{50}$

Eggs $\frac{1}{50}$

SOURCES: U.S. Bureau of Labor Statistics; Consumer Price Index; *The Hoosier Farmer*, Summer 2008

91. $\frac{11}{50}$

92. $\frac{9}{50}$

93. $\frac{3}{25}$

94. $\frac{1}{25}$

95. $\frac{3}{20}$

96. $\frac{13}{100}$

Find fraction notation. Simplify, if possible.

97. 85%

98. 55%

99. 62.5%

100. 12.5%

101. $33\frac{1}{3}\%$

102. $83\frac{1}{3}\%$

103. $16.\overline{6}\%$

104. $66.\overline{6}\%$

105. 7.25%

106. 4.85%

107. 0.8%

108. 0.2%

109. $25\frac{3}{8}\%$

110. $48\frac{7}{8}\%$

111. $78\frac{2}{9}\%$

112. $16\frac{5}{9}\%$

113. $64\frac{7}{11}\%$

114. $73\frac{3}{11}\%$

115. 150%

116. 110%

117. 0.0325%

118. 0.419%

119. $33.\overline{3}\%$

120. $83.\overline{3}\%$

In Exercises 121–126, find fraction notation for the percent notations given in the table below.

U.S. POPULATION BY SELECTED AGE CATEGORIES
(Data have been rounded to the nearest percent.)

AGE CATEGORY	PERCENT OF POPULATION
5–17 years	18%
18–24 years	10
15–44 years	41
18 years and older	76
65 years and older	13
75 years and older	6

SOURCES: U.S. Census Bureau; 2010 American Community Survey

121. 6%

122. 18%

123. 13%

124. 41%

125. 76%

126. 10%

Find fraction notation for the percent notation in each sentence.

127. A $\frac{3}{4}$-cup serving of Post Selects Great Grains cereal with $\frac{1}{2}$ cup of fat-free milk satisfies 15% of the minimum daily requirement for calcium.

Source: Kraft Foods Global, Inc.

128. A 1.8-oz serving of Kellogg's Frosted Mini-Wheats Blueberry cereal with $\frac{1}{2}$ cup of fat-free milk satisfies 35% of the minimum daily requirement for vitamin B_{12}.

Source: Kellogg, Inc.

129. In 2011, 20.6% of Americans adults met the Physical Activity Guidelines published by the U.S. Department of Health and Human Services for both aerobic and muscle-strengthening physical activity.

Source: U.S. Centers for Disease Control and Prevention

130. In 2010, 11.9% of Californians age 18 and older smoked cigarettes.

Source: California Department of Public Health

Complete each table.

131.

Fraction Notation	Decimal Notation	Percent Notation
$\frac{1}{8}$		12.5%, or $12\frac{1}{2}\%$
$\frac{1}{6}$		
		20%
	0.25	
		$33.\overline{3}\%$, or $33\frac{1}{3}\%$
		37.5%, or $37\frac{1}{2}\%$
		40%
$\frac{1}{2}$		

132.

Fraction Notation	Decimal Notation	Percent Notation
$\frac{3}{5}$		
	0.625	
$\frac{2}{3}$		
	0.75	75%
$\frac{4}{5}$		
$\frac{5}{6}$		$83.\overline{3}\%$, or $83\frac{1}{3}\%$
$\frac{7}{8}$		87.5%, or $87\frac{1}{2}\%$
		100%

133.

Fraction Notation	Decimal Notation	Percent Notation
	0.5	
$\frac{1}{3}$		
		25%
		$16.\overline{6}\%$, or $16\frac{2}{3}\%$
	0.125	
$\frac{3}{4}$		
	$0.8\overline{3}$	
$\frac{3}{8}$		

134.

Fraction Notation	Decimal Notation	Percent Notation
		40%
		62.5%, or $62\frac{1}{2}\%$
	0.875	
$\frac{1}{1}$		
	0.6	
	$0.\overline{6}$	
$\frac{1}{5}$		

Skill Maintenance

Solve.

135. $13 \cdot x = 910$ [1.7b]   **136.** $15 \cdot y = 75$ [1.7b]   **137.** $0.05 \times b = 20$ [5.7a]   **138.** $3 = 0.16 \times b$ [5.7a]

139. $\dfrac{24}{37} = \dfrac{15}{x}$ [7.3b]   **140.** $\dfrac{17}{18} = \dfrac{x}{27}$ [7.3b]   **141.** $\dfrac{9}{10} = \dfrac{x}{5}$ [7.3b]   **142.** $\dfrac{7}{x} = \dfrac{4}{5}$ [7.3b]

Synthesis

Find percent notation for each shaded area.

143.

144.

Write percent notation.

145. ▦ $\dfrac{41}{369}$

146. $2.5\overline{74631}$

Write decimal notation.

147. $\dfrac{14}{9}\%$

148. $\dfrac{19}{12}\%$

To draw a pie chart, or circle graph, think of a pie cut into 100 equal-size pieces. Then to represent, say, 20%, draw a wedge equal in size to 20 of these pieces. Use the given information for Exercises 149 and 150 to complete a circle graph.

149. *How Teens Spend Their Money*

Clothing	34%
Entertainment	22%
Food	22%
Other	22%

Source: Rand Youth Poll, eMarketer

150. *Reasons for Drinking Coffee*

To get going in the morning	32%
Like the taste	33%
Not sure	2%
To relax	4%
As a pick-me-up	10%
It's a habit	19%

Source: LMK Associates survey for Au Bon Pain Co., Inc.

Solving Percent Problems Using Percent Equations

a   TRANSLATING TO EQUATIONS

To solve a problem involving percents, it is helpful to translate first to an equation. To distinguish the method discussed in this section from that of Section 8.3, we will call these *percent equations*.

KEY WORDS IN PERCENT TRANSLATIONS

"**Of**" translates to "·" or "×".

"**What**" translates to any letter.

"**Is**" translates to "=".

"**%**" translates to "× $\frac{1}{100}$" or "× 0.01".

OBJECTIVES

a Translate percent problems to percent equations.

b Solve basic percent problems using percent equations.

SKILL TO REVIEW

Objective 5.7a: Solve equations containing decimals and one variable term.

Solve.

1. $0.05 \cdot x = 830$

2. $8 \cdot y = 40.648$

EXAMPLES   Translate each of the following.

1. 23%   of   5   is   what?

$0.23 \quad \cdot \quad 5 \quad = \quad a$ This is a *percent equation*.

2. What   is   11%   of   49?

$a \quad = \quad 0.11 \quad \cdot \quad 49$ Any letter can be used.

Do Margin Exercises 1 and 2. ▶

EXAMPLES   Translate each of the following.

3. 3   is   10%   of   what?

$3 \quad = \quad 0.10 \quad \cdot \quad b$

4. 45%   of   what   is   23?

$0.45 \quad \times \quad b \quad = \quad 23$

············· **Caution!** ···············

Don't forget to translate percent notation to decimal notation!

···

Do Exercises 3 and 4. ▶

EXAMPLES   Translate each of the following.

5. 10   is   what percent   of   20?

$10 \quad = \quad p \quad \times \quad 20$

6. What percent   of   50   is   7?

$p \quad \cdot \quad 50 \quad = \quad 7$

Do Exercises 5 and 6. ▶

Translate to an equation. Do not solve.

1. 12% of 50 is what?

2. What is 40% of 60?

Translate to an equation. Do not solve.

3. 45 is 20% of what?

4. 120% of what is 60?

Translate to an equation. Do not solve.

5. 16 is what percent of 40?

6. What percent of 84 is 10.5?

Answers

Skill to Review:
1. 16,600   **2.** 5.081

Margin Exercises:
1. $0.12 \cdot 50 = a$ **2.** $a = 0.40 \cdot 60$
3. $45 = 0.20 \cdot b$ **4.** $1.20 \times b = 60$
5. $16 = p \times 40$ **6.** $p \cdot 84 = 10.5$

Each percent equation in Examples 1–6 can be written in this form:

Amount = Percent number × Base.

In each case, one of the three pieces of information is missing.

b SOLVING PERCENT PROBLEMS

In solving percent problems, we use the *Translate* and *Solve* steps of the five-step problem-solving strategy.

Percent problems are actually of three different types. Although the method we present does *not* require that you be able to identify which type you are solving, it is helpful to know them. Each of the three types of percent problems depends on which of the three pieces of information is missing.

1. Finding the *amount* (the result of taking the percent)

   Example: **What** is 25% of 60?

   Translation: a = 0.25 · 60

2. Finding the *base* (the number you are taking the percent of)

   Example: 15 is 25% of **what?**

   Translation: 15 = 0.25 · b

3. Finding the *percent number* (the percent itself)

   Example: 15 is **what percent** of 60?

   Translation: 15 = p · 60

Finding the Amount

EXAMPLE 7   What is 17.9% of $14,660,000,000,000?

Translate: a = 0.179 · 14,660,000,000,000.

Solve: The letter is by itself. To solve the equation, we multiply:

a = 0.179 · 14,660,000,000,000 = 2,624,140,000,000.

Thus, $2,624,140,000,000 is 17.9% of $14,660,000,000,000. The answer is $2,624,140,000,000, or about $2.6 trillion.

◀ **Do Exercise 7.**

EXAMPLE 8   120% of 42 is what?

Translate: 1.20 · 42 = a.

Solve: The letter is by itself. To solve the equation, we carry out the calculation:

a = 1.20 · 42   120% = 1.20

a = 50.4.

Thus, 120% of 42 is 50.4. The answer is 50.4.

◀ **Do Exercise 8.**

The U.S. gross domestic product for 2010 was approximately $14.66 trillion. Of that amount, Americans spent 17.9% on health care. What was spent on health care? (See Example 7.)

Sources: *EconPost,* April 25, 2011; kaiseredu.org

7. Solve:

   What is 12% of $50?

8. Solve:

   64% of 55 is what?

Answers

7. $6   8. 35.2

Finding the Base

EXAMPLE 9   8% of what is 32?

Translate: $0.08 \cdot b = 32$.

Solve: This time the letter is *not* by itself. To solve the equation, we divide both sides by 0.08:

$$\frac{0.08 \cdot b}{0.08} = \frac{32}{0.08} \qquad \text{\color{red}{Dividing both sides by 0.08}}$$

$$b = \frac{32}{0.08}$$

$$b = 400.$$

Thus, 8% of 400 is 32. The answer is 400.

A survey of a group of people found that 8% of the group, or 32 people, chose cookies and cream as their favorite ice cream flavor. How many people were surveyed? (See Example 9.)

Source: Rasmussen Reports Survey

EXAMPLE 10 $3 is 16% of what?

Translate:

$$
\begin{array}{ccccc}
\$3 & \text{is} & 16\% & \text{of} & \text{what?} \\
\downarrow & \downarrow & \downarrow & \downarrow & \downarrow \\
3 & = & 0.16 & \cdot & b
\end{array}
$$

Solve: To solve the equation, we divide both sides by 0.16:

$$\frac{3}{0.16} = \frac{0.16 \cdot b}{0.16} \qquad \text{\color{red}{Dividing both sides by 0.16}}$$

$$\frac{3}{0.16} = b \qquad \text{\color{red}{16\% = 0.16}}$$

$$18.75 = b.$$

Thus, $3 is 16% of $18.75. The answer is $18.75.

Do Exercises 9 and 10. ▶

Solve.

9. 20% of what is 45?

$$
\begin{array}{ccccc}
20\% & \text{of} & \text{what} & \text{is} & 45? \\
\downarrow & \downarrow & \downarrow & \downarrow & \downarrow \\
\boxed{} & \cdot & b & = & \boxed{}
\end{array}
$$

$$\frac{0.20 \cdot b}{0.20} = \frac{45}{\boxed{}}$$

$$\boxed{} = \frac{45}{0.20}$$

$$b = \boxed{}$$

10. $60 is 120% of what?

Finding the Percent Number

When finding the percent number, you *must* remember to convert to percent notation after you have solved the equation.

EXAMPLE 11   414,000 is what percent of 621,000?

Translate:

$$
\begin{array}{ccccc}
414{,}000 & \text{is} & \text{what percent} & \text{of} & 621{,}000? \\
\downarrow & \downarrow & \downarrow & \downarrow & \downarrow \\
414{,}000 & = & p & \cdot & 621{,}000
\end{array}
$$

Solve: To solve the equation, we divide both sides by 621,000 and convert the result to percent notation:

$$414{,}000 = p \cdot 621{,}000$$

$$\frac{414{,}000}{621{,}000} = \frac{p \cdot 621{,}000}{621{,}000} \qquad \text{\color{red}{Dividing}}$$

$$0.66\overline{6} = p \qquad \text{\color{red}{Converting to decimal notation}}$$

$$66\tfrac{2}{3}\% = p. \qquad \text{\color{red}{Converting to percent notation}}$$

Thus, 414,000 is $66\tfrac{2}{3}\%$ of 621,000. The answer is $66\tfrac{2}{3}\%$.

In 2011, 621,000 new housing permits were issued in the United States. Of those new houses, 414,000 have porches. What percent of the houses have porches? (See Example 11.)

Source: U.S. Census Bureau

Solve.

11. 16 is what percent of 40?

16 is $\underbrace{\text{what percent}}$ of 40?

↓ ↓ ↓ ↓ ↓

16 ☐ p · ☐

$$\dfrac{16}{\boxed{}} = \dfrac{p \cdot \boxed{}}{40}$$

$$\dfrac{16}{40} = p$$

$$0.4 = p$$

$$\boxed{}\% = p$$

12. What percent of $84 is $10.50?

EXAMPLE 12 What percent of $50 is $16?

Translate: What percent of $50 is $16?

$\underbrace{\phantom{\text{What percent}}}$ ↓ ↓ ↓ ↓

p · 50 = 16

Solve: To solve the equation, we divide both sides by 50 and convert the answer to percent notation:

$$\dfrac{p \cdot 50}{50} = \dfrac{16}{50} \qquad \text{Dividing both sides by 50}$$

$$p = \dfrac{16}{50}$$

$$p = 0.32$$

$$p = 32\%. \qquad \text{Converting to percent notation}$$

Thus, 32% of $50 is $16. The answer is 32%.

◀ **Do Exercises 11 and 12.**

.... **Caution!** ....

When a question asks "what percent?", be sure to give the answer in percent notation.

....

CALCULATOR CORNER

Using Percents in Computations Many calculators have a ☐% key that can be used in computations. For example, to find 11% of 49, we press ☐1 ☐1 ☐2nd ☐% ☐× ☐4 ☐9 ☐=, or ☐4 ☐9 ☐× ☐1 ☐1 ☐SHIFT ☐%. The display reads ☐5.39☐, so 11% of 49 is 5.39.

In Example 9, we performed the computation 32/8%. To use the ☐% key in this computation, we press ☐3 ☐2 ☐÷ ☐8 ☐2nd ☐% ☐=, or ☐3 ☐2 ☐÷ ☐8 ☐SHIFT ☐%. The result is 400.

We can also use the ☐% key to find the percent number in a problem. In Example 11, for instance, we answered the question "414,000 is what percent of 621,000?" On a calculator, we press ☐4 ☐1 ☐4 ☐0 ☐0 ☐0 ☐÷ ☐6 ☐2 ☐1 ☐0 ☐0 ☐0 ☐2nd ☐% ☐=, or ☐4 ☐1 ☐4 ☐0 ☐0 ☐0 ☐÷ ☐6 ☐2 ☐1 ☐0 ☐0 ☐0 ☐SHIFT ☐%. The result is 66.$\overline{6}$, so 414,000 is 66$\frac{2}{3}$% of 621,000.

EXERCISES Use a calculator to find each of the following.

1. What is 12.6% of $40?

2. 0.04% of 28 is what?

3. 8% of what is 36?

4. $45 is 4.5% of what?

5. 23 is what percent of 920?

6. What percent of $442 is $53.04?

Answers

11. 40% **12.** 12.5%

Guided Solution:

11. =, 40; 40, 40, 40

✓ Reading Check

Match each question with the correct translation from the list at the right.

RC1. 18 is 40% of what? _____

RC2. What percent of 45 is 18? _____

RC3. What is 40% of 45? _____

RC4. 0.5% of 1200 is what? _____

RC5. 6 is what percent of 1200? _____

RC6. 6 is 0.5% of what? _____

a) $6 = 0.005 \cdot b$

b) $6 = p \cdot 1200$

c) $18 = 0.40 \cdot b$

d) $0.005 \cdot 1200 = a$

e) $p \cdot 45 = 18$

f) $a = 0.40 \cdot 45$

a   Translate to a percent equation. Do not solve.

1. What is 32% of 78?

2. 98% of 57 is what?

3. 89 is what percent of 99?

4. What percent of 25 is 8?

5. 13 is 25% of what?

6. 21.4% of what is 20?

b   Translate to a percent equation and solve.

7. What is 85% of 276?

8. What is 74% of 53?

9. 150% of 30 is what?

10. 100% of 13 is what?

11. What is 6% of $300?

12. What is 4% of $45?

13. 3.8% of 50 is what?

14. $33\frac{1}{3}\%$ of 480 is what?
(*Hint:* $33\frac{1}{3}\% = \frac{1}{3}$.)

15. $39 is what percent of $50?

16. $16 is what percent of $90?

17. 20 is what percent of 10?

18. 60 is what percent of 20?

19. What percent of $300 is $150?

20. What percent of $50 is $40?

21. What percent of 80 is 100?

22. What percent of 60 is 15?

23. 20 is 50% of what?

24. 57 is 20% of what?

25. 40% of what is $16?

26. 100% of what is $74?

27. 56.32 is 64% of what?

28. 71.04 is 96% of what?

29. 70% of what is 14?

30. 70% of what is 35?

31. What is $62\frac{1}{2}$% of 10?

32. What is $35\frac{1}{4}$% of 1200?

33. What is 8.3% of $10,200?

34. What is 9.2% of $5600?

35. 2.5% of what is 30.4?

36. 8.2% of what is 328?

Skill Maintenance

Write fraction notation.   [5.1b]

37. 0.09

38. 1.79

39. 0.875

40. 0.125

41. 0.9375

42. 0.6875

Write decimal notation.   [5.1b]

43. $\dfrac{89}{100}$

44. $\dfrac{7}{100}$

45. $\dfrac{3}{10}$

46. $\dfrac{17}{1000}$

Synthesis

Solve.

47. What is 7.75% of $10,880?
Estimate _____
Calculate _____

48. 50,951.775 is what percent of 78,995?
Estimate _____
Calculate _____

49. *Recyclables.* It is estimated that 40% to 50% of all trash is recyclable. If a community produces 270 tons of trash, how much of this trash is recyclable?

50. *Batting.* An all-star baseball player gets a hit in 30% to 35% of his at-bats. If an all-star had 520 to 580 at-bats, how many hits would he have had?

51. 40% of $18\frac{3}{4}$% of $25,000 is what?

a TRANSLATING TO PROPORTIONS

OBJECTIVES

a Translate percent problems to proportions.

b Solve basic percent problems using proportions.

SKILL TO REVIEW

Objective 7.3b: Solve proportions.

Solve.

1. $\dfrac{3}{100} = \dfrac{27}{b}$

2. $\dfrac{4.3}{20} = \dfrac{N}{100}$

A percent is a ratio of some number to 100. For example, 20% is the ratio $\dfrac{20}{100}$.

The numbers 30,800,000 and 154,000,000 have the same ratio as 20 and 100.

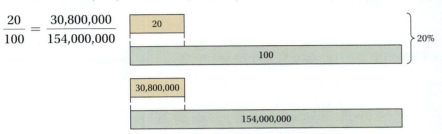

$$\frac{20}{100} = \frac{30,800,000}{154,000,000}$$

To solve a percent problem using a proportion, we translate as follows:

$$\text{Number} \longrightarrow \frac{N}{100} \longleftarrow 100 \qquad = \qquad \frac{a}{b} \longleftarrow \text{Amount} \atop \longleftarrow \text{Base}$$

> You might find it helpful to read this as "part is to whole as part is to whole."

For example, "60% of 25 is 15" translates to

$$\frac{60}{100} = \frac{15}{25} \begin{array}{l} \longleftarrow \text{Amount} \\ \longleftarrow \text{Base} \end{array}$$

A clue for translating is that the base, b, corresponds to 100 and usually follows the wording "percent of." Also, $N\%$ always translates to $N/100$. Another aid in translating is to make a comparison drawing. To do this, we start with the percent side and list 0% at the top and 100% near the bottom. Then we estimate where the specified percent—in this case, 60%—is located. The corresponding quantities are then filled in. The base—in this case, 25—always corresponds to 100%, and the amount—in this case, 15—corresponds to the specified percent.

In the United States, 20% of the labor force is age 55 or older. In 2012, there were approximately 154,000,000 people in the labor force. This means that about 30,800,000 workers were age 55 or older.

Sources: U.S. Department of Labor; U.S. Bureau of Labor Statistics

Percents	Quantities		Percents	Quantities		Percents	Quantities
0%	0		0%	0		0%	0
			60%			60%	15
100%			100%			100%	25

The proportion can then be read easily from the drawing: $\dfrac{60}{100} = \dfrac{15}{25}$.

---

*Note: This section presents an alternative method for solving basic percent problems. You can use either equations or proportions to solve percent problems, but you might prefer one method over the other, or your instructor may direct you to use one method over the other.

Answers

Skill to Review:
1. 900   2. 21.5

In the examples that follow, we use comparison drawings as an aid for visualization. It is not necessary to always make a drawing to solve problems of this sort.

EXAMPLE 1   Translate to a proportion.

23% of 5 is what?

number of hundredths   base   amount, a

$$\frac{23}{100} = \frac{a}{5}$$

EXAMPLE 2   Translate to a proportion.

What is 124% of 49?

amount, a number of hundredths   base

$$\frac{124}{100} = \frac{a}{49}$$

Translate to a proportion. Do not solve.

1. 12% of 50 is what?

2. What is 40% of 60?

3. 130% of 72 is what?

◀ **Do Exercises 1–3.**

EXAMPLE 3   Translate to a proportion.

3 is 10% of what?

amount   number of hundredths   base, b

$$\frac{10}{100} = \frac{3}{b}$$

EXAMPLE 4   Translate to a proportion.

45% of what is 23?

number of hundredths   base, b amount

$$\frac{45}{100} = \frac{23}{b}$$

Translate to a proportion. Do not solve.

4. 45 is 20% of what?

5. 120% of what is 60?

◀ **Do Exercises 4 and 5.**

EXAMPLE 5   Translate to a proportion.

10 is what percent of 20?

amount   number of hundredths, N base

$$\frac{N}{100} = \frac{10}{20}$$

Answers

1. $\frac{12}{100} = \frac{a}{50}$ **2.** $\frac{40}{100} = \frac{a}{60}$ **3.** $\frac{130}{100} = \frac{a}{72}$

4. $\frac{20}{100} = \frac{45}{b}$ **5.** $\frac{120}{100} = \frac{60}{b}$

EXAMPLE 6   Translate to a proportion.

What percent of 50 is 7?

number of   base   amount
hundredths, N

$$\frac{N}{100} = \frac{7}{50}$$

Do Exercises 6 and 7. ▶

Translate to a proportion. Do not solve.

6. 16 is what percent of 40?

7. What percent of 84 is 10.5?

b SOLVING PERCENT PROBLEMS

After a percent problem has been translated to a proportion, we solve as in Section 7.3.

EXAMPLE 7   5% of what is $20?

Translate: $\dfrac{5}{100} = \dfrac{20}{b}$

Solve: $5 \cdot b = 100 \cdot 20$ Equating cross products

$\dfrac{5 \cdot b}{5} = \dfrac{100 \cdot 20}{5}$ Dividing both sides by 5

$b = \dfrac{2000}{5}$

$b = 400$ Simplifying

Thus, 5% of $400 is $20. The answer is $400.

Do Exercise 8. ▶

EXAMPLE 8   120% of 42 is what?

Translate: $\dfrac{120}{100} = \dfrac{a}{42}$

Solve: $120 \cdot 42 = 100 \cdot a$ Equating cross products

$\dfrac{120 \cdot 42}{100} = \dfrac{100 \cdot a}{100}$ Dividing both sides by 100

$\dfrac{5040}{100} = a$

$50.4 = a$ Simplifying

Thus, 120% of 42 is 50.4. The answer is 50.4.

Do Exercises 9 and 10. ▶

GS **8.** Solve: 20% of what is $45?

$$\frac{20}{\boxed{}} = \frac{\boxed{}}{b}$$

$$20 \cdot b = 100 \cdot \boxed{}$$

$$\frac{20 \cdot b}{20} = \frac{100 \cdot 45}{20}$$

$$b = \frac{4500}{20}$$

$$b = \boxed{}$$

Thus, 20% of $ $\boxed{}$ is $45.

Solve.

GS **9.** 64% of 55 is what?

$$\frac{64}{100} = \frac{a}{\boxed{}}$$

$$\boxed{} \cdot 55 = 100 \cdot a$$

$$\frac{64 \cdot 55}{100} = \frac{100 \cdot a}{\boxed{}}$$

$$\frac{\boxed{}}{100} = a$$

$$\boxed{} = a$$

Thus, 64% of 55 is $\boxed{}$.

10. What is 12% of 50?

EXAMPLE 9 210 is $10\frac{1}{2}\%$ of what?

Translate: $\dfrac{210}{b} = \dfrac{10.5}{100}$ $\qquad 10\frac{1}{2}\% = 10.5\%$

Solve: $210 \cdot 100 = b \cdot 10.5$ $\qquad$ Equating cross products

$\dfrac{210 \cdot 100}{10.5} = \dfrac{b \cdot 10.5}{10.5}$ $\qquad$ Dividing both sides by 10.5

$\dfrac{21{,}000}{10.5} = b$ $\qquad$ Multiplying and simplifying

$2000 = b$ $\qquad$ Dividing

Thus, 210 is $10\frac{1}{2}\%$ of 2000. The answer is 2000.

◀ **Do Exercise 11.**

Percents	Quantities
0%	0
$10\frac{1}{2}\%$	210
100%	b

11. Solve:

60 is 120% of what?

12. Solve:

$12 is what percent of $40?

$\dfrac{12}{40} = \dfrac{N}{\boxed{}}$

$\boxed{} \cdot 100 = 40 \cdot N$

$\dfrac{12 \cdot 100}{\boxed{}} = \dfrac{40 \cdot N}{40}$

$\dfrac{\boxed{}}{40} = N$

$30 = N$

Thus, $12 is 30 $\boxed{}$ of $40.

EXAMPLE 10 $10 is what percent of $20?

Translate: $\dfrac{10}{20} = \dfrac{N}{100}$

Solve: $10 \cdot 100 = 20 \cdot N$ $\qquad$ Equating cross products

$\dfrac{10 \cdot 100}{20} = \dfrac{20 \cdot N}{20}$ $\qquad$ Dividing both sides by 20

$\dfrac{1000}{20} = N$ $\qquad$ Multiplying and simplifying

$50 = N$ $\qquad$ Dividing

Thus, $10 is 50% of $20. The answer is 50%.

Percents	Quantities
0%	0
$N\%$	$10
100%	$20

> Always "look before you leap." Many students can solve this problem mentally: $10 is half, or 50%, of $20.

◀ **Do Exercise 12.**

EXAMPLE 11 What percent of 50 is 16?

Translate: $\dfrac{N}{100} = \dfrac{16}{50}$

Solve: $50 \cdot N = 100 \cdot 16$ $\qquad$ Equating cross products

$\dfrac{50 \cdot N}{50} = \dfrac{100 \cdot 16}{50}$ $\qquad$ Dividing both sides by 50

$N = \dfrac{1600}{50}$ $\qquad$ Multiplying and simplifying

$N = 32$ $\qquad$ Dividing

Thus, 32% of 50 is 16. The answer is 32%.

Percents	Quantities
0%	0
$N\%$	16
100%	50

> Note when solving percent problems using proportions that N is a percent and needs only a % sign.

13. Solve:

What percent of 84 is 10.5?

◀ **Do Exercise 13.**

.... **Caution!** ....

Don't forget to add the % sign when the problem asks "What percent . . . ?"

✓ Reading Check

Match each question with the correct translation from the list at the right.

RC1. 70 is 35% of what? _____

RC2. 70 is what percent of 200? _____

RC3. What is 35% of 200? _____

RC4. 74.8 is 110% of what? _____

RC5. What percent of 68 is 74.8? _____

RC6. 110% of 68 is what? _____

a) $\dfrac{110}{100} = \dfrac{a}{68}$

b) $\dfrac{70}{b} = \dfrac{35}{100}$

c) $\dfrac{a}{200} = \dfrac{35}{100}$

d) $\dfrac{74.8}{68} = \dfrac{N}{100}$

e) $\dfrac{70}{200} = \dfrac{N}{100}$

f) $\dfrac{74.8}{b} = \dfrac{110}{100}$

a Translate to a proportion. Do not solve.

1. What is 37% of 74?

2. 66% of 74 is what?

3. 4.3 is what percent of 5.9?

4. What percent of 6.8 is 5.3?

5. 14 is 25% of what?

6. 133% of what is 40?

b Translate to a proportion and solve.

7. What is 76% of 90?

8. What is 32% of 70?

9. 70% of 660 is what?

10. 80% of 920 is what?

11. What is 4% of 1000?

12. What is 6% of 2000?

13. 4.8% of 60 is what?

14. 63.1% of 80 is what?

15. $24 is what percent of $96?

16. $14 is what percent of $70?

17. 102 is what percent of 100?

18. 103 is what percent of 100?

19. What percent of $480 is $120?

20. What percent of $80 is $60?

21. What percent of 160 is 150?

22. What percent of 33 is 11?

23. $18 is 25% of what?

24. $75 is 20% of what?

25. 60% of what is 54?

26. 80% of what is 96?

27. 65.12 is 74% of what?

28. 63.7 is 65% of what?

29. 80% of what is 16?

30. 80% of what is 10?

31. What is $62\frac{1}{2}$% of 40?

32. What is $43\frac{1}{4}$% of 2600?

33. What is 9.4% of $8300?

34. What is 8.7% of $76,000?

35. 80.8 is $40\frac{2}{5}$% of what?

36. 66.3 is $10\frac{1}{5}$% of what?

Skill Maintenance

Solve.   [7.3b]

37. $\dfrac{x}{188} = \dfrac{2}{47}$

38. $\dfrac{15}{x} = \dfrac{3}{800}$

39. $\dfrac{75}{100} = \dfrac{n}{20}$

40. $\dfrac{612}{t} = \dfrac{72}{244}$

Solve.

41. A recipe for muffins calls for $\frac{1}{2}$ qt of buttermilk, $\frac{1}{3}$ qt of skim milk, and $\frac{1}{16}$ qt of oil. How many quarts of liquid ingredients does the recipe call for?   [4.2d]

42. The Ferristown School District purchased $\frac{3}{4}$ ton (T) of clay. If the clay is to be shared equally among the district's 6 art departments, how much will each art department receive?   [3.8b]

Synthesis

Solve.

43. ⊞ What is 8.85% of $12,640?

Estimate _____
Calculate _____

44. ⊞ 78.8% of what is 9809.024?

Estimate _____
Calculate _____

45. 30% of 80 is what percent of 120?

46. 40% of what is the same as 30% of 200?

47. ⊞ What percent of 90 is the same as 26% of 135?

48. ⊞ What percent of 80 is the same as 76% of 150?

Mid-Chapter Review

Concept Reinforcement

Determine whether each statement is true or false.

_____ **1.** When converting decimal notation to percent notation, move the decimal point two places to the right and write a percent symbol. [8.1b]

_____ **2.** The symbol % is equivalent to $\times$ 0.10. [8.1a]

_____ **3.** Of the numbers $\frac{1}{10}$, 1%, 0.1%, 10%, and $\frac{1}{100}$, the smallest number is 0.1%. [8.1b, c]

Guided Solutions

 Fill in each blank with the number that creates a correct statement or solution. [8.1b, c]

4. $\dfrac{1}{2}\% = \dfrac{1}{2} \cdot \dfrac{1}{\boxed{}} = \dfrac{1}{\boxed{}}$

5. $\dfrac{80}{1000} = \dfrac{\boxed{}}{100} = \boxed{}\%$

6. $5.5\% = \dfrac{\boxed{}}{100} = \dfrac{\boxed{}}{1000} = \dfrac{11}{\boxed{}}$

7. $0.375 = \dfrac{\boxed{}}{1000} = \dfrac{\boxed{}}{100} = \boxed{}\%$

8. Solve: 15 is what percent of 80? [8.2b]

$15 = p \cdot \boxed{}$ Translating

$\dfrac{15}{\boxed{}} = \dfrac{p \cdot \boxed{}}{\boxed{}}$ Dividing on both sides

$\dfrac{15}{\boxed{}} = p$ Simplifying

$\boxed{} = p$ Dividing

$\boxed{}\% = p$ Converting to percent notation

Mixed Review

Find decimal notation. [8.1b]

9. 28%

10. 0.15%

11. $5\frac{3}{8}\%$

12. 240%

Find percent notation. [8.1b, c]

13. 0.71

14. $\dfrac{9}{100}$

15. 0.3891

16. $\dfrac{3}{16}$

17. 0.005

18. $\dfrac{37}{50}$

19. 6

20. $\dfrac{5}{6}$

Find fraction notation. Simplify.   [8.1c]

21. 85%   **22.** 0.048%   **23.** $22\frac{3}{4}\%$ **24.** $16.\overline{6}\%$

Write percent notation for the shaded area.   [8.1c]

25.

26.

Solve.   [8.2b], [8.3b]

27. 25% of what is 14.5?

28. 220 is what percent of 1320?

29. What is 3.2% of 80,000?

30. $17.50 is 35% of what?

31. What percent of $800 is $160?

32. 130% of $350 is what?

33. Arrange the following numbers from smallest to largest.   [8.1b, c]

$$\frac{1}{2}\%,\ 5\%,\ 0.275,\ \frac{13}{100},\ 1\%,\ 0.1\%,\ 0.05\%,\ \frac{3}{10},\ \frac{7}{20},\ 10\%$$

34. Solve: 8.5 is $2\frac{1}{2}\%$ of what?   [8.2b], [8.3b]

   A. 3.4   **B.** 21.25

   C. 0.2125   **D.** 340

35. Solve: $102,000 is what percent of $3.6 million?   [8.2b], [8.3b]

   A. $2.8\overline{3}$ million   **B.** $2\frac{5}{6}\%$

   C. $0.028\overline{3}\%$ **D.** $28.\overline{3}$

Understanding Through Discussion and Writing

36. Is it always best to convert from fraction notation to percent notation by first finding decimal notation? Why or why not?   [8.1c]

37. Suppose we know that 40% of 92 is 36.8. What is a quick way to find 4% of 92? 400% of 92? Explain.   [8.2b], [8.3b]

38. In solving Example 10 in Section 8.3, a student simplifies $\frac{10}{20}$ before solving. Is this a good idea? Why or why not?   [8.3b]

39. What do the following have in common? Explain.   [8.1b, c]

$$\frac{23}{16},\ 1\frac{875}{2000},\ 1.4375,\ \frac{207}{144},\ 1\frac{7}{16},\ 143.75\%,\ 1\frac{4375}{10,000}$$

Applications of Percent

8.4

a   APPLIED PROBLEMS INVOLVING PERCENT

Applied problems involving percent are not always stated in a manner easily translated to an equation. In such cases, it is helpful to rephrase the problem before translating. Sometimes it also helps to make a drawing.

EXAMPLE 1   *Transportation to Work.*   In the United States, there are about 154,000,000 workers who are 16 years old or older. Approximately 76.1% of these workers drive to work alone. How many workers drive to work alone?

Transportation to Work in the United States

Drive alone 76.1%
Taxi 0.1%   Motorcycle 0.2%
Bicycle 0.6%
Other means 0.8%
Walk 2.9%
Work from home 4.3%
Public transportation 5.0%
Carpool 10.0%

SOURCE: U.S. Bureau of Labor Statistics

OBJECTIVES

a   Solve applied problems involving percent.

SKILL TO REVIEW

Objective 5.4a: Divide using decimal notation.

Divide.
1. $345 \div 57.5$
2. $111.87 \div 9.9$

1. **Familiarize.**   We can simplify the pie chart shown above to help familiarize ourselves with the problem. We let $a =$ the total number of workers who drive to work alone.

Transportation to Work in the United States

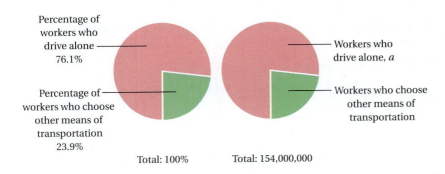

Percentage of workers who drive alone 76.1%

Percentage of workers who choose other means of transportation 23.9%

Total: 100%

Workers who drive alone, a

Workers who choose other means of transportation

Total: 154,000,000

Answers

Skill to Review:
1. 6   **2.** 11.3

2. Translate. There are two ways in which we can translate this problem.

Percent equation:

$$\underbrace{\text{What number}}_{a} \quad \underbrace{\text{is}}_{=} \quad \underbrace{76.1\%}_{0.761} \quad \underbrace{\text{of}}_{\cdot} \quad \underbrace{154{,}000{,}000?}_{154{,}000{,}000}$$

Proportion:

$$\frac{76.1}{100} = \frac{a}{154{,}000{,}000}$$

3. Solve. We now have two ways in which to solve the problem.

Percent equation:

$$a = 0.761 \cdot 154{,}000{,}000 = 117{,}194{,}000$$

Proportion:

$$\frac{76.1}{100} = \frac{a}{154{,}000{,}000}$$

$$76.1 \cdot 154{,}000{,}000 = 100 \cdot a \qquad \text{Equating cross products}$$

$$\frac{76.1 \cdot 154{,}000{,}000}{100} = \frac{100 \cdot a}{100} \qquad \text{Dividing both sides by 100}$$

$$\frac{11{,}719{,}400{,}000}{100} = a$$

$$117{,}194{,}000 = a \qquad \text{Simplifying}$$

4. Check. To check, we can repeat the calculations. We also can do a partial check by estimating. Since 76.1% is about 75%, or $\frac{3}{4}$, and $\frac{3}{4}$ of 154,000,000 is 115,500,000, which is close to 117,194,000, our answer is reasonable.

5. State. The number of workers who drive to work alone is 117,194,000.

◀ **Do Exercise 1.**

EXAMPLE 2 *Extinction of Mammals.* According to a study conducted for the International Union for the Conservation of Nature (IUCN), the world's mammals are facing an extinction crisis. Of the 5501 species of mammals on Earth, 1139 are on IUCN "vulnerable," "endangered," or "critically endangered" lists. What percent of all mammals are threatened with extinction?

1. Familiarize. The question asks for a percent of the world's mammals that are in danger of extinction. We note that 5501 is approximately 5500 and 1139 is approximately 1100. Since 1100 is $\frac{1100}{5500}$, or $\frac{1}{5}$, or 20%, of 5500, our answer should be close to 20%. We let p = the percent of mammals that are in danger of extinction.

2. Translate. There are two ways in which we can translate this problem.

Percent equation:

$$\underbrace{1139}_{1139} \quad \underbrace{\text{is}}_{=} \quad \underbrace{\text{what percent}}_{p} \quad \underbrace{\text{of}}_{\cdot} \quad \underbrace{5501?}_{5501}$$

Percents **Quantities**

1. *Transportation to Work.* There are about 154,000,000 workers 16 years old or older in the United States. Approximately 10.0% of them carpool to work. How many workers carpool to work?

Sources: U.S. Census Bureau; American Community Survey

Answer

1. 15,400,000

Mammals that are in danger of extinction include those shown here: Barbary macaque, black rhino, Galápagos seal, Malayan tapir, Cuvier's gazelle, Darwin's fox, and indri (a lemur).

Proportion:

$$\frac{N}{100} = \frac{1139}{5501}$$

For proportions, $N\% = p$.

Percents	Quantities
0%	0
$N\%$	1139
100%	5501

3. **Solve.** We now have two ways in which to solve the problem.

Percent equation:

$$1139 = p \cdot 5501$$

$$\frac{1139}{5501} = \frac{p \cdot 5501}{5501} \qquad \text{Dividing both sides by 5501}$$

$$\frac{1139}{5501} = p$$

$$0.207 \approx p \qquad \text{Finding decimal notation and rounding to the nearest thousandth}$$

$$20.7\% = p \qquad \text{Remember to find percent notation.}$$

Note that the solution, p, includes the % symbol.

Proportion:

$$\frac{N}{100} = \frac{1139}{5501}$$

$$N \cdot 5501 = 100 \cdot 1139 \qquad \text{Equating cross products}$$

$$\frac{N \cdot 5501}{5501} = \frac{113{,}900}{5501} \qquad \text{Dividing both sides by 5501}$$

$$N = \frac{113{,}900}{5501}$$

$$N \approx 20.7 \qquad \text{Dividing and rounding to the nearest tenth}$$

We use the solution of the proportion to express the answer to the problem as 20.7%. Note that in the proportion method, $N\% = p$.

4. **Check.** To check, we note that the answer 20.7% is close to 20%, as estimated in the *Familiarize* step.

5. **State.** About 20.7% of the world's mammals are threatened with extinction.

Do Exercise 2. ▶

2. *Presidential Assassinations in Office.* Of the 43 different U.S. presidents, 4 have been assassinated while in office. These were James A. Garfield, William McKinley, Abraham Lincoln, and John F. Kennedy. What percent have been assassinated in office?

Answer

2. About 9.3%

✓ Reading Check

Determine whether each statement is true or false.

_____ **RC1.** Applied problems involving percent can be translated using either equations or proportions.

_____ **RC2.** We can estimate to familiarize ourselves with a problem and to check to see if our answer is reasonable.

_____ **RC3.** Drawings such as pie charts are helpful when familiarizing ourselves with applied problems involving percent.

_____ **RC4.** When solving a percent problem using a proportion, N is already written in percent notation.

 a Solve.

1. *Facebook Daily Active Users.* In March 2014, there were, on average, 802 million daily active Facebook users worldwide. Approximately 35.2% of those users were in Europe, and approximately 19.8% were in the United States or Canada. About how many daily active users were in Europe, and about how many were in the United States and Canada?

Source: Facebook

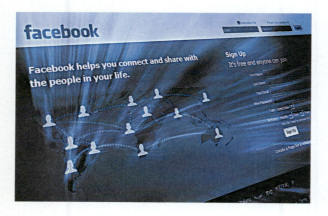

2. *Mississippi River.* The Mississippi River, which extends from its source, at Lake Itasca in Minnesota, to the Gulf of Mexico, is 2348 mi long. Approximately 77% of the river is navigable. How many miles of the river are navigable?

Source: National Oceanic and Atmospheric Administration

Mississippi River

3. Carrie earns $43,200 one year and then receives an 8% raise in salary. What is Carrie's new salary?

4. Chad earns $28,600 one year and then receives a 5% raise in salary. What is Chad's new salary?

5. *Test Results.* On a test, Juan got 85%, or 119, of the items correct. How many items were on the test?

6. *Test Results.* On a test, Maj Ling got 86%, or 81.7, of the items correct. (There was partial credit on some items.) How many items were on the test?

7. *Farmland.* In Kansas, 47,000,000 acres are farmland. About 5% of all the farm acreage in the United States is in Kansas. What is the total number of acres of farmland in the United States?

Sources: U.S. Department of Agriculture; National Agricultural Statistics Service

8. ▦ *World Population.* World population is increasing by 1.2% each year. In 2014, it was 7.2 billion. What will the population be in 2018?

Sources: U.S. Census Bureau; International Data Base

9. Car Depreciation. A car generally depreciates 25% of its original value in the first year. Lonnie's car is worth $27,300 after the first year. What was its original cost?

10. Car Depreciation. Given normal use, an American-made car will depreciate 25% of its original value the first year and 14% of its remaining value in the second year. What is the value of a car at the end of the second year if its original cost was $36,400? $28,400? $26,800?

11. Test Results. On a test of 80 items, Pedro got 93% correct. (There was partial credit on some items.) How many items did he get correct? incorrect?

12. Test Results. On a test of 40 items, Christina got 91% correct. (There was partial credit on some items.) How many items did she get correct? incorrect?

13. Olympic Team. The 2012 U.S. Summer Olympics team consisted of 529 members. Approximately 49.3% of the athletes were men. How many men were on the team?

Source: United States Olympic Committee

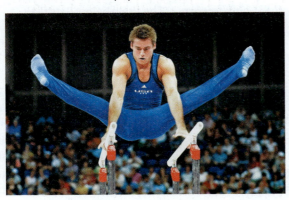

14. Murder Case Costs. The cost (for both trial and incarceration) of a life-without-parole case is about 30% of the cost of a death-penalty case. The average cost of a death-penalty case is $505,773. What is the average cost of a life-without-parole case?

Source: Indiana Legislative Services Agency, 2010 *Study of Murder Trials*

15. Spending on Pets. In 2011, Americans spent approximately $50.8 billion on their pets. Of this amount, $14.1 billion was for veterinarian bills. What percent of the total was spent on veterinary care?

Source: American Pet Products Association

16. Health Workers' Salaries. In 2010, the median annual salary of registered nurses was $64,690. The median annual salary of physicians and surgeons was $111,570. What percent of the physician and surgeon's median annual salary is the nurses' median annual salary?

Source: Bureau of Labor Statistics

17. Tipping. For a party of 8 or more, some restaurants add an 18% tip to the bill. What is the total amount charged for a party of 10 if the cost of the meal, without tip, is $195?

18. Tipping. Elaine adds a 15% tip when charging a meal to a credit card. What is the total amount charged to her card if the cost of the meal, without tip, is $18? $34? $49?

19. A lab technician has 540 mL of a solution of alcohol and water; 8% of the solution is alcohol. How many milliliters are alcohol? water?

20. A lab technician has 680 mL of a solution of water and acid; 3% of the solution is acid. How many milliliters are acid? water?

21. Wasting Food. As world population increases and farmland acreage decreases, food production and packaging must be improved and less food wasted. In the United States, consumers waste 39 lb of every 131 lb of fruit purchased. What percent is wasted?

Sources: United States Department of Agriculture; *National Geographic*, July 2011

22. Credit-Card Debt. Michael has disposable monthly income of $3400. Each month, he pays $470 toward his credit-card debt. What percent of his disposable income is allotted to paying off credit-card debt?

Monthly Disposable Income: $3400

Credit-card payments: $470

23. U.S. Volunteerism. The table below lists the numbers of Americans who did volunteer work in several areas in 2013. What percent of the total does each area represent? Round the answers to the nearest tenth of a percent.

VOLUNTEERS IN AMERICA: 2013

TOTAL	64,500,000
CIVIC	3,547,500
EDUCATIONAL	17,092,500
HEALTH	5,160,000
RELIGIOUS	22,059,000
SOCIAL SERVICE	9,288,000
SPORTS/ARTS	2,386,500

SOURCE: Corporation for National and Community Service

24. U.S. Volunteerism. The table below lists the population of the United States in 2013 and the number of volunteers by age group. What percent of the population in each age group volunteered? Round the answers to the nearest tenth of a percent.

AGE GROUP	POPULATION	NUMBER OF VOLUNTEERS
16–19 years old	17,000,000	4,500,000
20–24 years old	21,900,000	4,100,000
25–34 years old	41,200,000	9,400,000
35–44 years old	39,900,000	12,700,000
45–54 years old	44,000,000	13,200,000
55–64 years old	38,000,000	10,500,000
65–74 years old	23,400,000	6,300,000
75 years old and older	18,100,000	3,700,000

SOURCES: U.S. Census Bureau; Corporation for National and Community Service

Skill Maintenance

Convert to decimal notation.   [5.1b], [5.5a]

25. $\dfrac{25}{11}$

26. $\dfrac{11}{25}$

27. $\dfrac{27}{8}$

28. $\dfrac{43}{9}$

29. $\dfrac{23}{25}$

30. $\dfrac{20}{24}$

31. $\dfrac{14}{32}$

32. $\dfrac{2317}{1000}$

33. $\dfrac{34,809}{10,000}$

34. $\dfrac{27}{40}$

Synthesis

35. A couple uses a coupon to get $10 off a restaurant meal. Before subtracting $10, however, the restaurant adds a 20% tip. If the amount of the check is $40.40, how much would the meal alone (without tip) have cost without the coupon?

36. If p is 120% of q, then q is what percent of p?

37. Adult Height. At the age of 10, a girl has reached 84.4% of her final adult height. Cynthia is 4 ft 8 in. at the age of 10. What will be her adult height?

38. Adult Height. At the age of 15, a boy has reached 96.1% of his final adult height. Claude is 6 ft 4 in. at the age of 15. What will be his adult height?

a APPLIED PROBLEMS INVOLVING PERCENT INCREASE OR PERCENT DECREASE

OBJECTIVE

a Solve applied problems involving percent increase or percent decrease.

Percent is often used to state increase or decrease. Let's consider an example of each, using the price of a car as the original number.

Percent Increase

One year a car sold for $20,000. The manufacturer decides to raise the price of the following year's model by 5%. The increase is 0.05 × $20,000, or $1000. The new price is $20,000 + $1000, or $21,000. Note that the new price is 105% of the *former* price.

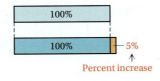

The increase, $1000, is 5% of the *former* price, $20,000. The *percent increase* is 5%.

New price: $21,000

Increase ↑

Former price: $20,000

Percent Decrease

Abigail buys a new car for $20,000. After one year, the car depreciates in value by 25%. The decrease is 0.25 × $20,000, or $5000. This lowers the value of the car to $20,000 − $5000, or $15,000. Note that the new value is 75% of the original price. If Abigail decides to sell the car after one year, $15,000 might be the most she could expect to get for it.

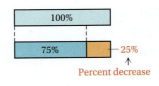

The decrease, $5000, is 25% of the *original* price, $20,000. The *percent decrease* is 25%.

Do Exercises 1 and 2. ▶

When a quantity is decreased by a certain percent, we say that there has been a **percent decrease**.

EXAMPLE 1 *Dow Jones Industrial Average.* The Dow Jones Industrial Average (DJIA) plunged from 11,143 to 10,365 on September 29, 2008. This was the largest one-day drop in its history. What was the percent decrease?

Sources: *Nightly Business Reports,* September 29, 2008; DJIA

Original price: $20,000

Decrease ↓

New value: $15,000

1. **Percent Increase.** The price of a car is $36,875. The price is increased by 4%.

   a) How much is the increase?

   b) What is the new price?

2. **Percent Decrease.** The value of a car is $36,875. The car depreciates in value by 25% after one year.

   a) How much is the decrease?

   b) What is the depreciated value of the car?

Answers

1. (a) $1475; (b) $38,350
2. (a) $9218.75; (b) $27,656.25

1. Familiarize. We first determine the amount of decrease and then make a drawing.

$$
\begin{array}{rl}
11{,}143 & \text{Opening average} \\
-10{,}365 & \text{Closing average} \\
\hline
778 & \text{Decrease}
\end{array}
$$

We are asking this question: The decrease is what percent of the opening average? We let p = the percent decrease.

2. Translate. There are two ways in which we can translate this problem.

Percent equation:

778	is	what percent	of	11,143?
778	=	p	·	11,143

Proportion:

$$\frac{N}{100} = \frac{778}{11{,}143}$$

For proportions, $N\% = p$.

3. Solve. We have two ways in which to solve the problem.

Percent equation:

$$778 = p \cdot 11{,}143$$

$$\frac{778}{11{,}143} = \frac{p \cdot 11{,}143}{11{,}143} \qquad \text{Dividing both sides by 11,143}$$

$$\frac{778}{11{,}143} = p$$

$$0.07 \approx p$$

$$7\% = p \qquad \text{Converting to percent notation}$$

Proportion:

$$\frac{N}{100} = \frac{778}{11{,}143}$$

$$11{,}143 \times N = 100 \times 778 \qquad \text{Equating cross products}$$

$$\frac{11{,}143 \times N}{11{,}143} = \frac{100 \times 778}{11{,}143} \qquad \text{Dividing both sides by 11,143}$$

$$N = \frac{77{,}800}{11{,}143}$$

$$N \approx 7$$

We use the solution of the proportion to express the answer to the problem as 7%.

Percents **Quantities**

0% ———— 0
N% ———— 778

100% ———— 11,143

CALCULATOR CORNER

Percent Increase or Decrease
On many calculators, there is a fast way to increase or decrease a number by any given percentage. For example, the result of taking 5% of $20,000 and adding it to $20,000 can be found by pressing

| 2 | 0 | 0 | 0 | 0 | + |

| 5 | SHIFT | % | = |

The displayed result will be

| 21000 |

EXERCISES

1. Use a calculator with a | % | key to confirm your answers to Margin Exercises 1 and 2.

2. The selling price of Lisa's $87,000 condominium was reduced by 8%. Find the new price.

---

542 CHAPTER 8 Percent Notation

4. Check. To check, we note that, with a 7% decrease, the closing Dow average should be 93% of the opening average. Since

$$93\% \cdot 11{,}143 = 0.93 \cdot 11{,}143 \approx 10{,}363,$$

and 10,363 is close to 10,365, our answer checks. (Remember that we rounded to get 7%.)

5. State. The percent decrease in the DJIA was approximately 7%.

Do Exercise 3. ▶

When a quantity is increased by a certain percent, we say that this is a **percent increase**.

EXAMPLE 2 *Costs for Moviegoers.* The average cost of a movie ticket was $5.80 in 2002. The cost rose to $8.12 in 2012. What was the percent increase in the cost of a movie ticket?

Sources: National Association of Theatre Owners; theaterseatstore.com

1. Familiarize. We first determine the increase in the cost and then make a drawing.

$$
\begin{array}{rl}
\$8.12 & \text{Cost in 2012} \\
-\ 5.80 & \text{Cost in 2002} \\
\hline
\$2.32 & \text{Increase}
\end{array}
$$

We are asking this question: The increase is what percent of the *original* cost? We let p = the percent increase.

2. Translate. There are two ways in which we can translate this problem.

Percent equation:

2.32	is	what percent	of	5.80?
↓	↓	↓	↓	↓
2.32	=	p	·	5.80

Proportion:

$$\frac{N}{100} = \frac{2.32}{5.80}$$

For proportions, $N\% = p$.

Percents	Quantities
0%	0
$N\%$	2.32
100%	5.80

3. *Volume of Mail.* The volume of U.S mail decreased from about 102,379 million pieces of mail in 2002 to 68,696 million pieces in 2012. What was the percent decrease?

Source: U.S. Postal Service

Answer

3. About 32.9%

3. **Solve.** We have two ways in which to solve the problem.

Percent equation:

$$2.32 = p \cdot 5.80$$

$$\frac{2.32}{5.80} = \frac{p \cdot 5.80}{5.80} \qquad \text{Dividing both sides by 5.80}$$

$$\frac{2.32}{5.80} = p$$

$$0.4 = p$$

$$40\% = p \qquad \text{Converting to percent notation}$$

Proportion:

$$\frac{N}{100} = \frac{2.32}{5.80}$$

$$5.80 \times N = 100 \times 2.32 \qquad \text{Equating cross products}$$

$$\frac{5.80 \times N}{5.80} = \frac{100 \times 2.32}{5.80} \qquad \text{Dividing both sides by 5.80}$$

$$N = \frac{232}{5.80}$$

$$N = 40$$

We use the solution of the proportion to express the answer to the problem as 40%.

4. **Check.** To check, we take 40% of 5.80:

$$40\% \times 5.80 = 0.40 \times 5.80 = 2.32.$$

This is the amount of the price increase, so the answer checks.

5. **State.** The percent increase in the cost of a movie ticket was 40%.

◀ **Do Exercise 4.**

The percent increase or decrease is *always* based on the original amount. To find a percent increase or decrease, we need to know (1) the amount of increase or decrease and (2) the original amount.

Percent increase or decrease $= \dfrac{\text{Amount of increase or decrease}}{\text{Original amount}}$

---

4. *Centenarians.* As of April 1, 2010, the number of centenarians in the United States was 53,364. It is projected that this number will increase to 601,000 by 2050. What is the projected percent increase?

Sources: *National Geographic,* November 2011; Population Reference Bureau, U.S. Census Bureau

Answer

4. About 1026%

Translating for Success

1. **Distance Walked.** After a knee replacement, Alex walked $\frac{1}{8}$ mi each morning and $\frac{1}{5}$ mi each afternoon. How much farther did he walk in the afternoon?

2. **Stock Prices.** A stock sold for $5 per share on Monday and only $2.125 per share on Friday. What was the percent decrease from Monday to Friday?

3. **SAT Score.** After attending a class titled "Improving Your SAT Scores," Jacob raised his total score from 884 to 1040. What was the percent increase?

4. **Change in Population.** The population of a small farming community decreased from 1040 to 884. What was the percent decrease?

5. **Lawn Mowing.** During the summer, brothers Steve and Rob earned money for college by mowing lawns. The largest lawn that they mowed was $2\frac{1}{8}$ acres. Steve can mow $\frac{1}{5}$ acre per hour, and Rob can mow only $\frac{1}{8}$ acre per hour. Working together, how many acres did they mow per hour?

The goal of these matching questions is to practice step (2), Translate, of the five-step problem-solving process. Translate each word problem to an equation and select a correct translation from equations A–O.

A. $x + \dfrac{1}{5} = \dfrac{1}{8}$

B. $250 = x \cdot 1040$

C. $884 = x \cdot 1040$

D. $\dfrac{250}{16.25} = \dfrac{1000}{x}$

E. $156 = x \cdot 1040$

F. $16.25 = 250 \cdot x$

G. $\dfrac{1}{5} + \dfrac{1}{8} = x$

H. $2\dfrac{1}{8} = x \cdot 5$

I. $5 = 2.875 \cdot x$

J. $\dfrac{1}{8} + x = \dfrac{1}{5}$

K. $1040 = x \cdot 884$

L. $\dfrac{250}{16.25} = \dfrac{x}{1000}$

M. $2.875 = x \cdot 5$

N. $x \cdot 884 = 156$

O. $x = 16.25 \cdot 250$

Answers on page A-18

6. **Land Sale.** Cole sold $2\frac{1}{8}$ acres of the 5 acres he inherited from his uncle. What percent of his land did he sell?

7. **Travel Expenses.** A magazine photographer is reimbursed 16.25¢ per mile for business travel, up to 1000 mi per week. In a recent week, he traveled 250 mi. What was the total reimbursement for travel?

8. **Trip Expenses.** The total expenses for Claire's recent business trip were $1040. She put $884 on her credit card and paid the balance in cash. What percent did she place on her credit card?

9. **Cost of Copies.** During the first summer session at a community college, the campus copy center advertised 250 copies for $16.25. At this rate, what is the cost of 1000 copies?

10. **Cost of Insurance.** Following a rise in the cost of health insurance, 250 of a company's 1040 employees canceled their insurance. What percent of the employees canceled their insurance?

✓ Reading Check

Complete the table by filling in the missing numbers.

	Original Price	New Price	Change	Percent Increase or Decrease
RC1.	$50	$40	$ _____	$\dfrac{\text{Change}}{\text{Original}} = \dfrac{\$}{\$} =$ _____ %
RC2.	$60	$75	$ _____	$\dfrac{\text{Change}}{\text{Original}} = \dfrac{\$}{\$} =$ _____ %
RC3.	$360	$480	$ _____	$\dfrac{\text{Change}}{\text{Original}} = \dfrac{\$}{\$} =$ _____ %
RC4.	$4000	$2400	$ _____	$\dfrac{\text{Change}}{\text{Original}} = \dfrac{\$}{\$} =$ _____ %

a Solve.

1. *Mortgage Payment Increase.* A monthly mortgage payment increases from $840 to $882. What is the percent increase?

2. *Savings Increase.* The amount in a savings account increased from $200 to $216. What was the percent increase?

3. A person on a diet goes from a weight of 160 lb to a weight of 136 lb. What is the percent decrease?

4. During a sale, a dress decreased in price from $90 to $72. What was the percent decrease?

5. *Overdraft Fees.* Consumers are paying record amounts of fees for overdrawing their bank accounts. In 2009, banks collected $38.5 billion in overdraft fees, which is $18.6 billion more than in 2000. What is the percent increase?

Source: CNNMoney.com

6. *Credit-Card Debt.* In 2013, the average credit-card debt per household in the United States was $15,162. In 1990, the average credit-card debt was $2966. What was the percent increase?

Sources: nerdwallet.com

7. Insulation. A roll of unfaced fiberglass insulation has a retail price of $23.43. For two weeks, it is on sale for $15.31. What is the percent decrease?

8. Set of Weights. A 300-lb weight set retails for $199.95. For its grand opening, a sporting goods store reduces the price to $154.95. What is the percent decrease?

9. Birds Killed. In 2009, approximately 440,000 birds were killed by wind turbines in the United States. It is estimated that this number will increase to 1,000,000 per year by 2020. What would the percent increase be?

Sources: National Wind Coordinating Collaborative; American Bird Conservancy; U.S. Fish and Wildlife Service

10. Miles of Railroad Track. The greatest combined length of U.S.-owned operating railroad track was 254,037 mi in 1916, when industrial activity increased during World War I. The total length has decreased ever since. By 2006, the number of miles of track had decreased to 140,490 mi. What was the percent decrease from 1916 to 2006?

U.S. Railroad Miles

NOTE: The lengths exclude yard tracks, sidings, and parallel tracks.
SOURCE: Association of American Railroads

11. Population over 1 Million. In 1950, 74 cities in the world had populations of 1 million or more. In 2010, 442 cities had populations of 1 million or more. What was the percent increase?

Sources: *World Cities*, George Modelski; United Nations

12. Prescription Drug Sales. Total spending on prescription drugs in the United States was $329.2 billion in 2011. This amount dropped to $325.8 billion in 2012. What was the percent decrease?

Source: IMS Institute for Healthcare Informatics

13. Two-by-Four. A cross-section of a standard or nominal "two-by-four" board actually measures $1\frac{1}{2}$ in. by $3\frac{1}{2}$ in. The rough board is 2 in. by 4 in. but is planed and dried to the finished size. What percent of the wood is removed in planing and drying?

14. Strike Zone. In baseball, the *strike zone* is normally a 17-in. by 30-in. rectangle. Some batters give the pitcher an advantage by swinging at pitches thrown out of the strike zone. By what percent is the area of the strike zone increased if a 2-in. border is added to the outside?

Source: Major League Baseball

Population Increase. The table below provides data showing how the populations of various states increased from 2000 to 2010. Complete the table by filling in the missing numbers. Round percents to the nearest tenth of a percent.

	State	Population in 2000	Population in 2010	Change	Percent Change
15.	Vermont	608,827	625,741		
16.	Wisconsin	5,363,675		323,311	
17.	Arizona		6,392,017	1,261,385	
18.	Virginia		8,001,024	922,509	
19.	Idaho	1,293,953		273,629	
20.	Georgia	8,186,453	9,535,483		

SOURCE: U.S. Census Bureau

21. ***Decrease in Population.*** Between 2010 and 2012, the population of Scranton, Pennsylvania, decreased from 76,089 to 75,809. What was the percent decrease?

Source: U.S. Census Bureau

22. ***Decrease in Population.*** Between 2010 and 2012, the population of Flint, Michigan, decreased from 102,434 to 100,515. What was the percent decrease?

Source: U.S. Census Bureau

Skill Maintenance

Find the perimeter of each polygon.   [2.7b]

23.
8 cm
4 cm   4 cm
4 cm   4 cm
8 cm

24.
10 ft
3 ft
5 ft
3 ft
12 ft

25.
6 in.
12 in.   12 in.
6 in.

26.
9 m   9 m
9 m   9 m

Synthesis

The U.S. Census Bureau reported the following household incomes by state. The numbers are given as averages of median incomes over 2-yr periods and have been standardized to 2012 dollars. Use the information in the table to answer Exercises 27 and 28.

27. Which state had the largest dollar amount of increase in median household income?

28. Which state had the largest dollar amount of decrease in median household income?

29. ▦ A worker receives raises of 3%, 6%, and then 9%. By what percent has the original salary increased?

Household Incomes 2009 to 2012

STATE	AVERAGE MEDIAN HOUSEHOLD INCOME, 2011–2012	CHANGE IN AVERAGE MEDIAN HOUSEHOLD INCOME, 2009–2010
Kentucky	$40,887	−8.0%
Michigan	49,957	2.1%
Ohio	44,978	−7.6%
Tennessee	43,079	2.6%
Utah	57,497	−6.0%
Wyoming	56,580	2.0%

Sales Tax, Commission, and Discount

a SALES TAX

Sales tax computations represent a special type of percent increase problem. For example, the sales tax rate in Colorado is 2.9%. This means that the tax is 2.9% of the purchase price. Suppose the purchase price of a canoe is $749.95. The sales tax is then 2.9% of $749.95, or 0.029 × $749.95, or $21.74855, which is about $21.75.

COLORADO

$749.95
+2.9% sales tax

BILL:	
Purchase price	= $749.95
Sales tax (2.9% of $749.95)	= + 21.75
Total price	$771.70

The total that you would pay for the canoe is the purchase price plus the sales tax:

$749.95 + $21.75, or $771.70.

SALES TAX

Sales tax = Sales tax rate × Purchase price

Total price = Purchase price + Sales tax

EXAMPLE 1 *Maine Sales Tax.* The sales tax rate in Maine is 5%. How much tax is charged on the purchase of 3 shrubs at $42.99 each? What is the total price?

a) We first find the purchase price of the 3 shrubs. It is

$3 × $42.99 = $128.97.$

b) The sales tax on items costing $128.97 is

$$\underbrace{5\%}_{\text{Sales tax rate}} \times \underbrace{128.97,}_{\text{Purchase price}}$$

or 0.05 × 128.97, or 6.4485. Thus the tax is $6.45 (rounded to the nearest cent).

Maine

$42.99
each
plus 5% sales tax

c) The total price is given by the purchase price plus the sales tax:

$128.97 + $6.45, or $135.42.

To check, note that the total price is the purchase price plus 5% of the purchase price. Thus the total price is 105% of the purchase price. Since 1.05 × 128.97 ≈ 135.42, we have a check. The sales tax is $6.45, and the total price is $135.42.

1. **Texas Sales Tax.** The sales tax rate in Texas is 6.25%. In Texas, how much tax is charged on the purchase of an ultrasound toothbrush that sells for $139.95? What is the total price?

2. **Wyoming Sales Tax.** In her hometown, Laramie, Wyoming, Samantha buys 4 copies of *It's All Good* by Gwyneth Paltrow for $18.95 each. The sales tax rate in Wyoming is 4%. How much sales tax will Samantha be charged? What is the total price?

Purchase price = 4 × $ [___]

            = $ [___]

Sales tax = [___] % × $ [___]

      = 0.04 × $75.80

      = $3.032

      ≈ $ [___]

Total price = $75.80 + $ [___]

          = $ [___]

3. The sales tax on the purchase of a set of holiday dishes that costs $449 is $26.94. What is the sales tax rate?

4. The sales tax on the purchase of a pair of designer jeans is $4.84 and the sales tax rate is 5.5%. Find the purchase price (the price before the tax is added).

◀ **Do Exercises 1 and 2.**

EXAMPLE 2 The sales tax on the purchase of an eReader that costs $199 is $13.93. What is the sales tax rate?

$199.00 plus $13.93 sales tax

We rephrase and translate as follows:

Rephrase: Sales tax   is   what percent   of   purchase price?

Translate: 13.93   = r ·   199.

To solve the equation, we divide both sides by 199:

$$\frac{13.93}{199} = \frac{r \cdot 199}{199}$$

$$\frac{13.93}{199} = r$$

$$0.07 = r$$

$$7\% = r.$$

The sales tax rate is 7%.

◀ **Do Exercise 3.**

EXAMPLE 3 The sales tax on the purchase of a stone-top firepit is $12.74 and the sales tax rate is 8%. Find the purchase price (the price before the tax is added).

We rephrase and translate as follows:

Rephrase: Sales tax   is   8%   of   what?

Translate: 12.74   =   8%   · b,

or   12.74 = 0.08 · b.

To solve, we divide both sides by 0.08:

$$\frac{12.74}{0.08} = \frac{0.08 \cdot b}{0.08}$$

$$\frac{12.74}{0.08} = b$$

$$159.25 = b.$$

The purchase price is $159.25.

Price: ?
$12.74 tax @ 8%

◀ **Do Exercise 4.**

b COMMISSION

When you work for a **salary**, you receive the same amount of money each week or month. When you work for a **commission**, you are paid a percentage of the total sales for which you are responsible.

COMMISSION

Commission = Commission rate × Sales

EXAMPLE 4 *Membership Sales.* A fitness club salesperson's commission rate is 3%. What is the commission on the sale of $8300 worth of fitness club memberships?

Commission rate **3%**

WHOLE FITNESS

$$Commission = Commission\ rate \times Sales$$
$$C = 3\% \times 8300$$
$$C = 0.03 \times 8300$$
$$C = 249$$

The commission is $249.

Do Exercise 5. ▶

5. Aniyah's commission rate is 15%. What commission does she earn on the sale of $9260 worth of exercise equipment?

EXAMPLE 5 *Earth-Moving Equipment Sales.* Gavin earns a commission of $20,800 for selling $320,000 worth of earth-moving equipment. What is the commission rate?

Commission $20,800

Total sales $320,000

$$Commission = Commission\ rate \times Sales$$
$$20,800 = r \times 320,000$$

Answer

5. $1389

To solve this equation, we divide both sides by 320,000:

$$\frac{20{,}800}{320{,}000} = \frac{r \times 320{,}000}{320{,}000}$$

$$0.065 = r$$

$$6.5\% = r.$$

The commission rate is 6.5%.

◀ **Do Exercise 6.**

6. Zion earns a commission of $2040 for selling $17,000 worth of concert tickets. What is the commission rate?

EXAMPLE 6 *Cruise Vacations.* Valentina's commission rate is 5.6%. She received a commission of $2457 on cruise vacation packages that she sold in November. How many dollars worth of cruise vacations did she sell?

$$\textit{Commission} = \textit{Commission rate} \times \textit{Sales}$$

$$2457 = 5.6\% \times S,$$

or

$$2457 = 0.056 \times S$$

To solve this equation, we divide both sides by 0.056:

$$\frac{2457}{0.056} = \frac{0.056 \times S}{0.056}$$

$$\frac{2457}{0.056} = S$$

$$43{,}875 = S.$$

Valentina sold $43,875 worth of cruise vacation packages.

◀ **Do Exercise 7.**

7. Nathan's commission rate is 7.5%. He receives a commission of $2970 from the sale of winter ski passes. How many dollars worth of ski passes did he sell?

$$\$2970 = \boxed{}\,\% \cdot S$$

$$\$2970 = 0.075 \cdot S$$

$$\frac{\$2970}{\boxed{}} = \frac{0.075 \cdot S}{0.075}$$

$$\$\boxed{} = S$$

c **DISCOUNT**

Suppose the regular price of a rug is $60, and the rug is on sale at 25% off. Since 25% of $60 is $15, the sale price is $60 − $15, or $45. We call $60 the **original**, or **marked**, **price**, 25% the **rate of discount**, $15 the **discount**, and $45 the **sale price**. Note that discount problems are a type of percent decrease problem.

> **DISCOUNT AND SALE PRICE**
>
> **Discount** = Rate of discount × Original price
> **Sale price** = Original price − Discount

EXAMPLE 7 A leather sofa marked $2379 is on sale at $33\frac{1}{3}$% off. What is the discount? the sale price?

Leather sofa
$**2379** original price
Save $33\frac{1}{3}$%

a)

Discount	=	Rate of discount	×	Original price

$$D = 33\frac{1}{3}\% \times 2379$$

$$D = \frac{1}{3} \times 2379$$

$$D = \frac{2379}{3} = 793$$

b)

Sale price	=	Original price	−	Discount

$$S = 2379 - 793$$

$$S = 1586$$

The discount is $793, and the sale price is $1586.

Do Exercise 8. ▶

8. A computer marked $660 is on sale at $16\frac{2}{3}$% off. What is the discount? the sale price?

EXAMPLE 8 The price of a snowblower is marked down from $950 to $779. What is the rate of discount?

We first find the discount by subtracting the sale price from the original price:

$$950 - 779 = 171.$$

The discount is $171.

Next, we use the equation for discount:

Discount	=	Rate of discount	×	Original price

$$171 = r \times 950.$$

To solve, we divide both sides by 950:

$$\frac{171}{950} = \frac{r \times 950}{950}$$

$$\frac{171}{950} = r$$

$$0.18 = r$$

$$18\% = r.$$

The rate of discount is 18%.

> To check, note that an 18% rate of discount means that the buyer pays 82% of the original price:
>
> $$0.82 \times \$950 = \$779.$$

Do Exercise 9. ▶

9. The price of a winter coat is reduced from $75 to $60. Find the rate of discount.

Answers
8. $110; $550 **9.** 20%

8.6 **Exercise Set**

For Extra Help

MyMathLab®   MathXL®
PRACTICE   WATCH   READ   REVIEW

☑ **Reading Check**

Complete each definition with the word *price*, *rate*, or *tax*.

RC1. Commission = Commission _____ × Sales

RC2. Discount = _____ of discount × Original price

RC3. Sale price = Original _____ − Discount

RC4. Sales tax = Sales _____ rate × Purchase price

RC5. Total price = Purchase price + Sales _____

a Solve.

1. *Wyoming Sales Tax.* The sales tax rate in Wyoming is 4%. How much sales tax would be charged on a fireplace screen with doors that costs $239?

2. *Kansas Sales Tax.* The sales tax rate in Kansas is 6.3%. How much sales tax would be charged on a fireplace screen with doors that costs $239?

3. *New Mexico Sales Tax.* The sales tax rate in New Mexico is 5.125%. How much sales tax is charged on a camp stove that sells for $129.95?

4. *Ohio Sales Tax.* The sales tax rate in Ohio is 5.5%. How much sales tax is charged on a pet carrier that sells for $39.99?

NEW MEXICO

OHIO

5. *California Sales Tax.* The sales tax rate in California is 7.5%. How much sales tax is charged on a purchase of 4 contour foam travel pillows at $39.95 each? What is the total price?

6. *Illinois Sales Tax.* The sales tax rate in Illinois is 6.25%. How much sales tax is charged on a purchase of 3 wet-dry vacs at $60.99 each? What is the total price?

7. The sales tax is $30 on the purchase of a diamond ring that sells for $750. What is the sales tax rate?

8. The sales tax is $48 on the purchase of a dining room set that sells for $960. What is the sales tax rate?

9. The sales tax is $9.12 on the purchase of a patio set that sells for $456. What is the sales tax rate?

10. The sales tax is $35.80 on the purchase of a refrigerator-freezer that sells for $895. What is the sales tax rate?

11. The sales tax on the purchase of a new fishing boat is $112 and the sales tax rate is 2%. What is the purchase price (the price before tax is added)?

12. The sales tax on the purchase of a used car is $100 and the sales tax rate is 5%. What is the purchase price?

13. The sales tax rate in New York City is 4.375% for the city plus 4% for the state. Find the total amount paid for 6 boxes of chocolates at $17.95 each.

14. The sales tax rate in Nashville, Tennessee, is 2.25% for Davidson County plus 7% for the state. Find the total amount paid for 2 ladders at $39 each.

15. The sales tax rate in Seattle, Washington, is 2.5% for King County plus 6.5% for the state. Find the total amount paid for 3 ceiling fans at $84.49 each.

16. The sales tax rate in Miami, Florida, is 1% for Dade County plus 6% for the state. Find the total amount paid for 2 tires at $49.95 each.

17. The sales tax rate in Atlanta, Georgia, is 1% for the city, 3% for Fulton County, and 4% for the state. Find the total amount paid for 6 basketballs at $29.95 each.

18. The sales tax rate in Dallas, Texas, is 1% for the city, 1% for Dallas County, and 6.25% for the state. Find the total amount paid for 5 flash drives at $19.95 each.

b Solve.

19. Benjamin's commission rate is 21%. What commission does he earn on the sale of $12,500 worth of windows?

20. Olivia's commission rate is 6%. What commission does she earn on the sale of $45,000 worth of lawn irrigation systems?

21. Alyssa earns $408 for selling $3400 worth of shoes. What is the commission rate?

22. Joshua earns $120 for selling $2400 worth of television sets. What is the commission rate?

23. *Real Estate Commission.* A real estate agent's commission rate is 7%. She receives a commission of $12,950 from the sale of a home. How much did the home sell for?

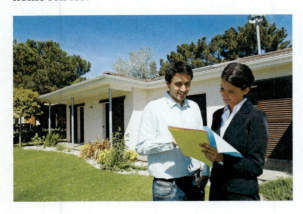

24. *Clothing Consignment Commission.* A clothing consignment shop's commission rate is 40%. The shop receives a commission of $552. How many dollars worth of clothing were sold?

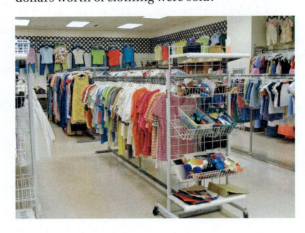

25. A real estate commission rate is 8%. What is the commission from the sale of a piece of land for $68,000?

26. A real estate commission rate is 6%. What is the commission from the sale of a $98,000 home?

27. David earns $1147.50 for selling $7650 worth of car parts. What is the commission rate?

28. Jayla earns $280.80 for selling $2340 worth of tee shirts. What is the commission rate?

29. Laila's commission is increased according to how much she sells. She receives a commission of 4% for the first $1000 of sales and 7% for the amount over $1000. What is her total commission on sales of $5500?

30. Malik's commission is increased according to how much he sells. He receives a commission of 5% for the first $2000 of sales and 8% for the amount over $2000. What is his total commission on sales of $6200?

C Complete the table below by filling in the missing numbers.

	Marked Price	Rate of Discount	Discount	Sale Price
31.	$300	10%		
32.	$2000	40%		
33.	$17	15%		
34.	$20	25%		
35.		10%	$12.50	
36.		15%	$65.70	
37.	$600		$240	
38.	$12,800		$1920	

39. Find the marked price and the rate of discount for the surfboard in this ad.

Save
$120.00

Surfboard Now
$180.00

40. Find the marked price and the rate of discount for the steel log rack in this ad.

Closeout
$35

Save
$15

41. Find the discount and the rate of discount for the pinball machine in this ad.

Best price of the year!
Now only
$3150
Was $3999

PINBALL

42. Find the discount and the rate of discount for the amaryllis in this ad.

Sale
3-in-1 Amaryllis
$37.95
Was $42.95

Skill Maintenance

Solve.   [7.3b]

43. $\dfrac{x}{12} = \dfrac{24}{16}$

44. $\dfrac{7}{2} = \dfrac{11}{x}$

Convert to standard notation.   [5.3b]

47. 4.03 trillion

48. 5.8 million

Solve.   [5.7a]

45. $0.64 \cdot x = 170$

46. $29.44 = 25.6 \times y$

Graph.   [6.4b]

49. $y = -\frac{1}{2}x$

50. $y = \frac{1}{2}x - 3$

Synthesis

51. 🖩 Sara receives a 10% commission on the first $5000 in sales and 15% on all sales beyond $5000. If Sara receives a commission of $2405, how much did she sell? Use a calculator and trial and error if you wish.

52. Elijah collects baseball memorabilia. He bought two autographed plaques, but then became short of funds and had to sell them quickly for $200 each. On one, he made a 20% profit, and on the other, he lost 20%. Did he make or lose money on the sale?

8.7

Simple Interest and Compound Interest; Credit Cards

OBJECTIVES

a Solve applied problems involving simple interest.

b Solve applied problems involving compound interest.

c Solve applied problems involving interest rates on credit cards.

SKILL TO REVIEW

Objective 8.1b: Convert between percent notation and decimal notation.

Find decimal notation.

1. $34\frac{5}{8}\%$ **2.** $5\frac{1}{4}\%$

1. What is the simple interest on $4300 invested at an interest rate of 4% for 1 year?

$$I = P \cdot r \cdot t$$
$$= \$4300 \cdot \boxed{}\% \cdot 1$$
$$= \$\boxed{} \cdot 0.04 \cdot 1$$
$$= \$\boxed{}$$

2. What is the simple interest on a principal of $4300 invested at an interest rate of 4% for 9 months?

a SIMPLE INTEREST

Suppose you put $1000 into an investment for 1 year. The $1000 is called the **principal**. If the **interest rate** is 5%, in addition to the principal, you get back 5% of the principal, which is

5% of $1000,   or   0.05 · $1000,   or $50.00.

The $50.00 is called **simple interest**. It is, in effect, the price that a financial institution pays for the use of the money over time.

SIMPLE INTEREST FORMULA

The **simple interest** I on principal P, invested for t years at interest rate r, is given by

$$I = P \cdot r \cdot t.$$

EXAMPLE 1   What is the simple interest on $2500 invested at an interest rate of 6% for 1 year?

We use the formula $I = P \cdot r \cdot t$:

$$I = P \cdot r \cdot t = \$2500 \cdot 6\% \cdot 1$$
$$= \$2500 \cdot 0.06$$
$$= \$150.$$

The simple interest for 1 year is $150.

◀ **Do Margin Exercise 1.**

EXAMPLE 2   What is the simple interest on a principal of $2500 invested at an interest rate of 6% for 3 months?

We use the formula $I = P \cdot r \cdot t$ and express 3 months as a fraction of a year:

$$I = P \cdot r \cdot t = \$2500 \cdot 6\% \cdot \frac{3}{12} = \$2500 \cdot 0.06 \cdot \frac{1}{4}$$

$$= \frac{\$2500 \cdot 0.06}{4} = \$37.50.$$

The simple interest for 3 months is $37.50.

◀ **Do Margin Exercise 2.**

When time is given in days, we generally divide it by 365 to express the time as a fractional part of a year.

EXAMPLE 3   To pay for a shipment of lawn furniture, Patio by Design borrows $8000 at $9\frac{3}{4}\%$ for 60 days. Find **(a)** the amount of simple interest that is due and **(b)** the total amount that must be paid after 60 days.

a) We express 60 days as a fractional part of a year:

$$I = P \cdot r \cdot t = \$8000 \cdot 9\tfrac{3}{4}\% \cdot \frac{60}{365}$$

$$= \$8000 \cdot 0.0975 \cdot \frac{60}{365}$$

$$\approx \$128.22.$$

The interest due for 60 days is $128.22.

b) The total amount to be paid after 60 days is the principal plus the interest:

$$\$8000 + \$128.22 = \$8128.22.$$

The total amount due is $8128.22.

Do Exercise 3. ▶

GS **3.** The Glass Nook borrows $4800 at $5\tfrac{1}{2}\%$ for 30 days. Find **(a)** the amount of simple interest due and **(b)** the total amount that must be paid after 30 days.

a) $I = P \cdot r \cdot t$

$$= \$4800 \cdot 5\tfrac{1}{2}\% \cdot \frac{\boxed{}}{365}$$

$$= \$4800 \cdot 0.055 \cdot \frac{30}{365}$$

$$\approx \$\boxed{}$$

b) Total amount

$$= \$4800 + \boxed{}$$

$$= \boxed{}$$

b COMPOUND INTEREST

When interest is paid *on interest*, we call it **compound interest**. This is the type of interest usually paid on investments. Suppose you have $5000 in a savings account at 6%. In 1 year, the account will contain the original $5000 plus 6% of $5000. Thus the total in the account after 1 year will be

106% of $5000,　or　1.06 · $5000,　or　$5300.

Now suppose that the total of $5300 remains in the account for another year. At the end of this second year, the account will contain the $5300 plus 6% of $5300. The total in the account will thus be

106% of $5300,　or　1.06 · $5300,　or　$5618.

Note that in the second year, interest is also earned on the first year's interest. When this happens, we say that interest is **compounded annually**.

EXAMPLE 4　Find the amount in an account if $2000 is invested at 8%, compounded annually, for 2 years.

a) After 1 year, the account will contain 108% of $2000:

1.08 · $2000 = $2160.

b) At the end of the second year, the account will contain 108% of $2160:

1.08 · $2160 = $2332.80.

The amount in the account after 2 years is $2332.80.

Do Exercise 4. ▶

4. Find the amount in an account if $2000 is invested at 2%, compounded annually, for 2 years.

Answers

3. (a) $21.70; **(b)** $4821.70　**4.** $2080.80

Guided Solution:
3. (a) 30, 21.70; **(b)** $21.70, $4821.70

Suppose the interest in Example 4 were **compounded semiannually**—that is, every half year. Interest would then be calculated twice a year at a rate of 8% ÷ 2, or 4% each time. The approach used in Example 4 can then be adapted, as follows.

After the first $\frac{1}{2}$ year, the account will contain 104% of $2000:

$$1.04 \cdot \$2000 = \$2080.$$

After a second $\frac{1}{2}$ year (1 full year), the account will contain 104% of $2080:

$$1.04 \cdot \$2080 = \$2163.20.$$

After a third $\frac{1}{2}$ year ($1\frac{1}{2}$ full years), the account will contain 104% of $2163.20:

$$1.04 \cdot \$2163.20 = \$2249.728$$
$$\approx \$2249.73. \qquad \text{Rounding to the nearest cent}$$

Finally, after a fourth $\frac{1}{2}$ year (2 full years), the account will contain 104% of $2249.73:

$$1.04 \cdot \$2249.73 = \$2339.7192$$
$$\approx \$2339.72. \qquad \text{Rounding to the nearest cent}$$

Let's summarize our results and look at them another way:

End of 1st $\frac{1}{2}$ year $\;\rightarrow\; 1.04 \cdot 2000 = 2000 \cdot (1.04)^1;$

End of 2nd $\frac{1}{2}$ year $\;\rightarrow\; 1.04 \cdot (1.04 \cdot 2000) = 2000 \cdot (1.04)^2;$

End of 3rd $\frac{1}{2}$ year $\;\rightarrow\; 1.04 \cdot (1.04 \cdot 1.04 \cdot 2000) = 2000 \cdot (1.04)^3;$

End of 4th $\frac{1}{2}$ year $\;\rightarrow\; 1.04 \cdot (1.04 \cdot 1.04 \cdot 1.04 \cdot 2000) = 2000 \cdot (1.04)^4.$

Note that each multiplication was by 1.04 and that

$$\$2000 \cdot 1.04^4 \approx \$2339.72. \qquad \text{Using a calculator and rounding to the nearest cent}$$

We have illustrated the following result.

COMPOUND INTEREST FORMULA

If a principal P has been invested at interest rate r, compounded n times a year, in t years it will grow to an amount A given by

$$A = P \cdot \left(1 + \frac{r}{n}\right)^{n \cdot t}.$$

In the compound interest formula, $n \cdot t$ is the total number of compounding periods, r is written in decimal notation, and $\dfrac{r}{n}$ is the interest rate for each period.

Let's apply this formula to confirm our preceding result, where the amount invested is $P = \$2000$, the interest rate is $r = 8\%$, the number of years is $t = 2$, and the number of compounding periods each year is $n = 2$. Substituting into the compound interest formula, we have

$$A = P \cdot \left(1 + \frac{r}{n}\right)^{n \cdot t} = 2000 \cdot \left(1 + \frac{8\%}{2}\right)^{2 \cdot 2}$$

$$= \$2000 \cdot \left(1 + \frac{0.08}{2}\right)^4 = \$2000(1.04)^4$$

$$= \$2000 \cdot 1.16985856 \approx \$2339.72.$$

If you were using a calculator, you could perform this computation in one step.

EXAMPLE 5 The Ibsens invest $4000 in an account paying $3\frac{5}{8}\%$, compounded quarterly. Find the amount in the account after $2\frac{1}{2}$ years.

The compounding is quarterly, so n is 4. We substitute $4000 for P, $3\frac{5}{8}\%$, or 0.03625, for r, 4 for n, and $2\frac{1}{2}$, or $\frac{5}{2}$, for t and compute A:

$$A = P \cdot \left(1 + \frac{r}{n}\right)^{n \cdot t} = \$4000 \cdot \left(1 + \frac{3\frac{5}{8}\%}{4}\right)^{4 \cdot 5/2}$$

$$= \$4000 \cdot \left(1 + \frac{0.03625}{4}\right)^{10}$$

$$= \$4000(1.0090625)^{10}$$

$$\approx \$4377.65.$$

The amount in the account after $2\frac{1}{2}$ years is $4377.65.

Do Exercise 5. ▶

5. A couple invests $7000 in an account paying $6\frac{3}{8}\%$, compounded semiannually. Find the amount in the account after $1\frac{1}{2}$ years.

CALCULATOR CORNER

Compound Interest A calculator is useful in computing compound interest. Not only does it perform computations quickly but it also eliminates the need to round until the computation is completed. This minimizes round-off errors that occur when rounding is done at each stage of the computation. We must keep order of operations in mind when computing compound interest.

To find the amount due on a $20,000 loan made for 25 days at 11% interest, compounded daily, we compute

$20,000\left(1 + \frac{0.11}{365}\right)^{25}$. To do this on a calculator, we press [2] [0] [0] [0] [0] [0] [×] [(] [1] [+] [.] [1] [1] [÷] [3] [6] [5] [)] [y^x]

[2] [5] [=]. Some calculators have a [^] key instead of [y^x]. The result is $20,151.23, rounded to the nearest cent.

Some calculators have business keys that allow such computations to be done more quickly.

EXERCISES

1. Find the amount due on a $16,000 loan made for 62 days at 13% interest, compounded daily.

2. An investment of $12,500 is made for 90 days at 8.5% interest, compounded daily. How much is the investment worth after 90 days?

C CREDIT CARDS

According to nerdwallet.com, the average credit-card debt among U.S. households with such debt was $15,162 per household in 2012. According to the Aggregate Revolving Consumer Debt Survey in 2012, the average credit-card debt carried by undergraduate college students is $3137.

The money you obtain through the use of a credit card is not "free" money. There is a price (interest) to be paid for the convenience of using a credit card. A balance carried on a credit card is a type of loan. Comparing interest rates is essential if one is to become financially responsible. A small change in an interest rate can make a large difference in the cost of a loan. When you make a payment on a credit-card balance, do you know how much of that payment is interest and how much is applied to reducing the principal?

Answer

5. $7690.94

EXAMPLE 6 *Credit Cards.* After the holidays, Addison has a balance of $3216.28 on a credit card with an annual percentage rate (APR) of 19.7%. She decides not to make additional purchases with this card until she has paid off the balance.

a) Many credit cards require a minimum monthly payment of 2% of the balance. At this rate, what is Addison's minimum payment on a balance of $3216.28? Round the answer to the nearest dollar.

b) Find the amount of interest and the amount applied to reducing the principal with the minimum payment found in part (a).

c) If Addison had transferred her balance to a card with an APR of 12.5%, how much of her first payment would be interest and how much would be applied to reducing the principal?

d) Compare the amounts for 12.5% from part (c) with the amounts for 19.7% from part (b).

We solve as follows.

a) We multiply the balance of $3216.28 by 2%:

$$0.02 \cdot \$3216.28 = \$64.3256.$$ Addison's minimum payment, rounded to the nearest dollar, is $64.

b) The amount of interest on $3216.28 at 19.7% for one month* is given by

$$I = P \cdot r \cdot t = \$3216.28 \cdot 0.197 \cdot \frac{1}{12} \approx \$52.80.$$

We subtract to find the portion of the first payment applied to reducing the principal:

$$\begin{array}{l}\text{Amount applied to} \\ \text{reducing the principal}\end{array} = \text{Minimum payment} - \begin{array}{l}\text{Interest for} \\ \text{the month}\end{array}$$

$$= \$64 - \$52.80$$
$$= \$11.20.$$

Thus the principal of $3216.28 is decreased by only $11.20 with the first payment. (Addison still owes $3205.08.)

c) The amount of interest on $3216.28 at 12.5% for one month is

$$I = P \cdot r \cdot t = \$3216.28 \cdot 0.125 \cdot \frac{1}{12} \approx \$33.50.$$

We subtract to find the portion of the first payment that is applied to reducing the principal:

$$\begin{array}{l}\text{Amount applied to} \\ \text{reducing the principal}\end{array} = \text{Minimum payment} - \begin{array}{l}\text{Interest for} \\ \text{the month}\end{array}$$

$$= \$64 - \$33.50$$
$$= \$30.50.$$

Thus the principal of $3216.28 would have decreased by $30.50 with the first payment. (Addison would still owe $3185.78.)

---

*Actually, the interest on a credit card is computed daily with a rate called a daily percentage rate (DPR). The DPR for Example 6 would be 19.7%/365 ≈ 0.054%. When no payments or additional purchases are made during a month, the difference in total interest for the month is minimal and we will not deal with it here.

d) Let's organize the information for both rates in the following table.

BALANCE BEFORE FIRST PAYMENT	FIRST MONTH'S PAYMENT	%APR	AMOUNT OF INTEREST	AMOUNT APPLIED TO PRINCIPAL	BALANCE AFTER FIRST PAYMENT
$3216.28	$64	19.7%	$52.80	$11.20	$3205.08
3216.28	64	12.5	33.50	30.50	3185.78

Difference in balance after first payment $\longrightarrow$ $19.30

At 19.7%, the interest is $52.80 and the principal is decreased by $11.20. At 12.5%, the interest is $33.50 and the principal is decreased by $30.50. Thus the interest at 19.7% is $52.80 $-$ $33.50, or $19.30, greater than the interest at 12.5%. Thus the principal is decreased by $30.50 $-$ $11.20, or $19.30, more with the 12.5% rate than with the 19.7% rate.

Do Exercise 6. ▶

Even though the mathematics of the information in the table below is beyond the scope of this text, it is interesting to compare how long it takes to pay off the balance of Example 6 if Addison continues to pay $64 each month, compared to how long it takes if she pays double that amount, or $128, each month. Financial consultants frequently tell clients that if they want to take control of their debt, they should pay double the minimum payment.

RATE	PAYMENT	NUMBER OF PAYMENTS TO PAY OFF DEBT	TOTAL PAID BACK	ADDITIONAL COST OF PURCHASES
19.7%	$64	107, or 8 yr 11 mo	$6848	$3631.72
19.7	128	33, or 2 yr 9 mo	4224	1007.72
12.5	64	72, or 6 yr	4608	1391.72
12.5	128	29, or 2 yr 5 mo	3712	495.72

As with most loans, paying an extra amount toward the principal with each payment greatly reduces the length of time it takes to pay off the loan completely. Note that at the rate of 19.7%, it will take Addison almost 9 years to pay off her debt if she pays only $64 per month and does not make additional purchases. If she transfers her balance to a card with a 12.5% rate and pays $128 per month, she can eliminate her debt in approximately $2\frac{1}{2}$ years. You can see how debt can get out of control if you continue to make purchases and pay only the minimum payment each month. The debt will never be eliminated.

6. *Credit Card.* After the holidays, Logan has a balance of $4867.59 on a credit card with an annual percentage rate (APR) of 21.3%. He decides not to make additional purchases with this card until he has paid off the balance.

a) Many credit cards require a minimum monthly payment of 2% of the balance. What is Logan's minimum payment on a balance of $4867.59? Round the answer to the nearest dollar.

b) Find the amount of interest and the amount applied to the principal with the minimum payment found in part (a).

c) If Logan had transferred his balance to a card with an APR of 13.6%, how much of his first payment would be interest and how much would be applied to the principal?

d) Compare the amounts for 13.6% from part (c) with the amounts for 21.3% from part (b).

✓ Reading Check

In the simple interest formula, $I = P \cdot r \cdot t$, the time t must be expressed in years. Convert each length of time to years. Use 1 year = 12 months = 365 days.

RC1. 6 months

RC2. 40 days

RC3. 285 days

RC4. 9 months

RC5. 3 months

RC6. 4 months

a   Find the simple interest.

	Principal	Rate of Interest	Time	Simple Interest
1.	$200	4%	1 year	
2.	$200	7.7%	1 year	
3.	$4300	10.56%	$\frac{1}{4}$ year	
4.	$80,000	$6\frac{3}{4}$%	$\frac{1}{12}$ year	
5.	$20,000	$4\frac{5}{8}$%	6 months	
6.	$8000	9.42%	2 months	
7.	$50,000	$5\frac{3}{8}$%	3 months	
8.	$100,000	$3\frac{1}{4}$%	9 months	

Solve. Assume that simple interest is being calculated in each case.

9. Mia's Boutique borrows $10,000 at 9% for 60 days. Find **(a)** the amount of interest due and **(b)** the total amount that must be paid after 60 days.

10. Mason's Drywall borrows $8000 at 10% for 90 days. Find **(a)** the amount of interest due and **(b)** the total amount that must be paid after 90 days.

11. Animal Instinct, a pet supply shop, borrows $6500 at $5\frac{1}{4}$% for 90 days. Find **(a)** the amount of interest due and **(b)** the total amount that must be paid after 90 days.

12. Andante's Cafe borrows $4500 at $12\frac{1}{2}$% for 60 days. Find **(a)** the amount of interest due and **(b)** the total amount that must be paid after 60 days.

13. Cameron's Garage borrows $5600 at 10% for 30 days. Find **(a)** the amount of interest due and **(b)** the total amount that must be paid after 30 days.

14. Shear Delights Hair Salon borrows $3600 at 4% for 30 days. Find **(a)** the amount of interest due and **(b)** the total amount that must be paid after 30 days.

b Interest is compounded annually. Find the amount in the account after the given length of time. Round to the nearest cent.

	Principal	Rate of Interest	Time	Amount in the Account
15.	$400	5%	2 years	
16.	$450	4%	2 years	
17.	$2000	8.8%	4 years	
18.	$4000	7.7%	4 years	
19.	$4300	10.56%	6 years	
20.	$8000	9.42%	6 years	
21.	$20,000	$6\frac{5}{8}\%$	25 years	
22.	$100,000	$5\frac{7}{8}\%$	30 years	

Interest is compounded semiannually. Find the amount in the account after the given length of time. Round to the nearest cent.

	Principal	Rate of Interest	Time	Amount in the Account
23.	$4000	6%	1 year	
24.	$1000	5%	1 year	
25.	$20,000	8.8%	4 years	
26.	$40,000	7.7%	4 years	
27.	$5000	10.56%	6 years	
28.	$8000	9.42%	8 years	
29.	$20,000	$7\frac{5}{8}\%$	25 years	
30.	$100,000	$4\frac{7}{8}\%$	30 years	

Solve.

31. The Aguilar family invests $4000 in an account paying 6%, compounded monthly. How much is in the account after 5 months?

32. Ryan and Alicia invest $2500 in an account paying 3%, compounded monthly. How much is in the account after 6 months?

33. Bryn and Andy invest $1200 in an account paying 10%, compounded quarterly. How much is in the account after 1 year?

34. The O'Hares invest $6000 in an account paying 8%, compounded quarterly. How much is in the account after 18 months?

35. Emilio loans his niece's business $20,000 for 50 days at 6% interest, compounded daily. How much is Emilio owed after 50 days?

36. Elsa loans her nephew's business $25,000 for 40 days at 5% interest, compounded daily. How much is Elsa owed after 40 days?

C Solve.

37. *Credit Cards.* Amelia has a balance of $1278.56 on a credit card with an annual percentage rate (APR) of 19.6%. The minimum payment required in the current statement is $25.57. Find the amount of interest and the amount applied to the principal with this payment and the balance after this payment.

38. *Credit Cards.* Lawson has a balance of $1834.90 on a credit card with an annual percentage rate (APR) of 22.4%. The minimum payment required in the current statement is $36.70. Find the amount of interest and the amount applied to the principal with this payment and the balance after this payment.

39. *Credit Cards.* Antonio has a balance of $4876.54 on a credit card with an annual percentage rate (APR) of 21.3%.

a) Many credit cards require a minimum monthly payment of 2% of the balance. What is Antonio's minimum payment on a balance of $4876.54? Round the answer to the nearest dollar.
b) Find the amount of interest and the amount applied to the principal with the minimum payment found in part (a).
c) If Antonio had transferred his balance to a card with an APR of 12.6%, how much of his payment would be interest and how much would be applied to the principal?
d) Compare the amounts for 12.6% from part (c) with the amounts for 21.3% from part (b).

40. *Credit Cards.* Becky has a balance of $5328.88 on a credit card with an annual percentage rate (APR) of 18.7%.

a) Many credit cards require a minimum monthly payment of 2% of the balance. What is Becky's minimum payment on a balance of $5328.88? Round the answer to the nearest dollar.
b) Find the amount of interest and the amount applied to the principal with the minimum payment found in part (a).
c) If Becky had transferred her balance to a card with an APR of 13.2%, how much of her payment would be interest and how much would be applied to the principal?
d) Compare the amounts for 13.2% from part (c) with the amounts for 18.7% from part (b).

Skill Maintenance

41. Find the LCM of 32 and 50. [4.1a]

42. Find the prime factorization of 228. [3.2c]

Divide and simplify. [3.7b]

43. $\dfrac{6}{125} \div \dfrac{8}{15}$

44. $\dfrac{16}{105} \div \dfrac{5}{14}$

Multiply and simplify. [3.6a]

45. $\dfrac{4}{15} \times \dfrac{3}{20}$

46. $\dfrac{8}{21} \times \dfrac{49}{800}$

47. Simplify: $4^3 - 6^2 \div 2^2$. [1.9c]

48. Solve: $x + \dfrac{2}{5} = \dfrac{9}{10}$. [4.3b]

Synthesis

Effective Yield. The *effective yield* is the yearly rate of simple interest that corresponds to a rate for which interest is compounded two or more times a year. For example, if P is invested at 12%, compounded quarterly, we multiply P by $(1 + 0.12/4)^4$, or 1.03^4. Since $1.03^4 \approx 1.126$, the 12% compounded quarterly corresponds to an effective yield of approximately 12.6%. In Exercises 49 and 50, find the effective yield for the indicated account.

49. ▦ The account pays 9% compounded monthly.

50. ▦ The account pays 10% compounded daily.

Vocabulary Reinforcement

Complete each statement with the appropriate word or phrase from the list at the right. Some of the choices will not be used.

1. When a quantity is decreased by a certain percent, we say that this is a _____ . [8.5a]

2. The _____ interest I on principal P, invested for t years at interest rate r, is given by $I = P \cdot r \cdot t$. [8.7a]

3. Sale price = Original price − _____ . [8.6c]

4. Commission = Commission rate × _____ . [8.6b]

5. Discount = _____ of discount × Original price. [8.6c]

6. When a quantity is increased by a certain percent, we say that this is a _____ . [8.5a]

discount

rate

sales

commission

price

principal

percent increase

percent decrease

simple

compound

Concept Reinforcement

Determine whether each statement is true or false.

_____ 1. A fixed principal invested for 4 years will earn more interest when interest is compounded quarterly than when interest is compounded semiannually. [8.7b]

_____ 2. Of the numbers 0.5%, $\frac{5}{1000}$%, $\frac{1}{2}$%, $\frac{1}{5}$, and $0.\overline{1}$, the largest number is $0.\overline{1}$. [8.1b, c]

_____ 3. If principal A equals principal B and principal A is invested for 2 years at 4%, compounded quarterly, while principal B is invested for 4 years at 2%, compounded semiannually, the interest earned from each investment is the same. [8.7b]

Study Guide

Objective 8.1b Convert between percent notation and decimal notation.

Example Find percent notation for 1.3.

We move the decimal point two places to the right and write a percent symbol:

$$1.3 = 130\%.$$

Example Find decimal notation for $12\frac{3}{4}$%.

We convert $12\frac{3}{4}$ to a decimal and move the decimal point two places to the left:

$$12\frac{3}{4}\% = 12.75\% = 0.1275.$$

Practice Exercise

1. Find percent notation for 0.082.

Practice Exercise

2. Find decimal notation for $62\frac{5}{8}$%.

Objective 8.1c   Convert between fraction notation and percent notation.

Example   Find percent notation for $\dfrac{5}{12}$.

$$12\overline{\smash{)}5.0\,0\,0}\;\;\;\begin{array}{r}0.4\,1\,6\end{array}\qquad \dfrac{5}{12}=0.41\overline{6}=41.\overline{6}\%,\text{ or }41\dfrac{2}{3}\%$$

$$\begin{array}{r}4\,8\\ \hline 2\,0\\ 1\,2\\ \hline 8\,0\\ 7\,2\\ \hline 8\end{array}$$

Practice Exercise

3. Find percent notation for $\dfrac{7}{11}$.

Example   Find fraction notation for 9.5%.

$$9.5\% = \dfrac{9.5}{100} = \dfrac{95}{1000}.$$

$$= \dfrac{5 \cdot 19}{5 \cdot 200}$$

$$= \dfrac{5}{5} \cdot \dfrac{19}{200}$$

$$= \dfrac{19}{200}$$

Practice Exercise

4. Find fraction notation for 6.8%.

Objective 8.2b   Solve basic percent problems using percent equations.

Example   165 is what percent of 3300?

We have

$$165 = p \cdot 3300 \qquad \text{Translating to a percent equation}$$

$$\dfrac{165}{3300} = \dfrac{p \cdot 3300}{3300}$$

$$\dfrac{165}{3300} = p$$

$$0.05 = p$$

$$5\% = p.$$

Thus, 165 is 5% of 3300.

Practice Exercise

5. 12 is what percent of 288?

Objective 8.3b   Solve basic percent problems using proportions.

Example   18% of what is 1296?

$$\dfrac{18}{100} = \dfrac{1296}{b} \qquad \text{Translating to a proportion}$$

$$18 \cdot b = 100 \cdot 1296$$

$$\dfrac{18 \cdot b}{18} = \dfrac{129{,}600}{18}$$

$$b = 7200.$$

Thus, 18% of 7200 is 1296.

Practice Exercise

6. 3% of what is 300?

Objective 8.5a Solve applied problems involving percent increase or percent decrease.

Example The total cost for 16 basic grocery items in March 2014 averaged $53.27 nationally. The total cost of these 16 items in March 2013 averaged $51.54. What was the percent increase?

Source: American Farm Bureau Federation

We first determine the amount of increase:

$$\$53.27 - \$51.54 = \$1.73$$

Then we translate to a percent equation or a proportion and solve.

Rewording: $1.73 is what percent of $51.54?

Percent Equation:

$$1.73 = p \cdot 51.54$$

$$\frac{1.73}{51.54} = \frac{p \cdot 51.54}{51.54}$$

$$\frac{1.73}{51.54} = p$$

$$0.034 \approx p$$

$$3.4\% \approx p$$

Proportion:

$$\frac{N}{100} = \frac{1.73}{51.54}$$

$$51.54 \cdot N = 100 \cdot 1.73$$

$$\frac{51.54 \cdot N}{51.54} = \frac{100 \cdot 1.73}{51.54}$$

$$N = \frac{173}{51.54}$$

$$N \approx 3.4$$

The percent increase was about 3.4%.

Practice Exercise

7. The number of concerts in the United States promoted by Live Nation Entertainment increased from 14,942 in 2012 to 15,582 in 2013. What was the percent increase?

Source: Live Nation Entertainment, Inc.

Objective 8.6a Solve applied problems involving sales tax and percent.

Example The sales tax is $34.23 on the purchase of a flat-screen high-definition television that costs $489. What is the sales tax rate?

Rephrase: Sales tax is what percent of purchase price?

Translate: $34.23 = r \cdot 489$

Solve:

$$\frac{34.23}{489} = \frac{r \cdot 489}{489}$$

$$\frac{34.23}{489} = r$$

$$0.07 = r$$

$$7\% = r$$

The sales tax rate is 7%.

Practice Exercise

8. The sales tax is $1102.20 on the purchase of a new car that costs $18,370. What is the sales tax rate?

Objective 8.6b Solve applied problems involving commission and percent.

Example A real estate agent's commission rate is $6\frac{1}{2}\%$. She received a commission of $17,160 on the sale of a home. For how much did the home sell?

   Rephrase: Commission is $6\frac{1}{2}\%$ of what selling price?

   Translate: $17,160 = 6\frac{1}{2}\% \cdot S$

   Solve: $17,160 = 0.065 \cdot S$

$$\frac{17,160}{0.065} = \frac{0.065 \cdot S}{0.065}$$

$$264,000 = S$$

The home sold for $264,000.

Practice Exercise

9. A real estate agent's commission rate is 7%. He received a commission of $12,950 on the sale of a home. For how much did the home sell?

Objective 8.7a Solve applied problems involving simple interest.

Example To meet its payroll, a business borrows $5200 at $4\frac{1}{4}\%$ for 90 days. Find the amount of simple interest that is due and the total amount that must be paid after 90 days.

$$I = P \cdot r \cdot t = \$5200 \cdot 4\frac{1}{4}\% \cdot \frac{90}{365}$$

$$= \$5200 \cdot 0.0425 \cdot \frac{90}{365}$$

$$\approx \$54.49$$

The interest due for 90 days $= \$54.49$.

The total amount due $= \$5200 + \$54.49 = \$5254.49$.

Practice Exercise

10. A student borrows $2500 for tuition at $5\frac{1}{2}\%$ for 60 days. Find the amount of simple interest that is due and the total amount that must be paid after 60 days.

Objective 8.7b Solve applied problems involving compound interest.

Example Find the amount in an account if $3200 is invested at 5%, compounded semiannually, for $1\frac{1}{2}$ years.

$$A = P \cdot \left(1 + \frac{r}{n}\right)^{n \cdot t}$$

$$= \$3200\left(1 + \frac{0.05}{2}\right)^{2 \cdot \frac{3}{2}}$$

$$= \$3200(1.025)^3$$

$$= \$3446.05$$

The amount in the account after $1\frac{1}{2}$ years is $3446.05.

Practice Exercise

11. Find the amount in an account if $6000 is invested at $4\frac{3}{4}\%$, compounded quarterly, for 2 years.

Review Exercises

Find percent notation.   [8.1b]

1. 1.7   **2.** 0.065

Find decimal notation for the percent notation(s) in each sentence.   [8.1b]

3. In the 2011–2012 school year, about 4% of the 15 million college students in the United States were foreign students. Approximately 13.1% of the foreign students were from India.

   Source: Institute of International Education

4. Poland is 60.84% urban; Sweden is 85.36% urban.

Source: The World Bank

Find percent notation.   [8.1c]

5. $\dfrac{3}{8}$ **6.** $\dfrac{1}{3}$

Find fraction notation.   [8.1c]

7. 24%   **8.** 6.3%

Translate to a percent equation. Then solve.   [8.2a, b]

9. 30.6 is what percent of 90?

10. 63 is 84% of what?

11. What is $38\frac{1}{2}$% of 168?

Translate to a proportion. Then solve.   [8.3a, b]

12. 24% of what is 16.8?

13. 42 is what percent of 30?

14. What is 10.5% of 84?

Solve.   [8.4a], [8.5a]

15. *Favorite Ice Creams.*   In one survey, 8.9% of those interviewed chose chocolate as their favorite ice cream flavor and 4.2% chose butter pecan. At this rate, of 2000 first-year students at a university, how many would choose chocolate as their favorite ice cream? How many would choose butter pecan?

Source: International Ice Cream Association

16. *Prescriptions.*   Of the 320 million people in the United States, about 224 million take at least one type of prescription drug. What percent take at least one type of prescription drug?

Source: R. Nauert (2013), "70 Percent of Americans Take Prescription Drugs," *Psych Central*

17. *Water Output.*   The average person loses 200 mL of water per day by sweating. This is 8% of the total output of water from the body. How much is the total output of water?

Source: Elaine N. Marieb, *Essentials of Human Anatomy and Physiology*, 6th ed. Boston: Addison Wesley Longman, Inc., 2000

18. *Test Scores.*   After Sheila got a 75 on a math test, she was allowed to go to the math lab and take a retest. She increased her score to 84. What was the percent increase?

19. *Test Scores.*   James got an 80 on a math test. By taking a retest in the math lab, he increased his score by 15%. What was his new score?

Solve.   [8.6a, b, c]

20. A state charges a meals tax of $7\frac{1}{2}$%. What is the meals tax charged on a dinner party costing $320?

21. In a certain state, a sales tax of $453.60 is collected on the purchase of a used car for $7560. What is the sales tax rate?

22. Kim earns $753.50 for selling $6850 worth of televisions. What is the commission rate?

23. What is the rate of discount for this stepladder?

SPECIAL VALUE!

Now **$67** Was $82

8' Aluminum Stepladder

24. An air conditioner has a marked price of $350. It is placed on sale at 12% off. What are the discount and the sale price?

25. The price of a fax machine is marked down from $305 to $262.30. What is the rate of discount?

26. An insurance salesperson receives a 7% commission. If $42,000 worth of life insurance is sold, what is the commission?

Solve.   [8.7a, b, c]

27. What is the simple interest on $1800 at 6% for $\frac{1}{3}$ year?

28. The Dress Shack borrows $24,000 at 10% simple interest for 60 days. Find **(a)** the amount of interest due and **(b)** the total amount that must be paid after 60 days.

29. What is the simple interest on a principal of $2200 at an interest rate of 5.5% for 1 year?

30. The Armstrongs invest $7500 in an investment account paying an annual interest rate of 4%, compounded monthly. How much is in the account after 3 months?

31. Find the amount in an investment account if $8000 is invested at 9%, compounded annually, for 2 years.

32. *Credit Cards.*   At the end of her junior year of college, Kasha has a balance of $6428.74 on a credit card with an annual percentage rate (APR) of 18.7%. She decides not to make additional purchases with this card until she has paid off the balance.

a) Many credit cards require a minimum payment of 2% of the balance. At this rate, what is Kasha's minimum payment on a balance of $6428.74? Round the answer to the nearest dollar.
b) Find the amount of interest and the amount applied to reducing the principal with the minimum payment found in part (a).
c) If Kasha had transferred her balance to a card with an APR of 13.2%, how much of her payment would be interest and how much would be applied to reducing the principal?
d) Compare the amounts for 13.2% from part (c) with the amounts for 18.7% from part (b).

33. A fishing boat listed at $16,500 is on sale at 15% off. What is the sale price?   [8.6c]

A. $14,025   **B.** $2475

C. 85%   **D.** $14,225

34. Find the amount in a money market account if $10,500 is invested at 6%, compounded semiannually, for $1\frac{1}{2}$ years.   [8.7b]

A. $11,139.45   **B.** $12,505.67

C. $11,473.63   **D.** $10,976.03

Synthesis

35. Mike's Bike Shop reduces the price of a bicycle by 40% during a sale. By what percent must the store increase the sale price, after the sale, to get back to the original price?   [8.6c]

Understanding Through Discussion and Writing

1. Which is the better deal for a consumer and why: a discount of 40% or a discount of 20% followed by another of 22%?   [8.6c]

2. Which is better for a wage earner, and why: a 10% raise followed by a 5% raise a year later, or a 5% raise followed by a 10% raise a year later?   [8.5a]

3. Ollie buys a microwave oven during a 10%-off sale. The sale price that Ollie paid was $162. To find the original price, Ollie calculates 10% of $162 and adds that to $162. Is this correct? Why or why not?   [8.6c]

4. You take 40% of 50% of a number. What percent of the number could you take to obtain the same result making only one multiplication? Explain your answer.   [8.4a]

5. A firm must choose between borrowing $5000 at 10% for 30 days and borrowing $10,000 at 8% for 60 days. Give arguments in favor of and against each option.   [8.7a]

6. On the basis of the mathematics presented in Section 8.7, discuss what you have learned about interest rates and credit cards.   [8.7c]

CHAPTER

8   **Test**

For Extra Help   For step-by-step test solutions, access the Chapter Test Prep Videos in MyMathLab® or on You[Tube] (search "BittingerPrealgebra" and click on Channels).

1. *Households Owning Pets.*   In 2011, about 61.3% of the households in Colorado owned pets. Find decimal notation for 61.3%.

   Source: American Veterinary Medical Association

2. *Gravity.*   The gravity of Mars is 0.38 as strong as Earth's. Find percent notation for 0.38.

   Source: www.marsinstitute.info/epo/mermarsfacts.html

3. Find percent notation for $\frac{11}{8}$.

4. Find fraction notation for 65%.

5. Translate to a percent equation. Then solve.

   What is 40% of 55?

6. Translate to a proportion. Then solve.

   What percent of 80 is 65?

Solve.

7. *Organ Transplants.*   In 2011, there were 28,535 organ transplants in the United States. The pie chart below shows the percentages for the main transplants. How many kidney transplants were there in 2011? liver transplants? heart transplants?

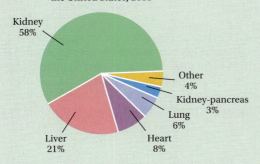

Organ Transplants in the United States, 2011

Kidney 58%
Other 4%
Kidney-pancreas 3%
Lung 6%
Heart 8%
Liver 21%

SOURCE: Milliman Research Report

8. *Batting Average.*   Miguel Cabrera of the Detroit Tigers got 193 hits during the 2013 baseball season. He got a hit in about 34.8% of his at-bats. How many at-bats did he have?

   Source: Major League Baseball

9. *Foreign Adoptions.*   The number of foreign children adopted by Americans declined from 8668 in 2012 to 7092 in 2013. Find the percent decrease.

   Source: U.S. State Department

10. There are about 7,200,000,000 people living in the world today, and approximately 4,300,000,000 live in Asia. What percent of people live in Asia?

    Source: Population Division/International Programs Center, U.S. Census Bureau, U.S. Dept. of Commerce

11. *Oklahoma Sales Tax.*   The sales tax rate in Oklahoma is 4.5%. How much tax is charged on a purchase of $560? What is the total price?

12. Noah's commission rate is 15%. What is his commission on the sale of $4200 worth of merchandise?

13. The marked price of a camera is $200, and it is on sale at 20% off. What are the discount and the sale price?

14. What is the simple interest on a principal of $120 at the interest rate of 7.1% for 1 year?

15. A city orchestra invests $5200 at 6% simple interest. How much is in the account after $\frac{1}{2}$ year?

16. Find the amount in an account if $1000 is invested at $5\frac{3}{8}$% compounded annually, for 2 years.

17. The Suarez family invests $10,000 at an annual interest rate of 4.9%, compounded monthly. How much is in the account after 3 years?

18. *Job Opportunities.* The table below lists job opportunities in 2006 and projected increases for 2016. Complete the table by filling in the missing numbers.

Occupation	Total Employment in 2006	Projected Employment in 2016	Change	Percent Increase
Dental assistant	280,000	362,000	82,000	29.3%
Plumber	705,000		52,000	
Veterinary assistant	71,000	100,000		
Motorcycle repair technician		24,000	3000	
Fitness professional		298,000		26.8%

SOURCE: EarnMyDegree.com

19. Find the discount and the rate of discount for the television in this ad.

19" LCD HDTV

$299⁹⁹

was $349⁹⁹

20. *Credit Cards.* Jayden has a balance of $2704.27 on a credit card with an annual percentage rate (APR) of 16.3%. The minimum payment required on the current statement is $54. Find the amount of interest and the amount applied to reducing the principal with this payment and the balance after this payment.

21. 0.75% of what number is 300?

    A. 2.25   **B.** 40,000   **C.** 400   **D.** 225

Synthesis

22. By selling a home without using a realtor, Juan and Marie can avoid paying a 7.5% commission. They receive an offer of $180,000 from a potential buyer. In order to give a comparable offer, for what price would a realtor need to sell the house? Round to the nearest hundred.

23. Karen's commission rate is 16%. She invests her commission from the sale of $15,000 worth of merchandise at an interest rate of 12%, compounded quarterly. How much is Karen's investment worth after 6 months?

1. *Passports.* In 2010, 14,794,604 U.S. passports and passport cards were issued. This number decreased to 12,613,153 in 2011 and then increased to 13,125,829 in 2012. What was the percent decrease from 2010 to 2011? What was the percent increase from 2011 to 2012? Round answers to the nearest tenth of a percent.

Sources: Travel.State.gov; Bureau of Consular Affairs, U.S. Department of State

2. Find percent notation: 0.269.

3. Find percent notation: $\dfrac{9}{8}$.

4. Find decimal notation: $\dfrac{13}{6}$.

5. Write fraction notation for the ratio 5 to 0.5.

6. Find the rate in kilometers per hour: 350 km, 15 hr.

Use $<$, $>$, or $=$ for ☐ to write a true sentence.

7. $\dfrac{5}{7}$ ☐ $\dfrac{6}{8}$

8. $\dfrac{6}{14}$ ☐ $\dfrac{15}{25}$

Estimate the sum or the difference by first rounding to the nearest hundred.

9. $263{,}961 + 32{,}090 + 127.89$

10. $73{,}510 - 23{,}450$

Calculate.

11. $46 - [4(6 + 4 \div 2) + 2 \times 3 - 5]$

12. Combine like terms: $5x - 9 - 7x - 5$.

Compute and simplify.

13. $\dfrac{6}{5} + 1\dfrac{5}{6}$

14. $-46.9 + 2.84$

15.
$$\begin{array}{r} 4\,8\,7{,}0\,9\,4 \\ 6{,}9\,3\,6 \\ +\ \ \ 2\,1{,}1\,2\,0 \\ \hline \end{array}$$

16. $35 - 34.98$

17. $3\dfrac{1}{3} - 2\dfrac{2}{3}$

18. $-\dfrac{8}{9} - \dfrac{6}{7}$

19. $\dfrac{7}{9} \cdot \dfrac{3}{14}$

20. $(-32)(-4)(-3)$

21.
$$\begin{array}{r} 4\,6.0\,1\,2 \\ \times\ \ \ \ \ 0.0\,3 \\ \hline \end{array}$$

22. $6\dfrac{3}{5} \div 4\dfrac{2}{5}$

23. $431.2 \div (-35.2)$

24. $15\,\overline{)\,1\,8\,5\,0}$

Solve.

25. $36 \cdot x = 3420$

26. $y + 142.87 = 151$

27. $\dfrac{3}{7}x - 5 = 16$

28. $\dfrac{3}{4} + x = \dfrac{5}{6}$

29. $3(x - 7) + 2 = 12x - 3$

30. $\dfrac{16}{n} = \dfrac{21}{11}$

31. Transportation. Boston has a higher percentage of workers who walk or bike to work than any other U.S. city. If 15.1% of Boston's 317,930 workers walk or bike to work, how many Boston workers walk or bike to work?

Source: American Community Survey Reports

32. Salaries of Medical Doctors. In 2014, the median salary of doctors in radiology was $149,298 higher than the median salary of doctors in neurology. If the median salary for radiologists was $377,459, what was the median salary for neurologists?

Source: www.salary.com

33. At one point during the regular NBA season, the Cleveland Cavaliers had won 39 out of 49 games. At this rate, how many games would they win in the entire season of 82 games?

Source: National Basketball Association

34. Unit Price. A 200-oz bottle of liquid laundry detergent costs $14.99. What is the unit price?

35. Patty walked $\frac{7}{10}$ mi to school and then $\frac{8}{10}$ mi to the library. How far did she walk?

36. Compound Interest. The Bakers invest $8500 in an account paying 8%, compounded monthly. How much is in the account after 5 years?

37. Ribbons. How many pieces of ribbon $1\frac{4}{5}$ yd long can be cut from a length of ribbon 9 yd long?

38. In what quadrant does the point $(-3, -1)$ lie?

39. Graph on a plane: $y = -\frac{3}{5}x$.

40. Find the mean: 19, 29, 34, 39, 45.

41. Find the median: 7, 7, 12, 15, 21.

42. Find the area of a 40-yd by 90-yd soccer field.

43. Subtract and simplify: $\frac{14}{25} - \frac{3}{20}$.

　　A. $\frac{11}{500}$　　　　　　**B.** $\frac{11}{5}$

　　C. $\frac{41}{100}$　　　　　　**D.** $\frac{205}{500}$

44. Find the perimeter of a 15-in. by 15-in. chessboard.

　　A. 60 in.　　　　　　**B.** 225 in.

　　C. 30 in.　　　　　　**D.** 90 in.

Synthesis

45. How many successive 10% discounts are necessary to lower the price of an item to below 50% of its original price?

On a trip through the mountains, a Dodge Neon traveled 240 mi on $7\frac{1}{2}$ gal of gasoline. Going across the plains, the same car averaged 36 miles per gallon.

46. What was the percent increase or decrease in miles per gallon when the car left the mountains for the plains?

47. How many miles per gallon did the Dodge average over the entire trip if it used 5 gal of gas to cross the plains?

CHAPTER
9

Geometry and Measurement

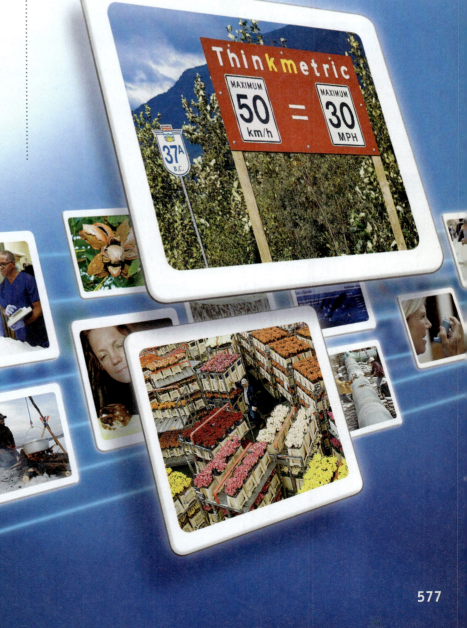

9.1 Systems of Linear Measurement

OBJECTIVES

a Convert from one American unit of length to another.

b Convert from one metric unit of length to another.

c Convert between American and metric units of length.

SKILL TO REVIEW

Objective 5.3a: Multiply using decimal notation.

Multiply.

1. $0.5603 \cdot 1000$

2. $1.87 \cdot 0.01$

Use the unit below to measure the length of each segment or object.

1.

2.

3.

4.

Answers

Skill to Review:
1. 560.3   **2.** 0.0187

Margin Exercises:
1. 2   **2.** 3   **3.** $1\frac{1}{2}$ **4.** $2\frac{1}{2}$

Length, or distance, is one kind of measure. To find lengths, we start with some **unit segment** and assign to it a measure of 1. Suppose $\overline{AB}$ below is a unit segment. Let's measure segment $\overline{CD}$, using $\overline{AB}$ as our unit segment.

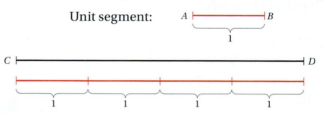

Since we can place 4 unit segments end to end along $\overline{CD}$, the measure of $\overline{CD}$ is 4.

Sometimes we need to use parts of units. For example, the measure of the segment $\overline{MN}$ below is $1\frac{1}{2}$. We place one unit segment and one half-unit segment end to end.

◀ **Do Margin Exercises 1–4.**

a AMERICAN MEASURES

American units of length are related as follows.

AMERICAN UNITS OF LENGTH	
12 inches (in.) = 1 foot (ft)	3 feet = 1 yard (yd)
36 inches = 1 yard	5280 feet = 1 mile (mi)

(Actual size, in inches)

We can visualize comparisons of the units as follows:

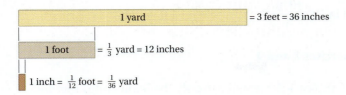

The symbols ′ and ″ are also used for inches and feet, as in 13 in. = 13″ and 27 ft = 27′. American units have also been called "English," or "British–American," because at one time they were used in both North America and Great Britain. Today, both Canada and England have officially converted to the metric system.

To change from certain American units to others, we make substitutions. Such a substitution is usually helpful when we are converting from a *larger* unit to a *smaller* one.

EXAMPLE 1 Complete: $7\frac{1}{3}$ yd = _____ in.

$$7\frac{1}{3}\text{ yd} = 7\frac{1}{3} \cdot 1\text{ yd}$$ We think of $7\frac{1}{3}$ yd as $7\frac{1}{3} \cdot$ yd, or $7\frac{1}{3} \cdot 1$ yd.

$$= 7\frac{1}{3} \cdot 36\text{ in.}$$ Substituting 36 in. for 1 yd

$$= \frac{22}{3} \cdot 36\text{ in.}$$

$$= 264\text{ in.}$$

Do Exercises 5–7. ▶

Sometimes it helps to use multiplying by 1 in making conversions. For example, 12 in. = 1 ft, so

$$\frac{12\text{ in.}}{1\text{ ft}} = 1 \quad \text{and} \quad \frac{1\text{ ft}}{12\text{ in.}} = 1.$$

EXAMPLE 2 Complete: 48 in. = _____ ft.

We want to convert from "in." to "ft." We multiply by 1 using a symbol for 1 with "in." on the bottom and "ft" on the top to eliminate inches and to convert to feet:

$$48\text{ in.} = \frac{48\text{ in.}}{1} \cdot \frac{1\text{ ft}}{12\text{ in.}}$$ Multiplying by 1 using $\frac{1\text{ ft}}{12\text{ in.}}$ to eliminate in.

$$= \frac{48\text{ in.}}{12\text{ in.}} \cdot 1\text{ ft}$$

$$= \frac{48}{12} \cdot \frac{\text{in.}}{\text{in.}} \cdot 1\text{ ft}$$

$$= 4 \cdot 1\text{ ft}$$ The $\frac{\text{in.}}{\text{in.}}$ acts like 1, so we can omit it.

$$= 4\text{ ft.}$$

Complete.

5. 8 yd = _____ in.

GS **6.** $2\frac{5}{6}$ yd = _____ ft

$$2\frac{5}{6}\text{ yd} = 2\frac{5}{6} \cdot 1\text{ yd}$$

$$= \frac{\boxed{}}{6} \cdot \boxed{}\text{ ft}$$

$$= \frac{17}{\boxed{}}\text{ ft}$$

$$= 8\frac{1}{2}\text{ ft}$$

7. 3.8 mi = _____ in.

Answers

5. 288   **6.** $8\frac{1}{2}$ **7.** 240,768

Guided Solution:
6. 17, 3, 2

Complete.

8. 72 in. = _____ ft

$$72 \text{ in.} = \frac{72 \text{ in.}}{1} \cdot \frac{\boxed{}}{12 \text{ in.}}$$

$$= \frac{\boxed{}}{12} \cdot 1 \text{ ft}$$

$$= \boxed{} \text{ ft}$$

9. 24 ft = _____ yd

Complete.

10. 35 ft = _____ yd

11. 26,400 ft = _____ mi

12. 2640 ft = _____ mi

Complete.

13. 6 mi = _____ ft

14. *Pedestrian Paths.* There are 23 mi of pedestrian paths in Central Park in New York City. Convert 23 miles to yards.

Answers

8. 6   **9.** 8   **10.** $11\frac{2}{3}$, or $11.\overline{6}$ **11.** 5

12. $\frac{1}{2}$, or 0.5   **13.** 31,680   **14.** 40,480 yd

Guided Solution:
8. 1 ft, 72, 6

We can also look at this conversion as "canceling" units:

$$48 \text{ in.} = \frac{48 \text{ in.}}{1} \cdot \frac{1 \text{ ft}}{12 \text{ in.}} = \frac{48}{12} \cdot 1 \text{ ft} = 4 \text{ ft.}$$

◀ **Do Exercises 8 and 9.**

In Examples 3–5, we will use only the "canceling" method.

EXAMPLE 3 Complete: 70 ft = _____ yd.

Since we are converting from "ft" to "yd," we choose a symbol for 1 with "yd" on the top and "ft" on the bottom:

$$70 \text{ ft} = 70 \text{ ft} \cdot \frac{1 \text{ yd}}{3 \text{ ft}} \qquad \text{If it helps, write 70 ft as } \frac{70 \text{ ft}}{1}.$$

$$= \frac{70}{3} \cdot 1 \text{ yd} = 23.\overline{3} \text{ yd, or } 23\frac{1}{3} \text{ yd.}$$

EXAMPLE 4 Complete: 23,760 ft = _____ mi.

We choose a symbol for 1 with "mi" on the top and "ft" on the bottom:

$$23,760 \text{ ft} = 23,760 \text{ ft} \cdot \frac{1 \text{ mi}}{5280 \text{ ft}} \qquad 5280 \text{ ft} = 1 \text{ mi, so } \frac{1 \text{ mi}}{5280 \text{ ft}} = 1$$

$$= \frac{23,760}{5280} \cdot 1 \text{ mi}$$

$$= 4.5 \text{ mi.}$$

◀ **Do Exercises 10–12.**

EXAMPLE 5 *Illuminated Bridge.* A computer-controlled light sculpture on the San Francisco–Oakland Bay Bridge was designed to be visible for a two-year period that began March 5, 2013. It consists of 25,000 white LED lights that run for 1.8 mi on the bridge's west span. The artist, Leo Villareal, wrote a program to ensure that the light patterns in the sculpture will never repeat. Convert 1.8 mi to yards.

Source: http://articles.latimes.com/2013/mar/05/news/
la-trb-california-san-francisco-bay-bridge-lighting-20130304

We have

$$1.8 \text{ mi} = 1.8 \text{ mi} \cdot \frac{5280 \text{ ft}}{1 \text{ mi}} \cdot \frac{1 \text{ yd}}{3 \text{ ft}}$$

$$= \frac{1.8 \cdot 5280}{1 \cdot 3} \cdot 1 \text{ yd}$$

$$= 3168 \text{ yd.}$$

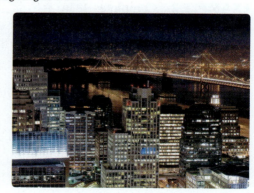

The illuminated lights run for 3168 yd on the west span of the bridge.

◀ **Do Exercises 13 and 14.**

b THE METRIC SYSTEM

Because the **metric system** is based on powers of 10, it allows for easy conversion between units. The metric system does not use inches, feet, pounds, and so on, but the units for time and electricity are the same as those in the American system.

The basic unit of length is the **meter.** It is just over a yard. In fact, 1 meter ≈ 1.1 yd.

(Comparative sizes are shown.)

1 Meter

1 Yard

The other units of length are multiples of the length of a meter:

10 times a meter,   100 times a meter,   1000 times a meter,   and so on,

or fractions of a meter:

$\frac{1}{10}$ of a meter, $\frac{1}{100}$ of a meter, $\frac{1}{1000}$ of a meter,   and so on.

You should memorize the names and abbreviations for metric units of length. Think of *kilo-* for 1000, *hecto-* for 100, *deka-* for 10, *deci-* for $\frac{1}{10}$, *centi-* for $\frac{1}{100}$, and *milli-* for $\frac{1}{1000}$. (The units dekameter and decimeter are not used often.) We will also use these prefixes when considering units of area, capacity, and mass.

Thinking Metric

To familiarize yourself with metric units, consider the following.

1 kilometer (1000 meters)	is a bit more than $\frac{1}{2}$ mile (≈ 0.6 mi).
1 meter	is just over a yard (≈ 1.1 yd).
1 centimeter (0.01 meter)	is a little more than the width of a jumbo paperclip (≈ 0.4 in.).
1 millimeter (0.001 meter)	is about the diameter of a paperclip wire.

> ### METRIC UNITS OF LENGTH
>
> 1 *kilo*meter (km) = 1000 meters (m)
>
> 1 *hecto*meter (hm) = 100 meters (m)
>
> 1 *deka*meter (dam) = 10 meters (m)
>
> 1 meter (m)
>
> 1 *deci*meter (dm) = $\frac{1}{10}$ meter (m)
>
> 1 *centi*meter (cm) = $\frac{1}{100}$ meter (m)
>
> 1 *milli*meter (mm) = $\frac{1}{1000}$ meter (m)

1 mm

1 cm

1 cm

1 inch is about 2.54 centimeters.

← 2.54 cm → (Actual size, in centimeters)

← 1 inch → (Actual size, in inches)

Using a centimeter ruler, measure each object.

15.

16.

17.

The millimeter (mm) is used to measure small distances, especially in industry.

1 mm

3 mm

Centimeters (cm) are used for body dimensions and clothing sizes.

163 cm
(64.2 in.)
(5 ft 4 in.)

RELAXED FIT
97 cm/81 cm
(38 in./32 in.)

◀ **Do Exercises 15–17.**

The meter (m) is used for expressing dimensions of large objects—say, the height of Hoover Dam, 221.4 m, or the distance around a standard athletic track, 400 m—and for expressing somewhat smaller dimensions like the length and width of an organic vegetable garden.

400 m around

6.2 m

4.5 m

Answers

15. 2 cm, or 20 mm
16. 2.3 cm, or 23 mm
17. 4.4 cm, or 44 mm

Kilometers (km) are used for longer distances, mostly in cases where miles are now being used. 1 mile is about 1.6 km.

1 km

1 mi

Do Exercises 18–23. ▶

Metric conversions from larger units to smaller ones are often most easily made using substitution.

EXAMPLE 6 Complete: 4 km = _____ m.

$$4 \text{ km} = 4 \cdot 1 \text{ km}$$
$$= 4 \cdot 1000 \text{ m} \quad \text{Substituting 1000 m for 1 km}$$
$$= 4000 \text{ m} \quad \text{Multiplying by 1000}$$

Do Exercises 24 and 25. ▶

Since

$$\frac{1}{10} \text{m} = 1 \text{ dm}, \quad \frac{1}{100} \text{m} = 1 \text{ cm}, \quad \text{and} \quad \frac{1}{1000} \text{m} = 1 \text{ mm},$$

we can use the following conversions.

METRIC CONVERSIONS

1 m = 10 dm,   1 m = 100 cm,   and   1 m = 1000 mm

Remembering these equations will help you to write forms of 1 when canceling to make conversions.

EXAMPLE 7 Complete: 93.4 m = _____ cm.

To convert from "m" to "cm," we multiply by 1 using a symbol for 1 with "m" on the bottom and "cm" on the top. This process introduces centimeters and at the same time eliminates meters.

$$93.4 \text{ m} = 93.4 \text{ m} \cdot \frac{100 \text{ cm}}{1 \text{ m}} \quad \text{Multiplying by 1 using } \frac{100 \text{ cm}}{1 \text{ m}}$$

$$= 93.4 \, \cancel{\text{m}} \cdot \frac{100 \text{ cm}}{1 \, \cancel{\text{m}}} = 93.4 \cdot 100 \cdot 1 \text{ cm} = 9340 \text{ cm}$$

Complete with mm, cm, m, or km.

18. A stick of gum is 7 _____ long.

19. New Orleans is 2941 _____ from San Diego.

20. A penny is 1 _____ thick.

21. The halfback ran 7 _____.

22. The book is 3 _____ thick.

23. The desk is 2 _____ long.

Complete.

GS **24.** 23 km = _____ m

$$23 \text{ km} = 23 \cdot 1 \text{ km}$$
$$= 23 \cdot \boxed{} \text{ m}$$
$$= \boxed{} \text{ m}$$

25. 4 hm = _____ m

Complete.

26. $1.78 \text{ m} = \underline{\hspace{2cm}} \text{ cm}$

27. $9.04 \text{ m} = \underline{\hspace{2cm}} \text{ mm}$

Complete.

28. $7814 \text{ m} = \underline{\hspace{2cm}} \text{ km}$

29. $7814 \text{ m} = \underline{\hspace{2cm}} \text{ dam}$ **GS**

$$7814 \text{ m} = 7814 \text{ m} \cdot \frac{1 \text{ dam}}{\boxed{}}$$

$$= \frac{7814}{\boxed{}} \cdot \frac{\text{m}}{\text{m}} \cdot 1 \text{ dam}$$

$$= \boxed{} \text{ dam}$$

30. $9.67 \text{ mm} = \underline{\hspace{2cm}} \text{ cm}$

31. $89 \text{ km} = \underline{\hspace{2cm}} \text{ cm}$

EXAMPLE 8 Complete: $0.248 \text{ m} = \underline{\hspace{2cm}} \text{ mm}$.

We are converting from "m" to "mm," so we choose a symbol for 1 with "mm" on the top and "m" on the bottom:

$$0.248 \text{ m} = 0.248 \,\cancel{\text{m}} \cdot \frac{1000 \text{ mm}}{1 \,\cancel{\text{m}}} = 0.248 \cdot 1000 \cdot 1 \text{ mm} = 248 \text{ mm}.$$

◀ **Do Exercises 26 and 27.**

EXAMPLE 9 Complete: $2347 \text{ m} = \underline{\hspace{2cm}} \text{ km}$.

We multiply by 1 using $\dfrac{1 \text{ km}}{1000 \text{ m}}$:

$$2347 \text{ m} = 2347 \,\cancel{\text{m}} \cdot \frac{1 \text{ km}}{1000 \,\cancel{\text{m}}} = \frac{2347}{1000} \cdot 1 \text{ km} = 2.347 \text{ km}.$$

Since $1 \text{ m} = 1000 \text{ mm}$ and $1 \text{ m} = 100 \text{ cm}$, we know that $1000 \text{ mm} = 100 \text{ cm}$. More simply, $10 \text{ mm} = 1 \text{ cm}$.

EXAMPLE 10 Complete: $8.42 \text{ mm} = \underline{\hspace{2cm}} \text{ cm}$.

We can multiply by 1 using either $\dfrac{1 \text{ cm}}{10 \text{ mm}}$ or $\dfrac{100 \text{ cm}}{1000 \text{ mm}}$. Both expressions for 1 will eliminate mm and leave cm:

$$8.42 \text{ mm} = 8.42 \,\cancel{\text{mm}} \cdot \frac{1 \text{ cm}}{10 \,\cancel{\text{mm}}} = \frac{8.42}{10} \cdot 1 \text{ cm} = 0.842 \text{ cm}.$$

◀ **Do Exercises 28–31.**

Mental Conversion

Changing from one unit to another in the metric system amounts to the movement of a decimal point. That is because the metric system is based on 10. Let's find a faster way to convert. Look at the following table.

1000 m	100 m	10 m	1 m	0.1 m	0.01 m	0.001 m
1 km	1 hm	1 dam	1 m	1 dm	1 cm	1 mm

Each place in the table has a value $\frac{1}{10}$ that of the place to the left or 10 times that of the place to the right. Thus, moving one place in the table corresponds to moving one decimal place.

EXAMPLE 11 Complete: $35.7 \text{ mm} = \underline{\hspace{2cm}} \text{ cm}$.

Think: Centimeters is the next larger unit after millimeters. Thus, we move the decimal point one place to the left.

$$35.7 \qquad 3.5.7 \qquad 35.7 \text{ mm} = 3.57 \text{ cm} \qquad \text{Converting to a larger unit shifts the decimal point to the left.}$$

Answers

26. 178 **27.** 9040 **28.** 7.814 **29.** 781.4
30. 0.967 **31.** 8,900,000

Guided Solution:
29. 10 m, 10, 781.4

EXAMPLE 12 Complete: 1.886 km = _____ cm.

Think: To go from km to cm in the table is a move of five places to the right. Thus, we move the decimal point five places to the right.

1000 m	100 m	10 m	1 m	0.1 m	0.01 m	0.001 m
1 km	1 hm	1 dam	1 m	1 dm	1 cm	1 mm

5 places to the right

1.886   1.88600.   1.886 km = 188,600 cm

EXAMPLE 13 Complete: 3 m = _____ cm.

Think: To go from m to cm in the table is a move of two places to the right. Thus, we move the decimal point two places to the right.

1000 m	100 m	10 m	1 m	0.1 m	0.01 m	0.001 m
1 km	1 hm	1 dam	1 m	1 dm	1 cm	1 mm

2 places to the right

3   3.00.   3 m = 300 cm

You should try to make metric conversions mentally as much as possible. The fact that conversions can be done so easily is an important advantage of the metric system.

> The most commonly used metric units of length are km, m, cm, and mm. We have purposely used these more often than the others in the exercises and examples.

Do Exercises 32–35. ▶

Complete. Try to do this mentally using the table.

32. 6780 m = _____ km

33. 9.74 cm = _____ mm

34. 1 mm = _____ cm

35. 845.1 mm = _____ dm

c ## CONVERTING BETWEEN AMERICAN UNITS AND METRIC UNITS

We can make conversions between American units and metric units by substituting based on the following table. The conversions shown are all approximations except for the first one; 1 in. is defined to be 2.54 cm.

AMERICAN	METRIC
1 in.	2.54 cm
1 ft	0.305 m
1 yd	0.914 m
1 mi	1.609 km
0.621 mi	1 km
1.094 yd	1 m
3.281 ft	1 m
39.370 in.	1 m

Answers

32. 6.78   **33.** 97.4   **34.** 0.1   **35.** 8.451

Complete.

36. 100 yd = _____ m
(The length of a football field excluding the end zones)

37. 2.5 mi = _____ km
(The length of the tri-oval track at Daytona International Speedway)

38. 2383 km = _____ mi
(The distance from St. Louis to Phoenix)

39. The Pacific Coast Highway, which is part of California State Route 1, is 655.8 mi long. Find this length in kilometers, rounded to the nearest tenth.

40. The height of the Stratosphere Tower in Las Vegas, Nevada, is 1149 ft. Find the height in meters.

41. Complete:
3.175 mm = _____ in.
(The thickness of a quarter)

Answers

36. 91.4 **37.** 4.0225 **38.** 1479.843
39. 1055.2 km **40.** 350.445 m **41.** 0.125

EXAMPLE 14 Complete: 11 in. = _____ cm.
(The wingspan of the world's largest butterfly, the Queen Alexandra)
Source: *Top 10 of Everything 2013*

$$11 \text{ in.} = 11 \cdot 1 \text{ in.}$$
$$= 11 \cdot 2.54 \text{ cm} \qquad \text{Substituting 2.54 cm for 1 in.}$$
$$= 27.94 \text{ cm}$$

This answer would probably be rounded to the nearest one: 28 cm.

EXAMPLE 15 Complete: 26.2 mi = _____ km.
(The length of the Olympic marathon)

Since we are given that 1 mi ≈ 1.609 km, we can convert using substitution.

$$26.2 \text{ mi} = 26.2 \cdot 1 \text{ mi}$$
$$\approx 26.2 \cdot 1.609 \text{ km} \qquad \text{Converting from mi to km}$$
$$\approx 42.16 \text{ km}$$

We can also convert by multiplying by 1.

$$26.2 \text{ mi} \approx \frac{26.2 \text{ mi}}{1} \cdot \frac{1 \text{ km}}{0.621 \text{ mi}}$$
$$\approx 42.19 \text{ km}$$

Note that, since the conversion factors given in the table are approximations, the answers found by the two methods differ slightly.

EXAMPLE 16 *Millau Viaduct.* The Millau viaduct is part of the E11 expressway connecting Paris, France, and Barcelona, Spain. The viaduct has the highest bridge piers ever constructed. The tallest pier is 804 ft high and the overall height including the pylon is 1122 ft, making this the highest bridge in the world. Convert 804 feet and 1122 feet to meters.
Source: www.abelard.org/france/viaduct-de-millau.php

We let P = the height of the pier and H = the overall height of the bridge. To convert feet to meters, we substitute 0.305 m for 1 ft.

$$P = 804 \text{ ft} \qquad\qquad H = 1122 \text{ ft}$$
$$= 804 \cdot 1 \text{ ft} \qquad\qquad = 1122 \cdot 1 \text{ ft}$$
$$\approx 804 \cdot 0.305 \text{ m} \qquad \approx 1122 \cdot 0.305 \text{ m}$$
$$= 245.22 \text{ m} \qquad\qquad = 342.21 \text{ m}$$

◀ **Do Exercises 36–40.**

EXAMPLE 17 Complete: 0.10414 mm = _____ in.
(The thickness of a $1 bill)

In this case, we must make two substitutions or multiply by two forms of 1 since the table on the preceding page does not provide a direct way to convert from millimeters to inches. Here we choose to multiply by forms of 1.

$$0.10414 \text{ mm} = 0.10414 \cdot 1 \text{ mm} \cdot \frac{1 \text{ cm}}{10 \text{ mm}} = 0.010414 \text{ cm}$$

$$= 0.010414 \cdot 1 \text{ cm} \cdot \frac{1 \text{ in.}}{2.54 \text{ cm}} = 0.0041 \text{ in.}$$

◀ **Do Exercise 41.**

9.1

Exercise Set

For Extra Help

MyMathLab®

MathXL®
PRACTICE   WATCH   READ   REVIEW

☑ **Reading Check**

Complete each symbol for 1 by choosing a number from the list at the right. Not every choice will be used.

RC1. $1 = \dfrac{1\text{ ft}}{\square\text{ in.}}$

RC2. $1 = \dfrac{\square\text{ ft}}{1\text{ mi}}$

RC3. $1 = \dfrac{1\text{ yd}}{\square\text{ in.}}$

RC4. $1 = \dfrac{\square\text{ cm}}{1\text{ m}}$

RC5. $1 = \dfrac{\square\text{ ft}}{1\text{ yd}}$

RC6. $1 = \dfrac{1\text{ cm}}{\square\text{ mm}}$

3	100
10	1000
12	1760
24	5280
36	

a   Complete.

1. 1 yd = _____ ft

2. 1 ft = _____ in.

3. 1 in. = _____ ft

4. 1 mi = _____ ft

5. 1 mi = _____ yd

6. 1 ft = _____ yd

7. 3 yd = _____ ft

8. 4 yd = _____ in.

9. 84 in. = _____ ft

10. 18 in. = _____ ft

11. 48 ft = _____ yd

12. 29 ft = _____ yd

13. 5 mi = _____ yd

14. 5 mi = _____ ft

15. 63 in. = _____ ft

16. 19 ft = _____ yd

17. 11,616 ft = _____ mi

18. 5.2 yd = _____ ft

19. 7.1 mi = _____ ft

20. 31,680 ft = _____ mi

21. $7\frac{1}{2}$ ft = _____ yd

22. 360 in. = _____ yd

23. 36 in. = _____ ft

24. 7.2 ft = _____ in.

25. 1760 yd = _____ mi

26. 3520 yd = _____ mi

27. 45 in. = _____ yd

28. $6\frac{1}{3}$ yd = _____ in.

29. 25 mi = _____ ft

30. 240 in. = _____ ft

31. 2 mi = _____ in.

32. 63,360 in. = _____ mi

b Complete. Do as much as possible mentally.

33. **a)** 1 km = _____ m
   b) 1 m = _____ km

34. **a)** 1 hm = _____ m
   b) 1 m = _____ hm

35. **a)** 1 dam = _____ m
   b) 1 m = _____ dam

36. **a)** 1 dm = _____ m
   b) 1 m = _____ dm

37. **a)** 1 cm = _____ m
   b) 1 m = _____ cm

38. **a)** 1 mm = _____ m
   b) 1 m = _____ mm

39. 8.3 km = _____ m

40. 27 km = _____ m

41. 98 cm = _____ m

42. 0.789 cm = _____ m

43. 8921 m = _____ km

44. 8664 m = _____ km

45. 32.17 m = _____ km

46. 4.733 m = _____ km

47. 289 m = _____ cm

48. 869 m = _____ cm

49. 477 cm = _____ m

50. 6.27 mm = _____ m

51. 6.88 m = _____ cm

52. 6.88 m = _____ dm

53. 1 mm = _____ cm

54. 1 cm = _____ km

55. 1 km = _____ cm

56. 2 km = _____ cm

57. 14.2 cm = _____ mm

58. 25.3 cm = _____ mm

59. 8.2 mm = _____ cm

60. 9.7 mm = _____ cm

61. 4500 mm = _____ cm

62. 8,000,000 m = _____ km

63. 0.024 mm = _____ m

64. 60,000 mm = _____ dam

65. 6.88 m = _____ dam

66. 7.44 m = _____ hm

67. 2.3 dam = _____ dm

68. 9 km = _____ hm

Complete the following table.

	Object	Millimeters (mm)	Centimeters (cm)	Meters (m)
69.	Width of a football field		4844	
70.	Length of a football field			109.09
71.	Width of a calculator			0.085
72.	Width of a credit card	56		
73.	Thickness of an index card	0.27		
74.	Thickness of a piece of cardboard		0.23	
75.	Height of One World Trade Center, New York, New York			541.3
76.	Height of The Gateway Arch, St. Louis, Missouri	192,000		

C Complete. Answers may vary slightly, depending on the conversion factor used.

77. 10 km = _____ mi
(A common running distance)

78. 5 mi = _____ km
(A common running distance)

79. 14 in. = _____ cm
(A common paper length)

80. 400 m = _____ yd
(A common race distance)

81. 65 mph = _____ km/h
(A common speed limit in the United States)

82. 100 km/h = _____ mph
(A common speed limit in Canada)

83. 94 ft = _____ m
(The length of an NCAA basketball court)

84. 1.91 m = _____ in.
(The height of Jeremy Lin of the NBA)

85. 180 cm = _____ in.
(A common snowboard length)

86. 0.25 in. = _____ mm
(The thickness of an eraser on a pencil)

87. 7.6 mm = _____ in.
(The thickness of an iPhone 5s)

88. 70 in. = _____ cm
(A common height for a man)

Complete the following table. Answers may vary, depending on the conversion factor used.

	Object	Yards (yd)	Centimeters (cm)	Inches (in.)	Meters (m)	Millimeters (mm)
89.	Width of a piece of typing paper			$8\frac{1}{2}$		
90.	Length of a piece of typing paper		27.94			
91.	Length of the Channel Tunnel connecting France and England				50,500	
92.	Height of Jin Mao Tower, Shanghai	460				

Skill Maintenance

Solve.

93. $-7x - 9x = 24$ [5.7b]

94. $-2a + 9 = 5a + 23$ [5.7b]

95. If 3 calculators cost \$43.50, how much would 7 calculators cost?   [7.4a]

96. A principal of \$500 is invested at a rate of 8.9% for 1 year. Find the simple interest.   [8.7a]

Convert to percent notation.

97. 0.47   [8.1b]

98. $\dfrac{7}{20}$ [8.1c]

99. A living room is 12 ft by 16 ft. Find the perimeter and area of the room.   [1.2b], [1.4b]

100. A bedroom measures 10 ft by 12 ft. Find the perimeter and area of the room.   [1.2b], [1.4b]

Synthesis

In Exercises 101–104, each sentence is incorrect. Insert or alter a decimal point to make the sentence correct.

101. When my right arm is extended, the distance from my left shoulder to the end of my right hand is 10 m.

102. The height of the Shanghai World Financial Center is 49.2 m.

103. A stack of ten quarters is 140 cm high.

104. The width of an adult's hand is 112 cm.

105. *Noah's Ark.*   It is thought that the biblical measure cubit was approximately 18 in. The dimensions of Noah's ark are given as follows: "The length of the ark shall be three hundred cubits, the breadth of it fifty cubits, and the height of it thirty cubits." What were the dimensions of Noah's ark in inches? in feet?

Source: *Holy Bible, King James Version*, Gen. 6:15

106. *Goliath's Height.*   In biblical measures, a span was considered to be half of a cubit (1 cubit ≈ 18 in.; see Exercise 105). The giant Goliath's height was "six cubits and a span." What was the height of Goliath in inches? in feet?

Source: *Holy Bible, King James Version*, 1 Sam. 17:4

Complete. Answers may vary, depending on the conversion factor used.

107. ▦ 2 mi = _____ cm

108. ▦ 10 km = _____ in.

109. ▦ As of May 2014, the women's world record for the 100-m dash was 10.49 sec, set by Florence Griffith-Joyner in Indianapolis, Indiana, on July 16, 1988. How fast is this in miles per hour? Round to the nearest tenth of a mile per hour.

Source: International Association of Athletics Federations

110. ▦ As of May 2014, the men's world record for the 100-m dash was 9.58 sec, set by Usain Bolt of Jamaica on August 16, 2009, in Berlin, Germany. How fast is this in miles per hour? Round to the nearest hundredth of a mile per hour.

Source: International Association of Athletics Federations

Use < or > to complete the following. Perform only approximate, mental calculations.

111. 59 in. ☐ 59 cm

112. 35 yd ☐ 35 m

113. 7 km ☐ 6 mi

114. 9 mi ☐ 18 km

115. 24 ft ☐ 6 m

116. 30 in. ☐ 90 cm

Converting Units of Area

9.2

a AMERICAN UNITS

OBJECTIVES

a Convert from one American unit of area to another.

b Convert from one metric unit of area to another.

It is often necessary to convert units of area. First we will convert from one American unit of area to another.

EXAMPLE 1 Complete: $1 \text{ yd}^2 =$ _____ ft^2.

We recall that 1 yd = 3 ft and make a sketch. Note that $1 \text{ yd}^2 = 9 \text{ ft}^2$. The same result can be found as follows:

$$1 \text{ yd}^2 = 1 \cdot (3 \text{ ft})^2 \qquad \textcolor{red}{\text{Substituting 3 ft for 1 yd}}$$
$$= 3 \text{ ft} \cdot 3 \text{ ft}$$
$$= 9 \text{ ft}^2. \qquad \textcolor{red}{\text{Note that ft} \cdot \text{ft} = \text{ft}^2.}$$

EXAMPLE 2 Complete: $2 \text{ ft}^2 =$ _____ in^2.

SKILL TO REVIEW

Objective 1.9b: Evaluate exponential notation.

Evaluate.

1. 9^2 **2.** 10^3

$$\textcolor{red}{\text{A sketch of } 2 \text{ ft}^2.}$$
$$\textcolor{red}{\text{Note that } 1 \text{ ft}^2 = 144 \text{ in}^2.}$$

$$2 \text{ ft}^2 = 2 \cdot (12 \text{ in.})^2 \qquad \textcolor{red}{\text{Substituting 12 in. for 1 ft}}$$
$$= 2 \cdot 12 \text{ in.} \cdot 12 \text{ in.} = 288 \text{ in}^2 \qquad \textcolor{red}{\text{Note that in.} \cdot \text{in.} = \text{in}^2.}$$

Do Margin Exercises 1–3. ▶

Complete.

1. $1 \text{ ft}^2 =$ _____ in^2

2. $10 \text{ ft}^2 =$ _____ in^2

3. $7 \text{ yd}^2 =$ _____ ft^2

American units of area are related as follows.

AMERICAN UNITS OF AREA

1 square yard (yd^2) = 9 square feet (ft^2)
1 square foot (ft^2) = 144 square inches (in^2)
1 square mile (mi^2) = 640 acres
1 acre = 43,560 ft^2

Caution!

Be aware of when you are converting units of length and when you are converting units of area.

$1 \text{ yd} = 3 \text{ ft} \Rightarrow 1 \text{ yd}^2 = 9 \text{ ft}^2$
$1 \text{ ft} = 12 \text{ in.} \Rightarrow 1 \text{ ft}^2 = 144 \text{ in}^2$

EXAMPLE 3 Complete: $36 \text{ ft}^2 =$ _____ yd^2.

To convert from "ft^2" to "yd^2" we write 1 with yd^2 on top and ft^2 on the bottom.

$$36 \text{ ft}^2 = 36 \text{ ft}^2 \cdot \frac{1 \text{ yd}^2}{9 \text{ ft}^2} \qquad \textcolor{red}{\text{Multiplying by 1 using } \frac{1 \text{ yd}^2}{9 \text{ ft}^2}}$$

$$= \frac{36}{9} \cdot \text{yd}^2 = 4 \text{ yd}^2$$

Answers

Skill to Review:
1. 81 **2.** 1000

Margin Exercises:
1. 144 **2.** 1440 **3.** 63

Complete.

4. $360 \text{ in}^2 = \underline{\hspace{1cm}} \text{ft}^2$

$360 \text{ in}^2 = 360 \text{ in}^2 \cdot \dfrac{1 \text{ ft}^2}{\boxed{}}$

$\phantom{360 \text{ in}^2} = \dfrac{360}{\boxed{}} \cdot \dfrac{\text{in}^2}{\text{in}^2} \cdot 1 \text{ ft}^2$

$\phantom{360 \text{ in}^2} = \boxed{} \text{ ft}^2$

5. $5 \text{ mi}^2 = \underline{\hspace{1cm}} \text{ acres}$

When converting from larger units to smaller units, we can substitute directly from the table.

(GS)

EXAMPLE 4 Complete: $7 \text{ mi}^2 = \underline{\hspace{1cm}} \text{ acres}.$

$7 \text{ mi}^2 = 7 \cdot \textcolor{red}{640 \text{ acres}} \qquad \textcolor{red}{\text{Substituting 640 acres for 1 mi}^2}$

$\phantom{7 \text{ mi}^2} = 4480 \text{ acres}$

◀ **Do Exercises 4 and 5.**

b METRIC UNITS

We next convert from one metric unit of area to another.

EXAMPLE 5 Complete: $1 \text{ cm}^2 = \underline{\hspace{1cm}} \text{ mm}^2.$

Since $1 \text{ cm} = 10 \text{ mm}$, a square centimeter will have sides of length 10 mm.

The area of the square can be written

$$(1 \text{ cm})(1 \text{ cm}) = 1 \text{ cm}^2$$

or

$$(10 \text{ mm})(10 \text{ mm}) = 100 \text{ mm}^2.$$

Thus, $1 \text{ cm}^2 = 100 \text{ mm}^2.$

EXAMPLE 6 Complete: $1 \text{ km}^2 = \underline{\hspace{1cm}} \text{ m}^2.$

$1 \text{ km}^2 = (1 \text{ km})(1 \text{ km})$

$\phantom{1 \text{ km}^2} = \textcolor{red}{(1000 \text{ m})(1000 \text{ m})} \qquad \textcolor{red}{\text{Substituting 1000 m for 1 km}}$

$\phantom{1 \text{ km}^2} = 1{,}000{,}000 \text{ m}^2 \qquad \textcolor{red}{\text{Note that m} \cdot \text{m} = \text{m}^2.}$

EXAMPLE 7 Complete: $1 \text{ cm}^2 = \underline{\hspace{1cm}} \text{ m}^2.$

$1 \text{ cm}^2 = (1 \text{ cm})(1 \text{ cm})$

$\phantom{1 \text{ cm}^2} = \textcolor{red}{\left(\dfrac{1}{100}\text{ m}\right)\left(\dfrac{1}{100}\text{ m}\right)} \qquad \textcolor{red}{\text{Substituting } \dfrac{1}{100}\text{ m for 1 cm}}$

$\phantom{1 \text{ cm}^2} = \dfrac{1}{10{,}000} \text{ m}^2 \qquad \textcolor{red}{\text{Note that m} \cdot \text{m} = \text{m}^2.}$

$\phantom{1 \text{ cm}^2} = 0.0001 \text{ m}^2$

Complete.

6. $1 \text{ m}^2 = \underline{\hspace{1cm}} \text{ mm}^2$

7. $1 \text{ mm}^2 = \underline{\hspace{1cm}} \text{ cm}^2$

◀ **Do Exercises 6 and 7.**

Mental Conversion

Let's compare conversions of units of length to the conversions of units of area in Examples 6 and 7.

Converting units of length

1 km → m
1 km = 1000 m
Move the decimal point
3 places to the right.

1 cm → m
1 cm = 0.01 m
Move the decimal point
2 places to the left.

Converting units of area

1 km² → m²
1 km² = 1,000,000 m²
Move the decimal point
6 places to the right.

1 cm² → m²
1 cm² = 0.0001 m²
Move the decimal point
4 places to the left.

In general, a metric area conversion requires moving the decimal point twice as many places as the corresponding length conversion. We can use a table as before and multiply the number of places we move by 2 to determine the number of places to move the decimal point.

EXAMPLE 8 Complete: $3.48 \text{ cm}^2 = \underline{\hspace{1cm}} \text{ mm}^2$.

Think: To go from cm to mm in the table is a move of 1 place to the right.

1000 m	100 m	10 m	1 m	0.1 m	0.01 m	0.001 m
1 km	1 hm	1 dam	1 m	1 dm	1 cm	1 mm

1 move to the right

So we move the decimal point 2 · 1, or 2, places to the right.

3.48   3.48. $3.48 \text{ cm}^2 = 348 \text{ mm}^2$

2 places to the right

EXAMPLE 9 Complete: $586.78 \text{ mm}^2 = \underline{\hspace{1cm}} \text{ m}^2$.

Think: To go from mm to m in the table is a move of 3 places to the left.

1000 m	100 m	10 m	1 m	0.1 m	0.01 m	0.001 m
1 km	1 hm	1 dam	1 m	1 dm	1 cm	1 mm

3 moves to the left

So we move the decimal point 2 · 3, or 6, places to the left.

586.78   0.000586.78 $586.78 \text{ mm}^2 = 0.00058678 \text{ m}^2$

6 places to the left

Do Exercises 8–10. ▶

Complete.

8. $2.88 \text{ m}^2 = \underline{\hspace{1cm}} \text{ cm}^2$

9. $4.3 \text{ mm}^2 = \underline{\hspace{1cm}} \text{ cm}^2$

10. $678,000 \text{ m}^2 = \underline{\hspace{1cm}} \text{ km}^2$

Answers
8. 28,800   **9.** 0.043   **10.** 0.678

9.2 Exercise Set

✓ Reading Check

Determine whether each equation is true or false.

RC1. _____ $9 \text{ ft}^2 = 1 \text{ yd}^2$

RC2. _____ $1 \text{ mi}^2 = 640 \text{ ft}^2$

RC3. _____ $1 \text{ acre} = 43{,}560 \text{ ft}^2$

RC4. _____ $36 \text{ in}^2 = 1 \text{ yd}^2$

RC5. _____ $1 \text{ km}^2 = 1{,}000{,}000 \text{ m}^2$

RC6. _____ $1 \text{ ft}^2 = 144 \text{ in}^2$

a Complete.

1. $1 \text{ ft}^2 = $ _____ in^2

2. $1 \text{ yd}^2 = $ _____ ft^2

3. $1 \text{ mi}^2 = $ _____ acres

4. $1 \text{ acre} = $ _____ ft^2

5. $1 \text{ in}^2 = $ _____ ft^2

6. $1 \text{ ft}^2 = $ _____ yd^2

7. $5 \text{ yd}^2 = $ _____ ft^2

8. $4 \text{ ft}^2 = $ _____ in^2

9. $7 \text{ ft}^2 = $ _____ in^2

10. $2 \text{ acres} = $ _____ ft^2

11. $432 \text{ in}^2 = $ _____ ft^2

12. $54 \text{ ft}^2 = $ _____ yd^2

13. $22 \text{ yd}^2 = $ _____ ft^2

14. $40 \text{ ft}^2 = $ _____ in^2

15. $15 \text{ ft}^2 = $ _____ in^2

16. $144 \text{ ft}^2 = $ _____ yd^2

17. $20 \text{ mi}^2 = $ _____ acres

18. $576 \text{ in}^2 = $ _____ ft^2

19. $1 \text{ mi}^2 = $ _____ ft^2

20. $1 \text{ mi}^2 = $ _____ yd^2

21. $720 \text{ in}^2 = $ _____ ft^2

22. $27 \text{ ft}^2 = $ _____ yd^2

23. $3 \text{ ft}^2 = $ _____ yd^2

24. $72 \text{ in}^2 = $ _____ ft^2

25. $1 \text{ acre} = $ _____ mi^2

26. $4 \text{ acres} = $ _____ ft^2

27. $40.3 \text{ mi}^2 = $ _____ acres

28. $1080 \text{ in}^2 = $ _____ ft^2

29. $216 \text{ in}^2 = $ _____ ft^2

30. $69 \text{ ft}^2 = $ _____ yd^2

b Complete.

31. $19 \text{ km}^2 =$ _____ m^2

32. $39 \text{ km}^2 =$ _____ m^2

33. $6.31 \text{ m}^2 =$ _____ cm^2

34. $2.7 \text{ m}^2 =$ _____ mm^2

35. $6.5432 \text{ mm}^2 =$ _____ cm^2

36. $8.38 \text{ cm}^2 =$ _____ mm^2

37. $349 \text{ cm}^2 =$ _____ m^2

38. $125 \text{ mm}^2 =$ _____ m^2

39. $250{,}000 \text{ mm}^2 =$ _____ cm^2

40. $5900 \text{ mm}^2 =$ _____ cm^2

41. $472{,}800 \text{ m}^2 =$ _____ km^2

42. $1.37 \text{ cm}^2 =$ _____ mm^2

Skill Maintenance

In Exercises 43 and 44, find the simple interest.   [8.7a]

43. On $2000 at an interest rate of 8% for 1.5 years

44. On $2000 at an interest rate of 5.3% for 2 years

In each of Exercises 45–48, find **(a)** the amount of simple interest due and **(b)** the total amount that must be paid back.   [8.7a]

45. A firm borrows $15,500 at 9.5% for 120 days.

46. A firm borrows $8500 at 10% for 90 days.

47. A firm borrows $6400 at 8.4% for 150 days.

48. A firm borrows $4200 at 11% for 30 days.

Synthesis

Complete.

49. ▦ $1 \text{ m}^2 =$ _____ ft^2

50. ▦ $1 \text{ in}^2 =$ _____ cm^2

51. ▦ $2 \text{ yd}^2 =$ _____ m^2

52. ▦ $1 \text{ acre} =$ _____ m^2

53. *Aalsmeer Flower Auction.*   The fourth-largest building in the world in terms of floor space houses the Aalsmeer Flower Auction in Aalsmeer, Netherlands. It covers approximately $990{,}000 \text{ m}^2$. Each day, over 20 million flowers are sold there. Convert $990{,}000 \text{ m}^2$ to square feet.

Sources: www.amsterdamlogue.com; www.youTube.com/Aalsmeer, Netherlands, Flower Auction

54. *The Palazzo.*   The largest building in the United States in terms of floor space is the Palazzo, a hotel and casino on the Las Vegas Strip in Paradise, Nevada. It contains approximately $6{,}953{,}000 \text{ ft}^2$. Convert $6{,}953{,}000 \text{ ft}^2$ to square meters. Round the answer to the nearest thousand.

Source: *Top 10 of Everything 2013*

55. Janie's Rubik's Cube has 54 cm² of area. Each side of Norm's cube is twice as wide as Janie's. Find the area of Norm's cube.

56. 🔲 In order to remodel an office, a carpenter needs to purchase carpeting, at $8.45 a square yard, and molding for the base of the walls, at $0.87 a foot. If the room is 9 ft by 12 ft, with a 3-ft-wide doorway, what will the materials cost?

Find the area of the shaded region of each figure. Give the answer in square feet. (Figures are not drawn to scale.)

57.

3 in.

4 ft

58.

3 in.

5 yd

59.

13 ft

4 in.

60.

16 ft

10 in.

Find the area of the shaded region of each figure.

61.

Each small triangle has a height and a base of 2 cm.

12 cm

12 cm

62.

2.8 m

18 mm

2.8 m

18 mm

63. 🔲 A 30-ft by 60-ft ballroom is to be remodeled by placing an 18-ft by 42-ft dance floor in the middle and carpeting the rest of the room. The new dance floor is to be laid in tiles that are 8-in. by 8-in. squares. How many such tiles are needed? What percent of the area is the dance floor?

Parallelograms, Trapezoids, and Circles

We have already discussed how to find the perimeter of polygons and the area of squares, rectangles, and triangles. In this section, we find the area of *parallelograms*, *trapezoids*, and *circles*. We also calculate the perimeter, or *circumference*, of a circle.

a PARALLELOGRAMS AND TRAPEZOIDS

A **parallelogram** is a four-sided figure with two pairs of parallel sides, as shown below.

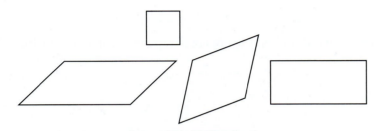

To understand how to find the area of a parallelogram, consider the one below.

If we cut off a piece and move it to the other end, we get a rectangle.

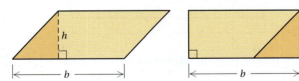

We can find the area by multiplying the length of a **base** b by h, the **height**.

AREA OF A PARALLELOGRAM

The **area of a parallelogram** is the product of the length of a base b and the height h:

$$A = b \cdot h.$$

EXAMPLE 1 Find the area of this parallelogram.

$$A = b \cdot h$$
$$= 7 \text{ km} \cdot 5 \text{ km}$$
$$= 35 \text{ km}^2$$

Answers

Skill to Review:
1. 52 ft   2. 300 cm

Find the area.

1.

6 cm

7.3 cm

2.

5.5 km

2.25 km

EXAMPLE 2 Find the area of this parallelogram.

$$A = b \cdot h$$
$$= (1.2\,\text{m}) \cdot (6\,\text{m})$$
$$= 7.2\,\text{m}^2$$

6 m

1.2 m

◀ **Do Exercises 1 and 2.**

A **trapezoid** is a polygon with four sides, two of which, the **bases**, are parallel to each other.*

To understand how to find the area of a trapezoid, consider the one below.

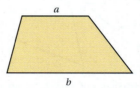

a

b

Think of cutting out another just like it and placing the second one like this:

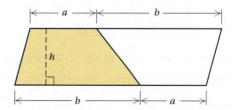

a   b

h

b   a

The resulting figure is a parallelogram with an area of

$$h \cdot (a + b).$$ The base of the parallelogram has length $a + b$.

The trapezoid we started with has half the area of the parallelogram, or

$$\frac{1}{2} \cdot h \cdot (a + b).$$

AREA OF A TRAPEZOID

The **area of a trapezoid** is half the product of the height and the sum of the lengths of the parallel sides, or the product of the height and the average length of the bases:

$$A = \frac{1}{2} \cdot h \cdot (a + b) = h \cdot \frac{a + b}{2}.$$

a

h

b

Answers

1. 43.8 cm² **2.** 12.375 km²

EXAMPLE 3 Find the area of this trapezoid.

$$A = \frac{1}{2} \cdot h \cdot (a + b)$$

$$= \frac{1}{2} \cdot 7 \text{ cm} \cdot (12 + 18) \text{ cm}$$

$$= \frac{7 \cdot 30}{2} \cdot \text{cm}^2 = \frac{7 \cdot 15 \cdot 2}{1 \cdot 2} \text{cm}^2$$

$$= 105 \text{ cm}^2 \qquad \text{Removing a factor equal to 1 } (\tfrac{2}{2} = 1) \text{ and multiplying}$$

Do Exercises 3 and 4. ▶

Find the area.

3.

4.

b CIRCLES

Radius and Diameter

At right is a circle with center O. Segment $\overline{AC}$ is a *diameter*. A **diameter** is a segment that passes through the center of the circle and has endpoints on the circle. Segment $\overline{OB}$ is called a *radius*. A **radius** is a segment with one endpoint on the center and the other endpoint on the circle. The words *radius* and *diameter* are also used to represent the lengths of a circle's radius and diameter, respectively.

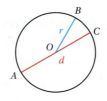

DIAMETER AND RADIUS

Suppose that d is the diameter of a circle and r is the radius. Then

$$d = 2 \cdot r \quad \text{and} \quad r = \frac{d}{2}.$$

EXAMPLE 4 Find the length of a radius of this circle.

$$r = \frac{d}{2}$$

$$= \frac{12 \text{ m}}{2}$$

$$= 6 \text{ m}$$

The radius is 6 m.

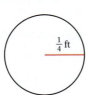

EXAMPLE 5 Find the length of a diameter of this circle.

$$d = 2 \cdot r$$

$$= 2 \cdot \frac{1}{4} \text{ ft}$$

$$= \frac{1}{2} \text{ ft}$$

The diameter is $\frac{1}{2}$ ft.

5. Find the length of a radius.

6. Find the length of a diameter.

Do Exercises 5 and 6. ▶

Answers

3. 100 m^2 **4.** 717.5 cm^2 **5.** 10 m   **6.** 5 ft

Circumference

The perimeter of a circle is called its **circumference**. The circumference of a circle depends on its diameter. Suppose that we take a dinner plate and measure the circumference C and diameter d with a tape measure. For the specific plate shown, we have

$$\frac{C}{d} = \frac{33.7 \text{ in.}}{10.75 \text{ in.}} \approx 3.1.$$

$d \approx 10.75$ in.

$C \approx 33.7$ in.

If we did this with plates and circles of several sizes, the result would always be a number close to 3.1. For any circle, if we divide the circumference C by the diameter d, we get the same number. We call this number π (pi). The number π is a nonterminating, nonrepeating decimal. It is impossible to precisely express π as a decimal or a fraction.

CIRCUMFERENCE AND DIAMETER

The circumference C of a circle of diameter d is given by

$$C = \pi \cdot d.$$

The number π is about 3.14, or about $\frac{22}{7}$.

EXAMPLE 6   Find the circumference of this circle. Use 3.14 for π.

$$\begin{aligned} C &= \pi \cdot d \\ &\approx 3.14 \cdot 6 \text{ cm} \\ &= 18.84 \text{ cm} \end{aligned}$$

The circumference is about 18.84 cm.

6 cm

◀ **Do Exercise 7.**

Since $d = 2 \cdot r$, where r is the length of a radius, it follows that

$$C = \pi \cdot d = \pi \cdot (2 \cdot r).$$

CIRCUMFERENCE AND RADIUS

The circumference C of a circle of radius r is given by

$$C = 2 \cdot \pi \cdot r.$$

EXAMPLE 7   Find the circumference of this circle. Use $\frac{22}{7}$ for π.

$$\begin{aligned} C &= 2 \cdot \pi \cdot r \\ &\approx 2 \cdot \frac{22}{7} \cdot 70 \text{ in.} \\ &= 2 \cdot 22 \cdot \frac{70}{7} \text{ in.} \\ &= 44 \cdot 10 \text{ in.} = 440 \text{ in.} \end{aligned}$$

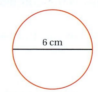

70 in.

The circumference is about 440 in.

---

7. Find the circumference of the circle. Use 3.14 for π.

18 in.

$$C = \pi \cdot d$$
$$\approx 3.14 \cdot \boxed{} \text{ in.}$$
$$= \boxed{} \text{ in.}$$

Answer

7. 56.52 in.

Guided Solution:
7. 18, 56.52

EXAMPLE 8   Find the perimeter of this figure. Use 3.14 for π.

3.2 km

8.5 km

We let $P=$ the perimeter. Note that the figure has three straight edges (line segments) and one curved edge. The curved edge is half the circumference of a circle of radius 3.2 km. The top and bottom straight edges have lengths of 8.5 km. The left edge is the same length as twice the radius of the semicircle on the right, or 6.4 km.

$$P = 2(8.5 \text{ km}) + 6.4 \text{ km} + \frac{1}{2} \cdot 2 \cdot \pi \cdot 3.2 \text{ km}$$

$$\approx 17 \text{ km} + 6.4 \text{ km} + 1 \cdot 3.14 \cdot 3.2 \text{ km}$$

$$= 23.4 \text{ km} + 10.048 \text{ km}$$

$$= 33.448 \text{ km}$$

The perimeter is about 33.448 km.

Do Exercises 8–10. ▶

Area

To understand how to find the area of a circle with radius r, think of cutting half a circular region into small slices and arranging them as shown below. Note that half of the circumference is half of $2\pi r$, or simply πr.

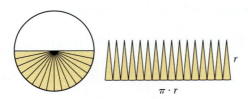

$\pi \cdot r$

Then imagine slicing the other half of the circular region and arranging the pieces in with the others as shown below.

$\pi \cdot r$

The thinner the slices, the closer this comes to being a rectangle. Its length is πr and its width is r, so its area is

$$(\pi \cdot r) \cdot r.$$

This is the area of a circle.

8. Find the circumference of this circle. Use 3.14 for π.

2.5 m

9. Find the circumference of this bicycle wheel. Use $\frac{22}{7}$ for π.

35 cm

10. Find the perimeter of this figure. Use 3.14 for π.

3.2 yd

7.1 yd

Answers

8. 15.7 m   **9.** 220 cm   **10.** 34.296 yd

AREA OF A CIRCLE

The **area of a circle** with radius of length r is given by

$$A = \pi \cdot r \cdot r, \quad \text{or} \quad A = \pi \cdot r^2.$$

11. Find the area of this circle. Use $\frac{22}{7}$ for π. **GS**

5 km

$A = \pi \cdot r \cdot r$
$\approx \frac{22}{7} \cdot 5 \text{ km} \cdot \boxed{} \text{ km}$
$= \frac{22}{7} \cdot 25 \text{ km}^2$
$= \frac{550}{7} \text{ km}^2$
$= \boxed{} \frac{4}{7} \text{ km}^2$

12. Find the area of this circle. Use 3.14 for π.

20.8 cm

EXAMPLE 9 Find the area of this circle. Use $\frac{22}{7}$ for π.

$A = \pi \cdot r^2 = \pi \cdot r \cdot r$

$\approx \frac{22}{7} \cdot 14 \text{ cm} \cdot 14 \text{ cm}$

$= \frac{22}{7} \cdot \frac{7 \cdot 2}{1} \text{ cm} \cdot 14 \text{ cm}$

$= 616 \text{ cm}^2 \quad \textit{Note: } r^2 \neq 2r.$

14 cm

The area is about 616 cm^2.

EXAMPLE 10 Find the area of this circle. Use 3.14 for π. Round to the nearest hundredth.

The diameter is 4.2 m; the radius is 4.2 m ÷ 2, or 2.1 m.

$A = \pi \cdot r \cdot r$

$\approx 3.14 \cdot 2.1 \text{ m} \cdot 2.1 \text{ m}$

$= 3.14 \cdot 4.41 \text{ m}^2$

$= 13.8474 \text{ m}^2 = 13.85 \text{ m}^2$

4.2 m

The area is about 13.85 m^2.

◀ **Do Exercises 11 and 12.**

We can often add or subtract areas to determine the area of a region.

EXAMPLE 11 Find the area of the shaded region.

1.4 ft
2.1 ft
4 ft

13. Find the area of the shaded region. Use 3.14 for π. (*Hint*: The figure consists of a square and two semicircles, or one complete circle.)

3 cm   3 cm

6 cm

The region consists of a triangle and a rectangle, as illustrated below. Note that the base of the triangle is the same as the length of the rectangle. We calculate the areas separately and add to find the area of the shaded region.

1.4 ft
4 ft
+
2.1 ft
4 ft

$\text{Area} = \frac{1}{2} \cdot 4 \text{ ft} \cdot 1.4 \text{ ft} + 4 \text{ ft} \cdot 2.1 \text{ ft}$

$= 2.8 \text{ ft}^2 + 8.4 \text{ ft}^2$

$= 11.2 \text{ ft}^2$

◀ **Do Exercise 13.**

Answers

11. $78\frac{4}{7} \text{ km}^2$ **12.** 339.6 cm^2 **13.** 64.26 cm^2

Guided Solution:

11. 5, 78

EXAMPLE 12 *Area of Cake Pans.* Tyler can make either a 9-in. round cake or a 9-in. square cake for a party. If he makes the square cake, how much more area will he have on the top for decorations?

9 in.

9 in.

9 in.

The area of the square is

$$A = s \cdot s$$
$$= 9 \text{ in.} \cdot 9 \text{ in.} = 81 \text{ in}^2.$$

The diameter of the circle is 9 in., so the radius is 9 in./2, or 4.5 in. The area of the circle is

$$A = \pi \cdot r \cdot r$$
$$\approx 3.14 \cdot 4.5 \text{ in.} \cdot 4.5 \text{ in.} = 63.585 \text{ in}^2.$$

The area of the top of the square cake is larger by about

$$81 \text{ in}^2 - 63.585 \text{ in}^2, \quad \text{or} \quad 17.415 \text{ in}^2.$$

Do Exercise 14. ▶

AREA AND CIRCUMFERENCE FORMULAS

Area of a parallelogram: $A = b \cdot h$

Area of a trapezoid: $A = \dfrac{1}{2} \cdot h \cdot (a + b)$

or $A = h \cdot \dfrac{a + b}{2}$

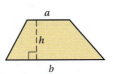

Circumference of a circle: $C = \pi \cdot d$

or $C = 2 \cdot \pi \cdot r$

Area of a circle: $A = \pi \cdot r^2$

Note: $d = 2r$ and $\pi \approx 3.14 \approx \frac{22}{7}$.

.... **Caution!** ....

Remember that circumference is always measured in linear units like ft, m, cm, yd, and so on. But area is measured in square units like ft², m², cm², yd², and so on.

....

14. Which is larger and by how much: a 10-ft-square flower bed or a 12-ft-diameter circular flower bed?

CALCULATOR CORNER

The $\boxed{\pi}$ Key   Many calculators have a $\boxed{\pi}$ key that can be used to enter the value of π in a computation. Results obtained using the $\boxed{\pi}$ key might be slightly different from those obtained when 3.14 or $\frac{22}{7}$ is used for the value of π in a computation.

When we use the $\boxed{\pi}$ key to find the circumference of the circle in Example 6, we get a result of approximately 18.85. Note that this is slightly different from the result found using 3.14 for the value of π.

EXERCISES

1. If you have a $\boxed{\pi}$ key on your calculator, to how many decimal places does this key give the value of π?

2. Find the circumference and the area of a circle with a radius of 225.68 in.

3. Find the area of a circle with a diameter of $46\frac{12}{13}$ in.

4. Find the area of a large irrigated farming circle with a diameter of 400 ft.

Answer

14. A 12-ft-diameter flower bed is 13.04 ft² larger.

✓ **Reading Check**

Complete each statement with the correct word from the following list. A word may be used more than once or not at all.

area   circumference   diameter   radius

RC1. The _____ of a circle is half the length of its diameter.

RC2. The _____ of a circle is found by multiplying its diameter by π.

RC3. The _____ of a circle is found by multiplying its radius by 2π.

RC4. The _____ of a circle is found by multiplying the square of its radius by π.

a Find the area of each parallelogram or trapezoid.

1.

5 cm · 10 cm

2.

4 cm · 4 cm

3.

6 ft · 8 ft · 20 ft

4.

5 yd · 4 yd · 10 yd

5.

11 m · 8 m · 15 m

6.
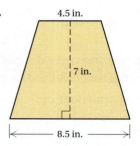
4.5 in. · 7 in. · 8.5 in.

7.

10.5 cm · 6.9 cm

8.

9 cm · 18 cm · 24 cm

9.

13 mi · 9 mi · 19 mi

10.

5.2 ft · 8 ft

11.

9 m · 7 m · 5 m

12.

14 in. · 17 in.

13.

$4\frac{1}{2}$ ft

$12\frac{1}{4}$ ft

14.

16 cm

35 cm

25 cm

15.

1.2 cm

1.6 cm

16.

6.2 m

9.1 m

17.

9.1 yd

8 yd

7.9 yd

18.

9 ft

13 ft

3 ft

b Find the length of a diameter of each circle.

19.

7 cm

20.

8 m

21.

$\frac{7}{8}$ in.

22.

$8\frac{2}{3}$ mi

Find the length of a radius of each circle.

23.

20 ft

24.

10 in.

25.

1.4 cm

26.

60.9 km

Find the circumference of each circle in Exercises 19–22. Use $\frac{22}{7}$ for π.

27. Exercise 19   **28.** Exercise 20   **29.** Exercise 21   **30.** Exercise 22

Find the circumference of each circle in Exercises 23–26. Use 3.14 for π.

31. Exercise 23   **32.** Exercise 24   **33.** Exercise 25   **34.** Exercise 26

Find the area of each circle in Exercises 19–22. Use $\frac{22}{7}$ for π.

35. Exercise 19   **36.** Exercise 20   **37.** Exercise 21   **38.** Exercise 22

Find the area of each circle in Exercises 23–26. Use 3.14 for π.

39. Exercise 23   **40.** Exercise 24   **41.** Exercise 25   **42.** Exercise 26

Solve. Use 3.14 for π unless otherwise specified.

43. *Pond Edging.*   Quiet Designs plans to incorporate a circular pond with a diameter of 30 ft in a landscape design. The pond will be edged using stone pavers. How many feet of pavers will be needed?

44. *Gypsy-Moth Tape.*   To protect an elm tree in your backyard, you decide to attach gypsy moth caterpillar tape around the trunk. The tree has a 1.1-ft diameter. What length of tape is needed?

45. *Areas of Pizza Pans.*   How much larger is a pizza made in a 16-in.-square pizza pan than a pizza made in a 16-in.-diameter circular pan?

16 in.

16 in.

16 in.

46. *Dimensions of a Penny.*   A penny has a 1-cm radius. What is its diameter? its circumference? its area?

1 cm

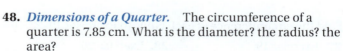

47. *Earth.*   The circumference of the earth at the equator is 24,901 mi. What is the diameter of the earth at the equator? the radius?

48. *Dimensions of a Quarter.*   The circumference of a quarter is 7.85 cm. What is the diameter? the radius? the area?

49. *Oil Spill.*   The radius of an oil slick is 150 mi. How much area is covered by the slick?

50. *Radio.*   A college radio station is allowed by the FCC to broadcast over an area with a radius of 7 mi. How much area is this?

51. *Masonry.* The Harris-Regency Hotel plans to install a 1-yd-wide walk around a circular swimming pool. The diameter of the pool is 20 yd. What is the area of the walk?

20 yd

1 yd

52. *Trampoline.* The standard backyard trampoline has a diameter of 14 ft. What is its area?

Source: International Trampoline Industry Association, Inc.

14 ft

Frame height: 36 in.

53. *Circumference of a Baseball Bat.* In Major League Baseball, the diameter of the barrel of a bat cannot be more than $2\frac{3}{4}$ in., and the diameter of the bat handle cannot be less than $\frac{16}{19}$ in. Find the maximum circumference of the barrel of a bat and the minimum circumference of the bat handle. Use $\frac{22}{7}$ for π.

Source: Major League Baseball

Maximum diameter of barrel of bat: $2\frac{3}{4}$ in.

Minimum diameter of bat handle: $\frac{16}{19}$ in.

54. *Roller-Rink Floor.* A roller-rink floor is shown below. Each end is a semicircle. What is the total area? If hardwood flooring costs $62.50 per square meter, how much will the flooring cost?

20 m

7 m

Find the perimeter of each figure. Use 3.14 for π.

55.

8 ft

8 ft

56.

8 in.   8 in.   8 in.   8 in.

57.

10 yd

10 yd

58.

12.8 cm

10.2 cm

Find the area of the shaded region in each figure. Use 3.14 for π.

59.

8 m

60.

10 yd

10 yd

61.

2.8 cm

2.8 cm

62.

8 km

8 km

63.

14.6 in.

11.4 in.

64.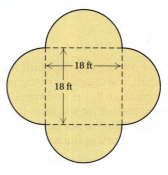

18 ft

18 ft

65.

15 cm

30 cm

30 cm

66.

3 in. 3 in. 2 in.

6 in.

4 in.

2 in.

12 in.

67.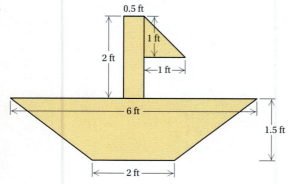

0.5 ft

1 ft

2 ft

1 ft

6 ft

1.5 ft

2 ft

68.

14 m

9 m

7 m

14 m

Skill Maintenance

Convert to fraction notation. [8.1c]

69. 9.25%

70. $87\frac{1}{2}\%$

Convert to percent notation. [8.1c]

71. $\frac{5}{4}$

72. $\frac{2}{3}$

73. The weight of a human brain is 2.5% of total body weight. A person weighs 200 lb. What does his brain weigh? [8.4a]

74. Jack's commission is increased according to how much he sells. He receives a commission of 6% for the first $3000 and 10% on the amount over $3000. What is the total commission on sales of $8500? [8.6b]

Synthesis

75. ▦ Calculate the surface area of an unopened steel can that has a height of 3.5 in. and a diameter of 2.5 in. (*Hint*: Make a sketch and "unroll" the sides of the can.) Use 3.14 for π.

76. ▦ The distance from Kansas City to Indianapolis is 500 mi. A car was driven this distance using tires with a radius of 14 in. How many revolutions of each tire occurred on the trip? Use $\frac{22}{7}$ for π.

77. *Sports Marketing.* Tennis balls are often packed vertically, three in a can, one on top of another. Without using a calculator, determine the larger measurement: the can's circumference or the can's height.

78. ▦ The sides of a cake box are trapezoidal, as shown in the figure. Determine the surface area of the box.

79. The radius of one circle is twice another circle's radius. If the area of the smaller circle is A, what is the area of the larger circle?

80. The radius of one circle is twice another circle's radius. If the circumference of the smaller circle is C, what is the circumference of the larger circle?

81. A cake recipe calls for a 9-inch round baking pan. If only square pans are available, what size pan should be used for the cake? Round to the nearest inch.

82. Peggy fenced a square garden with sides of 12 ft. If she decides to use the fence around a circular garden instead, what size garden can she fence? Which garden has the greater area?

Volume and Capacity

OBJECTIVES

a Find the volume of a rectangular solid, a cylinder, and a sphere.

b Convert from one unit of capacity to another.

c Solve applied problems involving volume and capacity.

SKILL TO REVIEW

Objective 9.3b: Find the circumference, area, radius, or diameter of a circle, given the length of a radius or a diameter.

1. Find the area of a circle whose radius is 21 in. Use $\frac{22}{7}$ for π.

2. Find the area of a circle whose diameter is $2\frac{4}{5}$ ft. Use $\frac{22}{7}$ for π.

1. Find the volume.

Answers

Skill to Review:

1. 1386 in^2 2. $6\frac{4}{25} \text{ ft}^2$, or 6.16 ft^2

Margin Exercise:

1. 12 cm^3

◀ **a** VOLUME

The **volume** of an object is the number of unit cubes needed to fill it.

Unit cube

Volume = 18 unit cubes

Two unit cubes commonly used to measure volume are shown below (actual size).

1 in.

1 in.

1 in.

1 cubic inch (in^3)

1 cm

1 cm

1 cm

1 cubic centimeter (cm^3)

EXAMPLE 1 Find the volume.

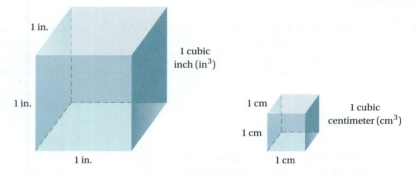

2 cm

3 cm

4 cm

The figure is made up of 2 layers of 12 cubes each, so its volume is 24 cubic centimeters (cm^3). Note that $24 = 4 \cdot 3 \cdot 2$.

◀ **Do Margin Exercise 1.**

VOLUME OF A RECTANGULAR SOLID

The **volume of a rectangular solid** is found by multiplying length by width by height:

$$V = \underbrace{l \cdot w}_{\text{Area of base}} \cdot \underbrace{h}_{\text{Height}}.$$

h

w

l

EXAMPLE 2 *Volume of a Safety Deposit Box.* Tricia rents a safety deposit box at her bank. The dimensions of the box are 18 in. × 10.5 in. × 5 in. Find the volume of this rectangular solid.

$$V = l \cdot w \cdot h$$
$$= 18 \text{ in.} \cdot 10.5 \text{ in.} \cdot 5 \text{ in.}$$
$$= 945 \text{ in}^3$$

Do Exercises 2 and 3. ▶

Volumes are described in units such as cubic centimeters (cm³) and cubic inches (in³). *Dimensional analysis* is an excellent way of determining the correct units for an answer.

- If measurements of length are added, use a one-dimensional unit of length: 3 ft + 2 ft + 7 ft = 12 ft.
- If two measurements of length are multiplied, use a two-dimensional unit of area: 8 ft · 7 ft = 56 ft².
- If three measurements of length are multiplied, use a three-dimensional unit of volume: 3 m · 2 m · 4 m = 24 m³.

A rectangular solid is shown below. Note that we can think of the volume as the product of the area of the base times the height:

$$V = l \cdot w \cdot h$$
$$= (l \cdot w) \cdot h$$
$$= (\text{Area of the base}) \cdot h$$
$$= B \cdot h,$$

Area of base $= B = l \cdot w$

where *B* represents the area of the base.

Like rectangular solids, **circular cylinders** have bases of equal area that lie in parallel planes. The bases of circular cylinders are circular regions.

The volume of a circular cylinder is found in a manner similar to the way the volume of a rectangular solid is found. The volume is the product of the area of the base times the height. The height is always measured perpendicular to the base.

2. Carry-on Luggage. A carry-on bag measures 23 in. by 10 in. by 13 in. Find the volume of this solid.

3. Cord of Wood. A cord of wood measures 4 ft by 4 ft by 8 ft. What is the volume of a cord of wood?

Answers

2. 2990 in³   **3.** 128 ft³

4. Find the volume of the cylinder. Use 3.14 for π.

10 ft

5 ft

$V = \pi \cdot r^2 \cdot h$

$\approx 3.14 \cdot 5\,\text{ft} \cdot 5\,\text{ft} \cdot \boxed{}\,\text{ft}$

$= 3.14 \cdot 250\,\text{ft}^3$

$= \boxed{}\,\text{ft}^3$

5. Find the volume of the cylinder. Use $\frac{22}{7}$ for π.

49 m

21 m

6. Find the volume of the sphere. Use $\frac{22}{7}$ for π.

28 ft

$V = \frac{4}{3} \cdot \pi \cdot r^3$

$\approx \frac{4}{3} \cdot \frac{22}{7} \cdot (\boxed{}\,\text{ft})^3$

$= \frac{4}{3} \cdot \frac{22}{7} \cdot \boxed{}\,\text{ft}^3$

$= \frac{275{,}968}{3}\,\text{ft}^3$

$= \boxed{}\tfrac{1}{3}\,\text{ft}^3$

7. The radius of a standard-sized golf ball is 2.1 cm. Find its volume. Use 3.14 for π.

Answers

4. 785 ft³   **5.** 67,914 m³   **6.** 91,989 $\frac{1}{3}$ ft³

7. 38.77272 cm³

Guided Solutions:
4. 10, 785   **6.** 28, 21,952, 91,989

VOLUME OF A CIRCULAR CYLINDER

The **volume of a circular cylinder** is the product of the area of the base B and the height h:

$$V = B \cdot h, \quad \text{or} \quad V = \pi \cdot r^2 \cdot h.$$

EXAMPLE 3   Find the volume of this circular cylinder. Use 3.14 for π.

$V = B \cdot h = \pi \cdot r^2 \cdot h$

$\approx 3.14 \cdot 4\,\text{cm} \cdot 4\,\text{cm} \cdot 12\,\text{cm}$

$= 602.88\,\text{cm}^3$

12 cm

4 cm

◀ **Do Exercises 4 and 5.**

A **sphere** is the three-dimensional counterpart of a circle. It is the set of all points in space that are a given distance (the radius) from a given point (the center). The volume of a sphere depends on its radius.

VOLUME OF A SPHERE

The **volume of a sphere** of radius r is given by

$$V = \frac{4}{3} \cdot \pi \cdot r^3.$$

r

EXAMPLE 4   *Bowling Ball.*   The radius of a standard-sized bowling ball is 4.2915 in. Find the volume of a standard-sized bowling ball (disregarding the finger holes). Round to the nearest hundredth of a cubic inch. Use 3.14 for π.

$r = 4.2915$ in.

9   10

We have

$V = \frac{4}{3} \cdot \pi \cdot r^3 \approx \frac{4}{3} \cdot 3.14 \cdot (4.2915\,\text{in.})^3$

$\approx 330.90\,\text{in}^3.$ Using a calculator

◀ **Do Exercises 6 and 7.**

b CAPACITY

American Units

To answer a question like "How much soda is in the bottle?" we need measures of **capacity**. American units of capacity are fluid ounces, cups, pints, quarts, and gallons. These units are related as follows.

AMERICAN UNITS OF CAPACITY	
1 gallon (gal) = 4 quarts (qt)	1 pt = 2 cups = 16 fluid ounces (fl oz)
1 qt = 2 pints (pt)	1 cup = 8 fluid oz

Fluid ounces, abbreviated fl oz, are often referred to as ounces, or oz.

EXAMPLE 5 Complete: 24 qt = _____ gal.

Since we are converting from a *smaller* unit to a *larger* unit, we multiply by 1 using 1 gal in the numerator and 4 qt in the denominator:

$$24 \text{ qt} = 24 \text{ qt} \cdot \frac{1 \text{ gal}}{4 \text{ qt}} = \frac{24}{4} \cdot 1 \text{ gal} = 6 \text{ gal.}$$

EXAMPLE 6 Complete: 9 gal = _____ oz.

The box above does not list how many ounces are in 1 gal. We convert gallons to quarts, quarts to pints, and pints to ounces, using the relationships given in the box.

Since we are converting from a *larger* unit to a *smaller* unit, we use substitution:

$9 \text{ gal} = 9 \cdot 1 \text{ gal} = 9 \cdot 4 \text{ qt} = 36 \text{ qt}$ Substituting 4 qt for 1 gal
$= 36 \cdot 1 \text{ qt} = 36 \cdot 2 \text{ pt} = 72 \text{ pt}$ Substituting 2 pt for 1 qt
$= 72 \cdot 1 \text{ pt} = 72 \cdot 16 \text{ oz} = 1152 \text{ oz.}$ Substituting 16 oz for 1 pt

Thus, 9 gal = 1152 oz.

Do Exercises 8 and 9. ▶

Complete.

GS **8.** 80 qt = _____ gal

$$80 \text{ qt} = 80 \text{ qt} \cdot \frac{1 \text{ gal}}{\boxed{}}$$

$$= \frac{80}{\boxed{}} \cdot 1 \text{ gal}$$

$$= \boxed{} \text{ gal}$$

9. 5 gal = _____ pt

Metric Units

One unit of capacity in the metric system is a **liter**. A liter is just a bit more than a quart. It is defined as follows.

1 liter ≈ 1.06 quarts

1 liter   1 quart

METRIC UNITS OF CAPACITY

> 1 liter (L) $=$ 1000 cubic centimeters (1000 cm^3)
>
> The script letter ℓ is also used for "liter."

The metric prefixes are also used with liters. The most common is **milli-**. The milliliter (mL) is, then, $\frac{1}{1000}$ liter.

> 1 L $=$ 1000 mL $=$ 1000 cm^3;
>
> 0.001 L $=$ 1 mL $=$ 1 cm^3.

A common unit for drug dosage is the milliliter (mL) or the cubic centimeter (cm^3). The notation "cc" is also used for cubic centimeter, especially in medicine. The milliliter and the cubic centimeter represent the same measure of capacity. A milliliter is about $\frac{1}{5}$ of a teaspoon.

5 mL

3 cm^3

> 1 mL $=$ 1 cm^3 $=$ 1 cc

Volumes for which quarts and gallons are used are expressed in liters. Large volumes are expressed using measures of cubic meters (m^3).

◀ **Do Exercises 10–13.**

EXAMPLE 7 Complete: 4.5 L $=$ _____ mL.

$$4.5 \text{ L} = 4.5 \cdot 1 \text{ L} = 4.5 \cdot 1000 \text{ mL} \qquad \text{Substituting 1000 mL for 1 L}$$
$$= 4500 \text{ mL}$$

We can also convert units of capacity by multiplying by a form of 1.

EXAMPLE 8 Complete: 280 mL $=$ _____ L.

$$280 \text{ mL} = 280 \text{ mL} \cdot \frac{1 \text{ L}}{1000 \text{ mL}}$$
$$= \frac{280}{1000} \text{ L}$$
$$= 0.28 \text{ L}$$

◀ **Do Exercises 14 and 15.**

Complete with mL or L.

10. The patient received an injection of 2 _____ of penicillin.

11. There are 250 _____ in a coffee cup.

12. The gas tank holds 80 _____.

13. Bring home 8 _____ of milk.

Complete.

14. 0.97 L $=$ _____ mL   **GS**

0.97 L $=$ 0.97 · 1 L

$= 0.97 \cdot$ ☐ mL

$=$ ☐ mL

15. 8990 mL $=$ _____ L

Answers

10. mL   **11.** mL   **12.** L   **13.** L   **14.** 970
15. 8.99

Guided Solution:
14. 1000, 970

c SOLVING APPLIED PROBLEMS

EXAMPLE 9 At a self-service gas station, 89-octane gasoline sells for 102.6¢ a liter. Estimate the price of 1 gal in dollars.

Since 1 liter is about 1 quart and there are 4 quarts in a gallon, the price of a gallon is about 4 times the price of a liter.

$$4 \cdot 102.6¢ = 410.4¢ = \$4.104$$

Thus, 89-octane gasoline sells for about $4.10 a gallon.

Do Exercise 16. ▶

EXAMPLE 10 *Propane Gas Tank.* A propane gas tank is shaped like a circular cylinder with half of a sphere at each end. Find the volume of the tank if the cylindrical section is 5 ft long with a 4-ft diameter. Use 3.14 for π.

1. Familiarize. We first make a drawing.

2. Translate. This is a two-step problem. We first find the volume of the cylindrical portion. Then we find the volume of the two ends and add. Note that the radius is 2 ft and that together the two ends make a sphere. We let $V =$ the total volume.

Total volume	is	Volume of the cylinder	plus	Volume of the sphere
		$\pi \cdot r^2 \cdot h$		$\frac{4}{3} \cdot \pi \cdot r^3$
V	$=$	$3.14 \cdot (2 \text{ ft})^2 \cdot 5 \text{ ft}$	$+$	$\frac{4}{3} \cdot 3.14 \cdot (2 \text{ ft})^3$

3. Solve. The volume of the cylinder is approximately

$$3.14 \cdot (2 \text{ ft})^2 \cdot 5 \text{ ft} = 3.14 \cdot 2 \text{ ft} \cdot 2 \text{ ft} \cdot 5 \text{ ft}$$
$$= 62.8 \text{ ft}^3.$$

The volume of the two ends is approximately

$$\frac{4}{3} \cdot 3.14 \cdot (2 \text{ ft})^3 = \frac{4}{3} \cdot 3.14 \cdot 2 \text{ ft} \cdot 2 \text{ ft} \cdot 2 \text{ ft}$$
$$\approx 33.5 \text{ ft}^3.$$

The total volume is approximately

$$62.8 \text{ ft}^3 + 33.5 \text{ ft}^3 = 96.3 \text{ ft}^3.$$

4. Check. We can repeat the calculations. The answer checks.

5. State. The volume of the tank is about 96.3 ft³.

Do Exercise 17. ▶

---

16. At a gas station, the price of 87-octane gasoline is 96.7 cents a liter. Estimate the price of 1 gal in dollars.

CALCULATOR CORNER

Volumes Using Pi Many calculators have a $\boxed{\pi}$ key that can be used to give a more precise value of π than 3.14.

We we use a $\boxed{\pi}$ key to find the volume of the circular cylinder in Example 3, the result is approximately 603.19. Note that this is slightly different from the result found using 3.14 for π.

EXERCISES

1. Use a calculator with a $\boxed{\pi}$ key to find the volume of a cylinder with radius 1.2 cm and height 0.25 cm.

2. Use a calculator with a $\boxed{\pi}$ key to find the volume of a sphere with radius $\frac{3}{8}$ in.

---

17. *Medicine Capsule.* A cold capsule is 8 mm long and 4 mm in diameter. Find the volume of the capsule. Use 3.14 for π. (*Hint*: First find the length of the cylindrical section.)

Answers

16. $3.87   **17.** 83.7$\overline{3}$ mm³

☑ **Reading Check**

Match each formula with the correct phrase from the list on the right.

RC1. _____ $V = l \cdot w \cdot h$

RC2. _____ $V = \pi \cdot r^2 \cdot h$

RC3. _____ $V = \dfrac{4}{3} \cdot \pi \cdot r^3$

a) the volume of a cylinder

b) the volume of a rectangular solid

c) the volume of a sphere

a   Find each volume. Use 3.14 for π in Exercises 9–12. Use $\dfrac{22}{7}$ for π in Exercises 13 and 14.

1.

10 cm   5 cm   5 cm

2.

4 ft   4 ft   4 ft

3.
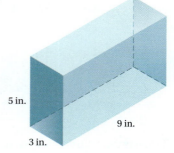
5 in.   9 in.   3 in.

4.

8 m   1.5 m   6 m

5.

10 m   1.5 m   5 m

6.

5 cm   5 cm   2.04 cm

7.

$5\frac{1}{2}$ yd $6\frac{1}{2}$ yd   10 yd

8.

$1\frac{1}{2}$ ft $\frac{1}{2}$ ft $2\frac{1}{4}$ ft

9.

13 ft   10 ft

10.
4 in.
8 in.

11.
7.5 cm
4 cm

12.
15.1 m
3 m

13.
300 yd
210 yd

14.
28 km
4 km

Find each volume. Use 3.14 for π in Exercises 15–18. Use $\frac{22}{7}$ for π in Exercises 19 and 20.

15.
$r = 100$ in.

16.
$r = 200$ ft

17.
$r = 3.1$ m

18.
$r = 15.2$ cm

19.
$r = 7$ km

20.
$r = 2.1$ m

b Complete.

21. 1 L = _____ mL = _____ cm^3

22. _____ L = 1 mL = _____ cm^3

23. 59 L = _____ mL

24. 714 L = _____ mL

25. 49 mL = _____ L

26. 43 mL = _____ L

27. 27.3 L = _____ cm^3

28. 49.2 L = _____ cm^3

29. 5 gal = _____ pt

30. 48 oz = _____ pt

31. 10 qt = _____ oz

32. 2 gal = _____ cups

33. 24 oz = _____ cups

34. 20 cups = _____ pt

35. 10 gal = _____ qt

36. 5 gal = _____ cups

37. 3 gal = _____ cups

38. 72 oz = _____ cups

39. 15 pt = _____ gal

40. 9 qt = _____ gal

c Solve.

41. *Oak Log.* An oak log has a diameter of 12 cm and a length (height) of 42 cm. Find the volume. Use 3.14 for π.

42. *Ladder Rung.* A rung of a ladder is 2 in. in diameter and 16 in. long. Find the volume. Use 3.14 for π.

43. *Architecture.* The largest sphere in the Oriental Pearl TV tower in Shanghai, China, measures 50 m in diameter. Find the volume of the sphere. Use 3.14 for π, and round to the nearest cubic meter.

Source: www.emporis.com

44. *Culinary Arts.* Raena often makes individual soufflés in cylindrical baking dishes called *ramekins*. The diameter of each ramekin is 3.5 in., and the height is 1.75 in. Find the approximate volume of a ramekin. Use 3.14 for π.

45. *Volume of a Trash Can.* The diameter of the base of a cylindrical trash can is 0.7 yd. The height is 1.1 yd. Find the volume. Use 3.14 for π and round the answer to the nearest hundredth.

46. *Tennis Ball.* The diameter of a tennis ball is 6.5 cm. Find the volume. Use 3.14 for π and round the answer to the nearest hundredth.

47. 🖩 *Volume of Earth.* The radius of the earth is about 3980 mi. Find the volume of the earth. Use 3.14 for π. Round to the nearest ten thousand cubic miles.

Source: *The Cambridge Factfinder,* 4th ed.

48. 🖩 *Astronomy.* The radius of the largest moon of Uranus is about 789 km. Find the volume of this satellite. Use $\frac{22}{7}$ for π. Round to the nearest ten thousand cubic kilometers.

49. *Oceanography.* The Deep-ESP (Environmental Sample Processor) is designed to study deep-sea environments. Its titanium pressure housing allows the ESP sphere to descend over 2 miles below the ocean surface. If the diameter of the sphere is 40 in., what is its volume? Use 3.14 for π and round to the nearest cubic inch.

Source: "Extreme Life," by Henry Bortman, 06/15/09, on www.astrobio.net

50. *Weather Forecasting.* Every day, the National Weather Service launches spherical weather balloons from 100 locations in the United States. Each balloon can rise up to 115,000 ft and travel up to 200 miles. Find the volume of a 6-ft-diameter weather balloon. Use 3.14 for π.

Source: Kaysam Worldwide, Inc.

51. 🔲 A sphere with diameter 1 m is circumscribed by a cube. How much greater is the volume of the cube than the volume of the sphere? Use 3.14 for π.

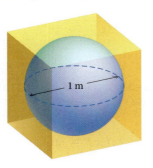

52. 🔲 *Golf-Ball Packaging.* The box shown is just big enough to hold 3 golf balls. If the radius of a golf ball is 2.1 cm, how much air surrounds the three balls? Use 3.14 for π.

53. *Architecture.* The Westhafen Tower in Frankfort, Germany, is a cylindrical building with a height of 110 m and a radius of 21 m. Find the volume of the building. Use 3.14 for π. Round to the nearest cubic meter.

54. *Gas Pipeline.* The 638-mi Rockies Express-East gas pipeline from Colorado to Ohio is constructed with 80-ft sections of steel pipe with a radius of 21 in., or $1\frac{3}{4}$ ft. Find the volume of one section. Use $\frac{22}{7}$ for π.

Source: Rockies Express Pipeline

55. *Metallurgy.* If all the gold in the world could be gathered together, it would form a cube 18 yd on a side. Find the volume of the world's gold.

56. 🔲 The width of a dollar bill is 2.3125 in., the length is 6.0625 in., and the thickness is 0.0041 in. Find the volume occupied by 1 million one-dollar bills.

57. *Toys.* Toys stores often sell capsules that dissolve in water allowing a toy inside the capsule to expand. One such capsule is 40 mm long with a diameter of 8 mm.

  a) What is the volume of the capsule? Use 3.14 for π.

  b) The manufacturer claims that the toy in the capsule will expand 600%. What is the volume of the toy after expansion?

58. *Silo.* A silo is a circular cylinder with half a sphere on the top of the cylinder. If the diameter of the cylinder is 6 m and the height of the cylinder is 13 m, find the volume of the silo. Use 3.14 for π.

Skill Maintenance

59. Find the simple interest on $600 at 8% for $\frac{1}{2}$ yr.   [8.7a]

60. Find the simple interest on $5000 at 7% for $\frac{1}{2}$ yr.   [8.7a]

61. If 9 pens cost $8.01, how much would 12 pens cost?   [7.4a]

62. Solve: $9(x - 1) = 3x + 5$.   [5.7b]

63. Solve: $-5y + 3 = -12y - 4$.   [5.7b]

64. A barge travels 320 km in 15 days. At this rate, how far will it travel in 21 days?   [7.4a]

65. Evaluate $\frac{9}{5}C + 32$ for $C = 15$.   [4.7c]

66. Evaluate $\frac{5}{9}(F - 32)$ for $F = 50$.   [4.7c]

Synthesis

67. 🖩 *Truck Rental.*   The storage compartment of a moving truck is 9.83 ft by 5.67 ft by 5.83 ft, with an "attic" measuring 1.5 ft by 5.67 ft by 2.5 ft. Find the total volume of the compartment.

68. *Tennis-Ball Packaging.*   Tennis balls are generally packaged in circular cylinders that hold 3 balls each. The diameter of a tennis ball is 6.5 cm. If the tennis balls were "liquid" and could be poured into the cylinder, how many tennis balls would the cylinder hold?

69. The volume of a ball is 36π cm³. Find the dimensions of a rectangular box that is just large enough to hold the ball.

70. 🖩 The volume of a basketball is 2304π cm³. Find the volume of a cube-shaped box that is just large enough to hold the ball.

71. 🖩 A 2-cm-wide stream of water passes through a 30-m-long garden hose. At the instant that the water is turned off, how many liters of water are in the hose? Use 3.141593 for π.

72. *Remarkable Feat.*   In 1982, Larry Walters captured the world's imagination by riding a lawn chair attached to 42 helium-filled weather balloons to an altitude of 16,000 ft, before using a BB gun to pop a few balloons and safely descend. Walters used balloons measuring approximately 7 ft in diameter. Find the total volume of the balloons used. Use $\frac{22}{7}$ for π.

Source: MarkBarry.com

73. *Conservation.*   Many people leave the water running while brushing their teeth. Suppose that one person wastes 32 oz of water in such a manner each day. How much water, in gallons, would that person waste in a week? in 30 days? in a year? If each of 310 million Americans wastes water this way, estimate how much water is wasted in a year.

74. The volumes occupied by all the world's gold and by a million dollar bills are described in Exercises 55 and 56. How many million dollar bills would fit into the volume of all the world's gold?

Mid-Chapter Review

Concept Reinforcement

Determine whether each statement is true or false.

_____ **1.** Distances that are measured in miles in the American system would probably be measured in meters in the metric system. [9.1c]

_____ **2.** One meter is slightly more than one yard. [9.1c]

_____ **3.** One kilometer is longer than one mile. [9.1c]

_____ **4.** The area of a parallelogram with base 8 cm and height 5 cm is the same as the area of a rectangle with length 8 cm and width 5 cm. [9.3a]

_____ **5.** The exact value of the ratio C/d is π. [9.3b]

Guided Solutions

 Fill in each blank with the number that creates a correct solution.

6. Complete: $16\frac{2}{3}$ yd = _____ ft. [9.1a]

$$16\frac{2}{3}\text{yd} = 16\frac{2}{3} \cdot 1\boxed{} = \frac{50}{3} \cdot 3\boxed{} = 50\boxed{}$$

7. Complete: 10,200 mm = _____ ft. [9.1c]

$$10{,}200\text{ mm} = 10{,}200\text{ mm} \cdot \frac{1}{1000\boxed{}} \approx 10.2\boxed{} \cdot \frac{3.281\boxed{}}{1} = 33.4662\boxed{}$$

8. Find the circumference and the area. Use 3.14 for π. [9.3b]

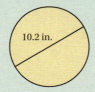

10.2 in.

$C = \pi \cdot d$

$C \approx \boxed{} \cdot \boxed{}$ in.

$C \approx \boxed{}$ in.

$A = \pi \cdot r \cdot r$

$A \approx \boxed{} \cdot \boxed{}$ in. $\cdot \boxed{}$ in.

$A \approx \boxed{}$ in$^{\boxed{}}$

Mixed Review

Complete.   [9.1a, b, c], [9.2a, b]

9. $5\frac{1}{2}$ mi = _____ yd

10. 840 in. = _____ ft

11. 24.05 cm = _____ dm

12. 0.15 m = _____ km

13. 3753 ft = _____ yd

14. 26,400 ft = _____ mi

15. 1800 m = _____ cm

16. 0.007 km = _____ cm

17. 2.5 yd = _____ m

18. 36 m = _____ ft

19. 6 in. = _____ cm

20. 100 mi = _____ km

21. 5 ft^2 = _____ in^2

22. 10 yd^2 = _____ ft^2

23. 60 ft^2 = _____ yd^2

24. 48 in^2 = _____ ft^2

25. 2 acres = _____ ft^2

26. 1.5 mi^2 = _____ acres

27. Arrange from smallest to largest:

100 in.,　430 ft,　$\frac{1}{100}$ mi,　3.5 ft,　6000 ft,　1000 in.,　2 yd.　[9.1a]

28. Arrange from largest to smallest:

3240 cm,　300 m,　250 dm,　150 hm,　33,000 mm,　310 dam,　13 km.　[9.1b]

Find the area.　[9.3a]

29.

20 in.
40 in.

30.

9 km
6 km
13 km

Find the circumference and the area. Use 3.14 for π.　[9.3b]

31.

7 in.

32.

8.6 cm

Find the volume. Use 3.14 for π in Exercises 34–36.　[9.4a]

33.

1.5 cm
5 cm
4 cm

34.

15 ft
6 ft

35.

3 in.
$\frac{1}{2}$ in.

36.

$r = 12$ m

Complete.　[9.4b]

37. 7.4 L = _____ mL

38. 62 cm^3 = _____ mL

39. 0.5 mL = _____ L

40. 3 gal = _____ qt

41. 18 cups = _____ pt

42. 32 oz = _____ cups

Understanding Through Discussion and Writing

43. A student makes the following error:

23 in. = 23 · (12 ft) = 276 ft.

Explain the error.　[9.1a]

44. Explain why a 16-in.-diameter pizza that costs $16.25 is a better buy than a 10-in.-diameter pizza that costs $7.85.　[9.3b]

45. Explain how the area of a triangle can be found by considering the area of a parallelogram.　[9.3a]

46. The radius of one circle is twice another circle's radius. Is the area of the first circle twice the area of the other circle? Why or why not?　[9.3b]

Angles and Triangles

9.5

a MEASURING ANGLES

We see a real-world application of *angles* of various types in the different back postures of the bicycle riders illustrated below.

Style of Biking Determines Cycling Posture

Road
About 180° flat

Riders prefer a more aerodynamic flat-back position.

Mountain
About 45°

Riders prefer a semi-upright position to help lift the front wheel over obstacles.

Comfort
About 90°

Riders prefer an upright position that lessens stress on the lower back and neck.

SOURCE: USA TODAY research

OBJECTIVES

a Name an angle in six different ways and measure an angle with a protractor.

b Classify an angle as right, straight, acute, or obtuse.

c Identify complementary, supplementary, and vertical angles and find the measure of a complement or a supplement of a given angle.

d Classify a triangle as equilateral, isosceles, or scalene, and as right, obtuse, or acute.

e Given two of the angle measures of a triangle, find the third.

An **angle** is a set of points consisting of two **rays**, or half-lines, with a common endpoint. The endpoint is called the **vertex** of the angle. The rays are called the **sides** of the angle.

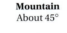

Ray *BA*, or $\overrightarrow{BA}$

A

Ray *BC*, or $\overrightarrow{BC}$

B

C

Vertex

The angle above can be named

angle *ABC*,   angle *CBA*,   angle *B*,   ∠*ABC*,   ∠*CBA*,   or   ∠*B*.

Note that the vertex is written in the middle of the name. If there is only one angle with a given vertex in a drawing, the angle may be named using simply its vertex.

Do Exercises 1 and 2. ▶

Name the angle in six different ways.

1.

D

E

F

2.

P

Q

R

Answers

1. Angle *DEF*, angle *FED*, angle *E*,
∠*DEF*, ∠*FED*, or ∠*E*   **2.** Angle *PQR*, angle *RQP*,
angle *Q*, ∠*PQR*, ∠*RQP*, or ∠*Q*

3. Give another name for ∠1.

Angles may also be numbered. In the following figure, another name for ∠ABC is ∠1, and another name for ∠CBD is ∠2.

◀ **Do Exercise 3.**

To measure angles, we start with some predetermined angle and assign to it a measure of 1. We call it a *unit angle*. Suppose ∠U, shown below, is a unit angle. Let's measure ∠DEF. If we made 3 copies of ∠U, they would "fill up" ∠DEF. Thus, the measure of ∠DEF would be 3 units.

The unit most commonly used for angle measure is the degree. Below is such a unit. Its measure is 1 degree, or 1°. There are 360 degrees in a circle.

A 1° angle:

Here are some other angles with their degree measures.

To indicate the *measure* of ∠XYZ, we write $m \angle XYZ = 90°$. The symbol ⌐ is sometimes drawn on a figure to indicate a 90° angle.

A device called a **protractor** is used to measure angles. Protractors often have two scales (inside and outside). In the center of the protractor is a vertex indicator such as ▲ or a small hole. To measure an angle like ∠Q below, we place the protractor's ▲ at the vertex and line up one of the angle's sides at 0°. Then we check where the angle's other side crosses the scale. In the figure below, the side $\overrightarrow{QR}$ lines up with 0° on the *inside* scale, so we check where the angle's other side crosses the *inside* scale. We see that $m \angle Q = 145°$.

4. Use a protractor to measure this angle.

145° is on the inside scale.

0° is on the inside scale.

◀ **Do Exercise 4.**

Let's find the measure of ∠ABC. This time we line up one of the angle's sides, $\overrightarrow{BC}$, with 0° on the *outside* scale. Then we check where the angle's other side, $\overrightarrow{BA}$, crosses the *outside* scale. We see that m ∠ABC = 42°.

Both 0° and 42° are on the outside scale.

Do Exercise 5.

A protractor can be used to draw a circle graph.

EXAMPLE 1 *Transportation.* According to a recent poll, 45% of adults believe that flying is the safest mode of transportation, 39% believe that cars are the safest, and 16% believe that trains are the safest. Draw a circle graph to represent these figures.

Source: *Marist Institute for Public Opinion*

Every circle graph contains a total of 360°. Thus,

45% of the circle is a 0.45(360°), or 162°, angle;

39% of the circle is a 0.39(360°), or 140.4°, angle;

16% of the circle is a 0.16(360°), or 57.6°, angle.

We begin by drawing a 162° angle. Beginning at the center of the circle, we draw a horizontal segment to the circle. That segment is one side of the angle. We use a protractor to mark off a 162° angle. From that mark, we draw a segment to the center of the circle to complete the angle. This section of the circle graph we label with both the percent (45%) and the type of transportation (Airplanes).

From the second segment drawn, we repeat the above procedure to draw a 140.4° angle. Since protractors are marked in units of 1°, we must approximate this angle. This section we label with 39% and Cars.

The remainder of the circle represents Trains and should be a 57.6° angle; we measure to confirm this, and label the section with 16% and Trains.

Finally, we give a title to the graph: Safest Mode of Transportation.

Safest Mode of Transportation

Airplanes: 45%

162°

57.6°

140.4°

Cars: 39%

Trains: 16%

Do Exercise 6.

5. Use a protractor to measure this angle.

6. Lengths of Engagement of Married Couples. The data below list the percentages of married couples who were engaged for a certain time period before marriage. Use this information to draw a circle graph.

Source: Bruskin Goldring Research

Less than 1 yr:	24%
1–2 yr:	21%
More than 2 yr:	35%
Never engaged:	20%

Answers

5. 33°

6. Lengths of Engagement of Married Couples

Greater than 0 but less than 1 year: 24%

More than 2 years: 35%

86.4°   126°

75.6°   72°

1–2 years: 21%

Never engaged: 20%

Arrangement of sections may vary.

b CLASSIFYING ANGLES

The following are ways in which we classify angles.

Classify each angle as right, straight, acute, or obtuse. Use a protractor if necessary.

7.

8.

9.

10.

> ### TYPES OF ANGLES
>
> **Right angle:** An angle that measures 90°.
>
> **Straight angle:** An angle that measures 180°.
>
> **Acute angle:** An angle that measures more than 0° but less than 90°.
>
> **Obtuse angle:** An angle that measures more than 90° but less than 180°.

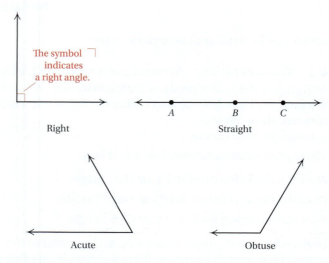

◀ **Do Exercises 7–10.**

c COMPLEMENTARY, SUPPLEMENTARY, AND VERTICAL ANGLES

Certain pairs of angles share special properties.

Complementary Angles

When the sum of the measures of two angles is 90°, the angles are said to be **complementary**. For example, in the figure below, $\angle 1$ and $\angle 2$ are complementary. If two angles are complementary, each is an acute angle. Note that when complementary angles are adjacent to each other (that is, they have a side in common), they form a right angle.

$$m \angle 1 + m \angle 2 = 90°$$
$$75° + 15° = 90°$$

> ### COMPLEMENTARY ANGLES
>
> Two angles are **complementary** if the sum of their measures is 90°. Each angle is called a **complement** of the other.

Answers

7. Right **8.** Acute **9.** Obtuse **10.** Straight

EXAMPLE 2 Identify each pair of complementary angles.

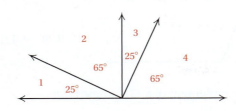

$\angle 1$ and $\angle 2$ $\quad 25° + 65° = 90°$ $\qquad$ $\angle 2$ and $\angle 3$
$\angle 1$ and $\angle 4$ $\qquad\qquad\qquad\qquad\quad$ $\angle 3$ and $\angle 4$

EXAMPLE 3 Find the measure of a complement of a 39° angle.

$90° - 39° = 51°$

The measure of a complement of a 39° angle is 51°.

Do Exercises 11–14. ▶

Supplementary Angles

Next, consider $\angle 1$ and $\angle 2$ as shown below. Because the sum of their measures is 180°, $\angle 1$ and $\angle 2$ are said to be **supplementary**. Note that when supplementary angles are adjacent, they form a straight angle.

$$m\angle 1 + m\angle 2 = 180°$$
$$30° + 150° \;\; = 180°$$

SUPPLEMENTARY ANGLES

Two angles are **supplementary** if the sum of their measures is 180°. Each angle is called a **supplement** of the other.

EXAMPLE 4 Identify each pair of supplementary angles.

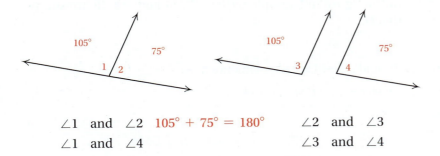

$\angle 1$ and $\angle 2$ $\quad 105° + 75° = 180°$ $\qquad$ $\angle 2$ and $\angle 3$
$\angle 1$ and $\angle 4$ $\qquad\qquad\qquad\qquad\quad$ $\angle 3$ and $\angle 4$

11. Identify each pair of complementary angles.

Find the measure of a complement of each angle.

12.

13.

GS **14.**

$\boxed{}° - 85° = \boxed{}°$

15. Identify each pair of supplementary angles.

Find the measure of a supplement of an angle with each given measure.

16. 38°   **17.** 157°

18. 90°

19. 71°   GS

$$\boxed{}° - 71° = \boxed{}°$$

20. Identify each pair of vertical angles.

EXAMPLE 5 Find the measure of a supplement of a 112° angle.

$$180° - 112° = 68°$$

The measure of a supplement of a 112° angle is 68°.

◀ **Do Exercises 15–19.**

Vertical Angles

When two lines intersect, four angles are formed. The pairs of angles that do not share any side in common are said to be **vertical** (or *opposite*) angles. Thus, in the drawing below, ∠1 and ∠3 are vertical angles, as are ∠4 and ∠2. Note that $m\angle 1 = m\angle 3$ and $m\angle 4 = m\angle 2$.

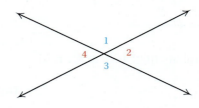

EXAMPLE 6 Identify each pair of vertical angles.

∠AMB and ∠CMD are vertical angles.
∠BMC and ∠DMA are vertical angles.

◀ **Do Exercise 20.**

Note in the figure in Example 6 that ∠DMA and ∠AMB are supplementary; that is,

$$m\angle DMA + m\angle AMB = 180°.$$

Also, ∠CMD and ∠DMA are supplementary, so

$$m\angle CMD + m\angle DMA = 180°.$$

Therefore,

$$m\angle DMA + m\angle AMB = m\angle CMD + m\angle DMA$$
$$m\angle AMB = m\angle CMD. \quad \text{Subtracting } m\angle DMA \text{ from both sides}$$

This shows that vertical angles have the same measure.

Answers

15. ∠1 and ∠2; ∠1 and ∠4; ∠2 and ∠3; ∠3 and ∠4 **16.** 142° **17.** 23° **18.** 90°
19. 109° **20.** ∠1 and ∠2; ∠3 and ∠4

Guided Solution:
19. 180, 109

VERTICAL ANGLES

Two angles are **vertical** if they are formed by two intersecting lines and have no side in common. Vertical angles have the same measure.

Do Exercise 21. ▶

If two angles have the same measure, we say that they are **congruent**, denoted by the symbol ≅ . In the figure below, the measures of angles *WVZ* and *XVY* are equal, and the angles are congruent:

$$m \angle WVZ = m \angle XVY$$
$$\angle WVZ \cong \angle XVY.$$

Note that we do not write that angles are equal: The *measures are equal* and the *angles are congruent*.

Do Exercise 22. ▶

d   TRIANGLES

A **triangle** is a polygon made up of three segments, or sides. Consider these triangles. The triangle with vertices *A*, *B*, and *C* can be named △*ABC*.

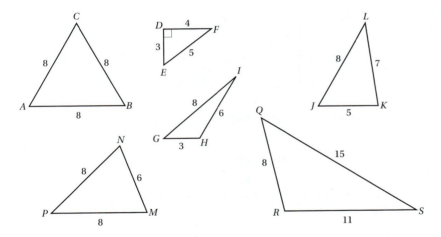

We can classify triangles according to sides and according to angles.

TYPES OF TRIANGLES

Equilateral triangle: All sides are the same length.
Isosceles triangle: Two or more sides are the same length.
Scalene triangle: All sides are of different lengths.
Right triangle: One angle is a right angle.
Obtuse triangle: One angle is an obtuse angle.
Acute triangle: All three angles are acute.

Do Exercises 23–26. ▶

21. Complete:
$$m \angle ANC = \underline{\qquad};$$
$$m \angle ANB = \underline{\qquad}.$$

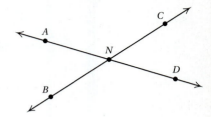

22. Complete:
$$\angle PMQ \cong \underline{\qquad};$$
$$\angle SMP \cong \underline{\qquad}.$$

23. Which triangles on this page are
a) equilateral?
b) isosceles?
c) scalene?

24. Are all equilateral triangles isosceles?

25. Are all isosceles triangles equilateral?

26. Which triangles on this page are
a) right triangles?
b) obtuse triangles?
c) acute triangles?

Answers
21. $m \angle BND$ or $m \angle DNB$; $m \angle CND$ or $m \angle DNC$
22. $\angle SMR$ or $\angle RMS$; $\angle QMR$ or $\angle RMQ$
23. (a) △*ABC*;   (b) △*ABC*, △*MPN*;
(c) △*DEF*, △*GHI*, △*JKL*, △*QRS*   **24.** Yes
25. No   **26.** (a) △*DEF*;   (b) △*GHI*, △*QRS*;
(c) △*ABC*, △*MPN*, △*JKL*

e SUM OF THE ANGLE MEASURES OF A TRIANGLE

The sum of the angle measures of every triangle is 180°. To see this, note that we can think of cutting apart a triangle as shown on the left below. If we reassemble the pieces, we see that a straight angle is formed.

$$64° + 32° + 84° = 180°$$

27. Find $m \angle P + m \angle Q + m \angle R$.

SUM OF THE ANGLE MEASURES OF A TRIANGLE

In any $\triangle ABC$, the sum of the measures of the angles is 180°:

$$m \angle A + m \angle B + m \angle C = 180°.$$

◀ **Do Exercise 27.**

If we know the measures of two angles of a triangle, we can calculate the measure of the third angle.

EXAMPLE 7 Find the missing angle measure.

$$
\begin{aligned}
m \angle A + m \angle B + m \angle C &= 180° \\
x + 65° + 24° &= 180° \\
x + 89° &= 180° \\
x &= 180° - 89° \qquad \text{Subtracting 89° from both} \\
& \qquad\qquad\qquad\qquad\quad \text{sides} \\
x &= 91°
\end{aligned}
$$

Thus, $m \angle A = 91°$.

28. Find the missing angle measure.

◀ **Do Exercise 28.**

✓ Reading Check

Match each definition with the correct term from the list on the right.

RC1. _____ An angle whose measure is 90°

RC2. _____ An angle whose measure is 180°

RC3. _____ An angle whose measure is greater than 0° and less than 90°

RC4. _____ An angle whose measure is greater than 90° and less than 180°

RC5. _____ A pair of angles whose measures add to 90°

RC6. _____ A pair of angles whose measures add to 180°

a) acute angle

b) complementary angles

c) straight angle

d) supplementary angles

e) obtuse angle

f) right angle

a Name each angle in six different ways.

1.

2.

Give another name for ∠1 in each figure.

3.

4.

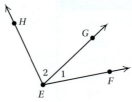

Use a protractor to measure each angle.

5.

6.

7.

8.

9.

10.

Use the given information and a protractor to draw a circle graph.

11. *Credit Score.* The FICO Credit Score is based on five types of information, as shown in the table below.

PERSONAL INFORMATION	PERCENT
Payment history	35%
Debt level	30%
Length of credit history	15%
Credit inquiries	10%
Mix of credit	10%

SOURCE: credit.about.com

12. *Snacking Habits.* Some adults snack heavily and some not at all. The table below lists the frequency of snacking of American adults.

FREQUENCY OF SNACKING	PERCENT
Never	10%
Occasionally	45%
Moderately	35%
Heavily	10%

SOURCE: Market Facts for Hershey Foods

13. *Wind Energy.* The table below shows how wind-turbine capacity is distributed among countries.

COUNTRY	PERCENT
China	26%
United States	20%
Germany	12%
Spain	9%
India	7%
Rest of the world	26%

SOURCE: Earth Policy Institute

14. *Causes of Spinal Cord Injuries.* The table below lists the causes of spinal cord injury.

CAUSES	PERCENT
Motor vehicle accidents	44%
Acts of violence	24%
Falls	22%
Sports	8%
Other	2%

SOURCE: National Spinal Cord Injury Association

b

15.–20. Classify each of the angles in Exercises 5–10 as right, straight, acute, or obtuse.

21.–24. Classify each of the angles in Margin Exercises 1, 2, 4, and 5 as right, straight, acute, or obtuse.

c Identify two pairs of vertical angles for each figure.

25.

26.

27.

28.

Complete.

29. Refer to Exercise 25.

$m\angle 2 = $ _____

$m\angle 3 = $ _____

30. Refer to Exercise 26.

$m\angle 4 = $ _____

$m\angle 2 = $ _____

31. Refer to Exercise 27.

$\angle AMC \cong$ _____

$\angle AMG \cong$ _____

32. Refer to Exercise 28.

$\angle RNS \cong$ _____

$\angle TNR \cong$ _____

Find the measure of a complement of an angle with the given measure.

33. 11°

34. 83°

35. 67°

36. 5°

37. 58°

38. 32°

39. 29°

40. 54°

Find the measure of a supplement of an angle with the given measure.

41. 3°

42. 54°

43. 139°

44. 13°

45. 75°

46. 128°

47. 104°

48. 49°

Classify each triangle as equilateral, isosceles, or scalene. Then classify it as right, obtuse, or acute.

49.

9, 7, 3

50.

7, 7, 5

51.

8, 6, 10

52.

8, 8, 12

53.

6, 6, 6

54.

12, 13, 5

55.

14, 8, 9

56.

5, 7, 8

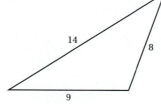

Find each missing angle measure.

57.

B 92°, A 42°, C x

58.

D 29°, E x, F 39°

59.

R 29°, Q 31°, S x

60.

S 60°, R x, T 60°

61.

M x, L 37°, N 85°

62.

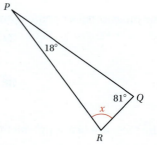

P 18°, Q 81°, R x

634 CHAPTER 9 Geometry and Measurement

Copyright © 2016 Pearson Education, Inc.

Skill Maintenance

Find the simple interest.   [8.7a]

	Principal	Rate of Interest	Time	Simple Interest
63.	$2000	8%	1 year	
64.	$750	6%	$\frac{1}{2}$ year	
65.	$4000	7.4%	$\frac{1}{2}$ year	
66.	$200,000	6.7%	$\frac{1}{12}$ year	

Interest is compounded semiannually. Find the amount in the account after the given length of time. Round to the nearest cent.   [8.7b]

	Principal	Rate of Interest	Time	Amount in the Account
67.	$25,000	6%	5 years	
68.	$150,000	$6\frac{7}{8}$%	15 years	
69.	$150,000	7.4%	20 years	
70.	$160,000	7.4%	20 years	

Synthesis

71. 🖩 In the figure, $m\angle 1 = 79.8°$ and $m\angle 3 = 33.07°$. Find $m\angle 2$, $m\angle 4$, $m\angle 5$, and $m\angle 6$.

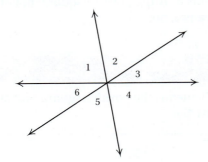

72. 🖩 In the figure, $m\angle 2 = 42.17°$ and $m\angle 6 = 81.9°$. Find $m\angle 1$, $m\angle 3$, $m\angle 4$, and $m\angle 5$.

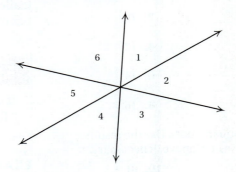

73. Find $m\angle ACB$, $m\angle CAB$, $m\angle EBC$, $m\angle EBA$, $m\angle AEB$, and $m\angle ADB$ in the rectangle shown below.

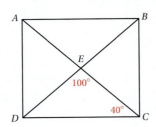

74. The angles in the figure are supplementary. Find the measure of each angle.

OBJECTIVES

a   Simplify square roots of squares, such as $\sqrt{25}$.

b   Approximate square roots.

c   Given the lengths of any two sides of a right triangle, find the length of the third side.

d   Solve applied problems involving right triangles.

SKILL TO REVIEW

Objective 1.9b: Evaluate exponential notation.

Evaluate.

1. 5^2 **2.** 8^2

Find each square.

1. 9^2 **2.** 10^2

3. 11^2 **4.** 12^2

5. 13^2 **6.** 14^2

7. 15^2 **8.** 16^2

Find all square roots. Use the results of Exercises 1–8 above, if necessary.

9. 100 **10.** 81

11. 49 **12.** 196

Simplify. Use the results of Exercises 1–8 above, if necessary.

13. $\sqrt{49}$ **14.** $\sqrt{16}$

15. $\sqrt{121}$ **16.** $\sqrt{100}$

17. $\sqrt{81}$ **18.** $\sqrt{64}$

19. $\sqrt{225}$ **20.** $\sqrt{169}$

21. $\sqrt{1}$ **22.** $\sqrt{0}$

Answers on p. 637

a   SQUARE ROOTS

SQUARE ROOT

If a number is a product of a factor times itself, then that factor is a **square root** of the number. (If $c^2 = a$, then c is a square root of a.)

The number 36 has two square roots, 6 and -6: we have $6 \cdot 6 = 36$ and $(-6) \cdot (-6) = 36$.

EXAMPLE 1   Find the square roots of 25.

The square roots of 25 are 5 and -5, because $5^2 = 25$ and $(-5)^2 = 25$. ◼

·· **Caution!** ··

To find the *square* of a number, multiply the number by itself. To find a *square root* of a number, find a number that, when squared, gives the original number.

◀ **Do Margin Exercises 1–12.**

Every positive number has two square roots. However, the symbol $\sqrt{}$ (called a **radical sign**) represents only the positive square root of the number underneath. Thus, $\sqrt{9}$ means 3, not -3.

RADICAL SIGN

If n is a positive number, $\sqrt{n}$ means the positive square root of n. "The" square root of n means $\sqrt{n}$.

EXAMPLES   Simplify.

2. $\sqrt{36} = 6$ The square root of 36 is 6 because $6^2 = 36$ and 6 is positive.

3. $\sqrt{144} = 12$ Note that $12^2 = 144$.

◀ **Do Exercises 13–22.**

b APPROXIMATING SQUARE ROOTS

Many square roots can't be written as whole numbers or fractions. For example, $\sqrt{2}$, $\sqrt{3}$, $\sqrt{39}$, and $\sqrt{70}$ cannot be precisely represented in decimal notation. To see this, consider the following decimal approximations for $\sqrt{2}$. Each gives a closer approximation, but none is exactly $\sqrt{2}$:

$$\sqrt{2} \approx 1.4 \qquad \text{because} \quad (1.4)^2 = 1.96;$$
$$\sqrt{2} \approx 1.41 \qquad \text{because} \quad (1.41)^2 = 1.9881;$$
$$\sqrt{2} \approx 1.414 \qquad \text{because} \quad (1.414)^2 = 1.999396;$$
$$\sqrt{2} \approx 1.4142 \quad \text{because} \quad (1.4142)^2 = 1.99996164.$$

Decimal approximations like these are commonly found by using a calculator.

EXAMPLE 4 Use a calculator to approximate $\sqrt{3}$, $\sqrt{27}$, and $\sqrt{180}$ to three decimal places.

We use a calculator to find each square root. Since the calculator displays more than three decimal places, we round to three places.

$$\sqrt{3} \approx 1.732, \quad \sqrt{27} \approx 5.196, \quad \sqrt{180} \approx 13.416$$

As a check, note that $1 \cdot 1 = 1$ and $2 \cdot 2 = 4$, so we expect $\sqrt{3}$ to be between 1 and 2. Similarly, we expect $\sqrt{27}$ to be between 5 and 6 and $\sqrt{180}$ to be between 13 and 14.

Do Exercises 23–26. ▶

c THE PYTHAGOREAN THEOREM

A **right triangle** is a triangle with a 90° angle, as shown here. In a right triangle, the longest side is called the **hypotenuse**. It is the side opposite the right angle. The other two sides are called **legs**. We generally use the letters a and b for the lengths of the legs and c for the length of the hypotenuse. They are related as follows.

THE PYTHAGOREAN THEOREM

In any right triangle, if a and b are the lengths of the legs and c is the length of the hypotenuse, then

$$a^2 + b^2 = c^2, \quad \text{or}$$

$$(\text{Leg})^2 + (\text{Other leg})^2 = (\text{Hypotenuse})^2.$$

The equation $a^2 + b^2 = c^2$ is called the **Pythagorean equation.**[*]

CALCULATOR CORNER

Finding Square Roots Many calculators have a square root key, $\boxed{\sqrt{}}$. On some calculators, we enter the number under the $\sqrt{}$ sign first and then press the $\boxed{\sqrt{}}$ key. On other calculators, we might press $\boxed{\sqrt{}}$ first.

It is always best to wait until calculations are complete before rounding. For example, to find $9 \cdot \sqrt{30}$ rounded to the nearest tenth, we do not first determine that $\sqrt{30} \approx 5.5$ and then multiply by 9 to get 49.5. Rather, we press $\boxed{9}\ \boxed{\times}\ \boxed{3}\ \boxed{0}\ \boxed{\text{2nd}}\ \boxed{\sqrt{}}$ $\boxed{=}$ or $\boxed{9}\ \boxed{\times}\ \boxed{\text{SHIFT}}\ \boxed{\sqrt{}}\ \boxed{3}\ \boxed{0}$ $\boxed{=}$. The result is 49.29503018, so $9 \cdot \sqrt{30} \approx 49.3$.

EXERCISES Use a calculator to find each of the following. Round to the nearest tenth.

1. $\sqrt{43}$
2. $7 \cdot \sqrt{8}$
3. $\sqrt{47} - 5$
4. $17 + \sqrt{57}$
5. $13\sqrt{68} + 14$
6. $7 \cdot \sqrt{90} + 3 \cdot \sqrt{40}$

Use a calculator to approximate to three decimal places.

23. $\sqrt{5}$ 24. $\sqrt{78}$

25. $\sqrt{168}$ 26. $\sqrt{321}$

Answers

Skill to Review:
1. 25   2. 64

Margin Exercises:
1. 81   2. 100   3. 121   4. 144   5. 169
6. 196   7. 225   8. 256   9. $-10, 10$
10. $-9, 9$ 11. $-7, 7$ 12. $-14, 14$ 13. 7
14. 4   15. 11   16. 10   17. 9   18. 8
19. 15   20. 13   21. 1   22. 0   23. 2.236
24. 8.832   25. 12.961   26. 17.916

---

[*]The *converse* of the Pythagorean theorem is also true. That is, if $a^2 + b^2 = c^2$, then the triangle is a right triangle.

The Pythagorean theorem is named for the Greek mathematician Pythagoras (569?–500? B.C.E.). We can think of this relationship as adding areas.

Hypotenuse: c
Area $= 5^2$

Leg: a
Area $= 3^2$

Leg: b
Area $= 4^2$

$$a^2 + b^2 = c^2$$
$$3^2 + 4^2 = 5^2$$
$$9 + 16 = 25$$

If we know the lengths of any two sides of a right triangle, we can use the Pythagorean equation to determine the length of the third side.

EXAMPLE 5 Find the length of the hypotenuse of this right triangle.

We substitute in the Pythagorean equation:

$$a^2 + b^2 = c^2$$
$$6^2 + 8^2 = c^2 \qquad \text{Substituting}$$
$$36 + 64 = c^2$$
$$100 = c^2.$$

The solution of this equation is the square root of 100, which is 10:

$$c = \sqrt{100} = 10.$$

◀ **Do Exercise 27.**

.... **Caution!** ....

Before applying the Pythagorean theorem, make sure you have a right triangle. Then determine which side of the triangle is the hypotenuse, or c. The hypotenuse is always the longest side of a right triangle and is always opposite the right angle.

....

EXAMPLE 6 Find the length b for the right triangle shown. Give an exact answer and an approximation to three decimal places.

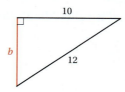

Recall that the leg opposite the right angle is the hypotenuse. Thus, for this triangle, $c = 12$. We substitute in the Pythagorean equation:

$$a^2 + b^2 = c^2$$
$$10^2 + b^2 = 12^2 \qquad \text{Substituting}$$
$$100 + b^2 = 144.$$

---

27. Find the length of the hypotenuse of this right triangle.

$$a^2 + b^2 = c^2$$
$$12^2 + \boxed{}^2 = c^2$$
$$144 + \boxed{} = c^2$$
$$\boxed{} = c^2$$
$$\boxed{} = c$$

Answer

27. $c = 13$

Guided Solution:
27. 5, 25, 169, 13

Next, we solve for b^2 and then b:

$$100 + b^2 - 100 = 144 - 100 \qquad \text{Subtracting 100 from both sides}$$
$$b^2 = 144 - 100$$
$$b^2 = 44 \qquad \text{Solving for } b^2$$
$$\textit{Exact answer:} \quad b = \sqrt{44} \qquad \text{Solving for } b$$
$$\textit{Approximation:} \quad b \approx 6.633. \qquad \text{Using a calculator}$$

Do Exercises 28–30. ▶

d APPLICATIONS

EXAMPLE 7 *Height of Ladder.* A 12-ft ladder leans against a building. The bottom of the ladder is 7 ft from the building. How high is the top of the ladder? Give an exact answer and an approximation to the nearest tenth of a foot.

1. **Familiarize.** We first make a drawing. In it we see a right triangle. We let $h =$ the unknown height. We also note that the ladder forms the hypotenuse of the triangle, so $c = 12$.

12 ft

h

7 ft

2. **Translate.** We substitute 7 for a, h for b, and 12 for c in the Pythagorean equation:

$$a^2 + b^2 = c^2 \qquad \text{Pythagorean equation}$$
$$7^2 + h^2 = 12^2.$$

3. **Solve.** We solve for h^2 and then h:

$$49 + h^2 = 144 \qquad 7^2 = 49 \text{ and } 12^2 = 144$$
$$49 + h^2 - 49 = 144 - 49 \qquad \text{Subtracting 49 from both sides}$$
$$h^2 = 144 - 49$$
$$h^2 = 95$$
$$\textit{Exact answer:} \quad h = \sqrt{95} \qquad \text{Solving for } h$$
$$\textit{Approximation:} \quad h \approx 9.7 \text{ ft.}$$

4. **Check.** $7^2 + (\sqrt{95})^2 = 49 + 95 = 144 = 12^2.$
5. **State.** The top of the ladder is $\sqrt{95}$ ft, or about 9.7 ft, from the ground.

Do Exercise 31. ▶

For each right triangle, find the length of the leg not given. Give an exact answer and an approximation to three decimal places.

28.

29.

30.

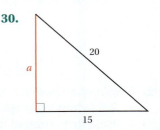

31. How long is a guy wire reaching from the top of an 18-ft pole to a point on the ground 10 ft from the pole? Give an exact answer and an approximation to the nearest tenth of a foot.

18 ft

$c = ?$

10 ft

Answers

28. $a = \sqrt{75}; a \approx 8.660$ 29. $b = \sqrt{120};$ $b \approx 10.954$ 30. $a = \sqrt{175}; a \approx 13.229$
31. $\sqrt{424}$ ft ≈ 20.6 ft

Translating for Success

1. **Servings of Pork.** An 8-lb pork roast contains 37 servings of meat. How many pounds of pork would be needed for 55 servings?

2. **Height of a Ladder.** A 14.5-ft ladder leans against a house. The bottom of the ladder is 9.4 ft from the house. How high is the top of the ladder?

3. **Cruise Cost.** A group of 6 college students pays $4608 for a spring break cruise. What is each person's share?

4. **Sales Tax Rate.** The sales tax is $14.95 on the purchase of a new ladder that costs $299. What is the sales tax rate?

5. **Volume of a Sphere.** Find the volume of a sphere whose radius is 7.2 cm.

The goal of these matching questions is to practice step (2), Translate, of the five-step problem-solving process. Translate each word problem to an equation and select a correct translation from equations A–O.

A. $x = 17.2 \text{ km} \cdot \dfrac{1000 \text{ m}}{1 \text{ km}}$

B. $6 \cdot x = \$4608$

C. $x = \dfrac{4}{3} \cdot \pi \cdot 6^2 \cdot (7.2)$

D. $x = \pi \cdot \left(5\dfrac{1}{2} \div 2\right)^2 \cdot 7$

E. $x = 6\% \times 5 \times \$14.95$

F. $x = \pi \cdot 5\dfrac{1}{2} \cdot 7$

G. $(9.4)^2 + x^2 = (14.5)^2$

H. $\$14.95 = x \cdot \$299$

I. $x = 2(14.5 + 9.4)$

J. $(9.4 + 14.5)^2 = x$

K. $\dfrac{8}{37} = \dfrac{x}{55}$

L. $x = 17.2 \text{ km} \cdot \dfrac{1 \text{ km}}{1000 \text{ m}}$

M. $x = 6 \cdot \$4608$

N. $8 \cdot 37 = 55 \cdot x$

O. $x = \dfrac{4}{3} \cdot \pi \cdot (7.2)^3$

Answers on page A-20

6. **Inheritance.** Six children each inherit $4608 from their mother's estate. What is the total inheritance?

7. **Sales Tax.** Erica buys 5 pairs of earrings at $14.95 each. The sales tax rate is 6%. How much sales tax will be charged?

8. **Bridge Length.** The Vasco Da Gama Bridge in Portugal is 17.2 km long. Convert this distance to meters.

9. **Volume of a Storage Tank.** The diameter of a cylindrical grain-storage tank is $5\frac{1}{2}$ yd. Its height is 7 yd. Find its volume.

10. **Perimeter of a Photo.** A rectangular photo is 14.5 cm by 9.4 cm. What is the perimeter of the photo?

☑ Reading Check

Determine whether each statement is true or false.

————————— **RC1.** 10 is a square root of 100.

————————— **RC2.** $\sqrt{3} = 9$.

————————— **RC3.** In a right triangle, the side opposite the right angle is the hypotenuse.

————————— **RC4.** In a right triangle, the sum of the lengths of the legs is the length of the hypotenuse.

a Find both square roots for each number listed.

1. 16

2. 9

3. 121

4. 49

5. 169

6. 144

7. 2500

8. 3600

Simplify.

9. $\sqrt{64}$

10. $\sqrt{4}$

11. $\sqrt{81}$

12. $\sqrt{49}$

13. $\sqrt{225}$

14. $\sqrt{121}$

15. $\sqrt{625}$

16. $\sqrt{900}$

17. $\sqrt{400}$

18. $\sqrt{169}$

19. $\sqrt{10,000}$

20. $\sqrt{1,000,000}$

b Approximate each number to the nearest thousandth. Use a calculator.

21. $\sqrt{48}$

22. $\sqrt{17}$

23. $\sqrt{8}$

24. $\sqrt{7}$

25. $\sqrt{3}$

26. $\sqrt{6}$

27. $\sqrt{12}$

28. $\sqrt{18}$

29. $\sqrt{19}$

30. $\sqrt{75}$

31. $\sqrt{110}$

32. $\sqrt{10}$

c Find the length of the third side of each right triangle. Give an exact answer and, when appropriate, an approximation to the nearest thousandth.

33.

34.

35.

36.

37.

38.

39.

40.

For each right triangle, find the length of the side not given. Assume that c represents the length of the hypotenuse. Give an exact answer and, when appropriate, an approximation to the nearest thousandth.

41. $a = 10, b = 24$

42. $a = 5, b = 12$

43. $a = 9, c = 15$

44. $a = 18, c = 30$

45. $a = 4, b = 5$

46. $a = 5, b = 6$

47. $a = 1, c = 32$

48. $b = 1, c = 20$

d   In Exercises 49–56, give an exact answer and an approximation to the nearest tenth.

49. A 30-ft string of lights reaches from the top of a pole to a point on the ground 16 ft from the base of the pole. How tall is the pole?

50. A 25-ft wire reaches from the top of a telephone pole to a point on the ground 18 ft from the base of the pole. How tall is the pole?

51. *Softball Diamond.* A slow-pitch softball diamond is actually a square 65 ft on a side. How far is it from home plate to second base?

52. *Baseball Diamond.* A baseball diamond is actually a square 90 ft on a side. How far is it from home plate to second base?

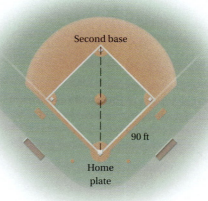

53. *Great Pyramid.* The Pyramid of Cheops is 146 m high. The distance at ground level from the center of the pyramid to the middle of one of the faces is 115 m, as shown below. What is the slant height of a side of the pyramid?

54. *Construction.* In order to support a masonry wall, Matthew erects braces at a height of 12 ft on the wall. The braces are anchored to the ground 15 ft from the base of the wall. How long are the braces?

55. An airplane is flying at an altitude of 4100 ft. The slanted distance directly to the airport is 15,100 ft. How far is the airplane horizontally from the airport?

56. A surveyor had poles located at points *P*, *Q*, and *R* around a lake. The distances that the surveyor was able to measure are marked on the drawing. What is the distance from *P* to *R* across the lake?

Skill Maintenance

Solve.

57. Food expenses account for 26% of the average family's budget. A family makes $1800 one month. How much do they spend for food? [8.4a]

58. Blakely County has a population that is increasing by 4% each year. This year the population is 180,000. What will it be next year? [8.5a]

59. Dexter College has a student body of 1850 students. Of these, 17.5% are seniors. How many students are seniors? [8.4a]

60. The price of a cell phone was reduced from $70 to $61.60. Find the percent decrease in price. [8.5a]

Simplify. [1.9b]

61. 2^3

62. 5^3

63. 10^3

64. 10^4

Simplify. [1.9c, d]

65. $90 \div 15 \cdot 2 - (1 + 2)^2$

66. $10^3 - \{2 \times [5 \times 3 - (4 + 2)]\}$

Synthesis

67. Find the area of the trapezoid shown. Round to the nearest hundredth.

b
8 cm
12 cm
4 cm

68. Which of the triangles below has the larger area? If the areas are the same, state so.

5   5   5   5
8   6

69. Caiden's new TV has a screen that measures $31\frac{3}{4}$ in. by $56\frac{1}{2}$ in. Determine the diagonal measurement of the screen. Round your answer to the nearest tenth of an inch.

$56\frac{1}{2}$ in.
$31\frac{3}{4}$ in.
?

70. A Philips 42-in. plasma television has a rectangular screen that measures 42 in. diagonally. The ratio of width to height is 16 to 9. Find the width and the height of the screen.

42 in.

71. A conventional 19-in. television set has a rectangular screen that measures 19 in. diagonally. The ratio of width to height in a conventional television set is 4 to 3. Find the width and the height of the screen.

19 in.

72. A cube is circumscribed by a sphere with a 1-m diameter. How much more volume is in the sphere?

1 m

Weight, Mass, and Temperature

a   WEIGHT: THE AMERICAN SYSTEM

The American units of weight are as follows.

> **AMERICAN UNITS OF WEIGHT**
>
> $$1 \text{ lb} = 16 \text{ ounces (oz)}$$
> $$1 \text{ ton (T)} = 2000 \text{ pounds (lb)}$$

OBJECTIVES

a   Convert from one American unit of weight to another.

b   Convert from one metric unit of mass to another.

c   Convert temperatures from Celsius to Fahrenheit and from Fahrenheit to Celsius.

The term "ounce" used here for weight is different from the "ounce" we used for capacity in Section 9.4.

EXAMPLE 1   A well-known hamburger is called a "quarter-pounder." Find its name in ounces: a " _____ ouncer."

$$\frac{1}{4} \text{ lb} = \frac{1}{4} \cdot 1 \text{ lb}$$

$$= \frac{1}{4} \cdot 16 \text{ oz} \qquad \text{Substituting 16 oz for 1 lb}$$

$$= 4 \text{ oz}$$

A "quarter-pounder" could also be called a "four-ouncer."

SKILL TO REVIEW

Objective 3.4a: Multiply an integer and a fraction.

Multiply.

1. $\dfrac{3}{4} \cdot 16$ **2.** $2480 \times \dfrac{1}{2000}$

EXAMPLE 2   Complete: $15,360 \text{ lb} = $ _____ T.

$$15,360 \text{ lb} = 15,360 \text{ lb} \cdot \frac{1 \text{ T}}{2000 \text{ lb}} \qquad \text{Multiplying by 1}$$

$$= \frac{15,360}{2000} \text{ T}$$

$$= 7.68 \text{ T} \qquad \text{Dividing by 2000}$$

Do Margin Exercises 1–3. ▶

Complete.

1. $5 \text{ lb} = $ _____ oz

2. $8640 \text{ lb} = $ _____ T

3. $1 \text{ T} = $ _____ oz

b   MASS: THE METRIC SYSTEM

There is a difference between **mass** and **weight**, but the terms are often used interchangeably. People sometimes use the word "weight" when, technically, they are referring to "mass." Weight is related to the force of the earth's gravity. The farther you are from the center of the earth, the less you weigh. Your mass, on the other hand, stays the same no matter where you are.

Answers

Skill to Review:

1. 12   **2.** $\dfrac{31}{25}$, or 1.24

Margin Exercises:

1. 80   **2.** 4.32   **3.** 32,000

METRIC UNITS OF MASS

$$1 \text{ metric ton (t)} = 1000 \text{ kilograms (kg)}$$
$$1 \textit{kilo}\text{gram (kg)} = 1000 \text{ grams (g)}$$
$$1 \textit{hecto}\text{gram (hg)} = 100 \text{ grams (g)}$$
$$1 \textit{deka}\text{gram (dag)} = 10 \text{ grams (g)}$$
$$1 \text{ gram (g)}$$
$$1 \textit{deci}\text{gram (dg)} = \frac{1}{10} \text{ gram (g)}$$
$$1 \textit{centi}\text{gram (cg)} = \frac{1}{100} \text{ gram (g)}$$
$$1 \textit{milli}\text{gram (mg)} = \frac{1}{1000} \text{ gram (g)}$$

The basic unit of mass is the **gram** (g), which is the mass of 1 cubic centimeter (1 cm^3) of water. Since a cubic centimeter is small, a gram is a small unit of mass.

$$1 \text{ g} = 1 \text{ gram} = \text{ the mass of 1 cm}^3 \text{ of water}$$

1 g = 1 cm^3
of water

The metric units of mass are listed to the left. The prefixes are the same as those for length.

Thinking Metric

One gram is about the mass of 1 raisin or 1 package of artificial sweetener. Since 1 kg is about 2.2 lb, 1000 kg is about 2200 lb, or 1 metric ton (t), which is about 10% more than 1 American ton (T), which is 2000 lb.

1 gram

1 kilogram of grapes   1 pound of grapes

Small masses, such as dosages of medicine and vitamins, may be measured in milligrams (mg). The gram (g) is used for objects ordinarily measured in ounces, such as the mass of a letter, a piece of candy, or a coin.

15 g

125 kg

Each 2.5 mg

1 kg

2 g

The kilogram (kg) is used for larger food packages and for body masses. The metric ton (t) is used for very large masses, such as the mass of an automobile, a truckload of gravel, or an airplane.

◀ **Do Exercises 4–8.**

Complete with mg, g, kg, or t.

4. A laptop computer has a mass of 6 _____.

5. Eric has a body mass of 85.4 _____.

6. This is a 3- _____ vitamin.

7. A pen has a mass of 12 _____.

8. A sport utility vehicle has a mass of 3 _____.

Answers

4. kg   **5.** kg   **6.** mg   **7.** g   **8.** t

Changing Units Mentally

As before, changing from one metric unit to another requires only the movement of a decimal point. We use this table.

1000 g	100 g	10 g	1 g	0.1 g	0.01 g	0.001 g
1 kg	1 hg	1 dag	1 g	1 dg	1 cg	1 mg

EXAMPLE 3 Complete: 8 kg = _____ g.

Think: A kilogram is 1000 times the mass of a gram. To go from kg to g in the table is a move of three places to the right. Thus, we move the decimal point three places to the right.

1000 g	100 g	10 g	1 g	0.1 g	0.01 g	0.001 g
1 kg	1 hg	1 dag	1 g	1 dg	1 cg	1 mg

3 places to the right

8.0   8.000.   8 kg = 8000 g

EXAMPLE 4 Complete: 4235 g = _____ kg.

Think: There are 1000 grams in 1 kilogram. To go from g to kg in the table is a move of three places to the left. Thus, we move the decimal point three places to the left.

1000 g	100 g	10 g	1 g	0.1 g	0.01 g	0.001 g
1 kg	1 hg	1 dag	1 g	1 dg	1 cg	1 mg

3 places to the left

4235.0   4.235.0   4235 g = 4.235 kg

EXAMPLE 5 Complete: 6.98 cg = _____ mg.

Think: One centigram has the mass of 10 milligrams. To go from cg to mg is a move of one place to the right. Thus, we move the decimal point one place to the right.

1000 g	100 g	10 g	1 g	0.1 g	0.01 g	0.001 g
1 kg	1 hg	1 dag	1 g	1 dg	1 cg	1 mg

1 place to the right

6.98   6.9.8   6.98 cg = 69.8 mg

> The most commonly used metric units of mass are kg, g, and mg. We have intentionally used those more often than the others in the exercises.

Do Exercises 9–12. ▶

Complete.

9. 6.2 kg = _____ g

10. 304.8 cg = _____ g

11. 7.7 cg = _____ mg

12. 2344 mg = _____ cg

Answers

9. 6200   10. 3.048   11. 77   12. 234.4

C TEMPERATURE

Estimated Conversions

Below are two temperature scales: **Fahrenheit** for American measure and **Celsius**, used internationally and in science.

By laying a straightedge horizontally between the scales, we can make an approximate conversion from one measure of temperature to another and get an idea of how the temperature scales compare.

EXAMPLES Convert to Celsius using the scales shown above. Approximate to the nearest ten degrees.

6. 212°F (Boiling point of water)   100°C   This is exact.
7. 32°F (Freezing point of water)   0°C   This is exact.
8. 105°F   40°C   This is approximate.

Do Exercises 13–15.

EXAMPLES Make an approximate conversion to Fahrenheit using the scales shown above.

9. 44°C (Hot bath)   110°F   This is approximate.
10. 20°C (Room temperature)   68°F   This is exact.
11. 83°C   180°F   This is approximate.

Do Exercises 16–18.

Convert to Celsius. Use a straightedge and the scales shown on this page. Approximate to the nearest ten degrees.

13. 180°F (Brewing coffee)

14. 25°F (Cold day)

15. −10°F (Miserably cold day)

Convert to Fahrenheit. Use a straightedge and the scales shown on this page. Approximate to the nearest ten degrees.

16. 25°C (Nice day at the park)

17. 40°C (Temperature of a patient with a high fever)

18. 10°C (Cold bath)

Answers

13. 80°C   **14.** 0°C   **15.** −20°C
16. 80°F   **17.** 100°F   **18.** 50°F

Exact Conversions

A formula allows us to make exact conversions from Celsius to Fahrenheit.

CELSIUS TO FAHRENHEIT

$$F = \frac{9}{5} \cdot C + 32, \quad \text{or} \quad F = 1.8 \cdot C + 32$$

$\left(\text{Multiply the Celsius temperature by } \frac{9}{5}, \text{ or } 1.8, \text{ and add } 32.\right)$

EXAMPLES Convert to Fahrenheit.

12. 0°C (Freezing point of water)

$$F = \frac{9}{5} \cdot C + 32 = \frac{9}{5} \cdot 0 + 32 = 0 + 32 = 32$$

Thus, 0°C = 32°F.

13. 37°C (Normal body temperature)

$$F = 1.8 \cdot C + 32 = 1.8 \cdot 37 + 32 = 66.6 + 32 = 98.6$$

Thus, 37°C = 98.6°F.

Check the answers to Examples 12 and 13 using the scales on p. 648.

Do Exercises 19 and 20. ▶

A second formula allows us to make exact conversions from Fahrenheit to Celsius.

FAHRENHEIT TO CELSIUS

$$C = \frac{5}{9} \cdot (F - 32), \quad \text{or} \quad C = \frac{F - 32}{1.8}$$

$\left(\text{Subtract 32 from the Fahrenheit temperature and multiply by } \frac{5}{9} \text{ or divide by } 1.8.\right)$

EXAMPLES Convert to Celsius.

14. 212°F (Boiling point of water)

$$C = \frac{5}{9} \cdot (F - 32) = \frac{5}{9} \cdot (212 - 32) = \frac{5}{9} \cdot 180 = 100$$

Thus, 212°F = 100°C.

15. 77°F

$$C = \frac{F - 32}{1.8} = \frac{77 - 32}{1.8} = \frac{45}{1.8} = 25$$

Thus, 77°F = 25°C.

Check the answers to Examples 14 and 15 using the scales on p. 648.

Do Exercises 21 and 22. ▶

Convert to Fahrenheit.

19. 80°C

20. 35°C

CALCULATOR CORNER

Temperature Conversions
Temperature conversions can be done quickly using a calculator. To convert 37°C to Fahrenheit, for example, we press 1 . 8 × 3 7 + 3 2 = . The calculator displays 98.6 , so 37°C = 98.6°F. We can convert 212°F to Celsius by pressing (2 1 2 − 3 2) ÷ 1 . 8 = . The display reads 100 , so 212°F = 100°C. Note that we must use parentheses when converting from Fahrenheit to Celsius in order to get the correct result.

EXERCISES Use a calculator to convert each temperature to Fahrenheit.

1. 5°C

2. 50°C

Use a calculator to convert each temperature to Celsius.

3. 68°F

4. 113°F

Convert to Celsius.

21. 95°F

$$C = \frac{5}{9}(F - 32)$$

$$= \frac{5}{9}(\boxed{} - 32)$$

$$= \frac{5}{9} \cdot \boxed{} = \boxed{}$$

Thus, 95°F = $\boxed{}$ °C.

22. 113°F

Answers

19. 176°F   **20.** 95°F   **21.** 35°C   **22.** 45°C

Guided Solution:
21. 95, 63, 35; 35

✓ Reading Check

Determine whether each statement is true or false.

RC1. _____ 400 g > 40 dg

RC2. _____ 5 hg < 400 g

RC3. _____ 0.5 kg = 500 g

RC4. _____ 48 oz = 4 lb

RC5. _____ 7500 lb < 3.5 T

RC6. _____ 800 cg > 6 g

a Complete.

1. 1 T = _____ lb

2. 1 lb = _____ oz

3. 6000 lb = _____ T

4. 8 T = _____ lb

5. 4 lb = _____ oz

6. 10 lb = _____ oz

7. 6.32 T = _____ lb

8. 8.07 T = _____ lb

9. 3200 oz = _____ T

10. 6400 oz = _____ T

11. 80 oz = _____ lb

12. 960 oz = _____ lb

13. *Pecans.* In 2011, U.S. farmers produced 269,700,000 pounds of pecans. How many tons of pecans were produced?

Source: National Agricultural Statistics Service, U.S. Department of Agriculture

14. *Peaches.* In 2011, U.S. farmers produced 978,260 tons of peaches. How many pounds of peaches were produced?

Source: National Agricultural Statistics Service, U.S. Department of Agriculture

b Complete.

15. 1 kg = _____ g

16. 1 hg = _____ g

17. 1 g = _____ kg

18. 1 dg = _____ g

19. 1 cg = _____ g

20. 1 mg = _____ g

21. 1 g = _____ mg

22. 1 g = _____ cg

23. 1 g = _____ dg

24. 25 kg = _____ g

25. 234 kg = _____ g

26. 9403 g = _____ kg

27. 5200 g = _____ kg

28. 1.506 kg = _____ g

29. 897 mg = _____ kg

30. 45 cg = _____ g

31. 7.32 kg = _____ g

32. 0.0025 cg = _____ mg

33. 8492 g = _____ kg

34. 9466 g = _____ kg

35. 585 mg = _____ cg

36. 96.1 mg = _____ cg

37. 8 kg = _____ cg

38. 0.06 kg = _____ mg

39. 1 t = _____ kg

40. 2 t = _____ kg

41. 3.4 cg = _____ dag

42. 115 mg = _____ g

43. 60.3 kg = _____ t

44. 15.68 kg = _____ t

C Convert to Celsius. Round the answer to the nearest ten degrees. Use the scales on p. 648.

45. 178°F

46. 195°F

47. 140°F

48. 107°F

49. 68°F

50. 45°F

51. 10°F

52. 120°F

Convert to Fahrenheit. Round the answer to the nearest ten degrees. Use the scales on p. 648.

53. 80°C

54. 93°C

55. 58°C

56. 33°C

57. −10°C

58. −5°C

59. 5°C

60. 15°C

Convert to Fahrenheit. Use the formula $F = \frac{9}{5} \cdot C + 32$.

61. 30°C   **62.** 85°C   **63.** 40°C   **64.** 90°C

65. 2°C   **66.** 8°C   **67.** −1°C   **68.** −15°C

69. 3000°C (The melting point of iron)   **70.** 1000°C (The melting point of gold)

Convert to Celsius. Use the formula $C = \frac{5}{9} \cdot (F - 32)$ or $C = \dfrac{F - 32}{1.8}$.

71. 77°F   **72.** 59°F   **73.** 131°F   **74.** 140°F

75. 178°F   **76.** 110°F   **77.** 5°F   **78.** −4°F

79. 98.6°F (Normal body temperature)   **80.** 104°F (High-fevered body temperature)

81. *Highest Temperatures.* The highest temperature ever recorded in the world is 136°F in the desert of Libya in 1922. The highest temperature ever recorded in the United States is $56\frac{2}{3}$ °C in California's Death Valley in 1913.

Source: *The Handy Geography Answer Book*

a) Convert each temperature to the other scale.
b) How much higher in degrees Fahrenheit was the world record than the U.S. record?

82. *Boiling Point and Altitude.* The boiling point of water actually changes with altitude. The boiling point is 212°F at sea level, but lowers about 1°F for every 500 ft of increase in altitude above sea level.

Sources: *The Handy Geography Answer Book; The New York Times Almanac*

a) What is the boiling point at an elevation of 1500 ft above sea level?
b) The elevation of Tucson is 2564 ft above sea level and that of Phoenix is 1117 ft. What is the boiling point in each city?
c) How much lower is the boiling point in Denver, whose elevation is 5280 ft, than in Tucson?
d) What is the boiling point at the top of Mt. McKinley (Denali) in Alaska, the highest point in the United States, at 20,320 ft?

California's Death Valley

Libyan desert

Skill Maintenance

Graph.　[6.4b]

83. $y = -3x$

84. $y = \frac{1}{3}x$

85. $y = x + 1$

86. $y = 3x - 4$

Solve.　[1.8a]

87. *Morel Mushrooms.* During the spring, Kate's Country Market sold 43 pounds of fresh morel mushrooms. The mushrooms sold for $22 a pound. Find the total amount Kate took in from the sale of the mushrooms.

88. Sandy can type 62 words per minute. How long will it take her to type 12,462 words?

Synthesis

89. A box of gelatin-mix packages weighs $15\frac{3}{4}$ lb. Each package weighs $1\frac{3}{4}$ oz. How many packages are in the box?

90. At $1.59 a dozen, the cost of eggs is $1.06 per pound. How much does an egg weigh?

91. Use the formula $F = \frac{9}{5} \cdot C + 32$ to find the temperature that is the same for both the Fahrenheit and Celsius scales.

92. Which represents a bigger change in temperature: a drop of 5°F or a drop of 5°C? Why?

93. *Chemistry.* Another temperature scale often used is the **Kelvin** scale. Conversions from Celsius to Kelvin can be carried out using the formula

$$K = C + 273.$$

A chemistry textbook describes an experiment in which a reaction takes place at a temperature of 400 Kelvin. A student wishes to perform the experiment, but has only a Fahrenheit thermometer. At what Fahrenheit temperature will the reaction take place?

94. 🖩 A large egg is about $5\frac{1}{2}$ cm tall with a diameter of 4 cm. Estimate the mass of such an egg by averaging the volumes of two spheres. (*Hint*: 1 cc of water has a mass of 1 g.)

Complete. Use 453.6 g = 1 lb. Round to four decimal places.

95. 🖩 1 lb = _____ kg

96. 🖩 1 g = _____ lb

97. 🖩 *Track and Field.* In shot put, a woman's shot weighs 8.8 lb and has a 4.5-in. diameter. Find its mass per cubic centimeter, given that 1 lb = 453.6 g.

Source: National Collegiate Athletic Association

98. 🖩 *Track and Field.* In shot put, a man's shot weighs 16 lb and has a 5-in. diameter. Find its mass per cubic centimeter, given that 1 lb = 453.6 g.

99. *Large Diamonds.* A **carat** (also spelled **karat**) is a unit of weight for precious stones; 1 carat = 200 mg. The Golden Jubilee Diamond weighs 545.67 carats and is the largest cut diamond in the world. The Hope Diamond, located at the Smithsonian National Museum of Natural History, weighs 45.52 carats.

Source: *National Geographic*, February 2001

a) How many grams does the Golden Jubilee Diamond weigh?
b) How many grams does the Hope Diamond weigh?
c) 🖩 Given that 1 lb = 453.6 g, how many ounces does each diamond weigh?

a MEASUREMENTS AND MEDICINE

Measurements play a critical role in health care. Doctors, nurses, aides, technicians, and others all need to use the proper units and perform the proper calculations to assure the best possible care of patients.

Because of the ease with which conversions can be made and its extensive use in science—among other reasons—the metric system is the primary system of measurement in medicine.

EXAMPLE 1 *Medical Dosage.* A physician ordered 3.5 L of 5% dextrose in water (abbreviated D5W) to be administered over a 24-hr period. How many milliliters were ordered?

We convert 3.5 L to milliliters:

$$3.5 \text{ L} = 3.5 \cdot 1 \text{ L}$$
$$= 3.5 \cdot 1000 \text{ mL} \qquad \text{Substituting}$$
$$= 3500 \text{ mL}.$$

The physician ordered 3500 mL of D5W.

Do Margin Exercise 1. ▶

Liquids at a pharmacy are often labeled in liters or milliliters. Thus, if a physician's prescription is given in ounces, it must be converted. For conversion, a pharmacist knows that 1 oz ≈ 29.57 mL.*

EXAMPLE 2 *Prescription Size.* A prescription calls for 3 oz of theophylline, a drug commonly used for children with asthma. For how many milliliters is the prescription?

We convert as follows:

$$3 \text{ oz} = 3 \cdot 1 \text{ oz}$$
$$\approx 3 \cdot 29.57 \text{ mL} \qquad \text{Substituting}$$
$$= 88.71 \text{ mL}.$$

The prescription calls for 88.71 mL of theophylline.

Do Exercise 2. ▶

---

*In practice, most pharmacists use 30 mL as an approximation of 1 oz.

OBJECTIVE

 Make conversions and solve applied problems concerning medical dosages.

SKILL TO REVIEW

Objective 9.4b: Convert from one unit of capacity to another.

Complete.
1. 1.07 L = _____ mL
2. 350 mL = _____ L

1. *Medical Dosage.* A physician ordered 2400 mL of 0.9% saline solution to be administered intravenously over a 24-hr period. How many liters were ordered?

2. *Prescription Size.* A prescription calls for 2 oz of theophylline.
   a) For how many milliliters is the prescription?
   b) For how many liters is the prescription?

Answers

Skill to Review:
1. 1070 2. 0.35

Margin Exercises:
1. 2.4 L 2. (a) About 59.14 mL;
(b) about 0.059 L

EXAMPLE 3 *Pill Splitting.* Chlorthalidone is a commonly prescribed drug used to treat hypertension. Tania's physician directs her to reduce her dosage from 25 mg to 12.5 mg. Tania's original prescription contained 30 tablets, each 25 mg.

a) How many milligrams of chlorthalidone, in total, were in the original prescription?

b) How many 12.5-mg doses can Tania obtain from the original prescription?

a) The original prescription contained 30 tablets, each containing 25 mg of chlorthalidone. To find the total amount in the prescription, we multiply:

$$30 \text{ tablets} \cdot 25 \text{ mg/tablet} = 750 \text{ mg}.$$

The original prescription contained 750 mg of chlorthalidone.

b) Since 12.5 mg is half of 25 mg ($25 \div 2 = 12.5$), Tania's original 30 doses can each be split in half, yielding $30 \cdot 2$, or 60, doses at 12.5 mg per dose.

◀ **Do Exercise 3.**

3. *Pill Splitting.* If Tania's physician originally prescribed 14 tablets that were each 25 mg, how many milligrams were in the original prescription? How many 12.5-mg doses could be obtained from the original prescription?

Another metric unit that is used in medicine is the microgram (mcg). It is defined as follows.

MICROGRAM

$$1 \text{ microgram} = 1 \text{ mcg} = \frac{1}{1,000,000} \text{ g} = 0.000001 \text{ g}$$

$$1,000,000 \text{ mcg} = 1 \text{ g}$$

One microgram is one-millionth of a gram, so one million micrograms is one gram. A microgram is also one-thousandth of a milligram, so one thousand micrograms is one milligram.

EXAMPLE 4 Complete: $1 \text{ mg} = \underline{\hspace{1cm}}$ mcg.

We convert to grams and then to micrograms:

$$1 \text{ mg} = 0.001 \text{ g}$$
$$= 0.001 \cdot 1 \text{ g}$$
$$= 0.001 \cdot 1,000,000 \text{ mcg} \qquad \text{Substituting 1,000,000 mcg for 1 g}$$
$$= 1000 \text{ mcg}.$$

◀ **Do Exercise 4.**

4. Complete:

$1 \text{ mcg} = \underline{\hspace{1cm}}$ mg.

$1 \text{ mcg} = \boxed{}$ g

$\quad = 0.000001 \cdot 1 \text{ g}$

$\quad = 0.000001 \cdot \boxed{}$ mg

$\quad = \boxed{}$ mg

EXAMPLE 5 *Medical Dosage.* Sublingual nitroglycerin comes in 0.4-mg tablets. How many micrograms are in each tablet?
Source: Steven R. Smith, M.D.

We are to complete $0.4 \text{ mg} = \underline{\hspace{1cm}}$ mcg. Thus,

$$0.4 \text{ mg} = 0.4 \cdot 1 \text{ mg}$$
$$= 0.4 \cdot 1000 \text{ mcg} \qquad \text{Substituting 1000 mcg for 1 mg (from Example 4)}$$
$$= 400 \text{ mcg}.$$

5. *Medical Dosage.* A physician prescribes 500 mcg of alprazolam, an antianxiety medication. How many milligrams is this dosage?
Source: Steven R. Smith, M.D.

Answers

3. 350 mg; 28 doses   **4.** 0.001 mg   **5.** 0.5 mg

Guided Solution:
4. 0.000001, 1000, 0.001

We can also do this problem in a manner similar to that used in Example 4.

◀ **Do Exercise 5.**

✓ Reading Check

Determine whether each statement is true or false.

_____ **RC1.** For medicines in liquid form, 1 oz is about 30 mL.

_____ **RC2.** One microgram is one-thousandth of a milligram.

_____ **RC3.** One microgram is one million grams.

_____ **RC4.** There are 1000 micrograms in a milligram and 1000 milligrams in a gram.

a *Medical Dosage.* Solve each of the following. (None of these medications should be taken without consulting your own physician.)

1. An emergency-room physician orders 2.0 L of Ringer's lactate to be administered over 2 hr for a patient in shock. How many milliliters is this?

2. Ingrid received 84 mL per hour of normal saline solution. How many liters did Ingrid receive in a 24-hr period?

3. To battle hypertension and prostate enlargement, Rick is directed to take 4 mg of doxazosin each day for 30 days. How many grams is this?

4. To battle high cholesterol, Kit is directed to take 40 mg of atorvastatin for 60 days. How many grams is this?

5. Cephalexin is an antibiotic that frequently is prescribed in 500-mg tablets. Dr. Bouvier prescribes 2 g of cephalexin per day for a patient with a skin abscess. How many 500-mg tablets would have to be taken in order to achieve this daily dosage?

6. Quinidine gluconate is a liquid mixture, part medicine and part water, which is administered intravenously. There are 80 mg of quinidine gluconate in each cubic centimeter (cc) of the liquid mixture. Dr. Nassat orders 500 mg of quinidine gluconate to be administered daily to a patient with malaria. How much of the solution would have to be administered in order to achieve the recommended daily dosage?

7. Albuterol is a medication used for the treatment of asthma. It comes in an inhaler that contains 17 mg of albuterol mixed with a liquid. One actuation (inhalation) from the mouthpiece delivers a 90-mcg dose of albuterol.
   a) Dr. Martinez orders 2 inhalations 4 times per day. How many micrograms of albuterol does the patient inhale per day?
   b) How many actuations/inhalations are contained in one inhaler?
   c) Delia is going away for 4 months and wants to take enough albuterol to last for that time. Her physician has prescribed 2 inhalations 4 times per day. Estimate how many inhalers Delia will need to take with her for the 4-month period.

8. Amoxicillin is a common antibiotic prescribed for children. It is a liquid suspension composed of part amoxicillin and part water. In one formulation of amoxicillin suspension, there are 250 mg of amoxicillin in 5 cc of the liquid suspension. Dr. Scarlotti prescribes 400 mg per day for a 2-year-old child with an ear infection. How much of the amoxicillin liquid suspension would the child's parent need to administer in order to achieve the recommended daily dosage of amoxicillin?

9. Dr. Norris tells a patient to purchase 0.5 L of hydrogen peroxide. Commercially, hydrogen peroxide is found on the shelves in bottles that hold 4 oz, 8 oz, and 16 oz. Which bottle comes closest to filling the prescription?

10. Dr. Lopez wants a patient to receive 3 L of a normal glucose solution in a 24-hr period. How many milliliters per hour should the patient receive?

11. Mirtazapine is a commonly prescribed antianxiety drug. Joanne has a 14-tablet supply of 30-mg tablets and her physician now directs her to reduce her dosage to 15 mg.
    a) How many grams were originally prescribed?
    b) How many 15-mg doses can Joanne obtain from the original prescription?

12. Nefazodone is a commonly prescribed antidepressant. Chad has a 90-tablet supply of 200-mg tablets when his physician directs him to cut his dosage to 100 mg.
    a) How many grams are in Chad's current supply?
    b) How many 100-mg doses can Chad obtain from his supply?

13. Amoxicillin is an antibiotic obtainable in a liquid suspension form, part medication and part water, and is frequently used to treat infections in infants. One formulation of the drug contains 125 mg of amoxicillin per 5 mL of liquid. A pediatrician orders 150 mg per day for a 4-month-old child with an ear infection. How much of the amoxicillin suspension would the parent need to administer to the infant in order to achieve the recommended daily dose?

14. Diphenhydramine HCL is an antihistamine available in liquid form, part medication and part water. One formulation contains 25 mg of medication in 5 mL of liquid. An allergist orders 40-mg doses for a high school student. How many milliliters should be in each dose?

Complete.

15. 1 mg = _____ mcg

16. 1 mcg = _____ mg

17. 325 mcg = _____ mg

18. 0.45 mg = _____ mcg

19. Dr. Djihn prescribes 0.25 mg of alprazolam, an antianxiety medication. How many micrograms are in this dose?

20. Dr. Kramer prescribes 0.4 mg of alprazolam, an antianxiety medication. How many micrograms are in this dose?

21. Digoxin is a medication used to treat heart problems. A cardiologist orders 0.125 mg of digoxin to be taken once daily. How many micrograms of digoxin are there in the daily dosage?

22. Digoxin is a medication used to treat heart problems. An internist orders 0.25 mg of digoxin to be taken once a day. How many micrograms of digoxin are there in the daily dosage?

23. Triazolam is a medication used for the short-term treatment of insomnia. A physician advises her patient to take one of the 0.125-mg tablets each night for 7 nights. How many milligrams of triazolam will the patient have ingested over that 7-day period? How many micrograms?

24. Clonidine is a medication used to treat high blood pressure. The usual starting dose of clonidine is one 0.1-mg tablet twice a day. If a patient is started on this dose by his physician, how many total milligrams of clonidine will the patient have taken before he returns to see his physician 14 days later? How many micrograms?

Skill Maintenance

Subtract.   [1.3a]

25.   5 7 8 9
        − 2 4 3 1

26.   8 4 2 9
        − 1 0 1 5

27.   4 0 9 7
        − 3 2 4 3

28.   8 3 9 0
        − 2 0 5 6

Simplify.   [2.7a]

29. $7x + 9 - 2x - 1$

30. $8x + 12 - 2x - 7$

31. $8t - 5 - t - 4$

32. $9r - 6 - r - 4$

Synthesis

33. ▦ A patient is directed to take 200 mg of nefazodone three times a day for one week, then 200 mg twice a day for a week, and then 100 mg three times a day for a week.

    a) How many grams of medication are used altogether?
    b) What is the average dosage size?

34. ▦ A patient is directed to take 200 mg of bupropion twice a day for a week, then 200 mg in the evening and 100 mg in the morning for a week, and then 100 mg twice a day for a week.

    a) How many grams of medication are used altogether?
    b) What is the average dosage size?

35. Naproxen sodium is a painkiller that lasts approximately 12 hours. A typical dose is one 220-mg tablet. Ibuprofen is a similar painkiller that lasts about 6 hours and has a typical dosage of two 200-mg tablets. Which is the better buy: a bottle containing 44 g of naproxen sodium costing $11 or a bottle containing 72 g of ibuprofen costing $11.24? Why?

Conversions and Formulas

American Units of Length: 12 in. = 1 ft; 3 ft = 1 yd; 36 in. = 1 yd; 5280 ft = 1 mi

Metric Units of Length: 1 km = 1000 m; 1 hm = 100 m; 1 dam = 10 m; 1 dm = 0.1 m; 1 cm = 0.01 m; 1 mm = 0.001 m

American–Metric Conversion: 1 m = 39.370 in.; 1 m = 3.281 ft; 1 ft = 0.305 m; 1 in. = 2.54 cm; 1 km = 0.621 mi; 1 mi = 1.609 km; 1 m = 1.094 yd; 1 yd = 0.914 m

American Units of Area: $1 \text{ yd}^2 = 9 \text{ ft}^2$; $1 \text{ ft}^2 = 144 \text{ in}^2$; $1 \text{ mi}^2 = 640$ acres; 1 acre = 43,560 ft^2

American Units of Capacity: 1 gal = 4 qt; 1 qt = 2 pt; 1 pt = 16 oz; 1 pt = 2 cups; 1 cup = 8 oz

Metric Units of Capacity: $1 \text{ L} = 1000 \text{ mL} = 1000 \text{ cm}^3 = 1000 \text{ cc}$

American–Metric Conversion: 1 oz = 29.57 mL

American Units of Weight: 1 T = 2000 lb; 1 lb = 16 oz

Metric Units of Mass: 1 t = 1000 kg; 1 kg = 1000 g; 1 hg = 100 g; 1 dag = 10 g; 1 dg = 0.1 g; 1 cg = 0.01 g; 1 mg = 0.001 g; 1 mcg = 0.000001 g

Temperature Conversion: $F = \dfrac{9}{5} \cdot C + 32$, or $F = 1.8 \cdot C + 32$;

$C = \dfrac{5}{9} \cdot (F - 32)$, or $C = \dfrac{F - 32}{1.8}$

Area of a Parallelogram: $A = b \cdot h$

Area of a Trapezoid: $A = \dfrac{1}{2} \cdot h \cdot (a + b)$

Radius and Diameter of a Circle: $d = 2 \cdot r$, or $r = \dfrac{d}{2}$

Circumference of a Circle: $C = \pi \cdot d$, or $C = 2 \cdot \pi \cdot r$

Area of a Circle: $A = \pi \cdot r \cdot r$, or $A = \pi \cdot r^2$

Volume of a Rectangular Solid: $V = l \cdot w \cdot h$

Volume of a Circular Cylinder: $V = \pi \cdot r^2 \cdot h$

Volume of a Sphere: $V = \dfrac{4}{3} \cdot \pi \cdot r^3$

Sum of Angle Measures of a Triangle: $m \angle A + m \angle B + m \angle C = 180°$

Pythagorean Equation: $a^2 + b^2 = c^2$

Vocabulary Reinforcement

Complete each statement with the correct word from the list at the right. Some of the choices may not be used and some may be used more than once.

1. A parallelogram is a four-sided figure with two pairs of _____ sides.   [9.3a]

2. A(n) _____ is the set of all points in space that are a given distance from a given point.   [9.4a]

3. The _____ of a circle is half the length of its diameter.   [9.3b]

4. Two angles are _____ if the sum of their measures is 180°.   [9.5c]

5. A(n) _____ triangle has all sides of different lengths.   [9.5d]

6. The _____ of a right triangle is the side opposite the right angle.   [9.6c]

circumference

radius

isosceles

scalene

parallel

sphere

cylinder

hypotenuse

leg

complementary

supplementary

Concept Reinforcement

Determine whether each statement is true or false.

_____ 1. To convert mm² to cm², move the decimal point 2 places to the left. [9.2b]

_____ 2. Since 1 yd = 3 ft, we multiply by 3 to convert square yards to square feet. [9.2a]

_____ 3. The number π is greater than 3.14 and $\frac{22}{7}$. [9.3b]

_____ 4. The acute angles of a right triangle are complementary. [9.5b, c, e]

_____ 5. The length of the hypotenuse of a right triangle is greater than the length of either of its legs. [9.6c]

_____ 6. You would probably use your furnace when the temperature was 40°C. [9.7c]

Study Guide

Objective 9.1a   Convert from one American unit of length to another.

Example   Complete: 126 in. = _____ yd.

$$126 \text{ in.} = \frac{126 \text{ in.}}{1} \cdot \frac{1 \text{ yd}}{36 \text{ in.}} = 3.5 \text{ yd}$$

Practice Exercise

1. Complete: 7 ft = _____ yd.

Objective 9.1b   Convert from one metric unit of length to another.

Example   Complete: 38 km = _____ cm.

To go from km to cm, we move the decimal point 5 places to the right.

38   38.00000.   38 km = 3,800,000 cm

Practice Exercise

2. Complete: 4.6 cm = _____ km.

Objective 9.1c   Convert between American and metric units of length.

Example   Complete: 42 ft = _____ m.
(Note: 1 ft $\approx$ 0.305 m.)

$$42 \text{ ft} = 42 \cdot 1 \text{ ft} \approx 42 \cdot 0.305 \text{ m} = 12.81 \text{ m}$$

Practice Exercise

3. Complete: 10 m = _____ yd.
(Note: 1 m $\approx$ 1.094 yd.)

Objective 9.2a   Convert from one American unit of area to another.

Example   Complete: 14,400 in² = _____ ft².

$$14,400 \text{ in}^2 = \frac{14,400 \text{ in}^2}{1} \cdot \frac{1 \text{ ft}^2}{144 \text{ in}^2} = 100 \text{ ft}^2$$

Practice Exercise

4. Complete: 81 ft² = _____ yd².

Objective 9.2b   Convert from one metric unit of area to another.

Example   Complete: 9.6 m² = _____ cm².

To go from m² to cm², we move the decimal point 2 × 2, or 4, places to the right.

9.6   9.6000.   9.6 m² = 96,000 cm²

Practice Exercise

5. Complete: 52.4 cm² = _____ mm².

Objective 9.3a Find the area of a parallelogram or trapezoid.

Examples Find the area of this parallelogram.

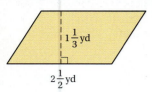

$$A = b \cdot h$$

$$= 2\frac{1}{2}\text{yd} \cdot 1\frac{1}{3}\text{yd}$$

$$= \frac{5}{2} \cdot \frac{4}{3} \cdot \text{yd} \cdot \text{yd}$$

$$= \frac{20}{6}\text{yd}^2 = \frac{10}{3}\text{yd}^2,$$

$$\text{or } 3\frac{1}{3}\text{ yd}^2$$

Find the area of this trapezoid.

$$A = \frac{1}{2} \cdot h \cdot (a + b)$$

$$= \frac{1}{2} \cdot 12 \text{ in.} \cdot (6 \text{ in.} + 18 \text{ in.})$$

$$= \frac{12 \cdot 24}{2}\text{in}^2 = 144 \text{ in}^2$$

Practice Exercises

6. Find the area of this parallelogram.

7. Find the area of this trapezoid.

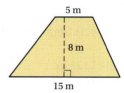

Objective 9.3b Find the circumference, area, radius, or diameter of a circle, given the length of a radius or diameter.

Examples Find the circumference of this circle. Use 3.14 for π.

$$C = \pi \cdot d, \quad \text{or} \quad 2 \cdot \pi \cdot r$$

$$\approx 2 \cdot 3.14 \cdot 4 \text{ ft}$$

$$= 25.12 \text{ ft}$$

Find the area of this circle. Use $\frac{22}{7}$ for π.

$$A = \pi \cdot r \cdot r, \quad \text{or} \quad \pi \cdot r^2$$

$$\approx \frac{22}{7} \cdot 21 \text{ mm} \cdot 21 \text{ mm}$$

$$= \frac{22 \cdot 21 \cdot 21}{7}\text{mm}^2 = 1386 \text{ mm}^2$$

Practice Exercises

8. Find the circumference of this circle. Use 3.14 for π.

9. Find the area of this circle. Use $\frac{22}{7}$ for π.

Objective 9.4a Find the volume of a rectangular solid, a cylinder, and a sphere.

Examples Find the volume of this rectangular solid.

$$V = l \cdot w \cdot h$$

$$= 3 \text{ yd} \cdot 2.6 \text{ yd} \cdot 7.1 \text{ yd}$$

$$= 3 \cdot 2.6 \cdot 7.1 \text{ yd}^3$$

$$= 55.38 \text{ yd}^3$$

Practice Exercises

10. Find the volume of this rectangular solid.

Objective 9.4a *(continued)*

Find the volume of this circular cylinder. Use 3.14 for π.

$$V = B \cdot h, \quad \text{or} \quad \pi \cdot r^2 \cdot h$$
$$\approx 3.14 \cdot 1.4\,\text{m} \cdot 1.4\,\text{m} \cdot 2.7\,\text{m}$$
$$= 16.61688\,\text{m}^3$$

2.7 m

1.4 m

11. Find the volume of this circular cylinder. Use $\dfrac{22}{7}$ for π.

$5\frac{2}{5}$ ft

$1\frac{1}{3}$ ft

Find the volume of this sphere. Use $\dfrac{22}{7}$ for π.

7 in.

$$V = \frac{4}{3} \cdot \pi \cdot r^3$$
$$\approx \frac{4}{3} \cdot \frac{22}{7} \cdot 7\,\text{in.} \cdot 7\,\text{in.} \cdot 7\,\text{in.}$$
$$= 1437\frac{1}{3}\,\text{in}^3$$

12. Find the volume of this sphere. Use 3.14 for π.

7.4 cm

Objective 9.4b   Convert from one unit of capacity to another.

Example   Complete: 6 gal = _____ pt.

$$6\,\text{gal} = 6 \cdot 1\,\text{gal}$$
$$= 6 \cdot 4\,\text{qt}$$
$$= 24 \cdot 1\,\text{qt}$$
$$= 24 \cdot 2\,\text{pt} = 48\,\text{pt}$$

Practice Exercise

13. Complete: 16 qt = _____ cups.

Objective 9.5c   Find the measure of a complement or a supplement of a given angle.

Example   Find the measure of a complement and a supplement of an angle that measures 65°.

The measure of the complement of an angle of 65° is 90° − 65°, or 25°.

The measure of the supplement of an angle of 65° is 180° − 65°, or 115°.

Practice Exercise

14. Find the measure of a complement and a supplement of an angle that measures 38°.

Objective 9.5e   Given two of the angle measures of a triangle, find the third.

Example   Find the missing angle measure.

$$m\angle A + m\angle B + m\angle C = 180°$$
$$x + 130° + 28° = 180°$$
$$x + 158° = 180°$$
$$x = 180° - 158°$$
$$x = 22°$$

The measure of $\angle A$ is 22°.

Practice Exercise

15. Find the missing angle measure.

72°   21° x

Objective 9.6c   Given the lengths of any two sides of a right triangle, find the length of the third side.

Example   Find the length of the third side of this triangle. Give an exact answer and an approximation to three decimal places.

$$a^2 + b^2 = c^2 \quad \text{Pythagorean equation}$$
$$a^2 + 11^2 = 15^2$$
$$a^2 + 121 = 225$$
$$a^2 = 225 - 121$$
$$a^2 = 104$$
$$a = \sqrt{104} \approx 10.198$$

Practice Exercise

16. Find the length of the third side of this right triangle. Give an exact answer and an approximation to three decimal places.

Objective 9.7a   Convert from one American unit of weight to another.

Example   Complete: $4020 \text{ oz} = $ _____ lb.

$$4020 \text{ oz} = 4020 \text{ o\!z} \times \frac{1 \text{ lb}}{16 \text{ o\!z}} = 251.25 \text{ lb}$$

Practice Exercise

17. Complete: $10{,}280 \text{ lb} = $ _____ T.

Objective 9.7b   Convert from one metric unit of mass to another.

Example   Complete: $5.62 \text{ cg} = $ _____ g.

To go from cg to g, we move the decimal point 2 places to the left.

$$5.62 \quad 0.05.62 \quad 5.62 \text{ cg} = 0.0562 \text{ g}$$

Practice Exercise

18. Complete: $9.78 \text{ mg} = $ _____ g.

Objective 9.7c   Convert temperatures from Celsius to Fahrenheit and from Fahrenheit to Celsius.

Examples   Convert 18°C to Fahrenheit and 95°F to Celsius.

$$F = \frac{9}{5}C + 32 = 1.8 \cdot 18 + 32$$
$$= 32.4 + 32 = 64.4$$

Thus, 18°C = 64.4°F.

$$C = \frac{5}{9} \cdot (F - 32) = \frac{5}{9} \cdot (95 - 32)$$
$$= \frac{5}{9} \cdot 63 = 35$$

Thus, 95°F = 35°C.

Practice Exercises

19. Convert 68°C to Fahrenheit.

20. Convert 104°F to Celsius.

Objective 9.8a   Make conversions and solve applied problems concerning medical dosages.

Example   A physician ordered 320 mL of 5% dextrose in water (D5W) to be administered intravenously over 4 hr. How many liters of D5W is this?

We convert 320 mL to liters:

$$320 \text{ mL} = 320 \cdot 1 \text{ mL}$$
$$= 320 \cdot 0.001 \text{ L}$$
$$= 0.32 \text{ L}.$$

The physician ordered 0.32 L of D5W.

Practice Exercise

21. A physician orders 0.5 L of normal saline solution. How many milliliters are ordered?

Review Exercises

Complete.

1. 10 ft = _____ yd
[9.1a]

2. $\frac{5}{6}$ yd = _____ in.
[9.1a]

3. 1.7 mm = _____ cm
[9.1b]

4. 2 yd = _____ in.
[9.1a]

5. 4 km = _____ cm
[9.1b]

6. 14 in. = _____ ft
[9.1a]

7. 200 m = _____ yd
[9.1c]

8. 20 mi = _____ km
[9.1c]

9. 5 lb = _____ oz
[9.7a]

10. 3 g = _____ kg
[9.7b]

11. 50 qt = _____ gal
[9.4b]

12. 28 gal = _____ pt
[9.4b]

13. 60 mL = _____ L
[9.4b]

14. 0.4 L = _____ mL
[9.4b]

15. 0.7 T = _____ lb
[9.7a]

16. 0.2 g = _____ mg
[9.7b]

17. 4.7 kg = _____ g
[9.7b]

18. 4 cg = _____ g
[9.7b]

19. 4 yd^2 = _____ ft^2
[9.2a]

20. 0.7 km^2 = _____ m^2
[9.2b]

21. 1008 in^2 = _____ ft^2
[9.2a]

22. 570 cm^2 = _____ m^2
[9.2b]

23. Find the circumference of a circle of radius 5 m. Use 3.14 for π. [9.3b]

24. Find the length of a radius of the circle. [9.3b]

$\frac{28}{11}$ in.

25. Find the length of a diameter of the circle. [9.3b]

12 m

26. *Track and Field.* Track meets are held on a track similar to the one shown below. Find the shortest distance around the track. Use 3.14 for π. [9.3b]

85.56 yd

85.56 yd

Find the area of each figure in Exercises 27–32. [9.3a, b]

27.

7 in.

3 in.

5 in.

28.

4.2 ft

7.1 ft

29.

5 cm

|← 12 cm →|

30.

4 mm

5 mm

|← 10 mm →|

31. Use $\frac{22}{7}$ for π.   **32.** Use 3.14 for π.

7 ft

20 cm

33. Find the area of the shaded region. Use 3.14 for π.   [9.3b]

21 ft

34. A "Norman" window is designed with dimensions as shown. Find its area. Use 3.14 for π.   [9.3b]

2 ft

5 ft

Use a protractor to measure each angle.   [9.5a]

35.

36.

P
Q
R

37.

38.

39–42. Classify each of the angles in Exercises 35–38 as right, straight, acute, or obtuse.   [9.5b]

43. Find the measure of a complement of $\angle BAC$.   [9.5c]

C

41°   120°   D

B   A

44. Find the measure of a supplement of a 44° angle.   [9.5c]

Use the following triangle for Exercises 45–47.

3.5

30°

4

2

x

45. Find the missing angle measure.   [9.5e]

46. Classify the triangle as equilateral, isosceles, or scalene.   [9.5d]

47. Classify the triangle as right, obtuse, or acute.   [9.5d]

Find the volume of each figure. Use 3.14 for π.   [9.4a]

48.

2.6 m

12 m

3 m

49.

14 ft

3 ft   4.6 ft

50.

90 cm

10 cm

51.

$r = 2$ yd

52.

12 cm

5 cm

53. Simplify: $\sqrt{64}$.   [9.6a]

54. ▣ Use a calculator to approximate $\sqrt{83}$ to three decimal places.   [9.6b]

For each right triangle, find the length of the side not given. Find an exact answer and an approximation to three decimal places. Assume that c represents the length of the hypotenuse.   [9.6c]

55. $a = 15, b = 25$ **56.** $a = 4, c = 10$

57.

c

5 ft

8 ft

58.

20 cm

a

18 cm

Solve.   [9.6d]

59. A wire 24 ft long reaches from the top of a pole to a point on the ground 16 ft from the base of the pole. How tall is the pole? Round to the nearest tenth of a foot.

60. *Construction.*   Chloe is designing rafters for a house. The rise of each rafter will be 6 ft and the run 12 ft. What is the rafter length? Round to the nearest hundredth of a foot.

Rafter length

Rise = 6 ft

Run = 12 ft

61. Convert 45°C to Fahrenheit.   [9.7c]

62. Convert 68°F to Celsius.   [9.7c]

Medical Dosage. Solve. [9.8a]

63. Amoxicillin is an antibiotic obtainable in a liquid suspension form, part medication and part water, and is frequently used to treat infections in infants. One formulation of the drug contains 125 mg of amoxicillin per 5 mL of liquid. A pediatrician orders 200 mg per day for a child with an ear infection. How much of the amoxicillin suspension would the parent need to administer to the child in order to achieve the recommended daily dose?

64. An emergency-room physician orders 3 L of Ringer's lactate to be administered over 4 hr for a patient suffering from shock and severe low blood pressure. How many milliliters is this?

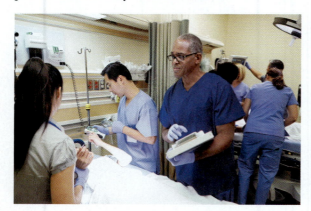

65. A physician prescribes 0.25 mg of alprazolam, an antianxiety medication. How many micrograms are in this dose?

66. Find the measure of a supplement of a $20\frac{3}{4}^{\circ}$ angle. [9.5c]

A. $339\frac{1}{4}^{\circ}$

B. $159\frac{1}{4}^{\circ}$

C. $69\frac{1}{4}^{\circ}$

D. $70\frac{1}{4}^{\circ}$

67. Complete: 0.16 gal = _____ cups. [9.4b]

A. 1.28

B. 2.56

C. 0.64

D. 160

Synthesis

68. *Running Record.* The world record for running the 200-m dash is 19.19 sec, set by Usain Bolt of Jamaica in the 2009 World Championships in Berlin. What would his record time be if he had run at the same rate in a 200-yd dash? [9.1c]

    Source: wikipedia.com

69. It is known that 1 gal of water weighs 8.3453 lb. Which weighs more: an ounce of pennies or an ounce (as capacity) of water? Explain. [9.4b], [9.7a]

70. A community center has a rectangular swimming pool that is 50 ft wide, 100 ft long, and 10 ft deep. The pool is filled with water to a line that is 1 ft from the top. Water costs $2.25 per 1000 ft³. How much does it cost to fill the pool? [9.4c]

71. 🔲 One lap around a standard running track is 440 yd. A marathon is 26 mi, 385 yd long. How many laps around a track does a marathon require? [9.1a]

72. 🔲 Find the area of the largest round pizza that can be baked in a 35-cm by 50-cm pan. Use 3.14 for π. [9.3b]

Understanding Through Discussion and Writing

1. Give at least two reasons why someone might prefer the use of grams to the use of ounces. [9.7a, b]

2. Is it possible for a triangle to contain two 90° angles? Why or why not? [9.5e]

3. Which is larger and why: one square meter or one square yard? [9.1c], [9.2a, b]

4. Explain a procedure that could be used to determine the measure of an angle's supplement from the measure of the angle's complement. [9.5c]

5. Explain how the Pythagorean theorem can be used to prove that a triangle is a *right* triangle. [9.6c]

6. Which occupies more volume: two spheres, each with radius *r*, or one sphere with radius 2*r*? Explain why. [9.4c]

CHAPTER

9 Test

For Extra Help

For step-by-step test solutions, access the Chapter Test Prep Videos in MyMathLab® or on YouTube (search "BittingerPrealgebra" and click on Channels).

Complete.

1. 8 ft = _____ in.

2. 280 cm = _____ m

3. 2 yd^2 = _____ ft^2

4. 5 km = _____ m

5. 9.1 mm = _____ cm

6. 4520 m^2 = _____ km^2

7. 2983 mL = _____ L

8. 3.8 kg = _____ g

9. 10 gal = _____ oz

10. 0.69 L = _____ mL

11. 9 lb = _____ oz

12. 4.11 T = _____ lb

13. Find the length of a radius of this circle.

14. Find the area of a circle of radius 4 m. Use 3.14 for π.

16 cm

15. Find the circumference of a circle of radius 14 ft. Use $\frac{22}{7}$ for π.

Find each area. Use 3.14 for π.

16.

2.5 cm

10 cm

17.

4 ft

3 ft

8 ft

18.

18.6 km

9.0 km

19. Find the measure of a complement and a supplement of $\angle CAD$.

C

D

65°

75°   30°

B   A   E

Use a protractor to measure each angle.

20.

21.

22. Classify the angle in Exercise 20 as right, straight, acute, or obtuse.

23. Classify the angle in Exercise 21 as right, straight, acute, or obtuse.

Use the following triangle for Exercises 24–26.

H

110°

10   10

35°   x

A   F

24. Find the missing angle measure.

25. Classify the triangle as equilateral, isosceles, or scalene.

26. Classify the triangle as right, obtuse, or acute.

27. A twelve-box rectangular carton of 12-oz juice boxes measures $10\frac{1}{2}$ in. by 8 in. by 5 in. What is the volume of the carton?

In Exercises 28–30, find the volume in each figure. Use 3.14 for π.

28.

5 ft

8 ft

29.

$r = 10$ yd

30.

5 m

2 m   3 m

31. How long must a wire be in order to reach from the top of a 13-m antenna to a point on the ground 9 m from the base of the antenna? Round to the nearest tenth of a meter.

For each right triangle, find the length of the side not given. Find an exact answer and an approximation to three decimal places.

32.

c 1

1

33.

10
7

b

34. Find $\sqrt{121}$.

35. Convert 32°F to Celsius.

36. An antihistamine solution contains 25 mg of medication in 5 mL of liquid. Pat is directed to take a 30-mg dose of the antihistamine. How many mL's of the solution should Pat take?

    A. 6 mL   **B.** 55 mL

    C. 125 mL   **D.** 150 mL

Synthesis

37. The measure of $\angle SMC$ is three times that of its complement. Find the measure of the supplement of $\angle SMC$.

38. ▦ A *board foot* is the amount of wood in a piece 12 in. by 12 in. by 1 in. A carpenter places the following order for a certain kind of lumber:

        25 pieces: 2 in. by 4 in. by 8 ft;

        32 pieces: 2 in. by 6 in. by 10 ft;

        24 pieces: 2 in. by 8 in. by 12 ft.

The price of this type of lumber is $225 per thousand board feet. What is the total cost of the carpenter's order?

Find the volume of the solid. (Note that the solids are not drawn in perfect proportion.) Give the answer in cubic feet. Use 3.14 for π.

39.

12 ft

2.6 in.
3 in.

40.

18 ft
$\frac{3}{4}$ in.

Solve.

1. *Energy.* The Indianapolis Resource Recovery Facility produces steam from trash. This steam is used to generate electricity. The facility produces at least 9.8 million tons of steam a day, enough to power the downtown Indianapolis heating loop. Find standard notation for 9.8 million.

   Source: covantaenergy.com

2. *Firefighting.* During a fire, the firefighters get a 1-ft layer of water on the 25-ft by 60-ft first floor of a 5-floor building. Water weighs $62\frac{1}{2}$ lb per cubic foot. What is the total weight of the water on the floor?

Calculate.

3. $1\frac{1}{2} + 2\frac{2}{3}$

4. $120.5 - 32.98$

5. $-27{,}148 \div 22$

6. $8^3 + 45 \cdot 24 - 9^2 \div 3$

7. $\left(\frac{1}{4}\right)^2 \div \left(\frac{1}{2}\right)^3 \times 2^4 + (10.3)(4)$

8. $14 \div [33 \div 11 + 8 \times 2 - (15 - 3)]$

Find fraction notation.

9. 1.209

10. 17%

Use $<$, $>$, or $=$ for $\square$ to write a true sentence.

11. $\dfrac{5}{6} \square \dfrac{7}{8}$

12. $\dfrac{15}{18} \square \dfrac{10}{12}$

Complete.

13. $2.5\,\text{yd} = \underline{\hspace{1cm}} \text{in.}$

14. $6\,\text{oz} = \underline{\hspace{1cm}} \text{lb}$

15. $15°C = \underline{\hspace{1cm}} °F$

16. $0.087\,\text{L} = \underline{\hspace{1cm}} \text{mL}$

17. $3\,\text{yd}^2 = \underline{\hspace{1cm}} \text{ft}^2$

18. $17\,\text{cm} = \underline{\hspace{1cm}} \text{m}$

Solve.

19. $x + \dfrac{3}{4} = \dfrac{7}{8}$

20. $\dfrac{3}{x} = \dfrac{7}{10}$

21. $1 - 7x = 4 - (x + 9)$

22. $-15x = 240$

23. Find the perimeter and the area.

24. Find the diameter, the circumference, and the area of this circle. Use $\frac{22}{7}$ for π.

25. Find the volume of this sphere. Use $\frac{22}{7}$ for π.

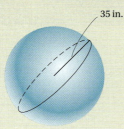

35 in.

26. Find the mean: 49, 53, 60, 62, 69.

Solve.

27. What is the simple interest on $8000 at 4.2% for $\frac{1}{4}$ year?

28. What is the amount in an account after 25 years if $8000 is invested at 4.2%, compounded annually?

29. How long must a rope be in order to reach from the top of an 8-m tree to a point on the ground 15 m from the bottom of the tree?

30. The sales tax on an office supply purchase of $5.50 is $0.33. What is the sales tax rate?

31. A bolt of fabric in a fabric store has $10\frac{3}{4}$ yd on it. A customer purchases $8\frac{5}{8}$ yd. How many yards remain on the bolt?

32. What is the cost, in dollars, of 15.6 gal of gasoline at 329.9¢ per gallon? Round to the nearest cent.

33. A box of powdered milk that makes 20 qt costs $4.99. A box that makes 8 qt costs $1.99. Which size has the lower unit price?

34. It is $\frac{7}{10}$ km from Maria's dormitory to the library. Maria started to walk from the dorm to the library, changed her mind after going $\frac{1}{4}$ of the distance, and returned to the dorm. How far did she walk?

35. Find the missing angle measure.

130°   20°

x

36. Combine like terms: $12a - 7 - 3a - 9$.

37. Graph: $y = -\dfrac{1}{3}x + 2$.

Synthesis

Find the volume in cubic feet. Use 3.14 for π.

38.

100 yd

10 ft

39.

14 ft

3 in.   4.6 in.

CHAPTER
10

Exponents and Polynomials

10.1

Integers as Exponents

OBJECTIVES

a Evaluate algebraic expressions containing whole-number exponents.

b Express exponential expressions involving negative exponents as equivalent expressions containing positive exponents.

SKILL TO REVIEW

Objective 1.9b: Evaluate exponential notation.

Evaluate.

1. 4^3 **2.** 12^2

Caution!

Don't confuse the powers 1 and 0. Be careful: $8^0 = 1$, but $8^1 = 8$.

Evaluate.

1. 16^1 **2.** -16^1

3. $(-35)^1$ **4.** 7^0

5. $(-9)^0$ **6.** -8^0

Answers

Skill to Review:
1. 64   **2.** 144

Margin Exercises:
1. 16   **2.** −16   **3.** −35   **4.** 1   **5.** 1
6. −1

We have already used the numbers 2, 3, . . ., as exponents. In this section, we consider 1, 0, and negative integers as exponents.

a   ONE AND ZERO AS EXPONENTS

Look for a pattern in the following:

$$8 \cdot 8 \cdot 8 \cdot 8 = 8^4 \quad \text{We divide by 8 each time.}$$
$$8 \cdot 8 \cdot 8 = 8^3$$
$$8 \cdot 8 = 8^2$$
$$8 = 8^?$$
$$1 = 8^?.$$

The exponents decrease by 1 each time. Continuing the pattern, we have

$$8 = 8^1 \quad \text{and} \quad 1 = 8^0.$$

We make the following definitions.

> **THE EXPONENTS 1 AND 0**
>
> $b^1 = b$, for any number b.
> $b^0 = 1$, for any nonzero number b.

EXAMPLE 1   Evaluate 23^1, -23^1, and $(-23)^1$.

$$23^1 = 23;$$
$$-23^1 = -23; \quad \text{We read } -23^1 \text{ as "the opposite of } 23^1\text{."}$$
$$(-23)^1 = -23 \quad \text{We read } (-23)^1 \text{ as "negative 23 to the first."}$$

EXAMPLE 2   Evaluate 3^0, $(-3)^0$, and -3^0.

$$3^0 = 1; \qquad\qquad\qquad\qquad 3 \text{ is the base; 0 is the exponent.}$$
$$(-3)^0 = 1; \qquad\qquad\qquad -3 \text{ is the base; 0 is the exponent.}$$
$$-3^0 = -1 \cdot 3^0 = -1 \cdot 1 = -1 \qquad \text{Note that } -3^0 \neq (-3)^0.$$

◀ **Do Margin Exercises 1–6.**

EXAMPLE 3   Evaluate $m^0 + 5$ for $m = 9$.

$$m^0 + 5 = 9^0 + 5 = 1 + 5 = 6$$

EXAMPLE 4 Evaluate $(3x + 2)^0$ for $x = -5$.

We substitute -5 for x and follow the rules for order of operations:

$$(3x + 2)^0 = (3(-5) + 2)^0 \quad \text{Substituting}$$
$$= (-15 + 2)^0 \quad \text{Multiplying}$$
$$= (-13)^0 = 1.$$

Do Exercises 7 and 8. ▶

7. Evaluate $(2x - 9)^0$ for $x = 3$.

8. Evaluate $t^0 - 4$ for $t = 7$.

b NEGATIVE INTEGERS AS EXPONENTS

The pattern used to help define the exponents 1 and 0 can also be used to define negative-integer exponents:

$$8 \cdot 8 \cdot 8 = 8^3 \quad | \quad \text{We divide by 8 each time.}$$
$$8 \cdot 8 = 8^2$$
$$8 = 8^1$$
$$1 = 8^0$$
$$\frac{1}{8} = 8^?$$
$$\frac{1}{8 \cdot 8} = 8^?.$$

The exponents decrease by 1 each time. To continue the pattern, we would say that

$$\frac{1}{8} = 8^{-1} \quad \text{and} \quad \frac{1}{8 \cdot 8} = 8^{-2}.$$

Thus, if we are to preserve the above pattern, we must have

$$\frac{1}{8^1} = 8^{-1} \quad \text{and} \quad \frac{1}{8^2} = 8^{-2}.$$

This leads to our definition of negative exponents.

NEGATIVE EXPONENTS

For any nonzero numbers a and b, and any integer n,

$$a^{-n} = \frac{1}{a^n} \quad \text{and} \quad \left(\frac{a}{b}\right)^{-n} = \left(\frac{b}{a}\right)^n.$$

(A base raised to a negative exponent is equal to the reciprocal of the base raised to a positive exponent.)

EXAMPLES Write an equivalent expression using a positive exponent. Then simplify.

5. $4^{-2} = \dfrac{1}{4^2} = \dfrac{1}{16}$ Note that 4^{-2} represents a *positive* number.

6. $(-3)^{-2} = \dfrac{1}{(-3)^2} = \dfrac{1}{(-3)(-3)} = \dfrac{1}{9}$

7. $m^{-3} = \dfrac{1}{m^3}$

8. $ab^{-1} = a \cdot b^{-1} = a\left(\dfrac{1}{b^1}\right) = a\left(\dfrac{1}{b}\right) = \dfrac{a}{b}$ Think of a as $\dfrac{a}{1}$ if you wish.

CALCULATOR CORNER

Expressions with Negative Exponents Most calculators can evaluate expressions like

$$\left(-\frac{2}{5}\right)^{-6}$$

with just a few keystrokes. On most calculators with a $\boxed{+/-}$ key, the following keystrokes will work.

$$\boxed{2} \; \boxed{\div} \; \boxed{5} \; \boxed{=} \; \boxed{+/-}$$
$$\boxed{x^y} \; \boxed{6} \; \boxed{+/-} \; \boxed{=}$$

On most calculators with a $\boxed{(-)}$ key, the following keystrokes will work.

$$\boxed{(} \; \boxed{(-)} \; \boxed{2} \; \boxed{\div} \; \boxed{5} \; \boxed{)}$$
$$\boxed{\wedge} \; \boxed{(-)} \; \boxed{6} \; \boxed{\text{ENTER}}$$

The same result, 244.140625, is found when $\left(-\frac{5}{2}\right)^6$ is evaluated.

EXERCISES Calculate the value of each pair of expressions.

1. $\left(\dfrac{3}{4}\right)^{-5}$; $\left(\dfrac{4}{3}\right)^5$

2. $\left(\dfrac{4}{5}\right)^{-7}$; $\left(\dfrac{5}{4}\right)^7$

3. $\left(-\dfrac{5}{4}\right)^{-7}$; $\left(-\dfrac{4}{5}\right)^7$

4. $\left(-\dfrac{7}{5}\right)^{-9}$; $\left(-\dfrac{5}{7}\right)^9$

5. $\left(-\dfrac{5}{6}\right)^{-4}$; $\left(-\dfrac{6}{5}\right)^4$

6. $\left(-\dfrac{9}{8}\right)^{-6}$; $\left(-\dfrac{8}{9}\right)^6$

Answers

7. 1   **8.** −3

Write an equivalent expression with positive exponents. Then simplify.

9. 4^{-3} **10.** 5^{-2}

11. 2^{-4} **12.** $(-2)^{-3}$

13. $\left(\dfrac{5}{3}\right)^{-2}$

14. $xy^{-2} = x \cdot y^{-2}$

$= x \cdot \dfrac{1}{\boxed{}}$

$= \dfrac{x}{\boxed{}}$

Write an equivalent expression with negative exponents.

15. $\dfrac{1}{9^2}$ **16.** $\dfrac{7}{x^4}$

Write an equivalent expression with positive exponents.

17. $\dfrac{m^3}{n^{-5}}$ **18.** $\dfrac{ab}{c^{-1}}$

Write an equivalent expression with positive exponents.

19. $\dfrac{a^4}{b^{-6}}$ **20.** $\dfrac{x^7 y}{z^{-4}}$

21. $\dfrac{a^4 b^{-7}}{c^{-3}} = \dfrac{a^4 c^{\boxed{}}}{b^{\boxed{}}}$

Answers

9. $\dfrac{1}{4^3}$; $\dfrac{1}{64}$ 10. $\dfrac{1}{5^2}$; $\dfrac{1}{25}$ 11. $\dfrac{1}{2^4}$; $\dfrac{1}{16}$
12. $\dfrac{1}{(-2)^3}$; $-\dfrac{1}{8}$ 13. $\left(\dfrac{3}{5}\right)^2$; $\dfrac{9}{25}$ 14. $x \cdot \dfrac{1}{y^2}$; $\dfrac{x}{y^2}$
15. 9^{-2} 16. $7x^{-4}$ 17. $m^3 n^5$
18. abc 19. $a^4 b^6$ 20. $x^7 y z^4$ 21. $\dfrac{a^4 c^3}{b^7}$

Guided Solutions:
14. y^2, y^2 21. $3, 7$

9. $\left(\dfrac{5}{6}\right)^{-2} = \left(\dfrac{6}{5}\right)^2$ $\left(\dfrac{a}{b}\right)^{-n} = \left(\dfrac{b}{a}\right)^n$

$= \dfrac{6}{5} \cdot \dfrac{6}{5} = \dfrac{36}{25}$

·············· **Caution!** ··············

Note in Example 5 that

$$4^{-2} \neq 4(-2) \quad \text{and} \quad 4^{-2} \neq \dfrac{1}{-4^2}.$$

In general, $a^{-n} \neq a(-n)$. The negative exponent also does not indicate that a negative number is in the result.

···

◀ **Do Exercises 9–14.**

EXAMPLES Write an equivalent expression using a negative exponent.

10. $\dfrac{1}{7^2} = 7^{-2}$ Reading $a^{-n} = \dfrac{1}{a^n}$ from right to left: $\dfrac{1}{a^n} = a^{-n}$

11. $\dfrac{5}{x^8} = 5 \cdot \dfrac{1}{x^8} = 5x^{-8}$

◀ **Do Exercises 15 and 16.**

Consider an expression like

$$\dfrac{a^2}{b^{-3}},$$

in which the denominator is a negative power. We can simplify as follows:

$$\dfrac{a^2}{b^{-3}} = \dfrac{a^2}{\dfrac{1}{b^3}} \qquad \text{Rewriting } b^{-3} \text{ as } \dfrac{1}{b^3}$$

$$= a^2 \cdot \dfrac{b^3}{1} \qquad \text{To divide by a fraction, we multiply by its reciprocal.}$$

$$= a^2 b^3.$$

◀ **Do Exercises 17 and 18.**

Our work above indicates that to divide by a base raised to a negative power, we can instead multiply by the opposite power of the same base. This will shorten our work.

EXAMPLES Write an equivalent expression using positive exponents.

12. $\dfrac{x^3}{y^{-2}} = x^3 y^2$ Instead of dividing by y^{-2}, multiply by y^2.

13. $\dfrac{a^2 b^5}{c^{-6}} = a^2 b^5 c^6$ Instead of dividing by c^{-6}, multiply by c^6.

14. $\dfrac{x^{-2} y}{z^{-3}} = x^{-2} y z^3 = \dfrac{y z^3}{x^2}$

◀ **Do Exercises 19–21.**

✓ Reading Check

Choose the number from the following list that makes each statement true. Not every number will be used.

$$-2 \qquad -1 \qquad 0 \qquad 1 \qquad 2$$

RC1. $7^{\square} = 1$

RC2. $7^{\square} = 7$

RC3. $7^{-2} = \dfrac{1}{7^{\square}}$

RC4. $\dfrac{1}{7} = 7^{\square}$

a   Evaluate.

1. 4^0

2. 17^0

3. 3.14^1

4. 2.67^1

5. $(-19.57)^1$

6. $(-34.6)^0$

7. $(-98.6)^0$

8. $(-98.6)^1$

9. $x^0, x \neq 0$

10. $a^0, a \neq 0$

11. x^1, for $x = 97$

12. a^1, for $a = -10$

13. $(3x - 17)^0$, for $x = 10$

14. $(7x - 45)^0$, for $x = 8$

15. $(5x - 3)^1$, for $x = 4$

16. $(35 - 4x)^1$, for $x = 8$

17. $(4m - 19)^0$, for $m = 3$

18. $(9 - 2x)^0$, for $x = 5$

19. $3x^0 + 4$, for $x = -2$

20. $7x^0 + 6$, for $x = -3$

21. $(3x)^0 + 4$, for $x = -2$

22. $(7x)^0 + 6$, for $x = -3$

23. $(5 - 3x^0)^1$, for $x = 19$

24. $(5x^1 - 29)^0$, for $x = 4$

b   Write an equivalent expression with a positive exponent. Then simplify, if possible.

25. 3^{-2}

26. 2^{-3}

27. 10^{-4}

28. 5^{-6}

29. t^{-4}

30. x^{-2}

31. 7^{-1}

32. 10^{-1}

33. $(-5)^{-2}$

34. $(-4)^{-3}$

35. $(-x)^{-2}$

36. $(-n)^{-2}$

Simplify.

37. $3x^{-7}$

38. $-6y^{-2}$

39. mn^{-5}

40. cd^{-1}

41. xy^{-1}

42. pw^{-3}

43. $-7a^{-9}$

44. $9p^{-4}$

45. $\dfrac{1}{x^{-3}}$

46. $\dfrac{1}{x^{-1}}$

47. $\dfrac{x}{y^{-4}}$

48. $\dfrac{r}{t^{-7}}$

49. $\dfrac{r^5}{t^{-3}}$

50. $\dfrac{x^7}{y^{-5}}$

51. $\dfrac{xy}{z^{-1}}$

52. $\dfrac{ab}{c^{-6}}$

53. $\dfrac{a^8 b}{c^{-2}}$

54. $\dfrac{x^7 y}{t^{-6}}$

55. $\dfrac{x^2 y^3}{z^{-8}}$

56. $\dfrac{x^8 y^{10}}{z^{-1}}$

57. $\dfrac{x^{-4} y^3}{t^{-2}}$

58. $\dfrac{ab^{-6}}{c^{-2}}$

59. $\dfrac{cd^{-4}}{a^{-1}}$

60. $\dfrac{x^3 y^{-8}}{z^{-4}}$

61. $\left(\dfrac{2}{5}\right)^{-2}$

62. $\left(\dfrac{3}{7}\right)^{-2}$

63. $\left(\dfrac{5}{a}\right)^{-3}$

64. $\left(\dfrac{x}{3}\right)^{-4}$

Write an equivalent expression using a negative exponent.

65. $\dfrac{1}{7^3}$

66. $\dfrac{1}{5^2}$

67. $\dfrac{9}{x^3}$

68. $\dfrac{4}{y^2}$

Skill Maintenance

69. When used for a singles match, a regulation tennis court is 27 ft by 78 ft. Find its perimeter.   [2.7b]

70. The Floral Doctor's delivery van traveled 147 mi on 10.5 gal of gas. How many miles per gallon did the van get?   [7.2a]

71. Ramon's new truck gets 21 mpg. This is 20% more than the mileage his old truck got. What mileage did the old truck get?   [8.5a]

72. A 5% sales tax is added to the price of a two-speed washing machine. If the machine is priced at $399, find the total amount paid.   [8.6a]

73. Of the 8 fish Mac caught, 3 were trout. What percentage were not trout?   [8.4a]

74. The diameter of a compact disc is 12 cm. What is its circumference? (Use 3.14 for π.)   [9.3b]

75. Multiply: $-57 \cdot 48$.   [2.4a]

76. Multiply: $(-72)(-46)$.   [2.4a]

Synthesis

Simplify.

77. $3 + 4^{-1} \cdot 10^2$

78. $(2 + 3)^{-2} \cdot 10^2 \div 2^{-1}$

79. $4^2 \div 2^{-1} \cdot 5^1 - (12 \cdot 10)^0$

80. $(42 - 6 \cdot 7)^8 \cdot (13 - 5 \cdot 4)^{-3}$

81. ▦ Evaluate $\dfrac{3^x}{3^{x-1}}$ for $x = -4$ and then for $x = -40$.

82. ▦ Evaluate $\dfrac{5^x}{5^{x+1}}$ for $x = -3$ and then for $x = -30$.

There are several rules that we can use to simplify exponential expressions. We first consider multiplying powers with like bases. Throughout this section, we assume that a variable in a denominator is not zero.

a MULTIPLYING POWERS WITH LIKE BASES

When we multiply expressions like a^3 and a^2, we can use the associative law to combine the factors:

$$a^3 \cdot a^2 = \underbrace{(a \cdot a \cdot a)}_{\text{3 factors}}\underbrace{(a \cdot a)}_{\text{2 factors}} = \underbrace{a \cdot a \cdot a \cdot a \cdot a}_{\text{5 factors}} = a^5.$$

Note that the exponent in a^5 is the sum of the exponents in a^3 and a^2. Similarly,

$$b^4 \cdot b^3 = (b \cdot b \cdot b \cdot b)(b \cdot b \cdot b) = b^7, \quad \text{where} \quad 4 + 3 = 7.$$

Adding the exponents gives the correct result.

THE PRODUCT RULE

For any nonzero number a and any integers m and n,

$$a^m \cdot a^n = a^{m+n}.$$

(To multiply powers with the same base, keep the base and add the exponents.)

EXAMPLES Multiply and simplify. Here "simplify" means to express the product as one base with a positive exponent whenever possible.

1. $3^6 \cdot 3^2 = 3^{6+2}$ Adding exponents: $a^m \cdot a^n = a^{m+n}$

     $= 3^8$

2. $x^4 \cdot x^5 = x^{4+5} = x^9$

3. $5 \cdot 5^6 \cdot 5^2 = 5^1 \cdot 5^6 \cdot 5^2$ Writing 5 as 5^1

     $= 5^{1+6+2}$

     $= 5^9$

4. $x^3 \cdot x^{-8} = x^{3+(-8)}$ The product rule works with negative exponents.

     $= x^{-5}$

     $= \dfrac{1}{x^5}$ Writing with a positive exponent

5. $(a^3b^2)(a^3b^5) = (a^3a^3)(b^2b^5)$ Using the associative law and the commutative law

     $= a^6b^7$ Adding exponents

Do Margin Exercises 1–5. ▶

OBJECTIVES

a   Use the product rule to multiply exponential expressions with like bases.

b   Use the quotient rule to divide exponential expressions with like bases.

c   Use the power rule to raise powers to powers.

d   Raise a product to a power and a quotient to a power.

SKILL TO REVIEW

Objective 2.3a: Subtract integers, and simplify combinations of additions and subtractions.

Simplify.

1. $2 - (-4)$

2. $-10 - (-6)$

············· **Caution!** ·············

Use the product rule correctly.

$2^4 \cdot 2^{10} = 2^{14}$

$2^4 \cdot 2^{10} \neq 4^{14}$ Incorrect!

$2^4 \cdot 2^{10} \neq 2^{40}$ Incorrect!

Simplify.

1. $x^4 \cdot x^9$ **2.** $y^{10} \cdot y$

3. $2^4 \cdot 2^2 \cdot 2^3$ **4.** $7^{-1} \cdot 7^4$

5. $(xy^3)(x^4y^5)$

Answers

Skill to Review:
1. 6   **2.** −4

Margin Exercises:
1. x^{13} **2.** y^{11} **3.** 2^9 **4.** 7^3 **5.** x^5y^8

---

b DIVIDING POWERS WITH LIKE BASES

Recall that any expression that is divided or multiplied by 1 is unchanged. This, together with the fact that anything (besides 0) divided by itself is 1, leads to a rule for division:

$$\frac{a^5}{a^2} = \frac{a \cdot a \cdot a \cdot a \cdot a}{a \cdot a} = \frac{a \cdot a \cdot a}{1} \cdot \frac{a \cdot a}{a \cdot a} = \frac{a \cdot a \cdot a}{1} \cdot 1 = a^3.$$

Note that the exponent in a^3 is the difference of the exponents in $\frac{a^5}{a^2}$; that is, $5 - 2 = 3$. Similarly,

$$\frac{x^4}{x^3} = \frac{x \cdot x \cdot x \cdot x}{x \cdot x \cdot x} = \frac{x}{1} \cdot \frac{x \cdot x \cdot x}{x \cdot x \cdot x} = \frac{x}{1} \cdot 1 = x^1, \text{ or } x.$$

Subtracting the exponents gives the correct result: $4 - 3 = 1$.

··········· **Caution!** ···············

Use the quotient rule correctly.

$$\frac{7^{10}}{7^2} = 7^8$$

$$\frac{7^{10}}{7^2} \neq 1^8 \quad \text{Incorrect!}$$

$$\frac{7^{10}}{7^2} \neq 7^5 \quad \text{Incorrect!}$$

> ### THE QUOTIENT RULE
> ....
> For any nonzero number a and any integers m and n,
>
> $$\frac{a^m}{a^n} = a^{m-n}.$$
>
> (To divide powers with the same base, keep the base and subtract the exponent of the denominator from the exponent of the numerator.)

EXAMPLES Divide and simplify. Here "simplify" means to write the expression as one base with a positive exponent whenever possible.

6. $\dfrac{4^7}{4^2} = 4^{7-2}$ Subtracting exponents

     $= 4^5$

7. $\dfrac{x^9}{x^{-3}} = x^{9-(-3)} = x^{12}$ Use parentheses when subtracting negative exponents.

8. $\dfrac{p}{p^5} = \dfrac{p^1}{p^5}$ Writing p as p^1

     $= p^{1-5}$

     $= p^{-4}$

     $= \dfrac{1}{p^4}$ Writing with a positive exponent

9. $\dfrac{3^{10}}{3^{10}} = 3^{10-10} = 3^0 = 1$ $a^0 = 1$

10. $\dfrac{c^8 d^{12}}{c^3 d^{10}} = \dfrac{c^8}{c^3} \cdot \dfrac{d^{12}}{d^{10}}$

     $= c^{8-3} d^{12-10}$

     $= c^5 d^2$

◀ **Do Exercises 6–10.**

Simplify.

6. $\dfrac{x^5}{x^3}$ **7.** $\dfrac{a^8}{a}$

8. $\dfrac{x^{11}}{x^{11}}$ **9.** $\dfrac{x^2 y^5}{x y^4}$

10. $\dfrac{4^{-3}}{4^{10}} = 4^{\square - \square}$ **GS**

     $= 4^{\square}$

     $= \dfrac{1}{4^{\square}}$

Answers

6. x^2 **7.** a^7 **8.** 1   **9.** xy **10.** $\dfrac{1}{4^{13}}$

Guided Solution:
10. $-3, 10, -13, 13$

c RAISING A POWER TO A POWER

Consider an expression like $(7^2)^4$:

$$(7^2)^4 = (7^2)(7^2)(7^2)(7^2) \qquad \text{There are four factors of } 7^2.$$
$$= (7 \cdot 7)(7 \cdot 7)(7 \cdot 7)(7 \cdot 7)$$
$$= 7 \cdot 7 \cdot 7 \cdot 7 \cdot 7 \cdot 7 \cdot 7 \cdot 7 \qquad \text{Using an associative law}$$
$$= 7^8.$$

Note that the exponent in 7^8 is the product of the exponents in $(7^2)^4$. Similarly,

$$(y^5)^3 = y^5 \cdot y^5 \cdot y^5 \qquad \text{There are three factors of } y^5.$$
$$= (y \cdot y \cdot y \cdot y \cdot y)(y \cdot y \cdot y \cdot y \cdot y)(y \cdot y \cdot y \cdot y \cdot y)$$
$$= y^{15};$$

or $\qquad (y^5)^3 = y^{5 \cdot 3} = y^{15}.$

> ### THE POWER RULE
> ....
> For any nonzero number a and any integers m and n,
> $$(a^m)^n = a^{mn}.$$
> (To raise a power to a power, multiply the exponents and leave the base unchanged.)

EXAMPLES Simplify. Express the answer using positive exponents.

11. $(2^4)^3 = 2^{4 \cdot 3} \qquad$ Multiplying exponents
$$ = 2^{12}$$

12. $(y^{-3})^8 = y^{-3 \cdot 8} = y^{-24} = \dfrac{1}{y^{24}} \qquad$ **13.** $(x^{-2})^{-10} = x^{(-2)(-10)} = x^{20}$

Do Exercises 11–13. ▶

Simplify.

11. $(3^4)^5$

12. $(5^{-2})^8$

13. $(t^{-8})^{-5}$

d RAISING A PRODUCT OR A QUOTIENT TO A POWER

When an expression inside parentheses is raised to a power, the inside expression is the base. Let's compare $2a^3$ and $(2a)^3$.

$2a^3 = 2 \cdot a \cdot a \cdot a \quad$ The base is a. $\qquad (2a)^3 = (2a)(2a)(2a) \quad$ The base is $2a$.
$$ = (2 \cdot 2 \cdot 2)(a \cdot a \cdot a)$$
$$ = 2^3 a^3$$
$$ = 8a^3$$

We see that $2a^3$ and $(2a)^3$ are *not* equivalent. Note too that $(2a)^3$ can be simplified by cubing each factor in $2a$. This leads to the following rule for raising a product to a power.

> ### RAISING A PRODUCT TO A POWER
> ....
> For any nonzero numbers a and b and any integer n,
> $$(ab)^n = a^n b^n.$$
> (To raise a product to a power, raise each factor to that power.)

EXAMPLES Simplify.

14. $(3a^4)^2 = (3^1a^4)^2$ Writing 3 as 3^1

$= (3^1)^2 \cdot (a^4)^2$ Raising each factor to the second power

$= 3^2 \cdot a^8 = 9a^8$ Multiplying exponents and simplifying

15. $(-5x^5y^8)^3 = (-5)^3(x^5)^3(y^8)^3$ Raising each factor to the third power

$= -125x^{15}y^{24}$

◀ **Do Exercises 14 and 15.**

There is a similar rule for raising a quotient to a power.

RAISING A QUOTIENT TO A POWER

For any nonzero numbers a and b and any integer n,

$$\left(\frac{a}{b}\right)^n = \frac{a^n}{b^n}.$$

(To raise a quotient to a power, raise the numerator to the power and divide by the denominator raised to the power.)

EXAMPLES

16. $\left(\frac{x}{7}\right)^2 = \frac{x^2}{7^2} = \frac{x^2}{49}$ Squaring the numerator and the denominator

17. $\left(\frac{5}{a^4}\right)^3 = \frac{5^3}{(a^4)^3}$ Raising a quotient to a power

$= \frac{5^3}{a^{12}} = \frac{125}{a^{12}}$ Using the power rule (multiplying exponents) and simplifying

Do Exercises 16 and 17.

In the following summary of definitions and rules, we assume that no denominators are 0 and 0^0 is not considered.

DEFINITIONS AND PROPERTIES OF EXPONENTS

For any integers m and n,

1 as an exponent:	$a^1 = a$
0 as an exponent:	$a^0 = 1$
Negative exponents:	$a^{-n} = \dfrac{1}{a^n}$
The product rule:	$a^m \cdot a^n = a^{m+n}$
The quotient rule:	$\dfrac{a^m}{a^n} = a^{m-n}$
The power rule:	$(a^m)^n = a^{mn}$
Raising a product to a power:	$(ab)^n = a^n b^n$
Raising a quotient to a power:	$\left(\dfrac{a}{b}\right)^n = \dfrac{a^n}{b^n}$

Simplify.

14. $(x^3y^8)^5$

15. $(-2ab^4)^2$

$= (-2)\ \square\ (a)\ \square\ (b^4)\ \square$

$= \square\ a\ \square\ b\ \square$

GS

Simplify.

16. $\left(\dfrac{a}{b}\right)^8$ **17.** $\left(\dfrac{x^4}{y^5}\right)^2$

Answers

14. $x^{15}y^{40}$ **15.** $4a^2b^8$ **16.** $\dfrac{a^8}{b^8}$ **17.** $\dfrac{x^8}{y^{10}}$

Guided Solution:

15. 2, 2, 2, 4, 2, 8

☑ Reading Check

Match each expression with an equivalent expression from the column on the right.

RC1. $4^5 \cdot 4^3$ _____

RC2. $(4^5)^3$ _____

RC3. $\dfrac{4^5}{4^3}$ _____

RC4. $\left(\dfrac{4}{5}\right)^3$ _____

RC5. $(4 \cdot 5)^3$ _____

a) 4^{5-3}

b) $4^{5 \cdot 3}$

c) 4^{5+3}

d) $4^3 \cdot 5^3$

e) $\dfrac{4^3}{5^3}$

a Simplify. Assume that no denominator is 0.

1. $3^4 \cdot 3^2$

2. $4^5 \cdot 4^7$

3. $x^2 \cdot x^{12}$

4. $y^3 \cdot y^9$

5. $a^6 \cdot a$

6. $x \cdot x^4$

7. $8 \cdot 8^5$

8. $2^{10} \cdot 2$

9. $5^{17} \cdot 5^{21}$

10. $4^{32} \cdot 4^{16}$

11. $x^2 \cdot x \cdot x^5$

12. $y \cdot y^4 \cdot y^8$

13. $a^{-5} \cdot a^8$

14. $c^{12} \cdot c^{-9}$

15. $p^6 \cdot p^{-5}$

16. $t^{-4} \cdot t^5$

17. $x^{-5} \cdot x^{-7}$

18. $y^{-3} \cdot y^{-4}$

19. $5^4 \cdot 5^{-10}$

20. $6^{-12} \cdot 6^2$

21. $(a^3 b^5)(ab^9)$

22. $(x^2 y^{10})(x^5 y)$

23. $a^{-4} \cdot a^9 \cdot a^{-6}$

24. $x^{-11} \cdot x^4 \cdot x^6$

b Simplify. Assume that no denominator is 0.

25. $\dfrac{8^9}{8^2}$

26. $\dfrac{3^{11}}{3^3}$

27. $\dfrac{x^4}{x^3}$

28. $\dfrac{x^{12}}{x^{11}}$

29. $\dfrac{a^{10}}{a}$

30. $\dfrac{c^{12}}{c}$

31. $\dfrac{x^2}{x^5}$

32. $\dfrac{y^4}{y^{12}}$

33. $\dfrac{7^8}{7^8}$

34. $\dfrac{10^6}{10^6}$

35. $\dfrac{t^2}{t^{-1}}$

36. $\dfrac{n^{-5}}{n^8}$

37. $\dfrac{c^{-2}}{c^{-6}}$

38. $\dfrac{t^{-2}}{t}$

39. $\dfrac{x^4 y^{12}}{xy^7}$

40. $\dfrac{u^9 v^{15}}{u^8 v}$

c Simplify. Assume that no denominator is 0.

41. $(x^2)^5$

42. $(y^5)^7$

43. $(2^{-3})^6$

44. $(3^{-2})^4$

45. $(7^{-1})^{-6}$

46. $(9^{-4})^{-3}$

47. $(y^{12})^{-1}$

48. $(x^{-18})^{-1}$

d Simplify. Assume that no denominator is 0.

49. $(x^2y^5)^2$

50. $(a^3b^4)^2$

51. $(2x^2y)^3$

52. $(3xy^3)^2$

53. $(-2a^2b^5)^3$

54. $(-5a^4b^2)^3$

55. $(-9x^2y^2)^2$

56. $(-10x^5y)^2$

57. $\left(\dfrac{a}{b}\right)^7$

58. $\left(\dfrac{x}{y}\right)^{10}$

59. $\left(\dfrac{a^2}{2}\right)^3$

60. $\left(\dfrac{x^4}{5}\right)^2$

61. $\left(\dfrac{x^4}{y^5}\right)^3$

62. $\left(\dfrac{a^6}{b^{10}}\right)^4$

63. $\left(\dfrac{2^4}{3^5}\right)^2$

64. $\left(\dfrac{5^9}{3^5}\right)^4$

Skill Maintenance

65. A sidewalk of uniform width is built around three sides of a store, as shown in the figure. What is the area of the sidewalk?   [1.8a]

66. A real estate agent's commission rate is 6%. The agent receives a commission of $7380 on the sale of a home. For how much did the home sell?   [8.6b]

Add.   [2.2a]

67. $-5 + (-12)$

68. $-19 + 12$

69. $17 + (-24)$

70. $-15 + (-2)$

Synthesis

Simplify. Assume that no denominator is 0.

71. $(3y)^{10}(3y)^6$

72. $(x^2y^3z^8)^0$

73. $\left(\dfrac{x}{y}\right)^{-1}$

74. $a^x \cdot a^{2x}$

75. $\dfrac{x^{3y}}{x^{2y}}$

76. $(a^{-3}b^2)^{-4}$

77. $\dfrac{a^{33}}{a^{3(11)}}$

78. $\left(\dfrac{1}{a}\right)^{-n}$

684   CHAPTER 10   Exponents and Polynomials

Copyright © 2016 Pearson Education, Inc.

Scientific Notation

a WRITING SCIENTIFIC NOTATION

We can write numbers using different types of notation, such as fraction notation, decimal notation, and percent notation. Now we study another, **scientific notation,** which makes use of exponential notation. Scientific notation is especially useful when calculations involve very large or very small numbers. The following are examples of scientific notation:

- The number of flamingos in Africa's Great Rift Valley:
  $4 \times 10^6 = 4{,}000{,}000$

- The length of an *E.coli* bacterium: 2×10^{-6} m $= 0.000002$ m

OBJECTIVES

a Convert between scientific notation and decimal notation.

b Multiply and divide using scientific notation.

c Solve applied problems using scientific notation.

SCIENTIFIC NOTATION

Scientific notation for a number is an expression of the type

$$M \times 10^n,$$

where n is an integer, M is greater than or equal to 1 and less than 10 $(1 \leq M < 10)$, and M is expressed in decimal notation. The notation 10^n is also considered to be scientific notation when $M = 1$.

You should try to make conversions to scientific notation mentally as much as possible. To do so, note the following.

In scientific notation, a positive power of 10 indicates a large number (greater than or equal to 10) and a negative power of 10 indicates a small number (between 0 and 1).

Converting from scientific notation to decimal notation involves multiplying by a power of 10. To convert $M \times 10^n$ to decimal notation, we move the decimal point.

- When n is positive, we move the decimal point right n places.
- When n is negative, we move the decimal point left $|n|$ places.

EXAMPLES Convert each number to decimal notation.

1. $7.893 \times 10^5 = 789,300$ The decimal point moves right 5 places.

   Think: Positive exponent indicates large number.

   Check: $7.893 \times 10^5 = 7.893 \times 100,000 = 789,300$

2. $4.7 \times 10^{-8} = 0.000000047$ The decimal point moves left 8 places.

   Think: Negative exponent indicates small number.

   Check: $4.7 \times 10^{-8} = 4.7 \times 0.00000001 = 0.000000047$

◀ **Do Exercises 1 and 2.**

We reverse the process when converting from decimal notation to scientific notation. To convert a number to scientific notation, we write it in the form $M \times 10^n$, where M has exactly one nonzero digit to the immediate left of the decimal point. Compare the following.

Scientific Notation	*NOT Scientific Notation*	
1.98×10^{12}	19.8×10^{11}	19.8 is greater than 10.
2.007×10^{-4}	0.2007×10^{-3}	0.2007 is less than 1.

◀ **Do Exercises 3–6.**

EXAMPLES Convert each number to scientific notation.

3. $78,000 = 78,000. = 7.8000 \times 10^4 = 7.8 \times 10^4$

                  4 places

   Check: $7.8 \times 10^4 = 7.8 \times 10,000 = 78,000$

4. $0.0000923 = 000009.23 \times 10^{-5} = 9.23 \times 10^{-5}$

              5 places

   Check: $9.23 \times 10^{-5} = 9.23 \times 0.00001 = 0.0000923$

◀ **Do Exercises 7 and 8.**

b **MULTIPLYING AND DIVIDING USING SCIENTIFIC NOTATION**

Multiplying

Consider the product $400 \cdot 2000 = 800,000$. In scientific notation, this is

$$(4 \times 10^2) \cdot (2 \times 10^3) = (4 \cdot 2)(10^2 \cdot 10^3) = 8 \times 10^5.$$

By applying the commutative and associative laws, we found this product by multiplying $4 \cdot 2$, to get 8, and $10^2 \cdot 10^3$, to get 10^5 (adding exponents).

Convert each number to decimal notation.

1. 6.893×10^{11}

2. 5.67×10^{-5}

Determine whether each number is written in scientific notation.

3. 8.06×10^3

4. 0.49×10^{12}

5. 10.608×10^{-3}

6. 1.008×10^{-6}

Convert each number to scientific notation.

7. 0.000517

8. $523,000,000$

Answers

1. 689,300,000,000   2. 0.0000567   3. Yes
4. No   5. No   6. Yes   7. 5.17×10^{-4}
8. 5.23×10^8

EXAMPLE 5 Multiply: $(1.8 \times 10^6) \cdot (2.3 \times 10^{-4})$.

We apply the commutative and associative laws to get

$$(1.8 \times 10^6) \cdot (2.3 \times 10^{-4}) = (1.8 \cdot 2.3) \times (10^6 \cdot 10^{-4})$$
$$= 4.14 \times 10^{6+(-4)}$$
$$= 4.14 \times 10^2.$$

EXAMPLE 6 Multiply: $(3.1 \times 10^5) \cdot (4.5 \times 10^{-3})$.

$$(3.1 \times 10^5) \cdot (4.5 \times 10^{-3}) = (3.1 \times 4.5)(10^5 \cdot 10^{-3})$$

$$= 13.95 \times 10^2 \qquad \text{Not scientific notation;} \atop \text{13.95 is greater than 10.}$$

$$= (1.395 \times 10^1) \times 10^2 \qquad \text{Substituting} \atop 1.395 \times 10^1 \text{ for } 13.95$$

$$= 1.395 \times (10^1 \times 10^2) \qquad \text{Using the associative law}$$

$$= 1.395 \times 10^3 \qquad \text{Adding exponents;} \atop \text{the answer is now in} \atop \text{scientific notation.}$$

Do Exercises 9 and 10. ▶

Multiply and write scientific notation for the result.

9. $(1.12 \times 10^{-8})(5 \times 10^{-7})$

GS **10.** $(9.1 \times 10^{-17})(8.2 \times 10^3)$

$$= (9.1 \times 8.2)(10^{-17} \cdot 10^3)$$
$$= \boxed{} \times 10^{-14}$$
$$= (\boxed{} \times 10^1) \times 10^{-14}$$
$$= 7.462 \times 10^{\boxed{}}$$

Dividing

Consider the quotient $800{,}000 \div 400 = 2000$. In scientific notation, this is

$$(8 \times 10^5) \div (4 \times 10^2) = \frac{8 \times 10^5}{4 \times 10^2} = \frac{8}{4} \times 10^{5-2} = 2 \times 10^3.$$

EXAMPLE 7 Divide: $(3.41 \times 10^5) \div (1.1 \times 10^{-3})$.

$$(3.41 \times 10^5) \div (1.1 \times 10^{-3}) = \frac{3.41 \times 10^5}{1.1 \times 10^{-3}}$$

$$= \frac{3.41}{1.1} \times \frac{10^5}{10^{-3}}$$

$$= 3.1 \times 10^{5-(-3)}$$

$$= 3.1 \times 10^8$$

Divide and write scientific notation for each result.

11. $\dfrac{4.2 \times 10^5}{2.1 \times 10^2}$

EXAMPLE 8 Divide: $(3.2 \times 10^{-7}) \div (8.0 \times 10^6)$.

$$(3.2 \times 10^{-7}) \div (8.0 \times 10^6) = \frac{3.2 \times 10^{-7}}{8.0 \times 10^6}$$

$$= \frac{3.2}{8.0} \times \frac{10^{-7}}{10^6}$$

$$= 0.4 \times 10^{-13} \qquad \text{Not scientific} \atop \text{notation; 0.4 is less} \atop \text{than 1.}$$

$$= (4.0 \times 10^{-1}) \times 10^{-13} \qquad \text{Substituting} \atop 4.0 \times 10^{-1} \text{ for } 0.4$$

$$= 4.0 \times (10^{-1} \times 10^{-13}) \qquad \text{Using the} \atop \text{associative law}$$

$$= 4.0 \times 10^{-14} \qquad \text{Adding exponents}$$

GS **12.** $\dfrac{1.1 \times 10^{-4}}{2.0 \times 10^{-7}}$

$$= \frac{1.1}{2.0} \times \frac{10^{-4}}{10^{-7}}$$
$$= 0.55 \times 10^{-4-(\boxed{})}$$
$$= 0.55 \times 10^{\boxed{}}$$
$$= (\boxed{} \times 10^{-1}) \times 10^3$$
$$= 5.5 \times 10^{\boxed{}}$$

Answers

9. 5.6×10^{-15} **10.** 7.462×10^{-13}
11. 2.0×10^3 **12.** 5.5×10^2

Guided Solutions:
10. $74.62, 7.462, -13$ **12.** $-7, 3, 5.5, 2$

Do Exercises 11 and 12. ▶

CALCULATOR CORNER

Scientific Notation

Most scientific calculators have a key labeled $\boxed{\text{EE}}$ which is used for scientific notation. We simply enter the decimal portion of the number, press $\boxed{\text{EE}}$, and then enter the exponent. Addition, subtraction, multiplication, and division are performed as usual. Many calculators abbreviate scientific notation. A calculator display like $\boxed{5\text{E}16}$ means 5×10^{16}.

EXERCISES Multiply or divide and express each answer in scientific notation.

1. $(3.15 \times 10^7)(4.3 \times 10^{-12})$

2. $(4 \times 10^4)(9 \times 10^7)$

3. $\dfrac{4 \times 10^{-9}}{5 \times 10^{16}}$

4. $\dfrac{9 \times 10^{11}}{3 \times 10^{-2}}$

C   APPLICATIONS WITH SCIENTIFIC NOTATION

EXAMPLE 9   *Distance from the Sun to Earth.*   Light from the sun traveling at a rate of 300,000 km/s (kilometers per second) reaches Earth in 499 sec. Find the distance, expressed in scientific notation, from the sun to Earth.

The time it takes light to reach Earth from the sun is 4.99×10^2 sec. The speed is 3.0×10^5 km/s. Recall that distance can be expressed in terms of speed and time as

$$\text{Distance} = \text{Speed} \times \text{Time}.$$

We substitute 3.0×10^5 for the speed and 4.99×10^2 for the time:

$$
\begin{aligned}
\text{Distance} &= (3.0 \times 10^5)(4.99 \times 10^2) && \text{Substituting}\\
&= 14.97 \times 10^7 && \text{Note that 14.97 is greater than 10.}\\
&= (1.497 \times 10^1) \times 10^7 &&\\
&= 1.497 \times (10^1 \times 10^7) && \text{Converting to scientific notation}\\
&= 1.497 \times 10^8 \text{ km.} &&
\end{aligned}
$$

Thus the distance from the sun to Earth is 1.497×10^8 km.

EXAMPLE 10   *Media Usage.*   In January 2013, the 800 million active YouTube users viewed 120 billion videos on the site. On average, how many videos did each user view?

Source: YouTube

In order to find the average number of YouTube videos that each user viewed, we divide the total number viewed by the number of users. We first write each number using scientific notation:

$$800 \text{ million} = 800{,}000{,}000 = 8 \times 10^8,$$
$$120 \text{ billion} = 120{,}000{,}000{,}000 = 1.2 \times 10^{11}.$$

We then divide 1.2×10^{11} by 8×10^8:

$$
\begin{aligned}
\frac{1.2 \times 10^{11}}{8 \times 10^8} &= \frac{1.2}{8} \times \frac{10^{11}}{10^8}\\
&= 0.15 \times 10^3 = (1.5 \times 10^{-1}) \times 10^3 = 1.5 \times 10^2.
\end{aligned}
$$

On average, each user viewed 1.5×10^2 videos.

◀ **Do Exercises 13 and 14.**

---

13. *Niagara Falls Water Flow.* On the Canadian side, the amount of water that spills over Niagara Falls in 1 min during the summer is about

$$1.3088 \times 10^8 \text{ L.}$$

How much water spills over the falls in one day? Express the answer in scientific notation.

14. *DNA.* The width of a DNA (deoxyribonucleic acid) double helix is about 2×10^{-9} m. If its length, fully stretched, is 5×10^{-2} m, how many times longer is the helix than it is wide?

Answers

13. 1.884672×10^{11} L
14. The length of the helix is 2.5×10^7 times its width.

Translating for Success

1. **Test Items.** On a test of 90 items, Sally got 80% correct. How many items did she get correct?

2. **Suspension Bridge.** The San Francisco/Oakland Bay suspension bridge is 0.4375 mi long. Convert this distance to yards.

3. **Population Growth.** Brookdale's growth rate per year is 0.9%. If the population was 1,500,000 in 2015, what is the population in 2016?

4. **Roller Coaster Drop.** The Manhattan Express Roller Coaster at the New York–New York Hotel and Casino, Las Vegas, Nevada, has a 144-ft drop. The California Screamin' Roller Coaster at Disney California Adventure, Anaheim, California, has a 32.635-m drop. How much larger in meters is the drop of the Manhattan Express than the drop of the California Screamin'?

5. **Driving Distance.** Nate drives the company car 675 mi in 15 days. At this rate, how far will he drive in 20 days?

The goal of these matching questions is to practice step (2), Translate, of the five-step problem-solving process. Translate each word problem to an equation and select a correct translation from equations A–O.

A. $32.635 \text{ m} + x = 144 \text{ ft} \cdot \dfrac{0.305 \text{ m}}{1 \text{ ft}}$

B. $x = 0.4375 \text{ mi} \times \dfrac{5280 \text{ ft}}{1 \text{ mi}} \times \dfrac{12 \text{ in.}}{1 \text{ ft}}$

C. $80\% \cdot x = 90$

D. $x = 0.89 \text{ km} \cdot \dfrac{1000 \text{ m}}{1 \text{ km}} \cdot \dfrac{1 \text{ m}}{3.281 \text{ ft}}$

E. $x = 80\% \cdot 90$

F. $x = 420 \text{ m} + 75 \text{ ft} \cdot \dfrac{0.305 \text{ m}}{1 \text{ ft}}$

G. $\dfrac{x}{20} = \dfrac{675}{15}$

H. $x = 0.4375 \text{ mi} \times \dfrac{5280 \text{ ft}}{1 \text{ mi}} \times \dfrac{1 \text{ yd}}{3 \text{ ft}}$

I. $x = 0.89 \text{ km} \cdot \dfrac{0.621 \text{ mi}}{1 \text{ km}} \cdot \dfrac{5280 \text{ ft}}{1 \text{ mi}}$

J. $\dfrac{x}{15} = \dfrac{675}{20}$

K. $x = 420 \text{ m} \cdot \dfrac{3.281 \text{ ft}}{1 \text{ m}} + 75 \text{ ft}$

L. $144 \text{ ft} + x = 32.635 \text{ m} \cdot \dfrac{1 \text{ ft}}{0.305 \text{ m}}$

M. $x = 1,500,000 - 0.9\% \,(1,500,000)$

N. $20 \cdot x = 675$

O. $x = 1,500,000 + 0.9\% \,(1,500,000)$

Answers on page A-22

6. **Test Items.** Jason answered correctly 90 items on a recent test. These items represented 80% of the total number of questions. How many items were on the test?

7. **Population Decline.** Flintville's growth rate per year is −0.9%. If the population was 1,500,000 in 2015, what is the population in 2016?

8. **Bridge Length.** The Tatara Bridge in Onomichi-Imabari, Japan, is 0.89 km long. Convert this distance to feet.

9. **Height of Tower.** The Willis Tower in Chicago is 75 ft taller than the Jin Mao Building in Shanghai. The height of the Jin Mao Building is 420 m. What is the height of the Willis Tower in feet?

10. **Gasoline Usage.** Nate's company car gets 20 miles to the gallon in city driving. How many gallons will it use in 675 mi of city driving?

Reading Check

Determine whether each statement is true or false.

_____ **RC1.** To convert a number less than 1 to scientific notation, move the decimal point to the right.

_____ **RC2.** A positive exponent in scientific notation indicates a number greater than or equal to 10.

_____ **RC3.** The number 12.856×10^{16} is written in scientific notation.

_____ **RC4.** The notations $\frac{1}{2}$, 0.5, 2^{-1}, and 5×10^{-1} all represent the same number.

a Convert each number to scientific notation.

1. 28,000,000,000

2. 4,900,000,000,000

3. 907,000,000,000,000,000

4. 168,000,000,000,000

5. 0.00000304

6. 0.000000000865

7. 0.000000018

8. 0.00000000002

9. 100,000,000,000

10. 0.0000001

11. *Population of the United States.* It is estimated that the population of the United States will be 419,854,000 in 2050. Convert 419,854,000 to scientific notation.
Source: U.S. Census Bureau

12. *Microprocessors.* The minimum feature size of a microprocessor is the transistor gate length. In 2011, the transistor gate length for a new microprocessor was about 0.000000028 m. Convert 0.000000028 to scientific notation.

13. *Wavelength of Light.* The wavelength of red light is 0.00000068 m. Convert 0.00000068 to scientific notation.

14. *Olympics.* Great Britain spent about $15,000,000,000 to stage the 2012 Summer Olympics. Convert 15,000,000,000 to scientific notation.
Source: cnn.com

Convert each number to decimal notation.

15. 8.74×10^7

16. 1.85×10^8

17. 5.704×10^{-8}

18. 8.043×10^{-4}

19. 10^7

20. 10^6

21. 10^{-5}

22. 10^{-8}

b   Multiply or divide and write scientific notation for each result.

23. $(3 \times 10^4)(2 \times 10^5)$

24. $(3.9 \times 10^8)(1.2 \times 10^{-3})$

25. $(5.2 \times 10^5)(6.5 \times 10^{-2})$

26. $(7.1 \times 10^{-7})(8.6 \times 10^{-5})$

27. $(9.9 \times 10^{-6})(8.23 \times 10^{-8})$

28. $(1.123 \times 10^4) \times 10^{-9}$

29. $\dfrac{8.5 \times 10^8}{3.4 \times 10^{-5}}$

30. $\dfrac{5.6 \times 10^{-2}}{2.5 \times 10^5}$

31. $(3.0 \times 10^6) \div (6.0 \times 10^9)$

32. $(1.5 \times 10^{-3}) \div (1.6 \times 10^{-6})$

33. $\dfrac{7.5 \times 10^{-9}}{2.5 \times 10^{12}}$

34. $\dfrac{4.0 \times 10^{-3}}{8.0 \times 10^{20}}$

c   Solve.

35. *Coral Reefs.*   There are 10 million bacteria per square centimeter of coral in a coral reef. The coral reefs near the Hawaiian Islands cover 14,000 km². How many bacteria are there in Hawaii's coral reef?

Sources: livescience.com; U.S. Geological Survey

36. *River Discharge.*   The average discharge at the mouths of the Amazon River is 4,200,000 cubic feet per second. How much water is discharged from the Amazon River in 1 yr? Express the answer in scientific notation.

Space Travel.   Use the following information for Exercises 37 and 38.

APPROXIMATE DISTANCE FROM EARTH TO:	
Moon	240,000 miles
Mars	35,000,000 miles
Pluto	2,670,000,000 miles

37. *Time to Reach Mars.*   Suppose it takes about 3 days for a space vehicle to travel from Earth to the moon. About how long would it take the same space vehicle traveling at the same speed to reach Mars? Express the answer in scientific notation.

38. *Time to Reach Pluto.*   Suppose it takes about 3 days for a space vehicle to travel from Earth to the moon. About how long would it take the same space vehicle traveling at the same speed to reach the dwarf planet Pluto? Express the answer in scientific notation.

39. *Information Technology.* In 2012, about 2.5×10^{18} bytes of information were generated each day by the worldwide online population of 2×10^9 people. Find the average amount of information generated per person per day in 2012.

Sources: IBM; internetworldstats.com

40. *Computer Technology.* Intel Corporation has developed silicon-based connections that use lasers to move data at a rate of 50 gigabytes per second. The printed collection of the U.S. Library of Congress contains 10 terabytes of information. How long would it take to copy this information using these connections? *Note:* 1 gigabyte $= 10^9$ bytes and 1 terabyte $= 10^{12}$ bytes.

Sources: spie.org; newworldencyclopedia.org

41. *Gold Leaf.* Gold can be milled into a very thin film called *gold leaf*. This film is so thin that it took only 43 oz of gold to cover the dome of Georgia's state capitol building. The gold leaf used was 5×10^{-6} m thick. In contrast, a U.S. penny is 1.55×10^{-3} m thick. How many sheets of gold leaf are in a stack that is the height of a penny?

Source: georgiaencyclopedia.org

42. *Relative Size.* An influenza virus is about 1.2×10^{-7} m in diameter. A staphylococcus bacterium is about 1.5×10^{-6} m in diameter. How many influenza viruses would it take, laid side by side, to equal the diameter of the bacterium?

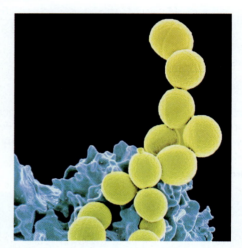

Skill Maintenance

Solve.

43. $2x - 4 - 5x + 8 = x - 3$ [5.7b]

44. $4(x - 3) + 5 = 6(x + 2) - 8$ [5.7b]

45. $2 + \dfrac{3}{4}t = \dfrac{1}{2}$ [4.4b]

46. $\dfrac{5}{16} = \dfrac{3}{x}$ [7.3b]

Synthesis

47. 🖩 Carry out the indicated operations. Express the result in scientific notation.

$$\frac{(5.2 \times 10^6)(6.1 \times 10^{-11})}{1.28 \times 10^{-3}}$$

48. Find the reciprocal and express it in scientific notation.
$$6.25 \times 10^{-3}$$

49. Find the LCM for 6.4×10^8 and 1.28×10^4.

50. *Stars.* It is estimated that there are 10 billion trillion stars in the known universe. Express the number of stars in scientific notation. *Note:* 1 billion $= 10^9$; trillion $= 10^{12}$.

Concept Reinforcement

Determine whether each statement is true or false.

_____ **1.** Raising a number to a negative power may not result in a negative number.   [10.1b]

_____ **2.** Any number raised to the zero power is zero.   [10.1a]

_____ **3.** In scientific notation, if the exponent is negative, the number is very large.   [10.3a]

Guided Solutions

 Fill in each blank with the number that creates a correct statement or solution.

4. $y^2 \cdot y^{10} = y^{\boxed{}+10} = y^{\boxed{}}$ [10.2a]

5. $\dfrac{y^2}{y^{10}} = y^{\boxed{}-10} = y^{\boxed{}} = \dfrac{1}{y^{\boxed{}}}$ [10.2b]

6. Convert to scientific notation.

$$4{,}032{,}000{,}000{,}000 = \boxed{} \times 10^{\boxed{}}$$ [10.3a]

7. Convert to scientific notation.

$$0.000038 = \boxed{} \times 10^{\boxed{}}$$ [10.3a]

Mixed Review

Evaluate.   [10.1a]

8. $(-12)^0$

9. $(-12)^1$

10. -12^0

11. $(12 - 6t)^1$ for $t = 5$

Write an equivalent expression with a positive exponent.   [10.1b]

12. x^{-7}

13. 3^{-1}

14. $9x^{-6}$

15. $\dfrac{1}{x^{-4}}$

Simplify. Assume that no denominator is 0.   [10.2a, b, c, d]

16. $3^4 \cdot 3^{11}$

17. $6 \cdot 6^9$

18. $x^{-6} \cdot x^{17}$

19. $y^{-2} \cdot y \cdot y^{-3}$

20. $(x^2y^8)(x^4y^3)$

21. $\dfrac{8^{11}}{8^2}$

22. $\dfrac{x^8}{x^9}$

23. $\dfrac{d^5}{d^5}$

24. $\dfrac{n^4}{n^{-1}}$

25. $\dfrac{a^{12}b^{11}}{a^4b^7}$

26. $(x^4)^8$

27. $(y^{-2})^{-6}$

28. $(x^{-3})^1$

29. $(5^4)^{-12}$

30. $(x^3y^4)^5$

31. $(3x^2y^4)^2$

32. $(-2ab^5)^3$

33. $\left(\dfrac{c}{d}\right)^{10}$

34. $\left(\dfrac{2}{x^3}\right)^2$

35. $\left(\dfrac{a^4}{b^8}\right)^7$

Convert each number to scientific notation. [10.3a]

36. 0.000907

37. 431,080,000,000

Convert each number to decimal notation. [10.3a]

38. 2.08×10^8

39. 4.1×10^{-6}

40. The wavelength of a radio wave is given by the velocity divided by the frequency. The velocity of radio waves is approximately 300,000,000 m/sec, and the frequency of Rick's ham radio repeater is 1,200,000,000 cycles per sec. What is the wavelength of Rick's repeater frequency? Express the answer in scientific notation. [10.3c]

41. A good length for a radio antenna is one fourth of the wavelength of the frequency it is designed to receive. What should the length of such a *quarterwave* antenna be for the frequency described in Exercise 40? [10.3c]

42. Multiply. Write scientific notation for the result: $(4.5 \times 10^{12})(5.2 \times 10^{20})$. [10.3b]
  A. 23.4×10^{240}
  B. 23.4×10^{32}
  C. 2.34×10^{33}
  D. 2.34×10^{31}

43. Divide. Write scientific notation for the result: $\dfrac{2.4 \times 10^{-6}}{4.8 \times 10^{-20}}$. [10.3b]
  A. 5.0×10^{13}
  B. 0.5×10^{14}
  C. 0.5×10^{-26}
  D. 5.0×10^{-15}

Understanding Through Discussion and Writing

44. Under what circumstances is $a^0 > a^1$? [10.1a]

45. Consider the expresssion x^{-3}. When evaluated, will the expression ever be negative? Explain. [10.1b]

46. Which number is larger and why: 5^{-8} or 6^{-8}? Do not use a calculator. [10.1b]

47. Emma can give a measurement using the unit km or the unit mm. Which measurement would require the larger exponent if scientific notation were used? Why? [10.3c]

Addition and Subtraction of Polynomials

10.4

A *term* is a number, a variable, a product of numbers and/or variables, or a quotient of numbers and/or variables. Thus, expressions like

$$5x^2, \quad -34, \quad \frac{3}{4}ab^2, \quad xy^3z^5, \quad \text{and} \quad \frac{7n}{m}$$

are terms. A term is called a **monomial** if there is no division by a variable expression. All of the terms above, except for $\frac{7n}{m}$, are monomials. Monomials are used to form **polynomials** like the following:

$$a^2b + c^3, \quad 5y + 3, \quad 3x^2 + 2x - 5, \quad -7a^3 + \tfrac{1}{2}a, \quad 37p^4, \quad x, \quad 0.$$

> ### POLYNOMIAL
>
> A **polynomial** is a monomial or a combination of sums and/or differences of monomials.

In a monomial, the number multiplied by the variable or variables is called the **coefficient**. The coefficient of $-6a^2b$ is -6. The coefficient of $\frac{1}{3}x^4$ is $\frac{1}{3}$. Since $x^2 = 1 \cdot x^2$, the coefficient of x^2 is 1.

Do Margin Exercises 1 and 2. ▶

a   ADDING POLYNOMIALS

Addition is commutative and associative; that is, we can reorder and regroup addends without changing the sum. We use these properties when adding polynomials in order to pair "like" terms. Recall that when two terms have the same variable(s) raised to the same power(s), they are like terms and can be combined.

EXAMPLE 1   Add: $(5x^3 + 4x^2 + 3x) + (2x^3 + 5x^2 - x)$.

$(5x^3 + 4x^2 + 3x) + (2x^3 + 5x^2 - x)$ $5x^3$ and $2x^3$ are like terms; $4x^2$ and $5x^2$ are like terms; $3x$ and $-x$ are like terms

$= (5x^3 + 2x^3) + (4x^2 + 5x^2) + (3x - x)$ Using the commutative and associative laws to pair like terms

$= 7x^3 + 9x^2 + 2x$ Combining like terms. Remember that $x = 1x$.

OBJECTIVES

a Add polynomials.

b Find the opposite of a polynomial.

c Subtract polynomials.

d Evaluate a polynomial.

SKILL TO REVIEW

Objective 2.7a: Combine like terms.

Combine like terms.
1. $8x^2 - 3x + 4x - x^2$
2. $2x^3 - 6 - 10 - 3x^3 + 9x$

---

1. What is the coefficient of $19x^3$?

GS **2.** What is the coefficient of $-a^2$?
Since $-a^2 = \boxed{} \cdot a^2$, the coefficient of $-a^2$ is $\boxed{}$.

Answers

Skill to Review:
1. $7x^2 + x$ 2. $-x^3 + 9x - 16$

Margin Exercises:
1. 19 2. -1

Guided Solution:
2. $-1, -1$

EXAMPLE 2 Add: $(3a^2 + 7a^2b) + (5a^2 - 6ab^2)$.

$$(3a^2 + 7a^2b) + (5a^2 - 6ab^2) = (3a^2 + 5a^2) + 7a^2b - 6ab^2$$
$$= 8a^2 + 7a^2b - 6ab^2 \qquad \text{Combining like terms}$$

EXAMPLE 3 Add: $(7x^2 + 5) + (5x^3 + 4x)$.

$$(7x^2 + 5) + (5x^3 + 4x) = 7x^2 + 5 + 5x^3 + 4x \qquad \text{There are no like terms here.}$$
$$= 5x^3 + 7x^2 + 4x + 5 \qquad \text{Rearranging the order}$$

Note in Example 3 that we wrote the answer so that the powers of x decrease as we read from left to right. This **descending order** is the traditional way of expressing an answer.

◀ **Do Exercises 3–5.**

Add.

3. $(7a^2 + 2a + 8) + (2a^2 + a - 9)$

4. $(5x^2y + 3x^2 + 4) + (2x^2y + 4x)$ **GS**
$$= (5x^2y + \boxed{}) + 3x^2 + 4 + 4x$$
$$= \boxed{} + 3x^2 + 4x + 4$$

5. $(2a^3 + 17) + (2a^2 - 9a)$

b OPPOSITES OF POLYNOMIALS

To subtract a number, we can add its opposite. We can similarly subtract a polynomial by adding its opposite. To check if two polynomials are opposites, recall that 5 and -5 are opposites because $5 + (-5) = 0$.

THE OPPOSITE OF A POLYNOMIAL

Two polynomials are **opposites**, or **additive inverses**, of each other if their sum is zero.

The opposite of $-9x^2 + x - 7$ is $9x^2 - x + 7$ because

$$(-9x^2 + x - 7) + (9x^2 - x + 7) = -9x^2 + 9x^2 + x + (-x) + (-7) + 7$$
$$= 0.$$

This can be said using algebraic symbolism:

$$\text{The opposite of} \quad (-9x^2 + x - 7) \quad \text{is} \quad 9x^2 - x + 7.$$

$$- \quad (-9x^2 + x - 7) \quad = \quad 9x^2 - x + 7$$

Note that in the opposite polynomials, the terms in each pair are opposites.

$-9x^2$ and $9x^2$ are opposites.

x and $-x$ are opposites.

-7 and 7 are opposites.

TO FIND THE OPPOSITE OF A POLYNOMIAL

We can find an equivalent polynomial for the opposite, or additive inverse, of a polynomial by replacing each term with its opposite— that is, by *changing the sign of every term*.

Answers

3. $9a^2 + 3a - 1$
4. $7x^2y + 3x^2 + 4x + 4$
5. $2a^3 + 2a^2 - 9a + 17$

Guided Solution:
4. $2x^2y, 7x^2y$

EXAMPLE 4 Find two equivalent expressions for the opposite of

$$4x^5 - 7x^3 - 8x + \tfrac{5}{6}.$$

a) $-\left(4x^5 - 7x^3 - 8x + \tfrac{5}{6}\right)$ This is one expression for the opposite of $4x^5 - 7x^3 - 8x + \tfrac{5}{6}$.

b) $-4x^5 + 7x^3 + 8x - \tfrac{5}{6}$ Changing the sign of every term

Thus, $-\left(4x^5 - 7x^3 - 8x + \tfrac{5}{6}\right)$ is equivalent to $-4x^5 + 7x^3 + 8x - \tfrac{5}{6}$, and each is the opposite of the original polynomial $4x^5 - 7x^3 - 8x + \tfrac{5}{6}$.

Do Exercises 6–9. ▶

EXAMPLE 5 Simplify: $-\left(-7x^4 - \tfrac{5}{9}x^3 + 8x^2 - x + 67\right)$.

$$-\left(-7x^4 - \tfrac{5}{9}x^3 + 8x^2 - x + 67\right) = 7x^4 + \tfrac{5}{9}x^3 - 8x^2 + x - 67$$

Do Exercises 10–12. ▶

c   SUBTRACTING POLYNOMIALS

We can now subtract a polynomial by adding the opposite of that polynomial. That is, for any polynomials P and Q, $P - Q = P + (-Q)$.

EXAMPLE 6 Subtract:

$$(9x^5 + x^3 - 2x^2 + 4) - (2x^5 + x^4 - 4x^3 - 3x^2).$$

We have

$(9x^5 + x^3 - 2x^2 + 4) - (2x^5 + x^4 - 4x^3 - 3x^2)$

$= (9x^5 + x^3 - 2x^2 + 4) + \left[-(2x^5 + x^4 - 4x^3 - 3x^2)\right]$ Adding the opposite

$= (9x^5 + x^3 - 2x^2 + 4) + \left[-2x^5 - x^4 + 4x^3 + 3x^2\right]$ Changing the sign of every term

$= 9x^5 + x^3 - 2x^2 + 4 - 2x^5 - x^4 + 4x^3 + 3x^2$

$= 7x^5 - x^4 + 5x^3 + x^2 + 4.$ Combining like terms

Do Exercises 13 and 14. ▶

To shorten our work, we often begin by changing the sign of each term in the polynomial being subtracted.

EXAMPLE 7 Subtract: $(5a^4 - 7a^3 + 5a^2b) - (-3a^4 + 4a^2b + 6)$.

We have

$(5a^4 - 7a^3 + 5a^2b) - (-3a^4 + 4a^2b + 6)$

$= 5a^4 - 7a^3 + 5a^2b + 3a^4 - 4a^2b - 6$ Removing all parentheses; changing the sign of every term in the second polynomial

$= 8a^4 - 7a^3 + a^2b - 6.$ Combining like terms

Do Exercise 15. ▶

Find two equivalent expressions for the opposite of each polynomial.

6. $12x^4 - 3x^2 + 4x$

7. $-4x^4 + 3x^2 - 4x$

8. $-13x^6 + 2x^4 - 3x^2 + x - \tfrac{5}{13}$

9. $-8a^3b + 5ab^2 - 2ab$

Simplify.

10. $-(4x^3 - 6x + 3)$

11. $-(5x^3y + 3x^2y^2 - 7xy^3)$

12. $-(14x^{10} - \tfrac{1}{2}x^5 + 5x^3 - x^2 + 3x)$

Subtract.

13. $(7x^3 + 2x + 4) - (5x^3 - 4)$

14. $(-3x^2 + 5x - 4) - (-4x^2 + 11x - 2)$

15. Subtract:
$(7x^3 + 3x^2 - xy) - (5x^3 + 3xy + 2)$.

Answers

6. $-(12x^4 - 3x^2 + 4x)$; $-12x^4 + 3x^2 - 4x$
7. $-(-4x^4 + 3x^2 - 4x)$; $4x^4 - 3x^2 + 4x$
8. $-\left(-13x^6 + 2x^4 - 3x^2 + x - \dfrac{5}{13}\right)$;

$13x^6 - 2x^4 + 3x^2 - x + \dfrac{5}{13}$
9. $-(-8a^3b + 5ab^2 - 2ab)$;
$8a^3b - 5ab^2 + 2ab$
10. $-4x^3 + 6x - 3$
11. $-5x^3y - 3x^2y^2 + 7xy^3$
12. $-14x^{10} + \dfrac{1}{2}x^5 - 5x^3 + x^2 - 3x$
13. $2x^3 + 2x + 8$
14. $x^2 - 6x - 2$
15. $2x^3 + 3x^2 - 4xy - 2$

d EVALUATING POLYNOMIALS AND APPLICATIONS

It is important to keep in mind that when we add or subtract polynomials, we are *not* solving an equation. Rather, we are finding an equivalent expression that is usually more concise. One reason we do this is to make it easier to evaluate.

EXAMPLE 8 Evaluate both $(5x^3 + 4x^2 + 3x) + (2x^3 + 5x^2 - x)$ and $7x^3 + 9x^2 + 2x$ for $x = 2$ (see Example 1).

a) When x is replaced by 2 in $(5x^3 + 4x^2 + 3x) + (2x^3 + 5x^2 - x)$, we have

$$5 \cdot 2^3 + 4 \cdot 2^2 + 3 \cdot 2 + 2 \cdot 2^3 + 5 \cdot 2^2 - 2,$$
or $5 \cdot 8 + 4 \cdot 4 + 3 \cdot 2 + 2 \cdot 8 + 5 \cdot 4 - 2,$
or $40 + 16 + 6 + 16 + 20 - 2$, which is 96.

b) Similarly, when x is replaced by 2 in $7x^3 + 9x^2 + 2x$, we have

$$7 \cdot 2^3 + 9 \cdot 2^2 + 2 \cdot 2,$$
or $7 \cdot 8 + 9 \cdot 4 + 4,$
or $56 + 36 + 4$. As expected, this is also 96.

Note how much easier it is to evaluate the simplified sum in part (b) rather than the original expression.

◀ **Do Exercise 16.**

Polynomials are frequently evaluated in real-world situations.

EXAMPLE 9 *Athletics.* In a sports league of n teams in which all teams play each other twice, the total number of games played is given by the polynomial

$$n^2 - n.$$

A women's softball league has 10 teams. If each team plays every other team twice, what is the total number of games played?

We evaluate the polynomial for $n = 10$:

$$n^2 - n = 10^2 - 10 = 100 - 10 = 90.$$

The league plays 90 games.

◀ **Do Exercises 17–19.**

---

16. Evaluate each expression for $a = 2$. (See Margin Exercise 3.)

a) $(7a^2 + 2a + 8) + (2a^2 + a - 9)$

b) $9a^2 + 3a - 1$

17. In the situation of Example 9, how many games are played in a league with 12 teams?

Wendy is pedaling down a hill. Her distance from the top of the hill, in meters, can be approximated by

$$\frac{1}{2}t^2 + 3t,$$

where t is the number of seconds she has been pedaling and $t < 30$.

18. How far has Wendy traveled in 4 sec?

19. How far has Wendy traveled in 10 sec?

Answers

16. (a) 41; (b) 41   **17.** 132 games
18. 20 m   **19.** 80 m

10.4 Exercise Set

MyMathLab®   MathXL°   PRACTICE   WATCH   READ   REVIEW

✓ Reading Check

Determine whether each statement is true or false.

_____ **RC1.** To find the opposite of a polynomial, we need change only the sign of the first term.

_____ **RC2.** We can subtract a polynomial by adding its opposite.

_____ **RC3.** The sum of two monomials is always a monomial.

_____ **RC4.** The coefficient of $3x^7$ is 7.

a Add.

1. $(3x + 7) + (-7x + 3)$

2. $(6x + 1) + (-7x + 2)$

3. $(-9x + 7) + (x^2 + x - 2)$

4. $(x^2 - 5x + 4) + (5x - 9)$

5. $(x^2 - 7) + (x^2 + 7)$

6. $(x^3 + x^2) + (2x^3 - 5x^2)$

7. $(6t^4 + 4t^3 - 1) + (5t^2 - t + 1)$

8. $(5t^2 - 3t + 12) + (2t^2 + 8t - 30)$

9. $(2 + 4x + 6x^2 + 7x^3) + (5 - 4x + 6x^2 - 7x^3)$

10. $(3x^4 - 6x - 5x^2 + 5) + (6x^2 - 4x^3 - 1 + 7x)$

11. $(9x^8 - 7x^4 + 2x^2 + 5) + (8x^7 + 4x^4 - 2x)$

12. $(4x^5 - 6x^3 - 9x + 1) + (6x^3 + 9x^2 + 9x)$

13. $(\frac{1}{2}t^3 - \frac{3}{4}t^2 + \frac{1}{3}) + (\frac{3}{2}t^3 - \frac{1}{4}t^2 + \frac{1}{6}t)$

14. $(\frac{2}{3}t^5 - t^4 + \frac{1}{2}t^2 + 4) + (\frac{7}{3}t^5 + \frac{1}{2}t^4 - \frac{1}{2}t^2 + \frac{1}{3}t)$

15. $(9x^3 - 3x + x^2 - 10) + (3x - x^2 - 9x^3 + 10)$

16. $(5 - x^4 + 3x^2 + 2x) + (x^4 - 3x^2 - 2x - 5)$

SECTION 10.4 Addition and Subtraction of Polynomials   **699**

17. $(8a^3b^2 + 5a^2b^2 + 6ab^2) + (5a^3b^2 - a^2b^2 - 4a^2b)$

18. $(6x^3y^3 - 4x^2y^2 + 3xy^2) + (x^3y^3 + 7x^3y^2 - 2xy^2)$

19. $(17.5abc^3 + 4.3a^2bc) + (-4.9a^2bc - 5.2abc)$

20. $(23.9x^3yz - 19.7x^2y^2z) + (-14.6x^3yz - 8x^2yz)$

b   Find two equivalent expressions for the opposite of each polynomial.

21. $-5x$

22. $x^2 - 3x$

23. $-x^2 + 13x - 7$

24. $-7x^3 - x^2 - x$

25. $12x^4 - 3x^3 + 3$

26. $4x^3 - 6x^2 - 8x + 1$

Simplify.

27. $-(3x - 5)$

28. $-(-2x + 4)$

29. $-(4x^2 - 3x + 2)$

30. $-(-6a^3 + 2a^2 - 9a + 1)$

31. $-(-4x^4 + 6x^2 + \frac{3}{4}x - 8)$

32. $-(-5x^4 + 4x^3 - x^2 + 0.9)$

c   Subtract.

33. $(3x + 2) - (-4x + 3)$

34. $(6x + 1) - (-7x + 2)$

35. $(9t^2 + 7t + 5) - (5t^2 + 7t - 1)$

36. $(8t^2 - 5t + 7) - (3t^2 - 5t - 8)$

37. $(-8x + 2) - (x^2 + x - 3)$

38. $(x^2 - 5x + 4) - (8x - 9)$

39. $(7a^2 + 5a - 9) - (8a^2 + 7)$

40. $(8a^2 - 6a + 5) - (2a^2 - 19a)$

41. $(8x^4 + 3x^3 - 1) - (4x^2 - 3x + 5)$

42. $(-4x^2 + 2x) - (3x^3 - 5x^2 + 3)$

43. $(1.2x^3 + 4.5x^2 - 3.8x) - (-3.4x^3 - 4.7x^2 + 23)$

44. $(0.5x^4 - 0.6x^2 + 0.7) - (2.3x^4 + 1.8x - 3.9)$

45. $\left(\frac{5}{8}x^3 - \frac{1}{4}x - \frac{1}{3}\right) - \left(-\frac{1}{8}x^3 + \frac{1}{4}x - \frac{1}{3}\right)$

46. $\left(\frac{1}{5}x^3 + 2x^2 - 0.1\right) - \left(-\frac{2}{5}x^3 + 2x^2 + 0.01\right)$

47. $(9x^3y^3 + 8x^2y^2 + 7xy) - (3x^3y^3 - 2x^2y + 6xy)$

48. $(3x^4y + 2x^3y - 7x^2y) - (5x^4y + 2x^2y^2 - 8x^2y)$

d Evaluate each polynomial for $x = 4$.

49. $-7x + 5$

50. $-3x + 1$

51. $2x^2 - 5x + 7$

52. $3x^2 + x + 7$

53. $x^3 - 5x^2 + x$

54. $7 - x + 3x^2$

Evaluate each polynomial for $x = -1$.

55. $2x + 9$

56. $6 - 2x$

57. $x^2 - 2x + 1$

58. $5x - 6 + x^2$

59. $-3x^3 + 7x^2 - 3x - 2$

60. $-2x^3 - 5x^2 + 4x + 3$

61. *Skydiving.* During the first 13 sec of a jump, the distance S, in feet, that a skydiver falls in t seconds can be approximated by the polynomial equation

$$S = 11.12t^2.$$

In 2009, 108 U.S. skydivers fell headfirst in formation from a height of 18,000 ft. How far had they fallen 10 sec after having jumped from the plane?

Source: www.telegraph.co.uk

62. *SCAD Thrill Ride.* The distance s, in feet, traveled by a body falling freely from rest in t seconds is approximated by the polynomial equation

$$s = 16t^2.$$

The SCAD thrill ride is a 2.5-sec free fall into a net. How far does the person fall?

Source: "What is SCAD?", www.scadfreefall.co.uk

Daily Accidents. The average number of accidents per day involving drivers who are a years old is approximated by the polynomial

$$0.4a^2 - 40a + 1039.$$

63. Evaluate the polynomial for $a = 18$ to find the daily number of accidents involving 18-year-old drivers.

64. Evaluate the polynomial for $a = 20$ to find the daily number of accidents involving 20-year-old drivers.

Total Revenue. Cutting Edge Electronics is marketing a new kind of phone. *Total revenue* is the total amount of money taken in. The firm determines that when it sells x phones, it takes in

$$280x - 0.4x^2 \text{ dollars.}$$

65. What is the total revenue from the sale of 75 phones?

66. What is the total revenue from the sale of 100 phones?

Total Cost. Cutting Edge Electronics determines that the total cost of producing x phones is given by

$$5000 + 0.6x^2 \text{ dollars.}$$

67. What is the total cost of producing 500 phones?

68. What is the total cost of producing 650 phones?

Minutes of Daylight. The number of minutes of daylight in Chicago, on a date n days after December 21, can be approximated by

$$-0.01096n^2 + 4n + 548.$$

69. ▦ Determine the number of minutes of daylight in Chicago 92 days after December 21.

70. ▦ Determine the number of minutes of daylight in Chicago 123 days after December 21.

Medicine. Ibuprofen is a medication used to relieve pain. The polynomial

$$0.5t^4 + 3.45t^3 - 96.65t^2 + 347.7t$$

can be used to estimate the number of milligrams of ibuprofen in the bloodstream t hours after 400 mg of the medication has been swallowed, where $0 \le t \le 6$.

Source: Based on data from Dr. P. Carey, Burlington, VT

71. ▦ Determine the number of milligrams of ibuprofen in the bloodstream 1 hr after the medication has been swallowed.

72. ▦ Determine the number of milligrams of ibuprofen in the bloodstream 5 hr after the medication has been swallowed.

Skill Maintenance

73. A 10-lb fish serves 7 people. What is the ratio of servings to pounds? [7.1a]

74. A bicycle salesperson's commission rate is 22%. A commission of $783.20 is received. How many dollars' worth of bicycles were sold? [8.6b]

75. The sales tax rate in Pennsylvania is 6%. How much sales tax would be paid in Pennsylvania on the purchase of a laptop computer for $1350? [8.6a]

76. Find the area of a rectangle that is 6.5 m by 4 m. [5.8b]

77. Find the area of a circle with radius 20 cm. Use 3.14 for π. [9.3b]

78. Melanie earned $4740 for working 12 weeks. What was the rate of pay? [7.2a]

Write the prime factorization for each number. [3.2c]

79. 168

80. 192

81. 735

82. 117

Synthesis

Minutes of Daylight.   The number of minutes of daylight in Los Angeles, on a date n days after December 21, can be approximated by

$$-0.0085n^2 + 3.1014n + 593.$$

83. ▦ Determine the number of minutes of daylight in Los Angeles on Ground Hog Day (February 2).

84. ▦ How much more daylight is available in Chicago than in Los Angeles on July 4? (See Exercises 69 and 70.)

85. ▦ *Medicine.*   The polynomial

$$0.5t^4 + 3.45t^3 - 96.65t^2 + 347.7t$$

used in Exercises 71 and 72 describes the number of milligrams of ibuprofen in the bloodstream t hours after 400 mg of the medication has been swallowed. To visualize this, make a line graph, with the number of hours on the horizontal axis and the amount of ibuprofen in the bloodstream on the vertical axis. Plot points for $t = 0, 1, 2, 3, 4, 5,$ and 6.

86. ▦ *Daily Accidents.*   The polynomial

$$0.4a^2 - 40a + 1039$$

from Exercises 63 and 64 can be used to estimate the average number of accidents per day involving drivers who are a years old. To visualize this, make a vertical bar graph, with the age of drivers on the horizontal axis and the average number of accidents per day on the vertical axis. Show bars for $a = 20, 30, 40, 50, 60, 70,$ and 80.

87. *Total Profit.*   Total profit is defined as total revenue minus total cost. Find a polynomial giving the total profit for Cutting Edge Electronics, described in Exercises 65–68, when x phones are produced and sold.

88. *Total Profit.*   Use the polynomial found in Exercise 87 to find **(a)** the total profit when 200 phones are produced and sold and **(b)** the total profit when 500 phones are produced and sold.

Perform the indicated operations and simplify.

89. $(7y^2 - 5y + 6) - (3y^2 + 8y - 12) + (8y^2 - 10y + 3)$

90. $(3x^2 - 4x + 6) - (-2x^2 + 4) + (-5x - 3)$

91. $(-y^4 - 7y^3 + y^2) + (-2y^4 + 5y - 2) - (-6y^3 + y^2)$

92. $(-4 + x^2 + 2x^3) - (-6 - x + 3x^3) - (-x^2 - 5x^3)$

93. Complete: $9x^4 + \underline{\quad} + 5x^2 - 7x^3 + \underline{\quad} - 9 + \underline{\quad} = 12x^4 - 5x^3 + 5x^2 - 16.$

94. Complete: $8t^4 + \underline{\quad} - 2t^3 + \underline{\quad} - 2t^2 + t - \underline{\quad} - 3 + \underline{\quad} = 8t^4 + 7t^3 - 3t + 4.$

OBJECTIVES

a Multiply monomials.

b Multiply a monomial and any polynomial.

c Use the distributive law to factor.

SKILL TO REVIEW

Objective 10.2a: Use the product rule to multiply exponential expressions with like bases.

Multiply.

1. $x^2 \cdot x^3$ **2.** $x \cdot x^4$

a **MULTIPLYING MONOMIALS**

Recall that the area of a square with sides of length x is x^2.

Area $= x^2$

If a rectangle is 3 times as long as it is wide, we can represent its width by x and its length by $3x$.

Area $= 3x^2$

The area, $3x^2$, is the product of $3x$ and x. This product can be found using an associative law:

$$(3x)x = 3(xx) = 3x^2.$$

To find other products of monomials, we may need to use a commutative law as well.

EXAMPLE 1 Multiply: $(4x)(5x)$.

$$
\begin{aligned}
(4x)(5x) &= 4 \cdot x \cdot 5 \cdot x && \text{Using an associative law} \\
&= 4 \cdot 5 \cdot x \cdot x && \text{Using a commutative law} \\
&= (4 \cdot 5)(x \cdot x) && \text{Using an associative law} \\
&= 20x^2
\end{aligned}
$$

The multiplication in Example 1 can be regarded as finding the area of a rectangle of width $4x$ and length $5x$. Note that the area consists of 20 squares, each of which has area x^2.

Multiply.

1. $(6a)(3a)$

2. $(-7x)(2x)$ GS
$$
\begin{aligned}
&= -7 \cdot x \cdot 2 \cdot x \\
&= -7 \cdot \boxed{} \cdot x \cdot x \\
&= (-7 \cdot 2)(x \cdot x) \\
&= \boxed{}
\end{aligned}
$$

Area $= 20x^2$

◄ **Do Margin Exercises 1 and 2.**

Usually the steps in Example 1 are combined: We multiply coefficients and we multiply variables.

Answers

Skill to Review:
1. x^5 **2.** x^5

Margin Exercises:
1. $18a^2$ **2.** $-14x^2$

Guided Solution:
2. $2, -14x^2$

EXAMPLES Multiply.

2. $(5x)(6x) = (5 \cdot 6)(x \cdot x)$ Multiplying the coefficients and the variables

$= 30x^2$ Simplifying

3. $(3x)(-y) = (3x)(-1y)$ Rewriting $-y$ as $-1y$

$= (3)(-1)(x \cdot y)$

$= -3xy$

Do Exercises 3–5. ▶

Multiply.

3. $(4a)(12a)$

4. $(-m)(5m)$

5. $(-6a)(-7b)$

We can use the product rule when multiplying monomials.

EXAMPLES Multiply and simplify.

4. $x^2 \cdot x^5 = x^{2+5}$ Adding exponents

$= x^7$

5. $(3a^4)(5a^2) = (3 \cdot 5)(a^4 \cdot a^2)$ Multiplying coefficients;

$= 15a^6$ adding exponents

6. $(-4x^2y^3)(3x^6y^7) = (-4 \cdot 3)(x^2 \cdot x^6)(y^3 \cdot y^7)$

$= -12x^8y^{10}$

Do Exercises 6–9. ▶

Multiply.

6. $a^5 \cdot a^4$

7. $(2x^8)(4x^5)$

8. $(-7m^4)(-5m^7)$

9. $(3a^5b^4)(5a^2b^8)$

b   MULTIPLYING A MONOMIAL AND ANY POLYNOMIAL

The product of x and $x + 2$ can be visualized as the area of a rectangle with width x and length $x + 2$, as illustrated in the figure at right.

The distributive law is used to find products of polynomials algebraically.

EXAMPLE 7 Multiply: $2x$ and $5x + 3$.

$2x(5x + 3) = 2x \cdot 5x + 2x \cdot 3$ Using the distributive law

$= 10x^2 + 6x$ Multiplying each pair of monomials

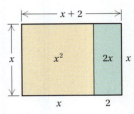

Area $= x(x + 2)$

$= x \cdot x + x \cdot 2$

$= x^2 + 2x$

EXAMPLE 8 Multiply: $5x(2x^2 - 3x + 4)$.

$5x(2x^2 - 3x + 4) = 5x \cdot 2x^2 - 5x \cdot 3x + 5x \cdot 4$ Note that $x \cdot x^2 = x^1 \cdot x^2 = x^3$.

$= 10x^3 - 15x^2 + 20x$

EXAMPLE 9 Multiply: $-3r^2s(2r^3s^2 - 5rs)$.

$-3r^2s(2r^3s^2 - 5rs) = -3r^2s \cdot 2r^3s^2 - (-3r^2s)5rs$

$= -6r^5s^3 + 15r^3s^2$

Do Exercises 10–12. ▶

Multiply.

10. $4x$ and $3x + 5$

11. $3a(2a^2 - 5a + 7)$

12. $4a^3b^2(2a^2 + 5b^4)$

Answers

3. $48a^2$ **4.** $-5m^2$ **5.** $42ab$ **6.** a^9

7. $8x^{13}$ **8.** $35m^{11}$ **9.** $15a^7b^{12}$ **10.** $12x^2 + 20x$

11. $6a^3 - 15a^2 + 21a$ **12.** $8a^5b^2 + 20a^3b^6$

C FACTORING

Factoring is the reverse of multiplying. We use the distributive law, beginning with a sum or a difference of terms that contain a common factor:

$$ab + ac = a(b + c) \quad \text{and} \quad rs - rt = r(s - t).$$

> ### FACTORING
>
> To **factor** an expression is to find an equivalent expression that is a product.

To *factor* an expression like $10y + 15$, we find an equivalent expression that is a product. To do this, we look to see if both terms have a factor in common. If there *is* a common factor, we can "factor it out" using the distributive law. Note the following:

The prime factorization of $10y$ is $2 \cdot 5 \cdot y$.

The prime factorization of 15 is $3 \cdot 5$.

We factor out the common factor, 5:

$$10y + 15 = 5 \cdot 2y + 5 \cdot 3 \qquad \text{Try to do this step mentally.}$$
$$= 5(2y + 3). \qquad \text{Using the distributive law}$$

We generally factor out the *largest* common factor. This is the product of all factors common to all terms.

EXAMPLE 10 Factor $12a - 30$.

The prime factorization of $12a$ is $2 \cdot 2 \cdot 3 \cdot a$.

The prime factorization of 30 is $2 \cdot 3 \cdot 5$.

Both factorizations include a factor of 2 and a factor of 3. Thus, 2 is a common factor, 3 is a common factor, and $2 \cdot 3$ is a common factor. The largest common factor is $2 \cdot 3$, or 6:

$$12a - 30 = 6(2a - 5). \qquad \text{Try to go directly to this step.}$$

EXAMPLE 11 Factor $9x + 27y - 9$.

The prime factorization of $9x$ is $3 \cdot 3 \cdot x$.

The prime factorization of $27y$ is $3 \cdot 3 \cdot 3 \cdot y$.

The prime factorization of 9 is $3 \cdot 3$.

$$9x + 27y - 9 = 9 \cdot x + 9 \cdot 3y - 9 \cdot 1 \qquad \begin{array}{l}\text{The largest common} \\ \text{factor is } 3 \cdot 3, \text{ or } 9.\end{array}$$
$$= 9(x + 3y - 1)$$

In Example 11, the 1 in the factorization is necessary. To see this, reverse the factorization process by multiplying. This provides a check for the answer.

$$9(x + 3y - 1) = 9 \cdot x + 9 \cdot 3y - 9 \cdot 1 = 9x + 27y - 9$$

> Factorizations can always be checked by multiplying.

.... **Caution!** ....

Note in Example 11 that although $3(3x + 9y - 3)$ is also equivalent to $9x + 27y - 9$, it is not factored "completely." However, we can complete the process by factoring out another factor of 3:

$$9x + 27y - 9 = 3(3x + 9y - 3) = 3 \cdot 3(x + 3y - 1) = 9(x + 3y - 1).$$

Remember to factor out the *largest common factor*.

....

EXAMPLES Factor. Try to write just the answer.

12. $-3x + 6y - 9z = -3(x - 2y + 3z)$

We generally factor out a negative factor when the first coefficient is negative. We might also factor as $-3x + 6y - 9z = 3(-x + 2y - 3z)$.

13. $18z - 12x - 24 = 6(3z - 2x - 4)$

The largest common factor is $2 \cdot 3.$ $\leftarrow$
$$\begin{cases} 18z = 2 \cdot 3 \cdot 3 \cdot z; \\ 12x = 2 \cdot 2 \cdot 3 \cdot x; \\ 24 = 2 \cdot 2 \cdot 2 \cdot 3 \end{cases}$$

Check: $6(3z - 2x - 4) = 6 \cdot 3z - 6 \cdot 2x - 6 \cdot 4 = 18z - 12x - 24$

Remember: An expression is factored when it is written as a product.

Do Exercises 13–16. ▶

EXAMPLE 14 Factor each of the following.

a) $10x^6 + 15x^2$

b) $8xy^3 - 6xy^2 + 4xy$

a) The prime factorization of $10x^6$ is $2 \cdot 5 \cdot x \cdot x \cdot x \cdot x \cdot x \cdot x$.
   The prime factorization of $15x^2$ is $3 \cdot 5 \cdot x \cdot x$.

$10x^6 + 15x^2 = 5x^2 \cdot 2x^4 + 5x^2 \cdot 3$ The largest common factor is $5x^2$.
$$= 5x^2(2x^4 + 3)$$

b) The prime factorization of $8xy^3$ is $2 \cdot 2 \cdot 2 \cdot x \cdot y \cdot y \cdot y$.
   The prime factorization of $6xy^2$ is $2 \cdot 3 \cdot x \cdot y \cdot y$.
   The prime factorization of $4xy$ is $2 \cdot 2 \cdot x \cdot y$.

$8xy^3 - 6xy^2 + 4xy = 2xy \cdot 4y^2 - 2xy \cdot 3y + 2xy \cdot 2$ The largest common factor is $2xy$.
$$= 2xy(4y^2 - 3y + 2)$$

The checks are left for the student.

The largest common factor can be determined by considering the coefficients and the variables separately. The largest common factor of the coefficients is found using prime factorizations. The largest common variable factors can be found by examining the exponents.

When a variable appears in every term of a polynomial, the *largest* common factor of that variable is the *smallest* of the powers of that variable in the polynomial.

Do Exercises 17–19. ▶

Factor.

13. $6z - 12$

14. $3x - 6y + 12$

15. $16a - 36b + 42$

16. $-12x + 32y - 16z$

Factor.

17. $5a^3 + 10a$

18. $14x^3 - 7x^2 + 21x$

GS **19.** $9a^2b - 6ab^2$
$9a^2b = 3 \cdot 3 \cdot a \cdot a \cdot \boxed{}$
$6ab^2 = 2 \cdot 3 \cdot a \cdot b \cdot \boxed{}$
The largest common factor is
   $3 \cdot a \cdot \boxed{}$.
$9a^2b - 6ab^2 = 3ab \cdot \boxed{}$
$ - 3ab \cdot \boxed{}$
$ = 3ab(\boxed{} - \boxed{})$

Answers

13. $6(z - 2)$ **14.** $3(x - 2y + 4)$
15. $2(8a - 18b + 21)$ **16.** $-4(3x - 8y + 4z)$
17. $5a(a^2 + 2)$ **18.** $7x(2x^2 - x + 3)$
19. $3ab(3a - 2b)$

Guided Solution:
19. b; b; b; $3a$, $2b$, $3a$, $2b$

✓ **Reading Check**

Match each expression with an equivalent expression from the column on the right. Choices may be used more than once or not at all.

RC1. $8x \cdot 2x$ _____

RC2. $(-16x)(-x)$ _____

RC3. $2x(8x - 1)$ _____

RC4. $8x(2x - 1)$ _____

a) $16x^2$
b) $-16x^2$
c) $16x^2 - 1$
d) $16x^2 - 2x$
e) $16x^2 - 8x$

a Multiply.

1. $(4a)(7a)$

2. $(7x)(6x)$

3. $(-4x)(15x)$

4. $(-9a)(10a)$

5. $(7x^5)(4x^3)$

6. $(10a^2)(3a^2)$

7. $(-0.1x^6)(0.7x^3)$

8. $(0.3x^3)(-0.4x^6)$

9. $(5x^2y^3)(7x^4y^9)$

10. $(9a^5b^4)(2a^4b^7)$

11. $(4a^3b^4c^2)(3a^5b^4)$

12. $(7x^3y^5z^2)(8x^3z^4)$

13. $(3x^2)(-4x^3)(2x^6)$

14. $(-2y^5)(10y^4)(-3y^3)$

b Multiply.

15. $3x(-x + 7)$

16. $2x(4x - 6)$

17. $-3x(x - 2)$

18. $-9x(-x - 1)$

19. $x^2(x^3 + 1)$

20. $-2x^3(x^2 - 1)$

21. $5x(2x^2 - 6x + 1)$

22. $-4x(2x^3 - 6x^2 - 5x + 1)$

23. $4xy(3x^2 + 2y)$

24. $7xy(3x^2 - 6y^2)$

25. $3a^2b(4a^5b^2 - 3a^2b^2)$

26. $4a^2b^2(2a^3b - 5ab^2)$

27. $2x + 8$

28. $3x + 12$

29. $7a - 35$

30. $9a - 18$

31. $28x + 21y$

32. $8x - 10y$

33. $9a - 27b + 81$

34. $5x + 10 + 15y$

35. $18 - 6m$

36. $28 - 4y$

37. $-16 - 8x + 40y$

38. $-35 + 14x - 21y$

39. $9x^5 + 9x$

40. $5x^6 + 5x$

41. $a^3 - 8a^2$

42. $a^5 - 9a^2$

43. $8x^3 - 6x^2 + 2x$

44. $9x^4 - 12x^3 + 3x$

45. $12a^4b^3 + 18a^5b^2$

46. $15a^5b^2 + 20a^2b^3$

Skill Maintenance

Graph. [6.4b, c]

47. $y = x - 5$

48. $y = -2x + 8$

49. $x = 2$

50. $y = \dfrac{2}{3}x$

Synthesis

Factor.

51. 🖩 $391x^{391} + 299x^{299}$

52. 🖩 $703a^{437} + 437a^{703}$

53. $84a^7b^9c^{11} - 42a^8b^6c^{10} + 49a^9b^7c^8$

54. Draw a figure similar to those preceding Examples 1 and 7 to show that $2x \cdot 3x = 6x^2$.

Vocabulary Reinforcement

Complete each statement with the correct word or number from the column at the right. Some of the choices may not be used.

1. Any nonzero number raised to the power 0 is equal to _____. [10.1a]

2. The _____ rule states that when multiplying with exponential notation, if the bases are the same, we keep the base and add the exponents. [10.2a]

3. The number 7.9×10^{-3} is written in _____ notation. [10.3a]

4. A base raised to a negative exponent is equal to the _____ of the base raised to a positive exponent. [10.1b]

5. The _____ rule states that when raising a power to a power, we multiply the exponents and leave the base unchanged. [10.2c]

6. The _____ of $3x^2$ is 3. [10.4a]

0
1
coefficient
exponent
fraction
scientific
power
product
quotient
reciprocal

Concept Reinforcement

_____ 1. If x^7 is divided by x^7, the result is the same as x^0. [10.1a]

_____ 2. We subtract a polynomial by adding its opposite. [10.4c]

_____ 3. If a polynomial is written as a product, it is factored. [10.5c]

Study Guide

Objective 10.1a Evaluate algebraic expressions containing whole-number exponents.

Examples Evaluate -6^1.
Since $6^1 = 6$, $-6^1 = -6$.
Evaluate $(-5)^0$.
Since $b^0 = 1$ for all $b \neq 0$, $(-5)^0 = 1$.

Practice Exercises
Evaluate.
1. $(-3)^1$
2. 12^0

Objective 10.1b Express exponential expressions involving negative exponents as equivalent expressions containing positive exponents.

Example Write an expression equivalent to x^{-10} using a positive exponent.

$$x^{-10} = \frac{1}{x^{10}}$$

Practice Exercise
3. Write an expression equivalent to 3^{-12} using a positive exponent.

Objective 10.2a Use the product rule to multiply exponential expressions with like bases.

Example Simplify $x^2 \cdot x^8$.

$$x^2 \cdot x^8 = x^{2+8} = x^{10}$$

Practice Exercise
4. Simplify $y \cdot y^{11}$.

Objective 10.2b Use the quotient rule to divide exponential expressions with like bases.

Example Simplify $\dfrac{7^{12}}{7^{10}}$.

$$\dfrac{7^{12}}{7^{10}} = 7^{12-10} = 7^2 = 49$$

Practice Exercise

5. Simplify $\dfrac{x^9}{x^2}$.

Objective 10.2c Use the power rule to raise powers to powers.

Example Simplify $(x^3)^{-5}$.

$$(x^3)^{-5} = x^{3(-5)} = x^{-15} = \dfrac{1}{x^{15}}$$

Practice Exercise

6. Simplify $(3^{-2})^{-6}$.

Objective 10.2d Raise a product to a power and a quotient to a power.

Example Simplify $(a^2 b)^9$.

$$(a^2 b)^9 = (a^2)^9 (b)^9 = a^{18} b^9$$

Practice Exercise

7. Simplify $(3x^7)^2$.

Objective 10.3a Convert between scientific notation and decimal notation.

Examples Convert 96,000,000,000 to scientific notation.

$$96{,}000{,}000{,}000 = 9.6 \times 10^{10}$$

Convert 6.02×10^{-5} to decimal notation.

$$6.02 \times 10^{-5} = 0.0000602$$

Practice Exercises

8. Convert 0.000803 to scientific notation.

9. Convert 3.48×10^3 to decimal notation.

Objective 10.4a Add polynomials.

Example Add: $(3x^3 - 7x^2 - 9) + (2x^3 + x^2 - 1)$.

$(3x^3 - 7x^2 - 9) + (2x^3 + x^2 - 1)$

$= 3x^3 + 2x^3 - 7x^2 + x^2 - 9 - 1$

$= 5x^3 - 6x^2 - 10$

Practice Exercise

10. Add: $(2a^2 - 5a + 3) + (a^3 + a - 5)$.

Objective 10.4c Subtract polynomials.

Example Subtract:
$(9y^3 - 8y^2 + 11y) - (4y^3 - 2y^2 - 12y)$.

$(9y^3 - 8y^2 + 11y) - (4y^3 - 2y^2 - 12y)$

$= 9y^3 - 8y^2 + 11y - 4y^3 + 2y^2 + 12y$

$= 5y^3 - 6y^2 + 23y$

Practice Exercise

11. Subtract: $(x^2 - x - 1) - (7x^2 - 6x + 2)$.

Objective 10.5a Multiply monomials.

Example Multiply: $(-3m^2)(9m^5)$.

$(-3m^2)(9m^5) = (-3 \cdot 9)(m^2 \cdot m^5)$

$= -27m^7$

Practice Exercise

12. Multiply: $(-2x^3)(-7x)$.

Objective 10.5b Multiply a monomial and any polynomial.

Example Multiply: $2x^3(3x^4 - 5)$. $\quad 2x^3(3x^4 - 5) = 2x^3 \cdot 3x^4 - 2x^3 \cdot 5$ $\qquad\qquad\qquad = 6x^7 - 10x^3$	**Practice Exercise** **13.** Multiply: $5x^2(x^3 + 2x^2 - 3x + 9)$.

Objective 10.5c Use the distributive law to factor.

Example Factor $6x^3 - 20x^2 + 2x$. $\quad 6x^3 - 20x^2 + 2x = 2x(3x^2 - 10x + 1)$	**Practice Exercise** **14.** Factor $20a^4 - 30a$.

Review Exercises

Evaluate. [10.1a]

1. $(-53)^0$ **2.** 46^1

3. $(5x + 7)^1$ for $x = -2$

4. $(3x - 2)^0$ for $x = 5$

Write an equivalent expression using positive exponents. Then simplify, if possible. [10.1b]

5. 12^{-2} **6.** $8a^{-7}$

7. $\dfrac{x^{-3}}{y^5 z^{-6}}$ **8.** $\left(\dfrac{4}{5}\right)^{-2}$

9. Write an expression equivalent to $\dfrac{1}{x^7}$ using a negative exponent. [10.1b]

Simplify. [10.2a, b, c, d]

10. $x^4 \cdot x^{11}$ **11.** $\dfrac{x^{16}}{x^4}$

12. $(3^4)^{11}$ **13.** $(x^2 y^4)^3$

14. $\dfrac{x^3}{x^{11}}$ **15.** $\dfrac{u^2 v^8}{uv^7}$

16. $x^{-5} \cdot x^{-6}$ **17.** $(3x^4)^2$

18. $a^{-2} \cdot a \cdot a^7$ **19.** $\left(\dfrac{y^2}{2}\right)^5$

20. Write scientific notation for 42,700,000. [10.3a]

21. Write scientific notation for 0.0001924. [10.3a]

Simplify. Write the answer in scientific notation. [10.3b]

22. $(5.1 \times 10^6)(2.3 \times 10^4)$

23. $\dfrac{1.6 \times 10^2}{6.4 \times 10^{18}}$

24. Every day about 12.4 billion spam e-mails are sent. If each spam e-mail wastes 4 sec, how many hours are wasted each day because of spam? Write the answer in scientific notation.

Source: spam-filter-review.toptenreviews.com   [10.3c]

Perform the indicated operation.   [10.4a, c]

25. $(-4x + 9) + (7x - 15)$

26. $(7x^4 - 5x^3 + 3x - 5) + (x^3 - 4x + 2)$

27. $(9a^5 + 8a^3 + 4a + 7) - (a^5 - 4a^3 + a^2 - 2)$

28. $(8a^3b^3 + 9a^2b^3) - (3a^3b^3 - 2a^2b^3 + 7)$

29. Find two equivalent expressions for the opposite of $12x^3 - 4x^2 + 9x - 3$.   [10.4b]

30. Evaluate $5t^3 + t$ for $t = -2$.   [10.4d]

31. The altitude, in feet, of a falling golf ball t seconds after it reaches the peak of its flight can be estimated by $-16t^2 + 200$. Find the ball's altitude 3 sec after it has reached its peak.   [10.4d]

Multiply.

32. $(5x^3)(6x^4)$ [10.5a]

33. $(-2xy)(9x^2y^4)$ [10.5a]

34. $3x(6x^3 - 4x - 1)$ [10.5b]

35. $2a^4b(7a^3b^3 + 5a^2b^3)$ [10.5b]

Factor.   [10.5c]

36. $45x^3 - 10x$

37. $7a - 35b - 49ac$

38. Evaluate $6 - 2x - x^2$ for $x = -1$.   [10.4d]

A. 3   B. 5
C. 7   D. 9

39. Factor out the largest common factor: $6x^3y - 9x^2y^5$.   [10.5c]

A. $3x^2y(2x - 3y^4)$ B. $3(2x^2y - 3x^2y^5)$
C. $3x^2(2xy - 3y^5)$ D. $y(6x^3 - 9x^2y^4)$

Synthesis

40. Simplify: $-3x^5 \cdot 3x^3 - x^6(2x)^2 + (3x^4)^2 + (2x^4)^2 - 40x^2(x^3)^2$.   [10.4a], [10.4c]

Factor.   [10.5c]

41. $39a^3b^7c^6 - 130a^2b^5c^8 + 52a^4b^6c^5$

42. $w^5x^6y^4z^5 - w^7x^3y^7z^3 + w^6x^2y^5z^6 - w^6x^7y^3z^4$

43. $10a^4b^{-5} + 12a^7b^{-3}$

Understanding Through Discussion and Writing

1. Can x^{-2} represent a negative number? Why or why not?   [10.1b]

2. Is it true that if $a > b$, then $a^{-1} < b^{-1}$? Explain your answer.   [10.1b]

3. Adi claims that $(3x^{-5})(-4x^{-2}) = -x^{10}$. What mistake(s) is she probably making?   [10.1b], [10.5a]

4. Using the quotient rule, explain why 7^0 is 1.   [10.2b]

5. Is every term a monomial? Why or why not?   [10.4a]

6. If all of a polynomial's coefficients are prime, is it still possible to factor the polynomial? Why or why not?   [10.5c]

For Extra Help   For step-by-step test solutions, access the Chapter Test Prep Videos in MyMathLab® or on You Tube (search "BittingerPrealgebra" and click on "Channels").

Evaluate.

1. 193^1

2. -9^0

3. $(3x - 7)^0$ for $x = 2$

Write an equivalent expression with positive exponents. Then simplify, if possible.

4. 5^{-3}

5. $\dfrac{5a^{-3}}{b^{-2}}$

6. $\left(\dfrac{3}{5}\right)^{-3}$

Simplify. Express the answer using positive exponents.

7. $a^{12} \cdot a^{13}$

8. $\dfrac{x}{x^2}$

9. $(5x^3y^4)^2$

10. $\left(\dfrac{3}{x^9}\right)^3$

11. $x^{-1} \cdot x^{-7} \cdot x$

12. $\dfrac{a^{10}b^{12}}{a^4b^{10}}$

13. Write scientific notation for 0.00047.

14. Write scientific notation for 8,250,000.

15. Find the product and write the answer using scientific notation:

$(3.2 \times 10^{-8})(5.7 \times 10^{-9})$.

16. Add: $(12a^3 - 9a^2 + 8) + (6a^3 + 4a^2 - a)$.

17. Find two equivalent expressions for the opposite of $-9a^4 + 7b^2 - ab + 3$.

18. Subtract: $(12x^4 + 7x^2 - 6) - (9x^4 + 8x^2 + 5)$.

19. The height, in meters, of a ball t sec after it has been thrown is approximated by $-4.9t^2 + 15t + 2$. How high is the ball 2 sec after it has been thrown?

Multiply.

20. $(-5x^4y^3)(2x^2y^5)$

21. $2a(5a^2 - 4a + 3)$

Factor.

22. $35x^6 - 25x^3 + 15x^2$

23. $6ab - 9bc + 12ac$

24. Americans throw away 2.5 million plastic beverage bottles every hour. How many bottles do they throw away in a week? Give your answer in scientific notation.

Source: www.grabstats.com

A. 6.0×10^7 bottles   **B.** 1.75×10^7 bottles
C. 2.52×10^9 bottles   **D.** 4.2×10^8 bottles

Synthesis

25. The polynomial

$$0.041h - 0.018A - 2.69$$

can be used to estimate the lung capacity, in liters, of a female of height h, in centimeters, and age A, in years. Find the lung capacity of a 30-yr-old woman who is 150 cm tall.

26. Simplify: $x^{8y} \cdot x^{2y} \cdot x^{-y}$.

This exam reviews the entire textbook. Many answers to exercises (and to real-world problems) can be written in several kinds of notation. For this exam, here is the guideline we follow: Use the notation given in the problem. That is, if the problem is given using mixed numerals, give the answer as a mixed numeral. If the problem is given in decimal notation, give the answer in decimal notation.

1. *Area.* There are 5.1 million km^2 of forest in the Amazon jungle. Find standard notation for 5.1 million.

    Source: www.outbackbrazil.com.

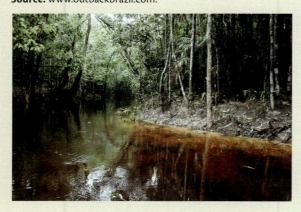

Add and, if possible, simplify.

2.   4 9 0 3
      5 2 7 8
      6 3 9 1
   + 4 5 1 3

3. $5\dfrac{4}{9}$
   $+ 3\dfrac{1}{3}$

4. $-29 + 53$

5. $-543 + (-219)$

6. $-34.56 + 2.783 + 0.433 + (-13.02)$

7. $(4x^5 + 7x^4 - 3x^2 + 9) + (6x^5 - 8x^4 + 2x^3 - 7)$

Subtract and, if possible, simplify.

8.   6 7 4
   − 4 3 1

9. $-4x - 13x$

10. $\dfrac{2}{5} - \dfrac{7}{8}$

11. $4\dfrac{1}{3}$
   $- 1\dfrac{5}{8}$

12.   2 0.0
   −   0.0 0 2 7

13. $(7x^3 + 2x^2 - x) - (5x^3 - 3x^2 - 8x)$

Multiply and, if possible, simplify.

14.   2 9 7
   ×   1 6

15. $349 \cdot (-213)$

16. $2\dfrac{3}{4} \cdot 1\dfrac{2}{3}$

17. $-\dfrac{9}{7} \cdot \dfrac{14}{15}$

18. $12 \cdot \dfrac{5}{6}$

19.   3 4.0 9
   ×   7.6

20. $3(8x - 5)$

21. $(9a^3b^2)(3a^5b)$

22. $7x^2(3x^3 - 2x + 8)$

Divide and simplify. State the answer using a remainder when appropriate.

23. $6\overline{)3438}$

24. $34\overline{)1914}$

Divide and, if possible, simplify.

25. $\dfrac{4}{5} \div \left(-\dfrac{8}{15}\right)$ **26.** $-2\dfrac{1}{3} \div (-30)$

27. $2.7\,\overline{)\,1\ 0\ 5.3}$

Simplify.

28. $10 \div 2 \times 20 - 5^2$ **29.** $\dfrac{|3^2 - 5^2|}{2 - 2 \cdot 5}$

30. Write exponential notation: $14 \cdot 14 \cdot 14$.

31. Round 68,489 to the nearest thousand.

32. Round $21.\overline{83}$ to the nearest hundredth.

33. Determine whether 1368 is divisible by 3.

34. Find all the factors of 15.

35. Find the LCM of 15 and 35.

36. Simplify $\dfrac{24}{33}$.

37. Convert to a mixed numeral: $-\dfrac{18}{5}$.

38. Use $<$ or $>$ for $\square$ to write a true sentence: $-17 \,\square\, -29$.

39. Use $<$ or $>$ for $\square$ to write a true sentence: $\dfrac{4}{7} \,\square\, \dfrac{3}{5}$.

40. Which number is greater: -1.001 or -0.9976?

41. Evaluate $\dfrac{a^2 - b}{3}$ for $a = -9$ and $b = -6$.

Factor.

42. $40 - 5t$

43. $18a^3 - 15a^2 + 6a$

44. What part is shaded?

Write decimal notation for each number.

45. $\dfrac{429}{10{,}000}$ **46.** $-\dfrac{13}{25}$

47. $\dfrac{8}{9}$ **48.** 7%

Write each number in fraction notation.

49. 6.71 **50.** $-7\dfrac{1}{4}$

51. 40%

Write each number in percent notation.

52. $\dfrac{17}{20}$ **53.** 1.5

54. Estimate the sum $9.389 + 4.2105$ to the nearest tenth.

Solve.

55. $234 + y = 789$

56. $3.9a = 249.6$

57. $\frac{2}{3} \cdot t = \frac{5}{6}$

58. $\frac{8}{17} = \frac{36}{x}$

59. $7x - 9 = 26$

60. $-2(x - 5) = 3x + 12$

Egg Consumption.   The line graph below shows egg consumption per person in the United States for recent years. Use the graph to solve Exercises 61–64.

Egg Consumption

SOURCE: American Egg Board

61. Find the lowest egg consumption and the year(s) in which it occurred.

62. Find the highest egg consumption and the year(s) in which it occurred.

63. Find the mean, the median, and the mode(s) of the egg consumptions.

64. What was the percent decrease in egg consumption from 2006 to 2010?

65. *Dead Sea.*   The lowest point in the world is the Dead Sea on the border of Israel and Jordan. It is 1312 ft below sea level. Convert 1312 ft to yards; to meters.

Source: *The Handy Geography Answer Book*

66. In Sam's writing lab, 3 of the 20 students are left-handed. If a student is randomly selected, what is the probability that he or she is left-handed?

67. Find the missing angle measure.

68. *Closest Star.*   Exluding the sun, the closest star to Earth is Proxima Centauri, which is 4.3 light-years away (one light-year $= 5.88 \times 10^{12}$ mi). How far, in miles, is Proxima Centauri from Earth? Express the answer in scientific notation.

69. On Monday morning, a bolt of fabric contained $8\frac{1}{4}$ yd. Madison sold $3\frac{5}{8}$ yd from the bolt. How much fabric remains?

70. From Indira's income of $32,000, amounts of $6400 and $1600 are paid for federal and state taxes. How much remains after these taxes have been paid?

71. A toddler walks $\frac{3}{5}$ km per hour. At this rate, how far would the child walk in $\frac{1}{2}$ hr?

72. Eight gallons of paint covers 2000 ft^2. How much paint is needed to cover 3250 ft^2?

73. What is the simple interest on $4000 principal at 8% for $\frac{3}{4}$ yr?

74. The population of Bridgeton is 29,000 this year and is increasing at a rate of 4% per year. What will the population be next year?

75. *Tree Height.* While visiting the Grove of the Patriarchs in Mt. Rainier National Park, Abi photographed one of the tallest trees in the grove. Standing 100 ft from the base of the tree, she estimated that the distance between her and the top of the tree was about 300 ft. About how tall is the tree? Round to the nearest foot.

76. *Medical Dosage.* A doctor suggests that a child who weighs 24 kg be given 42 mg phenytoin sodium. If the dosage is proportional to the child's weight, how much phenytoin sodium is recommended for a child who weighs 32 kg?

77. *Earth vs. Jupiter.* The mass of Earth is about 6×10^{21} metric tons. The mass of Jupiter is about 1.908×10^{24} metric tons. About how many times the mass of Earth is the mass of Jupiter? Express the answer in scientific notation.

78. At the start of a trip, the odometer on the Oquendos' minivan read 27,428.6 mi, and at the end of the trip, the reading was 27,914.5 mi. How long was the trip?

79. Eight identical dresses cost a total of $679.68. What is the cost of each dress?

80. Eighteen ounces of a fruit "smoothie" cost $3.06. Find the unit price in cents per ounce.

81. Luis paid $35 a day plus 15¢ a mile for a van rental. If his one-day van rental cost $68, how many miles did he drive?

82. Baldacci Real Estate received $5880 commission on the sale of an $84,000 home. What was the rate of commission?

83. *Falling Distance.* The distance, in feet, traveled by a body falling freely from rest in t seconds is approximated by the polynomial $16t^2$. A stone is dropped from a cliff and takes 8 sec to hit the ground. How high is the cliff?

Evaluate.

84. 18^2

85. 37^0

86. 42^1

87. $\sqrt{121}$

Write an equivalent expression with a positive exponent. Then simplify, if possible.

88. 4^{-3}

89. $\left(\dfrac{5}{4}\right)^{-2}$

Express each of the following in scientific notation.

90. 4,357,000

91. $(6.2 \times 10^{7})(4.3 \times 10^{-23})$

Complete.

92. $\dfrac{1}{3}$ yd = _____ in.

93. 5.8 km = _____ m

94. 10 lb = _____ oz

95. 8190 mL = _____ L

96. 3917 mm = _____ cm

97. 60,000 g = _____ kg

98. 2.3 g = _____ mg

99. 28 qt = _____ gal

100. 10 yd^2 = _____ ft^2

101. 200 cm^2 = _____ m^2

The data in the following table show the percent of people who eat salad a certain number of times per week.

NUMBER OF SALADS PER WEEK	PERCENT
None	3%
2 or fewer	37%
3–6	47%
At least one a day	13%

SOURCE: Market Facts for the Association of Dressings and Sauces

102. Make a bar graph of the data

103. Make a circle graph of the data.

104. Plot the following points:
$$(-5, 2), (4, 0), (3, -4), (0, 2)$$

105. Graph: $y = -\dfrac{1}{3}x$.

106. Graph: $y = 3$.

107. These triangles are similar. Find the missing lengths.

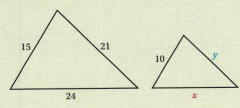

108. A rectangular mirror measures 20 in. by 24 in. Find its perimeter.

Find the area of each figure. Use 3.14 for π.

109.

4 cm
15.4 cm

110.

5 in.
10 in.

111.

10.8 yd
8.3 yd
20.2 yd

112.

2.5 m
10.3 m

113. Find the diameter, the circumference, and the area of this circle. Use 3.14 for π.

10.4 in.

Find the volume of each shape. Use 3.14 for π.

114.

2.3 m

2.3 m

10 m

115.

16 ft

4 ft

116. *Clouds.* Find the volume of a spherical cloud with a 1000-m diameter. Use 3.14 for π.

1000 m

117. Find the length of the third side of this right triangle. Give an exact answer and an approximation to three decimal places.

11 ft

a

6 ft

118. Find the measure of a complement of $\angle CBA$, if $m \angle CBA = 40°$.
  A. 140°   **B.** 40°
  C. 50°   **D.** 90°

119. Convert 100°C to Fahrenheit.
  A. $37\frac{7}{9}°\text{F}$ **B.** 237.6°F
  C. 180°F   **D.** 212°F

120. A month of the year is randomly selected. What is the probability that the name of the month contains an r in its spelling?
  A. $\frac{1}{2}$ **B.** $\frac{1}{3}$
  C. $\frac{2}{3}$ **D.** 8

Synthesis

121. A housing development is constructed on a dead-end road that runs along a river and ends in a cul-de-sac, as shown in the figure. The property owners agree to share the cost of maintaining the road in the following manner. The cost of the first fifth of the road in front of lot 1 is to be shared equally among all five lot owners. The cost of the second fifth in front of lot 2 is to be shared equally among the owners of lots 2–5, and so on. Assume that all five sections of the road cost the same to maintain.

  a) What fractional part of the cost is paid by each owner?

  b) What percent of the cost is paid by each owner?

  c) If lots 3, 4, and 5 were all owned by the same person, what percent of the cost of maintenance would this person pay?

Lot 1   Lot 2   Lot 3   Lot 4   Lot 5

Road

River

D

D

Developmental Units

A **ADDITION**

S **SUBTRACTION**

M **MULTIPLICATION**

D **DIVISION**

Addition

OBJECTIVES

a Add any two of the numbers 0, 1, 2, 3, 4, 5, 6, 7, 8, 9.

b Find certain sums of three numbers such as 1 + 7 + 9.

c Add two whole numbers when carrying is not necessary.

d Add two whole numbers when carrying is necessary.

a BASIC ADDITION

The sum 3 + 4 can be found by counting out a set of 3 objects and a separate set of 4 objects, putting them together, and counting all the objects.

A set of 3 + A set of 4 = A set of 7

The numbers to be added are called **addends**. The result is the **sum**.

$$3 + 4 = 7$$

Addend   Addend   Sum

EXAMPLES Add. Think of putting sets of objects together.

1. $5 + 6 = 11$

$$\begin{array}{r} 5 \\ + 6 \\ \hline 11 \end{array}$$

2. $8 + 5 = 13$

$$\begin{array}{r} 8 \\ + 5 \\ \hline 13 \end{array}$$

We can also do these problems by counting up from one of the numbers. For example, in Example 2, we start at 8 and count up 5 times: 9, 10, 11, 12, 13.

◀ **Do Exercises 1–6.**

What happens when we add 0? Think of a set of 5 objects. If we add 0 objects to it, we still have 5 objects. Similarly, if we have a set with 0 objects in it and add 5 objects to it, we have a set with 5 objects. Thus,

$$5 + 0 = 5 \quad \text{and} \quad 0 + 5 = 5.$$

ADDITION OF 0

Adding 0 to a number does not change the number:

$$a + 0 = 0 + a = a.$$

We say that 0 is the **additive identity**.

EXAMPLES Add.

3. $0 + 9 = 9$

$$\begin{array}{r} 0 \\ + 9 \\ \hline 9 \end{array}$$

4. $0 + 0 = 0$

$$\begin{array}{r} 0 \\ + 0 \\ \hline 0 \end{array}$$

5. $97 + 0 = 97$

$$\begin{array}{r} 97 \\ + 0 \\ \hline 97 \end{array}$$

◀ **Do Exercises 7–12.**

Add; think of joining sets of objects.

1. 4 + 5

2. 5 + 2

3. $\begin{array}{r} 9 \\ + 5 \end{array}$

4. $\begin{array}{r} 8 \\ + 8 \end{array}$

5. $\begin{array}{r} 9 \\ + 7 \end{array}$

6. $\begin{array}{r} 7 \\ + 9 \end{array}$

Add.

7. 8 + 0

8. 0 + 8

9. $\begin{array}{r} 7 \\ + 0 \end{array}$

10. $\begin{array}{r} 46 \\ + 0 \end{array}$

11. 0 + 13

12. 58 + 0

Answers

1. 9   2. 7   3. 14   4. 16
5. 16   6. 16   7. 8   8. 8
9. 7   10. 46   11. 13   12. 58

Your objective for this part of the section is to be able to add any two of the numbers 0, 1, 2, 3, 4, 5, 6, 7, 8, 9. Adding 0 is easy. The rest of the sums are listed in this table. Memorize the table by saying it to yourself over and over or by using flash cards.

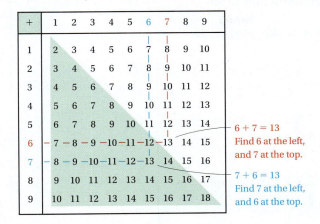

6 + 7 = 13
Find 6 at the left, and 7 at the top.

7 + 6 = 13
Find 7 at the left, and 6 at the top.

It is very important that you *memorize* the basic addition facts! If you do not, you will always have trouble with addition.

Note the following.

$$3 + 4 = 7 \qquad 7 + 6 = 13 \qquad 7 + 2 = 9$$
$$4 + 3 = 7 \qquad 6 + 7 = 13 \qquad 2 + 7 = 9$$

We can add whole numbers in any order. This is the *commutative law of addition*. Because of this law, you need to learn only about half the table above, as shown by the shading.

Do Exercises 13 and 14. ▶

b CERTAIN SUMS OF THREE NUMBERS

To add 3 + 5 + 4, we can add 3 and 5, then 4. We can also add 5 and 4, then add 3. Either way, we get 12.

$$3 + 5 + 4$$
$$8 + 4$$
$$12$$

$$3 + 5 + 4$$
$$3 + 9$$
$$12$$

EXAMPLE 6 Add from the top mentally.

1
7
+ 9

We first add 1 and 7, getting 8. Then we add 8 and 9, getting 17.

1
7 → 8
+ 9   9 → 17
17 ←

EXAMPLE 7 Add from the top mentally.

2
4 → 6
+ 8   8 → 14
14 ←

Do Exercises 15–18. ▶

Complete the table.

13.

+	1	2	3	4	5
1			4		
2					
3				7	
4					
5					

14.

+	6	5	7	4	9
7			14		
9					
5				9	
8					
4					

Add from the top mentally.

15.   1
    6
 + 9

16.   2
    3
 + 4

17.   6
    1
 + 4

18.   5
    2
 + 8

Answers

13.

+	1	2	3	4	5
1	2	3	4	5	6
2	3	4	5	6	7
3	4	5	6	7	8
4	5	6	7	8	9
5	6	7	8	9	10

14.

+	6	5	7	4	9
7	13	12	14	11	16
9	15	14	16	13	18
5	11	10	12	9	14
8	14	13	15	12	17
4	10	9	11	8	13

15. 16   **16.** 9   **17.** 11   **18.** 15

c ADDITION (NO CARRYING)

We now move to a gradual, conceptual development of addition of whole numbers. To add larger numbers, we can add the ones first, then the tens, then the hundreds, and so on.

EXAMPLE 8 Add: 5722 + 3234.

$$
\begin{array}{r}
5\ 7\ 2\ 2 \\
+\ 3\ 2\ 3\ 4 \\
\hline
6
\end{array}
$$
Add ones.

$$
\begin{array}{r}
5\ 7\ 2\ 2 \\
+\ 3\ 2\ 3\ 4 \\
\hline
5\ 6
\end{array}
$$
Add tens.

$$
\begin{array}{r}
5\ 7\ 2\ 2 \\
+\ 3\ 2\ 3\ 4 \\
\hline
9\ 5\ 6
\end{array}
$$
Add hundreds.

This is for explanation.

$$
\begin{array}{r}
5\ 7\ 2\ 2 \\
+\ 3\ 2\ 3\ 4 \\
\hline
8\ 9\ 5\ 6
\end{array}
$$
Add thousands.

$$
\begin{array}{r}
5\ 7\ 2\ 2 \\
+\ 3\ 2\ 3\ 4 \\
\hline
8\ 9\ 5\ 6
\end{array}
$$
You should write only this.

◀ Do Exercises 19–22.

Add.

19.
$$
\begin{array}{r}
2\ 4 \\
+\ 3\ 5 \\
\hline
\end{array}
$$

20.
$$
\begin{array}{r}
3\ 4\ 6 \\
+\ 2\ 0\ 3 \\
\hline
\end{array}
$$

21.
$$
\begin{array}{r}
8\ 3\ 2\ 7 \\
+\ 1\ 6\ 5\ 2 \\
\hline
\end{array}
$$

22.
$$
\begin{array}{r}
3\ 4\ 6\ 1 \\
+\ 2\ 0\ 3\ 5 \\
\hline
\end{array}
$$

d ADDITION (WITH CARRYING)

Carrying Tens, Hundreds, and Thousands

EXAMPLE 9 Add: 18 + 27.

$$
\begin{array}{r}
1\ 8 \\
+\ 2\ 7 \\
\hline
?
\end{array}
$$
Add ones. *Think:*
$$
\begin{array}{r}
8 \\
+\ 7 \\
\hline
1\ 5
\end{array}
$$
15 ones = 10 ones + 5 ones = 1 ten + 5 ones

$$
\begin{array}{r}
\overset{1}{1}\ 8 \\
+\ 2\ 7 \\
\hline
5
\end{array}
$$
Write 5 in the ones column. Then write 1 above the tens to indicate that we are *carrying* 1 ten from the ones column.

$$
\begin{array}{r}
\overset{1}{1}\ 8 \\
+\ 2\ 7 \\
\hline
4\ 5
\end{array}
$$
Add tens.

◀ Do Exercises 23 and 24.

We can use money to help explain Example 9.

18¢ → 1 dime and 8 pennies
+27¢ → 2 dimes and 7 pennies
        3 dimes and 15 pennies
              or
45¢ ← 4 dimes and 5 pennies

Add.

23.
$$
\begin{array}{r}
1\ 9 \\
+\ 3\ 7 \\
\hline
\end{array}
$$

24.
$$
\begin{array}{r}
4\ 6 \\
+\ 3\ 9 \\
\hline
\end{array}
$$

Answers

19. 59   20. 549   21. 9979
22. 5496   23. 56   24. 85

EXAMPLE 10 Add: $256 + 391$.

$$
\begin{array}{r}
2\;5\;6 \\
+\;3\;9\;1 \\
\hline
7
\end{array}
$$
Add ones.

$$
\begin{array}{r}
{}^{1} \\
2\;5\;6 \\
+\;3\;9\;1 \\
\hline
4\;7
\end{array}
$$
Add tens. We get 14 tens.
Now 14 tens = 10 tens + 4 tens = 1 hundred + 4 tens.
Write 4 in the tens column and a 1 above the hundreds.

$$
\begin{array}{r}
{}^{1} \\
2\;5\;6 \\
+\;3\;9\;1 \\
\hline
6\;4\;7
\end{array}
$$
Add hundreds.

Do Exercises 25 and 26. ▶

The carrying in Example 10 is like exchanging 14 dimes for 1 dollar bill and 4 dimes.

EXAMPLE 11 Add: $4803 + 3792$.

$$
\begin{array}{r}
4\;8\;0\;3 \\
+\;3\;7\;9\;2 \\
\hline
5
\end{array}
$$
Add ones.

$$
\begin{array}{r}
4\;8\;0\;3 \\
+\;3\;7\;9\;2 \\
\hline
9\;5
\end{array}
$$
Add tens.

$$
\begin{array}{r}
{}^{1} \\
4\;8\;0\;3 \\
+\;3\;7\;9\;2 \\
\hline
5\;9\;5
\end{array}
$$
Add hundreds. We get 15 hundreds. Now 15 hundreds = 10 hundreds + 5 hundreds = 1 thousand + 5 hundreds. Write 5 in the hundreds column and 1 above the thousands.

$$
\begin{array}{r}
{}^{1} \\
4\;8\;0\;3 \\
+\;3\;7\;9\;2 \\
\hline
8\;5\;9\;5
\end{array}
$$
Add thousands.

Do Exercise 27. ▶

TO THE STUDENT

If you had trouble with Section 1.2 and have studied Developmental Unit A, you should go back and work through Section 1.2 after completing Exercise Set A.

Carrying More Than Once

EXAMPLE 12 Add: $767 + 993$.

$$
\begin{array}{r}
{}^{1} \\
7\;6\;7 \\
+\;9\;9\;3 \\
\hline
0
\end{array}
$$
Add ones. We get 10 ones. Now 10 ones = 1 ten + 0 ones. Write 0 in the ones column and 1 above the tens.

$$
\begin{array}{r}
{}^{1}{}^{1} \\
7\;6\;7 \\
+\;9\;9\;3 \\
\hline
6\;0
\end{array}
$$
Add tens. We get 16 tens. Now 16 tens = 1 hundred + 6 tens. Write 6 in the tens column and 1 above the hundreds.

$$
\begin{array}{r}
{}^{1}{}^{1} \\
7\;6\;7 \\
+\;9\;9\;3 \\
\hline
1\;7\;6\;0
\end{array}
$$
Add hundreds. We get 17 hundreds.

Do Exercises 28 and 29. ▶

Add.

25.
$$
\begin{array}{r}
3\;4\;1 \\
+\;4\;8\;8
\end{array}
$$

26.
$$
\begin{array}{r}
7\;3\;0 \\
+\;2\;9\;6
\end{array}
$$

27. Add.
$$
\begin{array}{r}
7\;8\;5\;0 \\
+\;4\;8\;4\;8
\end{array}
$$

Add.

28.
$$
\begin{array}{r}
7\;9\;8\;9 \\
+\;5\;6\;7\;2
\end{array}
$$

29.
$$
\begin{array}{r}
5\;6,7\;8\;9 \\
+\;1\;4,5\;3\;9
\end{array}
$$

Answers

25. 829   **26.** 1026   **27.** 12,698
28. 13,661   **29.** 71,328

A   Exercise Set

For Extra Help

MyMathLab®

MathXL®
PRACTICE

WATCH

READ

REVIEW

a Add. Try to do these mentally. If you have trouble, think of putting sets of objects together.

1. 8 + 9	**2.** 8 + 7	**3.** 6 + 7	**4.** 9 + 5	**5.** 5 + 7	**6.** 5 + 6
7. 9 + 8	**8.** 9 + 7	**9.** 8 + 4	**10.** 9 + 1	**11.** 8 + 2	**12.** 3 + 8
13. 0 + 7	**14.** 4 + 3	**15.** 2 + 9	**16.** 0 + 0	**17.** 3 + 0	**18.** 9 + 9
19. 8 + 6	**20.** 3 + 7	**21.** 2 + 2	**22.** 7 + 7	**23.** 6 + 5	**24.** 7 + 8
25. 8 + 8	**26.** 8 + 1	**27.** 5 + 8	**28.** 5 + 9	**29.** 4 + 7	**30.** 6 + 1

31. 6 + 7   **32.** 7 + 7   **33.** 3 + 9   **34.** 6 + 0   **35.** 6 + 4

36. 9 + 3   **37.** 5 + 5   **38.** 5 + 3   **39.** 1 + 1   **40.** 4 + 5

41. 9 + 4   **42.** 0 + 8   **43.** 4 + 6   **44.** 2 + 7   **45.** 3 + 7

46. 3 + 3   **47.** 5 + 8   **48.** 3 + 6   **49.** 4 + 4   **50.** 4 + 7

b Add from the top mentally.

51. 1 8 + 3	**52.** 1 7 + 5	**53.** 3 2 + 5	**54.** 4 3 + 5	**55.** 1 7 + 9
56. 5 2 + 6	**57.** 4 5 + 1	**58.** 1 9 + 6	**59.** 1 8 + 7	**60.** 1 6 + 8

c Add.

61. 23 + 16	**62.** 54 + 35	**63.** 67 + 20	**64.** 496 + 503	**65.** 700 + 200
66. 801 + 67	**67.** 666 + 333	**68.** 523 + 325	**69.** 747 + 130	**70.** 8250 + 9430
71. 6552 + 4321	**72.** 3406 + 1293	**73.** 7340 + 3527	**74.** 4825 + 5070	**75.** 2073 + 1925
76. 9111 + 9111	**77.** 7889 + 9000	**78.** 52,433 + 12,056	**79.** 43,723 + 56,276	**80.** 51,670 + 26,107

d Add.

81. 38 + 8	**82.** 17 + 9	**83.** 17 + 38	**84.** 95 + 6	**85.** 862 + 781
86. 613 + 799	**87.** 355 + 491	**88.** 280 + 348	**89.** 814 + 390	**90.** 274 + 333
91. 9990 + 10	**92.** 999 + 11	**93.** 999 + 111	**94.** 839 + 388	**95.** 909 + 202
96. 808 + 909	**97.** 8718 + 1420	**98.** 3854 + 2700	**99.** 4828 + 1283	**100.** 6995 + 1432
101. 9889 + 1	**102.** 6889 + 4723	**103.** 9128 + 1997	**104.** 8898 + 6645	**105.** 9989 + 6785
106. 46,889 + 21,786	**107.** 23,448 + 10,989	**108.** 67,658 + 98,786	**109.** 77,548 + 23,767	**110.** 44,684 + 4,765

Subtraction

OBJECTIVES

a Find basic differences such as $5 - 3$, $13 - 8$, and so on.

b Subtract one whole number from another when borrowing is not necessary.

c Subtract one whole number from another when borrowing is necessary.

a BASIC SUBTRACTION

Subtraction can be explained by taking away part of a set.

EXAMPLE 1 Subtract: $7 - 3$.

We can do this by counting out 7 objects and then taking away 3 of them. Then we count the number that remain: $7 - 3 = 4$.

7 objects

Take away 3 objects 4 objects remain

We could also do this mentally by starting at 7 and counting down 3 times: 6, 5, 4.

EXAMPLE 2 Subtract: $11 - 6$. Think of "take away."

$11 - 6 = 5$ *Take away: "11 take away 6 is 5."*

$$\begin{array}{r} 11 \\ -\ 6 \\ \hline 5 \end{array}$$

◀ **Do Exercises 1–4.**

The addition table in Developmental Unit A will enable you to subtract also. First, recall how addition and subtraction are related.

An addition:

4   +   3   =   7

Two related subtractions:

a)

$7 - 3$ ⟵   =   4

b)

$7 - 4$ ⟵   =   3

Since we know that

$4 + 3 = 7$,   A basic addition fact

we also know the two subtraction facts:

$7 - 3 = 4$ and $7 - 4 = 3$.

Subtract.

1. $10 - 6$ **2.** $11 - 4$

3. $\begin{array}{r} 16 \\ -\ 8 \\ \hline \end{array}$ **4.** $\begin{array}{r} 10 \\ -\ 7 \\ \hline \end{array}$

Answers

1. 4   2. 7   3. 8   4. 3

EXAMPLE 3 From $8 + 9 = 17$, write two subtraction facts.

a) The addend 8 is subtracted from the sum 17.

$8 + 9 = 17.$ The related sentence is $17 - 8 = 9.$

b) The addend 9 is subtracted from the sum 17.

$8 + 9 = 17.$ The related sentence is $17 - 9 = 8.$

Do Exercises 5 and 6. ▶

We can use the idea that subtraction is defined in terms of addition to think of subtraction as "how much more."

EXAMPLE 4 Find: $13 - 6$.

To find $13 - 6$, we ask, "6 plus what number is 13?"

$6 + \square = 13$

Using the addition table at the right, we find 13 inside the table and 6 along the left side. Then we read the answer, 7, from the top. Thus we have

$13 - 6 = 7.$

Strive to do this kind of thinking as fast as you can mentally, without having to use the table.

Do Exercises 7–10. ▶

b SUBTRACTION (NO BORROWING)

We now move to a gradual, conceptual development of subtraction of whole numbers. To subtract larger numbers, we can subtract the ones first, then the tens, then the hundreds, and so on.

EXAMPLE 5 Subtract: $5787 - 3214$.

5 7 8 7
− 3 2 1 4
      3   Subtract ones.

5 7 8 7
− 3 2 1 4
    7 3   Subtract tens.

5 7 8 7
− 3 2 1 4
  5 7 3   Subtract hundreds.

5 7 8 7
− 3 2 1 4
2 5 7 3   Subtract thousands.

This is the explanation.

5 7 8 7
− 3 2 1 4
2 5 7 3   You should write only this.

Do Exercises 11–14. ▶

For each addition fact, write two subtraction facts.

5. $8 + 4 = 12$

6. $6 + 7 = 13$

+	1	2	3	4	5	6	7	8	9
1	2	3	4	5	6	7	8	9	10
2	3	4	5	6	7	8	9	10	11
3	4	5	6	7	8	9	10	11	12
4	5	6	7	8	9	10	11	12	13
5	6	7	8	9	10	11	12	13	14
6	7	8	9	10	11	12	13	14	15
7	8	9	10	11	12	13	14	15	16
8	9	10	11	12	13	14	15	16	17
9	10	11	12	13	14	15	16	17	18

Subtract. Try to do these mentally.

7. $14 - 6$ **8.** $12 - 5$

9.   1 3   **10.**   1 1
       − 4   − 7

Subtract.

11.   7 8   **12.**   2 9
        − 6 4   −   9

13.   5 4 2   **14.**   6 8 9 6
        − 3 0 1   − 4 8 7 1

Answers

5. $12 - 8 = 4$; $12 - 4 = 8$ **6.** $13 - 6 = 7$;
$13 - 7 = 6$ **7.** 8   **8.** 7   **9.** 9   **10.** 4
11. 14   **12.** 20   **13.** 241   **14.** 2025

C SUBTRACTION (WITH BORROWING)

We now consider subtraction when borrowing, or regrouping, is necessary.

Borrowing from the Tens Place

EXAMPLE 6 Subtract: $37 - 18$.

$$\begin{array}{r} 3\;7 \\ -\;1\;8 \\ \hline ? \end{array}$$

Try to subtract ones: $7 - 8$ is not a whole number.

$$\begin{array}{r} {\scriptstyle 2}\;{\scriptstyle 17} \\ 3\;7 \\ -\;1\;8 \\ \hline \end{array}$$

Borrow a ten. That is, 1 ten = 10 ones, and 10 ones + 7 ones = 17 ones. Write 2 above the tens column and 17 above the ones. We regard 37 as 20 + 17.

$$\begin{array}{r} {\scriptstyle 2}\;{\scriptstyle 17} \\ 3\;7 \\ -\;1\;8 \\ \hline 9 \end{array}$$

Subtract ones.

$$\begin{array}{r} {\scriptstyle 2}\;{\scriptstyle 17} \\ 3\;7 \\ -\;1\;8 \\ \hline 1\;9 \end{array}$$

Subtract tens.

> You should write only this.
>
> $$\begin{array}{r} {\scriptstyle 2}\;{\scriptstyle 17} \\ 3\;7 \\ -\;1\;8 \\ \hline 1\;9 \end{array}$$

◀ **Do Exercises 15 and 16.**

The borrowing in Example 6 is like exchanging 3 dimes and 7 pennies for 2 dimes and 17 pennies.

Subtract.

15.
$$\begin{array}{r} 4\;6 \\ -\;2\;9 \\ \hline \end{array}$$

16.
$$\begin{array}{r} 7\;4 \\ -\;3\;8 \\ \hline \end{array}$$

Borrowing Hundreds

EXAMPLE 7 Subtract: $538 - 275$.

$$\begin{array}{r} 5\;3\;8 \\ -\;2\;7\;5 \\ \hline 3 \end{array}$$

Subtract ones.

$$\begin{array}{r} 5\;3\;8 \\ -\;2\;7\;5 \\ \hline ?\;3 \end{array}$$

Try to subtract tens: 3 tens − 7 tens is not a whole number.

$$\begin{array}{r} {\scriptstyle 4}\;{\scriptstyle 13} \\ 5\;3\;8 \\ -\;2\;7\;5 \\ \hline 3 \end{array}$$

Borrow a hundred. That is, 1 hundred = 10 tens, and 10 tens + 3 tens = 13 tens. Write 4 above the hundreds column and 13 above the tens.

$$\begin{array}{r} {\scriptstyle 4}\;{\scriptstyle 13} \\ 5\;3\;8 \\ -\;2\;7\;5 \\ \hline 6\;3 \end{array}$$

Subtract tens.

$$\begin{array}{r} {\scriptstyle 4}\;{\scriptstyle 13} \\ 5\;3\;8 \\ -\;2\;7\;5 \\ \hline 2\;6\;3 \end{array}$$

Subtract hundreds.

> You should write only this.
>
> $$\begin{array}{r} {\scriptstyle 4}\;{\scriptstyle 13} \\ 5\;3\;8 \\ -\;2\;7\;5 \\ \hline 2\;6\;3 \end{array}$$

◀ **Do Exercises 17 and 18.**

The borrowing in Example 7 is like exchanging 5 dollars and 3 dimes for 4 dollars and 13 dimes.

Subtract.

17.
$$\begin{array}{r} 6\;4\;6 \\ -\;1\;9\;2 \\ \hline \end{array}$$

18.
$$\begin{array}{r} 7\;3\;3 \\ -\;4\;8\;3 \\ \hline \end{array}$$

Answers

15. 17 **16.** 36 **17.** 454 **18.** 250

Developmental Units

Borrowing More Than Once

Sometimes we must borrow more than once.

EXAMPLE 8   Subtract: $672 - 394$.

$$
\begin{array}{r}
6\ \overset{6}{7}\ \overset{12}{\cancel{2}} \\
-\ 3\ 9\ 4 \\
\hline
8
\end{array}
$$
Borrowing a ten to subtract ones

$$
\begin{array}{r}
\overset{5}{\cancel{6}}\ \overset{\overset{16}{\cancel{6}}}{}\ \overset{12}{\cancel{2}} \\
-\ 3\ 9\ 4 \\
\hline
2\ 7\ 8
\end{array}
$$
Borrowing a hundred to subtract tens

Do Exercises 19 and 20. ▶

EXAMPLE 9   Subtract: $6357 - 1769$.

$$
\begin{array}{r}
6\ 3\ \overset{4}{\cancel{5}}\ \overset{17}{\cancel{7}} \\
-\ 1\ 7\ 6\ 9 \\
\hline
8
\end{array}
$$
$7 - 9$ is not a whole number.
We borrow a ten.

$$
\begin{array}{r}
6\ \overset{2}{\cancel{3}}\ \overset{\overset{14}{\cancel{4}}}{}\ \overset{17}{\cancel{7}} \\
-\ 1\ 7\ 6\ 9 \\
\hline
8\ 8
\end{array}
$$
4 tens minus 6 tens is not a whole number.
We borrow a hundred.

$$
\begin{array}{r}
\overset{5}{\cancel{6}}\ \overset{\overset{12}{\cancel{3}}}{}\ \overset{14}{\cancel{4}}\ \overset{17}{\cancel{7}} \\
-\ 1\ 7\ 6\ 9 \\
\hline
4\ 5\ 8\ 8
\end{array}
$$
2 hundreds minus 7 hundreds is not a
whole number. We borrow a thousand.

We can always check by adding the answer to the number being subtracted.

EXAMPLE 10   Subtract: $8341 - 2673$. Check by adding.

We check by adding 5668 and 2673.

$Check:$
$$
\begin{array}{r}
\overset{1}{\ }\ \overset{1}{\ }\ \overset{1}{\ } \\
5\ 6\ 6\ 8 \\
+\ 2\ 6\ 7\ 3 \\
\hline
8\ 3\ 4\ 1
\end{array}
$$

Do Exercises 21 and 22. ▶

Zeros in Subtraction

Note the following:

50 is 5 tens;

70 is 7 tens.

Then

100 is 10 tens;

200 is 20 tens.

Do Exercises 23–26. ▶

Subtract.

19. $\begin{array}{r} 5\ 6\ 3 \\ -\ 1\ 8\ 7 \end{array}$ **20.** $\begin{array}{r} 7\ 3\ 3 \\ -\ 4\ 8\ 8 \end{array}$

Subtract. Check by adding.

21. $\begin{array}{r} 4\ 2\ 3\ 6 \\ -\ 1\ 6\ 7\ 9 \end{array}$ **22.** $\begin{array}{r} 7\ 5\ 4\ 1 \\ -\ 3\ 8\ 6\ 7 \end{array}$

Complete.

23. $80 =$ _____ tens

24. $60 =$ _____ tens

25. $300 =$ _____ tens

26. $900 =$ _____ tens

Answers

19. 376   **20.** 245   **21.** 2557   **22.** 3674
23. 8   **24.** 6   **25.** 30   **26.** 90

Complete.

27. 5000 = _____ tens

28. 9000 = _____ tens

29. 5380 = _____ tens

30. 6770 = _____ tens

Subtract.

31.   6 0
        − 1 8

32.   4 8 0
        − 2 5 6

Subtract.

33.   6 0 2
        − 4 6 4

34.   4 0 8
        − 3 6 4

Subtract.

35.   4 0 0 6
        − 1 2 3 8

36.   9 0 0 1
        − 7 8 0 4

Subtract.

37.   3 0 0 0
        − 1 7 5 4

38.   8 0 1 7
        − 3 2 8 9

Answers

27. 500 **28.** 900 **29.** 538 **30.** 677
31. 42 **32.** 224 **33.** 138 **34.** 44
35. 2768 **36.** 1197 **37.** 1246 **38.** 4728

Also,

230 is 2 hundreds + 3 tens
  or 20 tens + 3 tens
  or 23 tens.

Similarly,

1000 is 100 tens;

2000 is 200 tens;

4670 is 467 tens.

◀ **Do Exercises 27–30.**

EXAMPLE 11   Subtract: 50 − 37.

        ⁴ ¹⁰
        5̶ 0̶
      − 3 7
      _____
        1 3

We have 5 tens. We keep 4 of them in the tens column and put 1 ten, or 10 ones, with the ones.

◀ **Do Exercises 31 and 32.**

EXAMPLE 12   Subtract: 803 − 547.

        ⁷ ⁹ ¹³
        8̶ 0̶ 3̶
      − 5 4 7
      _____
        2 5 6

We have 8 hundreds, or 80 tens. We keep 79 tens and put 1 ten, or 10 ones, with the ones.

◀ **Do Exercises 33 and 34.**

EXAMPLE 13   Subtract: 9003 − 2789.

        ⁸ ⁹ ⁹ ¹³
        9̶ 0̶ 0̶ 3̶
      − 2 7 8 9
      _____
        6 2 1 4

We have 9 thousands, or 900 tens. We keep 899 tens and put 1 ten, or 10 ones, with the ones.

◀ **Do Exercises 35 and 36.**

EXAMPLES   Subtract.

14.
        ⁴ ⁹ ⁹ ¹⁰
        5̶ 0̶ 0̶ 0̶
      − 2 8 6 1
      _____
        2 1 3 9

15.
              ¹⁰
        ⁴ ⁹ 0̶ ¹³
        5̶ 0̶ 1̶ 3̶
      − 1 8 5 7
      _____
        3 1 5 6

We have 5 thousands, or 49 hundreds and 10 tens.

◀ **Do Exercises 37 and 38.**

TO THE STUDENT

If you had trouble with Section 1.3 and have studied Developmental Unit S, you should go back and work through Section 1.3 after completing Exercise Set S.

a Subtract. Try to do these mentally.

1. 7
   − 0

2. 8
   − 8

3. 7
   − 7

4. 8
   − 3

5. 5
   − 2

6. 1 6
   −   8

7. 1 7
   −   9

8. 1 2
   −   6

9. 1 1
   −   4

10. 1 2
    −   9

11. 1 4
    −   7

12. 1 8
    −   9

13. 1 3
    −   7

14. 1 5
    −   9

15. 9
    − 7

16. $7 - 3$

17. $4 - 1$

18. $2 - 0$

19. $3 - 3$

20. $6 - 3$

21. $7 - 6$

22. $9 - 8$

23. $10 - 3$

24. $6 - 6$

25. $11 - 7$

26. $12 - 8$

27. $5 - 0$

28. $4 - 0$

29. $13 - 9$

30. $14 - 9$

31. $11 - 2$

32. $12 - 3$

33. $16 - 9$

34. $18 - 9$

35. $11 - 5$

36. $10 - 4$

37. $10 - 8$

38. $14 - 8$

39. $15 - 8$

40. $10 - 2$

b Subtract.

41. 6 4
    − 3 1

42. 5 5
    − 3 4

43. 5 4 8
    − 3 0 1

44. 5 9 6
    − 4 0 3

45. 7 0 0
    − 2 0 0

46.
$$\begin{array}{r} 765 \\ -111 \\ \hline \end{array}$$

47.
$$\begin{array}{r} 525 \\ -323 \\ \hline \end{array}$$

48.
$$\begin{array}{r} 747 \\ -130 \\ \hline \end{array}$$

49.
$$\begin{array}{r} 988 \\ -700 \\ \hline \end{array}$$

50.
$$\begin{array}{r} 9450 \\ -8230 \\ \hline \end{array}$$

51.
$$\begin{array}{r} 6552 \\ -4321 \\ \hline \end{array}$$

52.
$$\begin{array}{r} 7547 \\ -3421 \\ \hline \end{array}$$

53.
$$\begin{array}{r} 5875 \\ -2111 \\ \hline \end{array}$$

54.
$$\begin{array}{r} 38,695 \\ -37,004 \\ \hline \end{array}$$

55.
$$\begin{array}{r} 67,899 \\ -66,673 \\ \hline \end{array}$$

56.
$$\begin{array}{r} 99,999 \\ -1 \\ \hline \end{array}$$

57.
$$\begin{array}{r} 56,780 \\ -56,770 \\ \hline \end{array}$$

58.
$$\begin{array}{r} 42,111 \\ -32,010 \\ \hline \end{array}$$

59.
$$\begin{array}{r} 77,654 \\ -66,611 \\ \hline \end{array}$$

60.
$$\begin{array}{r} 23,456 \\ -12,345 \\ \hline \end{array}$$

C Subtract.

61.
$$\begin{array}{r} 93 \\ -28 \\ \hline \end{array}$$

62.
$$\begin{array}{r} 42 \\ -13 \\ \hline \end{array}$$

63.
$$\begin{array}{r} 86 \\ -78 \\ \hline \end{array}$$

64.
$$\begin{array}{r} 98 \\ -89 \\ \hline \end{array}$$

65.
$$\begin{array}{r} 625 \\ -317 \\ \hline \end{array}$$

66.
$$\begin{array}{r} 735 \\ -609 \\ \hline \end{array}$$

67.
$$\begin{array}{r} 853 \\ -236 \\ \hline \end{array}$$

68.
$$\begin{array}{r} 961 \\ -747 \\ \hline \end{array}$$

69.
$$\begin{array}{r} 787 \\ -698 \\ \hline \end{array}$$

70.
$$\begin{array}{r} 6769 \\ -2367 \\ \hline \end{array}$$

71.
$$\begin{array}{r} 6431 \\ -2876 \\ \hline \end{array}$$

72.
$$\begin{array}{r} 7654 \\ -1765 \\ \hline \end{array}$$

73.
$$\begin{array}{r} 5246 \\ -2859 \\ \hline \end{array}$$

74.
$$\begin{array}{r} 6328 \\ -2679 \\ \hline \end{array}$$

75.
$$\begin{array}{r} 7641 \\ -3809 \\ \hline \end{array}$$

76.
$$\begin{array}{r} 8743 \\ -599 \\ \hline \end{array}$$

77.
$$\begin{array}{r} 12,647 \\ -4,897 \\ \hline \end{array}$$

78.
$$\begin{array}{r} 16,222 \\ -5,777 \\ \hline \end{array}$$

79.
$$\begin{array}{r} 46,781 \\ -12,988 \\ \hline \end{array}$$

80.
$$\begin{array}{r} 470 \\ -189 \\ \hline \end{array}$$

81.
$$\begin{array}{r} 690 \\ -235 \\ \hline \end{array}$$

82.
$$\begin{array}{r} 703 \\ -132 \\ \hline \end{array}$$

83.
$$\begin{array}{r} 6406 \\ -258 \\ \hline \end{array}$$

84.
$$\begin{array}{r} 2309 \\ -109 \\ \hline \end{array}$$

85.
$$\begin{array}{r} 3406 \\ -1293 \\ \hline \end{array}$$

86.
$$\begin{array}{r} 6807 \\ -3059 \\ \hline \end{array}$$

87.
$$\begin{array}{r} 8000 \\ -2794 \\ \hline \end{array}$$

88.
$$\begin{array}{r} 8002 \\ -6543 \\ \hline \end{array}$$

89.
$$\begin{array}{r} 38,000 \\ -37,695 \\ \hline \end{array}$$

90.
$$\begin{array}{r} 16,043 \\ -11,588 \\ \hline \end{array}$$

Multiplication

a   BASIC MULTIPLICATION

To multiply, we begin with two numbers, called **factors**, and get a third number, called a **product**. Multiplication of whole numbers can be explained by counting. The product 3×5 can be found by counting out 3 sets of 5 objects each, joining them (in a rectangular array if desired), and counting all the objects.

$$3 \times 5 = 15$$

Factor   Factor   Product

We can also think of multiplication as repeated addition.

$$3 \times 5 = \underbrace{5 + 5 + 5} = 15$$

3 addends of 5

OBJECTIVES

a   Multiply any two of the numbers 0, 1, 2, 3, 4, 5, 6, 7, 8, 9.

b   Multiply by multiples of 10, 100, and 1000.

c   Multiply larger numbers by 0, 1, 2, 3, 4, 5, 6, 7, 8, 9.

d   Multiply larger numbers by multiples of 10, 100, and 1000.

EXAMPLES   Multiply. Think either of putting sets of objects together in a rectangular array or of repeated addition.

1. $5 \times 6 = 30$

$$\begin{array}{r} 6 \\ \times\ 5 \\ \hline 30 \end{array}$$

2. $8 \times 4 = 32$

$$\begin{array}{r} 4 \\ \times\ 8 \\ \hline 32 \end{array}$$

Do Exercises 1–4.

Multiplying by 0

How do we multiply by 0? Consider $4 \cdot 0$. Using repeated addition, we see that

$$4 \cdot 0 = \underbrace{0 + 0 + 0 + 0} = 0.$$

4 addends of 0

We can also think of $4 \cdot 0$ as 4 sets with 0 objects in each set, so the total is 0.

Consider $0 \cdot 4$. Using repeated addition, this is 0 addends of 4, which is 0. Using sets, this is 0 sets with 4 objects in each set, which is 0. Thus we have the following.

MULTIPLICATION BY 0

Multiplying by 0 gives 0.

EXAMPLES   Multiply.

3. $13 \times 0 = 0$

$$\begin{array}{r} 0 \\ \times\ 13 \\ \hline 0 \end{array}$$

4. $0 \cdot 11 = 0$

$$\begin{array}{r} 11 \\ \times\ 0 \\ \hline 0 \end{array}$$

5. $0 \cdot 0 = 0$

$$\begin{array}{r} 0 \\ \times\ 0 \\ \hline 0 \end{array}$$

Do Exercises 5 and 6.

Multiply. Think of joining sets in a rectangular array or of repeated addition.

1. $7 \cdot 8$ (The dot, $\cdot$, means the same as $\times$.)

2.
$$\begin{array}{r} 9 \\ \times\ 4 \\ \hline \end{array}$$

3. $4 \cdot 7$

4.
$$\begin{array}{r} 7 \\ \times\ 6 \\ \hline \end{array}$$

Multiply.

5. $8 \cdot 0$

6.
$$\begin{array}{r} 17 \\ \times\ 0 \\ \hline \end{array}$$

Answers

1. 56   **2.** 36   **3.** 28   **4.** 42

5. 0   **6.** 0

Multiplying by 1

How do we multiply by 1? Consider $5 \cdot 1$. Using repeated addition, we see that

$$5 \cdot 1 = \underbrace{1 + 1 + 1 + 1 + 1}_{5 \text{ addends of } 1} = 5.$$

We can also think of $5 \cdot 1$ as 5 sets with 1 object in each set, for a total of 5 objects.

Consider $1 \cdot 5$. Using repeated addition, this is 1 addend of 5, which is 5. Using sets, this is 1 set of 5 objects, which is again 5 objects. Thus we have the following.

> **MULTIPLICATION BY 1**
>
> Multiplying a number by 1 does not change the number:
>
> $$a \cdot 1 = 1 \cdot a = a.$$
>
> We say that 1 is the **multiplicative identity**.

EXAMPLES   Multiply.

6. $13 \cdot 1 = 13$

$$\begin{array}{r} 1 \\ \times\ 13 \\ \hline 13 \end{array}$$

7. $1 \cdot 7 = 7$

$$\begin{array}{r} 7 \\ \times\ 1 \\ \hline 7 \end{array}$$

8. $1 \cdot 1 = 1$

$$\begin{array}{r} 1 \\ \times\ 1 \\ \hline 1 \end{array}$$

◀ **Do Exercises 7 and 8.**

You should be able to multiply any two of the numbers 0, 1, 2, 3, 4, 5, 6, 7, 8, 9. Multiplying by 0 and 1 is easy. The rest of the products are listed in the following table.

×	2	3	4	5	6	7	8	9
2	4	6	8	10	12	14	16	18
3	6	9	12	15	18	21	24	27
4	8	12	16	20	24	28	32	36
5	10	15	20	25	30	35	40	45
6	12	18	24	30	36	42	48	54
7	14	21	28	35	42	49	56	63
8	16	24	32	40	48	56	64	72
9	18	27	36	45	54	63	72	81

$5 \times 7 = 35$
Find 5 at the left, and 7 at the top.

$8 \cdot 4 = 32$
Find 8 at the left, and 4 at the top.

It is *very* important that you have the basic multiplication facts *memorized*. If you do not, you will always have trouble with multiplication.

The *commutative law of multiplication* says that we can multiply numbers in any order. Thus you need to learn only about half the table, as shown by the shading.

◀ **Do Exercises 9 and 10.**

Multiply.

7. $8 \cdot 1$

8.
$$\begin{array}{r} 2\ 3 \\ \times\quad 1 \\ \hline \end{array}$$

Complete the table.

9.

×	2	3	4	5
2				
3			12	
4				
5		15		
6				

10.

×	6	7	8	9
5				
6		48		
7				
8		56		
9				

Answers

7. 8   **8.** 23

9.

×	2	3	4	5
2	4	6	8	10
3	6	9	12	15
4	8	12	16	20
5	10	15	20	25
6	12	18	24	30

10.

×	6	7	8	9
5	30	35	40	45
6	36	42	48	54
7	42	49	56	63
8	48	56	64	72
9	54	63	72	81

b MULTIPLYING BY 10, 100, AND 1000

We now move to a gradual, conceptual development of multiplication of whole numbers. We begin by considering multiplication by 10, 100, and 1000.

Multiplying by 10

We know that

$$50 = 5 \text{ tens} \qquad 340 = 34 \text{ tens} \qquad \text{and} \qquad 2340 = 234 \text{ tens}$$
$$= 5 \cdot 10, \qquad = 34 \cdot 10, \qquad = 234 \cdot 10.$$

Turning this around, we see that to multiply any whole number by 10, all we need to do is write a 0 on the end of the number.

MULTIPLICATION BY 10

To multiply a whole number by 10, write 0 on the end of the number.

EXAMPLES Multiply.

9. $10 \cdot 6 = 60$

10. $10 \cdot 47 = 470$

11. $10 \cdot 583 = 5830$

Do Exercises 11–15. ▶

Let's find $4 \cdot 90$. This is $4 \cdot (9 \text{ tens})$, or 36 tens. The procedure is the same as multiplying 4 and 9 and writing a 0 on the end. Thus, $4 \cdot 90 = 360$.

EXAMPLES Multiply.

12. $5 \cdot 70 = 350$
     $5 \cdot 7$, then write a 0

13. $8 \cdot 80 = 640$

14. $5 \cdot 60 = 300$

Do Exercises 16 and 17. ▶

Multiplying by 100

Note the following:

$$300 = 3 \text{ hundreds} \quad 4700 = 47 \text{ hundreds} \quad \text{and} \quad 56{,}800 = 568 \text{ hundreds}$$
$$= 3 \cdot 100, \qquad = 47 \cdot 100, \qquad = 568 \cdot 100.$$

Turning this around, we see that to multiply any whole number by 100, all we need to do is write two 0's on the end of the number.

MULTIPLICATION BY 100

To multiply a whole number by 100, write two 0's on the end of the number.

Multiply.

11. $10 \cdot 7$ **12.** $10 \cdot 45$

13. $10 \cdot 273$ **14.** $10 \cdot 10$

15. $10 \cdot 100$

Multiply.

16. $\begin{array}{r} 7\,0 \\ \times\quad 8 \\ \hline \end{array}$ **17.** $\begin{array}{r} 6\,0 \\ \times\quad 6 \\ \hline \end{array}$

Answers

11. 70 **12.** 450 **13.** 2730 **14.** 100
15. 1000 **16.** 560 **17.** 360

Multiply.

18. $100 \cdot 7$ **19.** $100 \cdot 23$

20. $100 \cdot 723$ **21.** $100 \cdot 100$

22. $100 \cdot 1000$

Multiply.

23.
$$\begin{array}{r} 7\,0\,0 \\ \times \quad 8 \\ \hline \end{array}$$
     24.
$$\begin{array}{r} 4\,0\,0 \\ \times \quad 4 \\ \hline \end{array}$$

Multiply.

25. $1000 \cdot 9$ **26.** $1000 \cdot 852$

27. $1000 \cdot 10$ **28.** $3 \cdot 4000$

29. $9 \cdot 8000$

EXAMPLES Multiply.

15. $100 \cdot 6 = 600$

16. $100 \cdot 39 = 3900$

17. $100 \cdot 448 = 44{,}800$

◀ **Do Exercises 18–22.**

Let's find $4 \cdot 900$. This is $4 \cdot$ (9 hundreds), or 36 hundreds. The procedure is the same as multiplying 4 and 9 and writing two 0's on the end. Thus, $4 \cdot 900 = 3600$.

EXAMPLES Multiply.

18. $6 \cdot 800 = 4800$

         $6 \cdot 8$, then write 00

19. $9 \cdot 700 = 6300$
20. $5 \cdot 500 = 2500$

◀ **Do Exercises 23 and 24.**

Multiplying by 1000

Note the following:

$$6000 = 6 \text{ thousands} \quad \text{and} \quad 19{,}000 = 19 \text{ thousands}$$
$$= 6 \cdot 1000 \qquad\qquad\qquad = 19 \cdot 1000.$$

Turning this around, we see that to multiply any whole number by 1000, all we need to do is write three 0's on the end of the number.

> ### MULTIPLYING BY 1000
>
> To multiply a whole number by 1000, write three 0's on the end of the number.

EXAMPLES Multiply.

21. $1000 \cdot 8 = 8000$

22. $2000 \cdot 13 = 26{,}000$ $2 \cdot 13$, then write 000
23. $1000 \cdot 567 = 567{,}000$

◀ **Do Exercises 25–29.**

Multiplying Multiples by Multiples

Let's multiply 50 and 30. This is $50 \cdot$ (3 tens), or 150 tens, or 1500. The procedure is the same as multiplying 5 and 3 and writing two 0's on the end.

Answers

18. 700   **19.** 2300   **20.** 72,300
21. 10,000   **22.** 100,000   **23.** 5600
24. 1600   **25.** 9000   **26.** 852,000
27. 10,000   **28.** 12,000   **29.** 72,000

TO MULTIPLY MULTIPLES OF TENS, HUNDREDS, THOUSANDS, AND SO ON:

a) Multiply the one-digit numbers.

b) Count the number of zeros.

c) Write that many 0's on the end of the product.

EXAMPLES Multiply.

24.
$$\begin{array}{r} 80 \\ \times\ \ 60 \\ \hline 4800 \end{array}$$
1 zero at end
1 zero at end
↑—— 6 · 8, then write 00

25.
$$\begin{array}{r} 800 \\ \times\ \ 60 \\ \hline 48{,}000 \end{array}$$
2 zeros at end
1 zero at end
↑—— 6 · 8, then write 000

26.
$$\begin{array}{r} 800 \\ \times\ \ 600 \\ \hline 480{,}000 \end{array}$$
2 zeros at end
2 zeros at end
↑—— 6 · 8, then write 0,000

27.
$$\begin{array}{r} 800 \\ \times\ \ 50 \\ \hline 40{,}000 \end{array}$$
2 zeros at end
1 zero at end
↑—— 5 · 8, then write 000

Do Exercises 30–33. ▶

Multiply.

30.
$$\begin{array}{r} 9000 \\ \times\ \ \ \ \ 6 \\ \hline \end{array}$$

31.
$$\begin{array}{r} 80 \\ \times\ 70 \\ \hline \end{array}$$

32.
$$\begin{array}{r} 800 \\ \times\ \ 70 \\ \hline \end{array}$$

33.
$$\begin{array}{r} 600 \\ \times\ \ 30 \\ \hline \end{array}$$

C MULTIPLYING LARGER NUMBERS

The product 3×24 can be represented as

$$\begin{aligned} 3 \times (2 \text{ tens} + 4) &= (2 \text{ tens} + 4) + (2 \text{ tens} + 4) + (2 \text{ tens} + 4) \\ &= 6 \text{ tens} + 12 \\ &= 6 \text{ tens} + 1 \text{ ten} + 2 \\ &= 7 \text{ tens} + 2 \\ &= 72. \end{aligned}$$

We multiply the 4 ones by 3, getting   12
We multiply the 2 tens by 3, getting   + 60
          Then we add: 72

EXAMPLE 28 Multiply: 3×24.

$$\begin{array}{r} 24 \\ \times\ \ \ 3 \\ \hline 12 \\ 60 \\ \hline 72 \end{array}$$
We use the approach described above.

← Multiply the 4 ones by 3.
← Multiply the 2 tens by 3.
← Add.

Do Exercises 34–36. ▶

Multiply.

34.
$$\begin{array}{r} 14 \\ \times\ 2 \\ \hline \end{array}$$

35.
$$\begin{array}{r} 58 \\ \times\ 2 \\ \hline \end{array}$$

36.
$$\begin{array}{r} 37 \\ \times\ 4 \\ \hline \end{array}$$

EXAMPLE 29 Multiply: 5×734.

$$\begin{array}{r} 734 \\ \times\ \ \ \ 5 \\ \hline 20 \\ 150 \\ 3500 \\ \hline 3670 \end{array}$$

← Multiply the 4 ones by 5.
← Multiply the 3 tens by 5.
← Multiply the 7 hundreds by 5.
← Add.

Multiply.

37.
$$\begin{array}{r} 823 \\ \times\ \ \ \ 6 \\ \hline \end{array}$$

38.
$$\begin{array}{r} 1348 \\ \times\ \ \ \ \ \ 5 \\ \hline \end{array}$$

Answers

30. 54,000   **31.** 5600   **32.** 56,000
33. 18,000   **34.** 28   **35.** 116
36. 148   **37.** 4938   **38.** 6740

Do Exercises 37 and 38. ▶

Let's look at Example 29 again. Instead of writing each product on a separate line, we can use a shorter form.

EXAMPLE 30   Multiply: 5 × 734.

$$\begin{array}{r} 7 \overset{2}{3} 4 \\ \times \quad 5 \\ \hline 0 \end{array}$$

Multiply the ones by 5: 5 · (4 ones) = 20 ones = 2 tens + 0 ones. Write 0 in the ones column and 2 above the tens.

$$\begin{array}{r} 7 \overset{1}{\overset{}{3}} \overset{2}{4} \\ \times \quad 5 \\ \hline 7\ 0 \end{array}$$

Multiply 3 tens by 5 and add 2 tens: 5 · (3 tens) = 15 tens; 15 tens + 2 tens = 17 tens = 1 hundred + 7 tens. Write 7 in the tens column and 1 above the hundreds.

$$\begin{array}{r} \overset{1}{7} \overset{2}{3} 4 \\ \times \quad 5 \\ \hline 3\ 6\ 7\ 0 \end{array}$$

Multiply the 7 hundreds by 5 and add 1 hundred: 5 · (7 hundreds) = 35 hundreds; 35 hundreds + 1 hundred = 36 hundreds.

$$\begin{array}{r} \overset{1}{7} \overset{2}{3} 4 \\ \times \quad 5 \\ \hline 3\ 6\ 7\ 0 \end{array}$$ ← You should write only this.

◀ **Do Exercises 39–42.**

Multiply using the short form.

39.
$$\begin{array}{r} 5\ 8 \\ \times\ \ 2 \\ \hline \end{array}$$

40.
$$\begin{array}{r} 3\ 7 \\ \times\ \ 4 \\ \hline \end{array}$$

41.
$$\begin{array}{r} 8\ 2\ 3 \\ \times\ \ \ 6 \\ \hline \end{array}$$

42.
$$\begin{array}{r} 1\ 3\ 4\ 8 \\ \times\ \ \ \ \ 5 \\ \hline \end{array}$$

d   MULTIPLYING BY MULTIPLES OF 10, 100, AND 1000

To multiply 327 by 50, we multiply by 10 (write a 0) and then multiply 327 by 5.

$$\begin{array}{r} 3\ 2\ 7 \\ \times\qquad 5\boxed{0} \\ \hline 1\ 6,3\ 5\ 0 \end{array}$$

← Write a 0.
Multiply 5 · 327.

EXAMPLE 31   Multiply: 400 × 289.

$$\begin{array}{r} 2\ 8\ 9 \\ \times\ 4\boxed{0\ 0} \\ \hline 0\ 0 \end{array}$$ ← Write two 0's.

$$\begin{array}{r} 2\ 8\ 9 \\ \times\qquad 4\ 0\ 0 \\ \hline 1\ 1\ 5,6\ 0\ 0 \end{array}$$

Multiply 4 and 289:
$$\begin{array}{r} \overset{3}{}\overset{3}{} \\ 2\ 8\ 9 \\ \times\qquad 4 \\ \hline 1\ 1\ 5\ 6 \end{array}$$

$$\begin{array}{r} \overset{3}{}\overset{3}{} \\ 2\ 8\ 9 \\ \times\qquad 4\ 0\ 0 \\ \hline 1\ 1\ 5,6\ 0\ 0 \end{array}$$ ← Try to write only this.

Multiply.

43.
$$\begin{array}{r} 7\ 4\ 6 \\ \times\ \ \ \ 8 \\ \hline \end{array}$$

44.
$$\begin{array}{r} 7\ 4\ 6 \\ \times\ \ \ 8\ 0 \\ \hline \end{array}$$

45.
$$\begin{array}{r} 7\ 4\ 6 \\ \times\ 8\ 0\ 0 \\ \hline \end{array}$$

◀ **Do Exercises 43–45.**

TO THE STUDENT

If you had trouble with Section 1.4 and have studied Developmental Unit M, you should go back and work through Section 1.4 after completing Exercise Set M.

Answers

39. 116   **40.** 148   **41.** 4938   **42.** 6740
43. 5968   **44.** 59,680   **45.** 596,800

M **Exercise Set**

For Extra Help

MyMathLab®

MathXL®

PRACTICE

WATCH

READ

REVIEW

a Multiply. Try to do these mentally.

1. 3 × 4	**2.** 6 × 0	**3.** 7 × 1	**4.** 0 × 2	**5.** 10 × 1	**6.** 6 × 5
7. 5 × 2	**8.** 9 × 7	**9.** 9 × 6	**10.** 2 × 6	**11.** 7 × 0	**12.** 8 × 9
13. 1 × 8	**14.** 8 × 0	**15.** 4 × 7	**16.** 3 × 8	**17.** 5 × 9	**18.** 2 × 9
19. 0 × 7	**20.** 5 × 7	**21.** 9 × 5	**22.** 5 × 8	**23.** 0 × 0	**24.** 2 × 8

25. $5 \cdot 5$ **26.** $9 \cdot 9$ **27.** $1 \cdot 1$ **28.** $0 \cdot 0$ **29.** $2 \cdot 2$

30. $6 \cdot 6$ **31.** $1 \cdot 8$ **32.** $0 \cdot 1$ **33.** $3 \cdot 9$ **34.** $2 \cdot 9$

35. $6 \cdot 0$ **36.** $10 \cdot 1$ **37.** $6 \cdot 8$ **38.** $9 \cdot 6$ **39.** $8 \cdot 0$

40. $9 \cdot 8$ **41.** $3 \cdot 5$ **42.** $1 \cdot 8$ **43.** $1 \cdot 9$ **44.** $2 \cdot 1$

45. $8 \cdot 4$ **46.** $3 \cdot 2$ **47.** $5 \cdot 3$ **48.** $1 \cdot 6$ **49.** $4 \cdot 2$

50. $4 \cdot 5$ **51.** $5 \cdot 4$ **52.** $4 \cdot 4$ **53.** $5 \cdot 2$ **54.** $8 \cdot 0$

b Multiply.

55. $\begin{array}{r} 1\ 0 \\ \times\quad 8 \\ \hline \end{array}$

56. $\begin{array}{r} 7 \\ \times 1\ 0 \\ \hline \end{array}$

57. $\begin{array}{r} 2\ 0 \\ \times\quad 8 \\ \hline \end{array}$

58. $\begin{array}{r} 3\ 0 \\ \times\quad 7 \\ \hline \end{array}$

59. $\begin{array}{r} 4\ 5 \\ \times 1\ 0 \\ \hline \end{array}$

60. $\begin{array}{r} 7\ 8 \\ \times 1\ 0 \\ \hline \end{array}$

61. $\begin{array}{r} 8\ 0 \\ \times\quad 7 \\ \hline \end{array}$

62. $\begin{array}{r} 9\ 0 \\ \times\quad 4 \\ \hline \end{array}$

63. $\begin{array}{r} 1\ 0\ 0 \\ \times\quad 8 \\ \hline \end{array}$

64. $\begin{array}{r} 1\ 0\ 0 \\ \times\quad 3 \\ \hline \end{array}$

65. $\begin{array}{r} 1\ 0\ 0 \\ \times\quad 9 \\ \hline \end{array}$

66. $\begin{array}{r} 1\ 0\ 0 \\ \times\quad 1\ 0 \\ \hline \end{array}$

67. $\begin{array}{r} 3\ 4\ 5\ 7 \\ \times\quad 1\ 0\ 0 \\ \hline \end{array}$

68. $\begin{array}{r} 4\ 0\ 0 \\ \times\quad 3 \\ \hline \end{array}$

69. $\begin{array}{r} 7\ 0\ 0 \\ \times\quad 7 \\ \hline \end{array}$

70. $\begin{array}{r} 5\ 0\ 0 \\ \times\quad 8 \\ \hline \end{array}$

71. $\begin{array}{r} 1\ 0\ 0 \\ \times 1\ 0\ 0 \\ \hline \end{array}$

72. $\begin{array}{r} 1\ 0\ 0\ 0 \\ \times\quad 7 \\ \hline \end{array}$

73. $\begin{array}{r} 1\ 0\ 0\ 0 \\ \times\quad 9 \\ \hline \end{array}$

74. $\begin{array}{r} 1\ 0\ 0\ 0 \\ \times\quad 2 \\ \hline \end{array}$

75. $\begin{array}{r} 4\ 5\ 7 \\ \times 1\ 0\ 0\ 0 \\ \hline \end{array}$

76. $\begin{array}{r} 6\ 7\ 6\ 9 \\ \times 1\ 0\ 0\ 0 \\ \hline \end{array}$

77. $\begin{array}{r} 2\ 0\ 0\ 0 \\ \times\quad 9 \\ \hline \end{array}$

78. $\begin{array}{r} 5\ 0\ 0\ 0 \\ \times\quad 4 \\ \hline \end{array}$

79. $\begin{array}{r} 6\ 0\ 0\ 0 \\ \times\quad 8 \\ \hline \end{array}$

80. $\begin{array}{r} 8\ 0\ 0\ 0 \\ \times\quad 2 \\ \hline \end{array}$

81. $\begin{array}{r} 3\ 0\ 0\ 0 \\ \times\quad 2 \\ \hline \end{array}$

82. $\begin{array}{r} 1\ 0\ 0\ 0 \\ \times 1\ 0\ 0\ 0 \\ \hline \end{array}$

83. $\begin{array}{r} 4\ 0 \\ \times 3\ 0 \\ \hline \end{array}$

84. $\begin{array}{r} 2\ 0 \\ \times 1\ 0 \\ \hline \end{array}$

85. $\begin{array}{r} 8\ 0 \\ \times 5\ 0 \\ \hline \end{array}$

86. $\begin{array}{r} 5\ 0 \\ \times 5\ 0 \\ \hline \end{array}$

87. $\begin{array}{r} 4\ 0\ 0 \\ \times\quad 3\ 0 \\ \hline \end{array}$

88. $\begin{array}{r} 2\ 0\ 0 \\ \times\quad 3\ 0 \\ \hline \end{array}$

89. $\begin{array}{r} 7\ 0\ 0 \\ \times\quad 9\ 0 \\ \hline \end{array}$

90. $\begin{array}{r} 4\ 0\ 0 \\ \times 3\ 0\ 0 \\ \hline \end{array}$

91. $\begin{array}{r} 4\ 0\ 0\ 0 \\ \times\quad 2\ 0\ 0 \\ \hline \end{array}$

92. $\begin{array}{r} 6\ 0\ 0\ 0 \\ \times\quad 2\ 0 \\ \hline \end{array}$

93. $\begin{array}{r} 4\ 0\ 0\ 0 \\ \times 4\ 0\ 0\ 0 \\ \hline \end{array}$

94. $\begin{array}{r} 8\ 0\ 0\ 0 \\ \times\quad 1\ 0 \\ \hline \end{array}$

c Multiply.

95. $\begin{array}{r} 4\ 9 \\ \times\quad 3 \\ \hline \end{array}$

96. $\begin{array}{r} 7\ 4 \\ \times\quad 6 \\ \hline \end{array}$

97. $\begin{array}{r} 5\ 9\ 3 \\ \times\quad 5 \\ \hline \end{array}$

98. $\begin{array}{r} 6\ 0\ 9 \\ \times\quad 8 \\ \hline \end{array}$

99. $\begin{array}{r} 8\ 9\ 9 \\ \times\quad 7 \\ \hline \end{array}$

100. $\begin{array}{r} 8\ 6\ 5 \\ \times\quad 4 \\ \hline \end{array}$

101. $\begin{array}{r} 8\ 1\ 1\ 8 \\ \times\quad 2 \\ \hline \end{array}$

102. $\begin{array}{r} 6\ 7\ 5\ 4 \\ \times\quad 2 \\ \hline \end{array}$

103. $\begin{array}{r} 4\ 3,7\ 7\ 7 \\ \times\quad 2 \\ \hline \end{array}$

104. $\begin{array}{r} 3\ 2,5\ 6\ 4 \\ \times\quad 6 \\ \hline \end{array}$

d Multiply.

105. $\begin{array}{r} 5\ 8 \\ \times 6\ 0 \\ \hline \end{array}$

106. $\begin{array}{r} 9\ 3 \\ \times 3\ 0 \\ \hline \end{array}$

107. $\begin{array}{r} 4\ 2 \\ \times 8\ 0 \\ \hline \end{array}$

108. $\begin{array}{r} 7\ 8 \\ \times 9\ 0 \\ \hline \end{array}$

109. $\begin{array}{r} 3\ 4\ 6 \\ \times\quad 6\ 0 \\ \hline \end{array}$

110. $\begin{array}{r} 2\ 6\ 7 \\ \times\quad 4\ 0 \\ \hline \end{array}$

111. $\begin{array}{r} 8\ 9\ 7 \\ \times 4\ 0\ 0 \\ \hline \end{array}$

112. $\begin{array}{r} 3\ 6\ 6 \\ \times 3\ 0\ 0 \\ \hline \end{array}$

113. $\begin{array}{r} 8\ 3\ 4 \\ \times 7\ 0\ 0 \\ \hline \end{array}$

114. $\begin{array}{r} 3\ 3\ 3 \\ \times 9\ 0\ 0 \\ \hline \end{array}$

115. $\begin{array}{r} 5\ 6\ 7\ 3 \\ \times 2\ 0\ 0\ 0 \\ \hline \end{array}$

116. $\begin{array}{r} 4\ 6\ 7\ 8 \\ \times 5\ 0\ 0\ 0 \\ \hline \end{array}$

117. $\begin{array}{r} 6\ 7\ 8\ 8 \\ \times 9\ 0\ 0\ 0 \\ \hline \end{array}$

118. $\begin{array}{r} 9\ 1\ 2\ 9 \\ \times 8\ 0\ 0\ 0 \\ \hline \end{array}$

Division

a BASIC DIVISION

Division can be explained by arranging a set of objects in a rectangular array. This can be done in two ways. Both methods below illustrate the division $18 \div 6$.

OBJECTIVES

a Find basic quotients such as $20 \div 5$, $56 \div 7$, and so on.

b Divide by estimating multiples of thousands, hundreds, tens, and ones.

Method 1 We take 18 objects and determine how many rows, each with 6 objects, we can form.

3 rows of 6 objects

Since there are 3 rows of 6 objects, we have $18 \div 6 = 3$.

Method 2 We take 18 objects and arrange them in 6 rows, then determine how many objects are in each row.

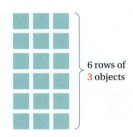

6 rows of 3 objects

Since there are 3 objects in each of the 6 rows, we have $18 \div 6 = 3$.

We can also use other notation for division:

$$18 \div 6 = 18/6 = \frac{18}{6} = 3 \quad \text{and} \quad 6\overline{)18}^{\,3}.$$

EXAMPLES Divide.

1. $9\overline{)3\ 6}^{\,4}$ *Think*: 36 objects: How many rows, each with 9 objects?
or: 36 objects: How many objects in each of 9 rows?

2. $42 \div 7 = 6$

3. $\dfrac{24}{3} = 8$

Do Exercises 1–4. ▶

Divide.

1. $24 \div 6$ **2.** $64 \div 8$

3. $\dfrac{63}{7}$ **4.** $\dfrac{27}{9}$

The multiplication table in Developmental Unit M will enable you to divide also. First, recall how multiplication and division are related.

A multiplication: $5 \cdot 4 = 20$

Two related divisions:

a)

4 rows of 5 objects

$20 \div 5 = 4$.

b)

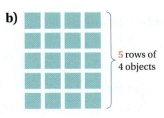

5 rows of 4 objects

$20 \div 4 = 5$.

Answers

1. 4   **2.** 8   **3.** 9   **4.** 3

Since we know that

$$5 \cdot 4 = 20, \qquad \text{A basic multiplication fact}$$

we also know two division facts:

$$20 \div 5 = 4 \quad \text{and} \quad 20 \div 4 = 5.$$

In the division $20 \div 4 = 5$, 20 is the *dividend*, 4 is the *divisor*, and 5 is the *quotient*.

EXAMPLE 4 From $7 \cdot 8 = 56$, write two division facts.

a) We have

$$7 \cdot 8 = 56 \qquad \text{Multiplication sentence}$$
$$7 = 56 \div 8. \qquad \text{Related division sentence}$$

b) We also have

$$7 \cdot 8 = 56 \qquad \text{Multiplication sentence}$$
$$8 = 56 \div 7. \qquad \text{Related division sentence}$$

◀ **Do Exercises 5 and 6.**

We can use the idea that division is defined in terms of multiplication to do basic divisions.

EXAMPLE 5 Find: $35 \div 5$.

To find $35 \div 5$, we ask, "5 times what number is 35?"

$$5 \cdot \square = 35$$

Using the multiplication table at the left, we find 35 inside the table and 5 along the left side. Then we read the answer, 7, from the top. Thus we have

$$35 \div 5 = 7.$$

Strive to do this kind of thinking as fast as you can mentally, without having to use the table.

◀ **Do Exercises 7–10.**

Division by 1

Note that $3 \div 1 = 3$ because $3 \cdot 1 = 3$.

DIVISION BY 1

Any number divided by 1 is that same number:

$$a \div 1 = \frac{a}{1} = a.$$

EXAMPLES Divide.

6. $\dfrac{8}{1} = 8$ **7.** $6 \div 1 = 6$ **8.** $34 \div 1 = 34$

◀ **Do Exercises 11–13.**

For each multiplication fact, write two division facts.

5. $6 \cdot 2 = 12$

6. $7 \times 6 = 42$

$\times$	2	3	4	5	6	7	8	9
2	4	6	8	10	12	14	16	18
3	6	9	12	15	18	21	24	27
4	8	12	16	20	24	28	32	36
5	10	15	20	25	30	35	40	45
6	12	18	24	30	36	42	48	54
7	14	21	28	35	42	49	56	63
8	16	24	32	40	48	56	64	72
9	18	27	36	45	54	63	72	81

Divide.

7. $28 \div 4$ **8.** $81 \div 9$

9. $\dfrac{16}{2}$ **10.** $\dfrac{54}{6}$

Divide.

11. $6 \div 1$ **12.** $\dfrac{13}{1}$

13. $1 \div 1$

Answers

5. $12 \div 2 = 6$; $12 \div 6 = 2$
6. $42 \div 6 = 7$; $42 \div 7 = 6$ **7.** 7   **8.** 9
9. 8   **10.** 9   **11.** 6   **12.** 13   **13.** 1

Division by 0

We cannot divide by 0. To see why, suppose the number 4 could be divided by 0, giving some number, $\square$, as the result. Because division is defined in terms of multiplication,

$$\text{if} \quad 4 \div 0 = \square, \quad \text{then} \quad 4 = \square \cdot 0 = 0. \qquad \text{False! } 4 \neq 0$$

Thus the only number that could possibly be divided by 0 would be 0 itself. But since 0 times any number is 0, this division could give us any number we wanted. We avoid these difficulties by agreeing not to divide *any* number by 0.

DIVISION BY 0

Division by 0 is not defined. (We agree not to divide by 0.)

Dividing 0 by Other Numbers

Note that $0 \div 3 = 0$ because $0 \cdot 3 = 0$.

DIVISION INTO 0

Zero divided by any number other than 0 is 0:

$$\frac{0}{a} = 0, \quad a \neq 0.$$

EXAMPLES Divide.

9. $0 \div 8 = 0$

10. $\dfrac{0}{9} = 0$

11. $\dfrac{137}{0}$ is undefined.

12. $12 \div 0$ is undefined.

Do Exercises 14–21. ▶

Dividing a Number by Itself

Note that $3 \div 3 = 1$ because $1 \cdot 3 = 3$.

DIVISION OF A NUMBER BY ITSELF

Any number other than 0 divided by itself is 1:

$$\frac{a}{a} = 1, \quad a \neq 0.$$

EXAMPLES Divide.

13. $8 \div 8 = 1 \qquad 1 \cdot 8 = 8$

14. $\dfrac{32}{32} = 1 \qquad 1 \cdot 32 = 32$

Do Exercises 22–27. ▶

Divide, if possible. If not possible, write "undefined."

14. $\dfrac{8}{4}$ **15.** $\dfrac{5}{0}$

16. $\dfrac{0}{5}$ **17.** $\dfrac{0}{0}$

18. $12 \div 0$ **19.** $100 \div 10$

20. $\dfrac{5}{3-3}$ **21.** $\dfrac{8-8}{4}$

Divide.

22. $23 \div 23$ **23.** $\dfrac{67}{67}$

24. $\dfrac{41}{41}$ **25.** $17 \div 17$

26. $17 \div 1$ **27.** $\dfrac{54}{54}$

Answers

14. 2   **15.** Undefined   **16.** 0
17. Undefined   **18.** Undefined   **19.** 10
20. Undefined   **21.** 0   **22.** 1   **23.** 1
24. 1   **25.** 1   **26.** 17   **27.** 1

b DIVIDING BY ESTIMATING MULTIPLES

We divide using a process of estimating multiples of 10, 100, 1000, and so on.

EXAMPLE 15 Divide: 743 ÷ 3.

a) Are there any hundreds in the quotient? To find how many hundreds, we find products of 3 and multiples of 100.

$3 \cdot 100 = 300$

$3 \cdot 200 = 600$ ← 743

$3 \cdot 300 = 900$

```
        2 0 0
  3 )‾7 4 3
      6 0 0  ← Multiply 3 × 200.
      1 4 3  ← Subtract.
```

b) Now look at 143 and go to the tens place. Are there any tens in the quotient?

$3 \cdot 10 = 30$

$3 \cdot 20 = 60$

$3 \cdot 30 = 90$

$3 \cdot 40 = 120$ ← 143

$3 \cdot 50 = 150$

```
          4 0
        2 0 0
  3 ) 7 4 3
      6 0 0
      1 4 3
      1 2 0  ← Multiply 3 × 40.
        2 3  ← Subtract.
```

c) Now look at 23 and go to the ones place. Are there any ones in the quotient?

$3 \cdot 1 = 3$

$3 \cdot 2 = 6$

$3 \cdot 3 = 9$

$3 \cdot 4 = 12$

$3 \cdot 5 = 15$

$3 \cdot 6 = 18$

$3 \cdot 7 = 21$ ← 23

$3 \cdot 8 = 24$

```
        2 4 7  ← Adding:
            7     200 + 40 + 7 = 247
          4 0
        2 0 0
  3 ) 7 4 3
      6 0 0
      1 4 3
      1 2 0
        2 3
        2 1  ← Multiply 3 × 7.
          2  ← Subtract. This is the
                remainder, since
                2 is less than 3.
```

We generally write only the *short form* when we divide.

```
      2 4 7
  3 ) 7 4 3
      6 0 0
      1 4 3
      1 2 0
        2 3
        2 1
          2
```

We write a 2 above the hundreds digit in the dividend to record 200.

We write a 4 to record 40.

We write a 7 to record 7.

The answer is 247 R 2.

◀ **Do Exercises 28–31.**

Divide.

28. 4)‾3 8 5   **29.** 7)‾8 8 4 6

Divide using the short form.

30. 2)‾6 4 8   **31.** 9)‾3 7 5 8

Answers

28. 96 R 1   **29.** 1263 R 5   **30.** 324
31. 417 R 5

a Divide, if possible.

1. $24 \div 8$ **2.** $72 \div 9$ **3.** $28 \div 7$ **4.** $22 \div 22$ **5.** $32 \div 1$

6. $45 \div 5$ **7.** $14 \div 2$ **8.** $40 \div 8$ **9.** $37 \div 1$ **10.** $10 \div 2$

11. $36 \div 4$ **12.** $12 \div 3$ **13.** $54 \div 9$ **14.** $18 \div 2$ **15.** $20 \div 4$

16. $16 \div 2$ **17.** $72 \div 8$ **18.** $42 \div 7$ **19.** $12 \div 4$ **20.** $8 \div 4$

21. $54 \div 6$ **22.** $18 \div 9$ **23.** $9 \div 3$ **24.** $28 \div 4$ **25.** $56 \div 7$

26. $24 \div 6$ **27.** $14 \div 2$ **28.** $14 \div 7$ **29.** $21 \div 7$ **30.** $36 \div 6$

31. $8 \div 8$ **32.** $32 \div 8$ **33.** $30 \div 5$ **34.** $18 \div 6$ **35.** $49 \div 7$

36. $81 \div 9$ **37.** $0 \div 7$ **38.** $9 \div 0$ **39.** $16 \div 0$ **40.** $42 \div 6$

41. $\dfrac{48}{6}$ **42.** $\dfrac{35}{5}$ **43.** $\dfrac{9}{9}$ **44.** $\dfrac{45}{9}$ **45.** $\dfrac{0}{5}$ **46.** $\dfrac{0}{8}$

47. $\dfrac{6}{2}$ **48.** $\dfrac{3}{3}$ **49.** $\dfrac{8}{2}$ **50.** $\dfrac{7}{1}$ **51.** $\dfrac{5}{5}$ **52.** $\dfrac{6}{1}$

53. $\dfrac{2}{2}$ **54.** $\dfrac{25}{5}$ **55.** $\dfrac{4}{2}$ **56.** $\dfrac{24}{3}$ **57.** $\dfrac{0}{9}$ **58.** $\dfrac{0}{4}$

59. $\dfrac{40}{5}$ **60.** $\dfrac{3}{1}$ **61.** $\dfrac{16}{4}$ **62.** $\dfrac{9}{0}$ **63.** $\dfrac{32}{8}$ **64.** $\dfrac{9}{9}$

b Divide.

65. 4) 2 7 7

66. 2) 3 9 9

67. 8) 7 3 7

68. 6) 8 3 1

69. 5) 1 0 5

70. 6) 7 0 8

71. 9) 8 2 0

72. 3) 9 6 5

73. 5) 8 6 1 9

74. 3) 8 7 7 5

75. 9) 7 7 7 7

76. 8) 4 1 7 9

77. 5) 4 8 2 3

78. 8) 5 4 3 7

79. 2) 5 7 9 4

80. 7) 9 2 9 8

81. 2 0) 8 7 5

82. 3 0) 9 8 7

83. 5 0) 6 8 9 3

84. 4 0) 9 9 4 6

85. 1 1) 4 1 5

86. 1 1) 8 3 6

87. 1 2) 6 3 1 2

88. 1 2) 1 9 7 0

Answers

CHAPTER 1

Exercise Set 1.1, p. 6
RC1. digit **RC2.** period **RC3.** expanded **RC4.** standard
1. 5 thousands **3.** 5 hundreds **5.** 1 **7.** 2
9. 2 thousands + 0 hundreds + 5 tens + 8 ones, or 2 thousands + 5 tens + 8 ones **11.** 1 thousand + 5 hundreds + 7 tens + 6 ones **13.** 5 thousands + 7 hundreds + 0 tens + 2 ones, or 5 thousands + 7 hundreds + 2 ones **15.** 9 ten thousands + 3 thousands + 9 hundreds + 8 tens + 6 ones
17. 1 billion + 3 hundred millions + 4 ten millions + 3 millions + 2 hundred thousands + 3 ten thousands + 9 thousands + 9 hundreds + 2 tens + 3 ones **19.** 2 hundred millions + 4 ten millions + 8 millions + 6 hundred thousands + 4 ten thousands + 5 thousands + 0 hundreds + 0 tens + 8 ones, or 2 hundred millions + 4 ten millions + 8 millions + 6 hundred thousands + 4 ten thousands + 5 thousands + 8 ones **21.** Eighty-five **23.** Eighty-eight thousand **25.** One hundred twenty-three thousand, seven hundred sixty-five **27.** Seven billion, seven hundred fifty-four million, two hundred eleven thousand, five hundred seventy-seven **29.** Seven hundred one thousand, seven hundred ninety-nine **31.** Two million, four hundred seventy-four thousand, two hundred eighty **33.** 632,896 **35.** 50,324 **37.** 2,233,812 **39.** 8,000,000,000 **41.** 40,000,000 **43.** 30,000,103 **45.** 64,186,000 **47.** 138

Calculator Corner, p. 10
1. 121 **2.** 1602 **3.** 1932 **4.** 864

Exercise Set 1.2, p. 12
RC1. addends **RC2.** sum **RC3.** 0 **RC4.** perimeter
1. 387 **3.** 164 **5.** 5198 **7.** 100 **9.** 8503 **11.** 5266
13. 4466 **15.** 6608 **17.** 34,432 **19.** 101,310 **21.** 230
23. 18,424 **25.** 31,685 **27.** 132 yd **29.** 1661 ft **31.** 570 ft
33. 8 ten thousands **34.** Nine billion, three hundred forty-six million, three hundred ninety-nine thousand, four hundred sixty-eight **35.** 1 + 99 = 100, 2 + 98 = 100, . . . , 49 + 51 = 100. Then 49 100's = 4900 and 4900 + 50 + 100 = 5050.

Calculator Corner, p. 15
1. 28 **2.** 47 **3.** 67 **4.** 119 **5.** 2128 **6.** 2593

Exercise Set 1.3, p. 17
RC1. minuend **RC2.** subtraction symbol **RC3.** subtrahend
RC4. difference
1. 44 **3.** 533 **5.** 39 **7.** 14 **9.** 369 **11.** 26 **13.** 234
15. 417 **17.** 5382 **19.** 2778 **21.** 3069 **23.** 1089
25. 7748 **27.** 4144 **29.** 56 **31.** 454 **33.** 3749
35. 2191 **37.** 43,028 **39.** 95,974 **41.** 4418 **43.** 1305
45. 9989 **47.** 48,017 **49.** 1345 **50.** 924 **51.** 22,692
52. 10,920 **53.** Six million, three hundred seventy-five thousand, six hundred two **54.** 9 thousands + 1 hundred + 0 tens + 3 ones, or 9 thousands + 1 hundred + 3 ones **55.** 3; 4

Calculator Corner, p. 21
1. 448 **2.** 21,970 **3.** 6380 **4.** 39,564 **5.** 180,480
6. 2,363,754

Exercise Set 1.4, p. 23
RC1. factors **RC2.** product **RC3.** 0 **RC4.** 1
1. 520 **3.** 564 **5.** 1527 **7.** 64,603 **9.** 4770 **11.** 3995
13. 870 **15.** 1920 **17.** 46,296 **19.** 14,652 **21.** 258,312
23. 798,408 **25.** 20,723,872 **27.** 362,128 **29.** 302,220
31. 49,101,136 **33.** 25,236,000 **35.** 20,064,048
37. 529,984 sq mi **39.** 8100 sq ft **41.** 12,685 **42.** 10,834
43. 8889 **44.** 254,119 **45.** 4 hundred thousands **46.** 0
47. 1 ten thousand + 2 thousands + 8 hundreds + 4 tens + 7 ones
48. Seven million, four hundred thirty-two thousand
49. 247,464 sq ft

Calculator Corner, p. 30
1. 28 **2.** 123 **3.** 323 **4.** 36

Exercise Set 1.5, p. 32
RC1. quotient **RC2.** dividend **RC3.** remainder
RC4. divisor
1. 12 **3.** 1 **5.** 22 **7.** 0 **9.** Not defined **11.** 6
13. 55 R 2 **15.** 108 **17.** 307 **19.** 753 R 3 **21.** 74 R 1
23. 92 R 2 **25.** 1703 **27.** 987 R 5 **29.** 12,700 **31.** 127
33. 52 R 52 **35.** 29 R 5 **37.** 40 R 12 **39.** 90 R 22 **41.** 29
43. 105 R 3 **45.** 1609 R 2 **47.** 1007 R 1 **49.** 23
51. 107 R 1 **53.** 370 **55.** 609 R 15 **57.** 304
59. 3508 R 219 **61.** 8070 **63.** 1241 **64.** 66,444
65. 19,800 **66.** 9380 **67.** 40 ft **68.** 99 sq ft
69. 54, 122; 33, 2772; 4, 8 **71.** 30 buses

Mid-Chapter Review: Chapter 1, p. 35
1. False **2.** True **3.** True **4.** False **5.** True
6. False **7.** Ninety-five million, four hundred six thousand, two hundred thirty-seven
8. $\begin{array}{r} \overset{5\ 9\ 14}{6\,\cancel{0}\,\cancel{4}} \\ -\ 4\ 9\ 7 \\ \hline 1\ 0\ 7 \end{array}$ **9.** 6 hundreds **10.** 6 ten thousands
11. 6 thousands **12.** 6 ones **13.** 2 **14.** 6 **15.** 5 **16.** 1
17. 5 thousands + 6 hundreds + 0 tens + 2 ones, or 5 thousands + 6 hundreds + 2 ones **18.** 6 ten thousands + 9 thousands + 3 hundreds + 4 tens + 5 ones **19.** One hundred thirty-six
20. Sixty-four thousand, three hundred twenty-five **21.** 308,716
22. 4,567,216 **23.** 798 **24.** 1030 **25.** 7922 **26.** 7534
27. 465 **28.** 339 **29.** 1854 **30.** 4328 **31.** 216
32. 15,876 **33.** 132,275 **34.** 5,679,870 **35.** 253

36. 112 R 5 **37.** 23 R 19 **38.** 144 R 31 **39.** 25 m
40. 8 sq in. **41.** When numbers are being added, it does not matter how they are grouped. **42.** Subtraction is not commutative. For example, $5 - 2 = 3$, but $2 - 5 \neq 3$.
43. Answers will vary. Suppose one coat costs $150. Then the multiplication $4 \cdot \$150$ gives the cost of four coats. Or, suppose one ream of copy paper costs $4. Then the multiplication $\$4 \cdot 150$ gives the cost of 150 reams. **44.** If we use the definition of division, $0 \div 0 = a$ such that $a \cdot 0 = 0$. We see that a could be *any* number since $a \cdot 0 = 0$ for any number a. Thus we cannot say that $0 \div 0 = 0$. This is why we agree not to allow division by 0.

Exercise Set 1.6, p. 43

RC1. True **RC2.** False **RC3.** False **RC4.** True
1. 50 **3.** 460 **5.** 730 **7.** 900 **9.** 100 **11.** 1000
13. 9100 **15.** 32,800 **17.** 6000 **19.** 8000 **21.** 45,000
23. 373,000 **25.** $80 + 90 = 170$ **27.** $8070 - 2350 = 5720$
29. 220; incorrect **31.** 890; incorrect
33. $7300 + 9200 = 16,500$ **35.** $6900 - 1700 = 5200$
37. 1600; correct **39.** 1500; correct
41. $10,000 + 5000 + 9000 + 7000 = 31,000$
43. $92,000 - 23,000 = 69,000$ **45.** $50 \cdot 70 = 3500$
47. $30 \cdot 30 = 900$ **49.** $900 \cdot 300 = 270,000$
51. $400 \cdot 200 = 80,000$ **53.** $350 \div 70 = 5$
55. $8450 \div 50 = 169$ **57.** $1200 \div 200 = 6$
59. $8400 \div 300 = 28$ **61.** $800 **63.** $1200; no
65. Answers will vary depending on the options chosen.
67. (a) $309,600; **(b)** $360,000 **69.** 90 people **71.** <
73. > **75.** < **77.** > **79.** > **81.** >
83. $1,335,475 < 4,134,519$, or $4,134,519 > 1,335,475$
85. $97,382 < 98,817$, or $98,817 > 97,382$ **87.** 86,754
88. 13,589 **89.** 48,824 **90.** 4415 **91.** 1702 **92.** 17,748
93. 54 R 4 **94.** 208 **95.** Left to the student **97.** Left to the student

Exercise Set 1.7, p. 52

RC1. (c) **RC2.** (a) **RC3.** (d) **RC4.** (b)
1. 14 **3.** 0 **5.** 90,900 **7.** 450 **9.** 352 **11.** 25 **13.** 29
15. 0 **17.** 79 **19.** 45 **21.** 8 **23.** 14 **25.** 32 **27.** 143
29. 17,603 **31.** 37 **33.** 1035 **35.** 66 **37.** 324 **39.** 335
41. 18,252 **43.** 104 **45.** 45 **47.** 4056 **49.** 2847 **51.** 15
53. 205 **55.** 457 **57.** 142 R 5 **58.** 142 **59.** 334
60. 334 R 11 **61.** < **62.** > **63.** > **64.** <
65. 6,376,000 **66.** 6,375,600 **67.** 347

Translating for Success, p. 62

1. E **2.** M **3.** D **4.** G **5.** A **6.** O **7.** F **8.** K
9. J **10.** H

Exercise Set 1.8, p. 63

RC1. Familiarize. **RC2.** Translate. **RC3.** Solve.
RC4. Check.
1. 1962 ft **3.** 1450 ft **5.** 95 milligrams **7.** 18 rows
9. 43 events **11.** 2054 mi **13.** 2,073,600 pixels **15.** 168 hr
17. $273 per month **19.** $197 **21.** $7092 **23.** 151,500
25. $78 **27.** $40 per month **29.** $24,456 **31.** 35 weeks;
2 episodes **33.** 21 columns **35.** 236 gal **37. (a)** 4200 sq ft;
(b) 268 ft **39.** 56 cartons **41.** 645 mi; 5 in. **43.** $247
45. 525 min, or 8 hr 45 min **47.** 168,300 jobs **49.** 104 seats
51. 32 $10 bills **53.** $400 **55.** 106 bones **57.** 8273
58. 7759 **59.** 806,985 **60.** 147 R 4 **61.** 34 m
62. 9706 sq ft **63.** $200 \times 600 = 120,000$ **64.** 66
65. 792,000 mi; 1,386,000 mi

Calculator Corner, p. 71

1. 243 **2.** 15,625 **3.** 20,736 **4.** 2048

Calculator Corner, p. 73

1. 49 **2.** 85 **3.** 36 **4.** 0 **5.** 73 **6.** 49

Exercise Set 1.9, p. 75

RC1. exponent **RC2.** squared **RC3.** multiplication
RC4. 3
1. 3^4 **3.** 5^2 **5.** 7^5 **7.** 10^3 **9.** 49 **11.** 729 **13.** 20,736
15. 243 **17.** 22 **19.** 20 **21.** 100 **23.** 1 **25.** 49 **27.** 5
29. 434 **31.** 41 **33.** 88 **35.** 4 **37.** 303 **39.** 20
41. 70 **43.** 295 **45.** 32 **47.** 906 **49.** 62 **51.** 102
53. 32 **55.** $94 **57.** 401 **59.** 110 **61.** 7 **63.** 544
65. 708 **67.** 27 **69.** 452 **70.** 835 **71.** 13 **72.** 37
73. 4898 **74.** 100 **75.** 104,286 sq mi **76.** 98 gal
77. $24; 1 + 5 \cdot (4 + 3) = 36$ **79.** $7; 12 \div (4 + 2) \cdot 3 - 2 = 4$

Summary and Review: Chapter 1, p. 78

Vocabulary Reinforcement

1. perimeter **2.** minuend **3.** digits; periods **4.** dividend
5. factors; product **6.** additive **7.** associative **8.** divisor;
remainder; dividend

Concept Reinforcement

1. True **2.** True **3.** False **4.** False **5.** True **6.** False

Important Concepts

1. 2 thousands **2.** 65,302 **3.** 3237 **4.** 225,036
5. 315 R 14 **6.** 36,000 **7.** < **8.** 36 **9.** 216

Review Exercises

1. 8 thousands **2.** 3 **3.** 2 thousands + 7 hundreds + 9 tens + 3 ones **4.** 5 ten thousands + 6 thousands + 0 hundreds + 7 tens + 8 ones, or 5 ten thousands + 6 thousands + 7 tens + 8 ones **5.** 4 millions + 0 hundred thousands + 0 ten thousands + 7 thousands + 1 hundred + 0 tens + 1 one, or 4 millions + 7 thousands + 1 hundred + 1 one
6. Sixty-seven thousand, eight hundred nineteen **7.** Two million, seven hundred eighty-one thousand, four hundred twenty-seven **8.** 476,588 **9.** 1,640,000,000 **10.** 14,272
11. 66,024 **12.** 21,788 **13.** 98,921 **14.** 5148 **15.** 1689
16. 2274 **17.** 17,757 **18.** 5,100,000 **19.** 6,276,800
20. 506,748 **21.** 27,589 **22.** 5,331,810 **23.** 12 R 3 **24.** 5
25. 913 R 3 **26.** 384 R 1 **27.** 4 R 46 **28.** 54 **29.** 452
30. 5008 **31.** 4389 **32.** 345,800 **33.** 345,760 **34.** 346,000
35. 300,000 **36.** > **37.** < **38.** $41,300 + 19,700 = 61,000$
39. $38,700 - 24,500 = 14,200$ **40.** $400 \cdot 700 = 280,000$
41. 8 **42.** 45 **43.** 58 **44.** 0 **45.** 4^3 **46.** 10,000
47. 36 **48.** 65 **49.** 233 **50.** 260 **51.** 165 **52.** $502
53. $484 **54.** 1982 **55.** 19 cartons **56.** $13,585
57. 14 beehives **58.** 98 sq ft; 42 ft **59.** 137 beakers filled;
13 mL left over **60.** $27,598 **61.** B **62.** A **63.** D **64.** 8
65. $a = 8, b = 4$ **66.** 6 days

Understanding Through Discussion and Writing

1. No; if subtraction were associative, then $a - (b - c) = (a - b) - c$ for any a, b, and c. But, for example,

$$12 - (8 - 4) = 12 - 4 = 8,$$

whereas

$$(12 - 8) - 4 = 4 - 4 = 0.$$

Since $8 \neq 0$, this example shows that subtraction is not associative.
2. By rounding prices and estimating their sum, a shopper can estimate the total grocery bill while shopping. This is particularly useful if the shopper wants to spend no more than a certain amount.

3. Answers will vary. Anthony is driving from Kansas City to Minneapolis, a distance of 512 mi. He stops for gas after driving 183 mi. How much farther must he drive?　**4.** The parentheses are not necessary in the expression $9 - (4 \cdot 2)$. Using the rules for order of operations, the multiplication would be performed before the subtraction even if the parentheses were not present. The parentheses are necessary in the expression $(3 \cdot 4)^2$; $(3 \cdot 4)^2 = 12^2 = 144$, but $3 \cdot 4^2 = 3 \cdot 16 = 48$.

Test: Chapter 1, p. 83

1. [1.1a] 5　**2.** [1.1b] 8 thousands + 8 hundreds + 4 tens + 3 ones　**3.** [1.1c] Thirty-eight million, four hundred three thousand, two hundred seventy-seven　**4.** [1.2a] 9989
5. [1.2a] 63,791　**6.** [1.2a] 3165　**7.** [1.2a] 10,515
8. [1.3a] 3630　**9.** [1.3a] 1039　**10.** [1.3a] 6848
11. [1.3a] 5175　**12.** [1.4a] 41,112　**13.** [1.4a] 5,325,600
14. [1.4a] 2405　**15.** [1.4a] 534,264　**16.** [1.5a] 3 R 3
17. [1.5a] 70　**18.** [1.5a] 97　**19.** [1.5a] 805 R 8
20. [1.6a] 35,000　**21.** [1.6a] 34,530　**22.** [1.6a] 34,500
23. [1.6b] 23,600 + 54,700 = 78,300
24. [1.6b] 54,800 − 23,600 = 31,200
25. [1.6b] 800 · 500 = 400,000　**26.** [1.6c] >　**27.** [1.6c] <
28. [1.7b] 46　**29.** [1.7b] 13　**30.** [1.7b] 14　**31.** [1.7b] 381
32. [1.8a] 83 calories　**33.** [1.8a] 20 staplers
34. [1.8a] 1,256,615 sq mi　**35. (a)** [1.2b], [1.4b] 300 in., 5000 sq in.; 264 in., 3872 sq in.; 228 in., 2888 sq in.;
(b) [1.8a] 2112 sq in.　**36.** [1.8a] 1852 12-packs; 7 cakes left over
37. [1.8a] $95　**38.** [1.9a] 12^4　**39.** [1.9b] 343
40. [1.9b] 100,000　**41.** [1.9c] 31　**42.** [1.9c] 98　**43.** [1.9c] 2
44. [1.9c] 18　**45.** [1.9d] 216　**46.** [1.9c] A
47. [1.4b], [1.8a] 336 sq in.　**48.** [1.9c] 9　**49.** [1.8a] 80 payments

CHAPTER 2

Exercise Set 2.1, p. 90

RC1. True　**RC2.** True　**RC3.** False　**RC4.** False
RC5. True　**RC6.** False　**RC7.** True　**RC8.** True
1. 820; −541　**3.** 950,000,000; −460　**5.** 1454; −55　**7.** <
9. >　**11.** >　**13.** <　**15.** <　**17.** <　**19.** >　**21.** 57
23. 0　**25.** 24　**27.** 53　**29.** 8　**31.** 7　**33.** −7　**35.** 0
37. 21　**39.** −53　**41.** 1　**43.** 7　**45.** −9　**47.** −17
49. 23　**51.** −1　**53.** 85　**55.** −345　**57.** 0　**59.** −8
61. 825　**62.** 125　**63.** 7106　**64.** 4　**65.** 42　**66.** 69
67. >　**69.** =　**71.** −1, 0, 1
73. $-100, -5, 0, |3|, 4, |-6|, 7^2, 10^2, 2^7, 2^{10}$

Calculator Corner, p. 94

1. 13　**2.** −8

Exercise Set 2.2, p. 95

RC1. right　**RC2.** left　**RC3.** left　**RC4.** right
1. −5　**3.** −4　**5.** 6　**7.** 0　**9.** −4　**11.** −12　**13.** −11
15. −15　**17.** 42　**19.** 5　**21.** −4　**23.** 0　**25.** 0
27. −9　**29.** 7　**31.** 0　**33.** 45　**35.** −3　**37.** 0
39. −10　**41.** −24　**43.** −5　**45.** −21　**47.** −30
49. 6　**51.** −21　**53.** 25　**55.** −17　**57.** 6　**59.** −65
61. −160　**63.** −62　**65.** −23　**67.** 6681　**68.** 73
69. 3 ten thousands + 9 thousands + 4 hundreds + 1 ten + 7 ones　**70.** 2352　**71.** 32　**72.** 3500　**73.** −40　**75.** −6483
77. All negative　**79.** Negative　**81.** Negative

Exercise Set 2.3, p. 100

RC1. (c)　**RC2.** (b)　**RC3.** (d)　**RC4.** (a)
1. −4　**3.** −7　**5.** −6　**7.** 0　**9.** −4　**11.** −7　**13.** −6
15. 0　**17.** 0　**19.** 14　**21.** 11　**23.** −14　**25.** 5　**27.** −7
29. −1　**31.** 18　**33.** −10　**35.** −3　**37.** −21　**39.** 5

41. −8　**43.** 12　**45.** −19　**47.** −68　**49.** −81　**51.** 116
53. 0　**55.** 55　**57.** 19　**59.** −62　**61.** −139　**63.** 6
65. 107　**67.** 219　**69.** Profit of $4300　**71.** 3780 m
73. 17 lb　**75.** −3°　**77.** 18 points　**79.** −10,411 ft
81. −$83, or $83 in debt　**83.** 64　**84.** 4896　**85.** 1
86. 4147　**87.** 8 cans　**88.** 288 oz　**89.** 35　**90.** 3
91. 32　**92.** 165　**93.** −309,882　**95.** False; $3 - 0 \neq 0 - 3$
97. True　**99.** True　**101.** 17　**103.** Up 15 points

Calculator Corner, p. 107

1. 148,035,889　**2.** −1,419,857　**3.** −1,124,864　**4.** 1,048,576
5. −531,441　**6.** −117,649　**7.** −7776　**8.** −19,683

Exercise Set 2.4, p. 107

RC1. negative　**RC2.** positive　**RC3.** positive　**RC4.** negative
1. −16　**3.** −60　**5.** −48　**7.** −30　**9.** 15　**11.** 18
13. 42　**15.** 20　**17.** −120　**19.** 0　**21.** 72　**23.** −340
25. 400　**27.** 0　**29.** 24　**31.** 420　**33.** −70　**35.** 30
37. 0　**39.** −294　**41.** 36　**43.** −125　**45.** 10,000
47. −16　**49.** −243　**51.** 1　**53.** −121　**55.** −64
57. 532,500　**58.** 60,000,000　**59.** 80　**60.** 2550　**61.** 5
62. 48　**63.** 40 sq ft　**64.** 240 cartons　**65.** 5 trips
66. 4 trips　**67.** 243　**69.** 0　**71.** 7　**73.** −2209
75. 130,321　**77.** −2197　**79.** 116,875　**81.** −$23
83. (a) Both m and n must be odd. **(b)** At least one of m and n must be even.

Calculator Corner, p. 112

1. −4　**2.** −2　**3.** 787

Exercise Set 2.5, p. 113

RC1. 0　**RC2.** undefined　**RC3.** undefined　**RC4.** 0
1. −6　**3.** −13　**5.** −2　**7.** 4　**9.** −9　**11.** 2　**13.** −12
15. −8　**17.** Undefined　**19.** −8　**21.** −23　**23.** 0
25. −19　**27.** −41　**29.** −7　**31.** −7　**33.** −334　**35.** 23
37. 8　**39.** 12　**41.** −1　**43.** 0　**45.** −10　**47.** −86
49. −9　**51.** 18　**53.** −10　**55.** −67　**57.** 10　**59.** −25
61. −7988　**63.** −3000　**65.** 60　**67.** 1　**69.** −37
71. −22　**73.** 2　**75.** 7　**77.** Undefined　**79.** 3　**81.** 2
83. 0　**85.** 28 sq in.　**86.** 248 rooms　**87.** 12 gal　**88.** 27 gal
89. 150 cal　**90.** 672 g　**91.** 4 pieces; 2 pieces　**92.** 4 lozenges; 4 lozenges　**93.** 0　**95.** 0　**97.** −2　**99.** 992
101. (1 5 x^2 − 5 x^y 3) ÷ (3 x^2 + 4 x^2)) =　**103.** 5　**105.** Positive　**107.** Negative
109. Positive

Mid-Chapter Review: Chapter 2, p. 116

1. False　**2.** False　**3.** True
4. $-x = -(-4) = 4$; $-(-x) = -(-(-4)) = -(4) = -4$
5. $5 - 13 = 5 + (-13) = -8$　**6.** $-6 - (-7) = -6 + 7 = 1$
7. 450; −79　**8.** −9　**9.** <　**10.** <　**11.** <　**12.** >
13. 38　**14.** 18　**15.** 0　**16.** 12　**17.** 56　**18.** −3
19. 0　**20.** 49　**21.** 19　**22.** 23　**23.** −2　**24.** −16
25. 0　**26.** −17　**27.** 1　**28.** 2　**29.** −26　**30.** −4
31. −13　**32.** 16　**33.** 6　**34.** −12　**35.** −36　**36.** −54
37. 26　**38.** 82　**39.** 81　**40.** −81　**41.** 25　**42.** −5
43. 75　**44.** 14　**45.** 42　**46.** −38　**47.** −13　**48.** 2
49. 33°C　**50.** $44　**51.** Answers may vary. The student may be confusing distance from 0 on the number line with position on the number line. Although −45 is farther from 0 than −21, it is less than −21 because it is to the left of −21 on the number line.
52. Answers may vary. Subtraction of integers is not associative, as can be illustrated by an example: Compare $3 - (9 - 10) = 4$ with $(3 - 9) - 10 = -16$.　**53.** Answers may vary. If we think of the addition on the number line, we start at a negative number

and move to the left. This always brings us to a point on the negative portion of the number line.   **54.** Yes: consider $m - (-n)$, where both m and n are positive. Then $m - (-n) = m + n$. Now $m + n$, the sum of two positive numbers, is positive.

Calculator Corner, p. 120

1. 243   **2.** 1024   **3.** -32 **4.** -3125

Exercise Set 2.6, p. 122

RC1. Division   **RC2.** Multiplication   **RC3.** Multiplication
RC4. Division
1. 20¢   **3.** -2 **5.** 1   **7.** 18 yr   **9.** -7
11. 14 ft   **13.** 14 ft   **15.** -21 **17.** -21 **19.** 400 ft
21. 22   **23.** 0   **25.** 36   **27.** -3 **29.** 11   **31.** -1100
33. $\dfrac{-5}{t}; \dfrac{5}{-t}$ **35.** $\dfrac{n}{-b}; -\dfrac{n}{b}$ **37.** $\dfrac{-9}{p}; -\dfrac{9}{p}$ **39.** $\dfrac{14}{-w}; -\dfrac{14}{w}$
41. $-5; -5; -5$ **43.** $-27; -27; -27$ **45.** $36; -12$ **47.** $45; 45$
49. $216; -216$ **51.** $1; 1$ **53.** $32; -32$ **55.** $5a + 5b$
57. $4x + 4$ **59.** $2b + 10$ **61.** $7 - 7t$ **63.** $30x - 12$
65. $8x + 56 + 48y$ **67.** $-7y + 14$ **69.** $3x + 6$
71. $-4x + 12y + 8z$ **73.** $8a - 24b + 8c$
75. $4x - 12y - 28z$ **77.** $20a - 25b + 5c - 10d$
79. $-3m - 2n$ **81.** $-2a + 3b - 4$ **83.** $-x + y + z$
85. Twenty-three million, forty-three thousand, nine hundred twenty-one   **86.** 901   **87.** $5280 - 2480 = 2800$ **88.** 994
89. 5 in.   **90.** \$63   **91.** 698°F   **93.** 4438   **95.** 279
97. 2   **99.** 2,560,000   **101.** $-32 \boxed{\times} (88 \boxed{-} 29) = -1888$
103. True   **105.** True

Translating for Success, p. 129

1. J   **2.** G   **3.** A   **4.** K   **5.** H   **6.** N   **7.** F
8. M   **9.** D   **10.** C

Exercise Set 2.7, p. 130

RC1. closed   **RC2.** perimeter   **RC3.** perimeter
RC4. square
1. $2a, 5b, -7c$ **3.** $mn, -6n, 8$ **5.** $3x^2y, -4y^2, -2z^3$ **7.** $14x$
9. $-5a$ **11.** $3x + 6y$ **13.** $-13a + 62$ **15.** $-4 + 4t + 6y$
17. $6a + 4b - 2$ **19.** $-1 + a - 12b$ **21.** $7x^2 + 3y$
23. $7x^4 + y^3$ **25.** $6a^2 - 3a$ **27.** $3x^3 - 8x^2 + 4$
29. $9x^3y - xy^3 + 3xy$ **31.** $-4a^6 - 3b^4 + 2a^6b^4$ **33.** 10 ft
35. 42 km   **37.** 8 m   **39.** 210 ft   **41.** 138 ft   **43.** 36 ft
45. 56 in.   **47.** 260 cm   **49.** 64 ft   **51.** 17 servings   **52.** 210
53. 16   **54.** 13   **55.** 15   **56.** 25   **57.** 375   **58.** 16
59. $7x + 1$ **61.** $-29 - 3a$ **63.** $-10 - x - 27y$ **65.** \$45
67. 912 mm

Exercise Set 2.8, p. 139

RC1. (c)   **RC2.** (a)   **RC3.** (d)   **RC4.** (b)
1. Equivalent equations   **3.** Equivalent expressions
5. Equivalent expressions   **7.** Equivalent equations
9. Equivalent expressions   **11.** Equivalent equations
13. -3 **15.** -8 **17.** 18   **19.** -15 **21.** 32   **23.** -17
25. 17   **27.** 0   **29.** 10   **31.** -14 **33.** 5   **35.** 0
37. -11 **39.** -56 **41.** 12   **43.** 390   **45.** 4   **47.** -9
49. -15 **51.** -26 **53.** -5 **55.** -4 **57.** -172 **59.** -4
61. 7   **63.** 3   **65.** -3 **67.** -7 **69.** -8 **71.** 8   **73.** 24
75. 6   **77.** 5   **79.** -8 **81.** 29   **82.** 7   **83.** 8   **84.** 27
85. 16   **86.** 26   **87.** 1   **88.** 24   **89.** 8   **91.** 0   **93.** -5
95. 29   **97.** -20 **99.** 1027   **101.** -343 **103.** 17

Summary and Review: Chapter 2, p. 142

Vocabulary Reinforcement

1. integers   **2.** absolute   **3.** opposite   **4.** not defined
5. substituting   **6.** constant   **7.** distributive law
8. addition principle   **9.** equivalent

Concept Reinforcement

1. True   **2.** True   **3.** False   **4.** False

Study Guide

1. **(a)** 17; **(b)** 300   **2.** 21   **3.** 14   **4.** -90 **5.** -11 **6.** -12
7. $30x - 40y - 5z$ **8.** $17a - 7b$ **9.** -6

Review Exercises

1. $-45; 72$ **2.** $>$ **3.** $<$ **4.** $>$ **5.** 39   **6.** 23   **7.** 0
8. 72   **9.** 59   **10.** -9 **11.** -11 **12.** 6   **13.** -24
14. -12 **15.** 23   **16.** -1 **17.** 0   **18.** -4 **19.** 12
20. 92   **21.** -84 **22.** -40 **23.** -3 **24.** -5 **25.** 0
26. -5 **27.** 2   **28.** -180 **29.** -20 **30.** -62 **31.** 7
32. $-6, -6, -6$ **33.** $20x + 36$ **34.** $6a - 12b + 15$
35. $-20x - 10y$ **36.** $17a$ **37.** $6x$ **38.** $-3m + 6$
39. 36 in.   **40.** 100 cm   **41.** -8 **42.** -9
43. -13 **44.** 11   **45.** 15   **46.** -7 **47.** 8-yd gain
48. $-\$130$ **49.** A   **50.** B   **51.** 403 and 397
52. **(a)** $-7 + (-6) + (-5) + (-4) + (-3) + (-2) +$
$(-1) + 0 + 1 + 2 + 3 + 4 + 5 + 6 + 7 + 8 = 8$; **(b)** 0
53. 662,582   **54.** $-88,174$ **55.** -240 **56.** $x < -2$
57. $x < 0$

Understanding Through Discussion and Writing

1. We know that the product of an even number of negative numbers is positive, and the product of an odd number of negative numbers is negative. Since $(-7)^8$ is equivalent to the product of eight negative numbers, it will be a positive number. Similarly, since $(-7)^{11}$ is equivalent to the product of eleven negative numbers, it will be a negative number.   **2.** The expression $-x$ does not always represent a negative number. When $x = 0$, $-x = 0$, and when x is negative, $-x$ is positive.   **3.** Jake is expecting the multiplication to be performed before the division.   **4.** The expression $-x^2$ represents a negative number, except when $x = 0$. For all other values of x, x^2 is positive, and the opposite of x^2 is negative.

Test: Chapter 2, p. 147

1. [2.1a] $-542; 307$ **2.** [2.1b] $>$ **3.** [2.1c] 739   **4.** [2.1d] -19
5. [2.2a] -11 **6.** [2.2a] -21 **7.** [2.2a] 9   **8.** [2.3a] -12
9. [2.3a] -15 **10.** [2.3a] -24 **11.** [2.3a] 19   **12.** [2.3a] 38
13. [2.4b] -64 **14.** [2.4a] -270 **15.** [2.4a] 0   **16.** [2.5a] 8
17. [2.5a] -8 **18.** [2.5b] -1 **19.** [2.5b] 25   **20.** [2.3b] 14°F higher   **21.** [2.6a] -3 **22.** [2.6b] $14x + 21y - 7$
23. [2.7a] $4x - 17$ **24.** [2.7b] 20 ft   **25.** [2.8b] 5
26. [2.8a] -12 **27.** [2.8b] -95 **28.** [2.8d] 4   **29.** [2.6b] C
30. [2.7b] 66 ft   **31.** [2.5b] $35x - 7$ **32.** [2.5b] $-24x - 57$
33. [2.5b] 103,097   **34.** [2.5b] 1086

Cumulative Review: Chapters 1–2, p. 149

1. [1.1c] 220,863   **2.** [1.1c] Five billion, three hundred eighty million, one thousand, four hundred thirty-seven
3. [1.2a] 18,827   **4.** [1.2a] 8857   **5.** [2.2a] 56   **6.** [2.2a] -102

7. [1.3a] 7846 **8.** [1.3a] 2428 **9.** [2.3a] 14 **10.** [2.3a] −43
11. [1.4a] 16,767 **12.** [1.4a] 8,266,500 **13.** [2.4a] −344
14. [2.4a] 72 **15.** [1.5a] 104 **16.** [1.5a] 62 **17.** [2.5a] 0
18. [2.5a] −5 **19.** [1.6a] 428,000 **20.** [1.6a] 5300
21. [1.6b] 749,600 + 301,400 = 1,051,000
22. [1.6b] 700 × 500 = 350,000 **23.** [2.1b] < **24.** [2.1c] 279
25. [1.9c], [2.5b] 36 **26.** [1.9d], [2.5b] 2 **27.** [1.9c], [2.5b] −86
28. [1.9b] 125 **29.** [2.6a] 3 **30.** [2.6a] 28
31. [2.6b] −2x − 10 **32.** [2.6b] 18x − 12y + 24
33. [2.2a] −2 **34.** [2.4b] −8 **35.** [2.3a] −15 **36.** [2.5a] 32
37. [2.3a] 13 **38.** [2.4b] −30 **39.** [2.3a] 16 **40.** [2.5b] −57
41. [1.7b], [2.8a] 27 **42.** [2.8b] −3 **43.** [2.8d] 15
44. [2.8d] −8 **45.** [1.8a] 104 yr **46.** [1.8a] 18,133 rooms
47. [2.3b] 155°C **48.** [1.8a] Westside Appliance
49. [2.7a] 10x − 14 **50.** [1.8a] Cases: 6; six-packs: 3;
loose cans: 4 **51.** [2.5b], [2.7a] 4a **52.** [2.5b] −1071
53. [2.1c], [2.8d] ±3

CHAPTER 3

Calculator Corner, p. 153

1. Yes **2.** No **3.** No **4.** Yes

Exercise Set 3.1, p. 156

RC1. (c) **RC2.** (a) **RC3.** (d) **RC4.** (f)
RC5. (b) **RC6.** (e) **1.** 7, 14, 21, 28, 35, 42, 49, 56, 63, 70
3. 20, 40, 60, 80, 100, 120, 140, 160, 180, 200 **5.** 3, 6, 9, 12, 15, 18,
21, 24, 27, 30 **7.** 12, 24, 36, 48, 60, 72, 84, 96, 108, 120
9. 10, 20, 30, 40, 50, 60, 70, 80, 90, 100 **11.** 25, 50, 75, 100, 125,
150, 175, 200, 225, 250 **13.** No **15.** Yes **17.** Yes
19. Yes; the sum of the digits is 12, which is divisible by 3.
21. No; the ones digit is not 0 or 5. **23.** Yes; the ones digit is 0.
25. Yes; the sum of the digits is 18, which is divisible by 9.
27. No; the ones digit is not even. **29.** No; the ones digit
is not even. **31.** 6825 is divisible by 3 and 5. **33.** 119,117 is
divisible by none of these numbers. **35.** 127,575 is divisible by 3,
5, and 9. **37.** 9360 is divisible by 2, 3, 5, 6, 9, and 10.
39. 555; 300; 36; 45,270; 711; 13,251; 8064 **41.** 300; 45,270
43. 300; 36; 45,270; 8064 **45.** 56; 324; 784; 200; 42; 812; 402
47. 55,555; 200; 75; 2345; 35; 1005 **49.** 324 **51.** 53 **52.** 5
53. −8 **54.** −24 **55.** 45 gal **56.** 4320 min **57.** 125
58. 16 **59.** 9³ **60.** 3⁶ **61.** 99,969 **63.** 30 **65.** 210
67. 840 **69.** (a) 999a + 99b + 9c = 9(111a +
11b + c) = 3(333a + 33b + 3c); therefore, 999a + 99b + 9c
is divisible by both 9 and 3; (b) if a + b + c + d is divisible
by 9, then a + b + c + d = 9n for some number n. Then
abcd = 999a + 99b + 9c + 9n = 9(111a + 11b + c + n),
so abcd is divisible by 9. A similar argument holds for 3.
71. 332,986,412 is divisible by 4 and 11.

Exercise Set 3.2, p. 163

RC1. True **RC2.** False **RC3.** True **RC4.** False
RC5. False **1.** No **3.** Yes **5.** 1, 2, 3, 6, 9, 18
7. 1, 2, 3, 6, 9, 18, 27, 54 **9.** 1, 3, 9 **11.** 1, 13
13. 1, 2, 7, 14, 49, 98 **15.** 1, 3, 5, 15, 17, 51, 85, 255 **17.** Prime
19. Composite **21.** Composite **23.** Prime **25.** Neither
27. Composite **29.** Prime **31.** Prime **33.** 3 · 3 · 3
35. 2 · 7 **37.** 2 · 2 · 2 · 2 · 5 **39.** 5 · 5 **41.** 2 · 31
43. 2 · 2 · 5 · 5 **45.** 11 · 13 **47.** 11 · 11 **49.** 3 · 7 · 13
51. 5 · 5 · 7 **53.** 11 · 19 **55.** 2 · 2 · 2 · 2 · 3 · 5 · 5
57. 3 · 3 · 7 · 11 **59.** 2 · 2 · 7 · 103 **61.** 2 · 3 · 11 · 17
63. −26 **64.** 256 **65.** 8 **66.** −23 **67.** 1 **68.** −98
69. 0 **70.** 0 **71.** 13 · 19 · 19 · 29 **73.** 23 · 31 · 61 · 73 · 149
75. Answers may vary. One arrangement is a three-dimensional
rectangular array consisting of 2 tiers of 12 objects each, where each
tier consists of a rectangular array of 4 rows with 3 objects each.

77.

Product	56	63	36	72	140	96
Factor	7	7	2	2	10	8
Factor	8	9	18	36	14	12
Sum	15	16	20	38	24	20

Product	48	168	110	90	432	63
Factor	6	21	10	9	24	3
Factor	8	8	11	10	18	21
Sum	14	29	21	19	42	24

Exercise Set 3.3, p. 170

RC1. (b) **RC2.** (a) **RC3.** (c) **RC4.** (e) **RC5.** (f)
RC6. (d) **1.** Numerator: 3; denominator: 4 **3.** Numerator: 7;
denominator: −9 **5.** Numerator: 2x; denominator: 3y

7. $\frac{6}{12}$ **9.** $\frac{1}{8}$ **11.** $\frac{3}{4}$ **13.** $\frac{4}{8}$ **15.** $\frac{12}{12}$ **17.** $\frac{9}{8}$

19. $\frac{4}{3}$ **21.** $\frac{11}{9}$ **23.** $\frac{5}{8}$ **25.** $\frac{4}{7}$ **27.** $\frac{12}{16}$ **29.** $\frac{38}{16}$

31. (a) $\frac{2}{8}$; (b) $\frac{6}{8}$ **33.** (a) $\frac{3}{8}$; (b) $\frac{5}{8}$ **35.** (a) $\frac{5}{7}$; (b) $\frac{5}{2}$;
(c) $\frac{2}{7}$; (d) $\frac{2}{5}$ **37.** (a) $\frac{4}{15}$; (b) $\frac{4}{11}$; (c) $\frac{11}{15}$
39. (a) $\frac{1060}{100,000}$; (b) $\frac{743}{100,000}$; (c) $\frac{865}{100,000}$; (d) $\frac{1026}{100,000}$; (e) $\frac{905}{100,000}$;
(f) $\frac{1728}{100,000}$ **41.** $\frac{4}{7}$ **43.** 0 **45.** 7 **47.** 1 **49.** 1 **51.** 0
53. 19x **55.** 1 **57.** −87 **59.** 0 **61.** Undefined
63. 7n **65.** Undefined **67.** −210 **68.** −322 **69.** 0
70. 0 **71.** 451 museums **72.** 71 gal **73.** $\frac{1}{6}$ **75.** $\frac{2}{16}$, or $\frac{1}{8}$

77.   **79.**   **81.** $\frac{52}{365}$ **83.** $\frac{3}{4}, \frac{1}{4}$

Exercise Set 3.4, p. 179

RC1. True **RC2.** True **RC3.** True **RC4.** True
1. $\frac{3}{8}$ **3.** $\frac{-5}{6}$, or $-\frac{5}{6}$ **5.** $\frac{14}{3}$ **7.** $-\frac{7}{9}$, or $\frac{-7}{9}$ **9.** $\frac{5x}{6}$
11. $\frac{-6}{5}$, or $-\frac{6}{5}$ **13.** $\frac{2a}{7}$ **15.** $\frac{-17m}{6}$, or $-\frac{17m}{6}$ **17.** $\frac{6}{5}$
19. $\frac{2x}{7}$ **21.** $\frac{4}{15}$ **23.** $\frac{-1}{40}$, or $-\frac{1}{40}$ **25.** $\frac{2}{15}$ **27.** $\frac{2x}{9y}$
29. $\frac{9}{16}$ **31.** $\frac{14}{39}$ **33.** $\frac{-3}{50}$, or $-\frac{3}{50}$ **35.** $\frac{7a}{64}$ **37.** $\frac{100}{y}$
39. $\frac{-147}{20}$, or $-\frac{147}{20}$ **41.** $\frac{40}{3}$ yd **43.** $\frac{1}{2625}$ **45.** $\frac{1}{24}$ **47.** $\frac{9}{20}$
49. 204 **50.** 700 **51.** 13 **52.** −90 **53.** 50 **54.** 6399
55. $\frac{71,269}{180,433}$ **57.** $\frac{-56}{1125}$, or $-\frac{56}{1125}$ **59.** $\frac{1}{80}$ gal

Calculator Corner, p. 184

1. $\frac{14}{15}$ **2.** $\frac{7}{8}$ **3.** $\frac{138}{167}$ **4.** $\frac{7}{25}$

Exercise Set 3.5, p. 186

RC1. Equivalent **RC2.** simplify **RC3.** common
RC4. cross **1.** $\frac{5}{10}$ **3.** $\frac{-36}{-48}$ **5.** $\frac{35}{50}$ **7.** $\frac{11t}{5t}$
9. $\frac{20}{4}$ **11.** $-\frac{51}{54}$ **13.** $\frac{15}{-40}$ **15.** $\frac{-42}{132}$ **17.** $\frac{x}{8x}$

19. $\dfrac{-10a}{7a}$ **21.** $\dfrac{4ab}{9ab}$ **23.** $\dfrac{12b}{27b}$ **25.** $\dfrac{1}{2}$ **27.** $-\dfrac{2}{3}$
29. $\dfrac{2}{5}$ **31.** 3 **33.** $\dfrac{3}{4}$ **35.** $-\dfrac{12}{7}$ **37.** $\dfrac{3}{4}$ **39.** $\dfrac{-1}{3}$
41. -5 **43.** $\dfrac{7}{8}$ **45.** $\dfrac{-2}{3}$ **47.** $\dfrac{2}{5}$ **49.** $\dfrac{9}{8}$ **51.** $\dfrac{12}{13}$
53. $\dfrac{17}{19}$ **55.** $\dfrac{3}{8}$ **57.** $\dfrac{3y}{2}$ **59.** $\dfrac{-9}{10b}$ **61.** $=$ **63.** $\ne$
65. $=$ **67.** $\ne$ **69.** $\ne$ **71.** $=$ **73.** $\ne$ **75.** $=$
77. $<$ **78.** $<$ **79.** $>$ **80.** $<$ **81.** 3520
82. 89 **83.** 6498 **84.** 85 **85.** $\dfrac{17}{29}$ **87.** $-\dfrac{29x}{15y}$
89. (a) $\dfrac{4}{10} = \dfrac{2}{5}$; (b) $\dfrac{6}{10} = \dfrac{3}{5}$ **91.** No; $\dfrac{162}{489} \ne \dfrac{193}{555}$ because $162 \cdot 555 \ne 489 \cdot 193$.

Mid-Chapter Review: Chapter 3, p. 188

1. True **2.** False **3.** False **4.** True
5. $\dfrac{25}{25} = 1$ **6.** $\dfrac{0}{9} = 0$ **7.** $\dfrac{8}{1} = 8$ **8.** $\dfrac{6}{13} = \dfrac{18}{39}$
9. $\dfrac{70}{225} = \dfrac{2 \cdot 5 \cdot 7}{3 \cdot 3 \cdot 5 \cdot 5} = \dfrac{5}{5} \cdot \dfrac{2 \cdot 7}{3 \cdot 3 \cdot 5} = 1 \cdot \dfrac{14}{45} = \dfrac{14}{45}$
10. 84; 17,576; 224; 132; 594; 504; 1632
11. 84; 300; 132; 500; 180 **12.** 17,576; 224; 500
13. 84; 300; 132; 120; 1632 **14.** 300; 180; 120
15. Prime **16.** Prime **17.** Composite **18.** Neither
19. 1, 2, 4, 5, 8, 10, 16, 20, 32, 40, 80, 160; $2 \cdot 2 \cdot 2 \cdot 2 \cdot 2 \cdot 5$
20. 1, 2, 3, 6, 37, 74, 111, 222; $2 \cdot 3 \cdot 37$ **21.** 1, 2, 7, 14, 49, 98; $2 \cdot 7 \cdot 7$ **22.** 1, 3, 5, 7, 9, 15, 21, 35, 45, 63, 105, 315; $3 \cdot 3 \cdot 5 \cdot 7$
23. $\dfrac{8}{24}$, or $\dfrac{1}{3}$ **24.** $\dfrac{5}{4}$ **25.** $\dfrac{7}{9}$ **26.** $\dfrac{8}{45}$ **27.** $\dfrac{-40}{11}$, or $-\dfrac{40}{11}$
28. $\dfrac{2}{5}$ **29.** $\dfrac{11}{3}$ **30.** 1 **31.** 0 **32.** $\dfrac{9}{31}$ **33.** $\dfrac{-9}{5}$
34. $\dfrac{5}{42}$ **35.** $\dfrac{21}{29}$ **36.** Undefined **37.** $=$ **38.** $\ne$
39. $\dfrac{25}{200}$, or $\dfrac{1}{8}$ **40.** $\dfrac{21}{10,000}$ mi^2 **41.** Find the product of two prime numbers. **42.** If we use the divisibility tests, it is quickly clear that none of the even-numbered years is a prime number. In addition, the divisibility tests for 5 and 3 show that 2001, 2005, 2007, 2013, 2015, and 2019 are not prime numbers. Then the years 2003, 2009, 2011, and 2017 can be divided by prime numbers to determine whether they are prime. When we do this, we find that 2003, 2011, and 2017 are prime numbers. If the divisibility tests are not used, each of the numbers from 2000 to 2020 can be divided by prime numbers to determine if it is prime. **43.** It is possible to cancel only when identical *factors* appear in the numerator and the denominator of a fraction. Situations in which it is not possible to cancel include the occurrence of identical *addends* or *digits* in the numerator and the denominator. **44.** No; since the only factors of a prime number are the number itself and 1, two different prime numbers cannot contain a common factor (other than 1).

Exercise Set 3.6, p. 194

RC1. products **RC2.** Factor **RC3.** 1 **RC4.** Carry out
1. $\dfrac{1}{3}$ **3.** $-\dfrac{1}{8}$ **5.** $\dfrac{4}{7}$ **7.** $\dfrac{1}{15}$ **9.** $-\dfrac{27}{10}$ **11.** $\dfrac{4x}{9}$ **13.** 1
15. 1 **17.** 3 **19.** 7 **21.** 12 **23.** $3a$ **25.** 1 **27.** 1
29. $\dfrac{1}{5}$ **31.** $\dfrac{9}{25}$ **33.** $60n$ **35.** $-\dfrac{10}{3}$ **37.** 1 **39.** 3
41. $\dfrac{119}{750}$ **43.** $-\dfrac{20}{187}$ **45.** $-\dfrac{42}{275}$ **47.** $-\dfrac{16}{5x}$ **49.** $-\dfrac{11}{40}$
51. $\dfrac{5a}{28b}$ **53.** $\dfrac{5}{8}$ in. **55.** 260 million ounces
57. 480 addresses **59.** $\dfrac{1}{3}$ cup **61.** \$115,500 **63.** 160 mi
65. Food: \$8400; housing: \$10,500; clothing: \$4200; savings: \$3000; taxes: \$8400; other expenses: \$7500 **67.** 60 in^2
69. $\dfrac{35}{4}$ mm^2 **71.** $\dfrac{63}{4}$ m^2 **73.** 92 mi^2 **75.** $\dfrac{7}{4}$ in^2 **77.** 35
78. 85 **79.** 125 **80.** 120 **81.** 4989 **82.** 8546
83. 8699 **84.** 6407 **85.** $\dfrac{129}{485}$ **87.** $\dfrac{1}{12}$ **89.** 13,380 mm^2

Calculator Corner, p. 200

1. $\dfrac{1}{6}$ **2.** $\dfrac{20}{9}$ **3.** $-\dfrac{9}{7}$ **4.** $\dfrac{3}{2}$

Exercise Set 3.7, p. 202

RC1. True **RC2.** False **RC3.** True **RC4.** True
1. $\dfrac{3}{7}$ **3.** $\dfrac{1}{9}$ **5.** 7 **7.** $-\dfrac{9}{8}$ **9.** $\dfrac{c}{a}$ **11.** $\dfrac{m}{-3n}$ **13.** $\dfrac{-15}{8}$
15. $\dfrac{1}{7m}$ **17.** $4a$ **19.** $-3z$ **21.** $\dfrac{4}{7}$ **23.** $\dfrac{4}{15}$ **25.** 4
27. -2 **29.** $\dfrac{25}{7}$ **31.** $\dfrac{1}{8}$ **33.** $\dfrac{3}{28}$ **35.** -8 **37.** $\dfrac{x}{2}$
39. $\dfrac{1}{9x}$ **41.** $35a$ **43.** 1 **45.** $-\dfrac{2}{3}$ **47.** $\dfrac{99}{224}$ **49.** $\dfrac{112a}{3}$
51. $\dfrac{14}{15}$ **53.** $\dfrac{7}{32}$ **55.** $-\dfrac{25}{12}$ **57.** -9 **58.** -81
59. 75 **60.** 81 **61.** 2 **62.** 2 **63.** $6x + 9y - 3$
64. $-4a + 12b - 8c$ **65.** $\dfrac{100}{9}$ **67.** 36 **69.** $\dfrac{121}{900}$
71. $\dfrac{9}{19}$ **73.** $\dfrac{220}{51}$

Translating for Success, p. 207

1. C **2.** H **3.** A **4.** N **5.** O **6.** F **7.** I
8. L **9.** D **10.** M

Exercise Set 3.8, p. 208

RC1. (b) **RC2.** (c) **RC3.** (d) **RC4.** (a) **1.** 15
3. 9 **5.** -45 **7.** $\dfrac{2}{17}$ **9.** $\dfrac{12}{5}$ **11.** $-\dfrac{16}{21}$ **13.** $-\dfrac{2}{25}$
15. $\dfrac{1}{6}$ **17.** -80 **19.** $-\dfrac{1}{6}$ **21.** $-\dfrac{7}{13}$ **23.** $\dfrac{27}{31}$
25. $\dfrac{6}{7}$ **27.** $\dfrac{12}{5}$ **29.** $-\dfrac{7}{15}$ **31.** $\dfrac{9}{5}$ **33.** 6 **35.** $\dfrac{10}{7}$
37. 960 extension cords **39.** 32 pairs **41.** 24 bowls
43. 16 L **45.** 20 packages **47.** $\dfrac{1}{8}$ T **49.** 9 bees
51. 2400 words **53.** 32 laps **55.** 288 km; 108 km
57. $\dfrac{1}{16}$ in. **59.** 26 **60.** -42 **61.** -67 **62.** -65 **63.** 20
64. 6 **65.** $17x$ **66.** $4a$ **67.** $7a + 3$ **68.** $4x - 7$
69. $\dfrac{2}{9}$ **71.** $\dfrac{7}{8}$ lb **73.** 103 slices **75.** 288 bees

Summary and Review: Chapter 3, p. 212

Vocabulary Reinforcement
1. multiplicative **2.** factors **3.** prime **4.** denominator
5. equivalent **6.** reciprocals **7.** factorization **8.** multiple

Concept Reinforcement
1. True **2.** False **3.** True **4.** True

Study Guide
1. 1, 2, 4, 8, 13, 26, 52, 104 **2.** $2 \cdot 2 \cdot 2 \cdot 13$ **3.** 0, 1, 18
4. $\dfrac{5}{14}$ **5.** $\ne$ **6.** $\dfrac{70}{9}$ **7.** $\dfrac{7}{10}$ **8.** $\dfrac{7}{3}$ cups

Review Exercises
1. 8, 16, 24, 32, 40, 48, 56, 64, 72, 80 **2.** No **3.** No **4.** No
5. Yes **6.** Yes **7.** 1, 2, 3, 4, 5, 6, 10, 12, 15, 20, 30, 60
8. 1, 2, 4, 8, 11, 16, 22, 44, 88, 176 **9.** Prime **10.** Neither
11. Composite **12.** $2 \cdot 5 \cdot 7$ **13.** $2 \cdot 2 \cdot 2 \cdot 3 \cdot 3$
14. $3 \cdot 3 \cdot 5$ **15.** $2 \cdot 3 \cdot 5 \cdot 5$ **16.** $2 \cdot 2 \cdot 2 \cdot 3 \cdot 3 \cdot 3 \cdot 3$
17. $2 \cdot 2 \cdot 2 \cdot 2 \cdot 3 \cdot 5 \cdot 5$ **18.** Numerator: 9; denominator: 7
19. $\dfrac{3}{5}$ **20.** $\dfrac{7}{6}$ **21.** (a) $\dfrac{3}{5}$; (b) $\dfrac{5}{3}$; (c) $\dfrac{3}{8}$ **22.** 0 **23.** 1
24. 48 **25.** 1 **26.** $-\dfrac{2}{3}$ **27.** $\dfrac{1}{4}$ **28.** -1 **29.** $\dfrac{3}{4}$
30. $\dfrac{2}{5}$ **31.** Undefined **32.** $\dfrac{2}{7}$ **33.** $\dfrac{32}{225}$ **34.** $\dfrac{15}{21}$ **35.** $\dfrac{-30}{55}$
36. $\dfrac{15}{100} = \dfrac{3}{20}; \dfrac{38}{100} = \dfrac{19}{50}; \dfrac{23}{100} = \dfrac{23}{100}; \dfrac{24}{100} = \dfrac{6}{25}$ **37.** $\ne$ **38.** $=$
39. $\ne$ **40.** $=$ **41.** $\dfrac{13}{2}$ **42.** $-\dfrac{1}{7}$ **43.** 8 **44.** $\dfrac{5y}{3x}$
45. $\dfrac{14}{45}$ **46.** $\dfrac{3y}{7x}$ **47.** $\dfrac{2}{3}$ **48.** -14 **49.** $\dfrac{10}{7}$ **50.** 1
51. $24x$ **52.** $\dfrac{1}{4}$ **53.** $\dfrac{80}{3}$ **54.** $\dfrac{1}{15}$ **55.** $6a$ **56.** -1
57. $\dfrac{3}{4}$ **58.** $\dfrac{4}{9}$ **59.** 240 **60.** $\dfrac{-3}{10}$ **61.** 28 **62.** $\dfrac{2}{3}$
63. 42 m^2 **64.** $\dfrac{35}{2}$ ft^2 **65.** 9 days **66.** \$32,085
67. 1000 km **68.** $\dfrac{1}{3}$ cup **69.** $\dfrac{1}{6}$ mi **70.** 60 bags
71. D **72.** B **73.** $\dfrac{17}{6}$ **74.** 2, 8
75. $a = 11,176$; $b = 9887$ **76.** 13, 11, 101, 37

Understanding Through Discussion and Writing

1. The student is probably multiplying the divisor by the reciprocal of the dividend rather than multiplying the dividend by the reciprocal of the divisor.
2. $9432 = 9 \cdot 1000 + 4 \cdot 100 + 3 \cdot 10 + 2 \cdot 1 = 9(999 + 1) + 4(99 + 1) + 3(9 + 1) + 2 \cdot 1 = 9 \cdot 999 + 9 \cdot 1 + 4 \cdot 99 + 4 \cdot 1 + 3 \cdot 9 + 3 \cdot 1 + 2 \cdot 1$. Since 999, 99, and 9 are each a multiple of 9, $9 \cdot 999$, $4 \cdot 99$, and $3 \cdot 9$ are multiples of 9. This leaves $9 \cdot 1 + 4 \cdot 1 + 3 \cdot 1 + 2 \cdot 1$, or $9 + 4 + 3 + 2$. If the sum of the digits, $9 + 4 + 3 + 2$, is divisible by 9, then 9432 is divisible by 9. **3.** Taking $\frac{1}{2}$ of a number is equivalent to multiplying the number by $\frac{1}{2}$. Dividing by $\frac{1}{2}$ is equivalent to multiplying by the reciprocal of $\frac{1}{2}$, or 2. Thus taking $\frac{1}{2}$ of a number is not the same as dividing by $\frac{1}{2}$.
4. We first consider an object and take $\frac{4}{7}$ of it. We divide the object into 7 parts and take 4 them, as shown by the shading below in the left figure.

Next, we take $\frac{2}{3}$ of the shaded area in the left figure. We divide it into 3 parts and take two of them, as shown in the right figure. The entire object has been divided into 21 parts, 8 of which have been shaded twice. Thus, $\frac{2}{3} \cdot \frac{4}{7} = \frac{8}{21}$. **5.** Since $\frac{1}{7}$ is a smaller number than $\frac{2}{3}$, there are more $\frac{1}{7}$'s in 5 than $\frac{2}{3}$'s. Thus, $5 \div \frac{1}{7}$ is a greater number than $5 \div \frac{2}{3}$. **6.** No; in order to simplify a fraction, we must be able to remove a factor of the type $\frac{n}{n}$, $n \neq 0$, where n is a factor that the numerator and the denominator have in common.

Test: Chapter 3, p. 217

1. [3.1b] Yes **2.** [3.1b] No **3.** [3.2a] 1, 2, 3, 5, 6, 9, 10, 15, 18, 30, 45, 90 **4.** [3.2b] Composite **5.** [3.2c] $2 \cdot 2 \cdot 3 \cdot 3$ **6.** [3.2c] $2 \cdot 2 \cdot 3 \cdot 5$ **7.** [3.3a] Numerator: 4; denominator: 9 **8.** [3.3a] $\frac{3}{4}$ **9.** [3.3a] $\frac{3}{7}$ **10.** [3.3a] (a) $\frac{180}{47}$; (b) $\frac{47}{93}$ **11.** [3.3b] 32 **12.** [3.3b] 1 **13.** [3.3b] 0 **14.** [3.5b] $\frac{-1}{3}$ **15.** [3.5b] 6 **16.** [3.5b] $\frac{1}{5}$ **17.** [3.3b] Undefined **18.** [3.5b] $\frac{2}{3}$ **19.** [3.5c] $=$ **20.** [3.5c] $\neq$ **21.** [3.5a] $\frac{15}{40}$ **22.** [3.7a] $\frac{42}{a}$ **23.** [3.7a] $-\frac{1}{9}$ **24.** [3.6a] $\frac{5}{2}$ **25.** [3.7b] $\frac{8}{33}$ **26.** [3.4a] $\frac{3x}{8}$ **27.** [3.7b] $-\frac{3}{14}$ **28.** [3.7b] 18 **29.** [3.6a] $\frac{2}{9}$ **30.** [3.8b] $\frac{3}{20}$ lb **31.** [3.6b] 125 lb **32.** [3.8a] 64 **33.** [3.8a] $-\frac{7}{4}$ **34.** [3.6b] $\frac{91}{2}$ m² **35.** [3.3a] C **36.** [3.6b] $\frac{7}{48}$ acre **37.** [3.6a], [3.7b] $-\frac{7}{960}$

Cumulative Review: Chapters 1–3, p. 219

1. [1.1c] Two million, fifty-six thousand, seven hundred eighty-three **2.** [1.2a] 10,982 **3.** [2.2a] -43 **4.** [2.2a] -33 **5.** [1.3a] 2129 **6.** [2.3a] -23 **7.** [2.3a] -8 **8.** [1.4a] 16,905 **9.** [2.4a] -312 **10.** [3.6a] $-30x$ **11.** [3.6a] $\frac{7}{15}$ **12.** [1.5a] 235 R 3 **13.** [2.5a] -17 **14.** [3.7b] -28 **15.** [3.7b] $\frac{2}{3}$ **16.** [1.6a] 4510 **17.** [1.6b] $900 \times 500 = 450,000$ **18.** [2.1c] 479 **19.** [2.5b] 8 **20.** [3.2b] Composite **21.** [2.6a] -21 **22.** [1.7b], [2.8a] 25 **23.** [1.7b], [2.8b] 7 **24.** [3.8a] -45 **25.** [2.8b] -12 **26.** [2.8d] -5 **27.** [2.8b] 10 **28.** [2.7a] $5x - 5$ **29.** [2.7a] $3x + 7y$ **30.** [3.3b] 1 **31.** [3.3b] 0 **32.** [3.3b] $63x$ **33.** [3.5b] $-\frac{5}{27}$ **34.** [3.7a] $\frac{5}{2}$ **35.** [3.7a] $\frac{1}{57}$ **36.** [3.5a] $\frac{21}{70}$ **37.** [1.8a] 8 oz **38.** [1.8a] 8 mpg **39.** [3.6b] 4375 students **40.** [3.8b] 5 qt **41.** [3.6b] $\frac{3}{5}$ mi **42.** [2.6a], [3.6a], [3.7b] $-\frac{54}{169}$ **43.** [2.1c], [2.6a], [3.6a] $-\frac{9}{100}$ **44.** [3.6b] $468

CHAPTER 4

Exercise Set 4.1, p. 227

RC1. True **RC2.** True **RC3.** True **RC4.** False
1. 4 **3.** 50 **5.** 40 **7.** 54 **9.** 150 **11.** 120 **13.** 72 **15.** 420 **17.** 144 **19.** 180 **21.** 42 **23.** 30 **25.** 72 **27.** 60 **29.** 36 **31.** 900 **33.** 48 **35.** 210 **37.** 300 **39.** 60 **41.** abc **43.** $9x^2$ **45.** $4x^3y$ **47.** $24r^3s^2t^4$ **49.** $a^3b^2c^2$ **51.** Every 60 yr **53.** Every 420 yr **55.** 14 **56.** -27 **57.** 7935 **58.** $\frac{2}{3}$ **59.** $-\frac{8}{7}$ **60.** -167 **61.** 2592 **63.** 54,033 **65.** 24 ft **67.** 24 strands **69.** (a) Not the LCM because a^2b^5 is not a factor of a^3b^3; (b) Not the LCM because a^3b^2 is not a factor of a^2b^5; (c) The LCM because both a^3b^2 and a^2b^5 are factors of a^3b^5 and it is the smallest such expression **71.** 2520 **73.** 27 and 2; 27 and 6; 27 and 18

Exercise Set 4.2, p. 235

RC1. True **RC2.** False **RC3.** False **RC4.** True
1. $\frac{5}{9}$ **3.** 1 **5.** $\frac{2}{5}$ **7.** $\frac{13}{a}$ **9.** $-\frac{1}{2}$ **11.** $\frac{7}{9}x$ **13.** $\frac{1}{2}t$ **15.** $-\frac{9}{x}$ **17.** $\frac{7}{24}$ **19.** $-\frac{1}{10}$ **21.** $\frac{23}{24}$ **23.** $\frac{83}{24}$ **25.** $\frac{5}{24}$ **27.** $\frac{37}{100}x$ **29.** $\frac{19}{20}$ **31.** $\frac{7}{8}$ **33.** $-\frac{99}{100}$ **35.** $-\frac{1}{30}x$ **37.** $-\frac{33}{7}t$ **39.** $-\frac{17}{24}$ **41.** $\frac{3}{4}$ **43.** $\frac{437}{1000}$ **45.** $\frac{5}{4}$ **47.** $\frac{239}{78}$ **49.** $\frac{59}{90}$ **51.** $-\frac{5}{4}$ **53.** $>$ **55.** $<$ **57.** $>$ **59.** $<$ **61.** $>$ **63.** $>$ **65.** $\frac{4}{15}; \frac{3}{10}; \frac{5}{12}$ **67.** $\frac{37}{12}$ mi **69.** $\frac{13}{16}$ lb **71.** $\frac{27}{32}''$ **73.** $\frac{13}{12}$ lb **75.** $\frac{7}{8}$ in. **77.** $\frac{33}{20}$ mi **79.** $\frac{4}{5}$ qt; $\frac{8}{5}$ qt; $\frac{2}{5}$ qt **81.** -13 **82.** 4 **83.** -8 **84.** -31 **85.** $\frac{10}{3}$ **86.** 42; 42 **87.** 35 **88.** 6407 **89.** 11 **90.** 3 **91.** $\frac{9}{7}$ **92.** $-\frac{2}{5}$ **93.** $\frac{13}{30}t + \frac{31}{35}$ **95.** $7t^2 + \frac{9}{a}t$ **97.** $>$ **99.** $\frac{4}{15}$; $320 **101.** $a = 2, b = 8$

Translating for Success, p. 243

1. J **2.** E **3.** D **4.** B **5.** I **6.** N **7.** A **8.** C **9.** L **10.** F

Exercise Set 4.3, p. 244

RC1. numerators; denominator **RC2.** denominators **RC3.** denominators **RC4.** denominators
1. $\frac{2}{3}$ **3.** $-\frac{1}{4}$ **5.** $\frac{2}{a}$ **7.** $-\frac{1}{2}$ **9.** $-\frac{4}{5a}$ **11.** $\frac{2}{t}$ **13.** $\frac{13}{16}$ **15.** $-\frac{1}{3}$ **17.** $\frac{7}{10}$ **19.** $-\frac{17}{60}$ **21.** $\frac{47}{100}$ **23.** $\frac{26}{75}$ **25.** $-\frac{71}{100}$ **27.** $\frac{13}{24}$ **29.** $-\frac{29}{50}$ **31.** $-\frac{1}{24}$ **33.** $-\frac{41}{72}$ **35.** $\frac{1}{360}$ **37.** $\frac{2}{9}x$ **39.** $-\frac{7}{20}a$ **41.** $\frac{7}{9}$ **43.** $\frac{4}{11}$ **45.** $\frac{1}{15}$ **47.** $\frac{9}{8}$ **49.** $\frac{2}{15}$ **51.** $-\frac{7}{24}$ **53.** $\frac{2}{15}$ **55.** $-\frac{5}{4}$ **57.** $\frac{5}{12}$ hr **59.** $\frac{1}{32}$ in. **61.** $\frac{11}{20}$ lb **63.** $\frac{7}{20}$ hr **65.** $\frac{3}{16}$ in. **67.** $\frac{7}{24}$ cup **69.** 1 **70.** Not defined **71.** Not defined **72.** 4 **73.** $\frac{4}{21}$ **74.** $\frac{3}{2}$ **75.** 21 **76.** $\frac{1}{32}$ **77.** 12 **78.** $\frac{5}{2}$ **79.** $-\frac{3}{10}$ **80.** $-\frac{9}{28}$ **81.** $\frac{1}{16}$ **83.** $-\frac{11}{10}$ **85.** $-\frac{64}{35}$ **87.** $-\frac{37}{1000}$ **89.** $\frac{1}{4}$ tub **91.** $\frac{43}{50}$ **93.** $\frac{1}{8}$ of the dealership **95.** $\frac{14}{3553}$ **97.** *Day 1:* Cut off $\frac{1}{7}$ of bar and pay the contractor.
  Day 2: Cut off $\frac{2}{7}$ of the bar's original length and trade it for the $\frac{1}{7}$.
  Day 3: Give the $\frac{1}{7}$ back to the contractor.
  Day 4: Trade the $\frac{4}{7}$ remaining for the contractor's $\frac{3}{7}$.
  Day 5: Give the contractor the $\frac{1}{7}$ again.
  Day 6: Trade the $\frac{2}{7}$ for the $\frac{1}{7}$.
  Day 7: Give the contractor the $\frac{1}{7}$ again. This assumes that the contractor does not spend parts of the gold bar immediately.

Exercise Set 4.4, p. 252

RC1. False **RC2.** False **RC3.** False **RC4.** True
1. 3 **3.** $\frac{3}{5}$ **5.** -1 **7.** $\frac{27}{2}$ **9.** $\frac{1}{2}$ **11.** $\frac{3}{2}$ **13.** $-\frac{4}{3}$ **15.** $\frac{1}{2}$ **17.** $\frac{8}{7}$ **19.** -6 **21.** $\frac{21}{5}$ **23.** 6 **25.** $-\frac{17}{4}$ **27.** $-\frac{3}{4}$ **29.** $\frac{3}{2}$ **31.** $\frac{9}{2}$ **33.** $-\frac{10}{3}$ **35.** $-\frac{1}{5}$ **37.** $-\frac{1}{6}$ **39.** $\frac{35}{12}$ **41.** -13 **42.** -8

43. 18 **44.** 27 **45.** The balance has decreased $150.
46. $1180 profit **47.** $\frac{5}{7m}$ **48.** $20n$ **49.** 4 **51.** $-\frac{290}{697}$
53. $\frac{436}{35}$ **55.** 2 cm

Calculator Corner, p. 256
1. $5\frac{4}{7}$ **2.** $8\frac{2}{5}$ **3.** $1476\frac{1}{6}$ **4.** $676\frac{4}{9}$ **5.** $134\frac{1}{15}$ **6.** $12\frac{169}{454}$

Exercise Set 4.5, p. 258
RC1. True **RC2.** True **RC3.** False **RC4.** True
1. $\frac{23}{3}$ **3.** $\frac{25}{4}$ **5.** $-\frac{161}{4}$ **7.** $\frac{51}{10}$ **9.** $\frac{103}{5}$ **11.** $-\frac{100}{9}$ **13.** $\frac{13}{8}$
15. $-\frac{51}{4}$ **17.** $\frac{57}{10}$ **19.** $-\frac{507}{100}$ **21.** $5\frac{1}{3}$ **23.** $7\frac{1}{2}$ **25.** $5\frac{7}{9}$ **27.** $7\frac{2}{9}$
29. $-5\frac{1}{2}$ **31.** $11\frac{1}{2}$ **33.** $-1\frac{1}{2}$ **35.** $61\frac{2}{3}$ **37.** $-8\frac{13}{50}$ **39.** $108\frac{5}{8}$ **41.** $906\frac{3}{7}$
43. $40\frac{4}{7}$ **45.** $-20\frac{2}{15}$ **47.** $-22\frac{3}{7}$ **49.** $275\frac{2}{3}$ backpacks
51. $5504\frac{1}{6}$ hr **53.** $18x + 6y - 24$ **54.** $-3a + 6b - 12c$
55. $237\frac{19}{541}$ **57.** $8\frac{2}{3}$ **59.** $3\frac{2}{3}$ **61.** $52\frac{1}{7}$

Mid-Chapter Review: Chapter 4, p. 260
1. True **2.** True **3.** False **4.** False
5.
$$\frac{11}{42} - \frac{3}{35} = \frac{11}{2 \cdot 3 \cdot 7} - \frac{3}{5 \cdot 7}$$
$$= \frac{11}{2 \cdot 3 \cdot 7} \cdot \left(\frac{5}{5}\right) - \frac{3}{5 \cdot 7} \cdot \left(\frac{2 \cdot 3}{2 \cdot 3}\right)$$
$$= \frac{11 \cdot 5}{2 \cdot 3 \cdot 7 \cdot 5} - \frac{3 \cdot 2 \cdot 3}{5 \cdot 7 \cdot 2 \cdot 3}$$
$$= \frac{55}{2 \cdot 3 \cdot 5 \cdot 7} - \frac{18}{2 \cdot 3 \cdot 5 \cdot 7}$$
$$= \frac{55 - 18}{2 \cdot 3 \cdot 5 \cdot 7} = \frac{37}{210}$$
6.
$$x + \tfrac{1}{8} = \tfrac{2}{3}$$
$$x + \tfrac{1}{8} - \tfrac{1}{8} = \tfrac{2}{3} - \tfrac{1}{8}$$
$$x + 0 = \tfrac{2}{3} \cdot \tfrac{8}{8} - \tfrac{1}{8} \cdot \tfrac{3}{3}$$
$$x = \tfrac{16}{24} - \tfrac{3}{24}$$
$$x = \tfrac{13}{24}$$
The solution is $\frac{13}{24}$.
7.
45 and 50 — 120
50 and 80 — 720
30 and 24 — 400
18, 24, and 80 — 450
30, 45, and 50 —
8. $\frac{16}{45}$ **9.** $\frac{25}{12}$ **10.** $\frac{1}{18}$ **11.** $-\frac{19}{90}$ **12.** $\frac{7}{240}$ **13.** $-\frac{7}{24}x$ **14.** $-\frac{17}{60}$
15. $\frac{6}{91}$ **16.** $\frac{22}{15}$ mi **17.** $\frac{101}{20}$ hr **18.** $\frac{1}{2}, \frac{3}{7}, \frac{3}{10}, \frac{4}{9}$ **19.** $\frac{13}{15}$ **20.** $-\frac{8}{9}$
21. $17\frac{8}{15}$ **22.** C **23.** C **24.** No; if one number is a multiple of the other, for example, the LCM is the larger of the numbers.
25. We multiply by 1, using the notation n/n, to express each fraction in terms of the least common denominator. **26.** Write $\frac{8}{5}$ as $\frac{16}{10}$ and $\frac{8}{2}$ as $\frac{40}{10}$ and since taking 40 tenths away from 16 tenths would give a result less than 0, the answer cannot possibly be $\frac{8}{3}$. You could also find the sum $\frac{8}{3} + \frac{8}{2}$ and show that it is not $\frac{8}{5}$. **27.** No; $2\frac{1}{3} = \frac{7}{3}$ but $2 \cdot \frac{1}{3} = \frac{2}{3}$.

Exercise Set 4.6, p. 268
RC1. (d) **RC2.** (a) **RC3.** (b) **RC4.** (c)
1. $11\frac{2}{5}$ **3.** $9\frac{1}{2}$ **5.** $5\frac{1}{3}$ **7.** $13\frac{7}{12}$ **9.** $12\frac{1}{10}$ **11.** $17\frac{5}{24}$ **13.** $21\frac{1}{2}$
15. $27\frac{7}{8}$ **17.** $7\frac{1}{5}$ **19.** $6\frac{1}{10}$ **21.** $1\frac{3}{5}$ **23.** $13\frac{1}{4}$ **25.** $15\frac{3}{8}$ **27.** $7\frac{5}{12}$
29. $11\frac{5}{18}$ **31.** $8\frac{3}{4}t$ **33.** $2x$ **35.** $8\frac{3}{10}y$ **37.** $11\frac{8}{9}t$ **39.** $10\frac{5}{24}x$
41. $2\frac{1}{3}a$ **43.** $23\frac{3}{20}$ ft **45.** $20\frac{1}{12}$ yd **47.** $\frac{15}{16}$ in. **49.** $134\frac{1}{4}$ in.
51. $28\frac{3}{4}$ yd **53.** $4\frac{5}{6}$ ft **55.** $7\frac{3}{8}$ ft **57.** $5\frac{7}{8}$ in. **59.** $20\frac{1}{8}$ in.
61. $3\frac{4}{5}$ hr **63.** $66\frac{5}{8}$ ft **65.** $-\frac{2}{5}$ **67.** $-3\frac{1}{4}$ **69.** $-3\frac{13}{15}$
71. $-7\frac{3}{5}$ **73.** $2\frac{7}{12}$ **75.** $-1\frac{8}{9}$ **77.** 3 ten thousands +

8 thousands + 1 hundred + 2 tens + 5 ones **78.** Two million, five thousand, six hundred eighty-nine **79.** 9^4 **80.** 81 **81.** Yes
82. No **83.** No **84.** Yes **85.** No **86.** Yes **87.** Yes
88. Yes **89.** $\frac{10}{13}$ **90.** $\frac{1}{10}$ **91.** $8568\frac{786}{1189}$ **93.** $10\frac{7}{12}$ **95.** $-28\frac{3}{8}$
97. $55\frac{3}{4}$ in.

Calculator Corner, p. 277
1. $\frac{7}{12}$ **2.** $\frac{11}{10}$ **3.** $\frac{35}{16}$ **4.** $-\frac{3}{10}$ **5.** $10\frac{2}{15}$ **6.** $1\frac{1}{28}$ **7.** $10\frac{11}{15}$
8. $2\frac{91}{115}$

Translating for Success, p. 278
1. O **2.** K **3.** F **4.** D **5.** H **6.** G **7.** L **8.** E
9. M **10.** J

Exercise Set 4.7, p. 279
RC1. True **RC2.** True **RC3.** True **RC4.** True
1. $22\frac{2}{3}$ **3.** $1\frac{2}{3}$ **5.** $-56\frac{2}{3}$ **7.** $16\frac{1}{3}$ **9.** $-10\frac{3}{25}$ **11.** $35\frac{91}{100}$ **13.** $6\frac{1}{4}$
15. $1\frac{1}{5}$ **17.** $1\frac{2}{3}$ **19.** $-1\frac{1}{8}$ **21.** $1\frac{8}{43}$ **23.** $\frac{9}{40}$ **25.** $23\frac{2}{5}$
27. $15\frac{5}{9}$ **29.** $-27\frac{2}{9}$ **31.** $-1\frac{1}{3}$ **33.** $12\frac{1}{4}$ **35.** $8\frac{3}{20}$ **37.** About $24 million **39.** About 4,800,000 **41.** About $42,000,000,000
43. 75 mph **45.** $12\frac{4}{5}$ tiles **47.** $5\frac{1}{2}$ cups of flour, $2\frac{2}{3}$ cups of sugar
49. 15 mpg **51.** $62\frac{1}{2}$ sq ft **53.** 400 cu ft **55.** $16\frac{1}{2}$ servings
57. $441\frac{1}{4}$ sq ft **59.** $76\frac{1}{4}$ sq ft **61.** $27\frac{5}{16}$ sq cm **63.** 68°F
65. $13\frac{1}{4}$ in. $\times$ $13\frac{1}{4}$ in.: perimeter = 53 in., area = $175\frac{9}{16}$ sq in.; $13\frac{1}{4}$ in. $\times$ $3\frac{1}{4}$ in.: perimeter = 33 in., area = $43\frac{1}{16}$ sq in.
67. 22 m **69.** $16\frac{25}{64}$ **71.** $\frac{4}{9}$ **73.** $r = \frac{240}{13}$, or $18\frac{6}{13}$ **75.** 88 gal

Exercise Set 4.8, p. 288
RC1. (c) **RC2.** (d) **RC3.** (a) **RC4.** (b)
1. $\frac{7}{24}$ **3.** $\frac{3}{2}$, or $1\frac{1}{2}$ **5.** $\frac{7}{8}$ **7.** $\frac{59}{30}$, or $1\frac{29}{30}$ **9.** $\frac{7}{16}$ **11.** $\frac{3}{20}$ **13.** -6
15. $\frac{1}{36}$ **17.** $-\frac{3}{100}$ **19.** -1 **21.** $\frac{19}{4}$, or $4\frac{3}{4}$ **23.** $\frac{3}{8}$ **25.** $\frac{2}{9}$
27. $\frac{3}{11}$ **29.** $-\frac{14}{3}$, or $-4\frac{2}{3}$ **31.** $-\frac{5}{4}$, or $-1\frac{1}{4}$ **33.** $\frac{1}{100}$ **35.** $-\frac{1}{6}$
37. $\frac{2x}{35}$ **39.** $-\frac{3n}{28}$ **41.** $\frac{6}{7x}$ **43.** $-\frac{5}{18}$ **45.** $-\frac{7}{12}$ **47.** $\frac{1}{3}$
49. $\frac{37}{48}$ **51.** $\frac{25}{72}$ **53.** $\frac{103}{16}$, or $6\frac{7}{16}$ **55.** $16\frac{7}{96}$ mi **57.** $9\frac{19}{40}$ lb
59. 20 **60.** 2 **61.** 84 **62.** 100 **63.** 1, 2, 3, 6, 7, 14, 21, 42
64. No **65.** Prime: 5, 7, 23, 43; composite: 9, 14; neither: 1
66. $2 \cdot 3 \cdot 5 \cdot 5$ **67.** $3\frac{11}{14}$ **69.** $\frac{8x}{147}$ **71.** 0 **73.** $\frac{1}{2}$ **75.** $\frac{1}{2}$

Summary and Review: Chapter 4, p. 291
Vocabulary Reinforcement
1. least common multiple **2.** mixed numeral **3.** fraction
4. complex fraction **5.** denominators **6.** least common multiple
7. greatest **8.** numerators

Concept Reinforcement
1. True **2.** True **3.** True **4.** False

Study Guide
1. 156 **2.** $\frac{28}{45}$ **3.** $<$ **4.** $\frac{4}{35}$ **5.** $-\frac{1}{12}$ **6.** $\frac{26}{3}$ **7.** $7\frac{5}{6}$
8. $7\frac{27}{28}$ **9.** $14\frac{14}{25}$ **10.** About 4,500,000 **11.** $\frac{9}{2}$, or $4\frac{1}{2}$

Review Exercises
1. 36 **2.** 90 **3.** 30 **4.** 1404 **5.** $\frac{7}{9}$ **6.** $\frac{9}{x}$ **7.** $-\frac{7}{15}$
8. $\frac{7}{16}$ **9.** $\frac{2}{9}$ **10.** $-\frac{1}{8}$ **11.** $\frac{4}{27}$ **12.** $\frac{1}{18}$ **13.** $>$ **14.** $<$
15. $\frac{19}{40}$ **16.** 11 **17.** $-\frac{5}{6}$ **18.** $\frac{12}{5}$ **19.** $\frac{2}{5}$ **20.** $\frac{15}{2}$ **21.** $\frac{67}{8}$
22. $\frac{13}{3}$ **23.** $-\frac{12}{7}$ **24.** $2\frac{1}{3}$ **25.** $-6\frac{3}{4}$ **26.** $12\frac{3}{5}$ **27.** $3\frac{1}{2}$
28. $-877\frac{1}{3}$ **29.** $82\frac{1}{3}$ **30.** $10\frac{2}{5}$ **31.** $11\frac{11}{15}$ **32.** -9 **33.** $1\frac{3}{4}$
34. $7\frac{7}{9}$ **35.** $4\frac{11}{15}$ **36.** $-5\frac{1}{8}$ **37.** $-14\frac{1}{4}$ **38.** $\frac{7}{9}x$ **39.** $3\frac{7}{40}a$

40. 16 **41.** $-3\frac{1}{2}$ **42.** $2\frac{21}{50}$ **43.** 6 **44.** -24 **45.** $-1\frac{7}{17}$
46. $\frac{1}{8}$ **47.** $\frac{9}{10}$ **48.** $13\frac{5}{7}$ **49.** $2\frac{8}{11}$ **50.** $4\frac{1}{4}$ yd **51.** $3\frac{1}{8}$ pizzas
52. 24 lb **53.** $63\frac{2}{3}$ pies; $19\frac{1}{3}$ pies **54.** $3\frac{3}{4}$ mi **55.** $177\frac{3}{4}$ sq in.
56. $50\frac{1}{4}$ sq in. **57.** $850 **58.** $4\frac{5}{6}$ ft **59.** 1 **60.** $\frac{39}{40}$ **61.** 3
62. $\frac{77}{240}$ **63.** $-\frac{8}{9}$ **64.** $-\dfrac{2x}{3}$ **65.** A **66.** D **67.** $\frac{600}{13}$, or $46\frac{2}{13}$
68. $\frac{6}{3}+\frac{5}{4}=3\frac{1}{4}$ **69. (a)** 6; **(b)** 10; **(c)** -28; **(d)** -1

Understanding Through Discussion and Writing

1. No; if the sum of the fractional parts of the mixed numerals is n/n, then the sum of the mixed numerals is an integer. For example, $1\frac{1}{5}+6\frac{4}{5}=7\frac{5}{5}=8$. **2.** A wheel makes $33\frac{1}{3}$ revolutions per minute. It rotates for $4\frac{1}{2}$ min. How many revolutions does it make? Answers may vary. **3.** The student is multiplying the whole numbers to get the whole-number portion of the answer and multiplying fractions to get the fraction part of the answer. The student should have converted each mixed numeral to fraction notation, multiplied, simplified, and then converted back to a mixed numeral. The correct answer is $4\frac{6}{7}$. **4.** It might be necessary to find the least common denominator before adding or subtracting. The least common denominator is the least common multiple of the denominators. **5.** Suppose that a room has dimensions $15\frac{3}{4}$ ft by $28\frac{5}{8}$ ft. The equation $2\cdot15\frac{3}{4}+2\cdot28\frac{5}{8}=88\frac{3}{4}$ gives the perimeter of the room, in feet. Answers may vary. **6.** Note that $5\cdot3\frac{2}{7}=5(3+\frac{2}{7})=5\cdot3+5\cdot\frac{2}{7}$. The products $5\cdot3$ and $5\cdot\frac{2}{7}$ should be added rather than multiplied together. The student could also have converted $3\frac{2}{7}$ to fraction notation, multiplied, simplified, and converted back to a mixed numeral. The correct answer is $16\frac{3}{7}$.

Test: Chapter 4, p. 297

1. [4.1a] 48 **2.** [4.2a] 3 **3.** [4.2b] $-\frac{5}{24}$ **4.** [4.3a] $\frac{2}{t}$ **5.** [4.3a] $\frac{1}{12}$
6. [4.3a] $-\frac{1}{12}$ **7.** [4.3b] $\frac{1}{4}$ **8.** [4.4a] $-\frac{12}{5}$ **9.** [4.4a] $\frac{3}{20}$
10. [4.2c] $>$ **11.** [4.5a] $\frac{7}{9}$ **12.** [4.5a] $-\frac{75}{8}$ **13.** [4.5a] $-8\frac{2}{9}$
14. [4.5b] $162\frac{7}{11}$ **15.** [4.6a] $14\frac{1}{5}$ **16.** [4.6a] $12\frac{5}{24}$ **17.** [4.6b] $4\frac{7}{24}$
18. [4.6d] $8\frac{4}{7}$ **19.** [4.6d] $-5\frac{7}{10}$ **20.** [4.3a] $-\frac{1}{8}x$ **21.** [4.6b] $1\frac{54}{55}a$
22. [4.7a] 39 **23.** [4.7a] -18 **24.** [4.7b] 6 **25.** [4.7b] 2
26. [4.7c] $19\frac{3}{5}$ **27.** [4.7c] $28\frac{1}{20}$ **28.** [4.7d] About 105 kg
29. [4.7d] 80 books **30.** [4.6c] **(a)** 3 in.; **(b)** $4\frac{1}{2}$ in. **31.** [4.3c] $\frac{1}{16}$ in.
32. [4.8b] $6\frac{11}{36}$ ft **33.** [4.8a] $3\frac{1}{2}$ **34.** [4.8a] $-\frac{9}{4}$, or $-2\frac{1}{4}$
35. [4.8b] $-\frac{2}{3}$ **36.** [4.1a] D **37.** [4.1a] **(a)** 24, 48, 72; **(b)** 24
38. [4.3c] Rebecca walks $\frac{17}{56}$ mi farther.

Cumulative Review: Chapters 1–4, p. 299

1. [4.6c] $23\frac{1}{4}$ gal **2.** [1.8a] 31 people **3.** [3.6b] $\frac{2}{5}$ tsp; 4 tsp
4. [4.7d] 16 pieces **5.** [1.8a] $108 **6.** [4.2d] $\frac{33}{20}$ mi **7.** [3.3a] $\frac{5}{16}$
8. [3.3a] $\frac{4}{5}$ **9.** [1.2a] 8982 **10.** [1.3a] 4518 **11.** [1.4a] 5004
12. [2.4a] -145 **13.** [4.2b] $\frac{5}{12}$ **14.** [4.6a] $8\frac{1}{4}$ **15.** [4.3a] $-\dfrac{5}{t}$
16. [4.6b] $1\frac{1}{6}$ **17.** [3.6a] $\frac{3}{2}$ **18.** [3.6a] -15 **19.** [2.3a] 16
20. [4.7b] $7\frac{1}{3}$ **21.** [1.5a] 715 **22.** [1.5a] 56 R 11 **23.** [4.5b] $56\frac{11}{45}$
24. [1.1a] 5 **25. (a)** [4.7d] $142\frac{1}{4}$ sq ft; **(b)** [4.6c] 54 ft
26. [1.6a] 38,500 **27.** [4.1a] 72 **28.** [4.8a] $\frac{1377}{100}$, or $13\frac{77}{100}$
29. [4.2c] $>$ **30.** [3.5c] $=$ **31.** [4.2c] $>$ **32.** [2.6a] 4
33. [3.5b] $\frac{4}{5}$ **34.** [3.3b] 0 **35.** [3.5b] -32 **36.** [4.5a] $\frac{37}{8}$
37. [4.5a] $-5\frac{2}{3}$ **38.** [1.7b] 93 **39.** [4.3b] $\frac{5}{9}$ **40.** [3.8a] $-\frac{12}{7}$
41. [4.4a] $\frac{2}{21}$ **42.** [3.1a, b], [3.2a, b, c]
Factors of 68: 1, 2, 4, 17, 34, 68
Factorization of 68: $2\cdot2\cdot17$, or $2\cdot34$
Prime factorization of 68: $2\cdot2\cdot17$
Numbers divisible by 6: 12, 54, 72, 300
Numbers divisible by 8: 8, 16, 24, 32, 40, 48, 64, 864
Numbers divisible by 5: 70, 95, 215
Prime numbers: 2, 3, 17, 19, 23, 31, 47, 101
43. [3.2b] 2003 **44.** [4.4a], [4.6b] $\frac{3}{7}$

CHAPTER 5

Exercise Set 5.1, p. 309

RC1. 5 **RC2.** 2 **RC3.** 3 **RC4.** 0 **RC5.** 8
RC6. 4 **RC7.** 1 **RC8.** 6
1. One hundred nineteen ten-thousandths
3. One hundred thirty-seven and six tenths
5. Five hundred nineteen and twenty-two hundredths
7. Three and seven hundred eighty-five thousandths
9. $\frac{73}{10}$; $7\frac{3}{10}$ **11.** $\frac{2167}{100}$; $21\frac{67}{100}$ **13.** $-\frac{2703}{1000}$; $-2\frac{703}{1000}$ **15.** $\frac{109}{10,000}$
17. $-\frac{40,003}{10,000}$; $-4\frac{3}{10,000}$ **19.** $-\frac{207}{10,000}$ **21.** $\frac{7,000,105}{100,000}$; $70\frac{105}{100,000}$
23. 0.3 **25.** -0.59 **27.** 3.798 **29.** 0.0078 **31.** -0.00018
33. 0.486197 **35.** 7.013 **37.** -8.431 **39.** 2.1739 **41.** 8.953073
43. 0.58 **45.** 0.410 **47.** -5.043 **49.** 235.07 **51.** $\frac{7}{100}$
53. -0.872 **55.** 0.2 **57.** -0.4 **59.** 3.0 **61.** -327.2
63. 0.89 **65.** -0.67 **67.** 1.00 **69.** -0.03 **71.** 0.572
73. 17.002 **75.** -20.202 **77.** 9.985 **79.** 809.5
81. 809.47 **83.** 830 **84.** $\frac{830}{1000}$, or $\frac{83}{100}$ **85.** 182
86. $\frac{182}{100}$, or $\frac{91}{50}$ **87.** $-\frac{12}{55}$ **88.** $-\frac{15}{34}$ **89.** 32,958 **90.** 10,726
91. $-1.09, -1.009, -0.989, -0.898, -0.098$ **93.** 6.78346
95. 99.99999 **97.** 1998, 2005, 2007, 2010, 2013 **99.** 1976

Calculator Corner, p. 315

1. 317.645 **2.** 49.08 **3.** 33.83 **4.** 0.99 **5.** 242.93
6. -11.692

Exercise Set 5.2, p. 316

RC1. 21.824; 23.7 **RC2.** 146.723; 40.9
1. 464.37 **3.** 1576.015 **5.** 132.56, or 132.560 **7.** 7.823
9. 50.7124 **11.** 10.06 **13.** 771.967 **15.** 20.8649
17. 227.468, or 227.4680 **19.** 41.381 **21.** 49.02 **23.** 3.564
25. 85.921 **27.** 1.6666 **29.** 4.0622 **31.** 29.999 **33.** 3.37
35. 1.045 **37.** 3.703 **39.** 0.9092 **41.** 605.21 **43.** 53.203
45. 161.62 **47.** 44.001 **49.** -3.29 **51.** -2.5 **53.** -7.2
55. 3.379 **57.** -16.6 **59.** 2.5 **61.** -3.519 **63.** 9.601 **65.** 75.5
67. 9.7 **69.** -10.292 **71.** -0.3 **73.** $5.7x$ **75.** $4.86a$
77. $21.1t+7.9$ **79.** $-2.917x$ **81.** $8.106y-7.1$ **83.** $-0.9x+3.1y$
85. $-1.1-8.4t$ **87.** $\frac{12}{35}$ **88.** $\frac{14}{45}$ **89.** $\frac{63}{1000}$ **90.** -10
91. -7 **92.** 31 **93.** $-12.001-12.2698a+10.366b$
95. $4.593a-10.996b-59.491$ **97.** -138.5 **99.** 2

Calculator Corner, p. 322

1. 142.803 **2.** -0.5076 **3.** 7916.4 **4.** 20.4153

Exercise Set 5.3, p. 326

RC1. (c) **RC2.** (e) **RC3.** (f) **RC4.** (a) **RC5.** (d) **RC6.** (b)
1. 47.6 **3.** 6.72 **5.** 0.252 **7.** 0.522 **9.** 426.3
11. $-783,686.852$ **13.** -780 **15.** 7.918 **17.** 0.09768
19. -0.287 **21.** 43.68 **23.** 0.030504 **25.** 89.76
27. -322.07 **29.** 55.68 **31.** 3487.5 **33.** 0.1155 **35.** -9420
37. 0.00953 **39.** 5706¢ **41.** 95¢ **43.** 1¢ **45.** $0.72
47. $0.02 **49.** $63.99 **51.** 3,480,000 **53.** 50,960,000,000; 13,410,000,000 **55.** 2,200,000 **57.** 11,000 **59.** 26.025
61. (a) 44 ft; **(b)** 118.75 sq ft **63. (a)** 37.8 m; **(b)** 88.2 m²
65. 2.7625 million nurses, or 2,762,500 nurses **67.** -27
68. 36 **69.** 69 **70.** -141 **71.** 1176 R 14 **72.** -27
73. $10^{21}=1$ sextillion **75.** $10^{24}=1$ septillion
77. 6,600,000,000,000 **79.** 366.5488175 **81.** $61.45

Calculator Corner, p. 330

1. 14.3 **2.** 2.56 **3.** -0.064 **4.** 75.8

Calculator Corner, p. 331

1. 28 R 2   **2.** 116 R 3   **3.** 74 R 10   **4.** 415 R 3

Exercise Set 5.4, p. 335

RC1. Subtract: $2 - 0.04$ **RC2.** Divide: $2.06 \div 0.01$
RC3. Evaluate: 8^3 **RC4.** Add: $4.1 + 6.9$ **RC5.** Divide: $9 \div 3$
RC6. Subtract: $10 - 5$
1. 2.99   **3.** 23.78   **5.** 7.08   **7.** 1.2   **9.** -0.9 **11.** -6000 **13.** 140
15. 40   **17.** -0.15 **19.** 3.2   **21.** -3.9 **23.** 0.625   **25.** 0.26
27. 2.34   **29.** -0.3045 **31.** 2.134567   **33.** -2.359 **35.** 1023.7
37. -9236 **39.** 0.08172   **41.** 9.7   **43.** -0.0527
45. $-75,300$ **47.** -0.0753 **49.** 2107   **51.** -302.997
53. -178.1 **55.** 206.0176   **57.** 0.5   **59.** -400.0108
61. 0.6725   **63.** 5.383   **65.** 10.5   **67.** 13,748.5 ft
69. 6.22 million stays   **71.** 31.24 mi   **73.** $\dfrac{19}{73}$ **74.** $\dfrac{1}{5}$

75. $2 \cdot 2 \cdot 3 \cdot 3 \cdot 19$, or $2^2 \cdot 3^2 \cdot 19$ **76.** $5 \cdot 401$ **77.** $15\dfrac{1}{8}$
78. $5\dfrac{7}{8}$ **79.** 343   **80.** 64   **81.** 47   **82.** 41   **83.** -56.6916
85. 6.254194585   **87.** 1000   **89.** 100   **91.** 46.7 points
93. 450 kWh

Mid-Chapter Review: Chapter 5, p. 339

1. False   **2.** True   **3.** True
4. $P(1 + r) = 5000(1 + 0.045)$
$= 5000(1.045)$
$= 5225$
5. $5.6 + 4.3 \times (6.5 - 0.25)^2 = 5.6 + 4.3 \times (6.25)^2$
$= 5.6 + 4.3 \times 39.0625$
$= 5.6 + 167.96875$
$= 173.56875$
6. Twenty-nine and forty-three hundredths   **7.** 9,400,000
8. $\frac{453}{100}$; $4\frac{53}{100}$ **9.** $\frac{287}{1000}$ **10.** 0.13   **11.** -5.09 **12.** 0.7
13. 6.39   **14.** -35.67 **15.** 8.002   **16.** 28.462   **17.** 28.46
18. 28.5   **19.** 28   **20.** 50.095   **21.** 1214.862   **22.** -10.23
23. 18.24   **24.** 272.19   **25.** 5.593   **26.** 15.55   **27.** -19.9
28. 4.14   **29.** 92.871   **30.** 8123.6   **31.** -0.0483 **32.** 5.06
33. 3.2   **34.** 763   **35.** 0.914036   **36.** 2045¢   **37.** $1.47
38. $-1.22x - 7.1$ **39.** 4.2   **40.** 59.774   **41.** 33.33
42. The student probably rounded over successively from the thousandths place as follows: $236.448 \approx 236.45 \approx 236.5 \approx 237$. The student should have considered only the tenths place and rounded down.   **43.** The decimal points were not lined up before the subtraction was carried out.   **44.** $10 \div 0.2 = \frac{10}{0.2} = \frac{10}{0.2} \cdot \frac{10}{10} = \frac{100}{2} = 100 \div 2$.   **45.** $0.247 \div 0.1 = \frac{247}{1000} \div \frac{1}{10} = \frac{247}{1000} \cdot \frac{10}{1} = \frac{247 \cdot 10}{10 \cdot 100} = \frac{247}{100} = 2.47 \neq 0.0247$; $0.247 \div 10 = \frac{247}{1000} \div 10 = \frac{247}{1000} \cdot \frac{1}{10} = \frac{247}{10,000} = 0.0247 \neq 2.47$

Calculator Corner, p. 343

1. $-0.1\overline{6}$ **2.** $0.6\overline{3}$ **3.** $6.\overline{3}$ **4.** $-57.\overline{1}$

Calculator Corner, p. 344

1. 123.150432   **2.** 52.59026102

Exercise Set 5.5, p. 346

RC1. Repeating   **RC2.** Terminating   **RC3.** Terminating
RC4. Repeating   **RC5.** Repeating   **RC6.** Terminating
1. 0.375   **3.** -0.5 **5.** 0.12   **7.** 0.225   **9.** 0.52   **11.** -0.85
13. -0.5625 **15.** 1.4   **17.** 1.12   **19.** -1.375 **21.** -0.975
23. 0.605   **25.** $0.5\overline{3}$ **27.** $0.\overline{3}$ **29.** $-1.\overline{3}$ **31.** $1.1\overline{6}$
33. $-1.\overline{27}$ **35.** $-0.41\overline{6}$ **37.** 0.254   **39.** $0.\overline{12}$ **41.** $-0.2\overline{18}$
43. $0.\overline{571428}$ **45.** 0.2; 0.18; 0.182   **47.** 0.3; 0.28; 0.278
49. 0.4; 0.36; 0.364   **51.** -1.7; -1.67; -1.667
53. -0.5; -0.47; -0.471 **55.** 0.6; 0.58; 0.583

57. -0.2; -0.19; -0.193 **59.** -0.8; -0.78; -0.778
61. **(a)** 0.571; **(b)** 1.333; **(c)** 0.429; **(d)** 0.75   **63.** 15.8 mpg
65. 17.8 mpg   **67.** 11.06   **69.** 2.736   **71.** $-417.51\overline{6}$
73. 0   **75.** 0.09705   **77.** -1.5275 **79.** 24.375   **81.** 1.08 m^2

83. 5.78 cm^2 **85.** 790.92 in^2 **87.** 21   **88.** $1\dfrac{1}{2}$ **89.** 10
90. $30\dfrac{7}{10}$ **91.** $1\dfrac{1}{24}$ cups   **92.** $1\dfrac{33}{100}$ in.   **93.** $0.\overline{142857}$,
$0.\overline{285714}$, $0.\overline{428571}$, $0.\overline{571428}$, $0.\overline{714285}$; $0.\overline{857142}$ **95.** 13.86 cm^2
97. 1.76625 ft^2 or 1.767145868 ft^2

Exercise Set 5.6, p. 352

RC1. (b)   **RC2.** (d)   **RC3.** (c)   **RC4.** (e)
RC5. (a)   **RC6.** (f)
1. (d)   **3.** (c)   **5.** (a)   **7.** (c)   **9.** 1.6   **11.** 6   **13.** 60
15. 2.3   **17.** 180   **19.** (a)   **21.** (c)   **23.** (b)   **25.** (b)
27. $1800 \div 9 = 200$ posts; answers may vary
29. $\$2 \cdot 12 = \$24$; answers may vary   **31.** 18,940,000 lb
32. 56 mm^2 **33.** 126 slices   **34.** 9 cups
35. $36,500 a year   **36.** 16 packages   **37.** Yes
39. No   **41.** **(a)** $+, \times$;   **(b)** $+, \times, -$

Exercise Set 5.7, p. 359

RC1. False   **RC2.** True   **RC3.** True   **RC4.** True
1. 5.4   **3.** 0.2   **5.** -6.24 **7.** -12.6 **9.** -2.55
11. 1.97   **13.** 8.38   **15.** -5.9 **17.** 6   **19.** 1.8
21. -3.7 **23.** -4.7 **25.** 1.7   **27.** 2.94   **29.** 4.5
31. 9   **33.** 3.2   **35.** -1.75 **37.** 30   **39.** 3.2
41. 9   **43.** 2.1   **45.** 4.5   **47.** 4.12   **49.** 5.6
51. -1.9 **53.** 13   **55.** 0   **57.** -1.5 **59.** 14 m^2
60. 27 cm^2 **61.** $\frac{25}{2}$ in^2 **62.** 24 ft^2 **63.** 5 ft^2
64. 12 m^2 **65.** $-\frac{29}{50}$ **66.** 0   **67.** -2 **68.** 8
69. 3.1   **71.** 36   **73.** 1.1212963

Translating for Success, p. 370

1. I   **2.** C   **3.** N   **4.** A   **5.** G   **6.** B   **7.** D   **8.** O
9. F   **10.** M

Exercise Set 5.8, p. 371

RC1. Familiarize   **RC2.** Translate   **RC3.** Solve
RC4. Check   **RC5.** State
1. Let $a = $ Ron's age: $a + 5$, or $5 + a$ **3.** $b + 6$, or $6 + b$
5. $c - 9$ **7.** Let $n = $ the number; $n - 16$
9. Let $s = $ Nate's speed; $8s$ **11.** $\dfrac{x}{17}$
13. Let $x = $ the number; $\frac{1}{2}x + 20$, or $20 + \frac{1}{2}x$
15. Let $x = $ the number; $4x - 20$
17. Let $l = $ the length and $w = $ the width; $l + w$, or $w + l$
19. Let $r = $ the rate and $t = $ the time; $rt + 10$, or $10 + rt$
21. Let $n = $ the number; $10n + n$, or $n + 10n$
23. Let x and $y = $ the numbers; $5(x - y)$
25. $60.2 billion   **27.** $66.28   **29.** $0.51
31. $218,666,666.67   **33.** 9.58 sec   **35.** 22,691.5 mi
37. 20.2 mpg   **39.** $24.33   **41.** 2.66 cc   **43.** 10.8¢
45. $26,900.21   **47.** Area: 268.96 sq ft; perimeter: 65.6 ft
49. 193.04 cm^2 **51.** 2.31 cm   **53.** 331.74 ft^2
55. 125 gigabytes   **57.** 296,183.55 sq ft   **59.** 17 bottles
61. 2.5 million protected acres, 2.9 million unprotected acres
63. $65.30 for food, $195.90 for lodging   **65.** $906.50
67. $316,987.20; $196,987.20   **69.** 0   **70.** Undefined
71. 1   **72.** $-\dfrac{1}{10}$ **73.** $-\dfrac{20}{33}$ **74.** $6\dfrac{5}{6}$ **75.** 3999.76 cm^2
77. $0.19   **79.** 25 cm^2. We assume that the figures are nested squares formed by connecting the midpoints of consecutive sides of the next larger square.

Summary and Review: Chapter 5, p. 377

Vocabulary Reinforcement

1. repeating   **2.** terminating   **3.** billion
4. million   **5.** trillion   **6.** rational numbers

Concept Reinforcement

1. True   **2.** False   **3.** True   **4.** False   **5.** True

Study Guide

1. $\frac{5093}{100}$ **2.** 81.7   **3.** 42.159   **4.** 153.35   **5.** 38.611
6. 207.848   **7.** 19.11   **8.** 0.176   **9.** 60,437   **10.** 7.4
11. 0.047   **12.** 15,690

Review Exercises

1. 6,590,000   **2.** 3,100,000,000
3. Three and forty-seven hundredths
4. Thirty-one thousandths
5. Twenty-seven and one ten-thousandth   **6.** Nine tenths
7. $\frac{9}{100}$ **8.** $-\frac{4561}{1000}$; $-4\frac{561}{1000}$ **9.** $-\frac{89}{1000}$ **10.** $\frac{30,227}{10,000}$; $3\frac{227}{10,000}$
11. 0.034   **12.** 4.2603   **13.** 27.91   **14.** -867.006
15. 0.034   **16.** -0.19 **17.** 0.741   **18.** 1.041
19. 17.4   **20.** 17.43   **21.** 17.429   **22.** 17   **23.** 499.829
24. 29.148   **25.** 229.1   **26.** 685.0519   **27.** -57.3 **28.** 2.37
29. 12.96   **30.** -1.073 **31.** 24,680   **32.** 3.2   **33.** -1.6
34. 0.2763   **35.** 0.003056   **36.** -4380 **37.** $2.2x - 9.1y$
38. $-2.84a + 12.57$ **39.** 925   **40.** 40.84   **41.** \$15.49
42. 248.27   **43.** 2.6   **44.** 1.28   **45.** 3.25   **46.** $-1.1\overline{6}$
47. 21.08   **48.** -3.2 **49.** -3 **50.** -7.5 **51.** 6.5
52. \$2.19 billion   **53.** \$15.52   **54.** 249.76 ft^2 **55.** \$784.47
56. 95 transactions   **57.** 195.7 lb of newspaper, 65.7 lb of glass
58. 14.5 mpg   **59.** 20.7 million books   **60.** \$55.50   **61.** \$1.33
62. 61.5 ft; 235.625 sq ft   **63.** B   **64.** A   **65.** (a) $+$; (b) $-$
66. $\frac{-13}{15}, \frac{-17}{20}, -\frac{11}{13}, -\frac{15}{19}, \frac{-5}{7}, -\frac{2}{3}$ **67.** 26,260 mi
68. $1 = 3 \cdot \frac{1}{3} = 3(0.33333333\ldots) = 0.99999999\ldots$, or $0.\overline{9}$
69. The rectangular pizza, at $\dfrac{4.4¢}{\text{in}^2}$, is a better buy than the round pizza, which costs $\dfrac{5.5¢}{\text{in}^2}$.

Understanding Through Discussion and Writing

1. Count the number of decimal places. Move the decimal point that many places to the right and write the result over a denominator of 1 followed by that many zeros.
2. $346.708 \times 0.1 = \frac{346,708}{1000} \times \frac{1}{10} = \frac{346,708}{10,000} = 34.6708 \neq 3467.08$
3. When the denominator of a fraction is a multiple of 10, long division is not the fastest way to convert the fraction to decimal notation. Many times this is also the case when the denominator has only 2's or 5's or both as factors.   **4.** Multiply by 1 to get a denominator that is a power of 10:
$$\frac{44}{125} = \frac{44}{125} \cdot \frac{8}{8} = \frac{352}{1000} = 0.352.$$
We can also divide to find that $\frac{44}{125} = 0.352$.

Test: Chapter 5, p. 382

1. [5.3b] 2,600,000,000
2. [5.1a] One hundred twenty-three and forty-seven thousandths
3. [5.1b] $-\frac{91}{100}$ **4.** [5.1b] $\frac{2769}{1000}$; $2\frac{769}{1000}$ **5.** [5.1b] 0.074
6. [5.1b] -3.7047 **7.** [5.1b] 756.09   **8.** [5.1b] 91.703
9. [5.1c] 0.162   **10.** [5.1c] 8.049   **11.** [5.1c] -0.09
12. [5.1d] 6   **13.** [5.1d] 5.68   **14.** [5.1d] 5.678   **15.** [5.1d] 5.7
16. [5.2a] 405.219   **17.** [5.3a] 0.03   **18.** [5.3a] 0.21345
19. [5.2b] 44.746   **20.** [5.2a] 356.37   **21.** [5.2c] -2.2
22. [5.2b] 1.9946   **23.** [5.3a] 73,962   **24.** [5.4a] 4.75
25. [5.4a] 30.4   **26.** [5.4a] -0.34682 **27.** [5.4a] 34,682
28. [5.3b] 17,982¢   **29.** [5.2d] $9.8x - 3.9y - 4.6$

30. [5.3c] 11.6   **31.** [5.4b] 7.6   **32.** [5.5c] 1.6   **33.** [5.5c] 5.25
34. [5.5a] -0.4375 **35.** [5.5a] $1.\overline{5}$ **36.** [5.5b] 1.56
37. [5.4b] -22.25 **38.** [5.4b] 1.8045   **39.** [5.5d] 9.72
40. [5.7a] -3.24 **41.** [5.7b] 10   **42.** [5.7b] 1.4
43. [5.8b] \$46.69   **44.** [5.8b] 28.3 mpg   **45.** [5.8b] \$592.45
46. [5.8b] \$293.93   **47.** [5.8b] 68.5 yr   **48.** [5.6a] C
49. [5.3a] (a) Always; (b) never; (c) sometimes; (d) sometimes
50. [5.8b] \$35   **51.** [5.8b] (a) Fly; (b) fly; (c) drive

Cumulative Review: Chapters 1–5, p. 385

1. [4.5a] $\frac{20}{9}$ **2.** [5.1b] $\frac{3051}{1000}$ **3.** [5.5a] -1.4 **4.** [5.5a] $0.\overline{54}$
5. [3.2b] Prime   **6.** [3.1b] No   **7.** [1.9c] 1754   **8.** [5.4b] 4.364
9. [5.1d] 584.97   **10.** [5.5b] 218.56   **11.** [5.6a] 160
12. [5.6a] 4   **13.** [2.6a] 12   **14.** [2.7a] $7p - 8$ **15.** [4.6a] $6\frac{1}{20}$
16. [1.2a] 139,116   **17.** [4.2b] $\frac{31}{18}$ **18.** [5.2a] 145.953
19. [1.3a] 710,137   **20.** [5.2b] -13.097 **21.** [4.6b] $\frac{5}{7}$
22. [4.3a] $-\frac{1}{110}$ **23.** [3.6a] $-\frac{1}{6}$ **24.** [1.4a] 5,317,200
25. [5.3a] 4.78   **26.** [5.3a] 0.0279431   **27.** [5.4a] 2.122
28. [1.5a] 1843   **29.** [5.4a] 13,862.1   **30.** [3.7b] $\frac{5}{6}$
31. [5.7b] -11.5 **32.** [2.8b] -28 **33.** [5.7a] 3.8125
34. [1.7b] 367,251   **35.** [4.3b] $\frac{1}{18}$ **36.** [3.8a] $-\frac{1}{2}$
37. [1.8a] 732 million   **38.** [1.8a] 885 million   **39.** [3.8b] \$1500
40. [3.6b] \$2400   **41.** [5.8b] \$258.77   **42.** [4.6c] $6\frac{1}{2}$ lb
43. [3.6b] 88 ft^2 **44.** [5.8b] 43.585 in^2 **45.** [4.8a] $\frac{9}{32}$
46. [5.4b] 527.04   **47.** [5.8b] \$2.39   **48.** [4.7d] 144 packages

CHAPTER 6

Exercise Set 6.1, p. 394

RC1. False   **RC2.** False   **RC3.** True   **RC4.** False
1. 92°   **3.** 108°   **5.** 85%, 60%; 90%, 40%; 100%, 10%
7. 90° and higher   **9.** 30% and higher
11. $90\% - 40\% = 50\%$ **13.** 483,612,200 mi   **15.** Neptune
17. 11 Earth diameters   **19.** 31,191.75 mi   **21.** 300 calories
23. Yes   **25.** 410 mg   **27.** White rhino
29. About 1350 rhinos   **31.** About 4100 rhinos
33. \$1270 per person   **35.** 1950 and 1970
37. \$1532.70 per person
39. (a) 15% less; (b) \$2003.51 more per person
41. About 24 games   **43.** 120–129

Exercise Set 6.2, p. 402

RC1. True   **RC2.** True   **RC3.** False   **RC4.** True
1. Miniature tall bearded   **3.** 16 in. to 26 in.   **5.** Tall bearded
7. 25 in.   **9.** 190 calories
11. 1 slice of chocolate cake with fudge frosting
13. 1 cup of premium chocolate ice cream
15. About 125 calories   **17.** 1950 and 1970
19. About 175,000 bachelor's degrees
21.

23. Denver and Indianapolis   **25.** 68.9   **27.** New York City
29. 16.6 min   **31.** \$42
33. June to July, September to October, and November to December
35. About \$900   **37.** 20 years   **39.** About \$440
41. About $120(\$1200) = \$144,000$

43.

45. 3.5 yr   **47.** 1.6 yr   **48.** 83   **49.** $\frac{4}{3}$ **50.** $\frac{7}{24}$ **51.** 2.26
52. 6.348   **53.** 8   **54.** 21   **55.** 17.26   **56.** $\frac{31}{60}$ **57.** 45

Exercise Set 6.3, p. 410

RC1. E   **RC2.** C   **RC3.** G   **RC4.** B   **RC5.** H

1.

3.

5. A: $(3, 3)$; B: $(0, -4)$; C: $(-5, 0)$; D: $(-1, -1)$; E: $(2, 0)$; F: $(-3, 5)$
7. A: $(5, 0)$; B: $(0, 5)$; C: $(-3, 4)$; D: $(2, -4)$; E: $(2, 3)$; F: $(-4, -2)$
9. II   **11.** IV   **13.** III   **15.** I   **17.** positive; negative
19. III   **21.** IV; positive   **23.** Yes   **25.** No   **27.** Yes
29. No   **31.** Yes   **33.** Yes   **35.** 7   **36.** 3   **37.** 0   **38.** $\frac{30}{17}$
39. $1\frac{28}{33}a$ **40.** $7x - 24$ **41.** Yes   **43.** I, IV   **45.** I, III
47. $(-1, -5)$

Calculator Corner, p. 414

1. $y = \frac{2}{3}x + 1$

2. $y = x + 1$

3. $y = -2x + 1$

4. $y = \frac{3}{5}x$

Exercise Set 6.4, p. 419

RC1. False   **RC2.** True   **RC3.** True   **RC4.** True
RC5. True   **RC6.** False
1. $(5, 3)$ **3.** $(3, 1)$ **5.** $(5, 14)$ **7.** $(10, -3)$ **9.** $(0, 3)$
11. $(2, -1)$ **13.** $(1, 3)$; $(-1, 5)$ **15.** $(7, 3)$; $(10, 6)$
17. $(0, 10)$; $(15, 0)$ **19.** $(3, 1)$; $(-2, 4)$ **21.** $(1, 4)$; $(-2, -8)$
23. $(0, \frac{3}{5})$; $(\frac{3}{2}, 0)$ **25.** $(0, 9)$, $(4, 5)$, $(10, -1)$
27. $(0, 0)$, $(1, 4)$, $(2, 8)$ **29.** $(0, 13)$, $(1, 10)$, $(2, 7)$
31. $(0, -1)$, $(2, 5)$, $(-1, -4)$ **33.** $(0, 0)$, $(1, -7)$, $(-2, 14)$
35. $(0, -4)$, $(4, 0)$, $(1, -3)$ **37.** $(0, 6)$, $(4, 0)$, $(1, \frac{9}{2})$
39. $(0, 2)$, $(3, 3)$, $(-3, 1)$

41.

43.

45.

47.

49.

51.

53.

55.

57.

59.

61.

63.

65.
$x = -3$

67.
$y = -4$

69. 9 min **70.** 319.75 pages **71.** $1\frac{7}{8}$ cups **72.** $-\frac{7}{11}$
73. 42 **74.** $-\frac{5}{8}$ **75.** $(0, 0.2), (-4, -1), (1, 0.5)$;
answers may vary

$21x - 70y = -14$

77. $(-3, 4.4), (-3.9, 5), (3, 0.4)$; answers may vary

$50x + 75y = 180$

79. $(0, 6), (1, 5), (2, 4), (3, 3), (4, 2), (5, 1), (6, 0)$
81. (a)
$y_1 = -0.63x + 2.8$

(b)
$y_1 = 2.3x - 4.1$
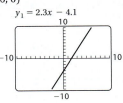

Mid-Chapter Review: Chapter 6, p. 424

1. True **2.** True **3.** False
4. $3 \cdot 2 + (-1); 6 + (-1); 5$
5. $1 - y = 6; -y = 5; y = -5$; Thus, $(1, -5)$ is a solution of
$x - y = 6$. **6.** 8 oz **7.** 6%
8. Hershey's Special Dark chocolate bar **9.** 7 oz
10. Nabisco Chips Ahoy cookies **11.** III **12.** II **13.** IV
14. No **15.** $(-10, -7)$ **16.** $(-1, -10)$ **17.** $(1, 3), (0, 5), (-1, 7)$
18.

$x + y = 3$
19.
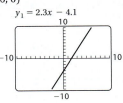
$y = x - 2$

20.
$y = 4x$

21.
$y = 2x - 4$

22.
$y = -3$

23.
$x = 1$

24. The points (a, b) and (b, a) will be in the same quadrant when both a and b are negative or when both are positive. **25.** The point $(5, 0)$ is not in any quadrant. It lies on the x-axis, and the axes are not considered part of any quadrant. **26.** The student should check each point carefully in the original equation. At least one of the points will not be a solution of the equation. After identifying any points that are not on the graph of the line, the student should calculate and plot new points to replace them. If these points do line up with each other, the student can then draw the line. **27.** A line can pass through at most three quadrants. For example, the line $y = x - 2$ passes through quadrants I, III, and IV. A straight line cannot pass through four quadrants.

Translating for Success, p. 431

1. F **2.** N **3.** A **4.** O **5.** G **6.** D **7.** C **8.** L
9. H **10.** E

Exercise Set 6.5, p. 432

RC1. statistic **RC2.** mean **RC3.** weight **RC4.** mode
1. Mean: 9.325 million visitors; median: 9.35 million visitors;
modes: 9.0 million visitors, 9.5 million visitors
3. Mean: 21; median: 18.5; mode: 29
5. Mean: 21; median: 20; modes: 5, 20
7. Mean: 5.38; median: 5.7; no mode exists
9. Mean: 239.5; median: 234; mode: 234 **11.** 36 mpg **13.** 2.7
15. Mean: $4.19; median: $3.99; mode: $3.99
17. 90 **19.** 263 days **21. (a)** Jefferson County: $133,987;
Hamilton County: $146,989; **(b)** Jefferson County
23. (a) 1.455 billion tickets; **(b)** 1.3775 billion tickets;
(c) 2002 to 2009 **25.** 225.05 **26.** 126.0516 **27.** $\frac{3}{35}$ **28.** $\frac{14}{15}$
29. 181 **31.** 10 home runs **33.** $6950

Exercise Set 6.6, p. 439

RC1. interpolation **RC2.** extrapolation **RC3.** probability
1. 83 **3.** About $6.2 billion **5.** About 233,000 students
7. About 10,300 kWh **9.** $\frac{1}{6}$, or $0.1\overline{6}$ **11.** $\frac{1}{2}$, or 0.5 **13.** $\frac{1}{52}$
15. $\frac{2}{13}$ **17.** $\frac{3}{26}$ **19.** $\frac{4}{39}$ **21.** $\frac{34}{39}$ **23.** About 1,075,000,000
24. $\frac{3}{64}$ acre **25.** $\frac{1}{4}$, or 0.25 **27.** $\frac{1}{36}$ **29.** 0

Summary and Review: Chapter 6, p. 442

Vocabulary Reinforcement
1. table **2.** origin **3.** pictograph **4.** mode **5.** mean

Concept Reinforcement
1. False   **2.** True   **3.** True

Study Guide
1. Quaker Organic Maple & Brown Sugar; $0.54 per serving
2. 12 g   **3.** Arrowhead Stadium   **4.** About $110 million more
5. No   **6.**

7. Mean: 8; median: 8; mode: 8   **8.** $\frac{1}{6}$

Review Exercises
1. 52%   **2.** Mexico, Germany, and the Philippines
3. United States   **4.** About 21 World Series
5. 7 games   **6.** About 16 more World Series
7. $70,000–$89,000   **8.** About 2 more governors
9. About 22 governors   **10.** April   **11.** About 150 more
tornadoes   **12.** In the spring   **13.** 1998   **14.** About 10,000
children   **15.** 2004 and 2010   **16.** Between 1986 and 1989,
between 1995 and 1998, and between 2004 and 2007
17. By about 4000 children
18.

19.

20. $(-5, -1)$ **21.** $(-2, 5)$ **22.** $(3, 0)$ **23.** $(4, -2)$
24.–26.   **27.** IV   **28.** III   **29.** I

30. $(1, 2)$; $(9, -2)$
31.   **32.**

33.   **34.**

35.

36. 38.5   **37.** 13.4   **38.** 1.55   **39.** 1840   **40.** $16.\overline{6}$
41. $321.\overline{6}$ **42.** 96   **43.** 28 mpg   **44.** 3.1   **45.** 38.5   **46.** 14
47. 1.8   **48.** 1900   **49.** $17   **50.** 375
51. Mean: $260; median: $228   **52.** 26   **53.** 11 and 17
54. 0.2   **55.** 700 and 800   **56.** $17   **57.** 20
58. About 22,000 children   **59.** $\frac{1}{52}$ **60.** $\frac{1}{2}$ **61.** D   **62.** A
63. $(0, \frac{100}{47})$, $(\frac{100}{34}, 0)$, $(2, \frac{32}{47})$ (Ordered pairs may vary.)

64. $11.52/hr   **65.** $a = 316, b = 349$
66.

67. 26 units   **68.** 65 square units

Understanding Through Discussion and Writing
1. The equation could represent a person's average income during
a 4-yr period. Answers may vary.   **2.** Bar graphs that show change
over time can be successfully converted to line graphs. Other bar
graphs cannot be successfully converted to line graphs.
3. It is possible for the mean of a set of numbers to be larger than all
but one number in the set. To see this, note that the mean of the set
$\{6, 8\}$ is 7, which is larger than all of the numbers in the set but one.
4. The median of a set of four numbers *can* be in the set. For
example, the median of the set $\{11, 15, 15, 17\}$ is 15, which is in the
set.   **5.** The median income is often used instead of the mean
income because it is not artificially raised by the small number of
people with very high incomes. The mean of the medians over a 3-yr
period would lessen the impact of a sudden drop or rise in income
for one year.   **6.** A company considering expansion would
probably be more interested in extrapolation than in interpolation
because it will be future sales or other activity that will influence
profit after expansion.

Test: Chapter 6, p. 449

1. [6.1a] Hiking at 3 mph with 20-lb load
2. [6.1a] Hiking at 3 mph with 10-lb load, hiking at 3 mph with 20-lb load   **3.** [6.1b] Spain   **4.** [6.1b] Norway and the United States
5. [6.1b] 900 lb   **6.** [6.1b] 1000 lb
7. [6.2b]

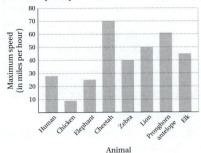

8. [6.1a] 61 mph   **9.** [6.1a] $2\frac{1}{2}$ times faster
10. [6.1a], [6.5a] 41 mph   **11.** [6.1a], [6.5b] 42.5 mph
12. [6.2c] 2005   **13.** [6.2c] 2009   **14.** [6.2c] 10 hurricanes
15. [6.2c] 10 more hurricanes   **16.** [6.2c] About 8 hurricanes/year
17. [6.6a] About 13 hurricanes   **18.** [6.3b] II   **19.** [6.3b] III
20. [6.3a] $(3, 4)$ **21.** [6.3a] $(0, -4)$ **22.** [6.3a] $(-4, 2)$
23. [6.4a] $(4, 2)$ **24.** [6.4b]

25. [6.4b]   **26.** [6.4c]

27. [6.5a] 50   **28.** [6.5a] 3   **29.** [6.5a] 15.5
30. [6.5b, c] Median: 50.5; mode: 54
31. [6.5b, c] Median: 3; no mode exists
32. [6.5b, c] Median: 17.5; modes: 17, 18   **33.** [6.5a] 76
34. [6.5a] 2.9   **35.** [6.6b] $\frac{2}{5}$
36. [6.4b]   **37.** [6.4b]

38. [6.3a] 56 sq units

Cumulative Review: Chapters 1–6, p. 453

1. [5.3b] 53,500,000,000   **2.** [6.5a] 24 mpg   **3.** [1.1a] 5 hundreds
4. [1.9c] 128   **5.** [3.2a] 1, 2, 3, 4, 5, 6, 10, 12, 15, 20, 30, 60
6. [5.1d] 52.0   **7.** [4.5a] $\frac{33}{10}$ **8.** [5.3b] \$2.10   **9.** [2.1d] 9
10. [2.6a] -2 **11.** [4.2b] $\frac{1}{6}$ **12.** [2.2a] -39 **13.** [4.3a] $\frac{1}{3}$
14. [5.2b] 325.43   **15.** [4.7a] 15   **16.** [5.3a] -42.282
17. [3.7b] $\frac{9}{10}$ **18.** [5.4a] 62.345   **19.** [4.3b] $-\frac{13}{15}$ **20.** [3.8a] $\frac{3}{4}$
21. [4.4a] 24   **22.** [5.7b] $-\frac{17}{4}$ **23.** [1.8a] 2572 billion kWh
24. [4.7d] $\frac{1}{4}$ yd   **25.** [3.6b] $\frac{3}{8}$ cup   **26.** [5.8b] 6.2 lb
27. [5.8b] \$9.55   **28.** [1.4b], [2.7b] 22 cm; 28 cm^2 **29.** [6.3b] II
30. [6.4b]

31. [2.7a] $-2x + y$ **32.** [2.6b] $10a - 15b + 5$
33. [6.5a, b] \$79.557; \$80.10
34. [6.2b]

35. [6.2d]

36. [4.8a] $\frac{9}{32}$ **37.** [4.6a] $7\frac{47}{1000}$
38. [6.3a] $(-2, -1), (-2, 7), (6, 7), (6, -1)$

CHAPTER 7

Exercise Set 7.1, p. 460

RC1. True   **RC2.** True   **RC3.** False   **RC4.** False

1. $\frac{4}{5}$ **3.** $\frac{178}{572}$ **5.** $\frac{0.4}{12}$ **7.** $\frac{3.8}{7.4}$ **9.** $\frac{56.78}{98.35}$ **11.** $\frac{8\frac{3}{4}}{9\frac{5}{6}}$ **13.** $\frac{29}{75}$

15. $\frac{11}{18}$ **17.** $\frac{107}{366}$ **19.** $\frac{104.8}{3100.7}$; $\frac{3100.7}{104.8}$ **21.** $\frac{60}{100}$; $\frac{100}{60}$ **23.** $\frac{2}{3}$ **25.** $\frac{3}{4}$
27. $\frac{12}{25}$ **29.** $\frac{7}{9}$ **31.** $\frac{2}{3}$ **33.** $\frac{14}{25}$ **35.** $\frac{1}{2}$ **37.** $\frac{3}{4}$ **39.** $\frac{478}{213}$; $\frac{213}{478}$
41. 408,550   **42.** 27.006   **43.** 397.27   **44.** $\frac{19}{30}$ **45.** $\frac{23}{24}$
46. $14\frac{1}{3}$ **47.** 942,219   **48.** 2.6   **49.** 94.375   **50.** $\frac{1}{16}$
51. $1\frac{13}{15}$ **52.** $4\frac{2}{3}$ **53.** $\frac{30}{47}$ **55.** $1:2:3$

Exercise Set 7.2, p. 466

RC1. (b)   **RC2.** (c)   **RC3.** (d)   **RC4.** (a)
1. 40 km/hr   **3.** 7.48 mi/sec   **5.** 25 mpg   **7.** 33 mpg
9. 47,236 people/sq mi   **11.** 25 mph; 0.04 hr/mi   **13.** About
393 performances/year   **15.** 186,000 mi/sec   **17.** 124 km/hr
19. 0.623 gal/ft^2 **21.** 28 beats/min   **23.** 26.188¢/oz;
26.450¢/oz; 16 oz   **25.** 13.111¢/oz; 11.453¢/oz; 75 oz
27. 18.222¢/oz; 17.464¢/oz; 28 oz   **29.** 10.187¢/oz; 10.346¢/oz;
10.7 oz   **31.** B: 18.719¢/oz; E: 14.563¢/oz; Brand E

33. A: 10.375¢/oz; B: 9.139¢/oz; H: 8.022¢/oz; Brand H
35. 11,550   **36.** 679.4928   **37.** 2.74568   **38.** $\frac{20}{3}$ **39.** $8\frac{11}{20}$
40. $13\frac{2}{3}$ **41.** 125   **42.** 9.5   **43.** 9.63   **44.** $\frac{125}{27}$ **45.** 150
46. $19\frac{3}{7}$ **47.** 6-oz: 10.833¢/oz; 5.5 oz: 10.909¢/oz

Exercise Set 7.3, p. 474

RC1. ratio   **RC2.** proportional   **RC3.** proportion
RC4. cross products   **1.** No   **3.** Yes   **5.** Yes   **7.** No   **9.** 45
11. 12   **13.** 10   **15.** 20   **17.** 5   **19.** 18   **21.** 22   **23.** 28
25. $\frac{28}{3}$, or $9\frac{1}{3}$ **27.** $\frac{26}{9}$, or $2\frac{8}{9}$ **29.** 5   **31.** 5   **33.** 0   **35.** 14
37. 2.7   **39.** 1.8   **41.** 0.06   **43.** 0.7   **45.** 12.5725   **47.** 1
49. $\frac{1}{20}$ **51.** $\frac{3}{8}$ **53.** $\frac{16}{75}$ **55.** $\frac{51}{16}$, or $3\frac{3}{16}$ **57.** $\frac{546}{185}$, or $2\frac{176}{185}$
59. 4063 theaters   **60.** 6 lb   **61.** $\frac{1}{4}$ lb   **62.** $\frac{3}{8}$ lb   **63.** $6\frac{7}{20}$ cm
64. $37\frac{1}{2}$ servings   **65.** $59.81   **66.** 21.5 mpg
67. Approximately 2731.4   **69.** Ruth: 1.863 strikeouts per home run; Schmidt: 3.436 strikeouts per home run

Mid-Chapter Review: Chapter 7, p. 477

1. True   **2.** True   **3.** False   **4.** False
5. $\frac{120\,\text{mi}}{2\,\text{hr}} = \frac{120}{2}\frac{\text{mi}}{\text{hr}} = 60\,\text{mi/hr}$
6.
$$\frac{x}{4} = \frac{3}{6}$$
$$x \cdot 6 = 4 \cdot 3$$
$$\frac{x \cdot 6}{6} = \frac{4 \cdot 3}{6}$$
$$x = 2$$
7. $\frac{4}{7}$ **8.** $\frac{313}{199}$ **9.** $\frac{35}{17}$ **10.** $\frac{59}{101}$ **11.** $\frac{2}{3}$ **12.** $\frac{1}{3}$ **13.** $\frac{8}{7}$ **14.** $\frac{25}{19}$
15. $\frac{2}{1}$ **16.** $\frac{5}{1}$ **17.** $\frac{2}{7}$ **18.** $\frac{3}{5}$ **19.** 60.75 mi/hr, or 60.75 mph
20. 48.67 km/hr   **21.** 13 m/sec   **22.** 16.17 ft/sec   **23.** 27 in./day
24. About 0.860 free throw made/attempt   **25.** 11.611¢/oz
26. 49.917¢/oz   **27.** Yes   **28.** No   **29.** No   **30.** Yes
31. 12   **32.** 40   **33.** 9   **34.** 35   **35.** 2.2   **36.** 4.32   **37.** $\frac{1}{2}$
38. $\frac{65}{4}$, or $16\frac{1}{4}$ **39.** Yes; every ratio $\frac{a}{b}$ can be written as $\frac{\frac{a}{b}}{1}$.
40. By making some sketches, we see that the rectangle's length must be twice the width.   **41.** The student's approach will work. However, when we use the approach of equating cross products, we eliminate the need to find the least common denominator.
42. The instructor thinks that the longer a student studies, the higher his or her grade will be. An example is the situation in which one student gets a test grade of 96 after studying for 8 hr while another student gets a grade of 78 after studying for $6\frac{1}{2}$ hr. This is represented by the proportion $\frac{96}{8} = \frac{78}{6\frac{1}{2}}$.

Translating for Success, p. 484

1. N   **2.** I   **3.** A   **4.** K   **5.** J   **6.** F   **7.** M   **8.** B
9. G   **10.** E

Exercise Set 7.4, p. 485

RC1. $\frac{m}{8}$ **RC2.** $\frac{8}{m}$ **RC3.** $\frac{6}{8}$ **RC4.** $\frac{8}{6}$ **1.** 11.04 hr
3. 7680 frames   **5. (a)** About 103 gal; **(b)** 3640 mi
7. 232.53 million, or 232,530,000   **9.** 175 bulbs
11. About 262 mg every 8 hr   **13.** 2975 ft^2 **15.** 450 pages
17. 13,500 mi   **19.** 880 calories   **21.** 60 students   **23.** 64 cans
25. 9.75 gal   **27.** 100 oz   **29.** 90 whales   **31.** 58.1 mi
33. (a) 38 euros; **(b)** $11,368.42   **35. (a)** 66 games;
(b) About 2504 points   **37.** 162   **38.** 4014   **39.** 4.45   **40.** 0.4
41. 14.3   **42.** 0.26   **43.** $\frac{1}{10}$ **44.** $\frac{5}{4}$ **45.** 17 positions
47. 2150 earned runs   **49.** CD player: $150; receiver: $450; speakers: $300

Exercise Set 7.5, p. 494

RC1. 5   **RC2.** 8   **RC3.** a **RC4.** 3
1. 25   **3.** $\frac{4}{3}$, or $1\frac{1}{3}$ **5.** $x = \frac{27}{4}$, or $6\frac{3}{4}$; $y = 9$ **7.** $x = 7.5$; $y = 7.2$
9. 1.25 m   **11.** 36 ft   **13.** 7 ft   **15.** 100 ft   **17.** 4   **19.** $10\frac{1}{2}$
21. $x = 6$; $y = 5.25$; $z = 3$ **23.** $x = 5\frac{1}{3}$, or $5.\overline{3}$; $y = 4\frac{2}{3}$, or $4.\overline{6}$;
$z = 5\frac{1}{3}$, or $5.\overline{3}$ **25.** 20 ft   **27.** 152 ft   **29.** Prime
30. Composite   **31.** $2 \cdot 2 \cdot 2 \cdot 101$, or $2^3 \cdot 101$ **32.** $3 \cdot 31$
33. $=$ **34.** $\neq$ **35.** $<$ **36.** $>$ **37.** $\frac{13}{4}$, or $3\frac{1}{4}$ **38.** 26.29   **39.** 127
40. 75   **41.** 13.75 ft   **43.** 1.25 cm   **45.** 3681.437
47. $x \approx 0.35$; $y = 0.4$

Summary and Review: Chapter 7, p. 498

Vocabulary Reinforcement

1. quotient   **2.** shape   **3.** cross products   **4.** rate
5. proportion   **6.** price

Concept Reinforcement

1. True   **2.** True   **3.** False   **4.** True

Study Guide

1. $\frac{17}{3}$ **2.** $\frac{8}{7}$ **3.** $7.50/hr   **4.** A: 9.964¢/oz; B: 10.281¢/oz;
Brand A   **5.** Yes   **6.** $\frac{27}{8}$ **7.** 175 mi   **8.** 21

Review Exercises

1. $\frac{47}{84}$ **2.** $\frac{46}{1.27}$ **3.** $\frac{83}{100}$ **4.** $\frac{0.72}{197}$ **5. (a)** $\frac{12{,}480}{16{,}640}$, or $\frac{3}{4}$; **(b)** $\frac{16{,}640}{29{,}120}$, or $\frac{4}{7}$
6. $\frac{3}{4}$ **7.** $\frac{9}{16}$ **8.** 28 mpg   **9.** 6300 revolutions/min
10. 0.638 gal/ft^2 **11.** 6.33¢/tablet   **12.** 19.542¢/oz
13. 14.969¢/oz; 12.479¢/oz; 15.609¢/oz; 48 oz   **14.** Yes   **15.** No
16. 32   **17.** 7   **18.** $\frac{1}{40}$ **19.** 24   **20.** 27 circuits
21. (a) 247.50 Canadian dollars; **(b)** 50.51 U.S. dollars
22. 832 mi   **23.** 133.5 billion gallons, or 133,500,000,000 gallons
24. About 3,774,811 lb   **25.** 320 calories   **26.** About 12,453
lawyers   **27.** $x = \frac{14}{3}$, or $4\frac{2}{3}$ **28.** $x = \frac{56}{5}$, or $11\frac{1}{5}$; $y = \frac{63}{5}$, or $12\frac{3}{5}$
29. 40 ft   **30.** $x = 3$; $y = \frac{21}{2}$, or $10\frac{1}{2}$; $z = \frac{15}{2}$, or $7\frac{1}{2}$ **31.** B
32. C   **33.** 1.806¢/sheet; 1.658¢/sheet; 1.409¢/sheet; 6 big rolls
34. $x = 4258.5$; $z \approx 10{,}094.3$ **35.** Finishing paint: 11 gal;
primer: 16.5 gal

Understanding Through Discussion and Writing

1. In terms of cost, a low faculty-to-student ratio is less expensive than a high faculty-to-student ratio. In terms of quality of education and student satisfaction, a high faculty-to-student ratio is more desirable. A college president must balance the cost and quality issues.   **2.** Yes; unit prices can be used to solve proportions involving money. In Example 3 of Section 7.4, for instance, we could have divided $90 by the unit price, or the price per ticket, to find the number of tickets that could be purchased for $90.   **3.** Leslie used 4 gal of gasoline to drive 92 mi. At the same rate, how many gallons would be needed to travel 368 mi?   **4.** Yes; consider the following pair of triangles.

Two pairs of sides are proportional, but we can see that x is shorter than y, so the ratio of x to y is clearly not the same as the ratio of 1 to 1 (or 2 to 2).

Test: Chapter 7, p. 503

1. $[7.1a] \frac{85}{97}$ **2.** $[7.1a] \frac{0.34}{124}$ **3.** $[7.1b] \frac{9}{10}$ **4.** $[7.1b] \frac{25}{32}$
5. $[7.2a]$ 0.625 ft/sec **6.** $[7.2a]$ $1\frac{1}{3}$ servings/lb **7.** $[7.2a]$ 32 mpg
8. $[7.2b]$ About 15.563¢/oz **9.** $[7.2b]$ 16.475¢/oz; 13.980¢/oz;
11.490¢/oz; 16.660¢/oz; 100 oz **10.** $[7.3a]$ Yes **11.** $[7.3a]$ No
12. $[7.3b]$ 12 **13.** $[7.3b]$ 360 **14.** $[7.3b]$ 42.1875 **17.** $[7.3b]$ 100
16. $[7.4a]$ 1512 km **17.** $[7.4a]$ 4.8 min **18.** $[7.4a]$ 525 mi
19. $[7.5a]$ 66 m **20.** $[7.4a]$ **(a)** 13 hats; **(b)** 30 packages
21. $[7.4a]$ About \$59.17 **22.** $[7.5a]$ $x = 8$; $y = 8.8$
23. $[7.5b]$ $x = \frac{24}{5}$, or 4.8; $y = \frac{32}{5}$, or 6.4; $z = 12$ **24.** $[7.2a]$ C
25. $[7.3b]$ $\frac{4}{3}$, or $1.\overline{3}$ **26.** $[7.4a]$ 5888 marbles

Cumulative Review: Chapters 1–7, p. 505

1. $[5.2a]$ 513.996 **2.** $[4.6a]$ $6\frac{3}{4}$ **3.** $[4.2b]$ $\frac{7}{20}$ **4.** $[5.2b]$ 30.491
5. $[2.3a]$ -17 **6.** $[4.3a]$ $-\frac{7}{60}$ **7.** $[5.3a]$ 222.076 **8.** $[2.4a]$ -645
9. $[4.7a]$ 3 **10.** $[5.4a]$ 43 **11.** $[2.5a]$ 51 **12.** $[3.7b]$ $\frac{3}{2}$
13. $[1.1b]$ 3 ten thousands + 0 thousands + 0 hundreds +
7 tens + 4 ones, or 3 ten thousands + 7 tens + 4 ones
14. $[5.1a]$ One hundred twenty and seven hundredths **15.** $[5.1c]$ 0.7
16. $[5.1c]$ -0.799 **17.** $[3.2c]$ $2 \cdot 2 \cdot 2 \cdot 3 \cdot 3$, or $2^4 \cdot 3^2$
18. $[4.1a]$ 90 **19.** $[3.3a]$ $\frac{5}{8}$ **20.** $[3.5b]$ $\frac{5}{8}$ **21.** $[5.5d]$ 5.718
22. $[5.4b]$ -25.56 **23.** $[2.6a]$ 5 **24.** $[7.1a]$ $\frac{0.3}{15}$
25. $[6.5a]$ 48.75 **26.** $[7.3a]$ yes **27.** $[7.2a]$ 55m/sec
28. $[7.2b]$ 8-oz can **29.** $[7.3b]$ 30.24 **30.** $[5.7a]$ -26.4375
31. $[3.8a]$ $\frac{8}{9}$ **32.** $[2.8d]$ -4 **33.** $[5.7a]$ 33.34
34. $[5.7b]$ 3 **35.** $[3.6b]$ 390 cal **36.** $[7.4a]$ 7 min
37. $[4.6c]$ $2\frac{1}{4}$ cups **38.** $[3.8b]$ 12 doors
39. $[5.8b]$, $[7.2a]$ 42.2025 mi **40.** $[5.8b]$ 132 orbits
41. $[3.2b]$ D **42.** $[7.2a]$ B **43.** $[7.5a]$ $10\frac{1}{2}$ ft

CHAPTER 8

Calculator Corner, p. 510

1. 0.14 **2.** 0.00069 **3.** 0.438 **4.** 1.25

Calculator Corner, p. 513

1. 52% **2.** 38.46% **3.** 110.26% **4.** 171.43% **5.** 59.62%
6. 28.31%

Exercise Set 8.1, p. 515

RC1. 43% **RC2.** 86% **RC3.** 19% **RC4.** 50%
1. $\frac{90}{100}$; $90 \times \frac{1}{100}$; 90×0.01 **3.** $\frac{12.5}{100}$; $12.5 \times \frac{1}{100}$; 12.5×0.01
5. 0.67 **7.** 0.456 **9.** 0.5901 **11.** 0.1 **13.** 0.01 **15.** 2
17. 0.001 **19.** 0.0009 **21.** 0.0018 **23.** 0.2319 **25.** 0.565
27. 0.14875 **29.** 0.13; 0.11 **31.** 0.07; 0.08 **33.** 0.139; 0.4
35. 47% **37.** 3% **39.** 870% **41.** 33.4% **43.** 1200%
45. 40% **47.** 0.6% **49.** 1.7% **51.** 27.18% **53.** 2.39%
55. 34% **57.** 65.1% **59.** 35.9%; 20% **61.** 41% **63.** 5%
65. 20% **67.** 30% **69.** 50% **71.** 87.5%, or $87\frac{1}{2}$% **73.** 80%
75. $66.\overline{6}$%, or $66\frac{2}{3}$% **77.** $16.\overline{6}$%, or $16\frac{2}{3}$% **79.** 18.75%, or $18\frac{3}{4}$%
81. 81.25%, or $81\frac{1}{4}$% **83.** 16% **85.** 5% **87.** 34%
89. 55%; 70% **91.** 22% **93.** 12% **95.** 15% **97.** $\frac{17}{20}$ **99.** $\frac{5}{8}$
101. $\frac{1}{3}$ **103.** $\frac{1}{6}$ **105.** $\frac{29}{400}$ **107.** $\frac{1}{125}$ **109.** $\frac{203}{800}$ **111.** $\frac{176}{225}$
113. $\frac{711}{1100}$ **115.** $\frac{3}{2}$ **117.** $\frac{13}{40,000}$ **119.** $\frac{1}{3}$ **121.** $\frac{3}{50}$ **123.** $\frac{13}{100}$
125. $\frac{19}{25}$ **127.** $\frac{3}{20}$ **129.** $\frac{103}{500}$

131.

Fraction Notation	Decimal Notation	Percent Notation
$\frac{1}{8}$	0.125	12.5%, or $12\frac{1}{2}$%
$\frac{1}{6}$	$0.1\overline{6}$	$16.\overline{6}$%, or $16\frac{2}{3}$%
$\frac{1}{5}$	0.2	20%
$\frac{1}{4}$	0.25	25%
$\frac{1}{3}$	$0.\overline{3}$	$33.\overline{3}$%, or $33\frac{1}{3}$%
$\frac{3}{8}$	0.375	37.5%, or $37\frac{1}{2}$%
$\frac{2}{5}$	0.4	40%
$\frac{1}{2}$	0.5	50%

133.

Fraction Notation	Decimal Notation	Percent Notation
$\frac{1}{2}$	0.5	50%
$\frac{1}{3}$	$0.\overline{3}$	$33.\overline{3}$%, or $33\frac{1}{3}$%
$\frac{1}{4}$	0.25	25%
$\frac{1}{6}$	$0.1\overline{6}$	$16.\overline{6}$%, or $16\frac{2}{3}$%
$\frac{1}{8}$	0.125	12.5%, or $12\frac{1}{2}$%
$\frac{3}{4}$	0.75	75%
$\frac{5}{6}$	$0.8\overline{3}$	$83.\overline{3}$%, or $83\frac{1}{3}$%
$\frac{3}{8}$	0.375	37.5%, or $37\frac{1}{2}$%

135. 70 **136.** 5 **137.** 400 **138.** 18.75 **139.** 23.125
140. 25.5 **141.** 4.5 **142.** 8.75 **143.** 20% **145.** $11.\overline{1}$%
147. $0.01\overline{5}$
149.

Food 22% ; Entertainment 22% ; Clothing 34% ; Other 22%

Calculator Corner, p. 524

1. \$5.04 **2.** 0.0112 **3.** 450 **4.** \$1000 **5.** 2.5% **6.** 12%

Exercise Set 8.2, p. 525

RC1. (c)   **RC2.** (e)   **RC3.** (f)   **RC4.** (d)   **RC5.** (b)
RC6. (a)

1. $a = 0.32 \cdot 78$ **3.** $89 = p \cdot 99$ **5.** $13 = 0.25 \cdot b$ **7.** 234.6
9. 45   **11.** $18   **13.** 1.9   **15.** 78%   **17.** 200%   **19.** 50%
21. 125%   **23.** 40   **25.** $40   **27.** 88   **29.** 20   **31.** 6.25
33. $846.60   **35.** 1216   **37.** $\frac{9}{100}$ **38.** $\frac{179}{100}$ **39.** $\frac{875}{1000}$, or $\frac{7}{8}$
40. $\frac{125}{1000}$, or $\frac{1}{8}$ **41.** $\frac{9375}{10,000}$, or $\frac{15}{16}$ **42.** $\frac{6875}{10,000}$, or $\frac{11}{16}$ **43.** 0.89
44. 0.07   **45.** 0.3   **46.** 0.017   **47.** $800 (can vary); $843.20
49. 108 to 135 tons   **51.** $1875

Exercise Set 8.3, p. 531

RC1. (b)   **RC2.** (e)   **RC3.** (c)   **RC4.** (f)   **RC5.** (d)
RC6. (a)

1. $\frac{37}{100} = \frac{a}{74}$ **3.** $\frac{N}{100} = \frac{4.3}{5.9}$ **5.** $\frac{25}{100} = \frac{14}{b}$ **7.** 68.4   **9.** 462
11. 40   **13.** 2.88   **15.** 25%   **17.** 102%   **19.** 25%
21. 93.75%, or $93\frac{3}{4}$%   **23.** $72   **25.** 90   **27.** 88   **29.** 20
31. 25   **33.** $780.20   **35.** 200   **37.** 8   **38.** 4000   **39.** 15
40. 2074   **41.** $\frac{43}{48}$ qt   **42.** $\frac{1}{8}$ T   **43.** $1170 (can vary); $1118.64
45. 20%   **47.** 39%

Mid-Chapter Review: Chapter 8, p. 533

1. True   **2.** False   **3.** True   **4.** $\frac{1}{2}$% = $\frac{1}{2} \cdot \frac{1}{100} = \frac{1}{200}$
5. $\frac{80}{1000} = \frac{8}{100} = 8\%$ **6.** $5.5\% = \frac{5.5}{100} = \frac{55}{1000} = \frac{11}{200}$
7. $0.375 = \frac{375}{1000} = \frac{37.5}{100} = 37.5\%$
8. $15 = p \cdot 80$
$\dfrac{15}{80} = \dfrac{p \cdot 80}{80}$
$\dfrac{15}{80} = p$
$0.1875 = p$
$18.75\% = p$
9. 0.28   **10.** 0.0015   **11.** 0.05375   **12.** 2.4   **13.** 71%
14. 9%   **15.** 38.91%   **16.** 18.75%, or $18\frac{3}{4}$%   **17.** 0.5%
18. 74%   **19.** 600%   **20.** 83.3%, or $83\frac{1}{3}$%   **21.** $\frac{17}{20}$ **22.** $\frac{3}{6250}$
23. $\frac{91}{400}$ **24.** $\frac{1}{6}$ **25.** 62.5%, or $62\frac{1}{2}$%   **26.** 45%   **27.** 58
28. 16.6%, or $16\frac{2}{3}$%   **29.** 2560   **30.** $50   **31.** 20%   **32.** $455
33. 0.05%, 0.1%, $\frac{1}{2}$%, 1%, 5%, 10%, $\frac{13}{100}$, 0.275, $\frac{3}{10}$, $\frac{7}{20}$ **34.** D   **35.** B
36. Some will say that the conversion will be done most accurately by first finding decimal notation. Others will say that it is more efficient to become familiar with some or all of the fraction and percent equivalents that appear inside the back cover and to make the conversion by going directly from fraction notation to percent notation.   **37.** Since $40\% \div 10 = 4\%$, we can divide 36.8 by 10, obtaining 3.68. Since $400\% = 40\% \times 10$, we can multiply 36.8 by 10, obtaining 368.   **38.** Answers may vary. Some will say this is a good idea since it makes the computations in the solution easier. Others will say it is a poor idea since it adds an extra step to the solution.   **39.** They all represent the same number.

Exercise Set 8.4, p. 538

RC1. True   **RC2.** True   **RC3.** True   **RC4.** False
1. Europe: about 282 million users; United States and Canada: about 159 million users   **3.** $46,656   **5.** 140 items
7. About 940,000,000 acres   **9.** $36,400   **11.** 74.4 items correct; 5.6 items incorrect   **13.** 261 men   **15.** About 27.8%
17. $230.10   **19.** Alcohol: 43.2 mL; water: 496.8 mL
21. About 29.8%   **23.** Civic: 5.5%; educational: 26.5%; health: 8%; religious: 34.2%; social service: 14.4%; sports/arts: 3.7%
25. $2.\overline{27}$ **26.** 0.44   **27.** 3.375   **28.** $4.\overline{7}$ **29.** 0.92   **30.** $0.8\overline{3}$
31. 0.4375   **32.** 2.317   **33.** 3.4809   **34.** 0.675   **35.** $42
37. About 5 ft 6 in.

Calculator Corner, p. 542

1. Left to the student   **2.** $80,040

Translating for Success, p. 545

1. J   **2.** M   **3.** N   **4.** E   **5.** G   **6.** H   **7.** O   **8.** C
9. D   **10.** B

Exercise Set 8.5, p. 546

RC1. $10, $\frac{\$10}{\$50}$, 20%   **RC2.** $15, $\frac{\$15}{\$60}$, 25%   **RC3.** $120, $\frac{\$120}{\$360}$, $33\frac{1}{3}$%
RC4. $1600, $\frac{\$1600}{\$4000}$, 40%
1. 5%   **3.** 15%   **5.** About 93.5%   **7.** About 34.7%
9. About 127%   **11.** About 497%   **13.** 34.375%, or $34\frac{3}{8}$%
15. 16,914; 2.8%   **17.** 5,130,632; 24.6%   **19.** 1,567,582; 21.1%
21. 0.4%   **23.** 32 cm   **24.** 33 ft   **25.** 36 in.   **26.** 36 m
27. Wyoming   **29.** About 19%

Exercise Set 8.6, p. 554

RC1. rate   **RC2.** rate   **RC3.** price   **RC4.** tax   **RC5.** tax
1. $9.56   **3.** $6.66   **5.** $11.99; $171.79   **7.** 4%   **9.** 2%
11. $5600   **13.** $116.72   **15.** $276.28   **17.** $194.08
19. $2625   **21.** 12%   **23.** $185,000   **25.** $5440   **27.** 15%
29. $355   **31.** $30; $270   **33.** $2.55; $14.45   **35.** $125; $112.50
37. 40%; $360   **39.** $300; 40%   **41.** $849; 21.2%   **43.** 18
44. $\frac{22}{7}$ **45.** 265.625   **46.** 1.15   **47.** 4,030,000,000,000
48. 5,800,000

49.   **50.**

51. $17,700

Calculator Corner, p. 561

1. $16,357.18   **2.** $12,764.72

Exercise Set 8.7, p. 564

RC1. $\frac{6}{12}$ year   **RC2.** $\frac{40}{365}$ year   **RC3.** $\frac{285}{365}$ year   **RC4.** $\frac{9}{12}$ year
RC5. $\frac{3}{12}$ year   **RC6.** $\frac{4}{12}$ year
1. $8   **3.** $113.52   **5.** $462.50   **7.** 671.88   **9.** (a) $147.95;
(b) $10,147.95   **11.** (a) $84.14; (b) $6584.14   **13.** (a) $46.03;
(b) $5646.03   **15.** $441   **17.** $2802.50   **19.** $7853.38
21. $99,427.40   **23.** $4243.60   **25.** $28,225.00   **27.** $9270.87
29. $129,871.09   **31.** $4101.01   **33.** $1324.58   **35.** $20,165.05
37. Interest: $20.88; amount applied to principal: $4.69; balance after the payment: $1273.87   **39.** (a) $98; (b) interest: $86.56; amount applied to principal: $11.44; (c) interest: $51.20; amount applied to principal: $46.80; (d) At 12.6%, the principal is reduced by $35.36 more than at the 21.3% rate. The interest at 12.6% is $35.36 less than at 21.3%.   **41.** 800   **42.** $2 \cdot 2 \cdot 3 \cdot 19$ **43.** $\frac{9}{100}$
44. $\frac{32}{75}$ **45.** $\frac{1}{25}$ **46.** $\frac{7}{300}$ **47.** 55   **48.** $\frac{1}{2}$ **49.** 9.38%

Summary and Review: Chapter 8, p. 567

Vocabulary Reinforcement

1. percent decrease   **2.** simple   **3.** discount   **4.** sales
5. rate   **6.** percent increase

Concept Reinforcement
1. True **2.** False **3.** True

Study Guide
1. 8.2% **2.** 0.62625 **3.** $63.\overline{63}$%, or $63\frac{7}{11}$% **4.** $\frac{17}{250}$
5. $4.1\overline{6}$%, or $4\frac{1}{6}$% **6.** 10,000 **7.** About 4.3% **8.** 6%
9. $185,000 **10.** Simple interest: $22.60; total amount due:
$2522.60 **11.** $6594.26

Review Exercises
1. 170% **2.** 6.5% **3.** 0.04; 0.131 **4.** 0.6084; 0.8536
5. 37.5%, or $37\frac{1}{2}$% **6.** $33.\overline{3}$%, or $33\frac{1}{3}$% **7.** $\frac{6}{25}$ **8.** $\frac{63}{1000}$
9. $30.6 = p \cdot 90$; 34% **10.** $63 = 0.84 \cdot b$; 75
11. $a = 0.385 \cdot 168$; 64.68 **12.** $\frac{24}{100} = \frac{16.8}{b}$; 70
13. $\frac{42}{30} = \frac{N}{100}$; 140% **14.** $\frac{10.5}{100} = \frac{a}{84}$; 8.82 **15.** 178 students;
84 students **16.** 70% **17.** 2500 mL **18.** 12% **19.** 92
20. $24 **21.** 6% **22.** 11% **23.** About 18.3% **24.** $42; $308
25. 14% **26.** $2940 **27.** $36 **28.** (a) $394.52; (b) $24,394.52
29. $121 **30.** $7575.25 **31.** $9504.80 **32.** (a) $129;
(b) interest: $100.18; amount applied to principal: $28.82;
(c) interest: $70.72; amount applied to principal: $58.28;
(d) At 13.2%, the principal is decreased by $29.46 more than at the
18.7% rate. The interest at 13.2% is $29.46 less than at 18.7%.
33. A **34.** C **35.** $66.\overline{6}$%, or $66\frac{2}{3}$%

Understanding Through Discussion and Writing
1. A 40% discount is better. When successive discounts are taken,
each is based on the previous discounted price rather than on the
original price. A 20% discount followed by a 22% discount is the
same as a 37.6% discount off the original price. **2.** Let $S =$ the
original salary. After both raises have been given, the two situations
yield the same salary: $1.05 \cdot 1.1S = 1.1 \cdot 1.05S$. However, the first
situation is better for the wage earner, because $1.1S$ is earned the
first year when a 10% raise is given while in the second situation
$1.05S$ is earned that year. **3.** No; the 10% discount was based on
the original price rather than on the sale price. **4.** For a number n,
40% of 50% of n is $0.4(0.5n)$, or $0.2n$, or 20% of n. Thus, taking 40% of
50% of a number is the same as taking 20% of the number.
5. The interest due on the 30-day loan will be $41.10 while that
due on the 60-day loan will be $131.51. This could be an argument
in favor of the 30-day loan. On the other hand, the 60-day loan puts
twice as much cash at the firm's disposal for twice as long as the 30-
day loan does. This could be an argument in favor of the 60-day loan.
6. Answers will vary.

Test: Chapter 8, p. 573
1. [8.1b] 0.613 **2.** [8.1b] 38% **3.** [8.1c] 137.5% **4.** [8.1c] $\frac{13}{20}$
5. [8.2a, b] $a = 0.40 \cdot 55$; 22 **6.** [8.3a, b] $\frac{N}{100} = \frac{65}{80}$; 81.25%
7. [8.4a] 16,550 kidney transplants; 5992 liver transplants;
2283 heart transplants **8.** [8.4a] 555 at-bats
9. [8.5a] About 18.2% **10.** [8.4a] About 59.7%
11. [8.6a] $25.20; $585.20 **12.** [8.6b] $630 **13.** [8.6c] $40; $160
14. [8.7a] $8.52 **15.** [8.7a] $5356 **16.** [8.7b] $1110.39
17. [8.7b] $11,580.07 **18.** [8.5a] Plumber: 757,000, 7.4%;
veterinary assistant: 29,000, 40.8%; motorcycle repair technician:
21,000, 14.3%; fitness professional: 235,000, 63,000 **19.** [8.6c] $50;
about 14.3% **20.** [8.7c] Interest: $36.73; amount applied to the
principal: $17.27; balance after payment: $2687 **21.** [8.2a, b],
[8.3a, b] B **22.** [8.6b] $194,600 **23.** [8.6b], [8.7b] $2546.16

Cumulative Review: Chapters 1–8, p. 575
1. [8.5a] 14.7%; 4.1% **2.** [8.1b] 26.9% **3.** [8.1c] 112.5%
4. [5.5a] $2.1\overline{6}$ **5.** [7.1a] $\frac{5}{0.5}$, or $\frac{10}{1}$ **6.** [7.2a] $\frac{70\text{ km}}{3\text{ hr}}$, or $23.\overline{3}$ km/hr,
or $23\frac{1}{3}$ km/hr **7.** [4.2c] < **8.** [4.2c] < **9.** [5.6a] 296,200
10. [1.6b] 50,000 **11.** [1.9d] 13 **12.** [2.7a] $-2x - 14$
13. [4.6a] $3\frac{1}{30}$ **14.** [5.2c] -44.06 **15.** [1.2a] 515,150
16. [5.2b] 0.02 **17.** [4.6b] $\frac{2}{3}$ **18.** [4.3a] $-\frac{110}{63}$ **19.** [3.6a] $\frac{1}{6}$
20. [2.4b] -384 **21.** [5.3a] 1.38036 **22.** [4.7b] $\frac{3}{2}$, or $1\frac{1}{2}$
23. [5.4a] -12.25 **24.** [1.5a] 123 R 5 **25.** [1.7b] 95
26. [5.7a] 8.13 **27.** [4.4a] 49 **28.** [4.3b] $\frac{1}{12}$ **29.** [5.7b] $-\frac{16}{9}$,
or $-1\frac{7}{9}$ **30.** [7.3b] $\frac{176}{21}$ **31.** [8.2a, b] About 48,000 workers
32. [1.8a] $228,161 **33.** [7.4a] About 65 games
34. [7.2b] 7.495 cents/oz **35.** [4.2d] $\frac{3}{4}$ mi, or $1\frac{1}{2}$ mi
36. [8.7b] $12,663.69 **37.** [4.7d] 5 pieces **38.** [6.3b] III
39. [6.4b]

40. [6.5a] 33.2 **41.** [6.5b] 12 **42.** [1.4b] 3600 yd^2
43. [4.3a] C **44.** [2.7b] A **45.** [8.6c] 7 discounts
46. [8.5a] 12.5% increase **47.** [4.7d], [6.5a] 33.6 mpg

CHAPTER 9

Exercise Set 9.1, p. 587
RC1. 12 **RC2.** 5280 **RC3.** 36 **RC4.** 100 **RC5.** 3
RC6. 10
1. 3 **3.** $\frac{1}{12}$ **5.** 1760 **7.** 9 **9.** 7 **11.** 16 **13.** 8800
15. $5\frac{1}{4}$, or 5.25 **17.** $2\frac{1}{5}$, or 2.2 **19.** 37,488 **21.** $2\frac{1}{2}$, or 2.5
23. 3 **25.** 1 **27.** $1\frac{1}{4}$, or 1.25 **29.** 132,000 **31.** 126,720
33. (a) 1000; (b) 0.001 **35.** (a) 10; (b) 0.1 **37.** (a) 0.01;
(b) 100 **39.** 8300 **41.** 0.98 **43.** 8.921 **45.** 0.03217
47. 28,900 **49.** 4.77 **51.** 688 **53.** 0.1 **55.** 100,000
57. 142 **59.** 0.82 **61.** 450 **63.** 0.000024 **65.** 0.688
67. 230 **69.** 48,440; 48.44 **71.** 85; 8.5 **73.** 0.027; 0.00027
75. 541,300; 54,130 **77.** 6.21 **79.** 35.56 **81.** 104.585
83. 28.67 **85.** 70.866 **87.** 0.299

	yd	cm	in.	m	mm
89.	0.2361	21.59	$8\frac{1}{2}$	0.2159	215.9
91.	55,247	5,050,000	1,988,892	50,500	50,500,000

93. $-\frac{3}{2}$ **94.** -2 **95.** $101.50 **96.** $44.50 **97.** 47%
98. 35% **99.** 56 ft; 192 ft^2 **100.** 44 ft; 120 ft^2 **101.** 1.0 m
103. 1.4 cm **105.** Length: 5400 in., or 450 ft; breadth: 900 in.,
or 75 ft; height: 540 in., or 45 ft **107.** 321,800 **109.** 21.3 mph
111. > **113.** < **115.** >

Exercise Set 9.2, p. 594
RC1. True **RC2.** False **RC3.** True **RC4.** False
RC5. True **RC6.** True
1. 144 **3.** 640 **5.** $\frac{1}{144}$ **7.** 45 **9.** 1008 **11.** 3 **13.** 198
15. 2160 **17.** 12,800 **19.** 27,878,400 **21.** 5 **23.** $\frac{1}{3}$
25. $\frac{1}{640}$ **27.** 25,792 **29.** $1\frac{1}{2}$ **31.** 19,000,000 **33.** 63,100
35. 0.065432 **37.** 0.0349 **39.** 2500 **41.** 0.4728 **43.** $240
44. $212 **45.** (a) $484.11; (b) $15,984.11 **46.** (a) $209.59;
(b) $8709.59 **47.** (a) $220.93; (b) $6620.93 **48.** (a) $37.97;

(b) $4237.97   **49.** 10.76   **51.** 1.67   **53.** 10,657,311 ft^2
55. 216 cm^2 **57.** 1 ft^2 **59.** $4\frac{1}{3}$ ft^2 **61.** 84 cm^2
63. 1701 tiles; 42%

Calculator Corner, p. 603

1. Answers will vary.   **2.** 1417.99 in.; 160,005.91 in^2
3. 1729.27 in^2 **4.** 125,663.71 ft^2

Exercise Set 9.3, p. 604

RC1. radius   **RC2.** circumference   **RC3.** circumference
RC4. area
1. 50 cm^2 **3.** 104 ft^2 **5.** 142.5 m^2 **7.** 72.45 cm^2 **9.** 144 mi^2
11. 49 m^2 **13.** $55\frac{1}{8}$ ft^2 **15.** 1.92 cm^2 **17.** 68 yd^2 **19.** 14 cm
21. $1\frac{3}{4}$ in.   **23.** 10 ft   **25.** 0.7 cm   **27.** 44 cm   **29.** $5\frac{1}{2}$ in.
31. 62.8 ft   **33.** 4.396 cm   **35.** 154 cm^2 **37.** $2\frac{13}{32}$ in^2
39. 314 ft^2 **41.** 1.5386 cm^2 **43.** 94.2 ft   **45.** About 55.04 in^2
larger   **47.** About 7930.25 mi; about 3965.13 mi   **49.** 70,650 mi^2
51. 65.94 yd^2 **53.** Maximum circumference of barrel: $8\frac{9}{14}$ in.;
minimum circumference of handle: $2\frac{86}{133}$ in.   **55.** 45.68 ft
57. 45.7 yd   **59.** 100.48 m^2 **61.** 6.9972 cm^2 **63.** 64.4214 in^2
65. 675 cm^2 **67.** 7.5 ft^2 **69.** $\frac{37}{400}$ **70.** $\frac{7}{8}$ **71.** 125%
72. $66.\overline{6}$%, or $66\frac{2}{3}$%   **73.** 5 lb   **74.** $730   **75.** 37.2875 in^2
77. Circumference   **79.** $4A$ **81.** An 8-inch square pan

Calculator Corner, p. 615

1. 1.13 cm^3 **2.** 0.22 in^3

Exercise Set 9.4, p. 616

RC1. (b)   **RC2.** (a)   **RC3.** (c)
1. 250 cm^3 **3.** 135 in^3 **5.** 75 m^3 **7.** $357\frac{1}{2}$ yd^3 **9.** 4082 ft^3
11. 376.8 cm^3 **13.** 41,580,000 yd^3 **15.** $4,186,666.\overline{6}$ in^3
17. Approximately 124.725 m^3 **19.** $1437\frac{1}{3}$ km^3 **21.** 1000; 1000
23. 59,000   **25.** 0.049   **27.** 27,300   **29.** 40   **31.** 320
33. 3   **35.** 40   **37.** 48   **39.** $1\frac{7}{8}$ **41.** 4747.68 cm^3
43. 65,417 m^3 **45.** 0.42 yd^3 **47.** 263,947,530,000 mi^3
49. 33,493 in^3 **51.** 0.477 m^3 **53.** 152,321 m^3 **55.** 5832 yd^3
57. **(a)** About 1875.63 mm^3; **(b)** about 11,253.78 mm^3
59. $24   **60.** $175   **61.** $10.68   **62.** $\frac{7}{3}$, or $2.\overline{3}$ **63.** -1
64. 448 km   **65.** 59   **66.** 10   **67.** About 346.2 ft^3
69. 6 cm by 6 cm by 6 cm   **71.** 9.424779 L   **73.** $1\frac{3}{4}$ gal; 7.5 gal;
$91\frac{1}{4}$ gal; about 28,000,000,000 gal

Mid-Chapter Review: Chapter 9, p. 621

1. False   **2.** True   **3.** False   **4.** True   **5.** True
6. $16\frac{2}{3}$ yd $= 16\frac{2}{3} \cdot 1$ yd $= \frac{50}{3} \cdot 3$ ft $= 50$ ft
7. 10,200 mm $= 10,200$ mm $\cdot \frac{1\,m}{1000\,mm} \approx 10.2$ m $\cdot \frac{3.281\,ft}{1\,m} =$
33.4662 ft
8. $C \approx 3.14 \cdot 10.2$ in.
$C \approx 32.028$ in.;
$A \approx 3.14 \cdot 5.1$ in. $\cdot 5.1$ in.
$A \approx 81.6714$ in^2
9. 9680   **10.** 70   **11.** 2.405   **12.** 0.00015   **13.** 1251   **14.** 5
15. 180,000   **16.** 700   **17.** 2.285   **18.** 118.116   **19.** 15.24
20. 160.9   **21.** 720   **22.** 90   **23.** $6\frac{2}{3}$ **24.** $\frac{1}{3}$ **25.** 87,120
26. 960   **27.** 3.5 ft, 2 yd, 100 in., $\frac{1}{100}$ mi, 1000 in., 430 ft, 6000 ft
28. 150 hm, 13 km, 310 dam, 300 m, 33,000 mm, 3240 cm, 250 dm
29. 800 in^2 **30.** 66 km^2 **31.** $C = 43.96$ in.; $A = 153.86$ in^2
32. $C = 27.004$ cm; $A = 58.0586$ cm^2 **33.** 30 cm^3
34. 1695.6 ft^3 **35.** 2.355 in^3 **36.** 7234.56 m^3 **37.** 7400
38. 62   **39.** 0.0005   **40.** 12   **41.** 9   **42.** 4
43. The student should have multiplied by $\frac{1}{12}$ (or divided by 12) to
convert inches to feet. The correct procedure is as follows:
$$23 \text{ in.} = 23 \text{ in.} \cdot \frac{1\,ft}{12\,in.} = \frac{23\,in.}{12\,in.} \cdot 1\,ft$$
$$= \frac{23}{12} \cdot \frac{in.}{in.} \cdot 1\,ft = \frac{23}{12} \cdot 1\,ft = \frac{23}{12}\,ft.$$

44. The area of a 16-in-diameter pizza is approximately
$3.14 \cdot 8$ in. $\cdot 8$ in., or 200.96 in^2. At $16.25, its unit price is
$\frac{\$16.25}{200.96\,\text{in}^2}$, or about $0.08/in^2. The area of 10-in.-diameter pizza
is approximately $3.14 \cdot 5$ in. $\cdot 5$ in., or 78.5 in^2. At $7.85, its unit
price is $\frac{\$7.85}{78.5\,\text{in}^2}$, or $0.10/in^2. Since the 16-in.-diameter pizza
has the lower unit price, it is a better buy.   **45.** A parallelogram
with base b and height h can be divided into two triangles, each with
base b and height h. Since the area of a parallelogram is given by
$A = bh$, the area of one of the triangles is given by $A = \frac{1}{2}bh$.
46. No; let $r =$ radius of the smaller circle. Then its area is
$\pi \cdot r \cdot r$, or πr^2. The radius of the larger circle is $2r$, and its area is
$\pi \cdot 2r \cdot 2r$, or $4\pi r^2$, or $4 \cdot \pi r^2$. Thus, the area of the larger circle is
4 times the area of the smaller circle.

Exercise Set 9.5, p. 631

RC1. (f)   **RC2.** (c)   **RC3.** (a)   **RC4.** (e)   **RC5.** (b)
RC6. (d)
1. Angle GHI, angle IHG, angle H, $\angle GHI$, $\angle IHG$, or $\angle H$
3. $\angle ADB$ or $\angle BDA$ **5.** 10°   **7.** 180°   **9.** 90°
11.

Arrangement of sections may vary.

13.

Arrangement of sections may vary.

15. Acute   **17.** Straight   **19.** Right   **21.** Acute   **23.** Obtuse
25. $\angle 1$ and $\angle 3$; $\angle 2$ and $\angle 4$ **27.** $\angle GME$ (or $\angle EMG$) and
$\angle AMC$ (or $\angle CMA$); $\angle AMG$ (or $\angle GMA$) and $\angle EMC$ (or $\angle CME$)
29. $m\angle 4$; $m\angle 1$ **31.** $\angle GME$ or $\angle EMG$; $\angle CME$ or $\angle EMC$
33. 79°   **35.** 23°   **37.** 32°   **39.** 61°   **41.** 177°   **43.** 41°
45. 105°   **47.** 76°   **49.** Scalene; obtuse   **51.** Scalene; right
53. Equilateral; acute   **55.** Scalene; obtuse   **57.** 46°
59. 120°   **61.** 58°   **63.** $160   **64.** $22.50   **65.** $148
66. $1116.67   **67.** $30,474.86   **68.** $413,458.31
69. $641,566.26   **70.** $684,337.34   **71.** $m\angle 2 = 67.13°$;
$m\angle 4 = 79.8°$; $m\angle 5 = 67.13°$; $m\angle 6 = 33.07°$
73. $m\angle ACB = 50°$; $m\angle CAB = 40°$; $m\angle EBC = 50°$;
$m\angle EBA = 40°$; $m\angle AEB = 100°$; $m\angle ADB = 50°$

Calculator Corner, p. 637

1. 6.6   **2.** 19.8   **3.** 1.9   **4.** 24.5   **5.** 121.2   **6.** 85.4

Translating for Success, p. 640

1. K   **2.** G   **3.** B   **4.** H   **5.** O   **6.** M   **7.** E   **8.** A
9. D   **10.** I

Exercise Set 9.6, p. 641

RC1. True **RC2.** False **RC3.** True **RC4.** False
1. $-4, 4$ **3.** $-11, 11$ **5.** $-13, 13$ **7.** $-50, 50$ **9.** 8
11. 9 **13.** 15 **15.** 25 **17.** 20 **19.** 100 **21.** 6.928
23. 2.828 **25.** 1.732 **27.** 3.464 **29.** 4.359 **31.** 10.488
33. $c = 15$ **35.** $c = \sqrt{98}; c \approx 9.899$ **37.** $a = 5$
39. $b = \sqrt{45}; b \approx 6.708$ **41.** $c = 26$ **43.** $b = 12$
45. $c = \sqrt{41}; c \approx 6.403$ **47.** $b = \sqrt{1023}; b \approx 31.984$
49. $\sqrt{644}$ ft ≈ 25.4 ft **51.** $\sqrt{8450}$ ft ≈ 91.9 ft
53. $\sqrt{34{,}541}$ m ≈ 185.9 m **55.** $\sqrt{211{,}200{,}000}$ ft $\approx 14{,}532.7$ ft
57. \$468 **58.** 187,200 **59.** About 324 students **60.** 12%
61. 8 **62.** 125 **63.** 1000 **64.** 10,000 **65.** 3 **66.** 982
67. 47.80 cm^2 **69.** 64.8 in. **71.** Width: 15.2 in.; height: 11.4 in.

Calculator Corner, p. 649

1. 41°F **2.** 122°F **3.** 20°C **4.** 45°C

Exercise Set 9.7, p. 650

RC1. True **RC2.** False **RC3.** True **RC4.** False
RC5. False **RC6.** True
1. 2000 **3.** 3 **5.** 64 **7.** 12,640 **9.** 0.1 **11.** 5
13. 134,850 tons **15.** 1000 **17.** $\frac{1}{1000}$, or 0.001 **19.** $\frac{1}{100}$, or 0.01
21. 1000 **23.** 10 **25.** 234,000 **27.** 5.2 **29.** 0.000897
31. 7320 **33.** 8.492 **35.** 58.5 **37.** 800,000 **39.** 1000
41. 0.0034 **43.** 0.0603 **45.** 80°C **47.** 60°C **49.** 20°C
51. -10°C **53.** 180°F **55.** 140°F **57.** 10°F **59.** 40°F
61. 86°F **63.** 104°F **65.** 35.6°F **67.** 30.2°F **69.** 5432°F
71. 25°C **73.** 55°C **75.** 81.1°C **77.** -15°C **79.** 37°C
81. (a) 136°F $= 57.\overline{7}$°C, $56\frac{2}{3}$°C $= 134$°F; **(b)** 2°F
83.

84.

85.

86.

87. \$946 **88.** 201 min, or 3 hr 21 min **89.** 144 packages
91. -40° **93.** 260.6°F **95.** 0.4536 **97.** About 5.1 g/cm^3
99. (a) 109.134 g; **(b)** 9.104 g; **(c)** Golden Jubilee: 3.85 oz; Hope: 0.321 oz

Exercise Set 9.8, p. 657

RC1. True **RC2.** True **RC3.** False **RC4.** True
1. 2000 mL **3.** 0.12 g **5.** 4 tablets **7. (a)** 720 mcg; **(b)** about
189 actuations; **(c)** 6 inhalers **9.** 16 oz **11. (a)** 0.42 g;
(b) 28 doses **13.** 6 mL **15.** 1000 **17.** 0.325 **19.** 250 mcg
21. 125 mcg **23.** 0.875 mg; 875 mcg **25.** 3358 **26.** 7414
27. 854 **28.** 6334 **29.** $5x + 8$ **30.** $6x + 5$ **31.** $7t - 9$
32. $8r - 10$ **33. (a)** 9.1 g; **(b)** 162.5 mg **35.** Naproxen
sodium is the better buy. It costs 11 cents/day and the ibuprofen
costs 25 cents/day.

Summary and Review: Chapter 9, p. 660

Vocabulary Reinforcement
1. parallel **2.** sphere **3.** radius **4.** supplementary
5. scalene **6.** hypotenuse

Concept Reinforcement
1. True **2.** False **3.** False **4.** True **5.** True **6.** False

Study Guide
1. $\frac{7}{3}$, or $2\frac{1}{3}$, or $2.\overline{3}$ **2.** 0.000046 **3.** 10.94 **4.** 9 **5.** 5240
6. 15.5 m^2 **7.** 80 m^2 **8.** 37.68 in. **9.** 616 cm^2
10. 1683.3 m^3 **11.** $30\frac{6}{35}$ ft^3 **12.** 1696.537813 cm^3 **13.** 64
14. Complement: 52°; supplement: 142° **15.** 87°
16. $\sqrt{208} \approx 14.422$ **17.** 5.14 **18.** 0.00978 **19.** 154.4°F
20. 40°C **21.** 500 mL

Review Exercises
1. $3.\overline{3}$ **2.** 30 **3.** 0.17 **4.** 72 **5.** 400,000 **6.** $1\frac{1}{6}$ **7.** 218.8
8. 32.18 **9.** 80 **10.** 0.003 **11.** 12.5 **12.** 224 **13.** 0.06
14. 400 **15.** 1400 **16.** 200 **17.** 4700 **18.** 0.04 **19.** 36
20. 700,000 **21.** 7 **22.** 0.057 **23.** 31.4 m **24.** $\frac{14}{11}$ in.
25. 24 m **26.** 439.7784 yd **27.** 18 in^2 **28.** 29.82 ft^2
29. 60 cm^2 **30.** 35 mm^2 **31.** 154 ft^2 **32.** 314 cm^2
33. 1038.555 ft^2 **34.** 26.28 ft^2 **35.** 54° **36.** 180° **37.** 140°
38. 90° **39.** Acute **40.** Straight **41.** Obtuse **42.** Right
43. 49° **44.** 136° **45.** 60° **46.** Scalene **47.** Right
48. 93.6 m^3 **49.** 193.2 ft^3 **50.** 28,260 cm^3 **51.** $33\frac{37}{75}$ yd^3
52. 942 cm^3 **53.** 8 **54.** 9.110 **55.** $c = \sqrt{850}; c \approx 29.155$
56. $b = \sqrt{84}; b \approx 9.165$ **57.** $c = \sqrt{89}$ ft; $c \approx 9.434$ ft
58. $a = \sqrt{76}$ cm; $a \approx 8.718$ cm **59.** About 17.9 ft
60. About 13.42 ft **61.** 113°F **62.** 20°C **63.** 8 mL
64. 3000 mL **65.** 250 mcg **66.** B **67.** B **68.** 17.54 sec
69. 1 gal $= 128$ oz, so 1 oz of water (as capacity) weighs $\frac{8.3453}{128}$ lb,
or about 0.0652 lb. An ounce of pennies weighs $\frac{1}{16}$ lb, or 0.0625 lb.
Thus, an ounce of water (as capacity) weighs more than an ounce of
pennies. **70.** \$101.25 **71.** 104.875 laps **72.** 961.625 cm^2

Understanding Through Discussion and Writing
1. Grams are more easily converted to other units of mass than
ounces. Since 1 gram is much smaller than 1 ounce, masses that
might be expressed using fractional or decimal parts of ounces can
often be expressed by whole numbers when grams are used.
2. No, the sum of the three angles of a triangle is 180°. If you have
two 90° angles, totaling 180°, the third angle would be 0°. A triangle
can't have an angle of 0°. **3.** Since 1 m is slightly longer than
1 yd, it follows that 1 m^2 is larger than 1 yd^2. **4.** Add 90° to the
measure of the angle's complement. **5.** Show that the sum of
the squares of the lengths of the legs is the same as the square of
the length of the hypotenuse. **6.** Volume of two spheres, each
with radius r: $2(\frac{4}{3}\pi r^3) = \frac{8}{3}\pi r^3$; volume of one sphere with radius
$2r$: $\frac{4}{3}\pi(2r)^3 = \frac{32}{3}\pi r^3 = 4 \cdot \frac{8}{3}\pi r^3$. The volume of the sphere with
radius $2r$ is four times the volume of the two spheres, each with
radius r.

Test: Chapter 9, p. 669

1. [9.1a] 96 **2.** [9.1b] 2.8 **3.** [9.2a] 18 **4.** [9.1b] 5000
5. [9.1b] 0.91 **6.** [9.2b] 0.00452 **7.** [9.4b] 2.983
8. [9.7b] 3800 **9.** [9.4b] 1280 **10.** [9.4b] 690 **11.** [9.7a] 144
12. [9.7a] 8220 **13.** [9.3b] 8 cm **14.** [9.3b] 50.24 m^2
15. [9.3b] 88 ft **16.** [9.3a] 25 cm^2 **17.** [9.3a] 18 ft^2
18. [9.3b] 103.815 km^2 **19.** [9.5c] Complement: 25°;
supplement: 115° **20.** [9.5a] 90° **21.** [9.5a] 35°
22. [9.5b] Right **23.** [9.5b] Acute **24.** [9.5e] 35°

25. [9.5d] Isosceles **26.** [9.5d] Obtuse **27.** [9.4c] 420 in^3
28. [9.4a] 628 ft^3 **29.** [9.4a] 4186.$\overline{6}$ yd^3 **30.** [9.4a] 30 m^3
31. [9.6d] About 15.8 m **32.** [9.6c] $c = \sqrt{2}$; $c \approx 1.414$
33. [9.6c] $b = \sqrt{51}$; $b \approx 7.141$ **34.** [9.6a] 11 **35.** [9.7c] 0°C
36. [9.8a] A **37.** [9.5c] 112.5° **38.** [9.4c] $188.40
39. [9.4a] 0.65 ft^3 **40.** [9.4a] 0.055 ft^3

Cumulative Review: Chapters 1–9, p. 671

1. [5.3b] 9,800,000 **2.** [9.4c] 93,750 lb **3.** [4.6a] $4\frac{1}{6}$
4. [5.2b] 87.52 **5.** [2.5a] −1234 **6.** [1.9c] 1565
7. [5.5d] 49.2 **8.** [1.9d] 2 **9.** [5.1b] $\frac{1209}{1000}$ **10.** [8.1c] $\frac{17}{100}$
11. [4.2c] < **12.** [3.5c] = **13.** [9.1a] 90 **14.** [9.7a] $\frac{3}{8}$
15. [9.7c] 59 **16.** [9.4b] 87 **17.** [9.2a] 27 **18.** [9.1b] 0.17
19. [4.3b] $\frac{1}{8}$ **20.** [7.3b] $4\frac{2}{7}$ **21.** [5.7b] 1 **22.** [2.8b], [3.8a] −16
23. [2.7b], [9.3a] 380 cm; 5500 cm^2 **24.** [9.3b] 70 in.; 220 in.; 3850 in^2
25. [9.4a] $179{,}666\frac{2}{3}$ in^3 **26.** [6.5a] 58.6 **27.** [8.7a] $84
28. [8.7b] $22,376.03 **29.** [9.6d] 17 m **30.** [8.6a] 6%
31. [4.6c] $2\frac{1}{8}$ yd **32.** [5.8b] $51.46 **33.** [7.2b] The 8-qt box
34. [3.6b] $\frac{7}{20}$ km **35.** [9.5e] 30° **36.** [2.7a] $9a - 16$
37. [6.4b]

38. [9.1a], [9.4a] 94,200 ft^3 **39.** [9.1a], [9.4a] 1.342 ft^3

CHAPTER 10

Calculator Corner, p. 675

1. 4.21399177; 4.21399177 **2.** 4.768371582; 4.768371582
3. −0.2097152; −0.2097152 **4.** −0.0484002582;
−0.0484002582 **5.** 2.0736; 2.0736 **6.** 0.4932701843;
0.4932701843

Exercise Set 10.1, p. 677

RC1. 0 **RC2.** 1 **RC3.** 2 **RC4.** −1
1. 1 **3.** 3.14 **5.** −19.57 **7.** 1 **9.** 1 **11.** 97 **13.** 1
15. 17 **17.** 1 **19.** 7 **21.** 5 **23.** 2 **25.** $\frac{1}{3^2}$; $\frac{1}{9}$
27. $\frac{1}{10^4}$; $\frac{1}{10{,}000}$ **29.** $\frac{1}{t^4}$ **31.** $\frac{1}{7^1}$; $\frac{1}{7}$ **33.** $\frac{1}{(-5)^2}$; $\frac{1}{25}$
35. $\frac{1}{(-x)^2}$; $\frac{1}{x^2}$ **37.** $\frac{3}{x^7}$ **39.** $\frac{m}{n^5}$ **41.** $\frac{x}{y}$ **43.** $\frac{-7}{a^9}$
45. x^3 **47.** xy^4 **49.** r^5t^3 **51.** xyz **53.** a^8bc^2
55. $x^2y^3z^8$ **57.** $\frac{t^2y^3}{x^4}$ **59.** $\frac{ac}{d^4}$ **61.** $\frac{25}{4}$ **63.** $\frac{a^3}{125}$
65. 7^{-3} **67.** $9x^{-3}$ **69.** 210 ft **70.** 14 mpg **71.** 17.5 mpg
72. $418.95 **73.** 62.5% **74.** 37.68 cm **75.** −2736
76. 3312 **77.** 28 **79.** 159 **81.** 3; 3

Exercise Set 10.2, p. 683

RC1. (c) **RC2.** (b) **RC3.** (a) **RC4.** (e) **RC5.** (d)
1. 3^6 **3.** x^{14} **5.** a^7 **7.** 8^6 **9.** 5^{38} **11.** x^8 **13.** a^3
15. p **17.** $\frac{1}{x^{12}}$ **19.** $\frac{1}{5^6}$ **21.** a^4b^{14} **23.** $\frac{1}{a}$ **25.** 8^7
27. x **29.** a^9 **31.** $\frac{1}{x^3}$ **33.** 1 **35.** t^3 **37.** c^4 **39.** x^3y^5
41. x^{10} **43.** $\frac{1}{2^{18}}$ **45.** 7^6 **47.** $\frac{1}{y^{12}}$ **49.** x^4y^{10} **51.** $8x^6y^3$

53. $-8a^6b^{15}$ **55.** $81x^4y^4$ **57.** $\frac{a^7}{b^7}$ **59.** $\frac{a^6}{8}$ **61.** $\frac{x^{12}}{y^{15}}$
63. $\frac{2^8}{3^{10}}$ **65.** 912 m^2 **66.** $123,000 **67.** −17 **68.** −7 **69.** −7
70. −17 **71.** $(3y)^{16}$ **73.** $\frac{y}{x}$ **75.** x^y **77.** 1

Calculator Corner, p. 688

1. 1.3545×10^{-4} **2.** 3.6×10^{12} **3.** 8×10^{-26} **4.** 3×10^{13}

Translating for Success, p. 689

1. E **2.** H **3.** O **4.** A **5.** G **6.** C **7.** M **8.** I
9. K **10.** N

Exercise Set 10.3, p. 690

RC1. True **RC2.** True **RC3.** False **RC4.** True
1. 2.8×10^{10} **3.** 9.07×10^{17} **5.** 3.04×10^{-6} **7.** 1.8×10^{-8}
9. 10^{11} **11.** 4.19854×10^8 **13.** 6.8×10^{-7} **15.** 87,400,000
17. 0.00000005704 **19.** 10,000,000 **21.** 0.00001 **23.** 6×10^9
25. 3.38×10^4 **27.** 8.1477×10^{-13} **29.** 2.5×10^{13}
31. 5.0×10^{-4} **33.** 3.0×10^{-21} **35.** 1.4×10^{21} bacteria
37. 4.375×10^2 days **39.** 1.25×10^9 bytes
41. 3.1×10^2 sheets **43.** $\frac{7}{4}$, or 1.75 **44.** $-\frac{11}{2}$, or −5.5
45. −2 **46.** 9.6 **47.** 2.478125×10^{-1} **49.** 6.4×10^8

Mid-Chapter Review: Chapter 10, p. 693

1. True **2.** False **3.** False **4.** $y^2 \cdot y^{10} = y^{2+10} = y^{12}$
5. $\frac{y^2}{y^{10}} = y^{2-10} = y^{-8} = \frac{1}{y^8}$ **6.** $4{,}032{,}000{,}000{,}000 = 4.032 \times 10^{12}$
7. $0.000038 = 3.8 \times 10^{-5}$ **8.** 1 **9.** −12 **10.** −1 **11.** −18
12. $\frac{1}{x^7}$ **13.** $\frac{1}{3}$ **14.** $\frac{9}{x^6}$ **15.** x^4 **16.** 3^{15} **17.** 6^{10} **18.** x^{11}
19. $\frac{1}{y^4}$ **20.** x^6y^{11} **21.** 8^9 **22.** $\frac{1}{x}$ **23.** 1 **24.** n^5 **25.** a^8b^4
26. x^{32} **27.** y^{12} **28.** $\frac{1}{x^3}$ **29.** $\frac{1}{5^{48}}$ **30.** $x^{15}y^{20}$ **31.** $9x^4y^8$
32. $-8a^3b^{15}$ **33.** $\frac{c^{10}}{d^{10}}$ **34.** $\frac{4}{x^6}$ **35.** $\frac{a^{28}}{b^{56}}$ **36.** 9.07×10^{-4}
37. 4.3108×10^{11} **38.** 208,000,000 **39.** 0.0000041
40. 2.5×10^{-1} m **41.** 6.25×10^{-2} m, or 6.25 cm **42.** C **43.** A
44. When a is not zero, $a^0 = 1$. Thus, $a^0 > a^1$ is equivalent to
$1 > a$. So $a^0 > a^1$ for all numbers less than 1. **45.** The
expression x^{-3} is equivalent to $\frac{1}{x^3}$. This expression will be negative
whenever x is negative. **46.** Since $5^{-8} = \frac{1}{5^8}$ and $6^{-8} = \frac{1}{6^8}$, we
know that $5^{-8} > 6^{-8}$ because $5^8 < 6^8$. **47.** An answer given in
mm would require a larger exponent in scientific notation because
mm is a smaller unit than km.

Exercise Set 10.4, p. 699

RC1. False **RC2.** True **RC3.** False **RC4.** False
1. $-4x + 10$ **3.** $x^2 - 8x + 5$ **5.** $2x^2$ **7.** $6t^4 + 4t^3 + 5t^2 - t$
9. $7 + 12x^2$ **11.** $9x^8 + 8x^7 - 3x^4 + 2x^2 - 2x + 5$
13. $2t^3 - t^2 + \frac{1}{6}t + \frac{1}{3}$ **15.** 0 **17.** $13a^3b^2 + 4a^2b^2 - 4a^2b + 6ab^2$
19. $-0.6a^2bc + 17.5abc^3 - 5.2abc$ **21.** $-(-5x)$; $5x$
23. $-(-x^2 + 13x - 7)$; $x^2 - 13x + 7$
25. $-(12x^4 - 3x^3 + 3)$; $-12x^4 + 3x^3 - 3$ **27.** $-3x + 5$
29. $-4x^2 + 3x - 2$ **31.** $4x^4 - 6x^2 - \frac{3}{4}x + 8$ **33.** $7x - 1$
35. $4t^2 + 6$ **37.** $-x^2 - 9x + 5$ **39.** $-a^2 + 5a - 16$
41. $8x^4 + 3x^3 - 4x^2 + 3x - 6$ **43.** $4.6x^3 + 9.2x^2 - 3.8x - 23$

45. $\frac{3}{4}x^3 - \frac{1}{2}x$ **47.** $6x^3y^3 + 8x^2y^2 + 2x^2y + xy$ **49.** -23
51. 19   **53.** -12 **55.** 7   **57.** 4   **59.** 11   **61.** 1112 ft
63. About 449 accidents   **65.** \$18,750   **67.** \$155,000
69. About 823 min   **71.** 255 mg   **73.** $\frac{7}{10}$ serving per pound
74. \$3560   **75.** \$81   **76.** 26 m^2 **77.** 1256 cm^2 **78.** \$395 per
week   **79.** $2 \cdot 2 \cdot 2 \cdot 3 \cdot 7$ **80.** $2 \cdot 2 \cdot 2 \cdot 2 \cdot 2 \cdot 3$
81. $3 \cdot 5 \cdot 7 \cdot 7$ **82.** $3 \cdot 3 \cdot 13$ **83.** About 711 min
85.

87. $-x^2 + 280x - 5000$ **89.** $12y^2 - 23y + 21$
91. $-3y^4 - y^3 + 5y - 2$ **93.** $3x^4, 2x^3, (-7)$; order of answers
may vary.

Exercise Set 10.5, p. 708

RC1. (a)   **RC2.** (a)   **RC3.** (d)   **RC4.** (e)
1. $28a^2$ **3.** $-60x^2$ **5.** $28x^8$ **7.** $-0.07x^9$ **9.** $35x^6y^{12}$ **11.** $12a^8b^8c^2$
13. $-24x^{11}$ **15.** $-3x^2 + 21x$ **17.** $-3x^2 + 6x$ **19.** $x^5 + x^2$
21. $10x^3 - 30x^2 + 5x$ **23.** $12x^3y + 8xy^2$ **25.** $12a^7b^3 - 9a^4b^3$
27. $2(x + 4)$ **29.** $7(a - 5)$ **31.** $7(4x + 3y)$ **33.** $9(a - 3b + 9)$
35. $6(3 - m)$ **37.** $-8(2 + x - 5y)$ **39.** $9x(x^4 + 1)$
41. $a^2(a - 8)$ **43.** $2x(4x^2 - 3x + 1)$ **45.** $6a^4b^2(2b + 3a)$
47.   **48.**

49.   **50.**

                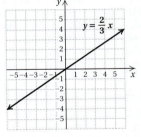

51. $23x^{299}(17x^{92} + 13)$ **53.** $7a^7b^6c^8(12b^3c^3 - 6ac^2 + 7a^2b)$

Summary and Review: Chapter 10, p. 710

Vocabulary Reinforcement
1. 1   **2.** product   **3.** scientific   **4.** reciprocal   **5.** power
6. coefficient

Concept Reinforcement
1. True   **2.** True   **3.** True

Study Guide
1. -3 **2.** 1   **3.** $\frac{1}{3^{12}}$ **4.** y^{12} **5.** x^7 **6.** 3^{12} **7.** $9x^{14}$
8. 8.03×10^{-4} **9.** 3480   **10.** $a^3 + 2a^2 - 4a - 2$
11. $-6x^2 + 5x - 3$ **12.** $14x^4$ **13.** $5x^5 + 10x^4 - 15x^3 + 45x^2$
14. $10a(2a^3 - 3)$

Review Exercises
1. 1   **2.** 46   **3.** -3 **4.** 1   **5.** $\frac{1}{12^2}; \frac{1}{144}$ **6.** $\frac{8}{a^7}$ **7.** $\frac{z^6}{x^3y^5}$
8. $\left(\frac{5}{4}\right)^2; \frac{25}{16}$ **9.** x^{-7} **10.** x^{15} **11.** x^{12} **12.** 3^{44} **13.** x^6y^{12}
14. $\frac{1}{x^8}$ **15.** uv **16.** $\frac{1}{x^{11}}$ **17.** $9x^8$ **18.** a^6 **19.** $\frac{y^{10}}{32}$
20. 4.27×10^7 **21.** 1.924×10^{-4} **22.** 1.173×10^{11}
23. 2.5×10^{-17} **24.** About 1.4×10^7 hr   **25.** $3x - 6$
26. $7x^4 - 4x^3 - x - 3$ **27.** $8a^5 + 12a^3 - a^2 + 4a + 9$
28. $5a^3b^3 + 11a^2b^3 - 7$ **29.** $-(12x^3 - 4x^2 + 9x - 3)$;
$-12x^3 + 4x^2 - 9x + 3$ **30.** -42 **31.** 56 ft   **32.** $30x^7$
33. $-18x^3y^5$ **34.** $18x^4 - 12x^2 - 3x$ **35.** $14a^7b^4 + 10a^6b^4$
36. $5x(9x^2 - 2)$ **37.** $7(a - 5b - 7ac)$ **38.** C   **39.** A
40. $-40x^8$ **41.** $13a^2b^5c^5(3ab^2c - 10c^3 + 4a^2b)$
42. $w^5x^2y^3z^3(x^4yz^2 - w^2xy^4 + wy^2z^3 - wx^5z)$
43. $2a^4b^{-5}(5 + 6a^3b^2)$

Understanding Through Discussion and Writing
1. Because $x^{-2} = \frac{1}{x^2}$ and because x^2 is never negative, it follows
that x^{-2} is never negative.   **2.** It is not true that if $a > b$, then
$a^{-1} < b^{-1}$. For example, $3 > -4$, but $\frac{1}{3} \not< -\frac{1}{4}$.   **3.** Adi is probably
adding coefficients and multiplying exponents instead of the other
way around.   **4.** Consider the expression $\frac{7^2}{7^2}$. Since this is
equivalent to $\frac{49}{49}, \frac{7^2}{7^2} = 1$. And, using the quotient rule, we have
$\frac{7^2}{7^2} = 7^{2-2} = 7^0$. Therefore, $7^0 = 1$.   **5.** Not every term is a
monomial. For example, $\frac{3}{x^2}$ is a term but not a monomial.
6. A polynomial whose coefficients are all prime may
still be factored. For example, $3x + 3y = 3(x + y)$, and
$5x^3 + 7x^2 + 13x = x(5x^2 + 7x + 13)$.

Test: Chapter 10, p. 714

1. [10.1a] 193   **2.** [10.1a] -1 **3.** [10.1a] 1   **4.** [10.1b] $\frac{1}{5^3}; \frac{1}{125}$
5. [10.1b] $\frac{5b^2}{a^3}$ **6.** [10.1b] $\left(\frac{5}{3}\right)^3; \frac{125}{27}$ **7.** [10.2a] a^{25}
8. [10.2b] $\frac{1}{x}$ **9.** [10.2d] $25x^6y^8$ **10.** [10.2d] $\frac{27}{x^{27}}$ **11.** [10.2a] $\frac{1}{x^7}$
12. [10.2b] a^6b^2 **13.** [10.3a] 4.7×10^{-4} **14.** [10.3a] 8.25×10^6
15. [10.3b] 1.824×10^{-16} **16.** [10.4a] $18a^3 - 5a^2 - a + 8$
17. [10.4b] $-(-9a^4 + 7b^2 - ab + 3); 9a^4 - 7b^2 + ab - 3$
18. [10.4c] $3x^4 - x^2 - 11$ **19.** [10.4d] 12.4 m   **20.** [10.5a] $-10x^6y^8$
21. [10.5b] $10a^3 - 8a^2 + 6a$ **22.** [10.5c] $5x^2(7x^4 - 5x + 3)$
23. [10.5c] $3(2ab - 3bc + 4ac)$ **24.** [10.3c] D
25. [10.4d] 2.92 L   **26.** [10.2a] x^{9y}

Cumulative Review/Final Examination: Chapters 1–10, p. 716

1. [5.3b] 5,100,000   **2.** [1.2a] 21,085   **3.** [4.6a] $8\frac{7}{9}$
4. [2.2a] 24   **5.** [2.2a] -762 **6.** [5.2c] -44.364
7. [10.4a] $10x^5 - x^4 + 2x^3 - 3x^2 + 2$ **8.** [1.3a] 243
9. [2.7a] $-17x$ **10.** [4.3a] $-\frac{19}{40}$ **11.** [4.6b] $2\frac{17}{24}$
12. [5.2b] 19.9973   **13.** [10.4c] $2x^3 + 5x^2 + 7x$ **14.** [1.4a] 4752
15. [2.4a] $-74,337$ **16.** [4.7a] $4\frac{7}{12}$ **17.** [3.6a] $-\frac{6}{5}$ **18.** [3.6a] 10
19. [5.3a] 259.084   **20.** [2.6b] $24x - 15$ **21.** [10.5a] $27a^8b^3$
22. [10.5b] $21x^5 - 14x^3 + 56x^2$ **23.** [1.5a] 573   **24.** [1.5a] 56 R 10

25. [3.7b] $-\dfrac{3}{2}$ **26.** [4.7b] $\dfrac{7}{90}$ **27.** [5.4a] 39   **28.** [1.9c] 75

29. [2.1c], [2.5b] -2 **30.** [1.9a] 14^3 **31.** [1.6a] 68,000

32. [5.5b] 21.84   **33.** [3.1b] Yes   **34.** [3.2a] 1, 3, 5, 15

35. [4.1a] 105   **36.** [3.5b] $\dfrac{8}{11}$ **37.** [4.5a] $-3\dfrac{3}{5}$ **38.** [2.1b] $>$

39. [4.2c] $<$ **40.** [5.1c] -0.9976 **41.** [2.6a] 29

42. [10.5c] $5(8 - t)$ **43.** [10.5c] $3a(6a^2 - 5a + 2)$ **44.** [3.3a] $\dfrac{3}{5}$

45. [5.1b] 0.0429   **46.** [5.5a] -0.52 **47.** [5.5a] $0.\overline{8}$ **48.** [8.1b] 0.07

49. [5.1b] $\dfrac{671}{100}$ **50.** [4.5a] $-\dfrac{29}{4}$ **51.** [8.1c] $\dfrac{2}{5}$ **52.** [8.1c] 85%

53. [8.1b] 150%   **54.** [5.6a] 13.6   **55.** [1.7b] 555   **56.** [5.7a] 64

57. [3.8a] $\dfrac{5}{4}$ **58.** [7.3b] 76.5   **59.** [2.8d] 5   **60.** [5.7b] $-\dfrac{2}{5}$

61. [6.2c] 248; 2010, 2011   **62.** [6.2c] 260; 2006   **63.** [6.5a, b, c] Mean: 251.75; median: 249.5; modes: 248, 249   **64.** [8.5a] About 4.6%

65. [9.1a], [9.1c] $437\dfrac{1}{3}$ yd; about 400 m   **66.** [6.6b] $\dfrac{3}{20}$, or 0.15

67. [9.5e] 118°   **68.** [10.3c] 2.5284×10^{13} mi   **69.** [4.6c] $4\dfrac{5}{8}$ yd

70. [1.8a] $24,000   **71.** [3.4c] $\dfrac{3}{10}$ km   **72.** [7.4a] 13 gal

73. [8.7a] $240   **74.** [8.5a] 30,160   **75.** [9.6d] About 283 ft

76. [9.8a] 56 mg   **77.** [10.3c] The mass of Jupiter is 3.18×10^2 times the mass of Earth.   **78.** [5.8b] 485.9 mi   **79.** [5.8b] $84.96

80. [7.2b] 17¢/oz   **81.** [5.8b] 220 mi   **82.** [8.6b] 7%

83. [10.4d] 1024 ft   **84.** [1.9b] 324   **85.** [10.1a] 1   **86.** [10.1a] 42

87. [9.6a] 11   **88.** [10.1b] $\dfrac{1}{4^3}$; $\dfrac{1}{64}$ **89.** [10.1b] $\left(\dfrac{4}{5}\right)^2$; $\dfrac{16}{25}$

90. [10.3a] 4.357×10^6 **91.** [10.3b] 2.666×10^{-15}

92. [9.1a] 12   **93.** [9.1b] 5800   **94.** [9.7a] 160   **95.** [9.4b] 8.19

96. [9.1b] 391.7   **97.** [9.7b] 60   **98.** [9.7b] 2300   **99.** [9.4b] 7

100. [9.2a] 90   **101.** [9.2b] 0.02

102. [6.2b]

103. [9.5a]

Number of Salads per Week

Arrangement of sections may vary.

104. [6.3a]

105. [6.4b]

106. [6.4c]

107. [7.5a] $x = 16, y = 14$ **108.** [2.7b] 88 in.   **109.** [9.3a] 61.6 cm²

110. [3.6b] 25 in²   **111.** [9.3a] 128.65 yd²   **112.** [9.3b] 67.390325 m²

113. [9.3b] Diameter: 20.8 in.; circumference: 65.312 in.; area: 339.6224 in²   **114.** [9.4a] 52.9 m³   **115.** [9.4a] 803.84 ft³

116. [9.4c] $523,333,333.\overline{3}$ m³   **117.** [9.6c] $\sqrt{85}$ ft; 9.220 ft

118. [9.5c] C   **119.** [9.7c] D   **120.** [6.6b] C

121. (a) [3.6a], [4.2d] Lot 1 pays $\dfrac{1}{25}$, lot 2 pays $\dfrac{9}{100}$, lot 3 pays $\dfrac{47}{300}$, lot 4 pays $\dfrac{77}{300}$, lot 5 pays $\dfrac{137}{300}$; **(b)** [8.1c], [8.4a] 4%, 9%, $15\dfrac{2}{3}$%, $25\dfrac{2}{3}$%, $45\dfrac{2}{3}$%; **(c)** [8.1c], [8.4a] 87%

DEVELOPMENTAL UNITS

Exercise Set A, p. 728

1. 17   **2.** 15   **3.** 13   **4.** 14   **5.** 12   **6.** 11   **7.** 17   **8.** 16

9. 12   **10.** 10   **11.** 10   **12.** 11   **13.** 7   **14.** 7   **15.** 11

16. 0   **17.** 3   **18.** 18   **19.** 14   **20.** 10   **21.** 4   **22.** 14

23. 11   **24.** 15   **25.** 16   **26.** 9   **27.** 13   **28.** 14   **29.** 11

30. 7   **31.** 13   **32.** 14   **33.** 12   **34.** 6   **35.** 10   **36.** 12

37. 10   **38.** 8   **39.** 2   **40.** 9   **41.** 13   **42.** 8   **43.** 10

44. 9   **45.** 10   **46.** 6   **47.** 13   **48.** 9   **49.** 8   **50.** 11

51. 12   **52.** 13   **53.** 10   **54.** 12   **55.** 17   **56.** 13   **57.** 10

58. 16   **59.** 16   **60.** 15   **61.** 39   **62.** 89   **63.** 87

64. 999   **65.** 900   **66.** 868   **67.** 999   **68.** 848   **69.** 877

70. 17,680   **71.** 10,873   **72.** 4699   **73.** 10,867   **74.** 9895

75. 3998   **76.** 18,222   **77.** 16,889   **78.** 64,489   **79.** 99,999

80. 77,777   **81.** 46   **82.** 26   **83.** 55   **84.** 101   **85.** 1643

86. 1412   **87.** 846   **88.** 628   **89.** 1204   **90.** 607

91. 10,000   **92.** 1010   **93.** 1110   **94.** 1227   **95.** 1111

96. 1717   **97.** 10,138   **98.** 6554   **99.** 6111   **100.** 8427

101. 9890   **102.** 11,612   **103.** 11,125   **104.** 15,543

105. 16,774   **106.** 68,675   **107.** 34,437   **108.** 166,444

109. 101,315   **110.** 49,449

Exercise Set S, p. 735

1. 7   **2.** 0   **3.** 0   **4.** 5   **5.** 3   **6.** 8   **7.** 8   **8.** 6   **9.** 7

10. 3   **11.** 7   **12.** 9   **13.** 6   **14.** 6   **15.** 2   **16.** 4

17. 3   **18.** 2   **19.** 0   **20.** 3   **21.** 1   **22.** 1   **23.** 7

24. 0   **25.** 4   **26.** 4   **27.** 5   **28.** 4   **29.** 4   **30.** 5

31. 9   **32.** 9   **33.** 7   **34.** 9   **35.** 6   **36.** 6   **37.** 2

38. 6   **39.** 7   **40.** 8   **41.** 33   **42.** 21   **43.** 247   **44.** 193

45. 500   **46.** 654   **47.** 202   **48.** 617   **49.** 288   **50.** 1220

51. 2231   **52.** 4126   **53.** 3764   **54.** 1691   **55.** 1226

56. 99,998   **57.** 10   **58.** 10,101   **59.** 11,043   **60.** 11,111

61. 65   **62.** 29   **63.** 8   **64.** 9   **65.** 308   **66.** 126

67. 617   **68.** 214   **69.** 89   **70.** 4402   **71.** 3555

72. 5889   **73.** 2387   **74.** 3649   **75.** 3832   **76.** 8144

77. 7750   **78.** 10,445   **79.** 33,793   **80.** 281   **81.** 455

82. 571   **83.** 6148   **84.** 2200   **85.** 2113   **86.** 3748

87. 5206   **88.** 1459   **89.** 305   **90.** 4455

Exercise Set M, p. 743

1. 12   **2.** 0   **3.** 7   **4.** 0   **5.** 10   **6.** 30   **7.** 10   **8.** 63

9. 54   **10.** 12   **11.** 0   **12.** 72   **13.** 8   **14.** 0   **15.** 28

16. 24   17. 45   18. 18   19. 0   20. 35   21. 45   22. 40
23. 0   24. 16   25. 25   26. 81   27. 1   28. 0   29. 4
30. 36   31. 8   32. 0   33. 27   34. 18   35. 0   36. 10
37. 48   38. 54   39. 0   40. 72   41. 15   42. 8   43. 9
44. 2   45. 32   46. 6   47. 15   48. 6   49. 8   50. 20
51. 20   52. 16   53. 10   54. 0   55. 80   56. 70   57. 160
58. 210   59. 450   60. 780   61. 560   62. 360   63. 800
64. 300   65. 900   66. 1000   67. 345,700   68. 1200
69. 4900   70. 4000   71. 10,000   72. 7000   73. 9000
74. 2000   75. 457,000   76. 6,769,000   77. 18,000
78. 20,000   79. 48,000   80. 16,000   81. 6000
82. 1,000,000   83. 1200   84. 200   85. 4000   86. 2500
87. 12,000   88. 6000   89. 63,000   90. 120,000
91. 800,000   92. 120,000   93. 16,000,000   94. 80,000
95. 147   96. 444   97. 2965   98. 4872   99. 6293
100. 3460   101. 16,236   102. 13,508   103. 87,554
104. 195,384   105. 3480   106. 2790   107. 3360
108. 7020   109. 20,760   110. 10,680   111. 358,800
112. 109,800   113. 583,800   114. 299,700   115. 11,346,000
116. 23,390,000   117. 61,092,000   118. 73,032,000

Exercise Set D, p. 749

1. 3   2. 8   3. 4   4. 1   5. 32   6. 9   7. 7   8. 5
9. 37   10. 5   11. 9   12. 4   13. 6   14. 9   15. 5
16. 8   17. 9   18. 6   19. 3   20. 2   21. 9   22. 2
23. 3   24. 7   25. 8   26. 4   27. 7   28. 2   29. 3
30. 6   31. 1   32. 4   33. 6   34. 3   35. 7   36. 9
37. 0   38. Undefined   39. Undefined   40. 7   41. 8
42. 7   43. 1   44. 5   45. 0   46. 0   47. 3   48. 1
49. 4   50. 7   51. 1   52. 6   53. 1   54. 5   55. 2
56. 8   57. 0   58. 0   59. 8   60. 3   61. 4
62. Undefined   63. 4   64. 1   65. 69 R 1   66. 199 R 1
67. 92 R 1   68. 138 R 3   69. 21   70. 118   71. 91 R 1
72. 321 R 2   73. 1723 R 4   74. 2925   75. 864 R 1
76. 522 R 3   77. 964 R 3   78. 679 R 5   79. 2897
80. 1328 R 2   81. 43 R 15   82. 32 R 27   83. 137 R 43
84. 248 R 26   85. 37 R 8   86. 76   87. 526   88. 164 R 2

Guided Solutions

CHAPTER 1

Section 1.1

8. $2718 = 2$ thousands $+ 7$ hundreds $+ 1$ ten $+ 8$ ones

17. One million, eight hundred seventy-nine thousand, two hundred four

Section 1.2

2.
$$
\begin{array}{r}
\overset{1\ \ 1\ \ 1}{7\ 9\ 6\ 8} \\
+\ 5\ 4\ 9\ 7 \\
\hline
1\ 3,4\ 6\ 5
\end{array}
$$

5. Perimeter $= 4$ in. $+ 5$ in. $+ 9$ in. $+ 6$ in. $+ 5$ in. $= 29$ in.

Section 1.3

1.
$$
\begin{array}{r}
7\ 8\ 9\ 3 \\
-\ 4\ 0\ 9\ 2 \\
\hline
3\ 8\ 0\ 1
\end{array}
$$
Check:
$$
\begin{array}{r}
3\ 8\ 0\ 1 \\
+\ 4\ 0\ 9\ 2 \\
\hline
7\ 8\ 9\ 3
\end{array}
$$

5.
$$
\begin{array}{r}
\overset{4\ \ 9\ 13}{5\ \cancel{0}\ \cancel{3}} \\
-\ 2\ 9\ 8 \\
\hline
2\ 0\ 5
\end{array}
$$

Section 1.4

4.
$$
\begin{array}{r}
\overset{1\ \ 2\ \ 4}{1\ 3\ 4\ 8} \\
\times\quad\quad 5 \\
\hline
6\ 7\ 4\ 0
\end{array}
$$

20. $A = l \cdot w$
$= 12 \text{ ft} \cdot 8 \text{ ft}$
$= 96 \text{ sq ft}$

Section 1.5

8. $0 \div 2$ means 0 divided by 2.
Since zero divided by any nonzero number is 0, $0 \div 2 = 0$.

9. $7 \div 0$ means 7 divided by 0.
Since division by 0 is not defined, $7 \div 0$ is not defined.

Section 1.6

26. Nearest ten:
$$
\begin{array}{r}
8\ 4\ 0 \\
\times\ 2\ 5\ 0 \\
\hline
4\ 2\ 0\ 0\ 0 \\
1\ 6\ 8\ 0\ 0\ 0 \\
\hline
2\ 1\ 0,0\ 0\ 0
\end{array}
$$

Nearest hundred:
$$
\begin{array}{r}
8\ 0\ 0 \\
\times\ 2\ 0\ 0 \\
\hline
1\ 6\ 0,0\ 0\ 0
\end{array}
$$

31. Since 8 is to the left of 12 on the number line, $8 < 12$.

Section 1.7

13.
$$
\begin{aligned}
x + 9 &= 17 \\
x + 9 - 9 &= 17 - 9 \\
x &= 8
\end{aligned}
$$
Check:
$$
\begin{array}{c}
x + 9 \stackrel{?}{=} 17 \\
8 + 9 \ ? \ 17 \\
17
\end{array}
$$
Since $17 = 17$ is true, the answer checks.
The solution is 8.

19.
$$
\frac{144}{9} = \frac{9 \cdot n}{9}
$$
$$
16 = n
$$
Check:
$$
\begin{array}{c}
144 \stackrel{?}{=} 9 \cdot n \\
144 \ ? \ 9 \cdot 16 \\
144
\end{array}
$$
Since $144 = 144$ is true, the answer checks.
The solution is 16.

Section 1.8

4. 1. Familiarize. Let $p =$ the number of pages William still has to read.

2. Translate.

Pages already read	plus	Number of pages to read	is	Total number of pages
↓	↓	↓	↓	↓
86	+	p	=	234

3. Solve.
$$
\begin{aligned}
86 + p &= 234 \\
86 + p - 86 &= 234 - 86 \\
p &= 148
\end{aligned}
$$

4. Check. If William reads 148 more pages, he will have read a total of $86 + 148$ pages, or 234 pages.

5. State. William has 148 more pages to read.

9. 1. Familiarize. Let $x =$ the number of hundreds in 3500. Let $t =$ the time it takes to lose one pound.

2. Translate.
$$
\begin{aligned}
100 \cdot x &= 3500 \\
x \cdot 2 &= t
\end{aligned}
$$

3. Solve. From Example 7, we know that $x = 35$.
$$
\begin{aligned}
x \cdot 2 &= t \\
35 \cdot 2 &= t \\
70 &= t
\end{aligned}
$$

4. Check. Since $70 \div 2 = 35$, there are 35 groups of 2 min in 70 min. Thus you will burn $35 \times 100 = 3500$ calories.

5. State. You must swim for 70 min, or 1 hr 10 min, in order to lose one pound.

Section 1.9

5. $10^4 = 10 \cdot 10 \cdot 10 \cdot 10 = 10{,}000$

15.
$$
\begin{aligned}
&9 \times 4 - (20 + 4) \div 8 - (6 - 2) \\
&= 9 \times 4 - 24 \div 8 - 4 \\
&= 36 - 24 \div 8 - 4 \\
&= 36 - 3 - 4 \\
&= 33 - 4 \\
&= 29
\end{aligned}
$$

25.
$$
\begin{aligned}
&[18 - (2 + 7) \div 3] - (31 - 10 \times 2) \\
&= [18 - 9 \div 3] - (31 - 10 \times 2) \\
&= [18 - 3] - (31 - 20) \\
&= 15 - 11 \\
&= 4
\end{aligned}
$$

CHAPTER 2

Section 2.1

10. The distance of 18 from 0 is 18, so $|18| = 18$.

22. $-(-x) = -(-(-2))$
$= -(2) = -2$

Section 2.2

19. -12 and 12 have the same absolute value. The answer is 0.

27. Add the positive numbers:

$$25 + 10 = 35.$$

Add the negative numbers:

$$-15 + (-5) + (-9) + (-14) = -43.$$

Finally, add the results:

$$35 + -43 = -8.$$

Section 2.3

11. $2 - 8 = 2 + (-8) = -6$

17. $-6 - (-2) - (-4) - 12 + 3 = -6 + 2 + 4 + (-12) + 3$
$= -6 + (-12) + 2 + 4 + 3$
$= -18 + 9$
$= -9$

Section 2.4

9. Multiply absolute values:

$$3 \cdot 6 = 18.$$

The signs are different, so the answer is negative.

$$3(-6) = -18$$

17. $(-4)(-5)(-2)(-3)(-1) = 20 \cdot 6 \cdot (-1)$
$= 120 \cdot (-1)$
$= -120$

Section 2.5

11. $(-2) \cdot |3 - 2^2| + 5 = (-2) \cdot |3 - 4| + 5$
$= (-2) \cdot |-1| + 5$
$= (-2) \cdot 1 + 5$
$= -2 + 5$
$= 3$

Section 2.6

4. $\dfrac{-6}{x} = -\dfrac{6}{x} = \dfrac{6}{-x}$

17. $6(x + y + z) = 6 \cdot x + 6 \cdot y + 6 \cdot z$
$= 6x + 6y + 6z$

Section 2.7

7. The like terms are

$$4m \text{ and } m,$$
$$-2n^2 \text{ and } n^2,$$
$$5 \text{ and } -9.$$

$4m + (-2n^2) + 5 + n^2 + m + (-9)$
$= 4m + m + (-2n^2) + n^2 + 5 + (-9)$
$= 5m + (-n^2) + (-4)$
$= 5m - n^2 - 4$

13. $P = 4s$
$= 4 \cdot 9$ in.
$= 36$ in.

Section 2.8

12. $-t = -3$
$-1 \cdot t = -3$
$\dfrac{-1 \cdot t}{-1} = \dfrac{-3}{-1}$
$t = 3$

18. $2x - 9 = 43$
$2x - 9 + 9 = 43 + 9$
$2x = 52$
$\dfrac{2x}{2} = \dfrac{52}{2}$
$x = 26$

CHAPTER 3

Section 3.1

4.
$$\begin{array}{r} 8 \\ 2\overline{)16} \\ \underline{16} \\ 0 \end{array}$$

Since the remainder is 0, 16 is divisible by 2.

22. Add the digits:

$$1 + 7 + 2 + 1 + 6 = 17.$$

Since 17 is not divisible by 3, the number 17,216 is not divisible by 3.

Section 3.2

4. 1 is a factor of 45. $1 \cdot 45$
2 is not a factor of 45.
3 is a factor of 45. $3 \cdot 15$
4 is not a factor of 45.
5 is a factor of 45. $5 \cdot 9$
6 is not a factor of 45.
7 is not a factor of 45.
8 is not a factor of 45.
Factors of 45: 1, 3, 5, 9, 15, 45.

9.
$$\begin{array}{r} 7 \\ 7\overline{)49} \\ 2\overline{)98} \end{array}$$
$98 = 2 \cdot 7 \cdot 7$

Section 3.3

12. Each gallon is divided into 4 equal parts.

The unit is $\dfrac{1}{4}$.

There are 7 equal units shaded.

The part that is shaded is $\dfrac{7}{4}$.

24. $\dfrac{4 - 4}{567} = \dfrac{0}{567} = 0$

Section 3.4

9. $\dfrac{3}{8} \cdot \dfrac{5}{7} = \dfrac{3 \cdot 5}{8 \cdot 7}$
$= \dfrac{15}{56}$

Section 3.5

5. $\dfrac{4}{3} = \dfrac{4}{3} \cdot \dfrac{5}{5}$
$= \dfrac{4 \cdot 5}{3 \cdot 5}$
$= \dfrac{20}{15}$

25. $2 \cdot 20 = 40 \qquad 3 \cdot 14 = 42$

$$\frac{2}{3} \; \square \; \frac{14}{20}$$

Since $40 \neq 42, \dfrac{2}{3} \neq \dfrac{14}{20}$.

Section 3.6

1. $\dfrac{2}{3} \cdot \dfrac{7}{8} = \dfrac{2 \cdot 7}{3 \cdot 8}$

$\qquad = \dfrac{2 \cdot 7}{3 \cdot 2 \cdot 2 \cdot 2}$

$\qquad = \dfrac{2}{2} \cdot \dfrac{7}{3 \cdot 2 \cdot 2}$

$\qquad = 1 \cdot \dfrac{7}{3 \cdot 2 \cdot 2}$

$\qquad = \dfrac{7}{12}$

7. $A = \dfrac{1}{2} \cdot b \cdot h$

$\qquad = \dfrac{1}{2} \cdot 11 \, \text{cm} \cdot \dfrac{12}{5} \, \text{cm}$

$\qquad = \dfrac{1 \cdot 11 \cdot 12}{2 \cdot 5} \, \text{cm}^2$

$\qquad = \dfrac{1 \cdot 11 \cdot 2 \cdot 2 \cdot 3}{2 \cdot 5} \, \text{cm}^2$

$\qquad = \dfrac{66}{5} \, \text{cm}^2$

Section 3.7

6. $\dfrac{6}{7} \div \dfrac{3}{4} = \dfrac{6}{7} \cdot \dfrac{4}{3}$

$\qquad = \dfrac{6 \cdot 4}{7 \cdot 3}$

$\qquad = \dfrac{2 \cdot 3 \cdot 2 \cdot 2}{7 \cdot 3}$

$\qquad = \dfrac{3}{3} \cdot \dfrac{2 \cdot 2 \cdot 2}{7}$

$\qquad = \dfrac{2 \cdot 2 \cdot 2}{7}$

$\qquad = \dfrac{8}{7}$

Section 3.8

1. $\dfrac{3}{2} \cdot \dfrac{2}{3}x = \dfrac{3}{2} \cdot 8$

$\qquad 1x = \dfrac{3 \cdot 8}{2}$

$\qquad x = \dfrac{3 \cdot 2 \cdot 4}{2}$

$\qquad x = 12$

4. $-\dfrac{7}{6} \cdot \left(-\dfrac{6}{7}a \right) = -\dfrac{7}{6} \cdot \dfrac{9}{14}$

$\qquad a = -\dfrac{7 \cdot 3 \cdot 3}{2 \cdot 3 \cdot 2 \cdot 7}$

$\qquad a = -\dfrac{3}{4}$

CHAPTER 4

Section 4.1

12. 1. $18 = 2 \cdot 3 \cdot 3$

$\qquad 40 = 2 \cdot 2 \cdot 2 \cdot 5$

2. Select the factorization of 40:

$\qquad 2 \cdot 2 \cdot 2 \cdot 5$.

This is not a multiple of 18. We need two factors of 3.

3. LCM $= 2 \cdot 2 \cdot 2 \cdot 5 \cdot 3 \cdot 3 = 360$

18. $5a^2 = 5 \cdot a^2$

$\qquad a^3 b = \quad a^3 \cdot b$

$\qquad$ LCM $= 5 \cdot a^3 \cdot b$, or $5a^3 b$

Section 4.2

9. The LCD is 24.

$$\frac{3}{8} + \frac{5}{6} = \frac{3}{8} \cdot 1 + \frac{5}{6} \cdot 1$$

$$= \frac{3}{8} \cdot \frac{3}{3} + \frac{5}{6} \cdot \frac{4}{4}$$

$$= \frac{9}{24} + \frac{20}{24}$$

$$= \frac{29}{24}$$

20. 1. Familiarize. Let $T =$ the total amount of berries in the salad.

2. Translate. To find the total amount, we add.

$$\frac{7}{8} + \frac{3}{4} + \frac{5}{16} = T$$

3. Solve. The LCD is 16.

$$\frac{7}{8} \cdot \frac{2}{2} + \frac{3}{4} \cdot \frac{4}{4} + \frac{5}{16} = T$$

$$\frac{14}{16} + \frac{12}{16} + \frac{5}{16} = T$$

$$\frac{31}{16} = T$$

4. Check. The answer is reasonable because it is larger than any of the individual amounts.

5. State. There are $\dfrac{31}{16}$ qt of berries in the salad.

Section 4.3

6. The LCD is 18.

$$\frac{5}{6} - \frac{1}{9} = \frac{5}{6} \cdot \frac{3}{3} - \frac{1}{9} \cdot \frac{2}{2}$$

$$= \frac{15}{18} - \frac{2}{18}$$

$$= \frac{13}{18}$$

13. $\dfrac{3}{5} + t = -\dfrac{7}{8}$

$\qquad \dfrac{3}{5} + t - \dfrac{3}{5} = -\dfrac{7}{8} - \dfrac{3}{5}$

$\qquad t + 0 = -\dfrac{7}{8} \cdot \dfrac{5}{5} - \dfrac{3}{5} \cdot \dfrac{8}{8}$

$\qquad t = -\dfrac{35}{40} - \dfrac{24}{40}$

$\qquad t = \dfrac{-35 - 24}{40}$

$\qquad t = \dfrac{-35 + (-24)}{40}$

$\qquad t = \dfrac{-59}{40} = -\dfrac{59}{40}$

Section 4.4

2.
$$\frac{1}{2}x - \frac{1}{5} = \frac{7}{10}$$
$$\frac{1}{2}x - \frac{1}{5} + \frac{1}{5} = \frac{7}{10} + \frac{1}{5}$$
$$\frac{1}{2}x = \frac{7}{10} + \frac{2}{10}$$
$$\frac{1}{2}x = \frac{9}{10}$$
$$2 \cdot \frac{1}{2}x = 2 \cdot \frac{9}{10}$$
$$1x = \frac{2 \cdot 3 \cdot 3}{2 \cdot 5}$$
$$x = \frac{9}{5}$$

6.
$$20 = 6 - \frac{2}{3}x$$
$$3(20) = 3\left(6 - \frac{2}{3}x\right)$$
$$60 = 3 \cdot 6 - \frac{3 \cdot 2}{3}x$$
$$60 = 18 - 2x$$
$$60 - 18 = 18 - 2x - 18$$
$$42 = -2x$$
$$\frac{42}{-2} = \frac{-2x}{-2}$$
$$-21 = x$$

Section 4.5

6.
$$4 \cdot 6 = 24$$
$$24 + 5 = 29$$
$$4\frac{5}{6} = \frac{29}{6}$$

16.
$$\begin{array}{r} 3 \\ 5)\overline{17} \\ \underline{15} \\ 2 \end{array}$$
$$\frac{17}{5} = 3\frac{2}{5}, \text{ so } \frac{-17}{5} = -3\frac{2}{5}$$

Section 4.6

5.
$$8\frac{2}{3} = 8\frac{4}{6}$$
$$\underline{-5\frac{1}{2}} = \underline{-5\frac{3}{6}}$$
$$3\frac{1}{6}$$

6.
$$5 = 4\frac{3}{3}$$
$$\underline{-1\frac{1}{3}} = \underline{-1\frac{1}{3}}$$
$$3\frac{2}{3}$$

Section 4.7

3.
$$-2 \cdot 6\frac{2}{5} = -\frac{2}{1} \cdot \frac{32}{5}$$
$$= -\frac{64}{5}$$
$$= -12\frac{4}{5}$$

6.
$$2\frac{1}{4} \div 1\frac{1}{5} = \frac{9}{4} \div \frac{6}{5}$$
$$= \frac{9}{4} \cdot \frac{5}{6}$$
$$= \frac{3 \cdot 3 \cdot 5}{2 \cdot 2 \cdot 2 \cdot 3}$$
$$= \frac{3}{3} \cdot \frac{3 \cdot 5}{2 \cdot 2 \cdot 2}$$
$$= \frac{15}{8}$$
$$= 1\frac{7}{8}$$

Section 4.8

2.
$$\frac{1}{3} \cdot \frac{3}{4} \div \frac{5}{8} - \frac{1}{10} = \frac{3}{12} \div \frac{5}{8} - \frac{1}{10}$$
$$= \frac{3}{12} \cdot \frac{8}{5} - \frac{1}{10}$$
$$= \frac{3 \cdot 2 \cdot 2 \cdot 2}{3 \cdot 2 \cdot 2 \cdot 5} - \frac{1}{10}$$
$$= \frac{2}{5} - \frac{1}{10}$$
$$= \frac{4}{10} - \frac{1}{10} = \frac{3}{10}$$

7.
$$\frac{\frac{10}{5}}{\frac{5}{8}} = 10 \div \frac{5}{8}$$
$$= 10 \cdot \frac{8}{5}$$
$$= \frac{10 \cdot 8}{5}$$
$$= \frac{2 \cdot 5 \cdot 8}{5 \cdot 1}$$
$$= 16$$

CHAPTER 5

Section 5.1

7. 0.896. 3 places
$$0.896 = \frac{896}{1000}$$

10. $\frac{743}{100}$ 7.43.

2 zeros 2 places
$$\frac{743}{100} = 7.43$$

Section 5.2

7.
$$\begin{array}{r} {\scriptstyle 1\ 1\ 1\ 1} \\ 4\ 5.7\ 8\ 0 \\ 2\ 4\ 6\ 7.0\ 0\ 0 \\ +\ \ \ \ \ 1.9\ 9\ 3 \\ \hline 2\ 5\ 1\ 4.7\ 7\ 3 \end{array}$$

8.
$$\begin{array}{r} {\scriptstyle 13} \\ {\scriptstyle 6\ 3\ 12} \\ 3\ 7.4\ 2\ 8 \\ -\ 2\ 6.6\ 7\ 4 \\ \hline 1\ 0.7\ 5\ 4 \end{array}$$

14.
$$\begin{array}{r} {\scriptstyle 4\ 9\ 9\ 9\ 10} \\ 5.0\ 0\ 0\ 0 \\ -\ 0.0\ 0\ 8\ 9 \\ \hline 4.9\ 9\ 1\ 1 \end{array}$$

Section 5.3

3.
$$\begin{array}{r} 4\ 2.6\ 5 \\ \times\ \ 0.8\ 0\ 4 \\ \hline 1\ 7\ 0\ 6\ 0 \\ 3\ 4\ 1\ 2\ 0\ 0\ 0 \\ \hline 3\ 4.2\ 9\ 0\ 6\ 0 \end{array}$$

15. $\$15.69 = 15.69 \times \$1$
$$= 15.69 \times 100¢$$
$$= 1569¢$$

Section 5.4

6.
$$\begin{array}{r} 0.0\ 2\ 5 \\ 8\ 6)\overline{2.1\ 5\ 0} \\ \underline{1\ 7\ 2} \\ 4\ 3\ 0 \\ \underline{4\ 3\ 0} \\ 0 \end{array}$$

7. $\dfrac{0.375}{0.25} = \dfrac{0.375}{0.25} \times \dfrac{100}{100}$

$= \dfrac{37.5}{25}$

$$0.2\,5\,)\overline{0.3\,7\,5} \quad \frac{1.5}{}$$

$$
\begin{array}{r}
1.5 \\
0.25\,\overline{)0.3\,7\,5} \\
\underline{2\,5} \\
1\,2\,5 \\
\underline{1\,2\,5} \\
0
\end{array}
$$

16. $625 \div 62.5 \times 25 \div 6250$

$= 10 \times 25 \div 6250$

$= 250 \div 6250$

$= 0.04$

Section 5.5

3. $\dfrac{1}{6} = 1 \div 6$

$$
\begin{array}{r}
0.1\,6\,6 \\
6\,\overline{)1.0\,0\,0} \\
\underline{6} \\
4\,0 \\
\underline{3\,6} \\
4\,0 \\
\underline{3\,6} \\
4
\end{array}
$$

$\dfrac{1}{6} = 0.1666\ldots = 0.1\overline{6}$

17. Method 1:

$\dfrac{3}{4} \times 0.62 = \dfrac{3}{4} \times \dfrac{0.62}{1}$

$= \dfrac{1.86}{4} = 0.465$

Method 2:

$\dfrac{3}{4} \times 0.62 = 0.75 \times 0.62$

$= 0.465$

Method 3:

$\dfrac{3}{4} \times 0.62 = \dfrac{3}{4} \cdot \dfrac{62}{100}$

$= \dfrac{186}{400}$

$= \dfrac{93}{200} = 0.465$

Section 5.7

6.
$$8 + 4x = 9x - 3$$
$$8 + 4x - 4x = 9x - 3 - 4x$$
$$8 = 5x - 3$$
$$8 + 3 = 5x - 3 + 3$$
$$11 = 5x$$
$$\dfrac{11}{5} = \dfrac{5x}{5}$$
$$2.2 = x$$

8.
$$3(x + 5) = 20 - x$$
$$3x + 15 = 20 - x$$
$$3x + 15 + x = 20 - x + x$$
$$4x + 15 = 20$$
$$4x + 15 - 15 = 20 - 15$$
$$4x = 5$$
$$\dfrac{4x}{4} = \dfrac{5}{4}$$
$$x = 1.25$$

CHAPTER 6

Section 6.1

3. The population density of Australia in 2008 was 7 people per square mile. The population density of Kenya in 2008 was 168 people per square mile.

$$7 \cdot n = 168$$
$$\dfrac{7 \cdot n}{7} = \dfrac{168}{7}$$
$$n = 24$$

In 2008, Kenya had 24 times more people per square mile than Australia did.

8. The graph shows $1\frac{1}{2}$ symbols for South America.

This represents 150 roller coasters.

The graph shows $\frac{1}{2}$ symbol for Africa.

This represents 50 roller coasters.
There are 100 more roller coasters in South America than in Africa.

Section 6.2

7. We look from left to right along a line at $400 per ounce. The points on the graph that are below this line correspond to the years 1970, 1975, 1985, 1990, 1995, and 2000.

Section 6.3

12. To plot the point $(5, 3)$, we locate 5 on the horizontal axis and go up 3 units. We are now in the first quadrant, or quadrant I.

16. We substitute 5 for x and 1 for y.

$$
\begin{array}{c|c}
y = 2x + 3 \\
\hline
1 \ ? \ 2 \cdot 5 + 3 \\
1 \ | \ 13
\end{array}
$$

Since $1 = 13$ is false, $(5, 1)$ is not a solution.

Section 6.4

1. Substitute 5 for y and solve for x.

$$x - y = 3$$
$$x - 5 = 3$$
$$x - 5 + 5 = 3 + 5$$
$$x = 8$$

One solution of $x - y = 3$ is $(8, 5)$.

Section 6.5

5. Course grade $= \dfrac{100 \cdot 15 + 92 \cdot 25 + 88 \cdot 40}{15 + 25 + 40}$

$= \dfrac{7320}{80} = 91.5$

Soha's course grade is 91.5%.

12. Rearrange the numbers in order from smallest to largest:

34, 34, 67, 68, 69, 70.

The middle numbers are 67 and 68.
The average of 67 and 68 is 67.5.
The median is 67.5.

15. Rearrange the numbers in order from smallest to largest.

13, 24, 27, 28, 67, 89.

Each number occurs one time.
There is no mode.

Section 6.6

3. Probability of landing on red or blue

$= \dfrac{\text{Number of ways to land on red or blue}}{\text{Number of ways to land on a space}}$

$= \dfrac{8}{12}$

$= \dfrac{2}{3}$

CHAPTER 7

Section 7.1

6. $\dfrac{\text{Length of shortest side}}{\text{Length of longest side}} = \dfrac{38.2}{55.5}$

9. Ratio of 3.6 to 12: $\dfrac{3.6}{12}$

Simplifying:

$\dfrac{3.6}{12} \cdot \dfrac{10}{10} = \dfrac{36}{120} = \dfrac{12 \cdot 3}{12 \cdot 10} = \dfrac{12}{12} \cdot \dfrac{3}{10} = \dfrac{3}{10}$

Section 7.2

4. $\dfrac{52\,\text{ft}}{13\,\text{sec}} = 4\,\text{ft/sec}$

6. Unit price $= \dfrac{\text{Price}}{\text{Number of units}}$

$\qquad = \dfrac{\$2.79}{26\,\text{oz}} = \dfrac{279\,\text{cents}}{26\,\text{oz}}$

$\qquad = \dfrac{279}{26}\dfrac{\text{cents}}{\text{oz}} \approx 10.731\text{¢/oz}$

Section 7.3

3. We compare cross products.

$1 \cdot 39 = 39 \qquad \dfrac{1}{2}\,?\,\dfrac{20}{39} \qquad 2 \cdot 20 = 40$

Since $39 \neq 40$, the numbers are not proportional.

8. $\dfrac{x}{9} = \dfrac{5}{4}$

$x \cdot 4 = 9 \cdot 5$

$\dfrac{x \cdot 4}{4} = \dfrac{9 \cdot 5}{4}$

$x = \dfrac{45}{4} = 11\dfrac{1}{4}$

Section 7.4

2. **1. Familiarize.** Let $p =$ the amount of paint needed, in gallons.

2. Translate. $\dfrac{4}{1600} = \dfrac{p}{6000}$

3. Solve.

$4 \cdot 6000 = 1600 \cdot p$

$15 = p$

4. Check. The cross products are the same.

5. State. For 6000 ft^2, they would need 15 gal of paint.

6. **1. Familiarize.** Let $D =$ the number of deer in the forest.

2. Translate. $\dfrac{153}{D} = \dfrac{18}{62}$

3. Solve.

$153 \cdot 62 = D \cdot 18$

$527 = D$

4. Check. The cross products are the same.

5. State. There are about 527 deer in the forest.

Section 7.5

1. The ratio of x to 20 is the same as the ratio of 9 to 12.

$\dfrac{x}{20} = \dfrac{9}{12}$

$x \cdot 12 = 20 \cdot 9$

$\dfrac{x \cdot 12}{12} = \dfrac{20 \cdot 9}{12}$

$x = \dfrac{180}{12} = 15$

5. Let $w =$ the width of an actual skylight.

$\dfrac{12}{52} = \dfrac{3}{w}$

$12 \cdot w = 52 \cdot 3$

$w = 13$

The width of an actual skylight will be 13 ft.

CHAPTER 8

Section 8.1

21. $\dfrac{19}{25} = \dfrac{19}{25} \cdot \dfrac{4}{4}$

$\qquad = \dfrac{76}{100} = 76\%$

25. $3.25\% = \dfrac{3.25}{100} = \dfrac{3.25}{100} \times \dfrac{100}{100}$

$\qquad = \dfrac{325}{10{,}000} = \dfrac{13 \times 25}{400 \times 25}$

$\qquad = \dfrac{13}{400} \times \dfrac{25}{25} = \dfrac{13}{400}$

Section 8.2

9. 20% of what is 45?

$0.20 \cdot b = 45$

$\dfrac{0.20 \cdot b}{0.20} = \dfrac{45}{0.20}$

$b = \dfrac{45}{0.2}$

$b = 225$

11. 16 is what percent of 40?

$16 = p \cdot 40$

$\dfrac{16}{40} = \dfrac{p \cdot 40}{40}$

$\dfrac{16}{40} = p$

$0.4 = p$

$40\% = p$

Section 8.3

8. $\dfrac{20}{100} = \dfrac{45}{b}$

$20 \cdot b = 100 \cdot 45$

$\dfrac{20 \cdot b}{20} = \dfrac{100 \cdot 45}{20}$

$b = \dfrac{4500}{20}$

$b = 225$

Thus, 20% of \$225 is \$45.

9. $\dfrac{64}{100} = \dfrac{a}{55}$

$64 \cdot 55 = 100 \cdot a$

$\dfrac{64 \cdot 55}{100} = \dfrac{100 \cdot a}{100}$

$\dfrac{3520}{100} = a$

$35.2 = a$

Thus, 64% of 55 is 35.2.

12. $\dfrac{12}{40} = \dfrac{N}{100}$

$12 \cdot 100 = 40 \cdot N$

$\dfrac{12 \cdot 100}{40} = \dfrac{40 \cdot N}{40}$

$\dfrac{1200}{40} = N$

$30 = N$

Thus, \$12 is 30% of \$40.

Section 8.6

2. Purchase price $= 4 \times \$18.95$

$\qquad\qquad = \$75.80$

Sales tax $= 4\% \times \$75.80$

$\qquad\quad = 0.04 \times \$75.80$

$\qquad\quad = \$3.032$

$\qquad\quad \approx \$3.03$

Total price $= \$75.80 + \$3.03$

$\qquad\qquad = \$78.83$

7. $\$2970 = 7.5\% \times S$

$\$2970 = 0.075 \cdot S$

$\dfrac{\$2970}{0.075} = \dfrac{0.075 \cdot S}{0.075}$

$\$39{,}600 = S$

Section 8.7

1. $I = P \cdot r \cdot t$
$= \$4300 \cdot 4\% \cdot 1$
$= \$4300 \cdot 0.04 \cdot 1$
$= \$172$

3. a) $I = P \cdot r \cdot t$
$= \$4800 \cdot 5\frac{1}{2}\% \cdot \dfrac{30}{365}$
$= \$4800 \cdot 0.055 \cdot \dfrac{30}{365}$
$\approx \$21.70$

b) Total amount
$= \$4800 + \$21.70$
$= \$4821.70$

CHAPTER 9

Section 9.1

6. $2\frac{5}{6}\,\text{yd} = 2\frac{5}{6} \cdot 1\,\text{yd}$
$= \dfrac{17}{6} \cdot 3\,\text{ft}$
$= \dfrac{17}{2}\,\text{ft}$
$= 8\frac{1}{2}\,\text{ft}$

8. $72\,\text{in.} = \dfrac{72\,\text{in.}}{1} \cdot \dfrac{1\,\text{ft}}{12\,\text{in.}}$
$= \dfrac{72}{12} \cdot 1\,\text{ft}$
$= 6\,\text{ft}$

24. $23\,\text{km} = 23 \cdot 1\,\text{km}$
$= 23 \cdot 1000\,\text{m}$
$= 23{,}000\,\text{m}$

29. $7814\,\text{m} = 7814\,\text{m} \cdot \dfrac{1\,\text{dam}}{10\,\text{m}}$
$= \dfrac{7814}{10} \cdot \dfrac{\text{m}}{\text{m}} \cdot 1\,\text{dam}$
$= 781.4\,\text{dam}$

Section 9.2

4. $360\,\text{in}^2 = 360\,\text{in}^2 \cdot \dfrac{1\,\text{ft}^2}{144\,\text{in}^2}$
$= \dfrac{360}{144} \cdot \dfrac{\text{in}^2}{\text{in}^2} \cdot 1\,\text{ft}^2$
$= 2.5\,\text{ft}^2$

Section 9.3

7. $C = \pi \cdot d$
$\approx 3.14 \cdot 18\,\text{in.}$
$= 56.52\,\text{in.}$

11. $A = \pi \cdot r \cdot r$
$\approx \frac{22}{7} \cdot 5\,\text{km} \cdot 5\,\text{km}$
$= \frac{22}{7} \cdot 25\,\text{km}^2$
$= \frac{550}{7}\,\text{km}^2$
$= 78\frac{4}{7}\,\text{km}^2$

Section 9.4

4. $V = \pi \cdot r^2 \cdot h$
$\approx 3.14 \cdot 5\,\text{ft} \cdot 5\,\text{ft} \cdot 10\,\text{ft}$
$= 3.14 \cdot 250\,\text{ft}^3$
$= 785\,\text{ft}^3$

6. $V = \frac{4}{3} \cdot \pi \cdot r^3$
$\approx \frac{4}{3} \cdot \frac{22}{7} \cdot (28\,\text{ft})^3$
$= \frac{4}{3} \cdot \frac{22}{7} \cdot 21{,}952\,\text{ft}^3$
$= \frac{275{,}968}{3}\,\text{ft}^3$
$= 91{,}989\frac{1}{3}\,\text{ft}^3$

8. $80\,\text{qt} = 80\,\text{qt} \cdot \dfrac{1\,\text{gal}}{4\,\text{qt}}$
$= \dfrac{80}{4} \cdot 1\,\text{gal}$
$= 20\,\text{gal}$

14. $0.97\,\text{L} = 0.97 \cdot 1\,\text{L}$
$= 0.97 \cdot 1000\,\text{mL}$
$= 970\,\text{mL}$

Section 9.5

14. $90° - 85° = 5°$
19. $180° - 71° = 109°$

Section 9.6

27. $a^2 + b^2 = c^2$
$12^2 + 5^2 = c^2$
$144 + 25 = c^2$
$169 = c^2$
$13 = c$

Section 9.7

21. $C = \frac{5}{9}(F - 32)$
$= \frac{5}{9}(95 - 32)$
$= \frac{5}{9} \cdot 63 = 35$
Thus, $95°\text{F} = 35°\text{C}$.

Section 9.8

4. $1\,\text{mcg} = 0.000001\text{g}$
$= 0.000001 \cdot 1\text{g}$
$= 0.000001 \cdot 1000\text{mg}$
$= 0.001\,\text{mg}$

CHAPTER 10

Section 10.1

14. $xy^{-2} = x \cdot y^{-2}$
$= x \cdot \dfrac{1}{y^2}$
$= \dfrac{x}{y^2}$

21. $\dfrac{a^4 b^{-7}}{c^{-3}} = \dfrac{a^4 c^3}{b^7}$

Section 10.2

10. $\dfrac{4^{-3}}{4^{10}} = 4^{-3-10}$
$= 4^{-13}$
$= \dfrac{1}{4^{13}}$

15. $(-2ab^4)^2 = (-2)^2(a)^2(b^4)^2$
$= 4a^2 b^8$

Section 10.3

10. $(9.1 \times 10^{-17})(8.2 \times 10^3)$
$= (9.1 \times 8.2)(10^{-17} \cdot 10^3)$
$= 74.62 \times 10^{-14}$
$= (7.462 \times 10^1) \times 10^{-14}$
$= 7.462 \times 10^{-13}$

12. $\dfrac{1.1 \times 10^{-4}}{2.0 \times 10^{-7}} = \dfrac{1.1}{2.0} \times \dfrac{10^{-4}}{10^{-7}}$
$= 0.55 \times 10^{-4-(-7)}$
$= 0.55 \times 10^3$
$= (5.5 \times 10^{-1}) \times 10^3$
$= 5.5 \times 10^2$

Section 10.4

2. Since $-a^2 = -1 \cdot a^2$, the coefficient of $-a^2$ is -1.

4. $(5x^2 y + 3x^2 + 4) + (2x^2 y + 4x)$
$= (5x^2 y + 2x^2 y) + 3x^2 + 4 + 4x$
$= 7x^2 y + 3x^2 + 4x + 4$

Section 10.5

2. $(-7x)(2x) = -7 \cdot x \cdot 2 \cdot x$
$= -7 \cdot 2 \cdot x \cdot x$
$= (-7 \cdot 2)(x \cdot x)$
$= -14x^2$

19. $9a^2 b = 3 \cdot 3 \cdot a \cdot a \cdot b$
$6ab^2 = 2 \cdot 3 \cdot a \cdot b \cdot b$
The largest common factor is $3 \cdot a \cdot b$.
$9a^2 b - 6ab^2 = 3ab \cdot 3a - 3ab \cdot 2b$
$= 3ab(3a - 2b)$

Glossary

A

Absolute value The distance that a number is from 0 on the number line

Acute angle An angle whose measure is greater than $0°$ and less than $90°$

Acute triangle A triangle in which all three angles are acute

Addends In addition, the numbers being added

Additive identity The number 0

Additive inverse A number's opposite; two numbers are additive inverses of each other if their sum is zero.

Additive inverse of a polynomial Two polynomials are additive inverses, or opposites, of each other if their sum is zero.

Algebraic expression A number or variable, or a collection of numbers and variables, on which operations are performed

Angle A set of points consisting of two rays (half-lines) with a common endpoint (vertex)

Area The number of square units that fill a plane region

Associative law of addition The statement that when three numbers are added, regrouping the addends gives the same sum

Associative law of multiplication The statement that when three numbers are multiplied, regrouping the factors gives the same product

Average A center point of a set of numbers found by adding the numbers and dividing by the number of items of data; also called the *mean*

Axes Two perpendicular number lines used to identify points in a plane

B

Bar graph A graphic display of data using bars proportional in length to the numbers represented

Base In exponential notation, the number being raised to a power

C

Celsius A temperature scale in which water freezes at $0°$ and boils at $100°$

Circumference The distance around a circle

Coefficient The numeric multiplier of a variable

Commission A percent of total sales paid to a salesperson

Commutative law of addition The statement that when two numbers are added, changing the order in which the numbers are added does not affect the sum

Commutative law of multiplication The statement that when two numbers are multiplied, changing the order in which the numbers are multiplied does not affect the product

Complementary angles Two angles for which the sum of their measures is $90°$

Complex fraction A fraction in which the numerator and/or denominator contains one or more fractions

Composite number A natural number, other than 1, that is not prime

Compound interest Interest computed on the sum of an original principal and the interest previously accrued by that principal

Congruent angles Two angles that have the same measure

Constant A number or letter that stands for just one number

Cross products Given an equation with a single fraction on each side, the products formed by multiplying the left numerator and the right denominator, and the left denominator and the right numerator

D

Decimal notation A representation of a number containing a decimal point

Denominator The number below the fraction bar in a fraction

Descending order When a polynomial is written with the powers of the variable decreasing as read from left to right, it is said to be in descending order.

Diameter A segment that passes through the center of a circle and has its endpoints on the circle, or the length of such a segment

Difference The result of subtracting one number from another

Digit A number 0, 1, 2, 3, 4, 5, 6, 7, 8, or 9 that names a place-value location

Discount The amount subtracted from the original price of an item to find the sale price

Distributive law The statement that multiplying a factor by the sum of two numbers gives the same result as multiplying the factor by each of the two numbers and then adding

Dividend In division, the number being divided

Divisible The number b is said to be divisible by another number a if b is a multiple of a.

Divisor In division, the number dividing another number

E

Equation A number sentence that says that the expressions on either side of the equals sign, $=$, represent the same number

Equilateral triangle A triangle in which all sides are the same length

Equivalent equations   Equations with the same solutions

Equivalent expressions   Expressions that have the same value for all allowable replacements

Equivalent fractions   Fractions that represent the same number

Evaluate   To substitute a value for each occurrence of a variable in an expression and calculate the result

Exponent   In expressions of the form a^n, the number n is an exponent.

Exponential notation   A representation of a number using a base raised to a power

Extrapolation   The process of estimating a value that goes beyond the given data

F

Factor   *Verb*: to write an equivalent expression that is a product. *Noun*: a multiplier

Factoring   Writing an expression as a product

Factorization   A number expressed as a product of two or more numbers

Fahrenheit   A temperature scale in which water freezes at 32° and boils at 212°

Fraction notation   A number written using a numerator and a denominator

G

Grade point average (GPA)   The average of the grade point values for each credit hour taken

H

Hypotenuse   In a right triangle, the side opposite the right angle

I

Inequality   A mathematical sentence using <, >, ≤, ≥, or ≠

Integers   The whole numbers and their opposites

Interest   A percentage of an amount that has been invested or borrowed

Interest rate   The percent at which interest is calculated on a principal

Interpolation   The process of estimating a value between given values

Isosceles triangle   A triangle in which two or more sides are the same length

L

Least common denominator (LCD)   The least common multiple of the denominators of two or more fractions

Least common multiple (LCM)   The smallest number that is a multiple of two or more numbers

Legs   In a right triangle, the two sides that form the right angle

Like terms   Terms that have exactly the same variable factors

Line graph   A graph in which quantities are represented as points connected by straight-line segments

Linear equation   Any equation that can be written in the form $Ax + By = C$, where x and y are variables

M

Marked price   The original price of an item

Mean   A center point of a set of numbers found by adding the numbers and dividing the sum of the numbers by the number of items in the set; also called the *average*

Median   In a set of data listed in order from smallest to largest, the middle number if there is an odd number of data items, or the average of the two middle numbers if there is an even number of data items

Minuend   The number from which another number is being subtracted

Mixed numeral   A number represented by a whole number and a fraction less than 1

Mode   The number or numbers that occur most often in a set of data

Monomial   A constant, a variable, or a product of a constant and one or more variables

Multiple of a number   The product of the number and an integer

Multiplicative identity   The number 1

N

Natural numbers   The counting numbers: 1, 2, 3, 4, 5, . . .

Negative integers   Integers to the left of zero on the number line

Numerator   The number above the fraction bar in a fraction

O

Obtuse angle   An angle whose measure is greater than 90° and less than 180°

Obtuse triangle   A triangle in which one angle is an obtuse angle

Opposite   The opposite, or additive inverse, of a number x is written $-x$. Opposites are the same distance from 0 on the number line but on different sides of 0.

Opposite of a polynomial   Two polynomials are opposites, or additive inverses, of each other if their sum is zero.

Ordered pair   A pair of numbers of the form (a, b) for which the order in which the numbers are listed is important

Origin   The point $(0, 0)$ on a graph where the two axes intersect

Original price   The price of an item before a discount is deducted

P

Palindrome prime   A prime number that remains a prime number when its digits are reversed

Parallelogram   A four-sided polygon with two pairs of parallel sides

Percent notation   A representation of a number as parts per 100; $n\%$

Perimeter   The distance around an object or the sum of the lengths of its sides

Periods   Groups of three digits, separated by commas

Pi (π)   The number that results when the circumference of a circle is divided by its diameter; $\pi \approx 3.14$, or $\frac{22}{7}$

Pictograph   A graphic means of displaying information using symbols to represent the amounts

Polygon   A closed geometric figure with three or more line segments as sides

Polynomial   A monomial or a sum of monomials

Positive integers   Integers to the right of zero on the number line

Prime factorization A factorization of a composite number as a product of prime numbers

Prime number A natural number that has exactly two different factors: itself and 1

Principal An amount of money that is invested or borrowed

Product The result when one number is multiplied by another

Proportion An equation stating that two ratios are equal

Protractor A device used to measure and draw angles

Purchase price The price of an item before sales tax is added

Pythagorean equation The equation $a^2 + b^2 = c^2$, where a and b are lengths of the legs of a right triangle and c is the length of the hypotenuse

Q

Quadrants The four regions into which the axes divide a plane

Quotient The result when one number is divided by another

R

Radical sign The symbol $\sqrt{}$

Radius A segment with one endpoint on the center of a circle and the other endpoint on the circle, or the length of such a segment

Rate A ratio used to compare two different kinds of measure

Ratio The quotient of two quantities; the ratio of a to b is $\dfrac{a}{b}$, also written $a:b$

Rational number Any number that can be written as the ratio of two integers $\dfrac{a}{b}$, where $b \neq 0$

Ray A part of a line consisting of one endpoint and all the points of the line on one side of the endpoint

Reciprocal A multiplicative inverse; two numbers are reciprocals if their product is 1

Rectangle A four-sided polygon with four 90° angles

Repeating decimal A decimal in which a block of digits repeats indefinitely

Right angle An angle whose measure is 90°

Right triangle A triangle that includes a right angle

Rounding Approximating the value of a number; used when estimating

S

Sale price The price of an item after a discount has been deducted

Sales tax A tax added to the purchase price of an item

Scalene triangle A triangle in which all sides are of different lengths

Scientific notation A representation of a number written in the form $M \times 10^n$, where n is an integer, $1 \leq M < 10$, and M is expressed in decimal notation

Similar triangles Triangles in which corresponding sides are proportional and corresponding angles are congruent

Simple interest A percentage of an amount P invested or borrowed for t years, computed by calculating principal $\times$ interest rate $\times$ time

Simplify To rewrite an expression in an equivalent, abbreviated form

Solution of an equation A replacement for the variable that makes the equation true

Solve To find all solutions of an equation, inequality, or problem

Sphere The set of all points in space that are a given distance (the radius) from a given point (the center)

Square A four-sided polygon with four right angles and all sides of equal length

Square root The number c is a square root of a if $c^2 = a$.

Standard form of a linear equation An equation written in the form $Ax + By = C$

Statistic A number that describes a set of data

Straight angle An angle whose measure is 180°

Substitute To replace a variable with a number

Subtrahend In subtraction, the number being subtracted

Sum The result in addition

Supplementary angles Two angles for which the sum of their measures is 180°

T

Table A representation of data in rows and columns

Term A number, a variable, or a product or a quotient of numbers and/or variables

Terminating decimal A decimal that can be written using a finite number of decimal places

Total price The sum of the purchase price of an item and the sales tax on the item

Trapezoid A four-sided polygon with exactly two parallel sides

Triangle A three-sided polygon

U

Unit price The ratio of price to the number of units

Unit rate The ratio of quantity to the number of units

V

Value The numerical result after a number has been substituted into an expression and the calculations have been carried out

Variable A letter that represents an unknown number

Vertex The common endpoint of the two rays that form an angle

Vertical angles Two angles that are formed by two intersecting lines and that have no side in common

Volume The number of unit cubes needed to fill an object

W

Whole numbers The natural numbers and 0: 0, 1, 2, 3, 4, 5, . . .

Index

Division
  basic, 745
  on a calculator, 30, 330
  checking, 29
  with decimal notation, 329–333
  definition, 27, 110
  dividend, 26
  divisor, 26
  by estimating multiples, 748
  estimating quotients, 40, 351
  with fraction notation, 200
  of integers, 110
  with mixed numerals, 274
  of a number by itself, 27, 747
  by one, 27, 746
  by a power of 10, 332, 333
  quotient, 26
    expressed as a mixed numeral, 257
  and reciprocals, 200
  and rectangular arrays, 26, 745
  related multiplication sentence, 745
  with remainders, 28
  as repeated subtraction, 26
  using scientific notation, 687
  undefined, 28, 110, 747
  of whole numbers, 26
  by zero, 28, 110, 747
  of zero by a nonzero number, 27,
    111, 747
  zeros in quotients, 31
Division principle, 135
Divisor, 26
Dollars, converting from/to cents, 324
DPR (daily percentage rate), 562

E

Eight, divisibility by, 158
Eleven, divisibility by, 158
English system of measures, *see*
    American system of measures
Equality of fractions, 184, 470
Equation(s), 42, 48. *See also* Solving
    equations.
  equivalent, 133
  false, 42
  graph of, 414, 417, 418
  linear, 413, 414
  percent, 521
  Pythagorean, 637, 660
  solution of, 48, 133, 409
  solving, *see* Solving equations
  true, 42
Equilateral triangle, 629
Equivalent
  equations, 133
  expressions, 118, 120
  fractions, 181
Estimating, 39. *See also* Rounding.
  in checking, 55
  differences, 39, 350
  products, 40, 351
  quotients, 40, 351
  sums, 39, 350

Evaluating
  algebraic expressions, 118, 275, 325
  exponential notation, 71
  polynomials, 698
Even number, 153
Expanded notation, 4
Exponent(s), 70
  adding, 679, 682
  on a calculator, 675
  multiplying, 681, 682
  negative integers as, 675
  one as, 674
  product rule, 679, 682
  properties of, 682
  quotient rule, 680, 682
  raising a power to a power, 681,
    682
  raising a product to a power, 681,
    682
  raising a quotient to a power, 682
  subtracting, 680, 682
  zero as, 674
Exponential notation, 70
  and dividing, 680
  evaluating, 71
  and least common multiples, 225
  and multiplying, 679
Expressions
  algebraic, 118
  equivalent, 118, 120
  evaluating, 118, 275, 325
  simplifying, 71
  terms, 126, 695
  value of, 118
Extrapolation, 436

F

Factor, 19, 159, 737
  common, 706
Factor tree, 162
Factoring algebraic expressions, 706
Factorization
  and LCMs, 223
  of natural numbers, 159
  prime, 161
Factorization method for finding LCM,
    223
Factors, *see* Factor
Fahrenheit temperature, 648, 649
False equation, 42
False inequality, 42
Familiarize, 54
Feet, 578
First coordinate, 407
Five, divisibility by, 154
Fluid ounce, 613
Foot, 578
Four, divisibility by, 158
Fraction, *see* Fractions
Fraction equivalent, 514
Fraction notation, 165. *See also*
    Fractions.
  addition, 230, 231

converting
  from/to decimal notation, 305,
    306, 341, 344
  from/to mixed numerals, 255, 256
  from/to percent notation, 511, 513
  and decimal notation together in
    calculations, 344
  denominator, 165
  of zero, 169
  in division, 200
  for integers, 169
  multiplication using, 176
  numerator, 165
  for one, 168
  percent notation as, 509
  for ratios, 167
  reciprocal, 199
  simplest, 182
  simplifying, 182
  subtraction, 239
  for zero, 169
Fractions, 165. *See also* Fraction
    notation.
  clearing, 250
  complex, 201, 285
  cross products, 185
  equality test for, 184, 470
  equivalent, 181
  improper, 167
  multiplication by an integer, 175
  multiplicative identity for, 181
  order, 233
  as partitions, 165
  as ratios, 167

G

Gallon, 613
Geometric shapes and proportions,
    490, 492
GPA, 427
Grade point average, 427
Gram, 646
Graph
  bar, 398
  circle, 625
  line, 400
  of a linear equation, 414, 417, 418
  picture, 390
  pie chart, 625
  of a point, 407
Graphing calculator, 414. *See also*
    Calculator.
Greater than (>), 42
Grid, 407
Grouping symbols on a calculator, 112

H

Hectogram, 646
Hectometer, 581
Height of a parallelogram, 597
Histogram, 393
Horizontal line, graph of, 417, 418
Hundred, 323

expanded notation, 4
integers, 86, 302
natural, 9
negative, on a calculator, 94
opposites, 86, 88
periods, 2
place-value, 2, 302
prime, 160
rational, 302
sign of, 89
standard notation, 4
whole, 9
word names for, 4
Numerals, mixed, 254. *See also* Mixed
    numerals.
Numerator, 165

O

Obtuse angle, 626
Obtuse triangle, 629
One
    division by, 27, 746
    as an exponent, 674
    fraction notation for, 168
    multiplying by, 181, 579, 583, 591, 738
    removing a factor equal to, 182, 184
One hundred, multiplication by, 739
One thousand, multiplication by, 740
Ones period, 2
Operations, order of, 71, 73, 111, 284,
    333
Opposite(s), 86, 88
    addition of, 93, 98
    of a polynomial, 696
    and subtraction, 98, 697
Opposite angles, 628
Order
    of addition, *see* Commutative law of
        addition
    in decimal notation, 307
    descending, 696
    in fraction notation, 233
    of multiplication, *see* Commutative
        law of multiplication
    on the number line, 87
    of operations, 71, 73, 111, 284, 333
        and calculators, 73
    of whole numbers, 42
Ordered pair, 407
    as the solution of an equation, 409
Origin, 407
Original price, 552
Ounce, 613, 645

P

Pair, ordered, 407
Palindrome prime, 216
Parallelogram, 597
    area, 597, 603, 660
Parentheses
    in multiplication, 19
    within parentheses, 74
Partition of an object, 165

Percent, *see* Percent notation
Percent decrease, 541
Percent equations, 521
Percent equivalent, 514
Percent increase, 541, 542
Percent key on a calculator, 524
Percent notation, 508, 509
    converting
        from/to decimal notation, 510
        from/to fraction notation, 511, 513
Percent symbol (%), 508
Perimeter, 11, 127
    of a rectangle, 128
    of a square, 128
Periods, 2
Pi (π), 600
    on a calculator, 344, 603, 615
Pictographs, 390
Picture graphs, 390
Pie chart, 625
Pint, 613
Pitch of a screw, 195
Place-value, 2, 302
Plane, 407
Plotting points, 407
Points, 407
Polygon, 127
Polynomials, 695
    addition of, 695
    additive inverse, 696
    descending order, 696
    evaluating, 696
    factoring, 706
    monomials, 695
    multiplication of monomials, 704
    opposite of, 696
    subtraction, 697
    terms, 695
Positive integers, 86
Pound, 645
Power(s). *See also* Exponent(s).
    of an integer, 106
Power rule, 681, 682
Predictions, 435
Price
    marked, 552
    original, 552
    purchase, 549
    sale, 552
    total, 549
    unit, 464
Primary principle of probability, 437
Prime, relatively, 164
Prime factorization, 161
    in finding LCMs, 223
Prime number, 160
    palindrome, 216
Principal, 558
Principle(s)
    addition, 134
    division, 135
    multiplication, 204
    primary, of probability, 437

Probability, 437
Problem solving, 54, 61. *See also* Index
    of Applications.
Product(s), 19, 737
    cross, 185
    estimating, 40, 351
    raising to a power, 681, 682
Product rule, 679, 682
Properties of exponents, 682
Proportional, 470
Proportions, 470
    and geometric figures, 490, 492
    and similar triangles, 490
    solving, 471
    used in solving percent problems, 527
Protractor, 624
Purchase price, 549
Pythagoras, 638
Pythagorean equation, 637, 660
Pythagorean theorem, 637

Q

Quadrants, 408
Quadrillion, 328
Quart, 613
Quintillion, 328
Quotient, 26
    estimating, 40, 351
    as a mixed numeral, 257
    raising to a power, 682
    zeros in, 31
Quotient rule, 680, 682

R

Radical sign $(\sqrt{\ })$, 636
Radius, 367, 599, 660
Raising a power to a power, 681, 682
Raising a product to a power, 681, 682
Raising a quotient to a power, 682
Rate
    commission, 551
    of discount, 552
    interest, 558
    ratio as, 463
    sales tax, 549
    unit, 464
Ratio, 167, 456
    percent as, 509
    and proportion, 470
    as a rate, 463
    simplifying, 459
Rational numbers, 302
Rays, 623
Reciprocal, 199
    and division, 200
    of zero, 199
Rectangle, 127
    area of, 22
    perimeter of, 128
Rectangular array
    in division, 26, 745
    in multiplication, 19, 737
Rectangular solid, volume of, 610, 660

Trillion, 323
Trillions period, 2
True equation, 42
True inequality, 42
Truncating, 311
Two, divisibility by, 153

U

Undefined division, 28, 110, 747
Unit angle, 624
Unit cube, 610
Unit price, 464
Unit rate, 464
Unit segment, 578

V

Value of an expression, 118
Variable, 118
   substituting for, 118
Vertex of an angle, 623
Vertical angles, 628, 629

Vertical line, graph of, 418
Volume, 610
   of a circular cylinder, 612, 660
   of a rectangular solid, 610, 660
   of a sphere, 612, 660

W

Weight, 645, 660. *See also*
     Mass.
Whole numbers, 9
   addition, 9, 10, 724
   division, 26, 30
   expanded notation, 4
   multiplication, 19, 21
   order, 42
   as remainders on a calculator,
     331
   rounding, 37
   standard notation, 4
   subtraction, 14, 15
   word names, 4
Window, 414

Word names
   for decimal notation,
     303
   for whole numbers, 4

Y

Yard, 578

Z

Zero
   addition of, 10, 724
   denominator of, 169
   divided by a nonzero number, 27,
     111, 747
   division by, 28, 110, 747
   as an exponent, 674
   fraction notation for, 169
   multiplication by, 105, 737
   in quotients, 31
   reciprocal, 199
   in subtraction, 733
Zoom menu, 414